# Genetic Engineering of Vegetable Crops

## CABI BIOTECHNOLOGY SERIES

Biotechnology, in particular the use of transgenic organisms, has a wide range of applications including agriculture, forestry, food and health. There is evidence that it could make a major impact in producing plants and animals that are able to resist stresses and diseases, thereby increasing food security. There is also potential to produce pharmaceuticals in plants through biotechnology, and provide foods that are nutritionally enhanced. Genetically modified organisms can also be used in cleaning up pollution and contamination. However, the application of biotechnology has raised concerns about biosafety, and it is vital to ensure that genetically modified organisms do not pose new risks to the environment or health. To understand the full potential of biotechnology and the issues that relate to it, scientists need access to information that not only provides an overview of and background to the field, but also keeps them up to date with the latest research findings.

This series, which extends the scope of CABI's successful "Biotechnology in Agriculture" series, addresses all topics relating to biotechnology including transgenic organisms, molecular analysis techniques, molecular pharming, *in vitro* culture, public opinion, economics, development and biosafety. Aimed at researchers, upper-level students and policy makers, titles in the series provide international coverage of topics related to biotechnology, including both a synthesis of facts and discussions of future research perspectives and possible solutions.

### Titles Available

1. Animal Nutrition with Transgenic Plants
   Edited by G. Flachowsky
2. Plant-derived Pharmaceuticals: Principles and Applications for Developing Countries
   Edited by K.L. Hefferon
3. Transgenic Insects: Techniques and Applications
   Edited by M.Q. Benedict
4. Bt Resistance: Characterization and Strategies for GM Crops Producing *Bacillus thuringiensis* Toxins
   Edited by Mario Soberón, Yulin Gao and Alejandra Bravo
5. Plant Gene Silencing: Mechanisms and Applications
   Edited by Tamas Dalmay
6. Ethical Tensions from New Technology: The Case of Agricultural Biotechnology
   Edited by Harvey James
7. GM Food Systems and their Economic Impact
   Tatjana Brankov and Koviljko Lovre
8. Endophyte Biotechnology: Potential for Agriculture and Pharmacology
   Edited by Alexander Schouten
9. Forest Genomics and Biotechnology
   Richard Meilan and Matias Kirst
10. Transgenic Insects: Techniques and Applications, 2nd Edition
    Edited by M.Q. Benedict and Maxwell J. Scott
11. Plant Omics: Advances in Big Data Biology
    Edited by Hajime Ohyanagi, Eiji Yamamoto, Ai Kitazumi and Kentaro Yano
12. Next-generation Sequencing and Agriculture
    Edited by Phillip Bayer and Dave Edwards
13. Aquaculture and Fisheries Biotechnology: Genetic Approaches
    Rex A. Dunham
14. Algal Biotechnology
    Edited by Qiang Wang
15. Genetic Engineering of Vegetable Crops
    Edited by Pritam Kalia

# Genetic Engineering of Vegetable Crops

*Edited by*

**Pritam Kalia**

CABI

**CABI is a trading name of CAB International**

CABI
Nosworthy Way
Wallingford
Oxfordshire OX10 8DE
UK

Tel: +44 (0)1491 832111
E-mail: info@cabi.org
Website: www.cabi.org

CABI
200 Portland Street
Boston
MA 02114
USA

Tel: +1 (617)682-9015
E-mail: cabi-nao@cabi.org



A catalogue record for this book is available from the British Library, London, UK.

ISBN-13: 9781800627109 (hardback)
9781800627116 (ePDF)
9781800627123 (ePub)

DOI: 10.1079/9781800627123.0000

Commissioning Editor: Rebecca Stubbs
Editorial Assistant: Helen Elliott
Production Editor: Shankari Wilford

Typeset by Straive, Pondicherry, India
**Printed and bound by CPI Group (UK) Ltd, Croydon, CR0 4YY**

# Contents

# Contributors

**K. Anjala**, Centre for Plant Biotechnology and Molecular Biology, Kerala Agricultural University, Thrissur, Kerala, India. Email: anjuhasee95@gmail.com

**Suresh H. Antre**, Department of Plant Biotechnology, College of Agriculture, University of Agricultural Sciences, Bangalore, Karnataka-560 065, India. Email: suresh.antre@gmail.com

**Donata Arena**, Dipartimento di Agricoltura, Alimentazione e Ambiente (Di3A), Università di Catania, Catania, Italy. Email: donata.arena@phd.unict.it

**T.S. Aruna**, Division of Vegetable Science, ICAR-Indian Agricultural Research Institute, New Delhi 110012, India. Email: arunts387@gmail.com

**Rehna Augustine**, Centre for Plant Biotechnology and Molecular Biology, Kerala Agricultural University, Thrissur, Kerala, India. Email: rehna.augustine@kau.in

**S. Backiyarani**, ICAR-National Research Center for Banana, Thogamalai Road, Thayanur Post, Trichy 620102, Tamil Nadu, India. Email: backiyarani@gmail.com

**Nikita Baliyan**, Division of Vegetable Science, ICAR-Indian Agricultural Research Institute, New Delhi 110012, India. Email: nikita.biotech@gmail.com

**Rafal Baranski**, Department of Plant Biology and Biotechnology, Faculty of Biotechnology and Horticulture, University of Agriculture in Krakow, Krakow, Poland. ORCID: 0000-0001-9276-3870. Email: rafal.baranski@urk.edu.pl

**Satyaprakash Barik**, Department of Agriculture and Allied Sciences, CV Raman Global University, Bhubaneswar, India. Email: satyabarik001@gmail.com

**D.R. Bhardwaj**, ICAR-Indian Institute Vegetable Research, Varanasi 221305, Uttar Pradesh, India. Email: dram_iivr@yahoo.com

**Vinay Bhardwaj**, ICAR-Central Potato Research Institute, Shimla 171001, Himachal Pradesh, India. Email: vinaycpri@gmail.com

**Ferdinando Branca**, Dipartimento di Agricoltura, Alimentazione e Ambiente (Di3A), Università di Catania, Catania, Italy. Email: fbranca@unict.it

**T.M. Brindha**, Centre for Plant Biotechnology and Molecular Biology, Kerala Agricultural University, Thrissur, Kerala, India. Email: brinititus@gmail.com

**S.K. Chakrabarti**, ICAR-Central Potato Research Institute, Shimla 171001, Himachal Pradesh, India. Email: skc_cpri@yahoo.co.in

**Supriya Chakraborty**, School of Life Sciences, Jawaharlal Nehru University, New Delhi-110064, India. Email: supriyachakrasls@yahoo.com

**K. Divya**, ICAR-Central Tuber Crops Research Institute, Thiruvananthapuram 695017, Kerala, India. Email: divyakariatt@gmail.com

**Megha Ganachari**, Division of Basic Sciences, ICAR-Indian Institute of Horticultural Research, Hesaraghatta, Bengaluru 560 089, Karnataka, India. Email: meghaganachari@gmail.com

**P.H. Ramanjini Gowda**, Department of Plant Biotechnology, College of Agriculture, University of Agricultural Sciences, Bangalore, Karnataka-560 065, India. Email: ramanjini@yahoo.com

**Rasmieh Hamid**, Department of Plant Breeding, Cotton Research Institute of Iran (CRII), Agricultural Research, Education and Extension Organization (AREEO), Gorgān, Iran. Email: r.hamid@areeo.ac.ir

**Vageeshbabu S. Hanur**, Division of Basic Sciences, ICAR-Indian Institute of Horticultural Research, Hesaraghatta Lake Post, Bangalore-560089, India. Email: vageeshbabu.hs@icar.gov.in

**Feba Jacob**, Centre for Plant Biotechnology and Molecular Biology, Kerala Agricultural University, Thrissur 680 656, Kerala, India. Email: febajacob53@gmail.com

**A. Jeevalatha**, ICAR-Central Potato Research Institute, Shimla 171001, Himachal Pradesh, India. Email: jeevalatha_a@yahoo.co.in

**Pritam Kalia**, Division of Vegetable Science, ICAR-Indian Agricultural Research Institute, New Delhi 110012, India. Email: pritam.kalia@gmail.com

**S. Kalpana**, ICAR-National Research Center for Banana, Thogamalai Road, Thayanur Post, Trichy 620102, Tamil Nadu, India. Email: ramajayamkalpana@gmail.com

**R. Karthic**, ICAR-National Research Center for Banana, Thogamalai Road, Thayanur Post, Trichy 620102, Tamil Nadu, India. Email: rajukarthic.dr@gmail.com

**Ashish Kaushal**, Graphic Era Hill University, School of Agriculture, Dehradun 284001, Uttarakhand, India; ORCID ID: 0000-0002-6029-2303. Email: akiihr04@gmail.com

**Magdalena Klimek-Chodacka**, Department of Plant Biology and Biotechnology, Faculty of Biotechnology and Horticulture, University of Agriculture in Krakow, Krakow, Poland. ORCID: 0000-0002-6254-9280. Email: m.chodacka@urk.edu.pl

**Manoj Kumar**, ICAR-Central Potato Research Institute, Shimla 171001, Himachal Pradesh, India. Email: manojkumar0105@gmail.com

**Polumetla Ananda Kumar**, ICAR-Indian Institute of Rice Research, Hyderabad-500030, India. Email: polumetla@gmail.com

**Ravinder Kumar**, ICAR-Central Potato Research Institute, Shimla 171001, Himachal Pradesh, India. Email: chauhanravinder97@gmail.com

**D.C. Lakshamanareddy**, Division of Basic Sciences, ICAR-Indian Institute of Horticultural Research, Bengaluru, India. Email: Lakshmanareddy.DC@icar.gov.in

**Aneta Lukasiewicz**, Department of Plant Biology and Biotechnology, Faculty of Biotechnology and Horticulture, University of Agriculture in Krakow, Krakow, Poland. ORCID: 0000-0003-4352-6398. Email: aneta.lukasiewicz1@gmail.com

**T. Makeshkumar**, ICAR-Central Tuber Crops Research Institute, Thiruvananthapuram 695017, Kerala, India. Email: makeshctcri@gmail.com or makeshkumar.t@icar.gov.in

**M. Malavika**, School of Life Sciences, Jawaharlal Nehru University, New Delhi-110064, India. Email: malsjune18@gmail.com

**Manisha Mangal**, Division of Vegetable Science, ICAR-Indian Agricultural Research Institute, New Delhi 110012, India. Email: manishamangal@rediffmail.com

**Deepu Mathew**, Centre for Plant Biotechnology and Molecular Biology, Kerala Agricultural University, Thrissur 680 656, Kerala, India. ORCiD: https://orcid.org/0000-0002-2941-1060. Email: deepu.mathew@kau.in

**Pallavi Mishra**, Division of Vegetable Improvement, ICAR-Indian Institute of Vegetable Research, Shahanshapur, PO-Jakhini, Varanasi-221305, Uttar Pradesh, India. Email: pallavimishra23@gmail.com

**Upasana Mohapatra**, Department of Plant Biotechnology, College of Agriculture, University of Agricultural Sciences, Bangalore, Karnataka-560 065, India. Email: upasanaprangya92@gmail.com

**N. Nagesha**, Department of Plant Biotechnology, College of Agriculture, University of Agricultural Sciences, Bangalore, Karnataka-560 065, India. Email: nagesha.n@uasbangalore.edu.in

**T.M. Ningaraju**, Department of Plant Biotechnology, College of Agriculture, University of Agricultural Sciences, Bangalore, Karnataka-560 065, India. Email: ningarajutm@gmail.com

**Saurabh Pandey**, National Institute of Plant Genome Research, Aruna Asaf Ali Marg, New Delhi 110067, India; Department of Agriculture, Guru Nanak Dev University, Amritsar, Punjab-143005, India. Email: saurabh_agri@gndu.ac.in

**Naresh Ponnam**, Division of Vegetable Crops, ICAR-Indian Institute of Horticultural Research, Bengaluru, India. Email: Ponnam.Naresh@icar.gov.in

**K.N. Poornima**, Division of Basic Sciences, ICAR-Indian Institute of Horticultural Research, Hesaraghatta, Bengaluru 560 089, Karnataka, India. Email: Poornima.N@icar.gov.in

**Roshini Pydi**, Division of Vegetable Crops, ICAR-Indian Institute of Horticultural Research, Bengaluru, India. Email: pydi.roshni1995@gmail.com

**Manoj Prasad**, National Institute of Plant Genome Research, Aruna Asaf Ali Marg, New Delhi 110067, India. Email: manoj_prasad@nipgr.ac.in

**K.V. Ravishankar**, ICAR-Indian Institute of Horticultural Research, Bengaluru 560089, Karnataka, India. ORCID ID: 0000-0001-9935-763X. Email: kv_ravishankar@yahoo.in

**K. Madhavi Reddy**, Division of Vegetable Crops, ICAR-Indian Institute of Horticultural Research, Bengaluru, India. Email: Madhavireddy.K@icar.gov.in

**Giulio Flavio Rizzo**, Dipartimento di Agricoltura, Alimentazione e Ambiente (Di3A), Università di Catania, Catania, Italy. Email: giulio.rizzo@phd.unict.it

**A.T. Sadashiva**, Nethra Crop Science, Bengaluru 562123, Karnataka, India. Email: atsbrs@gmail.com

**Vidya Sagar**, Division of Vegetable Improvement, ICAR-Indian Institute of Vegetable Research, Shahanshapur, PO-Jakhini, Varanasi-221305, Uttar Pradesh, India. Email: vidya.sagarkaushal@gmail.com

**Vinay Sagar**, ICAR-Central Potato Research Institute, Shimla 171001, Himachal Pradesh, India. Email: Vinay.Sagar@icar.gov.in

**Koushik Saha**, Division of Vegetable Crops, ICAR-Indian Institute of Horticultural Research, Bengaluru, India. Email: ksahaiari93@gmail.com

**B.G. Sangeetha**, ICAR-Central Tuber Crops Research Institute, Thiruvananthapuram 695017, Kerala, India. Email: sangeethageethu214@gmail.com

**Sanjay Sathian**, Centre for Plant Biotechnology and Molecular Biology, Kerala Agricultural University, Thrissur, Kerala, India. Email: sanjaysathian01@gmail.com

**K.M. Senthilkumar**, ICAR-Central Tuber Crops Research Institute, Thiruvananthapuram 695017, Kerala, India. Email: senthilmbb@gmail.com

**Sanjeev Sharma**, ICAR-Central Potato Research Institute, Shimla 171001, Himachal Pradesh, India. Email: sanjeevsharma.cpri@gmail.com

**M.N. Sheela**, ICAR-Central Tuber Crops Research Institute, Thiruvananthapuram 695017, Kerala, India. Email: sheelactcri@gmail.com

**H.K. Shreyas**, Division of Basic Sciences, ICAR-Indian Institute of Horticultural Research, Hesaraghatta, Bengaluru 560 089, Karnataka, India. Email: hkshreyas44@gmail.com

**B.P. Singh**, ICAR-Central Potato Research Institute, Shimla 171001, Himachal Pradesh, India. Email: birpals53@gmail.com

**Arpita Srivastava**, Division of Vegetable Science, ICAR-Indian Agricultural Research Institute, New Delhi 110012, India. Email: asrivastava45@gmail.com

**S. Sundaresha**, ICAR-Central Potato Research Institute, Shimla 171001, Himachal Pradesh, India. Email: sundaresha8@gmail.com

**K.R.M. Swamy**, (retired), Division of Vegetable Crops, ICAR-Indian Institute of Horticultural Research, Bengaluru-560089, India. Email: krmswamy@gmail.com

**Kamil Szymonik**, Department of Plant Biology and Biotechnology, Faculty of Biotechnology and Horticulture, University of Agriculture in Krakow, Krakow, Poland. ORCID: 0000-0002-0821-2396. Email: kamil.szymonik@urk.edu.pl

**Jagesh K. Tiwari**, ICAR-Central Potato Research Institute, Shimla 171001, Himachal Pradesh, India; current address: ICAR-Indian Institute of Vegetable Research, Varanasi 221305, Uttar Pradesh, India. Email: jageshtiwari@gmail.com

**Shailesh K. Tiwari**, Division of Vegetable Improvement, ICAR-Indian Institute of Vegetable Research, Shahanshapur, PO-Jakhini, Varanasi-221305, Uttar Pradesh, India. Email: tiwarishailu@gmail.com

**Simone Treccarichi**, Dipartimento di Agricoltura, Alimentazione e Ambiente (Di3A), Università di Catania, Catania, Italy. Email: simone.treccarichi@phd.unict.it

**S. Uma**, ICAR-National Research Center for Banana, Thogamalai Road, Thayanur Post, Trichy 620102, Tamil Nadu, India. Email: umabinit@yahoo.co.in

**Priti Upadhyay**, Division of Vegetable Science, ICAR-Indian Agricultural Research Institute, New Delhi 110012, India. Email: dr.pritiupadhyay84@gmail.com

**T.R. Usharani**, Division of Basic Sciences, ICAR-Indian Institute of Horticultural Research, Hesaraghatta, Bengaluru 560 089, Karnataka, India. Email: usharani.tr@icar.gov.in

**B. Varalakshmi**, Division of Vegetable Crops, ICAR-Indian Institute of Horticultural Research, Bengaluru 560089, India. Email: Varalakshmi.B@icar.gov.in

**K. Varsha**, Centre for Plant Biotechnology and Molecular Biology, Kerala Agricultural University, Thrissur, Kerala, India. Email: varshakvasudevan96@gmail.com

**Suresh Reddy Yerasu**, ICAR-Indian Institute Vegetable Research, Varanasi 221305, Uttar Pradesh, India. Email: yerasusureshreddy@yahoo.co.in

**Sneha Yogindran**, School of Life Sciences, Jawaharlal Nehru University, New Delhi-110064, India; Cochin University of Science and Technology, Cochin, Kerala-682022, India. Email: snehayogindran@gmail.com

# Preface

Over the last 60 years, plant breeders have been successful in developing and releasing new crop cultivars. These have provided the basis for the consistent supply of nutritious food in a changing environment, which has been a major contributing factor toward the alleviation of world hunger and malnutrition. Advances in yield performance, specific quality attributes, resistance to diseases and insect pests, and environmental adaptation are constantly demanded from plant breeders. In the recent history of modern plant breeding, "new technologies" have been used regularly to facilitate the development of new gene combinations during the breeding of new cultivars.

More recently, molecular biology has resulted in technology known as cloning and DNA sequencing of specific genes, the reassembly of specific DNA fragments into functional chimeric genes, and the transfer of such genes to single plant cells from which complete plants can be regenerated via cell and tissue culture. This approach involves genetic engineering of plants and offers immense opportunities for the incorporation of new genes into crop cultivars. Most genetically engineered crop plants currently on the market or under development involve the transfer of genes conferring resistance to insect pests, diseases, herbicides and environmental stress, as well as quality traits such as improved postharvest storage, nutrition, and color. The resulting novel germplasm is predicted to allow plant breeders to respond much more quickly to the market need for new and improved cultivars, and satisfy the increasing consumer demand for a consistent supply of high-quality fruits and vegetables with reduced pesticide residues.

Most of the characteristics targeted for improvement through genetic engineering are similar to those achieved by gene transfer from wild species via traditional breeding. Genetic engineering allows the repeated transfer of new genes directly into existing cultivars. This contributes to the more efficient development of new cultivars without many generations of hybridization. The transfer of single, discrete genes can readily be achieved by genetic engineering. This overcomes the problem of linkage drag in traditional plant breeding, which is associated with the transfer of many unwanted and undefined closely linked genes.

Crops developed by genetic engineering can be used not only to enhance yields and nutritional quality, but also to provide increased tolerance to various biotic and abiotic stresses. Although there have been some expressions of concern about biosafety and health hazards associated with genetically modified (GM) crops, there is no reason to hesitate in consuming genetically engineered food crops that have been thoughtfully developed and carefully tested. Integration of modern biotechnology with conventional agricultural practices in a sustainable manner can fulfill the goal of attaining food security for present as well as future generations.

To be healthy, our daily diet must include ample high-quality foods with all of the essential nutrients, above the basic nutrition. To maintain the per capita requirement, this will become increasingly hard because of the continued loss of arable lands and the prevalence of unfavorable environmental conditions including drought, salinity, floods and diseases. To ensure food security for future generations, the world must produce 50–100% more food than at present, in spite of the predicted adverse environmental conditions.

Conventional plant breeding alone can no longer sustain the ever-rising global food demand. This is the time to promote sustainable agricultural practices for boosting crop productivity with the utmost conservation of all available natural resources. GM crops can help us to meet the demand for high-yielding, nutritionally balanced, and biotic and abiotic stress-tolerant crop varieties. While the global area under GM crops continues to expand every year, concerns have been expressed regarding unintended and unpredictable pleiotropic effects of these crops on human health and the environment.

GM strategies are being employed to minimize yield losses due to various stresses (biotic and abiotic) and are being used extensively for value addition in food crops by enrichment with quality proteins, vitamins, iron, zinc, carotenoids, anthocyanins and so on. Other ongoing efforts include the enhancement of shelf life of perishable crops such as vegetables in order to significantly reduce postharvest losses. The global area under GM crops continues to expand every year, yet there have been no harmful effects encountered so far. Insect-resistant *Bacillus thuringiensis* (*Bt*) crops and/or herbicide-tolerant GM crops that are currently under commercial cultivation have benefited farmers through better insect and weed management, higher yields and reduced chemical pesticide use. Thus, modern biotechnology can enable the achievement of food and nutritional security for present and future generations. The performance of GM crops, however, needs to be closely monitored for several generations through rigorous biosafety assessments on a case-by-case basis before being released for commercial cultivation. One of the major goals of introducing genetic engineering in vegetables is to improve qualitative traits such as nutritional content, organoleptic qualities such as flavor and aroma, and the development of seedless vegetables.

Recent years have brought new possibilities of genome modification by utilizing novel tools based on the clustered regularly interspaced short palindromic repeats (CRISPR)-associated protein (Cas) system. This system has been used with success in numerous plant species, including vegetable crops, and can be used not only for basic research but also to provide benefits for commercial farming.

Using cutting-edge biotechnological and genome sequencing techniques, we can move closer to targeted breeding of vegetable crops for superior alleles and development of new cultivars fostered with precise genetic modification. In this regard, *Agrobacterium*-mediated gene transfer remains the foundation for most vegetable transformation protocols; however, new emerging technologies such as transcriptome engineering and genome editing, particularly CRISPR/Cas9, have had a significant impact on vegetable crop improvement.

A decade after the invention of genome editing, what we are seeing is application across a range of different crops and trait combinations that will bring benefits to producers and consumers. The first applications of genome editing in food and agriculture are now being commercialized, with many more in the late stages of research and development. Increasingly favorable regulatory decisions are unfolding globally that should facilitate the realized benefits of the technology for the health of people and the planet.

Based on this background, there are many reviews and books on genetic engineering focusing on cereal crops specifically, but the present one is the only comprehensive book devoted entirely to vegetable crops in relation to modern technologies in a single volume. The 20 chapters in this book represent a collection of recent and highly relevant reviews prepared by authors who are well-recognized experts in their respective areas. The first ten chapters focus on tomato, eggplant, peppers, amaranth, cauliflower, carrot, cucurbits, potato, tropical tubers and melons, with the remaining chapters exploring transgenic resistance to viral diseases, embryogenic cell suspension culture, genome editing, biotechnology to maintain biodiversity, genomics and molecular breeding, genetic engineering in vegetable crops, biotechnological interventions for vegetable crop improvement, molecular techniques for biofortification, RNA interference strategies, and designing futuristic vegetable crop varieties.

I remain immensely grateful to all the contributing authors for their highly useful contributions in developing excellent informative reviews on genetic transformation of vegetable crops. I hope that this book reaches all corners the world, benefiting students, teachers and researchers and any others who are interested.

Above all, I am grateful to the publisher for their support and decision to bring out this important publication.

**Pritam Kalia**
New Delhi, India

# Dedication

**In loving memory of the late Professor K.V. Peter.**

I am indebted to Professor Peter for inspiring me to bring out this book entitled *Genetic Engineering of Vegetable Crops*. He planned the initial topics to be included in the book but sadly died before completion. Professor Peter was a voracious writer and a dedicated vegetable scientist breeder to his core. He was appreciative of my breeding efforts in cauliflower, broccoli and carrot and in shaping the Division of Vegetable Science at the Indian Agricultural Research Institute, New Delhi, India.

Pritam Kalia
New Delhi, India

# 1 Recent Advances in Tomato Genetic Engineering for Virus Resistance: Achievements and Limitations

**Saurabh Pandey**[1,2] **and Manoj Prasad**[1,3*]

[1]*National Institute of Plant Genome Research, Aruna Asaf Ali Marg, New Delhi, India;* [2]*Kumari Devi Choubey College of Agriculture and Research Station, Bemetara, IGKV, Raipur, Chattisgarh, India;* [3]*Department of Genetics, University of Delhi, New Delhi, India*

**Abstract**

Tomato (*Solanum lycopersicum* L.) is the second most extensively grown vegetable crop worldwide used in raw, cooked, and processed form. Optimum tomato production is limited by biotic and abiotic factors. Viruses cause nearly 45% of the newly discovered plant diseases. Tomato is infected by about 140 virus species. Plants deploy different mechanisms such as dominant resistance, recessive resistance, RNA interference-mediated degradation of viral transcripts, and hormone-mediated resistance to resist virus infections. However, resistance source limitations and continuous virus evolution have highlighted the importance of harnessing host factors for virus resistance. Consequently, engineering these host factors could enhance virus tolerance and help us develop a new generation of virus-tolerant crops. In this chapter, we will discuss the recent advances, achievements, and limitations of genetic engineering for virus resistance in tomato.

**Keywords:** genetic engineering, marker-assisted selection, molecular markers, resistance genes, tomato, virus resistance

## 1.1 Introduction

Tomato (*Solanum lycopersicum*) is a key vegetable crop member of the Solanaceae family. Cultivated tomato has its origin in western parts of Central and South America (Kimura and Sinha, 2008). Its fruit is a rich source of minerals including sodium and potassium (129 mg 100 $g^{-1}$), and antioxidants such as lycopene (351 mg 100 $g^{-1}$), folic acid (30 mg 100 $g^{-1}$) and vitamin C (27 mg 100 $g^{-1}$). It is consumed raw, cooked, or processed as soup and ketchup. Tomato also possesses anti-cancer properties and protects the exposed skin against ultraviolet radiation (Story *et al.*, 2010). The tomato system has been utilized as a model to determine plant-pathogen molecular intricacies (Arie *et al.*, 2007). The International Solanaceae Genome Project (SOL) sequenced the full genome (950 Mb) of cultivated tomato, providing resources for studying disease-resistance methods against various pathogens (The Tomato Genome Consortium, 2012).

Since the late 1990s, outbreaks of various recombinant virus strains with their progressive evolution in the infection of plants have brought the focus of the scientific community on to virus diseases. Widespread dispersion of these viruses has resulted in crop losses globally. Viruses are

*Email: manoj_prasad@nipgr.ac.in


DOI: 10.1079/9781800627123.0001

important pathogens causing major yield losses in vital agronomical and horticultural crops. Viruses in the family Geminiviridae are an economically important group of plant viruses that affect horticultural crops in tropical and subtropical areas throughout the world (Moffat, 1999). Under field conditions, tomatoes are the preferred host for 55 different virus species, among which 26 are probable begomoviruses. Of these virus species, 49 are related to tomato leaf curl disease and 17 are responsible for tomato yellow leaf curl disease. Tomato leaf curl disease has severely damaged the tomato yield in South and South-east Asia (Chakraborty, 2008). Diseases caused by tomato spotted wilt virus (TSWV), tomato yellow leaf curl virus (TYLCV), tomato mosaic virus (ToMV), tomato chlorotic spot virus, tomato ringspot virus, tobacco mosaic virus (TMV), and tomato mottle virus are some of the economically important diseases.

Plant resistance/tolerance against geminiviruses can be linked broadly to resistance loci or quantitative trait loci (QTLs), epigenetic modulation and subsequent RNA silencing, or to varieties developed through genetic engineering to provide viral coat protein (CP)-mediated resistance (Lindbo and Falk, 2017). In natural RNA interference (RNAi), small-interfering RNAs (siRNAs) or microRNAs are generated to target the corresponding RNA target, leading to its degradation or translational inhibition (Chapman and Carrington, 2007). During the course of crop domestication, disease-resistant cultivars were selected from their ancestors and maintained by farmers through preservation of seeds. This preliminary breeding led to an understanding of the genetics of crop resistance through plant disease resistance genes (*R* genes) and other defense molecules. Progress in plant breeding in combination with modern molecular tools and genetic engineering has equipped plant scientists to devise defense strategies for crop survival. When a pathogen breaches the plant defense barrier, it activates a cascade of host-cell responses such as activation of host *R* genes (*R*-gene mediated) or hormonal signaling pathways. Transfer of *R* genes from wild resources and landraces into cultivated lines has been the most exploited strategy of plant resistance to date.

In this chapter, we will discuss the genetic resources and *R* genes available, the key constraints for breeding virus-tolerant tomato plants, and how genetic engineering could help to overcome these constraints.

## 1.2 Genetic Resources Available in Tomato

The sequencing revolution in the early 21st century led to increased availability of genomic information and helped us understand the evolution and genetic architecture of tomato. Sequencing of modern cultivated tomato was initiated by SOL, a consortium of ten countries, launched in the USA in 2003 (Mueller *et al.*, 2005). The tomato genome was chosen over other Solanaceae family members because of its high synteny among the 3000 species. After a decade of effort, the first draft genome sequence of tomato cultivar Heinz 1706 (*S. lycopersicum*) along with that of its closest wild relative accession, LA1589 (*Solanum pimpinellifolium*) were published in 2012 (The Tomato Genome Consortium, 2012). Since this first publication, the tomato genome has been corrected and reannotated with new available sequencing data and now is available as SL 3.0 (genome version) and ITAG 3.2 (RNA sequence data). The advent of next-generation sequencing (NGS) with reduced costs led to resequencing of several tomato accessions (Goodwin *et al.*, 2016).

Publication of the genome of a stress-tolerant species of tomato (*Solanum pennellii*) enhanced our understanding of its drought tolerance and morphology (Bolger *et al.*, 2014). Several stress-related candidate genes were mapped in this wild cultivar. Expression level variation among *S. pennellii* LA716 and *S. lycopersicum* 'M82' was observed owing to promoter and coding region sequence polymorphisms. These wild species of tomato were resequenced with other sequencing platforms for long-read sequencing (Usadel *et al.*, 2017). Similarly, a panel of tomato accessions composed of wild species was sequenced (The 100 Tomato Genome Sequencing Consortium, 2014). This project suggested diversity in terms of heterozygosity present in wild and cultivated accessions. In the same year, resequencing of 360 different tomato accession genomes revealed the breeding path from wild to cultivated tomato (Lin *et al.*, 2014). This resequencing also led to interesting findings about two independent QTLs controlling fruit size. To date, almost 900 different tomato accessions have been resequenced by different sequencing platforms, generating a huge amount of genetic information for tomato improvement freely available at https://solgenomics.net (accessed 13 June 2024) (Tieman *et al.*, 2017; Ye *et al.*, 2017; Tranchida-Lombardo *et al.*, 2018). In addition, more than 83,000 tomato accessions are preserved in various genetic resource centers and seed banks worldwide (Table 1.1) (FAO, 2015).

### 1.2.1 *R* genes

During the course of evolution, natural variations caused by mutations have resulted in enhanced adaptability of plants. Introgression of these mutations in modern cultivars for disease resistance from wild cultivars mainly involves *R* genes (Bauchet *et al.*, 2017a; Rothan *et al.*, 2019). Nevertheless, the number of cloned *R* genes with functional validation is inadequate. Induced mutations through the technique of targeting induced local lesions in genomes (TILLING) offers mutant collections in particular accessions for the discovery of novel alleles (Mazzucato *et al.*, 2015; Gauffier *et al.*, 2016). Several mutant collections in tomato have been generated through TILLING worldwide (Table 1.1). These collections have been helpful in identification of resistance loci through molecular markers. To date, a total of 26 major *R* genes from tomato have been identified by molecular marker association with QTLs or genes (Table 1.2).

Resistance loci in tomato are named according to the following rules: (i) they are written in italic; (ii) they are abbreviated to one to three letters consisting of uppercase or lowercase letters based on dominant and recessive alleles; (iii) they are separated by a dash followed by a number, where the number indicates the order of their discovery. In order to separate different alleles, the last number is followed by dot and a number, or by a superscript letter or number. The majority of these loci are dominant in nature, except for *ty-5*, *ol-2*, and *pot-1* (Ruffel *et al.*, 2005; Bai *et al.*, 2008; Lapidot *et al.*, 2015). Another recessive allele for disease resistance is *py-1* against corky root rot caused by *Pyrenochaeta lycopersici*, which has not been cloned to date (Doganlar *et al.*, 1998).

In tomato diseases, major *R* genes/QTLs have been identified from various pathogen variants and *R*-gene sources. For example, six dominant and three recessive *R* genes were identified for TSWV (Foolad and Panthee, 2012). Some of these genes have been used in marker-assisted selection (MAS), such as *Mi-1.2* and *Sw-5* for broad-spectrum resistance to plant diseases. Some of these genes correspond to an allelic series, for example *Ty-1* and *Ty-3* on chromosome 6, or *Tm-2* and *Tm-22* on chromosome 9 (Lanfermeijer *et al.*, 2005; Verlaan *et al.*, 2013). These genes are also present as tightly linked loci, such as *Prf* and *Pto* on chromosome 5 (Salmeron *et al.*, 1996). Additionally, they belong to clusters of major resistance loci/genes, for example *Cf-4* and *Cf-9*, or *Cf-2* and *Cf-5* present on chromosomes 1 and 6, respectively

**Table 1.1.** List of important tomato genomic information resources (all accessed 13 June 2024).

| Genetic resource | Country | Website/reference |
|---|---|---|
| Tomato Genetic Resources Center (TGRC) | USA | https://tgrc.ucdavis.edu/ |
| US Department of Agriculture (USDA) | USA | www.ars.usda.gov/ |
| The World Vegetable Center | Taiwan | https://avrdc.org/ |
| Centre for Genetic Resources | Netherlands | www.wur.nl/en/Research-Results/Statutoryresearch-tasks/Centre-for-Genetic-Resources-the-Netherlands-1.htm |
| UCD Genome Center TILLING laboratory | USA | https://comailab.org/tilling/ |
| The Microtom collection | | Okabe *et al.* (2011) |
| TOMATOMA database | Japan | http://tomatoma.nbrp.jp/ |
| The Repository of Tomato Genomics Resources | India | https://lifesciences.uohyd.ac.in/images/index.html |
| The Genes That Make Tomatoes | USA | http://zamir.sgn.cornell.edu/mutants/index.html |
| The Tilling Platform of Tomato, INRA | France | www-urgv.versailles.inra.fr/tilling/tomato.htm |
| LycoTILL database, Metapontum Agrobios | Italy | www.agrobios.it/tilling/ |

**Table 1.2.** List of important resistance genes/quantitative trait loci discovered in tomato.

| Resistance gene/locus | Chromosome no. | Resistance against | *Solanum* sp. | Reference |
|---|---|---|---|---|
| *Asc-1* | 3 | *Alternaria alternata* | *S. pennellii* | Brandwagt *et al.* (2000) |
| *Bs-4* | 4 | *Xanthomonas campestris* pv. *vesicatoria* | *S. lycopersicum* | Schornack *et al.* (2004) |
| *Cf-2* | 6 | *Cladosporium fulvum* | *S. pimpinellifolium* | Dixon *et al.* (1996) |
| *Cf-4* | 1 | *Cladosporium fulvum* | *S. habrochaites* | Takken *et al.* (1999) |
| *Cf-5* | 6 | *Cladosporium fulvum* | *S. lycopersicum* | Dixon *et al.* (1998) |
| *Cf-9* | 1 | *Cladosporium fulvum* | *S. pimpinellifolium* | Jones *et al.* (1994) |
| *Hero* | 4 | *Globodera rostochiensis* | *S. pimpinellifolium* | Ernst *et al.* (2002) |
| *I* (*I-1*) | 11 | *Fusarium oxysporum* | *S. pimpinellifolium* | Catanzariti *et al.* (2017) |
| *I-2* | 11 | *Fusarium oxysporum* f. sp. *lycopersici* | *S. pimpinellifolium* | Simons *et al.* (1998) |
| *I-3* | 3 | *Fusarium oxysporum* f. sp. *lycopersici* | *S. pennellii* | Catanzariti *et al.* (2015) |
| *I-7* | 8 | *Fusarium oxysporum* f. sp. *lycopersici* | *S. pennellii* | Gonzalez-Cendales *et al.* (2016) |
| *Mi-1.2/Mi/Meu* | 6 | *Meloidogyne incognita*, *M. arenaria*, *M. javanica* | *S. peruvianum* | Casteel *et al.* (2007) |
| *ol-2/SlMlo1* | 4 | *Oidium neolycopersici* | *S. lycopersicum* | Bai *et al.* (2008) |
| *Ph-3* | 9 | *Phytophthora infestans* | *S. pimpinellifolium* | Zhang *et al.* (2014) |
| *pot-1* | 3 | Potyviruses (potato virus Y and tobacco etch virus) | *S. habrochaites* | Piron *et al.* (2010) |
| *Prf* | 5 | *Pseudomonas syringae* pv. *tomato* | *S. pimpinellifolium* | Salmeron *et al.* (1996) |
| *Pto* | 5 | *Pseudomonas syringae* pv. *tomato* | *S. pimpinellifolium* | Martin *et al.* (1993) |
| *Tm-1* | 2 | Tomato mosaic virus | *S. habrochaites* | Ishibashi *et al.* (2007) |
| *Tm-2* | 9 | Tomato mosaic virus | *S. peruvianum* | Lanfermeijer *et al.* (2005) |
| *Tm-2²/Tm-2.2/Tm-2ª* | 9 | Tomato mosaic virus | *S. peruvianum* | Lanfermeijer *et al.* (2005) |
| *Ty-1* | 6 | Tomato yellow leaf curl virus | *S. chilense* | Verlaan *et al.* (2013) |
| *Ty-2* | 11 | Tomato yellow leaf curl virus | *S. habrochaites* | Yamaguchi *et al.* (2018) |
| *Ty-3* | 6 | Tomato yellow leaf curl virus | *S. chilense* | Verlaan *et al.* (2013) |
| *ty-5* | 4 | Tomato yellow leaf curl virus | *S. peruvianum* | Lapidot *et al.* (2015) |
| *Ve-1/Ve* | 9 | *Verticillium dahliae* | *S. lycopersicum* | Fradin *et al.* (2009) |
| *Sw-5* | 9 | Tomato spotted wilt virus | *S. peruvianum* | Brommonschenkel *et al.* (2000) |

(Dixon *et al.*, 1998; Takken *et al.*, 1999). In tomato, these *R* genes sometimes provide broader or non-specific pathogen resistance; for example, *Mi-1.2/Meu* provides resistance against root knot nematode as well as against whitefly (*Bemisia tabaci*), the psyllid *Bactericera cockerelli*, and the aphid *Macrosiphum euphorbiae* (Milligan *et al.*, 1998; Rossi *et al.*, 1998; Vos *et al.*, 1998; Nombela *et al.*, 2003).

Major *R* genes/qualitative resistance are often subjected to frequent breakdown by virulent pathogens and needed to be stacked for broader and longer-term protection. For example, the late blight resistance gene *Ph-1* became ineffective due to the emergence of more virulent strains of *Phytophthora infestans*. This led researchers to shift their focus to new genetic resources such as quantitative resistance/partial resistance (Cowger and Brown, 2019). The frequency of available resistance QTLs in natural genetic resources has also played an important role in this shift. Several resistance QTLs were mapped in the tomato genome (Table 1.3). A search for dominant susceptibility genes responsible for pathogen multiplication could be a useful strategy for resistance breeding through mutations performed by TILLING. For example, this approach led to resistance against potyviruses in tomato (Piron *et al.*, 2010). Similarly, using natural variation in tomato populations by ecoTILLING, a new *Sw-5* variant was detected providing resistance against TSWV (Belfanti *et al.*, 2015).

### R *genes and QTL architecture*

In the tomato genome, various hotspots for *R* genes have been observed by mapping of resistance loci. Mapping of the major *R* genes in tomato suggested their tandem and clustered organization (Foolad, 2007). Several resistance loci from wild relatives or the same identified genitor have been mapped on chromosomes 6 and 9. For example, chromosome 6 carries *R* genes for alfalfa mosaic virus (*Am*), *Cladosporium fulvum* (*Cf-2* and *Cf-5*), *Meloidogyne incognita* (*Mi-1.2*), *O. neolycopersici* (*Ol-1*, *Ol-3*, *Ol-4*, *Ol-5*, and *Ol-6*), and TYLCV (*Ty-1* and *Ty-3*), with resistance QTLs for ToMoV and *Ralstonia solanacearum* (Agrama and Scott, 2006). Similarly, *R*-gene clusters of *Tm-2* and *Tm-22* controlling ToMV, *Sw-5* controlling TSWV, *Ve* controlling *Verticillium dahlia*, *Ph-3* controlling *P. infestans*, and *Frl* controlling *Fusarium oxysporum* f. sp. *radicis-lycopersici* were found on chromosome 9 (Stevens *et al.*, 1995; Pillen *et al.*, 1996; Vakalounakis *et al.*, 1997; Kawchuk *et al.*, 2001; Chunwongse *et al.*, 2002).

### *Molecular basis of* R *genes and QTLs*

Major *R* genes often encode proteins to recognize avirulent proteins of pathogens directly/indirectly that trigger plant immune responses. Recessive resistance alleles evolved due to loss-of-function mutations that hampered the pathogen's ability to infect the plants, such as the single recessive

**Table 1.3.** Resistance QTLs identified against specific pathogens in tomato.

| Pathogen | *Solanum* sp. resistance derived from | Chromosome no. | Reference(s) |
|---|---|---|---|
| *Phytophthora infestans* | *S. habrochaites* (accession LA1777) | 6 | Arafa *et al.* (2017) |
| | *S. hirsutum* | 12 | Brouwer *et al.* (2004) |
| | *S. hirsutum* | 12 | Brouwer and St Clair (2004) |
| | *S. habrochaites* | 7, 8, 12 | Li *et al.* (2011) |
| | *S. pimpinellifolium* (accession PI 163245) | 2, 3, 10, 11 | Ohlson *et al.* (2018) |
| | *S. pimpinellifolium* | 9, 10 | Panthee *et al.* (2017) |
| *Oidium lycopersici* | *S. neorickii* | 6, 12 | Bai *et al.* (2003) |
| *Alternaria solani* | *S. hirsutum* | 8, 11 | Foolad *et al.* (2002) |
| *Alternaria alternata* | *S. cheesmaniae* | 2 | Robert *et al.* (2001) |
| *Xanthomonas* sp. | *S. pimpinellifolium* | 5, 11 | Sim *et al.* (2015) |
| *Clavibacter michiganensis* | *S. hirsutum* | 2, 5 | Coaker and Francis (2004); Kabelka *et al.* (2002) |
| *Ralstonia solanacearum* | *S. lycopersicum* 'Hawaii 7996' | 6, 12 | Wang *et al.* (2013) |
| *Botrytis cinerea* | *S. lycopersicoides* | 1 | Davis *et al.* (2009) |
| Cucumber mosaic virus | *S. chilense* | 12 | Stamova and Chetelat (2000) |
| Tomato yellow leaf curl Thailand virus Taiwan strain | *S. chilense* | 4, 6, 10, 11 | Kadirvel *et al.* (2013) |

genes *pot-1* and *ol-2*. These susceptibility factors can often be mutated using modern biotechnological tools. Cloning of major *R* genes has provided structural information about their proteins. These proteins encode, for example, conserved nucleotide-binding site–leucine-rich repeat (NBS-LRR), receptor-like protein or receptor-like kinase domains that have been helpful in identification of homologous genes in the same or related species. For example, *Mi* and *Sw-5* are homologs of the *Cf* series of genes. Similarly, using the RenSeq (resistance gene enrichment sequencing) strategy, 105 and 355 novel NBS-LRR genes were identified in *S. lycopersicum* Heinz 1706 and *S. pimpinellifolium* LA1589 (Andolfo *et al.*, 2014). As well as these major *R* genes, many other genes have been characterized for tomato disease responses that are involved in plant–pathogen interactions by triggering defense mechanisms through *R* genes, such as the lipase-like protein EDS1. Similarly, the *Bs-3*, *I-2* and *Prf* proteins interact with HSP90, SGT-1 and RAR1 proteins for their actions. Some of these proteins regulate phytohormones such as jasmonic acid and salicylic acid. *R* genes corresponding to QTLs in tomato have not yet been cloned. These QTLs are predicted to work through genes operating in pathogen-associated molecular pattern (PAMP) recognition for basal defense, signal transduction pathway genes, phytoalexin production genes, and other unknown genes directly or indirectly regulating plant defense (Poland *et al.*, 2009).

### 1.2.2 Novel means of *R*-gene/QTL identification

The battle between pathogens and plants needs constant identification of novel resources for *R* gene/QTL discovery. One such strategy is whole-genome sequencing of bulked $F_2$ populations obtained by crossing tolerant and susceptible parents. This strategy offers identification of single-nucleotide polymorphisms and mutations created with chemical mutagens such as ethyl methanesulfonate associated with targeted resistance phenotypes by comparison with the reference genome (Garcia *et al.*, 2016). Limitations of genetic diversity in two-parent crossings were overcome by multi-parent advanced generation intercross (MAGIC) populations (Pascual *et al.*, 2015). For example, a MAGIC population was recently used for identification of water-deficit and salinity QTLs (Diouf *et al.*, 2018).

### 1.2.3 Wild cultivars

Tomato belongs to the Solanaceae family within the Lycopersicon clade, which consists of modern domesticated tomato (*S. lycopersicum*) plus 12 different wild species (Peralta *et al.*, 2005). One important genetic resource for tomato breeding is wild germplasm for disease resistance and other stress-related traits (Rick and Chetelat, 1995; Sharma *et al.*, 2017). The major *R* genes (~40) have mainly been discovered from wild cultivars of tomato against a variety of pathogens. Of these discovered genes, 20 have been incorporated into modern cultivars (Ercolano *et al.*, 2012). The main sources of these genes were *Solanum habrochaites*, *S. peruvianum*, *S. chilense*, and *S. pimpinellifolium* (Laterrot, 2000). Screening of these germplasms, which are available from various genetic centers, will open up avenues for discovery of more resistance alleles/QTLs.

## 1.3 Breeding for Resistance to Viruses

The history of tomato breeding dates back to the 1940s for phenotypic selection of various traits for crop improvement and disease resistance. However, we are increasingly facing novel pathogen and pest challenges for tomato production from various parts of world. The advent of NGS, combined with molecular biology and breeding, will expedite the process of identification and introgression of disease-resistant loci through breeding. Incorporation of resistance in tomato cultivars using phenotypic, enzymatic, and DNA markers linked to disease-resistant loci using MAS has been practiced since the early 1990s (Laterrot and Moretti, 1989; Aarts *et al.*, 1991; Messeguer *et al.*, 1991). MAS increased the efficiency of recessive gene resistance or genes having low penetrance, or environmentally influenced genes. Breeding through the use of MAS also facilitated the introgression and pyramiding of resistance loci due to the clustered nature of these genes in the tomato genome. For example, in *Tm-22* tomato cultivars, the *Frl* gene was also expressed providing resistance against *Fusarium* crown and root rot (Foolad and Panthee, 2012). However, repulsion-phase linkage has hindered breeding selection, as shown for *Ph-3* and *Sw-5* association (Robbins *et al.*, 2010). MAS has proved to be instrumental for tomato improvement, particularly for disease resistance, over the decades and has potential for future generations of tomato by integrating with modern molecular methods.

Currently, molecular DNA markers are available for more than 30 resistance loci important for tomato breeding (Foolad and Panthee, 2012), although their availability and use is less because of their complex inheritance and quantitative nature. Successful examples of MAS are for major *R* genes such as *Tm-22*, *I*, *I-2*, *I-3*, *Asc*, *Pto*, *Mi-1.1/Mi1.2*, *Sw5*, and *Sm*, providing resistance against various pathogens. Similarly, the *Ty* series of marker genes has been used to produce cultivars resistant to TYLCV (Dhaliwal *et al.*, 2020). The markers identified for disease resistance are not polymorphic for breeding populations that originated from intraspecific crosses. The advent of NGS solved this problem with the discovery of a large number of trait-associated markers (single-nucleotide polymorphisms) in a population. NGS, along with allele mining, genome-wide association studies and ecoTILLING, has proved to be highly important for discovering resistance-related markers from available genetic resources. In conclusion, tomato breeders can use a combination of these techniques with superior genotype combinations for disease-resistance trait association and for designing elite modern cultivars of tomato.

## 1.4 Genetic-engineering Strategies for Virus Tolerance

Conventional techniques for virus control range from cultural practices to cultivation of resistant crop varieties for economic virus disease management (Kyle, 1993). Breeding for virus-resistant crops is limited because of the narrow range of directly associated *R* genes with their recessive, dominant or co-dominant, and quantitative nature (Whitham and Hajimorad, 2016). Continuous evolution of virus decreased the resistance of these limited resources integrated into cultivated tomato cultivars. This prompted scientists to look for other resources involved in virus replication and movement inside the host as a target. An understanding of plant virus biology, host and insect interactions, pathogenesis, and resistance-associated mechanisms has helped to determine new avenues for plant virus resistance (Cillo and Palukaitis, 2014).

Genetic engineering for virus resistance in plants is a highly attractive and successful approach with intact agronomic traits remaining intact in the cultivar. Plant virus resistance is broadly classified into two categories, plant derived and pathogen derived. Plant-derived virus resistance consists of resistance sources such as *R* genes, protease inhibitors, ribosome-inactivating proteins, RNA-modifying enzymes, dsRNases and single-chain variable fragment (scFv) antibodies. Pathogen-derived resistance results from viral CPs, movement proteins (MPs), replicase, non-coding RNA sequences, defective-interfering RNA and viral proteases (Cillo and Palukaitis, 2014). Along with these traditional approaches, new techniques such as genome editing using clustered regularly interspaced palindromic repeats (CRISPR)/CRISPR-associated protein 9 (Cas9) have also been used for transgenic plant development. We will discuss transgenic approaches used for virus resistance development in this section.

### 1.4.1 Plant-derived resistance

#### *Host* R *genes*

In the battle between plants and viruses, plants have evolved several molecular mechanisms to counter virus attacks. Most of these molecular responses act as a basal defense, although some are specific responses (Dangl and Jones, 2001). Genes regulating multiple pathways also play a role in plant virus resistance. The most common example of a host-encoding factor is the *R* gene responsible for the hypersensitive reaction (HR); this R protein interacts with the viral avirulence gene (*Avr*) product inducing programmed cell death. Further *R* genes were classified into three categories based on their inheritance as dominant or recessive genes and their structure

(Maule *et al.*, 2007; Palukaitis and Carr, 2008). *R* genes act via recognition of effector proteins from the pathogen following the gene-for-gene hypothesis. Structurally, *R* genes comprise a NBS-LRR domain with a coiled-coil or Toll–interleukin receptor (TIR) domain at their N-terminal end (Dangl and Jones, 2001; Chisholm *et al.*, 2006). The first *R* gene was isolated and cloned from *Nicotiana glutinosa* against TMV and named the *N* gene. *N* genes are composed of TIR-NBS-LRR domains. Overexpression of these *N* genes produces the HR phenotype in tomato plants and provides resistance against TMV and ToMV at the site of infection. Tomato varieties were also introgressed with some *R* genes such as *Sw-5* and *Tm-22* to provide resistance against TSWV and ToMV, respectively. Overexpression of non-viral *R* genes such as *Prf* can also provide resistance against TMV by activating systemic acquired resistance responses in tomato plants (Oldroyd and Staskawicz, 1998). Recessive *R* genes evolved due to mutations in susceptibility factors indispensable for successful virus infection. However, plants have to be homozygous to display such resistance. One of the classic examples of a recessive *R* gene is eukaryotic translation initiation factors (eIFs). These genes are effective against potyviruses by inhibiting factors involved in replication and translation, such as eIF4E and eIF4G (Robaglia and Caranta, 2006; Maule *et al.*, 2007). The third category of host genes for virus resistance includes various transcription factors that are known to be master controllers of gene modulation (Dai *et al.*, 2008). Their role in the modulation of defense responses against plant viruses has been explored (Pandey *et al.*, 2018). Overexpression of these transcription factors resulted in broad-spectrum resistance against bacteria and viruses (Cao *et al.*, 2005). These are the prominent candidates for further exploration for virus resistance in tomatoes.

Other host genes involved in defense pathways are related to virus replication, translation, and movement. For example, antisense inhibition of *S*-adenosylhomocysteine hydrolase (SAHH) critical for 5′-end capping of viral RNAs during replication provides resistance (Masuta *et al.*, 1995). Overexpression of transgenic SAHH results in abnormal phenotypes due to increased cytokinin activity. Similarly, overexpression of carbohydrate-binding lectin proteins provides resistance against plant viruses (Yamaji *et al.*, 2012). The presence of lectins in most plants makes them a good candidate for virus resistance. Lectins are known to be induced by biotic and abiotic cues with a role in plant defense (Lannoo and van Damme, 2014). Overexpression of lectins such as RTM1 and JAX1 (jacalin-type lectin) provided resistance against potyviruses and potexviruses (Chisholm *et al.*, 2000; Yamaji *et al.*, 2012). Recently, RNA helicases have been reported to provide tolerance against geminiviruses and potyviruses (Pandey *et al.*, 2019, 2020a). These chaperones are important for plant virus replication in the case of potyviruses, whereas the mechanism of tolerance against geminiviruses needs to be explored further.

### *Ribosome-inactivating proteins*

Another important host factor for virus resistance is ribosome-inactivating proteins (RIPs). RIPs are *N*-glycosides involved in removal of the purine base from 28S rRNA, rendering it non-functional for protein translation. Overexpression studies of one RIP, pokeweed antiviral protein (PAP), displayed broad-spectrum resistance against potato virus X (PVX) and potato virus Y (PVY) in tobacco and potato crops, and against turnip mosaic virus in brassicas (Lodge *et al.*, 1993; Zhang *et al.*, 1999). Because of detrimental phenotypic effects of this overexpression, various mutants of PAP proteins were used for the production of transgenic plants. In the same way, overexpression of RIP from *Phytolacca insularis* and of the RIP dianthin from *Dianthus caryophyllus* provided resistance against PVX, PVY and potato leaf roll virus (PLRV) in potato and against African cassava mosaic virus in *Nicotiana benthamiana*, respectively (Hong *et al.*, 1996; Moon *et al.*, 1997). These proteins are thought to have direct effects on viral RNAs, although the mechanism of the defense response is yet to be elucidated (Wang and Tumer, 2000). A major problem with this type of host resistance is that consistency at the level of resistance varies among plant species and different viruses, limiting their application at the commercial level.

### *Plantibodies*

Plants do not have an immune system to produce antibodies, but the idea is to use antibodies produced in animal systems for virus treatment. Specific antibodies overexpressed in plants have a high affinity that can disrupt viral proteins. ScFv antibodies and synthetic scFv libraries along with the phage display approach have helped identification and production of specific plantibodies (Ziegler and Torrance, 2002). Both monoclonal and polyclonal antibodies have been shown to be stably overexpressed in plants (Düring *et al.*, 1990; Firek *et al.*, 1993). The first report of this technique successfully demonstrated their use in *N. benthamiana* against artichoke mottled crinkle virus (Tavladoraki *et al.*, 1993). However, significant progress could not be made for a number of reasons including reduced levels of resistance, the stability of the overexpressed protein in the plant, ambiguity around selection of the target fragment in the virus, and level of scFv expression required (Nölke *et al.*, 2004). These problems were resolved with technological advancements in terms of their stability and targeting, as demonstrated against replication-associated proteins and replicase proteins of viruses (Nickel *et al.*, 2008; Gil *et al.*, 2011).

#### *Host-mediated resistance/systemic acquired resistance*

Pathogen infection in plants results in an HR response that initially restricts the pathogen at the site of infection and consequently becomes systemic. This type of response is collectively known as systemic acquired resistance, and peculiarly activates a cluster of genes at the same time in distal plant parts. Initiation of this type of response leads to accumulation of salicylic acid (SA), which in turn increases the transcription of pathogenesis-related (PR) proteins. This activation cascade leads to an increase in phenolic compounds through activation of the phenylpropanoid pathway and reactive oxygen species accumulation, which favors the deposition of lignin and suberin in the cell wall. Systemic acquired resistance has been shown to be initiated in the plant against virus attack. For example, tobacco overexpression of bacterial salicylate hydrolase makes plants susceptible to TMV because of decreased endogenous SA levels (Delaney *et al.*, 1994). SA acts by binding with catalase to enhance hydrogen peroxide levels in plants. Silencing of catalase genes in transgenic tobacco leads to chlorosis of the lower leaves of the plant (Takahashi *et al.*, 1997). These plants also showed higher accumulation of SA and PR expression, providing resistance against TMV.

### 1.4.2 Pathogen-derived resistance

#### *CP-mediated resistance*

Plant viruses are composed of nucleic acid and CP. The concept of pathogen-derived resistance originated from the possible role of CP in cross-protection (Sanford and Johnston, 1985). This concept was proven afterwards in tobacco, where overexpression of the CP gene provided resistance against TMV infection (Abel *et al.*, 1986). These transgenic plants showed reduced infection with delayed symptoms appearance, which was attributed to targeting of the virions rather than the uncoated nucleic acid of TMV (Nelson *et al.*, 1987; Register and Beachy, 1988). These results also suggested that uncoating, the initial step of virus infection, was affected by CP overexpression in the host plant. Further research in this direction confirmed the reduction of virion disassembly in transgenic plants. A mutated version of CP from TMV enhanced CP–CP aggregation, which led to enhanced resistance in transgenic plants (Asurmendi *et al.*, 2004). When the CP was overexpressed through protoplasts, MP expression was reduced along with formation of replication complexes from TMV that correlated with a delay in the appearance of virus infection symptoms (Asurmendi *et al.*, 2004; Bendahmane *et al.*, 2007). Other processes related to virus infection can also be affected by CP overexpression in host plants and provide resistance. All of these combined phenomena are known as 'coat protein-mediated resistance' (CPMR) and have led to the development of various virus-resistant transgenic plants (Beachy, 1999). Interestingly, these transgenic studies showed that the level of resistance is not directly correlated with viral CP expression (Lawson *et al.*, 1990; Quemada *et al.*, 1991).

The first commercialized transgenic plant for virus resistance in the USA was squash (*Cucurbita pepo*) resistant to potyvirus, which was produced using the CPMR strategy (Tricoll *et al.*, 1995). This transgenic is a successful example of transgenic application for multi-virus resistance and improved economic gains (Lindbo and Falk, 2017). A similar strategy was applied in potato crops for PVY resistance (Kaniewski *et al.*, 1990). Another crop where the development of trangenics was successful is papaya, where the CPMR strategy was used to develop virus-resistant transgenic plants (Fuchs and Gonsalves, 2007; Fitch, 2016). Virus-resistant transgenic plants are successfully grown in the USA, Brazil, and China (Tollefson, 2011; James, 2017).

#### *Resistance through heterologous virus gene expression*

VIRUS RESISTANCE THROUGH THE REPLICASE. Following the success of CPMR, scientists tried to explore other viral gene sources that could be used for pathogen-derived resistance against plant viruses. Thus, they overexpressed the RNA-dependent RNA polymerase or replicase protein of TMV in tobacco, which provided increased resistance (Golemboski *et al.*, 1990). This type of resistance was termed 'replicase-mediated resistance' and was used against pea early browning virus, cucumber mosaic virus, and PVY (Anderson *et al.*, 1992). The replicase gene used in this type of resistance could be full length, truncated, or other mutant versions, giving resistance against various viruses. This suggested that replicase-mediated resistance does not require functional protein but acts via post-transcriptional gene silencing via RNA-mediated resistance (Marano and Baulcombe, 1998). This resistance is narrow spectrum and less effective in comparison with CPMR. Some DNA virus families such as geminiviruses do not encode polymerase and depend on their replication-associated protein (Rep) interaction with host-encoded polymerases for genome replication. Here, transgenic overexpression of the Rep gene inhibits replication and decreases the viral DNA. For example, *Nicotiana* plants transgenic for Rep showed a direct correlation between Rep transcripts and resistance against African cassava mosaic virus (Hong and Stanley, 1996). Interestingly, a truncated version of the Rep protein was highly effective at inhibiting virus replication in protoplasts. However, when the NTP-binding domain was mutated and expressed, it did not affect the replication of tomato yellow leaf curl Sardinia virus but inhibited viral Rep expression by non-functional Rep protein complex formation (Noris *et al.*, 1996; Bruneti *et al.*, 1997; Lucioli *et al.*, 2003).

VIRUS RESISTANCE THROUGH MPS. In the infection cycle, plant viruses require cell-to-cell movement of virions or ribonucleoprotein complexes, which is accomplished through specific proteins known as MPs. These proteins interact with plasmodesmata and allow intracellular movement of virus particles. MPs have a nucleic acid-binding domain that binds with viral nucleic acid, and other domains required for plasmodesmata modification. Mutations in the nucleic acid-binding domain do not affect their interaction with plasmodesmata but do affect viral DNA/RNA trafficking through the plasmodesmata. This strategy has been used to achieve broad-spectrum resistance against plant viruses; for example, MP from TMV, when overexpressed in tobacco plants, provided resistance against TMV and other tobacco-infecting viruses (Lapidot *et al.*, 1993; Malyshenko *et al.*, 1993; Cooper *et al.*, 1995). Similar results were observed when MP from PVX and white clover mosaic virus was expressed in potato and *Nicotiana benthamiana* plants, respectively (Beck *et al.*, 1994; Cooper *et al.*, 1995). In some cases, this resistance could be narrow spectrum, for example MP expression of PVX provided resistance only for the targeted virus (Kobayashi *et al.*, 2001). Some virus families such as Caulimoviridae and Comoviridae do not encode plasmodesmata-modifying proteins; instead, they encode a protein able to form tubules for their movement to adjacent cells. Transgenic expression of MP from PLRV provides resistance for an array of virus infections including PVY, PVX, and PLRV (Tacke *et al.*, 1996). Truncated MPs from begomoviruses (family Geminiviridae) provide resistance against heterologous and homologous virus infections (Duan *et al.*, 1997; Hou *et al.*, 2000).

### 1.4.3 RNA-mediated virus resistance

Plant virus resistance in the last decade of the 20th century shifted focus from protein-based resistance to RNA-based resistance. Discovery of RNAi in this era also provided a much-needed impetus in the area of plant virus resistance. This kind of resistance is governed by transgenically expressing target virus sequences in a sense or antisense orientation. This was first demonstrated by overexpressing tobacco etch virus non-translatable CP sequence in tobacco (Lindbo and Dougherty, 1992a,b). Molecular analysis of these plants suggested a higher level of transgene RNAs with a reduction in tobacco etch virus replication (Lindbo *et al.*, 1993; Dougherty *et al.*, 1994). The post-transcriptional gene silencing suggested to govern this type of resistance is a sequence-specific RNA-targeting phenomenon (Dougherty and Parks, 1995). RNA-mediated resistance was also tried with artificial defective-interfering RNAs and satellite RNAs (Kim *et al.*, 1997; Patil and Dasgupta, 2006). As our understanding of the phenomenon of RNAi increased, small RNAs such as siRNAs, short-hairpin RNAs, artificial microRNAs and *trans*-acting siRNAs were also utilized for plant virus resistance through transcriptional or post-transcriptional gene silencing (Lindbo and Dougherty, 2005). An understanding of RNA-mediated resistance explained the CPMR in papaya and plum transgenic plants (Scorza *et al.*, 2001; Tennant *et al.*, 2001). Replicase-mediated resistance could also be explained through RNAi (Vazquez *et al.*, 2001; Prins *et al.*, 2008).

Several transgenic plants have been produced using RNAi by expression of hairpin RNA/dsRNA (Waterhouse and Fusaro, 2006; Mlotshwa *et al.*, 2008). Here, the host gene-silencing machinery activated and restricted the plant RNA or DNA virus infection. However, evolution of RNA-silencing suppressor protein in viruses counteracted the induced silencing (Bisaro, 2006). Therefore, dsRNA expression was found to be highly effective for RNAi induction because instant production of siRNA triggered RNAi in the absence of virus (Fusaro *et al.*, 2006). RNA-mediated resistance and RNAi are already used to produce transgenic plants resistant to watermelon mosaic virus, zucchini yellow mosaic virus, cucumber mosaic virus, PVX, PVY, PLRV, plum pox virus, papaya ring spot virus, and other geminiviruses (Tricoll *et al.*, 1995; Thomas *et al.*, 2000; Klas *et al.*, 2006; Kundu *et al.*, 2008; Krubphachaya *et al.*, 2007; Kumar *et al.*, 2017). Similarly, artificial microRNAs against cucumoviruses, begomoviruses, potexviruses, tospoviruses, and ipomoviruses have provided plant virus resistance (Qu *et al.*, 2007; Duan *et al.*, 2008; Zhang *et al.*, 2011; Ai *et al.*, 2013; Miter *et al.*, 2016; Wagaba *et al.*, 2016). Application of these dsRNA molecules in nanosheets coated by spraying presents an attractive strategy for virus control (Mitter *et al.*, 2017a,b). Several other gene-silencing phenomena such as spray-induced gene silencing are in development that can be used for virus resistance at the field level (Pandey *et al.*, 2020b).

### 1.4.4 Genome editing as a tool for engineering virus resistance

Genome editing is one of the most revolutionary molecular biology techniques of the modern era. In genome editing, alleles can be introduced for the desired mutations in plants (Hsu *et al.*, 2014). These techniques include the use of protein–DNA recognition-based zinc-finger nucleases, transcription activator-like effector nucleases (TALENs), and RNA/DNA base pairing with protospacer-adjacent motif sequence recognition-based CRISPR-Cas9 (Sood *et al.*, 2021). These editing tools edit the genome and introduce double-stranded breaks, which are repaired by non-homology end joining, which is error prone, or by homology-directed repair pathways. These repairs result in the introduction of insertions/deletions or base substitution, which in turn produces knockout or precise editing. Among these techniques, CRISPR-Cas9 is the most popular for crop genome editing for knockout mutations because of its recognition precision and minimal off-target effects. The first case of CRISPR-based editing in tomato was reported in 2014 (Brooks *et al.*, 2014). This tool has since been used to characterize several developmental traits. For example, in tomato, RIN (ripening inhibitor) is known to be important for ripening induction. CRISPR-based editing of this gene showed tomato edited lines with moderate red color and an inability to ripen (Ito *et al.*, 2017). Additionally, this study showed that this is a gain-of-function mutation for the ripening repressor protein in contrast to the popular notion of RIN being the null mutation.

In tomato, various diseases lack information on directly associated *R* genes and QTLs in popular cultivars. Following genome sequencing of wild and cultivated tomato, many such genes were discovered in wild germplasms, which are often difficult to breed due to interspecific barriers. Genetic engineering tools such as CRISPR can therefore be used to modulate resistance against plant viruses. For example, ectopic expression/ introgression of the *pvr1* gene from *Capsicum chinense* provides resistance against potyviruses in tomato (Kang *et al.*, 2007).CRISPR/Cas9 was recently used for targeting TYLCV with guide RNA for the CP and replicase genes, which resulted in resistant tomato plants (Tashkandi *et al.*, 2018). Plant virus resistance in the host is often governed by recessive *R* genes/host factors such as eIFs (Palukaitis and Carr, 2008). These factors help virus translation by binding to their 5′ cap or 3′ tail. In particular, the eIF4E protein helps potyvirus protein translation by binding at the 3′ viral mRNA tail (Robaglia and Caranta, 2006). Silencing of eIF4E in tomato plants provided broad-spectrum virus resistance (Chandrasekaran *et al.*, 2016). Genome editing has also been used to control DNA viruses (TYLCV) in tomato by CRISPR-Cas9 (Ali *et al.*, 2015). With the application of an array of guide RNAs targeting the whole viral genome, CRISPR-Cas9 could be used to control a diversity of plant viruses infecting tomato. With the fast evolution of CRISPR/Cas9 and other gene-editing tools, new functional genes could be engineered to provide virus resistance in tomato.

## 1.5 Current Status of Tomato Transgenic Plants for Virus Tolerance

Since the release of the first transgenic variety for cultivation 22 years ago, the area of transgenic crops has increased over 100-fold (International Service for the Acquisition of Agri-biotech Applications Report, 2017). The first commercial food variety released was a tomato crop from Calgene with the trade name FLAVR SAVR™. This variety was engineered by RNA silencing of the polygalacturonase enzyme responsible for fruit softening (Kramer and Redenbaugh, 1994). This variety showed increased shelf life and increased resistance to fungal infections. Due to some unexpected results for flavor, this variety was removed from the market in 1999. Many transgenic plants have been registered, but not a single variety has been approved for commercial production. There is no commercially cultivated tomato variety available in the world (www.isaaa.org/gmapprovaldatabase/default.asp; accessed 17 June 2024).

However, silencing via RNAi has been used to manipulate traits such as ripening, increased shelf life, allergen removal, and parthenocarpic tomato fruit (Le *et al.*, 2006; Schijlen *et al.*, 2007). Control of fruit ripening was also achieved by silencing of pectin methylesterase, expansin, β-galactoside, SlNAC1, and pectate lyase genes (Tieman and Handa, 1994; Brummell *et al.*, 1999; Smith *et al.*, 2002; Meng *et al.*, 2016; Uluisik *et al.*, 2016). Similarly, several overexpression studies on biotic and abiotic stresses in tomato have resulted in resistance. Various exogenous and endogenous genes have been overexpressed for plant virus resistance, as discussed in this chapter. Moreover, scientists are pushing these transgenic plants for trials; hopefully by the next decade we will have commercially cultivated tomato varieties enriched in various traits.

Virus-resistant tomato transgenics are possible due to the advent of next-generation genome-editing tools. However, these tools are still in an evolving phase and need further research to realize their full potential, particularly for editing tomato crops for virus resistance. The regulatory challenges of these edited plants also impede the pace of commercialization. Public health concerns have been raised about these crops as they are considered to be genetically modified organisms (GMOs). Scientists and breeders have suggested that genome-edited plants should be considered as mutagenesis events and exempt from the regulations operative on GMOs. In countries like the USA and Canada, genome-edited plants are exempted from GMO regulations and are treated equivalent to plants generated by conventional breeding. However, European countries face similar regulations for genome-edited crops as for GMOs (Callaway, 2018). These regulations may affect genome-edited tomato commercialization worldwide, especially for virus resistance.

## 1.6 Conclusion and Future Perspectives

Tomato is an important vegetable crop adapted across the world in different climatic conditions.

Recently, virus infestation has severely affected the production of tomato. In contrast to available chemical control measures for other pathogens, virus infestation can be controlled by virus-resistant genetic resources created through genetic improvement by plant breeding and genetic engineering. Genome-sequencing consortiums for tomato and wild relatives have helped to discover *R* genes/QTLs and other host factors responsible for providing resistance. These *R* genes are both dominant and recessive in nature, and can be introgressed through plant breeding in modern cultivars. However, only a limited number of such genotypes have resulted in virus resistance without a yield penalty in terms of quality and other agronomic traits. Therefore, genetic-engineering technologies such as RNAi and gene editing through CRISPR-Cas9 are imperative for virus management. Transformation protocols to implement these methods are evolving, particularly for tomato. Targeting multiple viruses for broad-spectrum resistance is another advanced field that is being tried by engineering plant genomes for viral resistance by producing several guide RNAs against viral genomes. Transfer of these technologies to raise virus-resistant crops could help to fulfill the demands of an ever-increasing world population. With the regulatory framework for gene-edited crops becoming more relaxed in countries like the USA and Canada, we can expect early releases of virus-resistant crops in the near future.

## 1.7 References


Aarts, J., Hontelez, J., Fischer, P., Verkerk, R., Vankammen, A., and Zabel, P. (1991) Acid phosphatase-11, a tightly linked molecular marker for root-knot nematode resistance in tomato—from protein to gene, using pcr and degenerate primers containing deoxyinosine. *Plant Molecular Biology* 16, 647–661.

Abel, P.P., Nelson, R.S., De, B., Hoffmann, N., Rogers, S.G., *et al.* (1986) Delay of disease development in transgenic plants that express the tobacco mosaic virus coat protein gene. *Science* 232, 738–743.

Agrama, H.A. and Scott, J.W. (2006) Quantitative trait loci for tomato yellow leaf curl virus and tomato mottle virus resistance in tomato. *Journal of the American Society for Horticultural Science* 131, 267–272.

Ali, Z., Abulfaraj, A., Idris, A., Ali, S., Tashkandi, M., and Mahfouz, M.M. (2015) CRISPR/Cas9-mediated viral interference in plants. *Genome Biology* 16: 238.

Anderson, J.M., Palukaitis, P., and Zaitlin, M. (1992) A defective replicase gene induces resistance to cucumber mosaic virus in transgenic tobacco plants. *Proceedings of the National Academy of Sciences USA* 89, 8759–8763.

Andolfo, G., Jupe, F., Witek, K., Etherington, G.J., Ercolano, M.R., and Jones, J.D.G. (2014) Defining the full tomato NB-LRR resistance gene repertoire using genomic and cDNA RenSeq. *BMC Plant Biology* 14: 120.

Arafa, R.A., Rakha, M.T., Soliman, N.E.K., Moussa, O.M., Kamel, S.M., and Shirasawa, K. (2017) Rapid identification of candidate genes for resistance to tomato late blight disease using next-generation sequencing technologies. *PLOS ONE* 12: e0189951.

Arie, T., Takahashi, H., Kodama, M., and Teraoka, T. (2007) Tomato as a model plant for plant–pathogen interactions. *Plant Biotechnology* 24, 135–147.

Asurmendi, S., Berg, R.H., Koo, J.C., and Beachy, R.N. (2004) Coat protein regulates formation of replication complexes during tobacco mosaic virus infection. *Proceedings of the National Academy of Sciences USA* 101, 1415–1420.

Bai, Y.L., Huang, C.C., van der Hulst, R., Meijer-Dekens, F., Bonnema, G., and Lindhout, P. (2003) QTLs for tomato powdery mildew resistance (*Oidium lycopersici*) in *Lycopersicon parviflorum* G1.1601 co-localize with two qualitative powdery mildew resistance genes. *Molecular Plant–Microbe Interactions* 16, 169–176.

Bai, Y.L., Pavan, S., Zheng, Z., Zappel, N.F., Reinstadler, A., *et al.* (2008) Naturally occurring broad-spectrum powdery mildew resistance in a central American tomato accession is caused by loss of *Mlo* function. *Molecular Plant–Microbe Interactions* 21, 30–39.

Bauchet, G., Grenier, S., Samson, N., Bonnet, J., Grivet, L., and Causse, M. (2017) Use of modern tomato breeding germplasm for deciphering the genetic control of agronomical traits by Genome Wide Association study. *Theoretical and Applied Genetics* 130, 875–889.

Beachy, R.N. (1999) Coat-protein-mediated resistance to tobacco mosaic virus: discovery mechanisms and exploitation. *Philosophical Transactions of the Royal Society of London B: Biological Sciences* 354, 659–664.

Beck, D.L., van Dolleweerd, C.J., Lough, T.J., Balmori, E., Voot, D.M., *et al.* (1994) Disruption of virus movement confers broad-spectrum resistance to systemic infection by plant viruses with a triple gene block. *Proceedings of the National Academy of Sciences USA* 91, 10310–10314.

Belfanti, E., Malatrasi, M., Orsi, I., and Boni, A.G. (2015) Isolated nucleotide sequence from *Solanum lycopersicum* for improved resistance to tomato spotted wilt virus, TSWV. Patent WO/2015/090468; International Application No. PCT/EP2013/077799.

Bendahmane, M., Chen, I., Asurmendi, S., Bazzini, A.A., Szecsi, J., and Beachy, R.N. (2007) Coat protein mediated resistance to TMV infection of *Nicotiana tabacum* involves multiple modes of interference by coat protein. *Virology* 366, 107–116.

Bisaro, D.M. (2006) Silencing suppression by geminivirus proteins. *Virology* 344, 158–168.

Bolger, A., Scossa, F., Bolger, M.E., Lanz, C., Maumus, F., *et al.* (2014) The genome of the stress-tolerant wild tomato species *Solanum pennellii*. *Nature Genetics* 46, 1034–1038.

Brandwagt, B.F., Mesbah, L.A., Takken, F.L.W., Laurent, P.L., Kneppers, T.J.A., *et al.* (2000) A longevity assurance gene homolog of tomato mediates resistance to *Alternaria alternata* f. sp. *lycopersici* toxins and fumonisin $B_1$. *Proceedings of the National Academy of Sciences USA* 97, 4961–4966.

Brommonschenkel, S.H., Frary, A. and Tanksley, S.D. (2000) The broad-spectrum tospovirus resistance gene *Sw-5* of tomato is a homolog of the root-knot nematode resistance gene *Mi*. *Molecular Plant–Microbe Interactions* 13, 1130–1138.

Brooks, C., Nekrasov, V., Lippman, Z.B., and VanEck, J. (2014) Efficient gene editing in tomato in the first generation using the clustered regularly interspaced short palindromic repeats/CRISPRassociated9 system. *Plant Physiology* 166(3), 1292–1297.

Brouwer, D.J. and St Clair, D.A. (2004) Fine mapping of three quantitative trait loci for late blight resistance in tomato using near isogenic lines (NILs) and sub-NILs. *Theoretical and Applied Genetics* 108, 628–638.

Brouwer, D.J., Jones, E.S., and St Clair, D.A. (2004) QTL analysis of quantitative resistance to *Phytophthora infestans* (late blight) in tomato and comparisons with potato. *Genome* 47, 475–492.

Brummell, D.A., Harpster, M.H., Civello, P.M., Palys, J.M., Bennett, A.B., and Dunsmuir, P. (1999) Modification of expansin protein abundance in tomato fruit alters softening and cell wall polymer metabolism during ripening. *Plant Cell* 11(11), 2203–2216.

Bruneti, A., Tavazza, M., Noris, E., Tavazza, R., Caciagli, P., *et al.* (1997) High expression of truncated viral Rep protein confers resistance to Tomato yellow leaf curl virus in transgenic tomato plants. *Molecular Plant–Microbe Interactions* 10, 571–579.

Callaway, E. (2018) CRISPR plants now subject to tough GM laws in European Union. *Nature* 560: 16.

Cao, Y., Wu, Y., Zheng, Z., and Song, F. (2005) Overexpression of the rice EREBP-like gene *OsBIERF3* enhances disease resistance and salt tolerance in transgenic tobacco. *Physiological and Molecular Plant Pathology* 67, 202–211.

Casteel, C.L., Walling, L.L., and Paine, T.D. (2007) Effect of *Mi-1.2* gene in natal host plants on behavior and biology of the tomato psyllid *Bactericerca cockerelli* (Sulc) (Hemiptera: Psyllidae). *Journal of Entomological Science* 42, 155–162.

Catanzariti, A.M., Lim, G.T.T., and Jones, D.A. (2015) The tomato *I-3* gene: a novel gene for resistance to *Fusarium* wilt disease. *New Phytologist* 207, 106–118.

Catanzariti, A.M., Do, H.T.T., Bru, P., de Sain, M., Thatcher, L.F., *et al.* (2017) The tomato *I* gene for *Fusarium* wilt resistance encodes an atypical leucine-rich repeat receptor-like protein whose function is nevertheless dependent on SOBIR1 and SERK3/BAK1. *Plant Journal* 89, 1195–1209.

Chakraborty, S. (2008) Tomato leaf curl viruses from India (Geminiviridae). In: Mahy, B.W.J. and van Regenmortel, M.H.V. (eds.) *Encyclopedia of Virology*. Elsevier, Oxford, pp. 124–133.

Chandrasekaran, J., Brumin, M., Wolf, D., Leibman, D., Klap, C., *et al.* (2016) Development of broad virus resistance in non-transgenic cucumber using CRISPR/Cas9 technology. *Molecular Plant Pathology* 17, 1140–1153.

Chapman, E.J. and Carrington, J.C. (2007) Specialization and evolution of endogenous small RNA pathways. *Nature Reviews Genetics* 8, 884–896.

Chisholm, S.T., Mahajan, S.K., Whitham, S.A., Yamamoto, M.L., and Carrington, J.C. (2000) Cloning of the *Arabidopsis RTM1* gene, which controls restriction of long-distance movement of tobacco etch virus. *Proceedings of the National Academy of Sciences USA* 97, 489–494.

Chisholm, S.T., Coaker, G., Day, B., and Staskawicz, B.J. (2006) Host–microbe interactions: shaping the evolution of the plant immune response. *Cell* 124, 803–814.

Chunwongse, J., Chunwongse, C., Black, L., and Hanson, P. (2002) Molecular mapping of the *Ph-3* gene for late blight resistance in tomato. *Journal of Horticultural Science and Biotechnology* 77, 281–286.

Cillo, F. and Palukaitis, P. (2014) Transgenic resistance. *Advances in Virus Research* 90, 35–146.

Coaker, G.L. and Francis, D.M. (2004) Mapping, genetic effects, and epistatic interaction of two bacterial canker resistance QTLs from *Lycopersicon hirsutum*. *Theoretical and Applied Genetics* 108, 1047–1055.

Cooper, B., Lapidot, M., Heick, J.A., Dodds, J.A., and Beachy, R.N. (1995) A defective movement protein of TMV in transgenic plants confers resistance to multiple viruses whereas the functional analog increases susceptibility. *Virology* 206, 307–313.

Cowger, C. and Brown, J.K.M. (2019) Durability of quantitative resistance in crops: greater than we know? *Annual Review of Phytopathology* 57, 253–277.

Dai, S., Wei, X., Alfonso, A.A., Pei, L., Duque, U.G., *et al.* (2008) Transgenic rice plants that overexpress transcription factors RF2a and RF2b are tolerant to rice tungro virus replication and disease. *Proceedings of the National Academy of Sciences USA* 105, 21012–21016.

Dangl, J.L. and Jones, J.D. (2001) Plant pathogens and integrated defence responses to infection. *Nature* 411, 826–833.

Davis, J., Yu, D.Z., Evans, W., Gokirmak, T., Chetelat, R.T., and Stotz, H.U. (2009) Mapping of loci from *Solanum lycopersicoides* conferring resistance or susceptibility to *Botrytis cinerea* in tomato. *Theoretical and Applied Genetics* 119, 305–314.

Delaney, T.P., Uknes, S., Vernooij, B., Friedrich, L., Weymann, K., *et al.* (1994) A central role of salicylic acid in plant disease resistance. *Science* 266, 1247–1250.

Dhaliwal, M.S., Jindal, S.K., Sharma, A., and Prasanna, H.C. (2020) Tomato yellow leaf curl virus disease of tomato and its management through resistance breeding: a review. *Journal of Horticultural Science and Biotechnology* 95, 425–444.

Diouf, I.A., Derivot, L., Bitton, F., Pascual, L., and Causse, M. (2018) Water deficit and salinity stress reveal many specific QTL for plant growth and fruit quality traits in tomato. *Frontiers in Plant Science* 9: 279.

Dixon, M.S., Jones, D.A., Keddie, J.S., Thomas, C.M., Harrison, K., and Jones, J.D. (1996) The tomato *Cf-2* disease resistance locus comprises two functional genes encoding leucine-rich repeat proteins. *Cell* 84, 451–459.

Dixon, M.S., Hatzixanthis, K., Jones, D.A., Harrison, K., and Jones, J.D.G. (1998) The tomato *Cf-5* disease resistance gene and six homologs show pronounced allelic variation in leucine-rich repeat copy number. *Plant Cell* 10, 1915–1925.

Doganlar, S., Dodson, J., Gabor, B., Beck-Bunn, T., Crossman, C., and Tanksley, S.D. (1998) Molecular mapping of the *py-1* gene for resistance to corky root rot (*Pyrenochaeta lycopersici*) in tomato. *Theoretical and Applied Genetics* 97, 784–788.

Dougherty, W.G. and Parks, T.D. (1995) Transgenes and gene suppression: telling us something new? *Current Opinion in Cell Biology* 7, 399–405.

Duan, Y.P., Powell, C.A., Webb, S.E., Purcifull, D.E., and Hiebert, E. (1997) Geminivirus resistance in transgenic tobacco expressing mutated BC1 protein. *Molecular Plant–Microbe Interactions* 10, 617–623.

Düring, K., Hippe, S., Kreuzaler, F., and Schell, J. (1990) Synthesis and self-assembly of a functional monoclonal antibody in transgenic *Nicotiana tabacum*. *Plant Molecular Biology* 15, 281–293.

Ercolano, M.R., Sanseverino, W., Carli, P., Ferriello, F., and Frusciante, L. (2012) Genetic and genomic approaches for R-gene mediated disease resistance in tomato: retrospects and prospects. *Plant Cell Reports* 31, 973–985.

Ernst, K., Kumar, A., Kriseleit, D., Kloos, D.U., Phillips, M.S., and Ganal, M.W. (2002) The broad-spectrum potato cyst nematode resistance gene (Hero) from tomato is the only member of a large gene family of NBS-LRR genes with an unusual amino acid repeat in the LRR region. *Plant Journal* 31, 127–136.

FAO (Food and Agriculture Organization of the United Nations) (2015) *Coping with Climate Change: The Roles of Genetic Resources for Food And Agriculture*. FAO, Rome.

Firek, S., Draper, J., Owen, M.R., Gandecha, A., Cockburn, B., and Whitelam, G.C. (1993) Secretion of a functional single-chain Fv protein in transgenic tobacco plants and cell suspension cultures. *Plant Molecular Biology* 23, 861–870.

Fitch, M.M.M. (2016) Update on gene transfer biotechnology of papaya. *Acta Horticulturae* 1111, 7–18.

Foolad, M.R. (2007) Genome mapping and molecular breeding of tomato. *International Journal of Plant Genomics* 2007: 64358.

Foolad, M.R. and Panthee, D.R. (2012) Marker-assisted selection in tomato breeding. *Critical Reviews in Plant Sciences* 31, 93–123.

Foolad, M.R., Zhang, L.P., Khan, A.A., Nino-Liu, D., and Lin, G.Y. (2002) Identification of QTLs for early blight (*Alternaria solani*) resistance in tomato using backcross populations of a *Lycopersicon esculentum* × *L. hirsutum* cross. *Theoretical and Applied Genetics* 104, 945–958.
Fradin, E.F., Zhang, Z., Juarez Ayala, J.C., Castroverde, C.D., Nazar, R.N., *et al.* (2009) Genetic dissection of *Verticillium* wilt resistance mediated by tomato Ve1. *Plant Physiology* 150, 320–332.
Fuchs, M. and Gonsalves, D. (2007) Safety of virus resistant transgenic plants two decades after their introduction: lessons from realistic field risk assessment studies. *Annual Review of Phytopathology* 45, 173–202.
Fusaro, A.F., Matthew, L., Smith, N.A., Curtin, S.J., Dedic-Hagan, J., *et al.* (2006) RNA interference-inducing hairpin RNAs in plants act through the viral defence pathway. *EMBO Reports* 7, 1168–1175.
Garcia, V., Bres, C., Just, D., Fernandez, L., Tai, F.W.J., *et al.* (2016) Rapid identification of causal mutations in tomato EMS populations via mapping-by-sequencing. *Nature Protocols* 11, 2401–2418.
Gauffier, C., Lebaron, C., Moretti, A., Constant, C., Moquet, F., *et al.* (2016) A TILLING approach to generate broad-spectrum resistance to potyviruses in tomato is hampered by *eIF4E* gene redundancy. *Plant Journal* 85, 717–729.
Gil, M., Esteban, O., Garcia, J.A., Pena, L., and Cambra, M. (2011) Resistance to *Plum pox virus* in plants expressing cytosolic and nuclear single-chain antibodies against viral NIb replicase. *Plant Pathology* 60, 967–976.
Golemboski, D.B., Lomonossoff, G.P., and Zaitlin, M. (1990) Plants transformed with a tobacco mosaic virus nonstructural gene sequence are resistant to the virus. *Proceedings of the National Academy of Sciences USA* 87, 6311–6315.
Gonzalez-Cendales, Y., Catanzariti, A.M., Baker, B., McGrath, D.J., and Jones, D.A. (2016) Identification of *I-7* expands the repertoire of genes for resistance to *Fusarium* wilt in tomato to three resistance gene classes. *Molecular Plant Pathology* 17, 448–463.
Goodwin, S., McPherson, J.D., and McCombie, W.R. (2016) Coming of age: ten years of next-generation sequencing technologies. *Nature Reviews Genetics* 17, 333–351.
Hong, Y. and Stanley, J. (1996) Virus resistance in *Nicotiana benthamiana* conferred by African cassava mosaic virus replication-associated protein (AC1) transgene. *Molecular Plant–Microbe Interactions* 9, 219–225.
Hong, Y., Saunders, K., Hartley, M.R., and Stanley, J. (1996) Resistance to geminivirus infection by virus-induced expression of dianthin in transgenic plants. *Virology* 220, 119–127.
Hou, Y.M., Sanders, R., Ursin, V.M., and Gilbertson, R.L. (2000) Transgenic plants expressing geminivirus movement proteins: abnormal phenotypes and delayed infection by *Tomato mottle virus* in transgenic tomatoes expressing the *Bean dwarf mosaic virus* BV1 or BC1 proteins. *Molecular Plant–Microbe Interactions* 13, 297–308.
Hsu, P.D., Lander, E.S., and Zhang, F. (2014) Development and applications of CRISPR-Cas9 for genome engineering. *Cell* 157, 1262–1278.
Ishibashi, K., Masuda, K., Naito, S., Meshi, T., and Ishikawa, M. (2007) An inhibitor of viral RNA replication is encoded by a plant resistance gene. *Proceedings of the National Academy of Sciences USA* 104, 13833–13838.
Ito, Y., Nishizawa-Yokoi, A., Endo, M., Mikami, M., Shima, Y., *et al.* (2017) Re-evaluation of the rin mutation and the role of RIN in the induction of tomato ripening. *Nature Plants* 3, 866–874.
James, C. (2017) Global status of commercialized biotech. In: *GM Crops 2016*. ISAAA, New York.
Jones, D.A., Thomas, C.M., Hammondkosack, K.E., Balintkurti, P.J., and Jones, J.D.G. (1994) Isolation of the tomato *Cf-9* gene for resistance to *Cladosporium fulvum* by transposon tagging. *Science* 266, 789–793.
Kabelka, E., Franchino, B., and Francis, D.M. (2002) Two loci from *Lycopersicon hirsutum* LA407 confer resistance to strains of *Clavibacter michiganensis* subsp. *michiganensis*. *Phytopathology* 92, 504–510.
Kadirvel, P., de la Peña, R., Schafleitner, R., Huang, S., Geethanjali, S., *et al.* (2013) Mapping of QTLs in tomato line FLA456 associated with resistance to a virus causing tomato yellow leaf curl disease. *Euphytica* 190, 297–308.
Kang, B.C., Yeam, I., Li, H.X., Perez, K.W., and Jahn, M.M. (2007) Ectopic expression of a recessive resistance gene generates dominant potyvirus resistance in plants. *Plant Biotechnology Journal* 5, 526–536.
Kaniewski, W., Lawson, C., Sammons, B., Haley, L., Hart, J., *et al.* (1990) Field resistance of transgenic Russet Burbank potato to effects of infection by potato virus X and potato virus Y. *Nature Biotechnology* 8, 750–754.

Kawchuk, L.M., Hachey, J., Lynch, D.R., Kulcsar, F., van Rooijen, G., *et al.* (2001) Tomato *Ve* disease resistance genes encode cell surface-like receptors. *Proceedings of the National Academy of Sciences USA* 98, 6511–6515.

Kim, S.J., Lee, S.J., Kim, B.-D., and Paek, K.-H. (1997) Satellite-RNA-mediated resistance to cucumber mosaic virus in transgenic plants of hot pepper (*Capsicum annuum* cv. Golden Tower). *Plant Cell Reports* 16, 825–830.

Kimura, S. and Sinha, N. (2008) Tomato (*Solanum lycopersicum*): a model fruit-bearing crop. *CSH Protocols* 2008: pdb.emo105.

Klas, F.E., Fuchs, M., and Gonsalves, D. (2006) Comparative spatial spread overtime of *Zucchini Yellow Mosaic Virus* (ZYMV) and *Watermelon Mosaic Virus* (WMV) in fields of transgenic squash expressing the coat protein genes of ZYMV and WMV, and in fields of non-transgenic squash. *Transgenic Research* 15, 527–541.

Kobayashi, K., Cabral, S., Calamante, G., Maldonado, S., and Mentaberry, A. (2001) Transgenic tobacco plants expressing the potato virus X open reading frame 3 gene develop specific resistance and necrotic ring symptoms after infection with the homologous virus. *Molecular Plant–Microbe Interactions* 14, 1274–1285.

Kramer, M.G. and Redenbaugh, K. (1994) Commercialization of a tomato with an antisense polygalacturonase gene: the FLAVR SAVR™ tomato story. *Euphytica* 79, 293–297.

Krubphachaya, P., Jurícek, M., and Kertbundit, S. (2007) Induction of RNA-mediated resistance to papaya ringspot virus type W. *Journal of Biochemistry and Molecular Biology* 40, 404–411.

Kumar, S., Tanti, B., Patil, B.L., Mukherjee, S.K., and Sahoo, L. (2017) RNAi-derived transgenic resistance to mungbean yellow mosaic India virus in cowpea. *PLOS ONE* 12: e0186786.

Kundu, J.K., Briard, P., Hily, J.M., Ravelonandro, M., and Scorza, R. (2008) Role of the 25–26 nt siRNA in the resistance of transgenic *Prunus domestica* graft inoculated with plum poxvirus. *Virus Genes* 36, 215–220.

Kyle, M. (1993) *Resistance to Viral Diseases of Vegetables*. Timber Press, Portland, Oregon.

Lanfermeijer, F.C., Warmink, J., and Hille, J. (2005) The products of the broken *Tm-2* and the durable *Tm-2*$^2$ resistance genes from tomato differ in four amino acids. *Journal of Experimental Botany* 56, 2925–2933.

Lannoo, N. and van Damme, E.J. (2014) Lectin domains at the frontiers of plant defense. *Frontiers in Plant Science* 5: 397.

Lapidot, M., Gafny, R., Ding, B., Wolf, S., Lucas, W.J., and Beachy, R.N. (1993) A dysfunctional movement protein of *Tobacco mosaic virus* that partially modifies the plasmodesmata and limits virus spread in transgenic plants. *Plant Journal* 4, 959–970.

Lapidot, M., Karniel, U., Gelbart, D., Fogel, D., Evenor, D., *et al.* (2015) A novel route controlling begomovirus resistance by the messenger RNA surveillance factor Pelota. *PLOS Genetics* 11: e1005538.

Laterrot, H. (2000) Disease resistance in tomato: Practical situation. *Acta Physiologiae Plantarum* 22, 328–331.

Laterrot, H. and Moretti, A. (1989) Linkage between Pto and susceptibility to fenthion. *Tomato Genetics Cooperative Report* 39, 21–22.

Lawson, C., Kaniewski, W., Haley, L., Rozman, R., Newell, C., *et al.* (1990) Engineering resistance to mixed virus infection in a commercial potato cultivar: resistance to potato virus X and potato virus Y in transgenic Russet Burbank. *Biotechnology* 8, 127–134.

Le, L.Q., Lorenz, Y., Scheurer, S., Fötisch, K., Enrique, E., *et al.* (2006) Design of tomato fruits with reduced allergenicity by dsRNAi-mediated inhibition of ns-LTP (Lyc e 3) expression. *Plant Biotechnology Journal* 4, 231–242.

Li, J., Liu, L., Bai, Y., Finkers, R., Wang, F., *et al.* (2011) Identification and mapping of quantitative resistance to late blight (*Phytophthora infestans*) in *Solanum habrochaites* LA1777. *Euphytica* 179, 427–438.

Lin, T., Zhu, G., Zhang, J., Xu, X., Yu, Q., *et al.* (2014) Genomic analyses provide insights into the history of tomato breeding. *Nature Genetics* 46, 1220–1226.

Lindbo, J.A. and Dougherty, W.G. (1992a) Pathogen derived resistance to a potyvirus: immune and resistant phenotypes in transgenic tobacco expressing altered forms of a potyvirus coat protein nucleotide sequence. *Molecular Plant–Microbe Interactions* 5, 144–153.

Lindbo, J.A. and Dougherty, W.G. (1992b) Untranslatable transcripts of the tobacco etch virus coat protein gene sequence can interfere with tobacco etch virus replication in transgenic plants and protoplasts. *Virology* 189, 725–733.

Lindbo, J.A. and Dougherty, W.G. (2005) Plant pathology and RNAi: a brief history. *Annual Review of Phytopathology* 43, 191–204.

Lindbo, J.A. and Falk, B.W. (2017) the impact of 'coat protein-mediated virus resistance' in applied plant pathology and basic research. *Phytopathology* 107, 624–634.

Lindbo, J.A., Silva-Rosales, L., Proebsting, W.M., and Dougherty, W.G. (1993) Induction of a highly specific antiviral state in transgenic plants: implications for regulation of gene expression and virus resistance. *Plant Cell* 5, 1749–1759.

Lodge, J.K., Kaniewski, W.K., and Tumer, N.E. (1993) Broad-spectrum virus resistance in transgenic plants expressing pokeweed antiviral protein. *Proceedings of the National Academy of Sciences USA* 90, 7089–7093.

Lucioli, A., Noris, E., Bruneti, A., Tavazza, R., Ruzza, V., *et al.* (2003) Tomato yellow leaf curl Sardinia virus Rep-derived resistance to homologous and heterologous geminiviruses occurs by different mechanisms and is overcome if virus-mediated transgene silencing is activated. *Journal of Virology* 77, 6785–6798.

Malyshenko, S.I., Kondakova, O.A., Nazarova, J.V., Kaplan, I.B., Taliansky, M.E., and Atabekov, J.G. (1993) Reduction of Tobacco mosaic virus accumulation in transgenic plants producing non-functional viral transport proteins. *Journal of General Virology* 74, 1149–1156.

Marano, M.R. and Baulcombe, D. (1998) Pathogen-derived resistance targeted against the negative-strand RNA of tobacco mosaic virus: RNA strand-specific gene silencing? *Plant Journal* 13, 537–546.

Martin, G.B., Brommonschenkel, S.H., Chunwongse, J., Frary, A., Ganal, M.W., *et al.* (1993) Map-based cloning of a protein kinase gene conferring disease resistance in tomato. *Science* 262, 1432–1436.

Masuta, C., Tanaka, H., Uehara, K., Kuwata, S., Koiwai, A., and Noma, M. (1995) Broad resistance to plant viruses in transgenic plants conferred by antisense inhibition of a host gene essential in *S*-adenosylmethionine-dependent transmethylation reactions. *Proceedings of the National Academy of Sciences USA* 92, 6117–6121.

Maule, A.J., Caranta, C., and Boulton, M.I. (2007) Sources of natural resistance to plant viruses: status and prospects. *Molecular Plant Pathology* 8, 223–231.

Mazzucato, A., Cellini, F., Bouzayen, M., Zouine, M., Mila, I., *et al.* (2015) A TILLING allele of the tomato *Aux/IAA9* gene offers new insights into fruit set mechanisms and perspectives for breeding seedless tomatoes. *Molecular Breeding* 35: 22.

Meng, C., Yang, D., Ma, X., Zhao, W., Liang, X., *et al.* (2016) Suppression of tomato *SlNAC1* transcription factor delays fruit ripening. *Journal of Plant Physiology* 193, 88–96.

Messeguer, R., Ganal, M., de Vicente, M.C., Young, N.D., Bolkan, H., and Tanksley, S.D. (1991) High resolution RFLP map around the root knot nematode resistance gene (*Mi*) in tomato. *Theoretical and Applied Genetics* 82, 529–536.

Milligan, S.B., Bodeau, J., Yaghoobi, J., Kaloshian, I., Zabel, P., and Williamson, V.M. (1998) The root knot nematode resistance gene *Mi* from tomato is a member of the leucine zipper, nucleotide binding, leucine-rich repeat family of plant genes. *Plant Cell* 10, 1307–1319.

Mitter, N., Worrall, E.A., Robinson, K.E., Li, P., Jain, R.G., *et al.* (2017a) Clay nanosheets for topical delivery of RNAi for sustained protection against plant viruses. *Nature Plants* 3: 16207.

Mitter, N., Worrall, E.A., Robinson, K.E., Xu, Z.P., and Carroll, B.J. (2017b) Induction of virus resistance by exogenous application of double-stranded RNA. *Current Opinion in Virology* 26, 49–55.

Mlotshwa, S., Pruss, G.J., and Vance, V. (2008) Small RNAs in viral infection and host defense. *Trends in Plant Science* 13, 375–382.

Moffat, A. (1999) Geminiviruses emerge as serious crop threat. *Science* 286, 1835–1835.

Moon, Y.H., Song, S.K., Choi, K.W., and Lee, J.S. (1997) Expression of a cDNA encoding *Phytolacca insularis* antiviral protein confers virus resistance on transgenic potato plants. *Molecules and Cells* 7, 807–815.

Mueller, L.A., Tanksley, S.D., Giovannoni, J.J., van Eck, J., Stack, S., *et al.* (2005) The tomato sequencing project, the first cornerstone of the International Solanaceae Project (SOL). *Comparative Functional Genomics* 6, 153–158.

Nelson, R.S., Abel, P.P., and Beachy, R.N. (1987) Lesions and virus accumulation in inoculated transgenic tobacco plants expressing the coat protein gene of tobacco mosaic virus. *Virology* 158, 126–132.

Nickel, H., Kawchuk, L., Twyman, R.M., Zimmermann, S., Junghans, H., *et al.* (2008) Plantibody-mediated inhibition of the *Potato leafroll virus* P1 protein reduces virus accumulation. *Virus Research* 136, 140–145. DOI: 10.1016/j. virusres.2008.05.001.

Nölke, G., Fischer, R., and Schillberg, S. (2004) Antibody based resistance in plants. *Journal of Plant Pathology* 86, 5–17.

Nombela, G., Williamson, V.M., and Muniz, M. (2003) The root-knot nematode resistance gene *Mi-1.2* of tomato is responsible for resistance against the whitefly *Bemisia tabaci*. *Molecular Plant–Microbe Interactions* 16, 645–649.

Noris, E., Accoto, G.P., Tavazza, R., Bruneti, A., Crespi, S., and Tavazza, M. (1996) Resistance to tomato yellow leaf curl geminivirus in *Nicotiana benthamiana* plants transformed with a truncated viral *C1* gene. *Virology* 224, 130–138.

Ohlson, E.W., Ashrafi, H., and Foolad, M.R. (2018) Identification and mapping of late blight resistance quantitative trait loci in tomato accession PI 163245. *Plant Genome* 11: 180007.

Okabe, Y., Asamizu, E., Saito, T., Matsukura, C., Ariizumi, T., *et al.* (2011) Tomato TILLING technology: Development of a reverse genetics tool for the efficient isolation of mutants from Micro-Tom mutant libraries. *Plant and Cell Physiology* 52, 1994–2005.

Oldroyd, G.E. and Staskawicz, B.J. (1998) Genetically engineered broad-spectrum disease resistance in tomato. *Proceedings of the National Academy of Sciences USA* 95, 10300–10305.

Palukaitis, P. and Carr, J.P. (2008) Plant resistance responses to viruses. *Journal of Plant Pathology* 90, 153–171.

Pandey, S., Sahu, P., Kulshreshtha, R., and Prasad, M. (2018) Role of host transcription factors in modulating defense response during plant virus interaction. In: Patil, B. (ed.) *Genes, Genetics and Transgenics for Virus Resistance in Plants*. Caister Academic Press, Poole, UK, pp. 25–54.

Pandey, S., Muthamilarasan, M., Sharma, N., Chaudhry, V., Dulani, P., *et al.* (2019) Characterization of DEAD-box family of RNA helicases in tomato provides insights into their roles in biotic and abiotic stresses. *Environmental and Experimental Botany* 158, 107–116.

Pandey, S., Prasad, A., Sharma, N., and Prasad, M. (2020a) Linking the plant stress responses with RNA helicases. *Plant Science* 299: 110607.

Pandey, S., Sharma, N., and Prasad, M. (2020b) Role of RNA-interacting proteins in modulating plant–microbe interactions. *Advanced Genetics* 105, 67–94.

Panthee, D.R., Piotrowski, A., and Ibrahem, R. (2017) Mapping quantitative trait loci (QTL) for resistance to late blight in tomato. *International Journal of Molecular Sciences* 18: E1589.

Pascual, L., Desplat, N., Huang, B.E., Desgroux, A., Bruguier, L., and Bouchet, J.P. (2015) Potential of a tomato MAGIC population to decipher the genetic control of quantitative traits and detect causal variants in the resequencing era. *Plant Biotechnology Journal* 13, 565–577.

Patil, B.L. and Dasgupta, I. (2006) Defective-interfering DNAs of plant viruses. *Critical Reviews in Plant Sciences* 25, 47–64.

Peralta, I.E., Knapp, S., and Spooner, D.M. (2005) New species of wild tomatoes (*Solanum* Section *Lycopersicon*: Solanaceae) from Northern Peru. *Systematic Botany* 30, 424–434.

Pillen, K., Ganal, M.W., and Tanksley, S.D. (1996) Construction of a high-resolution genetic map and YAC-contigs in the tomato Tm-2a region. *Theoretical and Applied Genetics* 93, 228–233.

Piron, F., Nicolai, M., Minoia, S., Piednoir, E., Moretti, A., *et al.* (2010) An induced mutation in tomato eIF4E leads to immunity to two potyviruses. *PLOS ONE* 5: e11313.

Poland, J.A., Balint-Kurti, P.J., Wisser, R.J., Pratt, R.C., and Nelson, R.J. (2009) Shades of gray: the world of quantitative disease resistance. *Trends in Plant Science* 14, 21–29.

Prins, M., Laimer, M., Noris, E., Schubert, J., Wassenegger, M., and Tepfer, M. (2008) Strategies for antiviral resistance in transgenic plants. *Molecular Plant Pathology* 9, 73–83.

Quemada, H.D., Gonsalves, D., and Slightom, J. (1991) Expression of coat protein gene from cucumber mosaic virus strain C in tobacco: protection against infections by CMV strains transmitted mechanically or by aphids. *Phytopathology* 81, 794–802.

Register, J.C. and Beachy, R.N. (1988) Resistance to TMV in transgenic plants results from interference with an early event in infection. *Virology* 166, 524–532.

Rick, C.M. and Chetelat, R.T. (1995) Utilization of related wild species for tomato improvement. *Acta Horticulturae* 412, 21–38.

Robaglia, C. and Caranta, C. (2006) Translation initiation factors: a weak link in plant RNA virus infection. *Trends in Plant Science* 11, 40–45.

Robbins, M.D., Masud, M.A.T., Panthee, D.R., Gardner, R.G., Francis, D.M., and Stevens, M.R. (2010) Marker assisted selection for coupling phase resistance to *Tomato spotted wilt virus* and *Phytophthora infestans* (late blight) in tomato. *HortScience* 45, 1424–1428.

Robert, V.J.M., West, M.A.L., Inai, S., Caines, A., Arntzen, L., *et al.* (2001) Marker assisted introgression of black mold resistance QTL alleles from wild *Lycopersicon cheesmanii* to cultivated tomato (*L. esculentum*) and evaluation of QTL phenotypic effects. *Molecular Breeding* 8, 217–233.

Rossi, M., Goggin, F.L., Milligan, S.B., Kaloshian, I., Ullman, D.E., and Williamson, V.M. (1998) The nematode resistance gene *Mi* of tomato confers resistance against the potato aphid. *Proceedings of the National Academy of Sciences USA* 95, 9750–9754.

Rothan, C., Diouf, I., and Causse, M. (2019) Trait discovery and editing in tomato. *Plant Journal* 97, 73–90.

Ruffel, S., Gallois, J.L., Lesage, M.L., and Caranta, C. (2005) The recessive potyvirus resistance gene *pot-1* is the tomato orthologue of the pepper *pvr2-eIF4E* gene. *Molecular Genetics and Genomics* 274, 346–353.

Salmeron, J.M., Oldroyd, G.E., Rommens, C.M., Scofield, S.R., Kim, H.-S., *et al.* (1996) Tomato *Prf* is a member of the leucine-rich repeat class of plant disease resistance genes and lies embedded within the *Pto* kinase gene cluster. *Cell* 86, 123–133.

Sanford, J.C. and Johnston, S.A. (1985) The concept of parasite-derived resistance-deriving resistance genes from the parasite's own genome. *Journal of Theoretical Biology* 113, 395–405.

Schijlen, E.G., de Vos, C.R., Martens, S., Jonker, H.H., Rosin, F.M., *et al.* (2007) RNA interference silencing of chalcone synthase, the first step in the flavonoid biosynthesis pathway, leads to parthenocarpic tomato fruits. *Plant Physiology* 144, 1520–1530.

Schornack, S., Ballvora, A., Gürlebeck, D., Peart, J., Baulcombe, D., *et al.* (2004) The tomato resistance protein Bs4 is a predicted non-nuclear TIR-NB-LRR protein that mediates defense responses to severely truncated derivatives of AvrBs4 and overexpressed AvrBs3. *Plant Journal* 37, 46–60.

Scorza, R., Callahan, A., Levy, L., Damsteegt, V., Webb, K., and Ravelonandro, M. (2001) Post-transcriptional gene silencing in plum pox virus resistant transgenic European plum containing the plum pox potyvirus coat protein gene. *Transgenic Research* 10, 201–209.

Sharma, S., Pandey, S., Muthamilarasan, M., Chaudhry, V., Dulani, P., and Prasad, M. (2017) Genomics resources for abiotic stress tolerance in Solanaceae crops. In: Chakrabarti, S.K., Xie, C., and Tiwari, J.K. (eds.) *The Potato Genome*. Springer.

Sim, S.C., Robbins, M.D., Wijeratne, S., Wang, H., Yang, W.C., and Francis, D.M. (2015) Association analysis for bacterial spot resistance in a directionally selected complex breeding population of tomato. *Phytopathology* 105, 1437–1445.

Simons, G., Groenendijk, J., Wijbrandi, J., Reijans, M., Groenen, J., *et al.* (1998) Dissection of the *Fusarium I2* gene cluster in tomato reveals six homologs and one active gene copy. *Plant Cell* 10, 1055–1068.

Smith, D.L., Abbott, J.A., and Gross, K.C. (2002) Down-regulation of tomato β-galactosidase 4 results in decreased fruit softening. *Plant Physiology* 129, 1755–1762.

Sood, P., Pandey, S., and Prasad, M. (2021) Biotechnological advances and advanced mutation breeding techniques for crop plants. In: Kumar, P. and Kumar, T.A. (eds.) *Crop Improvement Biotechnological Advances*. CRC Press, Boca Raton, Florida, pp. 139–165.

Stamova, B.S. and Chetelat, R.T. (2000) Inheritance and genetic mapping of cucumber mosaic virus resistance introgressed from *Lycopersicon chilense* into tomato. *Theoretical and Applied Genetics* 101, 527–537.

Stevens, M.R., Lamb, E.M., and Rhoads, D.D. (1995) Mapping the *Sw-5* locus for tomato spotted wilt virus resistance in tomatoes using RAPD and RFLP analyses. *Theoretical and Applied Genetics* 90, 451–456.

Story, E.N., Kopec, R.E., Schwartz, S.J., and Harris, G.K. (2010) An update on the health effects of tomato lycopene. *Annual Review of Food Science and Technology* 1, 189–210.

Tacke, E., Salamini, F., and Rohde, W. (1996) Genetic engineering of potato for broad spectrum protection against virus infection. *Nature Biotechnology* 14, 1597–1601.

Takahashi, H., Chen, Z., Du, H., Liu, Y., and Klessig, D.F. (1997) Development of necrosis and activation of disease resistance in transgenic tobacco plants with severely reduced catalase levels. *Plant Journal* 11, 993–1005.

Takken, F.L.W., Thomas, C.M., Joosten, M., Golstein, C., Westerink, N., *et al.* (1999) A second gene at the tomato *Cf-4* locus confers resistance to *Cladosporium fulvum* through recognition of a novel avirulence determinant. *Plant Journal* 20, 279–288.

Tashkandi, M., Ali, Z., Aljedaani, F., Shami, A., and Mahfouz, M.M. (2018) Engineering resistance against *Tomato yellow leaf curl virus* via the CRISPR/Cas9 system in tomato. *Plant Signaling & Behavior* 13: 1525996.

Tavladoraki, P., Benvenuto, E., Trinca, S., de Martinis, D., Cataneo, A., and Galef, P. (1993) Transgenic plants expressing a functional single-chain Fv antibody are specifically protected from virus attack. *Nature* 366, 469–472.

Tennant, P., Fermin, G., Fitch, M.M., Manshardt, R.M., Slightom, J., and Gonsalves, D. (2001) Papaya ringspot virus resistance of transgenic rainbow and SunUp is affected by gene dosage, plant development, and coat protein homology. *European Journal of Plant Pathology* 107, 645–653.

The 100 Tomato Genome Sequencing Consortium (2014) Exploring genetic variation in the tomato (*Solanum* Section *Lycopersicon*) clade by whole-genome sequencing. *Plant Journal* 80, 136–148.

The Tomato Sequencing Consortium (2012) The tomato sequence provides insights into fleshy fruit tomato. *Nature* 485, 635–641.

Thomas, P.E., Lawson, E.C., Zalewski, J.C., Reed, G.L., and Kaniewski, W.K. (2000) Extreme resistance to *Potato leafroll virus* in potato cv. Russet Burbank mediated by the viral replicase gene. *Virus Research* 71, 49–62.

Tieman, D.M. and Handa, A.K. (1994) Reduction in pectin methylesterase activity modifies tissue integrity and cation levels in ripening tomato (*Lycopersicon esculentum* Mill.) fruits. *Plant Physiology* 106, 429–36.

Tieman, D., Zhu, G., Resende, M.F.R., Lin, T., Nguyen, C., *et al.* (2017) A chemical genetic roadmap to improved tomato flavor. *Science* 355, 391–394.

Tollefson, J. (2011) Brazil cooks up transgenic bean. *Nature* 478: 168.

Tranchida-Lombardo, V., AieseCigliano, R., Anzar, I., Landi, S., Palombieri, S., *et al.* (2018) Whole-genome re-sequencing of two Italian tomato landraces reveals sequence variations in genes associated with stress tolerance, fruit quality and long shelf-life traits. *DNA Research* 25, 149–160.

Tricoll, D.M., Carney, K.J., Russell, P.F., McMaster, J.R., Groff, D.W., *et al.* (1995) Field evaluation of transgenic squash containing single or multiple virus coat protein gene constructs for resistance to cucumber mosaic virus, watermelon mosaic virus 2, and zucchini yellow mosaic virus. *Nature Biotechnology* 13, 1458–1465.

Uluisik, S., Chapman, N.H., Smith, R., Poole, M., Adams, G., *et al.* (2016) Genetic improvement of tomato by targeted control of fruit softening. *Nature Biotechnology* 34, 950–952.

Usadel, B., Chetelat, R., Koren, S., Maumus, F., Fernie, A.R., *et al.* (2017) *De novo* assembly of a new *Solanum pennellii* accession using nanopore sequencing. *Plant Cell* 29, 2336–2348.

Vakalounakis, D.J., Laterrot, H., Moretti, A., Ligoxigakis, E.K., and Smardas, K. (1997) Linkage between *Frl* (*Fusarium oxysporum* f. sp. *radicis-lycopersici* resistance) and *Tm-2* (tobacco mosaic virus resistance-2) loci in tomato (*Lycopersicon esculentum*). *Annals of Applied Biology* 130, 319–323.

Vazquez Rovere, C., Asurmendi, S., and Hopp, H.E. (2001) Transgenic resistance in potato plants expressing potato leaf roll virus (PLRV) replicase gene sequences is RNA-mediated and suggests the involvement of postranscriptional gene silencing. *Archives of Virology* 146, 1337–1353.

Verlaan, M.G., Hutton, S.F., Ibrahem, R.M., Kormelink, R., Visser, R.G.F., *et al.* (2013) The tomato yellow leaf curl virus resistance genes *Ty-1* and *Ty-3* are allelic and code for DFDGD-class RNA-dependent RNA polymerases. *PLOS Genetics* 9: e1003399.

Vos, P., Simons, G., Jesse, T., Wijbrandi, J., Heinen, L., *et al.* (1998) The tomato *Mi-1* gene confers resistance to both root-knot nematodes and potato aphids. *Nature Biotechnology* 16, 1365–1369.

Wang, P. and Tumer, N.E. (2000) Virus resistance mediated by ribosome inactivating proteins. *Advances in Virus Research* 55, 325–355.

Wang, J.F., Ho, F.I., Truong, H.T.H., Huang, S.M., Balatero, C.H., *et al.* (2013) Identification of major QTLs associated with stable resistance of tomato cultivar 'Hawaii 7996' to *Ralstonia solanacearum*. *Euphytica* 190, 241–252.

Waterhouse, P.M. and Fusaro, A.F. (2006) Viruses face a double defense by plant small RNAs. *Science* 313, 54–55.

Whitham, S.A. and Hajimorad, M.R. (2016) *Plant Genetic Resistance to Viruses*. In: Wang, A. and Zhou, X. (eds.) *Current Research Topics in Plant Virology*. Springer, Cham, pp. 87–111.

Yamaguchi, H., Ohnishi, J., Saito, A., Ohyama, A., Nunome, T., *et al.* (2018) An NB-LRR gene, *TYNBS1*, is responsible for resistance mediated by the *Ty-2 Begomovirus* resistance locus of tomato. *Theoretical and Applied Genetics* 131, 1345–1362.

Yamaji, Y., Maejima, K., Ozeki, J., Komatsu, K., Shiraishi, T., *et al.* (2012) Lectin-mediated resistance impairs plant virus infection at the cellular level. *Plant Cell* 24, 778–793.

Ye, J., Wang, X., Hu, T., Zhang, F., Wang, B., *et al.* (2017) An Indel in the promoter of *Al-ACTIVATED MALATE TRANSPORTER9* selected during tomato domestication determines fruit malate contents and aluminum tolerance. *Plant Cell* 29, 2249–2268.

Zhang, H., Tian, Y., Zhou, Y., Dang, B., Lan, H., *et al.* (1999) Introduction of pokeweed antiviral protein cDNA into *Brassica napus* and acquisition of transgenic plants resistant to viruses. *Chinese Science Bulletin* 44, 701–704.

Zhang, C., Liu, L., Wang, X., Vossen, J., Li, G., *et al.* (2014) The *Ph-3* gene from *Solanum pimpinellifolium* encodes CC-NBS-LRR protein conferring resistance to *Phytophthora infestans*. *Theoretical and Applied Genetics* 127, 1353–1364.

Ziegler, A. and Torrance, L. (2002) Applications of recombinant antibodies in plant pathology. *Molecular Plant Pathology* 3, 401–407.

# 2 Transgenic Eggplant

**Vageeshbabu S. Hanur[1]* and Polumetla Ananda Kumar[2]**
[1]*Division of Basic Sciences, ICAR-Indian Institute of Horticultural Research, Hesaraghatta Lake Post, Bangalore, India;* [2]*ICAR-Indian Institute of Rice Research, Rajendranagar, Hyderabad, India*

**Abstract**

Eggplant (*Solanum melongena* L.; also known as brinjal or aubergine) is an important globally cultivated vegetable crop. Eggplant's yield and marketable quality are severely affected by various biotic and abiotic stress factors. The major biotic stresses are eggplant shoot and fruit borer (ESFB), fruit borers, soil-borne pathogens such as wilt and viruses, and *Mycoplasma* infections. Transgenic eggplants have been developed, including *Bacillus thuringiensis* (*Bt*) transgenics, to mitigate many of these stresses. Different *Bt* insecticidal crystal protein genes (*cry* genes) have been used to address the most serious problem of ESFB infestation. Commercialization of *Bt*-transgenic eggplant is varied and is influenced by issues of sociopolitical and technological considerations. Case studies of commercialization efforts in India, Bangladesh, and the Philippines are guiding the future of genetic-engineering efforts in different crops for a variety of traits in developing countries. Future transgenic crop research may involve clustered regularly interspaced palindromic repeats (CRISPR)-CRISPR-associated protein 9 (Cas9)-based gene editing and other technologies with clear biosafety and societal acceptability.

**Keywords:** agricultural biotechnology, biosafety transgenics, brinjal, *Bt*, genetic modification, insect pest resistance

## 2.1 Introduction

The burgeoning world population is estimated to reach 10 billion by 2050 and may severely stress the availability of inputs, with an unprecedented 25–70% rise in the demand for food (Hunter *et al.*, 2017). Eggplant (*Solanum melongena* L., Solanaceae, $2n = 24$), also known as brinjal, aubergine, or guinea squash, is an important non-tuberous vegetable crop. In global vegetable production, eggplant ranks fifth after tomato, onion, cucumber and gherkin, and cabbage. It is an important vegetable cash crop of small and marginal farmers, especially in developing countries. The cultivated eggplant is of Indian origin (de Candolle, 1886; Vavilov, 1928, 1951). Eggplant is one of the most important vegetable crops of the Indian subcontinent, although whether India is the center of origin of *Bacillus thuringiensis* (*Bt*)-transgenic eggplant has been debated (Hanur, 2011a). It is grown over $5.5 \times 10^5$ ha, mostly dry land, by approximately 400,000–500,000 small and resource-poor farmers, producing about 8.4 million t. Worldwide, it is cultivated over an area of 0.72 million ha with production of 32 million t. India accounts for 72% of the world's acreage and 26% of the world's production. Globally, China, India, Egypt, Turkey, and Iran account for nearly 93% of production. According to the International

*Email: vageeshbabu.hs@icar.gov.in

 *Genetic Engineering of Vegetable Crops* (Pritam Kalia ed.)
DOI: 10.1079/9781800627123.0002

Treaty on Food and Agriculture's Plant Genetic Resources, it is an important vegetable crop for food security (FAO, 2009). It is a versatile and well-adapted crop suited to varied agroclimatic conditions.

## 2.2 Production Constraints

Eggplant shoot and fruit borer (ESFB; *Leucinodes orbonalis*, Pyralidae, Lepidoptera) is the most devastating insect pest in this plant, causing yield losses of up to 50–70%. Losses of up to 100% have also been recorded under unsprayed situations (Hanur, 2008, 2010a; Chandra Kumar *et al.*, 2008), although the occurrence of other pests and diseases is quite common. In India, farmers rely mainly on insecticides to manage ESFB, and spray up to 40 times per season for its control. Management of this notorious pest has not been satisfactory, even under integrated pest-management packages, due to the unique bionomics of the pest. Biochemical means of resistance sources have been proposed and tested (Kalloo, 1988; Jat and Pareek, 2003; Prabhu, *et al.*, 2009) in the wild forms of eggplant, although *S. macrocarpon* was tested as not resistant to ESFB (Hanur *et al.*, 2011). In experiments where cultivated and wild forms of eggplant have been challenged with ESFB, it showed a preference for cultivated forms, suggesting that wild forms possess some type of resistance. However, when cultivated hosts are unavailable, this pest will damage the wild forms equally seriously. None of the cultivated genotypes was found to be resistant to ESFB. Management of ESFB is a herculean task due to its peculiar and unique feeding behavioural biology (Hanur *et al.*, 2014). However, the wild accessions *Solanum gilo*, *Solanum anomalum*, and *Solanum macrocarpon* were reported as tolerant due to non-preference. Attempts were made to transfer ESFB resistance from *S. macrocarpon*, but the interspecific hybrid (*S. melongena* × *S. macrocarpon*) was found to be sterile due to ovule abortion; studies are under way to restore fertility in this interspecific hybrid.

Other equally important pests include the fruit borer (*Helicoverpa armigera* Hubner), stem borer (*Euzophera perticella*), hadda beetle (*Epilachna vigintioctopunctata*), Colorado potato beetle (*Leptinotarsa decemlineata*), aphids (*Aphis* spp.), and parasitic nematodes (*Meloidogyne* spp.). Aphid-transmitted viruses (potato virus Y) cause mosaic mottling of the leaves and stunting of the plants with deformed leaves. Other viral pathogens include cucumber mosaic virus, eggplant mottled crinkle virus and alfalfa mosaic virus. The equally serious littleleaf disease, with chlorotic young leaves and deformed profuse axillary bud proliferation with bushy/stunted growth and sterile phenotype, is caused by a *Mycoplasma* sp. and is sap transmissible via insect vectors.

Although eggplant is generally fairly tolerant to abiotic stresses, sensitivity to prolonged and stronger stresses affect the yield and quality of the crop. Eggplant is susceptible to frost, moisture stress, salinity, and pests and diseases. Global climatic changes aggravated by numerous anthropogenic activities have reduced the productivity of eggplant worldwide. Genetic and biotechnological improvements can be used to intervene to stabilize and improve productivity.

The importance of biotechnological research in eggplant has been highlighted. In concerted research activities, investigations have been directed toward biotechnology and the molecular biology of the crop for genetic improvements in an attempt to overcome the problems. *In vitro* tissue-culture techniques have been used profitably for regeneration of the crop through organogenesis, somatic embryogenesis, and *Agrobacterium*-mediated genetic transformation to introduce useful genes related to stress tolerance and other purposes such as parthenogenesis. Various molecular markers such as quantitative trait loci, random amplified polymorphic DNA, amplified fragment length polymorphisms, microsatellites, and simple sequence repeats have been identified for effective utilization for crop improvement. Genes have been mapped for various morphological and other traits for genetic improvement. Various transgenic lines have been developed for resistance to insects such as beetles, stem borer, and aphids, and for resistance to diseases such as *Fusarium* and bacterial wilt.

## 2.3 Development of Transgenic Plants

Eggplant offers an optimized system for studying the basic mechanisms of regeneration, including shoot bud induction, stem-cell regulation, fine control of signal transduction mechanisms, and crosstalk that operates at subcellular levels during developmental ontogeny, and for applying the technologies developed so far to produce a range of transgenics with varied modified traits (Hanur, 2007, 2010c). A considerable amount of research has been undertaken to understand the tissue-culture responsiveness of the crop in terms of regeneration and *in vitro* plant differentiation, which shows comparative *in vitro* synteny with other solanaceous crops such as tomato, chili, potato, and tobacco. Several groups have attempted and achieved genetic transformation of eggplant in a number of genotypic backgrounds over the last three decades (Magioli and Mansur, 2005; Collonnier *et al.*, 2001; Rajam *et al.*, 2008). The production of transgenic plants was first reported by Guri and Sink (1988) using leaf explants and the cointegrated vector pMON 200 carrying the *nptII* gene. They used the cointegrated/hybrid Ti plasmid system instead of the routinely used binary system. In the following year, Fillipone and Lurquin (1989) reported stable transformation of leaf and cotyledon explants using a binary vector through a wild super-virulent *Agrobacterium* strain, A281. Since then, transgenic plants have been widely obtained using various *in vitro* techniques such as explants, *Agrobacterium* vectors, genes, and other factors.

### 2.3.1 Factors affecting genetic transformation

Several factors including the type of explant, preconditioning of the explants, and selection using antibiotic-containing media have demonstrated their usefulness in genetic transformation (Fillipone and Lurquin, 1989; Arpaia *et al.*, 1997; Billings *et al.*, 1997; Iannacone *et al.*, 1997; Magioli *et al.*, 2000; Hanur, 2011b). Fari *et al.* (1995) attempted *Agrobacterium*-mediated genetic transformation via the somatic embryogenesis route. Kumar *et al.* (1998) used cotyledonary leaves to transform eggplant with the *Bt* insecticidal crystal protein (Cry) gene *cry1Ab*. Franklin and Lakshmi Sita (2003) used root explants to optimize stable transformation. Hypocotyl explants were mostly preferred over the commonly used explants (Kamat and Rao, 1978; Matsuoka and Hinata, 1979; Alicchio *et al.*, 1982; Ali *et al.*, 1991; Chen *et al.*, 1995; Sharma and Rajam, 1995; Arpaia *et al.*, 1997; Magioli *et al.*, 1998; Zhang, 1999; Hanur *et al.*, 2006; Solanki *et al.*, 2006; Prakash *et al.*, 2007). Cotyledonary leaf explants derived from axenic cultures of seedlings were found not to be responsive to regeneration compared with their hypocotyl counterparts (Iannacone *et al.*, 1997; Rotino *et al.*, 1997b; Kumar *et al.*, 1998; Prabhavathi *et al.*, 2002; Kumar and Rajam, 2005; Hanur *et al.*, 2006; Prakash *et al.*, 2007). The regeneration and concurrent transformation efficiencies reported in these systems were relatively low, except for a few cases (Sharma and Rajam, 1995a; Hanur *et al.*, 2006; Prakash *et al.*, 2007).

Part of the answer to the problem of empirical results obtained by the use of various explants may lie in the developmental ontogeny, control, and cell niches, as governed by stem cells (Hanur, 2007, 2010c). These stem cells are tightly controlled developmentally by an array of genes including *WUSCHEL (WUS)*, *CLAVATA (CLV)*, *SHOOTMERISTEMLESS (STM)* (all master regulators), *KNOTTEDLIKE HOMEOBOX (KNOX)*, *CUP-SHAPED COTYLEDONS (CUC)*, *SCARCROW (SCR)*, and other critical housekeeping genes. Various external and internal signals and cues determine the cell fate either *in vitro* or *in vivo*. The study of stem-cell biology has helped in understanding plant regeneration, totipotency, and cloning of genes related to regeneration. Exploiting this knowledge, Hanur *et al.* (2006) and Prakash *et al.* (2007) reported the development of *Bt*-transgenic eggplant (Fig. 2.1) and tomato. Poor regeneration is one of the limiting factors in the development of transgenic crops, as *Agrobacterium* as a plant pathogen can disturb the fragile *in vitro* conditions by wounding and infection regimes.

While incorporating transgenic resistance in eggplant against EFSB, the synergistic use of hypocotyl explants and higher levels of cytokinin, especially 6-benzylaminopurine (BAP), during the preconditioning period for higher regeneration of transformants was studied (Hanur *et al.*, 2006). Challenging with *Agrobacterium* definitively reduced the regeneration response, vindicating the suggestion that infection by a pathogen (*Agrobacterium*) can cause havoc in the regeneration responses (Billings *et al.*, 1997; Hanur 2004). Preconditioning noticeably produced the highest

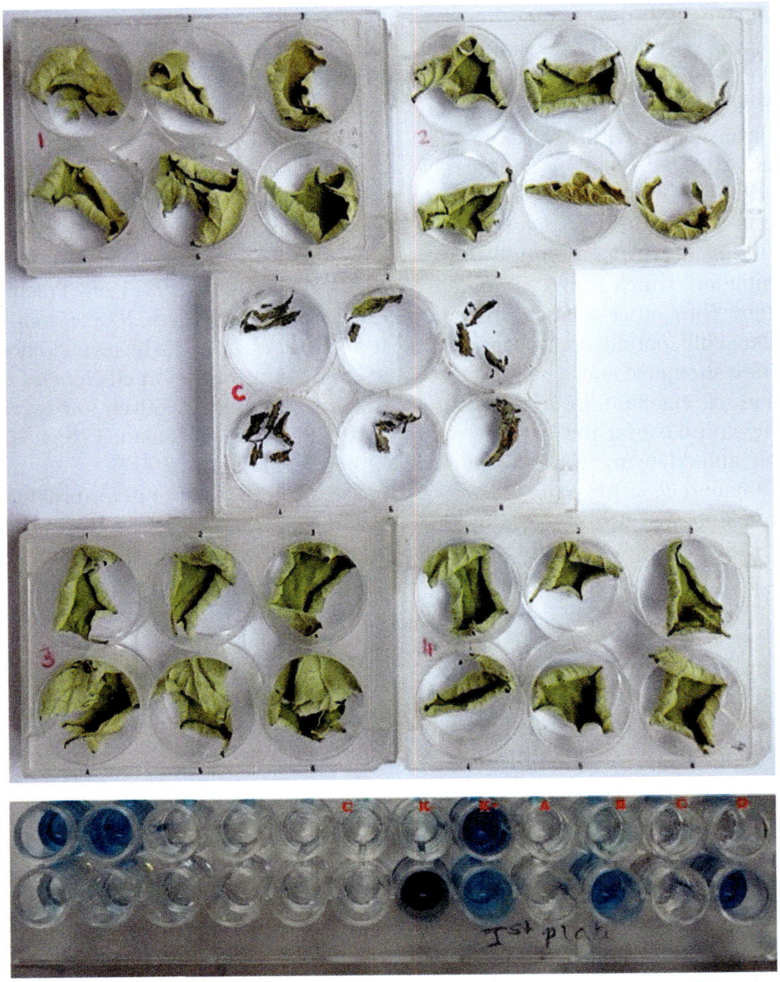

**Fig. 2.1.** Top: six-well plate leaf bioassays on control (central plate, C; non-transgenic eggplant leaves) and *Bacillus thuringiensis* (*Bt*)-*cry1Ab* eggplant lines exposed to eggplant shoot and fruit borer, *Leucinodes orbonalis* Guenee. Each leaf in each well was uniformly smeared with 100 µl of reference *Bt* culture followed by inoculation with two freshly hatched first-instar neonate larvae of *L. orbonalis* per leaf (plates 1–4: 2HA1-1 plants). Bottom: ELISA of *Bt*-transgenic eggplant lines to study expression levels of *Bt*-Cry1Ab protein. C: negative control; K: kit control; K+: positive kit standard control, A–D: different *Bt* lines of 2HA1-1 plants.

frequency of regeneration, and was sometimes better with a dark preculture period. Synergistic use of hypocotyl explants and high BAP preconditioning were found to increase the regeneration of transformed cells in eggplant.

The effects of growth regulators and the type of explant on transformation and *in vitro* morphogenetic responses of *S. melongena* 'Manjarigota' were studied (Prakash *et al.*, 2007, 2008). Both hypocotyl and cotyledonary explants showed a marked influence on *in vitro* morphogenetic responses after *Agrobacterium* co-cultivation. Hypocotyl explants showed callus initiation and regeneration responses earlier than cotyledonary leaves. Hypocotyl explants were found to be better than cotyledonary leaf explants in regenerating shoots after *Agrobacterium* co-cultivation. There was a delay and reduction in both callus and regeneration responses in *Agrobacterium* co-cultivated explants. Hypocotyl explants showed the highest regeneration response on Murashige and Skoog (MS) medium containing 2 mM BAP and 0.05 mM 1-naphthaleneacetic acid (NAA), while cotyledonary leaves did not show a regeneration response after *Agrobacterium* co-cultivation. However, they showed green buds on MS medium containing 10 mM BAP and 1 mM NAA, although these could not differentiate into shoots. Overall, hypocotyl explants were found to be better in regenerating shoots after *Agrobacterium* co-cultivation.

In transformation studies, once the target cells have been transformed, the transgenic cells or plants produced from them are identified on selection medium. A marker gene is necessary because only a low proportion of the cells exposed to the transformation processes subsequently become stably transformed (Klee *et al.*, 1987). The use of selection medium confers an advantage to those cells that stably incorporate the transgene construct and are therefore resistant to the specific antibiotic in the selective medium. The use of a marker gene in a transformation process thus aims to give a selective advantage to transformed cells by allowing them to grow. The selectable marker gene confers on the transformed cells an ability to metabolize compounds that are not usually metabolized by untransformed cells. One such widely used marker is the neomycin phosphotransferase (*nptII*) gene, which confers resistance to the aminoglycoside antibiotic kanamycin. Antibiotics are added to the culture medium to control *Agrobacterium*, because it may affect the plant-regeneration process as *Agrobacterium* is basically a plant pathogen and can cause defense reactions in the host (Rotino and Gleddie, 1990; Franklin and Sita, 2003; Hanur, 2004). Kanamycin-resistant transformed cells can be selected preferentially, followed by regeneration. Cefotaxime should also be added to the medium to eliminate *Agrobacterium* from the cultures. Cefotaxime has been shown to have both positive and negative effects on callus formation and regeneration in many crops. Various gelling agents differ in their affinity to bind kanamycin and inhibit the latter's activity in culture medium, which may indirectly affect transformation and recovery of transformants (Laine *et al.*, 2000).

Although the mechanism is not clear, the efficiency of kanamycin in inhibiting regeneration differs with various gelling agents (Chauvin *et al.*, 1999; Prakash *et al.*, 2007). In eggplant, the effect of the antibiotics cefotaxime and kanamycin has been studied to a lesser extent (Billings *et al.*, 1997). There have been many studies probing the roles of microbicidal antibiotics in affecting the responses of cultures *in vitro* in a variety of crops including eggplant. The presence of cefotaxime, ampicillin, or vancomycin did not affect embryogenic callus formation and development from leaf and cotyledonary explants of eggplant; however, antibiotic treatment reduced the number of embryos produced per callus using cefotaxime and vancomycin (Magioli *et al.*, 2001a). Similarly, negative effects of cefotaxime on callus formation and plant regeneration have been described in carrot (Okkels and Pederson, 1988). Cefotaxime was, in another case, found to enhance callus fresh weight, but also caused a decrease in the rate of embryo regeneration, whereas timentin showed no effect on embryo differentiation (Picoli *et al.*, 2000, 2002a,b). Stimulation of callus growth and regeneration due to cefotaxime has been reported in barley (Mathias and Mukasa, 1987), bread wheat (Mathias and Boyd, 1986), and many horticultural crops. Magioli *et al.* (2001b) also studied the effects of various morphological factors, antibiotics, and *Agrobacterium* co-cultivation on the efficiency of the somatic embryogenesis route.

A study was conducted to determine the effects of antibiotics and gelling agents during transformation in *S. melongena* 'Manjarigota' (Prakash *et al.*, 2007). Hypocotyl explants were found to be sensitive even to the lowest level of kanamycin tested (25 mg $l^{-1}$). Explants that showed increased callus initiation and regeneration response on co-cultivation with *Agrobacterium* and kanamycin at 100 mg $l^{-1}$ were selected, as this indicated a highly effective selection pressure. Cefotaxime did not affect the regeneration response and at 500 mg $l^{-1}$ effectively inhibited *Agrobacterium* overgrowth completely on *Agrobacterium* co-cultivated explants. In addition, marked differences were observed in the regeneration response in hypocotyl explants cultured on medium solidified with various gelling agents (agar, Agargel (Phytagel:agar 1:1), and Phytagel), indicating the influence of gelling agent on the activity of kanamycin in the culture medium, which indirectly affects selection and recovery of transformants. Both the callus initiation response and regeneration response differed when increasing levels of kanamycin were used. Potato as an alternate host to rear ESFB for use in the testing of *Bt* eggplant lines was also studied (Boopal *et al.*, 2013).

Subramanyam *et al.* (2013) studied a number of factors in eggplant transformation. They attempted *in planta* seed transformation. They studied genotype, preculture duration, acetosyringone concentration, the influence of surfactants (Tween-20, Silwet L-77, and Triton X-100), sonication, and the influence of vacuum infiltration on transformation efficiency. They also studied post-co-cultivation processing followed by selection of transformants. Inheritance of the β-glucuronidase (GUS) gene (*uidA*) and *bar* marker genes in the $T_1$ generation of the progeny plants was also studied.

Rajam and Kumar (2007) reviewed the work on regeneration and genetic engineering in eggplant, providing an overview of the progress in eggplant tissue culture (organogenesis, somatic embryogenesis, protoplast culture and somatic hybridization, and andropogenic haploids) and genetic engineering, and briefly discussed the future prospects for the genetic transformation of eggplant. More recently, efficient *in vitro* organogenesis in eggplant has been attempted (Garcia-Fortea *et al.*, 2020, 2021). Comprehensive reviews are available (Hanur, 2010a,b).

An efficient method for producing transgenic eggplant (*S. melongena* 'Haritha') *via Agrobacterium tumefaciens*-mediated genetic transformation using the *uidA* reporter gene was developed by Soniya *et al.* (2005). Leaf explants were transformed by co-cultivation with *A. tumefaciens* strain LBA4404 harboring the binary vector pCAMBIA 1301 carrying the reporter gene *uidA* intron and the marker gene hygromycin phosphotransferase (*hpt*). Transgenic calluses induced in medium containing 1 mg indole-3-acetic acid (IAA) $l^{-1}$ and 3 mg BAP $l^{-1}$, or 1 mg IAA $l^{-1}$ and 3 mg kinetin $l^{-1}$, subcultured in 0.1 mg thidiazuron (TDZ) $l^{-1}$ and 3.0 mg BAP $l^{-1}$, produced shoot buds. Direct shoot buds were also induced in medium containing 3 mg BAP $l^{-1}$ and 0.5 mg TDZ $l^{-1}$. The putative transgenic shoot buds elongated on basal medium and rooted efficiently on sterilized soil irrigated with water. Histochemical GUS assays and polymerase chain reaction (PCR) analysis of transgenic plants confirmed the presence of the *uidA* and *hpt* genes, respectively. Integration of T-DNA into the genome of putative transgenics was further confirmed by Southern blot analysis.

### 2.3.2 Transgenics for biotic stress tolerance

#### *Insect pest resistance*

Eggplant is affected by a number of insect pests. ESFB (*L. orbonalis* Guenee) is the most devastating pest of these. Other pests such as aphids and *Myllocerus* spp. weevil cause minor yield losses. Diseases caused by bacterial and fungal wilts, blights, nematodes and viruses also cause minor to significant economic losses. Transgenic eggplants expressing resistance to pests using several *Bt* genes have been developed (Table 2.1). The failure of conventional methods of pest management made the research option of transgenics the most sustainable and workable alternative. Among many potential genes, *Bt*-transgenic technology has become the front-runner with great potential and commercial success globally. India, as the main country for production and consumption of eggplant, as well as a leader in application of pesticides to manage this pest, has a need for *Bt*-transgenic technology in eggplant (Kumar and Sharma, 1994; Kumar *et al.*, 1996a,b).

EARLY EFFORTS. Rao *et al.* (1999) tested seven lepidopteran-specific *Bt* δ-endotoxins for their efficacy against the second-instar larvae of ESFB. The protein Cry2Aa was the most potent toxin, followed by Cry1C, Cry1Ac, Cry1Ab, and Cry1B. The first attempt to transform eggplant was made in the 1990s utilizing a codon-modified synthetic version of the *cry1Ab* gene (Fujimoto *et al.*, 1993). The gene was introduced into a popular variety of eggplant, *S. melongena* 'Pusa Purple Long,' and subjected to molecular and entomological studies (Kumar *et al.*, 1998). One of the resulting transgenic lines (BT6) resistant to ESFB was field tested at multiple locations with mixed results (P.A. Kumar, unpublished data).

EVENT EE1. Scientists at Mahyco, Jalna (Maharashtra, India), introduced a synthetic *cry1Ac*

**Table 2.1.** Selected *Bacillus thuringiensis* (*Bt*) genes transformed into eggplant.

| Gene/construct | Reference |
|---|---|
| *cry1Ab* | Kumar *et al.* (1998) |
| *cry1Ac* | Pal *et al.* (2009) |
| *cry1Aa3* | Rai *et al.* (2013) |
| *cry1Fa1* | Shrivastava *et al.* (2011) |
| *cry3A* | Jelenkovic *et al.* (1998) |
| *cry3B* | Arpaia *et al.* (1997, 1998) |
| *cry1F* | Jadhav *et al.* (2015) |
| *cry1F* | Shruti *et al.* (2015) |
| *cry3B* | Chen *et al.* (1995) |
| *cry3A* | Hamilton *et al.* (1997) |
| *cry1Ab* | Prakash *et al.* (2007, 2008, 2015, 2016) |
| *cry3B* | Rotino *et al.* (1990) |
| *cry1Ac* | Hautea *et al.* (2016) |
| *cy1Ac* | Prodhan *et al.* (2018, 2019) |
| *cry3B* | Iannacone *et al.* (1997) |
| *cry3B* | Billings *et al.* (1997) |

gene obtained from Monsanto (St. Louis, Missouri) in eggplant, and Event EE1, which exhibited good performance, was introgressed into four commercial hybrids (MHB 4, MHB 9, MHB 11, and MHB 39). The *Bt* hybrids were subjected to various field, biosafety, and environmental tests as prescribed by the Indian genetically modified organism regulatory system in the period 2002–2008 (Bhalla *et al.*, 2007; Choudhary and Gaur, 2009). The Event EE1 was also introgressed into four varieties at Tamil Nadu Agricultural University (TNAU, Coimbatore) and six varieties at the University of Agricultural Sciences, Dharwad (UAS-D). Two expert committees comprising the Genetic Engineering Approval Committee (GEAC) under the aegis of the Ministry of Environment and Forests (MoEF) analyzed the biosafety data. Based on observations made by the expert committees, GEAC approved the environmental safety of *Bt* eggplant Event EE1 on 14 October 2009. However, a moratorium on the environmental release of *Bt* eggplant was imposed by the MoEF on 9 February 2010 (Rao, 2010). Event EE1 was provided by Mahyco to the Bangladesh Agricultural Research Institute (BARI). Four *Bt* eggplant varieties developed and tested by BARI were approved by the Government of Bangladesh for cultivation in 2013 (Choudhary *et al.*, 2014; Shelton *et al.*, 2018). A socio-economic survey was performed and it was found that the *Bt* eggplant reduced labor and chemical costs and that 75% of farmers had increased yield, with 72% of them having enhanced income as a result of *Bt* eggplant cultivation. *Bt* eggplant also reduced pesticide use, as attested to by 97% of the farmers (Haque and Saha, 2020).

EVENT 142. The National Research Centre on Plant Biotechnology (NRCPB; currently NIPB, New Delhi) of the Indian Agricultural Research Institute (IARI) developed a *Bt* eggplant (Event 142) using a codon-modified synthetic version of the *Bt* insecticidal protein gene encoding Cry1Fa1 in *S. melongena* 'Pusa Purple Long' (Kumar, 2010; Indian Patent No. 242768; P.A. Kumar, unpublished data). Event 142 was introgressed into two commercial hybrids by Beej Sheetal (Jalna, Maharashtra, India). The relevant biosafety and environmental safety tests have been conducted except for Biosafety Research Level II (BRL II) (P.A. Kumar, unpublished data).

OTHER EVENTS. Genetic transformation of *S. melongena* 'Arka Keshav' and 'Manjarigota' with different *Bt* genes including *cry1Ab*, *cry1Aa3*, *cry2A*, *cry1Aabc*, and *vip3A* is under way at the Indian Institute of Horticultural Research (IIHR, Bangalore, India). Cry2A transgenics have been developed, and molecular analyses and resistance phenotyping against first-instar and neonate larvae of ESFB have been performed. Promising *Bt* eggplant lines are under field evaluation (Figs 2.1–2.3) (Hanur *et al.*, 2006; Hanur, 2008, 2010a).

Pal *et al.* (2009) reported *Agrobacterium*-mediated gene transfer of *Bt-cry1Ac* to an improved inbred line of eggplant (IVBL-9). PCR and Southern blot analyses confirmed the presence of a single-copy insertion of the *cry1Ac* gene in seven independent transgenic plants. The insertion of a single copy of the gene was also confirmed by segregation analysis of $T_1$ seed from $T_0$ plants. ELISA analyses revealed the presence of the Cry1Ac protein, and quantitative estimates confirmed significant levels of Cry1Ac protein (2.46–4.33 ng $ml^{-1}$ of leaf extract) expressed in all seven transformed plants. High levels of expression of this insecticidal protein resulted in significant larval mortality and in the reduced growth of any surviving larva on transformed

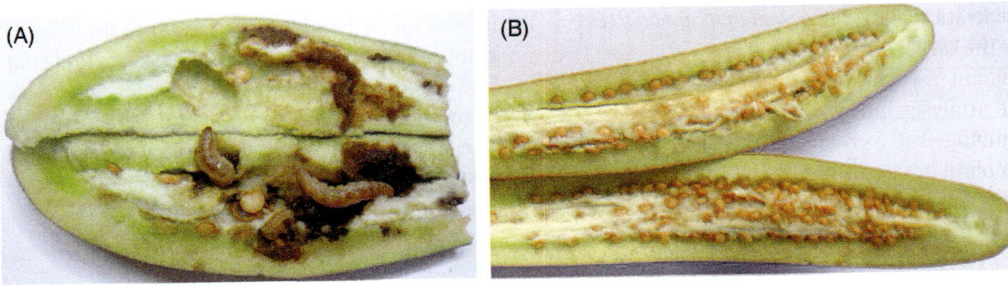


**Fig. 2.2.** Infested control (A) and Cry2A *Bacillus thuringiensis* (*Bt*)-transgenic (B) eggplant (*Solanum melongena* 'Arka Keshav') fruits, developed at ICAR-IIHR (Indian Council of Agricultural Research-Indian Institute of Horticultural Research), Bangalore, India.

**Fig. 2.3.** *Bacillus thuringiensis* (*Bt*) (A) and non-*Bt* control (B) shoots of *Solanum melongena* 'Arka Keshav' developed at ICAR-IIHR (Indian Council of Agricultural Research-Indian Institute of Horticultural Research), Bangalore, India.

eggplant tissue. Narendran *et al.* (2007) used *Agrobacterium*-mediated co-transformation for the generation of marker-free, insect-resistant transgenic plants for resistance to ESFB by expressing the *cry2Ab* gene. These plants were screened by double-antibody sandwich ELISA for the detection of Cry2Ab protein and GUS assays for expression of β-glucuronidase. The *cry-2Ab* and marker genes were found to have segregated independently in 44% of the primary transformants analyzed, indicating that the T-DNA insertions were unlinked. The marker-free status of the plants was confirmed by molecular analysis and germination tests *in vitro* on kanamycin-containing medium. Transformants that displayed normal Mendelian inheritance were chosen for further analysis. The shoots and fruits of a number of marker-free transgenic plants were screened using larvae of ESFB.

In Europe and the USA, the Colorado potato beetle (CPB; *Leptinotarsa decemlineata* Say) causes heavy damage to potatoes and eggplants, especially where it has developed resistance against many insecticides. As transgenic plants expressing *Bt* toxins are effective in controlling insect pests, the availability of insect-resistant transgenic eggplant hybrids provides new possibilities for the management of resistance to CPB. Transgenic eggplant for resistance to CPB using a coleopteran-specific *cry3B* gene was developed (Arpaia *et al.*, 1997, 1998). Arpaia *et al.* (2007) set up replicated field plots of *Bt*-expressing eggplants and near-isogenic untransformed eggplants as a control. They monitored the presence and abundance of herbivore and predator arthropods in weekly visual samplings of the plant canopy for three growing seasons (2001–2003). Insect species were pooled in organismal taxonomic units, and three multivariate methods were used to compare species assemblage as an estimate of insect biodiversity. This multistep statistical approach proved to be efficient in recognizing association patterns, as evidenced by the data for the target species, CPB, clearly showing a significant association with the control plots. All the analyses indicated a comparable species assemblage between transgenic and near-isogenic eggplant areas. Their results suggested that some taxa may warrant more specific study. For example, Alticinae beetles (Coleoptera: Chrysomelidae) were alternatively more abundant in either of the two treatments, and their overall abundance was significantly higher on transgenic eggplants. In light of these results and because of their taxonomic proximity to the target species, these herbivores may represent an important non-target group to be studied further. Moreover, some sap feeders (e.g. Homoptera: Cicadellidae) were more abundant on *Bt*-transgenic plants in some samples in all 3 years.

Transgenic eggplant lines carrying the *Bt-cry3B* gene were obtained and tested to assess their insect resistance capabilities by Mennella *et al.* (2004, 2005). The two best transgenic lines were selfed to obtain homozygous lines for CPB gene resistance, and were then used as parents for transgenic hybrids. Field trials were set up at two different locations for 2 years, evaluating the performances of two transgenic hybrids and their untransformed control. In the same experimental design, plots composed of 100% transgenic plants were compared with mix plots with 80% transgenic plants and 20% untransformed refuge plants. Data were recorded biweekly to assess the level of CPB infestation by monitoring the abundance of adults, egg masses, and larvae. At harvest, the total yield per plot was recorded. Field observations confirmed that *Bt*-transgenic eggplant hybrids were able to control CPB infestation. Significantly higher production was obtained from both transgenic hybrids compared with the control in the year with the heavier CPB attack. No statistical differences were evidenced (in yield or insect abundance) between the two methods of transgenic plant deployment, and thus the adoption of a mixed plot with refuge plants may represent a potentially effective and durable means of CPB control in eggplant cultivation.

The effect of a genetically modified eggplant line expressing a protease inhibitor, oryzacystatin, on the aphids *Myzus persicae* (Sulzer) and *Macrosiphum euphorbiae* (Thomas) was examined by Ribeiro *et al.* (2006). The transgenic eggplant reduced the net reproductive rate, the instantaneous rate of population increase, and the finite rate of population increase of both aphid species compared with a control eggplant line. The mean generation time of the aphids was unaffected by the transgenic plants. Age-specific mortality rates of *Myzus persicae* and *Macrosiphum euphorbiae* were higher on the transgenic plants. These results indicated that expression of oryzacystatin in eggplant has a negative impact on population growth and mortality rates of *Myzus persicae* and *Macrosiphum euphorbiae* and could be a source of plant resistance for pest management of these aphids.

An experiment was conducted by Ribeiro *et al.* (2005) to evaluate the resistance of transgenic eggplant carrying the oryzacystatin gene to *Mechanitis polymnia* and *Mechanitis lysimnia.* The leaf area consumed, and the mortality and duration of the larval and pupal phases of *M. polymnia* and *M. lysimnia* were evaluated, and it was found that the transgenic eggplant was not resistant to *M. polymnia* and *M. lysimnia.*

Non-*Bt* transgenics and transgenics for incorporating genetic resistance to biotic stresses have been attempted and are well documented (Table 2.2).

### *Resistance to bacterial and fungal diseases*

Xing and Chin (2000) attempted to engineer fungal resistance in eggplant by overexpression of the yeast Δ-9 desaturase gene, which resulted in higher concentrations of 16:1, 18:1, and 16:3 fatty acids. Such transgenics exhibited increased resistance to *Verticillium* wilt. Transgenic plants challenged by the pathogen also resulted in a marked increase in the content of cellular 16:1 and 16:3 fatty acids. The authors showed that *cis*-Δ-9 16:1 fatty acid was inhibitory to the growth of *Verticillium* pathogen. Transgenic eggplant resistant to *Verticillium dahliae* and *Fusarium oxysporum* wilts by overexpression of pathogenesis-related genes, such as alfalfa acidic glucanase, chitinase, and thaumatin (singly and in combination), was also produced (Singh, 2004; Singh *et al.*, 2014). Singh *et al.* (2015) developed transgenic eggplant expressing the class I rice endochitinase gene and showed that transgenic plants exhibited a delayed onset of disease by 5–7 days from *V. dahliae* and *F. oxysporum* infection.

**Table 2.2.** Transgenic eggplant genes/constructs for resistance to biotic stresses.

| Gene/construct | Trait | Reference |
|---|---|---|
| Oryzacystatin | Resistance to aphids | Ribeiro *et al.* (2005, 2006) |
| *Mi-1.2* | Resistance to root-knot nematode | Goggin *et al.* (2006) |
| Yeast $\Delta^9$ desaturase | Resistance to *Verticillium* wilt | Xing and Chin (2000) |
| Rice cystatin OC-I$\Delta$D86 | Resistance to nematode | Papolu *et al.* (2016) |
| *Mi-msp-1* | Resistance to nematode | Chaudhary *et al.* (2019) |
| Silencing pharyngeal gland genes | Resistance to nematode | Shivakumara *et al.* (2017) |
| *SmMYB44* (R2R3) transcription factor | Resistance to bacterial wilt | Qiu *et al.* (2019) |
| Alfalfa/rice acidic glucanase | Resistance to *Verticillium dahliae* and *Fusarium oxysporum* fungal wilt | Singh *et al.* (2014, 2015) |
| Wasabi defensin gene with R/RS site-specific recombination system | Resistance to *Alternaria solani* | Darwish *et al.* (2014) |
| NAC family transcription factor | Resistance to bacterial wilt | Na *et al.* (2016) |
| *SmMLO1* | Resistance/susceptibility to powdery mildew in Solanaceae | Bracuto *et al.* (2017) |
| Antifungal *Dm-AMP1* | Resistance to *Botrytis cinerea* and *Verticillium albo-atrum* | Turrini *et al.* (2004) |
| Hevein | Fungal resistance | Bhat *et al.* (2020) |
| *Escherichia coli mtlD* | Resistance to *Fusarium oxysporum*, *Verticillium dahliae*, and *Rhizoctonia solani* | Prabhavathi and Rajam (2007) |
| Coat protein gene | Resistance to cucumber mosaic virus | Pratap *et al.* (2011) |

Antimicrobial compounds synthesized by seeds or plant tissues constitute a powerful armory to enhance crop resistance against various fungal pathogens, as demonstrated in many crops. Many plants produce "morphogenic" and "non-morphogenic" defensins that inhibit pathogenic fungal growth by reducing hyphal elongation and increasing hyphal branching. For example, *Dahlia merckii* (Asteraceae) has Dm-AMP1, a typical non-morphogenic defensin that interacts with mannosylated sphingolipids occurring in the outer plasma membrane of hyphae. Turrini *et al.* (2004) developed transgenic eggplant expressing Dm-AMP1. The transgenic lines were found to be resistant to *Botrytis cinerea* and *Verticillium albo-atrum*. Bioassays with the arbuscular mycorrhizal symbiont *Glomus mosseae* indicated that transformed plants showed better growth, indicating enhanced growth of the symbiont.

Prabhavathi *et al.* (2002) reported that transgenic eggplant containing the *Escherichia coli* mannitol-1-phosphate dehydrogenase (*mtlD*) gene showed increased tolerance to abiotic stresses. The same group (Prabhavathi and Rajam, 2007) also showed that such transgenics expressing *mtlD* with mannitol accumulation exhibited increased resistance against three fungal wilts caused by *F. oxysporum*, *V. dahliae*, and *Rhizoctonia solani* under both *in vitro* and *in vivo* growth conditions. Mannitol levels could not be detected in the wild-type plants, but the presence of mannitol in the transgenics could be positively correlated with disease resistance. These results and previous data suggest that the *mtlD* gene could be used for engineering crop plants for both biotic and abiotic stress tolerance.

Darwish *et al.* (2014) cloned the wasabi defensin gene from *Wasabia japonica* (a Japanese horseradish plant with potential antimicrobial proteins) into an *ipt*-type multi-auto-transformation vector by gateway cloning technology and transformed eggplant using plant growth regulator- and antibiotic-free MS medium. *In vitro* whole plant and detached leaf assays of the marker-free transgenic plants exhibited improved resistance against *Alternaria solani*.

Polyamines, including putrescine, spermine, and spermidine, are natural aliphatic polycations found ubiquitously in organisms and modulate DNA–protein and protein–protein interactions and also affect RNA structure. They are involved in the regulation of plant growth, development, and specific responses to biotic and abiotic stressors. Their genes are under tight transcriptional regulation during stress responses. In eggplant, few resistance genes have been identified to address bacterial wilt caused by *Ralstonia solanacearum*, and therefore the underlying mechanism of wilt resistance is unclear. Qiu *et al.* (2019) speculated about the role of a spermidine synthase gene, *SmSPDS*. Using virus-induced gene silencing to create knockdown lines, a yeast one-hybrid assay, and a dual luciferase-based transactivation assay, they identified an R2R3-MYB transcription factor, SmMYB44, that directly binds to the *SmSPDS* promoter and activates its expression. Overexpression and *SmMYB44* RNA interference (RNAi) transgenic plants confirmed the enhancement of resistance by this candidate gene.

NAC family transcription factors possess an N terminus with a highly conserved NAC domain and originally were characterized due to their sequence similarities to the *Petunia* hybrid *NO APICAL MERISTEM* (*NAM*) and the *Arabidopsis thaliana ATAF1*, *ATAF2*, and *CUP-SHAPED COTYLEDON* (*CUC2*) genes. They perform a variety of roles in plant growth and development, responses to biotic and abiotic stresses, hormone signal transduction, flowering time, and fruit maturation. They are involved in coordinating responses to attacks by phytopathogens. Na *et al.* (2016) identified an NAC transcription factor (SmNAC) from eggplant based on RNA sequencing data and showed that SmNAC is involved in the resistance of eggplant to bacterial wilt caused by *R. solanacearum*. They studied its localization at the tissue and subcellular levels. They generated transgenic eggplant with either constitutive upregulation or suppression (using RNAi) of this gene, and demonstrated that, while overexpression of SmNAC decreased resistance to the pathogen by reduced accumulation of the signaling molecule salicylic acid together with reduced expression of the *ICS1* gene encoding isochorismate synthase 1 (involved in salicylic acid biosynthesis), RNAi-silenced *SmNAC* transgenic lines had higher salicylic acid levels. This suggested that SmNAC represses the accumulation of salicylic acid by inhibiting *ICS1* expression, thereby lowering resistance to the pathogen.

Similarly, another susceptibility gene was studied in eggplant. The Mildew Locus O (*MLO*) gene family encodes plant-specific proteins harboring seven transmembrane domains and a calmodulin-binding domain. Specific homologs of the *MLO* gene family act as susceptibility factors for powdery mildew fungal pathogens, as their loss-of-function alleles are associated with a recessive form of resistance, known as *mlo* resistance. Bracuto *et al.* (2017) transgenically expressed the eggplant powdery mildew susceptibility gene, *SmMLO1*, in a tomato *mlo* mutant with compromised resistance to the tomato powdery mildew pathogen (*Oidium neolycopersici*), while also restoring susceptibility by transgenic expression of a synthetic gene, named *s-SmMLO1*, stimulating powdery mildew resistance research in eggplant in particular and in other crops in general using reverse-genetics methods such as targeted induced local lesions in genomes (TILLING), RNAi, and genome editing.

In the search for and in breeding for resistance to the *R. solanacearum* species complex (RSSC), an interesting study by Morel *et al.* (2018) validated the contribution of the *GMI1000* allele of a type III-secreted effector and a rather highly prevalent effector protein, RipAX2, to control the disease by triggering specific resistance in eggplant accession AG91-25, which carries the major resistance genetic determinant locus *EBWR9*. RipAX2 is the second identified RSSC effector protein involved in triggering resistance in eggplant to *Ralstonia* spp., the first being RipP2 (formerly known as PopP2) recognized by *RE-bw* (Xi'ou *et al.*, 2014). It was shown that RipAX2 triggered resistance in AG91-25 resistant eggplant, whose recognition occurs in the early stages of infection, possibly mediated by a resistance (*R*) gene present in the *ERs1*/*EBWR9* locus. Additionally, another effector, Rip36, critically interacting with a putative zinc-binding motif, induced a hypersensitive response in the non-host wild eggplant, *Solanum torvum* (Nahar *et al.*, 2014), which needs further investigation.

### *Resistance to viral diseases*

Cucumber mosaic virus is a major constraint for yield and quality in eggplant. Pathogen-derived resistance breeding and transgenics are options in virus resistance breeding. Pratap *et al.* (2011)

developed cucumber mosaic virus coat protein transgenic eggplant that showed a significant level of tolerance against cucumber mosaic virus under glasshouse conditions.

### *Resistance to nematodes*

The *Mi-1.2* gene in tomato is a member of the nucleotide-binding site–leucine-rich repeat (NBS-LRR) class of plant resistance genes, and confers resistance against root-knot nematodes (*Meloidogyne* spp.), potato aphid (*Macrosiphum euphorbiae*), and sweet potato whitefly (*Bemisia tabaci*). *Mi-1.2* mediates a rapid, local defensive response at the site of infection, although the signaling and defensive pathways required for resistance are largely unknown. In a study by Goggin *et al.* (2006), eggplant was transformed with *Mi-1.2* to determine whether this gene could function in a genetic background other than tomato. Eggplants that carried *Mi-1.2* displayed resistance to the root-knot nematode *Meloidogyne javanica* but were fully susceptible to the potato aphid, whereas a susceptible tomato line transformed with the same transgene was resistant to both nematodes and aphids. This study showed that *Mi-1.2* can confer nematode resistance in another solanaceous species. It also indicates that the requirements for *Mi*-mediated aphid and nematode resistance differ. Potentially, aphid resistance requires additional genes that are not conserved between tomato and eggplant.

## 2.3.3 Transgenics for abiotic stress tolerance

Various genes have been deployed in transgenic eggplant to address the problem of abiotic stresses (Table 2.3). For example, by expressing the bacterial mannitol-1-phosphate dehydrogenase (*mtlD*) gene, which is involved in mannitol synthesis, Prabhavathi and Rajam (2007) achieved tolerance against osmotic stress induced by salt, drought, and chilling stress.

Interestingly, these transgenic plants also showed enhanced resistance to fungal wilts, as noted earlier in this chapter, caused by *V. dahliae* and *F. oxysporum*. Various transgenics have now been developed using potential genes such as arginine decarboxylase, ornithine decarboxylase, *S*-adenosylmethionine decarboxylase, and spermidine synthase, encoding enzymes in the polyamine metabolic pathway, and these transgenic plants have shown increased tolerance to multiple abiotic stresses (salinity, drought, extreme temperature, and heavy metals), as well as to biotic stresses. Gene-pyramiding (gene-stacking) approaches show potential to further increase the levels of tolerance in eggplant.

In an attempt to improve and enhance chilling tolerance in eggplant, Wan *et al.* (2014) transferred *Arabidopsis C-REPEAT BINDING FACTOR 3* (*AtCBF3*) and *COLD-REGULATED 15A* (*AtCOR15A*) genes, both driven by an effective *Arabidopsis RESPONSIVE TO DESSICATION 29A* (*AtRD29A*) promoter. In transgenic lines, the gene expression, proline content, and levels of catalase and peroxidase activities significantly increased, while relative electrical conductivity and malondialdehyde content were remarkably decreased compared with control plants, showing that ectopic expression of these two genes could promote the cold-adaptation process in eggplant from chilling stress. Similarly, Sagare and Mohanty (2012) developed transgenic eggplant using the *rd29A*::*DREB1A* promoter–gene combination. Analysis of relative water content,

**Table 2.3.** Transgenic eggplant genes/constructs for tolerance to abiotic stresses.

| Gene/construct | Trait | Reference |
|---|---|---|
| *Escherichia coli* mannitol-1-phosphate dehydrogenase (*mtlD*) | Abiotic stress tolerance | Prabhavathi *et al.* (2002) |
| *rd29A*::*DREB1A* | Moisture stress tolerance | Sagare and Mohanty (2012) |
| Wheat *TaNHX2* vacuolar $Na^+/H^+$ antiporter | Tolerance to salinity stress | Yarra and Kirti (2019) |
| Yeast *HAL1* gene | Salt tolerance | Kumar *et al.* (2014) |
| *Arabidopsis C-REPEAT BINDING FACTOR 3* (*AtCBF3*) and *COLD-REGULATED 15A* (*AtCOR15A*) | Improved chilling tolerance | Wan *et al.* (2014) |

proline content, photosynthetic pigments, and membrane stability index after a 15-day moisture stress induction indicated tolerance of the transgenic plants to moisture stress.

Vacuolar $Na^+/H^+$ antiporters provide an excellent mechanism for ionic homeostasis in plants under salt stress. Yarra and Kirti (2019) developed transgenic eggplant containing the wheat antiporter *TaNHX2* gene driven by a double cauliflower mosaic virus 35S promoter. Stress-inducible expression of *TaNHX2* significantly enhanced growth performance and the $Na^+$ and $K^+$ profiles from leaves and roots under salt stress. In addition, the transgenic plants displayed stable leaf relative water content, chlorophyll content, and proline accumulation, and had improved photosynthetic efficiency, transpiration rate, and stomatal conductivity under 200 mM NaCl stress. They also revealed a reduction in malondialdehyde content, and hydrogen peroxide and oxygen radical production, with increased antioxidant enzyme activity. Ectopic expression of the *HAL1* gene of yeast in transgenic eggplant, undertaken by Kumar *et al.* (2014), resulted in the development of salt tolerance up to 50 mM NaCl in the transgenic lines.

### 2.3.4 Transgenics for other plant traits

Eggplant has been an interesting model solanaceous plant to understand and improve plant traits (Table 2.4) and to answer important basic and strategic plant biology questions (Table 2.5) through the deployment of interesting genes. Through the introduction of the *iaaM* gene from *Pseudomonas syringae* pv. *savastanoi* under the control of the ovule-specific promoter *DefH9* from *Antirrhinum majus* and thereby engineering auxin biosynthesis, transgenic eggplants with parthenocarpic fruit were developed (Rotino *et al.*, 1997a,b). These transgenics produced seedless parthenocarpic fruit in the absence of pollination without the external application of the plant hormones, even at low temperature, which normally prohibit fruit production in untransformed lines. Furthermore, these transgenics exhibited significantly higher winter yields than the untransformed plants and a commercial hybrid under an unheated glasshouse trial (Donzella *et al.*, 2000). Trials in the open field for summer production and in the greenhouse for early spring production confirmed that transgenic parthenocarpic eggplant $F_1$ hybrids had higher production coupled with improved fruit quality with respect to the untransformed controls (Acciarri *et al.*, 2000, 2002). The quality parameters of the frozen parthenocarpic fruit did not show any significant changes when compared with their controls (Maestrelli *et al.*, 2003).

Du *et al.* (2016) showed that SmARF8 auxin response factor is an active transcriptional regulator involved in parthenocarpy in eggplant using a natural parthenocarpic mutant. Transgenic

**Table 2.4.** Transgenic eggplant genes/constructs for other improved traits.

| Gene/construct | Trait | Reference |
|---|---|---|
| General transcription factors | Artificial microRNA-mediated silencing and reversible male sterility | Toppino *et al.* (2011) |
| Barnase and Cre-*lox* system | Fertility restoration | Cao *et al.* (2010) |
| P450 *CYP76AD1*, *BvDoDA1* dioxygenase, and *cDOPA5GT* glycosyltransferase (betalain-related) genes | Improved antioxidant and fruit pigmentation by betaxanthin betalain | Polturak *et al.* (2017) |
| Gene editing of *SmelPPO4*, *SmelPPO5* and *SmelPPO6* polyphenoloxidase genes | Reduction of fruit flesh browning | Maioli *et al.* (2020) |
| *EEF48-crtB* | Enhanced carotenoid content in fruits | Mishiba *et al.* (2020) |
| *SmARF8* | Transgenic parthenocarpy | Du *et al.* (2016) |
| *DefH9-iaaM* | Transgenic parthenocarpy | Rotino *et al.* (1997b); Donzella *et al.* (2000); Acciarri *et al.* (2002) |
| *gfp*, *uidA*, *hpt*, *virE*, *lacZ* | Increased transformation | Kumar and Rajam (2005) |

**Table 2.5.** Transgenic eggplant genes/constructs for basic/strategic studies.

| Gene/construct | Trait | Reference |
|---|---|---|
| *nos-npt* | First successful genetic transformation using cointegrated/ hybrid Ti plasmid vector system | Guri and Sink (1988) |
| *gfp, uidA, hpt, virE, lacZ* | Increased transformation | Kumar and Rajam (2005) |
| *nptII* | Efficient and stable transformation | Franklin and Sita (2003) |
| Aminoglycoside 3′-adenyltransferase | Biolistic plastid transformation | Singh *et al.* (2010), Bansal and Singh (2014) |
| *AtGRP-5::uidA* | Optimizing transformation efficiency | Magioli *et al.* (2001a) |
| *luc* | Evaluating stability of luciferase marker expression | Hanyu *et al.* (1999) |
| *bar* and *uidA* | *In planta* vacuum infiltration transformation of seeds | Subramanyam *et al.* (2013) |

lines harboring RNAi for *SmARF8* exhibited parthenocarpy, indicating its negative role in parthenocarpic fruit development. Introduction of full-length cDNA *SmARF8* could partially complement the parthenocarpic phenotypes in *Arabidopsis arf8-1* and *arf8-4* mutants.

Eggplants are low in provitamin A carotenoid content, and any enhancement will go a long way toward addressing this vitamin A deficiency. Mishiba *et al.* (2020) implemented a metabolic engineering strategy to modify the plant carotenoid biosynthetic pathway by identifying endogenous phytoene synthase expression as the limiting step. They amplified the fruit-specific *EEF48* gene promoter by thermal asymmetric interlaced-PCR and transgenic lines were developed with an *EEF48*-driven *crtB* gene. One self-pollinated progeny line showed a 30-fold increase in β-carotene accumulation in the fruit.

Polturak *et al.* (2017) conducted metabolic engineering and biofortification in eggplant, along with tobacco, tomato, potato, and ornamental *Petunia* × *hybrida*. They simultaneously attempted engineering and heterologous production of betalains in eggplant and obtained strong red-violet coloration in the flesh and skin. Several betaxanthins were identified; vulgaxanthin I and III in potato and indicaxanthin in tomato. However, the major betaxanthin detected in eggplant was novel and remained unidentified. Transgenic tobacco also developed resistance to gray mold caused by *B. cinerea*.

Much work has been done to exploit biotechnology in heterosis breeding in crops, as exemplified using the barnase/barstar system. Cao *et al.* (2010) undertook the molecular regulation of pollen sterility/fertility using the barnase gene with Cre/*loxP*-mediated site-specific recombination system in transgenic eggplant. An eggplant inbred/pure line, E38, was transformed with the Cre gene, while another inbred/pure line was transformed with the barnase gene situated between *loxp* sites. In the transgenic lines, flower stamens were shorter and male sterile, but crossing male-sterile plants with Cre-expressing plants produced normal hybrid fruit with full restoration of fertility.

Toppino *et al.* (2011) induced male sterility by suppressing endogenous general transcription factor genes, TAFs, using anther-specific promoters combined with artificial microRNA technology. The system was made reversible by the ethanol-inducible expression of a microRNA-insensitive form of the target gene. They obtained completely male-sterile plants, and fertility could be restored by short treatments with ethanol. This, coupled with induced parthenocarpy, opens the possibility for heterosis breeding, transgene containment, and novel gene regulation.

Eggplant fruit contains high amounts of phenolic compounds, represented mainly by chlorogenic acid (5-*O*-caffeoylquinic acid), which has antioxidant, cardioprotective, antibacterial, anti-inflammatory, hepatoprotective, and antimicrobial therapeutic properties. However, phenolics are also responsible for the browning phenomenon and are the result of the action of polyphenol oxidases. Selective fruit-specific reduction of action of these enzymes has resulted in browning in tomato and potato. Maioli *et al.* (2020) selected three *PPO* genes, *SmelPPO4*, *SmelPPO5* and *SmelPPO6*, for the generation of knockout mutants using CRISPR-Cas9-based gene editing. The edited lines showed a reduction of the typical brown coloration in the flesh due to phenolics, without detectable off-target effects.

## 2.4 Commercialization

*Bt*-transgenic eggplant, popularly called *Bt* brinjal in India and *Bt* baingan in Bangladesh, has faced several hurdles for its commercialization in India. *Bt* eggplant can potentially decrease insecticide input use against EFSB by as much as 80%. The *Bt* transgene is effective on EFSB with 98% insect mortality compared with less than 30% mortality in non-*Bt* plants. The yield benefits on average increase to 116% in marketable fruits compared with non-Bt hybrids, while they increase to 166% over popular open-pollinated varieties. It is also estimated that *Bt* eggplant can bring about a 3% net increase (i.e. US$250 or INR 18,000) of average economic benefit per acre to the farmer, mainly through savings from insecticides. The national benefits may therefore extend to more than US$400 million year$^{-1}$ (Bennett *et al.*, 2008; Krishna and Qaim, 2008; Kumar *et al.*, 2011). This scenario can be applied on a global scale.

The Maharashtra Hybrid Seeds Company (Mahyco) began the development of commercial *Bt* eggplant in 2000. In late 2003, a partnership of Mahyco, Cornell University, Sathguru Management Consultants, and the US Agency for International Development (USAID) under the Agricultural Biotechnology Support Program (ABSP) II was formed. Three additional partners, TNAU, UAS-D, and ICAR-IIVR (Indian Council of Agricultural Research-Indian Institute of Vegetable Research), made arrangements for the development of *Bt* eggplant in open-pollinated varieties. Successful multilocation field trials of *Bt* lines were conducted. The GEAC of the MoEF of the Government of India, in its 97th meeting held on 14 October 2009, recommended the environmental release of the Mahyco-Monsanto *Bt* eggplant Event EE1. After seven rounds of nationwide consultative public meetings and numerous discussions, Mr. Jairam Ramesh, the then Minister of Environment and Forests, decided to impose a moratorium on the commercial cultivation of *Bt* eggplant (Event EE1). Currently, the fate of commercialization of *Bt* eggplant remains uncertain.

The same Event EE1 was provided to BARI (Bangladesh) under ABSP-II (now the South Asia Eggplant Improvement Partnership (SAEIP)) for introgression into local varieties. After all regulatory considerations, four *Bt* varieties were developed. BARI applied to the National Technical Committee on Crop Biotechnology (NTCCB) of Bangladesh for permission to commercially release the varieties. Following the recommendations of NTCCB and the National Technical Committee on Biosafety, the Bangladesh government granted approval for the commercial release and "limited cultivation" on 30 October 2013. Within 3 months, *Bt* seedlings were distributed to 20 farmers in four districts. In 2014–2015, BARI supplied planting material to 108 farmers in 19 districts. Over the next 2 years, demonstration trials were conducted in 250 farmers' fields in 25 districts and in 512 farmers' fields in 36 districts, respectively. By 2018, nearly 150,000 farmers cultivated *Bt* eggplant (Shelton *et al.*, 2018). *Bt* eggplant commercial cultivation brought about a significant change in the socio-economic status of vegetable farmers in Bangladesh (Haque and Saha, 2020).

Mahyco also provided the *Bt* eggplant Event EE1 to the Philippines. Scientists from the University of the Philippines, Institute of Plant Breeding, Los Banos, developed *Bt* eggplant, known as talong. After rigorous biosafety assessment, the Philippine Department of Agriculture approved *Bt* eggplant (Event EE1) for food, feed, and processing (ISAAA, 2021).

## 2.5 Future Perspectives

Conflicting interests and opinions prevail with regard to the acceptance and commercialization of genetically engineered eggplant in India. The generation of dependable and extensive data to convince all sections of society that the transgenic technology is safe and beneficial in the larger interest of society and the environment is key for the acceptance of transgenic or genetically modified crops. However, successful global cultivation of transgenic pest-resistant *Bt* crops such as cotton, maize, and eggplant should prompt the regulatory authorities to revisit the moratorium imposed on *Bt* eggplant in India. Overall, there is no doubt that transgenic eggplant offers a great opportunity for another revolution in vegetable crop cultivation, thereby addressing food and nutritional security and sustainable agriculture.

Next-generation speed breeding, along with transgenic technology, can involve newer technologies such as TILLING and other techniques through programmable endonucleases (e.g. zinc finger, transcription activator-like effector nucleases (TALENS)), RNAi, and the successful CRISPR-Cas9-based gene editing (El-Mounadi *et al.*, 2020) in solanaceous crops (Wang *et al.*, 2021) and with special reference to eggplant (Gianoglio *et al.*, 2017; Maioli *et al.*, 2020).

## 2.6 References


Acciarri, N., Vitelli, G., Arpaia, S., Mennella, G., Sunseri, F., and Rotino, G.L. (2000) Transgenic resistance to the Colorado potato beetle in Bt-expressing eggplant fields. *HortScience* 35, 722–725.

Acciarri, N., Restano, F., Vitelli, G., Perrone, D., Zottini, M., *et al.* (2002) Genetically modified parthenocarpic eggplants: improved fruit productivity under both greenhouse and open field cultivation. *BMC Biotechnology* 2: 4.

Ali, M., Okubo, H., and Fujieda, K. (1991) *In vitro* multiplication of intraspecific and interspecific *Solanum* hybrids through somatic embryogenesis and adventitious organogenesis. *Journal of the Japanese Society for Horticultural Science* 60, 601–612.

Alicchio, R., Del Grosso, E., and Boschueri, E. (1982) Tissue cultures and plant regeneration from different explants in six cultivars of *Solanum melongena*. *Experientia* 38, 449–450.

Arpaia, S., Mennella, G., Onofaro, V., Perri, E., Sunseri, E., and Rotino, G.L. (1997) Production of transgenic eggplant (*Solanum melongena* L.) resistant to Colorado potato beetle (*Leptinotarsa decemlineata* Say). *Theoretical and Applied Genetics* 95, 329–334.

Arpaia, S., Acciarri, N., Leo, G.M.D., Mennella, G., Sabino, G., *et al.* (1998) Field performance of *Bt* expressing transgenic eggplant lines resistant to Colorado potato beetle. In: *Proceedings of the Xth EUCARPIA Meeting on Genetics and Breeding of Capsicum and Eggplant, 7–11 September, Avignon, France*, pp. 191–194.

Arpaia, S., Leo, G.M., Fiore, M.C., Schmidt, J.E.U., and Scardi, M. (2007) Composition of arthropod species assemblages in *Bt*-expressing and near isogenic eggplants in experimental fields. *Environmental Entomology* 36, 213–227.

Bhalla, S., Gupta, K., and Khetarpal, R.K. (2007) Status of genetically engineered insect pest resistant crops and related biosafety concerns. *Indian Journal of Plant Protection* 35, 220–226.

Bansal, K.C., and Singh, A.K. (2014) Plastid transformation in eggplant. *Methods in Molecular Biology* 1132, 305–316.

Bennett, R., Shankar, B., and Srinivasan, C.S. (2008) The economics of agricultural biotechnology—an overview. *International Journal of Biotechnology* 10, 109–278.

Bhat, S.G., Arulananthu, G., Rajesh, G., and Ramesh, N. (2020) *Agrobacterium*-mediated transformation of brinjal (*Solanum melongena* L.) using fungal resistant gene. *Electronic Journal of Plant Breeding* 11, 160–168.

Billings, S., Jelenkovic, G., Chin, C.K., and Eberhardt, J. (1997) The effect of growth regulation and antibiotics on eggplant transformation. *Journal of the American Society for Horticultural Science* 122, 158–162.

Boopal, K., Arya, V.V., Rami Reddy, P.V., and Hanur, V. S. (2013) Potato, a suitable alternative host for brinjal shoot and fruit borer *Leucinodes orbonalis* Guenee (Lepidoptera: Pyralidae). *International Journal of Agricultural Sciences* 3, 179–184.

Bracuto, V., Appiano, M., Ricciardi, L., Gol, D., Visser, R.G.F., *et al.* (2017) Functional characterization of the powdery mildew susceptibility gene *SmMLO1* in eggplant (*Solanum melongena* L.). *Transgenic Research* 26, 323–330.

Cao, B., Huang, Z., Chen, G., and Lei, J. (2010) Restoring pollen fertility in transgenic male-sterile eggplant by *Cre*/*lox*p-mediated site-specific recombination system. *Genetics and Molecular Biology* 33, 298–307.

Chandra Kumar, H.L., Ashok Kumar, C.T., Kumar, N.G., Chakravarthy, A.K., and Putta Raju, T.B. (2008) Seasonal occurrence of major insect pests and their natural enemies on brinjal. *Current Biotica* 2, 63–73.

Chaudhary, S., Dutta, T.K., Tyagi, N., Shivakumara, T.N., Papolu, P.K., *et al.* (2019) Host-induced silencing of *Mi-msp-1* confers resistance to root-knot nematode *Meloidogyne incognita* in eggplant. *Transgenic Research* 28, 327–340.

Chauvin, E.J., Marhadour, S., Cohat, J., and Nard, L.M. (1999) Effects of gelling agents on *in vitro* regeneration and kanamycin efficiency as a selective agent in plant transformation procedures. *Plant Cell, Tissue and Organ Culture* 58, 213–217.

Chen, Q., Jelenkovic, G., Cheekok, C., Billings, S., Eberhardt, J., *et al.* (1995) Transfer and transcriptional expression of coleopteran *CryIIIB* endotoxin gene of *Bacillus thuringiensis* in eggplant. *Journal of the American Society for Horticultural Science* 120, 921–927.

Choudhary, B. and Gaur, K. (2009) *The Development and Regulation of* Bt *Brinjal in India*. International Service for the Acquisition of Agri-biotech Applications (ISAAA) Brief. No. 38. ISAAA, Ithaca, New York.

Choudhary, B., Nasiruddin, K.M., and Gaur, K. (2014) *The Status of Commercialized Bt Brinjal in Bangladesh*. International Service for the Acquisition of Agri-biotech Applications (ISAAA) Brief No. 47. ISAAA, Ithaca, New York.

Collonnier, C., Fock, I., Kashyap, V., Rotino, G.L., Daunay, M.C., *et al.* (2001) Applications of biotechnology in eggplant. *Plant Cell, Tissue and Organ Culture* 65, 91–107.

Darwish, N.A., Khan, R.S., Ntui, V.O., Nakamura, I., and Mii, M. (2014) Generation of selectable marker-free transgenic eggplant resistant to *Alternaria solani* using the R/RS site-specific recombination system. *Plant Cell Reports* 33, 411–421.

de Candolle, A. (1886) *Origin of Cultivated Plants*, 2nd edn (reprinted in 1959). Hafner Publishing Company, New York/London.

Donzella, G., Spena, A., and Rotino, G.L. (2000) Transgenic parthenocarpic eggplants: superior germplasm for increased winter production. *Molecular Breeding* 6, 79–86.

Du, L., Bao, C., Hu, T., Zhu, Q., Hu, H., *et al.* (2016) SmARF8, a transcription factor involved in parthenocarpy in eggplant. *Molecular Genetics and Genomics* 291, 93–105.

El-Mounadi, K., Morales-Floriano, M.L., and Garcia-Ruiz, H. (2020) Principles, applications and biosafety of plant genome editing using CRISPR-Cas9. *Frontiers in Plant Science* 11: 56.

FAO (Food and Agriculture Organization of the United Nations) (2009) *International Treaty on Plant Genetic Resources for Food and Agriculture*. FAO, Rome.

Fari, M., Nagy, I., Csanyi, M., Mityko, J., and Andrasfalvy, A. (1995) *Agrobacterium* mediated genetic transformation and plant regeneration via organogenesis and somatic embryogenesis from cotyledon leaves in eggplant (*Solanum melongena* L. cv. 'Kecskemeti lila'). *Plant Cell Reports* 15, 82–86.

Fillipone, E. and Lurquin, P.F. (1989) Stable transformation of eggplant (*Solanum melongena* L.) by cocultivation of tissues with *Agrobacterium tumefaciens* carrying a binary plasmid vector. *Plant Cell Reports* 8, 370–373.

Franklin, G., and Lakshmi Sita, G. (2003) *Agrobacterium tumefaciens*-mediated transformation of eggplant (*Solanum melongena* L.) using root explants. *Plant Cell Reports* 21, 549–554.

Fujimoto, H., Itoh, K., Yamamoto, M., Kyozuka, J., and Shimamoto, K. (1993) Insect resistant rice generated by introduction of a modified delta-endotoxin gene of *Bacillus thuringiensis*. *Bio/tech* 11, 1151–1155.

Garcia-Fortea, E., Lluch-Ruiz, A., Pineda-Chaza, B.J., Garcia-Perez, A., Bracho-Gil, J.P., *et al.* (2020) A highly efficient organogenesis protocol based on zeatin riboside for in vitro regenearation of eggplant. *BMC Plant Biology* 20, 1–16.

Garcia-Fortea, E., Garcia-Perez, A., Gimeno-Paez, E., Martinez-Lopez, M., Vilanova, S., *et al.* (2021) Ploidy modification for plant breeding using *in vitro* organogenesis: a case in eggplant. *Methods in Molecular Biology* 2264, 197–206.

Gianoglio, S., Comino, C., Moglia, A., Lanteri, S., Jaime, P., and Diego, O. (2017) CRISPR/Cas9-mediated knock-out of polyhenol oxidase genes in eggplant (*Solanum melongena* L.). In: *SolCuc2017: XIV Solanaceae and 3rd Cucurbitaceae Joint Conference, 3–6 September, València, Spain*, p. 159.

Goggin, F.L., Jia, L.L., Shah, G., Hebert, S., Williamson, V.M., and Ullman, D.E. (2006) Heterologous expression of the *Mi-1.2* gene from tomato confers resistance against nematodes but not aphids in eggplant. *Molecular Plant–Microbe Interactions* 19, 383–388.

Guri, A. and Sink, K.C. (1988) *Agrobacterium* transformation of eggplant. *Journal of Plant Physiology* 133, 52–55.

Hamilton, G.C., Jelenkovic, G.L., Lashomb, J.H., Ghidiu, G., Billings, S., and Patt, J.M. (1997) Effectiveness of transgenic eggplant (*Solanum melongena* L.) against the Colorado potato beetle. *Advances in Horticultural Science* 11, 189–192.

Hanur, V.S. (2004) Genetic transformation of human cells by a soil phytopathogen presents common molecular strategies. *Current Science* 87, 856–857.

Hanur, V.S. (2007) Exciting developments in plant stem cell research. *Current Science* 92, 1668–1670.

Hanur, V.S. (2008) *Bt* resistance and monophagous pests: handling with prudence. *Current Science* 95, 449–451.

Hanur, V.S. (2010a) Brinjal. In: Singh, H.P., Parthasarathy, V.A., and Nirmal Babu, K. (eds.) *Regeneration in Vegetable Crops. Advances in Horticultural Biotechnology Series*. Westville Publishing House, New Delhi, pp. 53–72.

Hanur, V.S. (2010b) Brinjal. In: Singh, H.P., Parthasarathy, V.A., and Nirmal Babu, K. (eds.) *Transgenics in Vegetable Crops. Advances in Horticultural Biotechnology Series*. Westville Publishing House, New Delhi, pp. 251–267.

Hanur, V.S. (2010c) Advances in stem cell research in plants. In: Malik, C.P. (ed.) *Genetic Engineering: A New Hope for Crop Production and Management*. Aavishkar Publishers, Jaipur, India, pp. 289–321.

Hanur, V.S. (2011a) GM crops and centres of origin—a case study of Bt brinjal in India. *Current Science* 100, 1285–1286.

Hanur, V.S. (2011b) Brinjal. In: Singh, H.P., Parthasarathy, V.A., and Nirmal Babu, K. (eds.) *Regeneration Systems – Vegetables, Ornamentals and Tuber Crops, Vol. I. Advances in Horticultural Biotechnology Series.* Westville Publishing House, New Delhi, pp. 53–72.

Hanur, V.S., Prakash, D.P., Deepali, B.S., Asokan, R., Ramachandra, Y.L., *et al.* (2006) Synergistic use of hypocotyl explants and high BAP preconditioning for enhanced transformation frequency in brinjal (*Solanum melongena* L.). *Journal of Horticultural Sciences* 1, 116–119.

Hanur, V.S., Reddy, P.V.R., Boopal, K., Srividya, K.N., Poovarasan, S., *et al.* (2011) *Solanum macrocarpon*, a wild brinjal, is not a source of resistance against brinjal shoot and fruit borer, *Leucinodes orbonalis* Guenee. *Current Science* 101, 1427–1429.

Hanur, V.S., Boopal, K., Arya, V.V., Srividya, K.N., and Saraswathi, M.S. (2014) Why is management of brinjal shoot and fruit borer, *Leucinodes orbonalis* Guenee, difficult? An examination into the pest's unique feeding behavioural biology. *Journal of Entomology and Zoology Studies* 2, 257–260.

Hanyu, H., Murata, A., Park, E.Y., Okabe, M., Billings, S., *et al.* (1999) Stability of luciferase gene expression in a long term period in transgenic eggplant, *Solanum melongena. Plant Biotechnology* 16, 403–407.

Haque, M.S. and Saha, N.R. (2020) Biosafety measures, socio-economic impacts and challenges of *Bt*-brinjal cultivation in Bangladesh. *Frontiers in Bioengineering and Biotechnology* 8: 337.

Hautea, D.M., Taylor, L.D., Masanga, A.P.L., Sison, M.L.J., Narciso, J.O., *et al.* (2016) Field performance of *Bt* eggplants (*Solanum melongena* L.) in the Philippines: Cry1Ac expression and control of the eggplant fruit and shoot borer (*Leucinodes orbonalis* Guenee). *PLOS ONE* 11: e0157498.

Hunter, M.C., Smith, R.G., Schipanski, M.E., Atwood, L.W., and Mortensen, D.A. (2017) Agriculture in 2050: recalibrating targets for sustainable intensification. *Bioscience* 67, 385–390.

Iannacone, R., Grieco, D., and Cellini, F. (1997) Specific sequence modifications of a *cry3B* endotoxin gene result in high levels of expression and insect resistance. *Plant Molecular Biology* 34, 485–496.

ISAAA (International Service for the Acquisition of Agri-biotech Applications) (2021) Philippines Approves Bt Eggplant for Direct Use as Food, Feed, or for Processing. ISAAA, Ithaca, New York. Available at: www.isaaa.org/blog/entry/default.asp?BlogDate=7/23/2021 (accessed 19 June 2024).

Jadhav, M., Jadhav, A., Pawar, B., Kale, A., and Kutes, N. (2015) *Agrobacterium*-mediated genetic transformation of brinjal with *cry1F* gene for resistance against shoot and fruit borer. *Journal of Crop Improvement* 29, 518–527.

Jat, K.L., and Pareek, B.L. (2003) Biophysical and biochemical factors of resistance in brinjal against *Leucinodes orbonalis*. *Indian Journal of Otolaryngology* 65, 252–258.

Jelenkovic, G., Billings, S., Qi, C., Lashomb, J., Hamilton, G., and Ghidiu, G. (1998) Transformation of eggplant with synthetic *cryIIIA* gene produces a high level of resistance to the Colorado potato beetle. *Journal of the American Society for Horticultural Science* 123, 9–25.

Kalloo, G. (1988) Biochemical basis of insect resistance in vegetables. In: *Vegetable Breeding*, Vol. II, CRC Press, Boca Raton, Florida.

Kamat, M.G., and Rao, T.S. (1978) Vegetative multiplication of eggplants (*Solanum melongena*) using tissue culture techniques. *Plant Science Letters* 13, 57–65.

Klee, H.J., Horsch, R.B., and Rogers, S.G. (1987) *Agrobacterium*-mediated plant transformation and its further applications to plant biology. *Annual Review of Plant Biology* 38, 467–486.

Krishna, V.V. and Qaim, M. (2008) Potential impacts of *Bt* eggplant on economic surplus and farmers' health in India. *Agricultural Economics* 38, 167–180.

Kumar, P.A. (2010) Synthetic gene encoding Cry1Fa1 delta-endotoxin of *Bacillus turingiensis*. India Patent No. 242768. Indian Patent Office, New Delhi.

Kumar, P.A. and Rajam, M.V. (2005) Enhanced induction of *Vir* genes results in the improvement of *Agrobacterium*-mediated transformation of eggplant. *Journal of Plant Biochemistry and Biotechnology* 14, 89–94.

Kumar, P.A. and Sharma, R.P. (1994) Genetic engineering of crop plants with *Bacillus thuringiensis* δ-endotoxin genes. *Journal of Plant Biochemistry and Biotechnology* 3, 3–9.

Kumar, P.A., Mohapatra, T., Archak, S., and Sharma, R.P. (1996a) Expression of *Bacillus thuringiensis* δ-endotoxin in transgenic tobacco and cauliflower. In: Chopra, V.L., Sharma, R.P., and Swaminathan, M.S. (eds.) *Agricultural Biotechnology*. Oxford-IBH, New Delhi, pp. 191–197.

Kumar, P.A., Sharma, R.P., and Malik, V. (1996b) Insecticidal proteins of *Bacillus thuringiensis*. *Advances in Applied Microbiology* 42, 1–43.

Kumar, P.A., Mandaokar, A., Sreenivasu, K., Chakrabarti, S.K., Bisaria, S., *et al.* (1998) Insect resistant transgenic brinjal plants. *Molecular Breeding* 4, 33–37.

Kumar, S., Lakshmi Prasanna, P.A., and Shwetal, W. (2011) Potential benefits of *Bt* brinjal in India—an economic assessment. *Agricultural Economics Research Review* 24, 83–90.

Kumar, S.K., Sivanesan, I., Murugesan, K., Jeong, B.R., and Hwang, S.J. (2014) Enhancing salt tolerance in eggplant by introduction of foreign halotolerance gene, *HAL1* isolated from yeast. *Horticulture, Environment, and Biotechnology* 55, 222–229.

Laine, E., Lamblin, F., LaCoux, P.D., Roger, D., Sihachakr, D., and David, A. (2000) Gelling agent influences the detrimental effect of kanamycin on adventitious budding in flax. *Plant Cell, Tissue and Organ Culture* 63, 77–80.

Maestrelli, A., Lo Scalzo, R., Rotino, G.L., Acciarri, N., Spena, A., *et al.* (2003) Freezing effect on some quality parameters of transgenic parthenocarpic eggplants. *Journal of Food Engineering* 56, 285–287.

Magioli, C. and Mansur, E. (2005) Eggplant (*Solanum melongena* L.): tissue culture, genetic transformation and use as an alternative model plant. *Acta Botanica Brasilica* 19, 139–148.

Magioli, C., Rocha, A.P.M., Oliveira, D.E., and Mansur, E. (1998) Efficient shoot organogenesis of eggplant (*Solanum melongena* L.) induced by thidiazuron. *Plant Cell Reports* 17, 661–663.

Magioli, C., Rocha, A.P.M., Margis-Pinheiro, M., Sachetto-Martins, G., and Mansur, E. (2000) Establishment of an efficient *Agrobacterium*-mediated transformation system for eggplant and study of a potential biotechnologically useful promoter. *Journal of Plant Biotechnology* 2, 43–49.

Magioli, C., Barroco, R.M., Rocha, C.A.B., Santiago-Fernandes, L.D., Mansur, E., *et al.* (2001a) Somatic embryo formation in *Arabidopsis* and eggplant is associated with expression of a glycin-rich protein gene (*Atgrp-5*). *Plant Science* 161, 559–567.

Magioli, C., Rocha, A.P.M., Tarre, E., Santiago-Fernandes, L.D., Oliveira, D.E., *et al.* (2001b) Effect of morphological factors, antibiotics and *Agrobacterium* co-cultivation in the efficiency of somatic embryogenesis of eggplant (*Solanum melongena* L.). *Journal of Plant Biotechnology* 3, 19–25.

Maioli, A., Gianoglio, S., Moglia, A., Acquadro, A., Valentino, D., *et al.* (2020) Simultaneous CRISPR/Cas9 editing of three PPO genes reduces fruit flesh browning in *Solanum melongena* L. *Frontiers in Plant Science* 11, 1–13.

Mathias, R.J. and Boyd, L.A. (1986) Cefotaxime stimulates callus growth, embryogenesis and regeneration in hexaploid bread wheat (*Triticum aestivum* L em. Thell). *Plant Science* 46, 217–223.

Mathias, R.J. and Mukasa, C. (1987) The effect of cefotaxime on the growth and regeneration of callus from four varieties of barley (*Hordeum vulgare* L.). *Plant Cell Reports* 6, 454–457.

Matsuoka, H. and Hinata, K. (1979) NAA-induced organogenesis and embryogenesis in hypocotyl callus of *Solanum melongena* L. *Journal of Experimental Botany* 30, 363–370.

Mennella, G., d'Alessandro, A., Arpaia, S., Acciarri, N., Sunseri, F., and Rotino, G.L. (2004) Management of resistance to Coleoptera (*Leptinotarsa decemlineata* Say) in transgenic eggplant (*Solanum melongena* L.) hybrids. In: *Proceedings of the XIIth EUCARPIA Meeting on Genetics and Breeding of Capsicum and Eggplant, 17–19 May, Noordwijkerhout, Netherlands*. Plant Research International, Wageningen, Netherlands, pp. 196–202.

Mennella, G., Acciarri, N., d'Alessandro, A., Perrone, D., Arpaia, S., *et al.* (2005) Mixed deployment of *Bt*-expressing transgenic eggplant hybrids as a reliable method to manage resistance to Colorado potato beetle. *Scientia Horticulturae* 104, 127–135.

Mishiba, K.-I., Nishida, K., Inoue, N., Fujiwara, T., Teranishi, S., *et al.* (2020) Genetic engineering of eggplant accumulating β-carotene in fruit. *Plant Cell Reports* 39, 1029–1039.

Morel, A., Guinard, J., Lonjon, F., Sujeeun, L., Barberis, P., *et al.* (2018) The eggplant AG91-25 recognizes the Type III-secreted effector RipAX2 to trigger resistance to bacterial wilt (*Ralstonia solanacearum* species complex). *Molecular Plant Pathology* 19, 2459–2472.

Na, C., Shuanghua, W., Jinglong, F., Bihao, C., Jianjun, L., *et al.* (2016) Overexpression of the eggplant (*Solanum melongena* L.) NAC family transcription factor SmNAC suppresses resistance to bacterial wilt. *Scientific Reports* 6: 31568.

Nahar, K., Matsumoto, I., Taguchi, F., Inagaki, Y., Yamamoto, M., *et al.* (2014) *Ralstonia solanacearum* type III secretion system effector Rip36 induces a hypersensitive response in the nonhost wild eggplant *Solanum torvum. Molecular Plant Pathology* 15, 297–303.

Narendran, M., Lalwani, S., Girhepuje, P., Shirale, D., Mustafa, G., *et al.* (2007) Production of marker-free insect resistant transgenic brinjal (*Solanum melongena* L.) carrying *cry2Ab* gene. *Acta Horticulturae* 752, 535–538.

Okkels, F.T. and Pedersen, M.G. (1988) The toxicity of plant tissue and *Agrobacterium tumefaciens* of some antibiotics. *Acta Horticulturae* 225, 199–207.

Pal, J.K., Singh, M., Rai, M., Satpathy, S., Singh, D.V., and Kumar, S. (2009) Development and bioassay of *Cry1Ac* transgenic eggplant (*Solanum melongena* L.) resistant to shoot and fruit borer. *Journal of Horticultural Sciences and Biotechnology* 84, 434–438.

Papolu, P.K., Dutta, T.K., Tyagi, N., Urwin, P.E., Lilley, C.J., and Rao, U. (2016) Expression of a cystatin transgene in eggplant provides resistance to root-knot nematode, *Meloidogyne incognita*. *Frontiers in Plant Science* 7, 1122–1130.

Picoli, E.A.T., Otoni, W.C., Cecon, P.R., and Fari, M. (2000) Influence of antibiotics on NAA-induced somatic embryogenesis in eggplant (*Solanum melongena* L. cv. Embil). *International Journal of Horticultural Science* 6, 88–95.

Picoli, E.A.T., Brommonschenkel, S.H., Cecon, P.R., da Silva, D.J.H., Fari, M.G., and Otoni, W.C. (2002a) Study of genetic transformation efficiency via organogenesis and embryogenesis in eggplant (*Solanum melongena* L. cv. Embli): effects of co-culture, temperature and kanamycin and hygromycin-based selection procedures. *International Journal of Horticultural Science* 8, 3–4.

Picoli, E.A.T., Cecon, P.R., Fari, M.G., and Otoni, W.C. (2002b) Organogenesis in eggplant as affected by antibiotics and growth regulators. *International Journal of Horticultural Science* 8, 76–82.

Polturak, G., Grossman, N., Vela-Corcia, D., Dong, Y., Nudel, A., *et al.* (2017) Engineered gray mold resistance, antioxidant capacity and pigmentation in betalain-producing crops and ornamentals. *Proceedings of the National Academy of Sciences of the USA* 114, 9062–9067.

Prabhavathi, V. and Rajam, M.V. (2007) Mannitol-accumulating transgenic eggplants exhibit enhanced resistance to fungal wilts. *Plant Science* 173, 50–54.

Prabhavathi, V., Yadav, J.S., Kumar, P.A., and Rajam, M.V. (2002) Abiotic stress tolerance in transgenic eggplant (*Solanum melongena* L.) by introduction of bacterial mannitol phosphate dehydrogenase gene. *Molecular Breeding* 9, 137–147.

Prabhu, M., Natarajan, S., Veeraragavathatham, D., and Pugalendhi, L. (2009) The biochemical basis of shoot and fruit borer resistance in interspecific progenies of brinjal (*Solanum melongena*). *EurAsian Journal of BioSciences* 3, 50–57.

Prakash, D.P., Deepali, B.S., Asokan, R., Ramachandra, Y.L., Anand, L., and Hanur, V.S. (2007) Effects of growth regulators and explant-type on *Agrobacterium*-mediated transformation in brinjal (*Solanum melongena* L.) cv. Manjarigota. *Journal of Horticultural Sciences* 2, 94–98.

Prakash, D.P., Deepali, B.S., Asokan, R., Ramachandra, Y.L., Shetti, D.L., *et al.* (2008) Effect of growth regulators on *in vitro* complete plant regeneration in brinjal (*Solanum melongena* L.). *Indian Journal of Horticulture* 65, 371–376.

Prakash, D.P., Ramachandra, Y.L., and Hanur, V.S. (2015) Factors affecting *in vitro* shoot regeneration in hypocotyls of brinjal (*Solanum melongena* L.) in the early steps of *Agrobacterium*-mediated transformation. *Journal of Horticultural Sciences* 10, 136–142.

Prakash, D.P., Ramachandra, Y.L., and Hanur, V.S. (2016) Effect of *Agrobacterium* infection time, co-cultivation and cell density on *in vitro* response in hypocotyls of eggplant (*Solanum melongena* L.). *Journal of Horticultural Sciences* 11, 21–26.

Pratap, D., Kumar, S., Raj, S.K., and Sharma, A.K. (2011) *Agrobacterium*-mediated transformation of eggplant (*Solanum melongena* L.) using cotyledon explants and coat protein gene of cucumber mosaic virus. *Indian Journal of Biotechnology* 10, 19–24.

Prodhan, M.Z.H., Hasan, M.T., Chowdhury, M.M.I., Alam, M.S., Rahman, M.L., *et al.* (2018) *Bt* eggplant (*Solanum melongena* L.) in Bangladesh: fruit production and control of eggplant fruit and shoot borer (*Leucinodes orbonalis* Guenee), effects on non-target arthropods and economic returns. *PLOS ONE* 13: e0205713.

Prodhan, M.Z.H., Shirale, D.K., Islam, Z., Hossain, J., Paranjape, V., and Shelton, A.M. (2019) Susceptibility of field populations of eggplant fruit and shoot borer (*Leucinodes orbonalis* Guenee) to Cry1Ac, the protein expressed in Bt eggplant (*Solanum melongena* L.) in Bangladesh. *Insects* 10: 198.

Qiu, Z., Yan, S., Xia, B., Jiang, J., Yu, B., *et al.* (2019) The eggplant transcription factor MYB44 enhances resistance to bacterial wilt by activating the expression of spermidine synthase. *Journal of Experimental Botany* 70, 5343–5354.

Rai, N.P., Rai, G.K., Kumar, S., Kumari, N., and Singh, M. (2013) Shoot and fruit borer resistant transgenic eggplant (*Solanum melongena* L.) expressing *cry1Aa3* gene: development and bioassay. *Crop Protect* 53, 37–45.

Rajam, M.V. and Kumar, S.V. (2007) Eggplant. In: Pua, E.-C. and Davey, M.R. (eds.) *Biotechnology in Agriculture and Forestry. Transgenic Crops IV*. Springer, Berlin, pp. 201–219.

Rajam, M.V., Rotino, G.L., Sihachakr, D., Souvannavong, V., Mansur, E., and Kumar, P.A. (2008) Eggplant. In: Kole, C., Hall T.C. (eds.) *Compendium of Transgenic Crops: Transgenic Vegetable Crops, Vol.* 6. Wiley-Blackwell, Chichester, UK, pp. 47–72.

Rao, C.K. (2010) *Moratorium on Bt Brinjal: A Review of the Order of the Minister of Environment and Forests, Government of India.* Foundation for Biotechnology Awareness and Education, Bangalore.

Rao, N.G.V., Majumdar, A., Mandaokar, A.D., Nimbalkar, S.A., and Kumar, P.A. (1999) Susceptibility of brinjal shoot and fruit borer to the δ-endotoxins of *Bacillus thuringiensis*. *Current Science* 77, 336–337.

Ribeiro, A.P.O., Otoni, W.C., Picanco, M.C., Galvan, T.L., Pereira, E.J.G., *et al.* (2005) Resistance of transgenic eggplant carrying the oryzacystatin gene to *Mechanitis polymnia* and *Mechanitis lysimnia*. *Acta Scientiarum Agronomy* 27, 647–650.

Ribeiro, A.P.O., Pereira, E.J.G., Galvan, T.L., Picanco, M.C., Picoli, E.A.T., *et al.* (2006) Effect of eggplant transformed with oryzacystatin gene on *Myzus persicae* and *Macrosiphum euphorbiae*. *Journal of Applied Entomology* 130, 84–90.

Rotino, G.L. and Gleddie, S. (1990) Transformation of eggplant (*Solanum melongena* L.) using a binary *Agrobacterium tumefaciens* vector. *Plant Cell Reports* 9, 26–29.

Rotino, G.L., Perri, E., Acciarri, N., Sunseri, F., and Arpaia, S. (1997a) Development of eggplant varietal resistance to insects and diseases via plant breeding. *Advances in Horticultural Science* 11, 193–201.

Rotino, G.L., Perri, E., Zottini, M., Sommer, H., and Spena, A. (1997b) Genetic engineering of parthenocarpic plants. *Nature Biotechnology* 15, 1398–1401.

Sagare, D.B. and Mohanty, I. (2012) Development of moisture stress tolerant brinjal cv. Utkal Anushree (*Solanum melongena* L.) using *Agrobacterium* mediated gene transformation. *Journal of Agricultural Science* 4, 141–148.

Sharma, P. and Rajam, M.V. (1995) Genotype, explants and position effects on organogenesis and embryogenesis in eggplant (*Solanum melongena*). *Journal of Experimental Botany* 46, 131–141.

Shelton, A.M., Hossain, M.J., Paranjape, V., Azad, A.K., Rahman, M.L., *et al.* (2018) Bt eggplant project in Bangladesh: History, present status and future direction. *Frontiers in Bioengineering and Biotechnology* 6: 106.

Shivakumara, T.N., Chaudhary, S., Kamaraju, D., Dutta, T.K., Papolu, P.K., *et al.* (2017) Host-induced silencing of two pharyngeal gland genes conferred transcriptional alteration of cell wall-modifying enzymes of *Meloidogyne incognita vis-à-vis* perturbed nematode infectivity in eggplant. *Frontiers in Plant Science* 8, 473–479.

Shrivastava, D., Dalal, M., Nain, V., Sharma, P., and Kumar, P.A. (2011) Targeted integration of *Bacillus thuringiensis* δ-endotoxin *cry1Fa1* in brinjal (*Solanum melongena* L.). *Current Trends in Biotechnology and Pharmacy* 5, 1149–1156.

Shruti, B.K., Roy, R., Kumar, P.A., and Roy, S.P. (2015) Gene integration of *cry1F* in brinjal for resistance against fruit and shoot borer. *Journal of Advanced Zoology* 36, 83–89.

Singh, D. (2004) Genetic engineering of eggplant for resistance to fungal pathogens. PhD thesis, University of Delhi, New Delhi.

Singh, A.K., Verma, S.S., and Bansal, K.C. (2010) Plastid transformation in eggplant (*Solanum melongena* L.). *Transgenic Research* 19, 113–119.

Singh, D., Ambroise, A., Haicour, R., Sihachakr, D., and Rajam, M.V. (2014) Increased resistance to fungal wilts in transgenic eggplant expressing alfalfa glucanase gene. *Physiology and Molecular Biology of Plants* 20, 143–150.

Singh, D., Haicour, R., Sihachakr, D., and Rajam, M.V. (2015) Expression of rice chitinase gene in transgenic eggplant confers resistance to fungal wilts. *Indian Journal of Biotechnology* 14, 230–234.

Solanki, J.J., Pawar, P.K., and Maheshwari, V.L. (2006) Efficient plant regeneration in *Solanum melongena* L. *Physiology and Molecular Biology of Plants* 12, 307–311.

Soniya, E.V., Kannapiran, P., Unni, S.C., Joy, N., and Radhakrishnan, E.K. (2005) Transformation and regeneration of *Solanum melongena* using *Agrobacterium tumefaciens*. *Journal of Plant Biology* 32, 29–32.

Subramanyam, K., Rajesh, M., Jaganath, B., Vasuki, A., Theboral, J., *et al.* (2013) Assessment of factors influencing the *Agrobacterium*-mediated *in planta* seed transformation of brinjal (*Solanum melongena* L.) *Applied Biochemistry and Biotechnology* 171, 450–468.

Toppino, L., Kooiker, M., Lindner, M., Dreni, L., Rotino, G.L., and Kater, M.M. (2011) Reversible male sterility in eggplant (*Solanum melongena* L.) by artificial microRNa-mediated silencing of general transcription factor genes. *Plant Biotechnology* 9, 684–692.

Turrini, A., Sbrana, C., Pitto, L., Castiglione, M.R., Giorgetti, L., *et al.* (2004) The antifungal Dm-AMP1 protein from *Dahlia merckii* expressed in *Solanum melongena* is released in root exudates and differentially affects pathogenic fungi and mycorrhizal symbiosis. *New Phytologist* 163, 393–403.

Vavilov, N.I. (1928) Geographical centers of our cultivated plants. In: *Proceedings of 5th International Congress of Genetics, New York*, pp. 342–369.

Vavilov, N.I. (1951) *The Origin, Variation, Immunity and Breeding of Cultivated Plants*. Chronica Botanica, Waltham, Massachusetts.

Wan, F., Pan, Y., Li, J., Chen, X., Pan, Y., *et al.* (2014) Heterologous expression of *Arabidopsis C-repeat binding factor 3* (*AtCBF3*) and *cold-regulated 15A* (*AtCOR15A*) enhanced chilling tolerance in transgenic eggplant (*Solanum melongena* L.). *Plant Cell Reports* 33, 1951–1961.

Wang, W., Liu, J., Wang, H., Li, T., and Zhao, H. (2021) A highly efficient regeneration, genetic transformation system and induction of targeted mutations using CRISPR/Cas9 in *Lycium ruthenicum*. *Plant Methods* 17: 71.

Xi'ou, X., Bihao, C., Guannan, L., Jianjun, L., Qinghua, C., *et al.* (2014) Functional characterization of a putative bacterial wilt resistance gene (*RE-bw*) in eggplant. *Plant Molecular Biology* 33, 1058–1073.

Xing, J. and Chin, C.K. (2000) Modification of fatty acids in eggplant affects its resistance to *Verticillium dahliae*. *Physiological and Molecular Plant Pathology* 56, 217–225.

Yarra, R. and Kirti, P.B. (2019) Expressing class I wheat NHX (*TaNHX2*) gene in eggplant (*Solanum melongena* L.) improves plant performance under saline condition. *Functional & Integrative Genomics* 19, 541–554.

Zhang, Y.H. (1999) Effects of different explants and phytohormones on organogenesis in scarlet eggplant (*Solanum aethiopicum* L.). *Journal of Yunnan Agricultural University* 14, 279–283.

# 3 Pepper Genetic Engineering

**K. Madhavi Reddy[1]*, Satyaprakash Barik[2], Naresh Ponnam[1], Koushik Saha[1], Roshini Pydi[1] and D.C. Lakshamanareddy[3]**

*[1]Division of Vegetable Crops, ICAR-Indian Institute of Horticultural Research, Bengaluru, India; [2]Department of Agriculture and Allied Sciences, CV Raman Global University, Bhubaneswar, India; [3]Division of Basic Sciences, ICAR-Indian Institute of Horticultural Research, Bengaluru, India*

**Abstract**

Fresh hot and sweet peppers are commercially grown as vegetables and dry pepper, an important spice crop that exhibits potential applications in various industries (food, phytofeed, cosmeceutical, and pharmaceutical). Conventional breeding coupled with molecular marker approaches has been well explored in pepper crop-improvement programs. However, genetic engineering has added new momentum in breeding crops for resistance and improved quality. Poor or low regeneration potential is a major hindrance in exploring genetic engineering in peppers. Genetic transformation and genetic engineering through transgenics and RNA interference technologies have been attempted successfully in peppers against various biotic and abiotic stresses. Recently, advanced breeding technology by genome editing using clustered regularly interspaced palindromic repeats (CRISPR)-CRISPR-associated protein 9 (Cas9) has been explored in peppers, mainly against anthracnose fruit rot disease. Functional characterization of susceptible genes responsible for various biotic and abiotic stresses will enable us to further explore genome editing in pepper crop-improvement programs.

**Keywords:** abiotic stresses, biotic stresses, *Capsicum* spp., conventional breeding, genetic engineering, molecular breeding

## 3.1 Introduction

Hot or sweet pepper belongs to the genus *Capsicum*, which consists of 38 species (USDA-ARS, 2011). Pepper originated in Central and South America (Basu and De, 2003), and *C. annuum* var. *glabriusculum* is considered the wild progenitor of cultivated chili (Aguilar-Melendez *et al.*, 2009). Hot pepper was one of the first plants to be selected for domestication in the Mesoamerican subcontinent. It has wide application as a vegetable, spice, and natural colorant, and in medicine in ancient times. Columbus discovered and named this plant red pepper, as the fruit mimicked the pungency of black pepper. It was introduced into Europe in 1493 and classified as *Capsicum*. To date, 38 *Capsicum* spp. have been reported (USDA-ARS, 2011); however, only five species (*C. annuum*, *C. chinense*, *C. frutescens* (*C. annuum* var. *glabriusculum*), *C. pubescens and C. baccatum*) are domesticated. Very recently, a distinct domesticated species, *C. assamicum*, was identified from the north-eastern part of India (Purukayastha *et al.*, 2012); it is closely related to *C. frutescens* and *C. chinense* but can be distinguished by its distinct morphological features.

*Email: Madhavireddy.K@icar.gov.in

 *Genetic Engineering of Vegetable Crops* (Pritam Kalia ed.)
DOI: 10.1079/9781800627123.0003

*C. annuum* is the most widely farmed of these domesticated species due to its greater resilience to shifting climates, although *C. chinense* is popular worldwide because of its intense pungency (Bosland and Baral, 2007). Pepper is rich in carotenoids, vitamins (A, C, and E) that help to destroy free radicals, flavonoids, and capsaicinoids such as capsaicin, dihydrocapsaicin, and nordihydrocapsaicin, which are the main pungency components (Maga, 1975; Simonne *et al.*, 1997). These are used in the pharmaceutical industry because of their substantial nutraceutical characteristics, in addition to acting as antioxidants. Capsaicin is used as a pain reliever during chemotherapy and diabetic neuropathy (Berger *et al.*, 1995). Pepper is also commonly used to prevent pancreatic and prostate cancer by inducing apoptosis, and to treat toothache, arthritis, and diarrhea, among other conditions (Surh, 2002; Ravishankar *et al.*, 2003). The area under pepper expanded by 33% from 2009 to 2019 (FAOSTAT). During 2021, around 4.84 million t and 36.29 million t of dry and green peppers were produced, respectively, across the world. India is the largest producer of dry pepper, while China dominates green pepper production (FAOSTAT, 2021).

Biotechnological tools have increased the efficiency of conventional methods of crop improvement involving breeding leading to enhanced yield and earliness, resistance to biotic and abiotic stresses, and other horticultural traits. Recently, biotechnology has been utilized in numerous crop-improvement programs, including the production of disease-free planting material and enhancement of the shelf life of vegetables, as well as the production of organic fertilizers and pesticides. The application of genetic engineering has added new momentum to breeding of horticultural crops. The use of effective protocols to transfer a desired targeted gene into plant cells and regeneration of whole plants that carry this foreign gene is known as genetic transformation (Ochoa-Alejo and Ramírez-Malagón, 2001). Delivery of the foreign gene during transformation is accomplished using a vector to carry the gene into a viable host cell (Razzaq *et al.*, 2004). To transfer foreign genes into the host plant, *Agrobacterium tumefaciens*-mediated (for dicotyledonous species) and biolistic bombardment-mediated (for both monocotyledonous and dicotyledonous species) genetic transformation is used. Genetic transformation is a potential approach for improving a variety of horticulturally desirable features in pepper.

## 3.2 Floral Biology and Botany

*Capsicum* spp. have a chromosome number of $2n = 24$. Anthesis occurs early in the morning from 05:00 to 08:00 and anther dehiscence between 07:00 and 08:00, with the highest stigma receptivity fruit set observed on the day of anthesis. After 24 h of anthesis, the stigmatic surface starts to dry (Dhall *et al.*, 2011). Natural cross-pollination of 58–62% has been reported by insect pollinators (Murthy and Murthy, 1962). In *C. frutescens*, maximum anthesis was recorded between 09:00 and 10:00 am, with the stigma being receptive for 24 h before and after anthesis (Sheela, 1998).

### 3.2.1 *Capsicum annuum* L.

This is an annual or perennial branched herb or shrub, growing to a height of 1–1.5 m. The tap root system has numerous lateral roots. It has simple alternate leaves, with ovate leaf blades, acuminate at the apex, and light to dark green in color. It has solitary flowers, with the calyx having five conspicuous teeth; the corolla is white to purple, and there are five to seven stamens with pale blue to purplish anthers. The ovary has two to four locules and the stigma is capitate. The fruit is a berry varying in shape, size, and color from green/yellow/purple to red/brown/orange at the fully ripe stage. A wide range of pungency levels are observed. The seeds are orbicular, flattened, and pale yellow in color.

### 3.2.2 *Capsicum frutescens* L.

According to the Integrated Taxonomic Information System (ITIS), *C. frutescens* is considered a variety of *C. annuum* (*C. annuum* var. *glabriusculum* Heiser & Pickersgill (TSN 30493); ITIS, 2022). It is a perennial shrub, with ovate-lanceolate leaves; there are two to three flowers per node, and the calyx is triangular and acute, with five lobes. There are five exerted stamens with oblong anthers and a capitate stigma. The fruit taper gradually toward the apex, with the color varying from green to white.

### 3.2.3 *Capsicum baccatum* L.

*C. baccatum* is a perennial herb or shrub, with densely pubescent branches and solitary leaves and flowers (double whorled). The calyx is five ribbed with five dentate teeth. The corolla is greenish white to straw white, with a yellow throat. It has a yellow anther and glabrous ovary, with glossy and rounded distal fruits.

### 3.2.4 *Capsicum chinense* L.

*C. chinense* is perennial in nature. It is a subshrub with a woody base, and is erect and upright, growing up to 2 m high. The stout taproot has many lateral roots. The leaves are crinkled, medium green to light green ovate with an acute leaf apex. The corolla is white with a tinge of light brown, and is clustered (three to six). The anthers are a blue or dark green color with purple. Anthesis occurs at 06.00–09.00 am. The ovary is perigynous.

### 3.2.5 *Capsicum pubescens* L.

*C. pubescens* is a perennial shrub, growing up to 4 m high. It is a woody plant that lives for up to 15 years and hence has a tree-like appearance. The leaves are wedge-shaped, large, and ovate, 5–12 cm long and 2.5–4 cm wide, with petioles 5–12 mm long. The flowers are solitary or in pairs. The calyx has five triangular, pointed teeth. The corolla is blue/violet in color, and the anthers are purple and white. The seeds are dark and rugose.

## 3.3 Hybridization and Crossability

Various intraspecific and interspecific hybridizations have been conducted to date (Krishna Kumari and Peter 1984; Mini, 1989; Pradeepkumar *et al.*, 1993; Sunil and Rasheed, 1998; Naresh *et al.*, 2016). It was reported that fertile hybrids could be obtained by interspecific crosses within the *C. annuum* complex to varying degrees (Baral and Bosland, 2004; Jarret and Dang, 2004; Panda *et al.*, 2004). However, no fertile hybrids were produced in crosses between *C. annuum* and *C. pubescens*. Varying rates of successful fertile hybrids were obtained by crossing wild and semi-domesticated *C. annuum* var. *glabriusculum* and *C. annuum* var. *annuum* (Hernández-Verdugo *et al.*, 2001; Jarret and Dang, 2004; Guzmán *et al.*, 2005).

## 3.4 Genetic Resources

The World Vegetable Center (WVC) in Taiwan holds a large collection of pepper germplasm, as well as other international organizations such as the US Department of Agriculture and New Mexico State University. The National Bureau of Plant Genetic Resources (NBPGR) in New Delhi acts as an essential nodal center for germplasm exploration, collection, and conservation. The All India Coordinated Research Project (AICRP) (Vegetable Crops), national institutes such as ICAR-IARI (Indian Council of Agricultural Research-Indian Agricultural Research Institute), ICAR-IIVR (ICAR-Indian Institute of Vegetable Research), ICAR-IIHR (ICAR-Indian Institute of Horticultural Research), and horticultural research stations such as Lam farm, Devihosur, India, along with state agricultural universities such as Dr. Y.S.R. Horticultural University and Punjab Agricultural University are active centers for chili germplasm collection, conservation, and maintenance.

India is considered the secondary center of origin for *Capsicum* spp. owing to its wide range of diversity with respect to fruit color and pungency. A few land races have been recognized with a geographical indication (GI) tag, such as Byadagi Dabbi, named after the village Byadagi in the Haveri district, Karnataka, India, in September 2010. It exhibits a high export value and is popular in the international market for its deep red color (150–200 ASTA color value) and low pungency. Its fruits are of medium length (8–10 cm), wrinkled with slight bulge at the calyx and a deep red color.

## 3.5 Conventional and Molecular Breeding for Major Diseases and Insect Pests of Pepper

In *Capsicum* spp., molecular markers have been applied to map quantitative trait loci (QTLs) related to important characteristics (Ramchiary *et al.*, 2013). However, marker development and its application in breeding are still in the early phase, mainly due to the large genome size and the lack of a complete genome sequence. An annotated genome of *C. annuum* is available at Solgenomics (https://solgenomics.net/organism/Capsicum_annuum/genome; accessed 19 June 2024). Molecular markers have been applied primarily to assess genetic diversity and for DNA fingerprinting (Singh *et al.*, 2008; Thul *et al.*, 2012). Some studies on the application of molecular markers on cytoplasmic male sterility have been reported (Kumar *et al.*, 2007, 2009).

The QTL *Bw1* is associated with bacterial wilt resistance, and sequence analysis of the amplified products of the corresponding marker CAMS451 showed that it corresponded to open reading frame 1 (ORF1) of the *RRP41* gene, and a 2 Mbp genomic region flanking CAMS451 harbored 44 defense-related genes (Mathew *et al.*, 2020). *Phytopthora* root rot resistance in chili was found to be governed by a single dominant gene along with some minor genes against a virulent Indian *Phytopthora* isolate; the authors used IHR3575, a resistant derivative of 'Criollo de Morelos 334' ('CM334'), and IHR3476, a highly susceptible genotype, and their derived populations to study the genetics and identification of markers. Two new simple sequence repeat (SSR) markers, HpmsE034 and HpmsE062, linked to *Phytopthora* root rot resistance against the Indian isolate were identified through a bulked segregant analysis using $F_2$ and $F_{2:3}$ populations (Kumar *et al.*, 2021).

Several viruses such as chili leaf curl virus (ChiLCV), tomato leaf curl Joydebpur virus, tomato leaf curl New Delhi virus, and chili leaf curl Vellanad virus have been reported to be associated with chili leaf curl disease. Two SSR markers (CA 516044 and PAU-LC-343-1) were reported as linked markers for chili leaf curl disease resistance, and are located on chromosome 6 of the pepper genome at a genetic distance of 6.8 and 8.9 cM, respectively, from the resistance gene (Thakur *et al.*, 2020). Earlier reports demonstrated recessive control of resistance (Kumar *et al.*, 2009; Anandhi and Khader, 2011; Rai, 2010; Rai *et al.*, 2014), with a different number of genes (reviewed by Thakur *et al.*, 2018), which may be due to utilization of different germplasms in these studies.

Progress in the development of insect resistance is limited due to the population pressure of insects on the host, the evolution of new biotypes, the breakdown of resistance, and the presence of one or more undesirable characteristics associated with them, and hence has not gained popularity (Gopal *et al.*, 2018). None the less, several genotypes have been identified exhibiting high tolerance to sucking pests (Tatagar *et al.*, 2001; Jayewar *et al.*, 2018; Kurbett *et al.*, 2018).

## 3.6 Genetic Transformation in Pepper

The first attempt at genetic transformation in *Capsicum* spp. was made by Liu *et al.* (1990) in wild pepper, *C. annuum* var. *glabriusculum*, as well as in six cultivated varieties of bell peppers. Individual suspensions of wild tumorigenic strains of *A. tumefaciens* (strains A281 and C58) and a disarmed strain containing the pGV3850 plasmid with the β-glucuronidase (GUS) gene (*uidA*) and neomycin phosphotransferase gene (*nptII*) under the control of the 35S promoter from cauliflower mosaic virus (CaMV) were co-cultured with explants (hypocotyls) in the presence of the 35S promoter from CaMV. Leaves and cotyledons formed callus and leaf-like structures from the explants in the presence of 200 mg $l^{-1}$ of kanamycin. GUS activity was observed in the leaf-like structures and shoot buds but not in explants derived from shoot buds.

Most transgenic pepper lines have been developed via *Agrobacterium*-mediated transformation with cotyledons and/or hypocotyls as explants (Table 3.1) (Shin *et al.*, 2002; Wang *et al.*, 2013b; Seo *et al.*, 2014; Mythili *et al.*, 2015; Kim *et al.*, 2016; Ko *et al.*, 2016; Bagga *et al.*, 2019). The *uidA* gene has been employed as a reporter gene, using the CaMV 35S promoter and nopaline synthase (*nos*) as a terminator gene in *Capsicum* transformation (Liu *et al.*, 1990; Zhu *et al.*, 1996; Manoharan *et al.*, 1998; Kim *et al.*,

**Table 3.1.** Transgenic peppers developed for various desired traits.

| Transgenic cultivar | Gene | Source of gene | Method of genetic transformation | Trait of interest | Achievement | Reference |
|---|---|---|---|---|---|---|
| *Capsicum annuum* 'A3' | *CanPOD* | Pepper | *Agrobacterium*-mediated transformation | Resistance to *Phytophtora capsici* | Enhanced resistance to *P. capsici* and tolerance of abiotic stresses, including salt, drought, and methyl jasmonate | Wang *et al.* (2013a) |
| *Capsicum annuum* 'Mesilla Cayenne' (NuMex 6-4) | *RB* gene | *Solanum bulbocastanum* | *Agrobacterium*-mediated transformation | Resistant to *P. capsici* | Resistance to *P. capsici* | Bagga *et al.* (2019) |
| *Capsicum annuum* 'Nokkwang' | Defensin, *J1-1* | Wild pepper | *Agrobacterium tumefaciens*-mediated gene transfer | Resistance to anthracnose disease | Strong resistance to the anthracnose fungus that was associated with highly reduced lesion formation and fungal colonization | Seo *et al.* (2014) |
| *Capsicum annuum* 'Nokkang' | *PepEST* | Pepper | *A. tumefaciens*-mediated gene transfer | Resistance to anthracnose disease | Inhibition of disease incidence of up to 30% in the transgenic plants compared with the wild-type | Ko *et al.* (2016) |
| *Capsicum annuum* 'Pusa Jwala' | Baculovirus chitinase | Pepper | *Agrobacterium*-mediated transformation | Resistance to anthracnose disease | Higher tolerance to anthracnose disease | Mythili *et al.* (2015) |
| *Capsicum annuum* 'Bukang' | *CaWRKY1* | Pepper | Virus-induced gene silencing, *Agrobacterium*-mediated transformation | Resistance to *Xanthomonas axonopodis* pv. *vesicatoria* | Enhanced resistance to *X. axonopodis* pv. *vesicatoria* | Oh *et al.* (2007) |
| *Capsicum annuum* | *Tsi1* | Tobacco | *A. tumefaciens*-mediated gene transfer | Pepper mild mottle virus (PMMV) and cucumber mosaic virus (CMV) | Resistance to PMMV and CMV. Enhanced resistance to *Xanthomonas campestris* pv. *vesicatoria* and *Phytophthora capsici* | Shin *et al.* (2002) |
| *Capsicum annuum* var. Longum 'H15' | CMV-CP | Coat protein gene | *Agrobacterium*-mediated transformation | Virus resistance | Increased resistance to cucumber mosaic virus | Kim *et al.* (2016) |

2001a,b; Shin *et al.*, 2002a; Li *et al.*, 2003; Mihalka *et al.*, 2003; Lee *et al.*, 2004a). Explants such as leaves, hypocotyls, and cotyledons have mostly been used for transformation in pepper (Wang, 1991; Lee *et al.*, 1993). Transient GUS activity in transformed pepper plants was measured using explants of leaves, hypocotyls, and cotyledons by Wang (1991). Cotyledons as explants were successfully used for transformation using *A. tumefaciens* strain LBA4404 and the binary vector pRok1/105 in *C. annuum* 'Golden Tower' with I17N satellite RNA from cucumber mosaic virus (CMV) (Lee *et al.*, 1993) and the frequency of transformation using cotyledons as explants was reported to be 4%.

Transgenic pepper was developed using *A. tumefaciens* strain GV311-SE for the CMV coat protein (CP) gene (Zhu *et al.*, 1996). Later, a reliable and efficient transformation protocol using the LBA4404 strain of *A. tumefaciens* carrying the plasmid pBI121 was developed (Li *et al.*, 2003). *A. tumefaciens* strain EHA105 was reported to obtain a higher frequency of shoot regeneration (Sobhakumari and Lalithakumari, 2005). Kumar *et al.* (2009) developed an alternate tissue-culture-independent *A. tumefaciens*-mediated *in planta* transformation procedure using two sweet pepper cultivars, 'Arka Gaurav' and 'Arka Mohini'. They used strain EHA105 harboring the binary vector pCAMBIA1301, which carries the genes for GUS (*uidA*) and hygromycin phosphotransferase II (*hpt II*) for transformation. GUS histochemical analysis of $T_0$ and $T_1$ plants at various stages of growth was followed by molecular analysis using polymerase chain reaction (PCR), Southern blot analysis, and reverse transcriptase (RT)-PCR, resulting in identification of 35.0% and 29.7% transgenic plants of 'Arka Gaurav' and 'Arka Mohini', respectively. Stable transformation in pepper is difficult to achieve due to its recalcitrant nature, and in recent decades, several researchers have worked to increase transformation efficacy in this species (Phillips *et al.*, 2000; Harpster *et al.*, 2002; Shin *et al.*, 2002; Lee *et al.*, 2004b; Delis *et al.*, 2005; Sharma *et al.*, 2008; Lee *et al.*, 2009). Genetic transformation efficiency was enhanced using the *WUSCHEL* gene from *Arabidopsis thaliana* in Habanero chili pepper, which efficiently promoted transition from the vegetative state to embryogenesis (Sols-Ramos *et al.*, 2009). Ectopic expression of the *Brassica napus* BABY BOOM AP2/ERF transcription factor was used to efficiently regenerate transgenic plants from the recalcitrant sweet pepper (*C. annuum*) varieties 'Fiesta', 'Ferrari,' and 'Spirit'. Transient activation of BABY BOOM in the progeny plants induced prolific cell regeneration and led to the production of a significant number of somatic embryos exhibiting immense potential to be raised as seedlings (Heidmann *et al.*, 2011).

## 3.7 Genetic Engineering in Pepper for Resistance to Biotic and Abiotic Stresses

### 3.7.1 Resistance to viral diseases

The traditional breeding system and approaches were previously successful, and they continue to be the main strategy for developing virus-resistant chili genotypes. However, a few studies have also explored the potential of genetic engineering in pepper for developing resistance to viral diseases. A detailed understanding of the method of virus accumulation in host cells, as well as preventive techniques to restrict virus entry into plant cells and the activation of plant defense mechanisms, was reported by Ishibashi *et al.* (2007) and Hofmann *et al.* (2009). Similarly, the role of viral DNA methylation and activation of sequence silencing machinery, as well as ubiquitylation-mediated protection against begomoviruses, was highlighted by Lozano-Duran *et al.* (2011), Marino *et al.* (2012) and Sahu *et al.* (2014a,b). Compared with susceptible cultivars, ChiLCV-resistant pepper cultivars showed upregulation of numerous defense genes such as polyphenol oxidase, thionin, nucleotide-binding site leucine-rich repeats and certain ATP/ADP transporter proteins. Initially, a 10-day delay in the appearance of symptoms, an inhibited rate of CMV development, and reduced disease severity after *A. tumefaciens*-mediated transfer of the CMV satellite RNA gene to transgenic pepper plants was reported (Dong *et al.*, 1991). Subsequently, an attempt to develop resistance to CMV by transfer of satellite RNA in hot pepper 'Golden Tower' was also reported (Kim *et al.*, 1997). Transgenic pepper plants with resistance to CMV through *A. tumefaciens*-mediated transfer of CMV CP genes using cotyledons as explants have also been reported (Zhang *et al.*, 1994).

Various studies describing resistance to virus diseases by utilizing the viral CP genes have been reported. Transgenic plants of *C. annuum* var. Longum were developed containing CMV and tobacco mosaic virus (TMV) CP genes using hypocotyls as explants through modified *A. tumefaciens*-mediated transformation. These transgenic lines exhibited high levels of resistance to homologous strains of both viruses (Cai *et al.*, 2003). CP gene-mediated resistance of transgenic peppers has shown promise against CMV (Kim *et al.*, 2016). Transgenic chili pepper plants containing the CP gene of the pathotype CMVP0 were successful in exhibiting tolerance to both CMVP0 and a new pathotype (Lee *et al.*, 2009). Genetically modified (GM) chili pepper plants of the inbred line P915 containing the CMVP0 CP gene inserted using *A. tumefaciens*-mediated transformation were found to be highly resistant to CMV (Kim *et al.*, 2009a,b). Transgenic hot pepper plants resistant to the CMV pathotypes CMVP0 and CMV-Ca-P1, developed through introgression of the CMVP0 CP gene using *Agrobacterium*-mediated transformation were patented in South Korea in 2008 (patent no. KR100804766B1). Gene flow from a GM variety containing the CMVP0 CP gene to a non-GM variety 'P915' and two commercial $F_1$ hybrids, 'Manidda' and 'Taesan', was assessed. The gene flow frequency was reported to be 17.89% between GM and non-GM 'Taesan' chili pepper at the closest distance, indicating that the recommended isolation distances are sufficient to avoid escape of the gene from GM varieties.

Virus-induced silencing of the *C. annuum WRKY1* gene led to inhibited proliferation of *Xanthomonas axonopodis* pv. *vesicatoria* and a hypersensitive cell death response to *Pseudomonas syringe* pv. *tabaci* as well as TMV infection in leaves of pepper (Oh *et al.*, 2008). Recently, RNA interference (RNAi) via antiviral RNA-silencing pathways has been suggested as a potential tool for creating virus-resistant pepper (Yang *et al.*, 2020). RNAi or RNA silencing can also be used to create transgenic plants with defense-related genes associated with jasmonic acid signaling and antiviral RNA-silencing pathways, which regulate virus resistance (Yang *et al.*, 2020). In a different study, multiplex editing of a begomovirus was used to restrict the formation of escape mutants and disease development (Roy *et al.*, 2019). A novel, multiplexed guide RNA-based clustered regularly interspaced palindromic repeats (CRISPR)-CRISPR-associated protein 9 (Cas9) approach that targets the viral genome at two or more sites simultaneously was used to eliminate the ChiLCV genome without the recurrence of functional escape mutants.

### 3.7.2 Resistance to fungal diseases

#### *Anthracnose*

*Colletotrichum* sp. causes chili anthracnose, a serious disease of *Capsicum annuum* L., leading to substantial pre- and postharvest yield losses of up to 54–80% in tropical and subtropical areas around the world (Than *et al.*, 2008; Garg *et al.*, 2014; Saxena *et al.*, 2016). Initially, anthracnose fruit rot-resistant pepper was developed by overexpressing the *PepEST* (pepper esterase) gene (Ko *et al.*, 2005). Recently, a host-induced gene-silencing strategy targeting the *Colletotrichum gloeosporioides* COM1 (*CgCOM1*) developmental gene, responsible for the generation of fungal conidia and appressorium, was used to develop transgenic plants with disease resistance (Mahto *et al.*, 2020). Small-interfering RNA-mediated silencing of *CgCOM1* led to resistance via repression of conidial germination, germ-tube development, appressoria formation and mycelial growth of *C. gloeosporioides*.

Recently, genome editing by the Clustered regularly interspaced short palindromic repeats (CRISPR)/CRISPR-associated protein (Cas) system has been used to develop disease resistance in vegetable crops including pepper. A single transcript unit CRISPR/Cas9-mediated alteration in the *CaERF28* gene, which determines susceptibility to anthracnose disease, enhanced disease resistance (Mishra *et al.*, 2021). The use of a construct with a single polymerase II promoter-driven expression of Cas9, single guide RNA, and a hammerhead ribozyme resulted in multiple insertions/deletions in the susceptible genotype 'Arka Lohit'. Segregation allowed identification of T-DNA-free and marker-free C-ERF28 mutant lines. The homozygous mutants showed enhanced resistance to anthracnose.

### Phytopthora *fruit rot*

*Phytophthora capsici*, an oomycete that causes pepper blight affecting all parts of pepper (roots, stem, and leaves) is one of the most damaging diseases in the world (Oelke *et al.*, 2003; Lee *et al.*, 2012). The disease manifests as black, rapidly developing, water-soaked lesions. The pathogen infects the pepper crop irrespective of its growth stage, leading to damping off and seedling and foliar blight, as well as wilt (Baysal *et al.*, 2005). The *CaRGA2* gene from *C. annuum* 'CM334' exhibiting resistance to *P. capsici* was extracted and characterized; it has a full-length cDNA of 3018 bp with a 2874 bp ORF encoding a 957 amino acid protein. On silencing of the *CaRGA2* gene, the resistance level was clearly suppressed, as evidenced by detached leaf inoculation, as well as semi-quantitative RT-PCR (Zhang *et al.*, 2013).

Recently, the broad-spectrum host resistance *RB* gene from *Solanum bulbocastanum*, which notably confers resistance to *Phytophthora infestans* infection in tomato and potato (*RB* gene construct pCLD04541; Song *et al.*, 2003), along with its native promoter were introduced into chili accession NuMex 6-4 by *Agrobacterium tumefaciens*-mediated transformation (Bagga *et al.*, 2019). Transformed $F_1$ progeny, in addition to leaf blight resistance, also showed some level of resistance to root rot. These transformants with an introgressed *RB* gene will be of potential use in breeding peppers with *Phytopthora* resistance.

### 3.7.3 Resistance to bacterial diseases

The *C. annuum* calmodulin 1 gene (*CaCaM1*), a pathogen-responsive gene, has been identified and functionally described from hot pepper (Choi *et al.*, 2009). *Agrobacterium* spp.-mediated transient expression in pepper followed by overexpression of transgenes in *Arabidopsis thaliana* was used to confirm the cellular activity of *CaCaM1*. Local acquired resistance against *Xanthomonas campestris* pv. *vesicatoria* was achieved by transient expression of *CaCaM1*, which stimulated nitric oxide production, reactive oxygen species, and hypersensitive response-like apoptosis in leaves. The overexpression of CaCaM1 in *A. thaliana* and pepper was inhibited by treatment with a calcium-channel blocker, indicating the high degree of association between CaCaM1-mediated defensive responses in plants and calcium influx.

### 3.7.4 Abiotic stress tolerance

*Agrobacterium*-mediated transformation with the wheat (*Triticum aestivum*) $Na^+/H^+$ antiporter gene (*TaNHX2*) under the control of the CaMV 35S promoter and *nptII* as a selectable marker resulted in salinity-tolerant transgenic pepper plants that showed increased levels of relative water content, ascorbate peroxidase, superoxide dismutase, chlorophyll, and proline, and reduced levels of hydrogen peroxide (Bulle *et al.*, 2016). Transgenic plants with salinity tolerance were also developed by ectopic expression of the osmotin gene derived from *Nicotiana tabacum* in pepper, under the control of the CaMV 35S promoter and *nptII* as a selectable marker (Subramanyam *et al.*, 2011). Constitutive coexpression of *Arabidopsis thaliana* homeodomain leucine zipper transcription factor, HOMEODOMAIN GLABROUS11 (AtEDT1/HDG11), enhanced drought tolerance in pepper without compromising fruit yield (Zhu *et al.*, 2015). Enhanced drought tolerance of transgenic plants was associated with reduced stomatal density and an improved root system. The transgenic pepper plants also exhibited high reactive oxygen species scavenging capacity, and high abscisic acid, soluble sugar, proline, chlorophyll fluorescence ($F_m/F_v$), and chlorophyll content under moisture stress compared with the control plants. The G4 variety was successfully transformed with the transcription factor gene *dreb1A* with the desiccation-inducible promoter *rd29A* using the pCAMBIA 2301 vector. Some transformants were found that showed a good level of drought tolerance (Maligeppagol *et al.*, 2016). These studies demonstrated genetic engineering as a potential tool for enhancing abiotic stress tolerance in peppers.

### 3.7.5 Horticultural traits

Apart from disease and abiotic stress resistance, the applications of genetic engineering are very limited in pepper. Transgenic pepper suppressed ripening-related endo-1,4-β-glucanase, as reported by Harpester *et al.* (2002). Transforming a dwarf transgenic pepper with the *OsMADS1* gene from rice (*Oryza sativa*) resulted in the production of a dwarf transgenic pepper (Kim *et al.*, 2001a,b). A non-pungency-causing novel gene in pepper, ketoacyl-ACP reductase (*CaKR1*), was discovered using an RNA-silencing technique (Koeda *et al.*, 2019).

## 3.8 Problems Associated with Capsicum Genetic Engineering

Various inherent problems are associated with transformation in pepper, including drastic recalcitrant morphogenic characteristics, the emergence of rosette shoots or ill-defined shoot buds, and genotype dependence. These problems alone or in conjunction severely limit transformation in pepper via genetic engineering.

### 3.8.1 Extreme recalcitrant morphogenic characteristics

The failure to respond to *in vitro* cultivation by explants is known as recalcitrance. Three key aspects influence tissue-culture responses: the physiology of the donor plant, manipulation, and stress physiology in *in vitro* conditions (Benson, 2000). Pepper exhibits poor regeneration potential in comparison with other solanaceous crops such as potato, tomato, and tobacco, which are widely utilized as model systems due to their high ability for regeneration *in vitro*. Although there exist some positive responses of peppers to shoot morphogenesis (Ramírez-Malagón and Ochoa-Alejo, 1996; Husain *et al.*, 1999; Venkataiah *et al.*, 2003), genetic engineering has had limited success in transforming pepper due to its low regeneration potential (Steinitz *et al.*, 1999; Ochoa-Alejo and Ramírez-Malagón, 2001). Hence, selection of proper explants exhibiting response potential to the nutritional media and other growth enhancers will be essential to alleviate recalcitrance.

### 3.8.2 Production of rosette shoots or ill-defined shoot buds

The production of rosette shoots or ill-defined leafy shoots lacking elongation is one of the components restricting regeneration of tissue-cultured transgenic pepper plants. Several approaches for overcoming the challenges of shoot elongation have been investigated. The rosette of buds was directly rooted by Arroyo and Revilla (1991), and elongation occurred in the field. To mitigate this problem, other studies have used growth adjuvants such as silver nitrate (Hyde and Phillips, 1996), phenylacetic acid (Husain *et al.*, 1999), and a lactone, 24-epibrassinolide (Franck-Duchenne *et al.*, 1998).

### 3.8.3 Genotypic dependency

Another important criterion that modulates organogenesis in pepper *in vitro* programs is genotypic dependency. Different cultivars exhibit different regeneration potentials and this is a significant constraint that necessitates the establishment of standard regeneration techniques for individual cultivars.

## 3.9 References

Aguilar-Meléndez, A., Morrell, P.L., Roose, M.L., and Kim, S.C. (2009) Genetic diversity and structure in semiwild and domesticated chiles (*Capsicum annuum*; Solanaceae) from Mexico. *American Journal of Botany* 96, 1190–1202.

Anandhi, K., and Abdul Khader, K. (2011). Gene effects of fruit yield and leaf curl virus resistance in interspecific crosses of chilli (*Capsicum annuum* L. and *C. frutescens* L.). *Journal of Tropical Agriculture* 49, 107–109.

Arroyo, R., and Revilla, M. (1991) *In vitro* plant regeneration from cotyledon and hypocotyl segments in two bell pepper cultivars. *Plant Cell Reports* 10, 414–416.

Bagga, S., Lucero, Y., Apodaca, K., Rajapakse, W., Lujan, P., *et al.* (2019) Chile (*Capsicum annuum*) plants transformed with the *RB* gene from *Solanum bulbocastanum* are resistant to *Phytophthora capsici*. *PLOS ONE* 14: e0223213.

Baral, J.B. and Bosland, P.W. (2004) Unraveling the species dilemma in *Capsicum frutescens* and *C. chinense* (Solanaceae): a multiple evidence approach using morphology, molecular analysis, and sexual compatibility. *Journal of the American Society for Horticultural Science* 129, 826–832.

Basu, S.K. and De, A.K. (2003) *Capsicum*: historical and botanical perspectives. In: De, A.K. (ed.) *Capsicum: The Genus Capsicum*. Taylor and Francis, London/New York, pp. 1–15.

Baysal, O., Turgut, C., and Mao, G. (2005) Acibenzolar-S-methyl induced resistance to *Phytophthora capsici* in pepper leaves. *Biologia Plantarum* 49, 599–604.

Benson, E.E. (2000) *In vitro* plant recalcitrance: an introduction. *In Vitro Cellular & Developmental Biology Plant* 36, 141–148.

Berger, A., Henderson, M., Nadoolman, W., Duffy, V., Cooper, D., *et al.* (1995) Oral capsaicin provides temporary relief for oral mucositis pain secondary to chemotherapy/radiation therapy. *Journal of Pain and Symptom Management* 10, 243–248.

Bosland, P.W., and Baral, J.B. (2007) 'Bhut Jolokia'—The world's hottest known chile pepper is a putative naturally occurring interspecific hybrid. *HortScience* 42, 222–224.

Bulle, M., Yarra, R., and Abbagani, S. (2016) Enhanced salinity stress tolerance in transgenic chilli pepper (*Capsicum annuum* L.) plants overexpressing the wheat antiporter (*TaNHX2*) gene. *Molecular Breeding* 36: 36.

Cai, W.Q., Fang, R.X., Shang, H.S., Wang, X., Zhang, F.L., *et al.* (2003) Development of CMV-and TMV-resistant chilli pepper: field performance and biosafety assessment. *Molecular Breeding* 11, 25–35.

Carrizo García, C., Sehr, E.M., Barfuss, M.H.J., Barboza, G.E., Samuel, R., *et al.* (2016) Phylogenetic relationships, diversification and expansion of chili peppers (*Capsicum*, Solanaceae). *Annals of Botany* 118, 35–51.

Choi, H.W., Lee, D.H., and Hwang, B.K. (2009) The pepper calmodulin gene *CaCaM1* is involved in reactive oxygen species and nitric oxide generation required for cell death and the defense response. *Molecular Plant–Microbe Interactions* 22, 1389–1400.

Dhall, R.K., Hundal, J.S., and Saxena, A. (2011) Floral biology studies in chilli (*Capsicum annum* L). *Vegetable Science* 38, 221–224.

Delis, M., Garbaczewska, G.Y., and Niemirowicz-Szczytt, K. (2005) Differentiation of adventitious buds from *Capsicum annuum* L. hypocotyls after co-culture. *Acta Biologica Cracoviensia Series Botanica* 47, 193–198.

Dong, X., Mindrinos, M., Davis, K.R., and Ausubel, F.M. (1991) Induction of *Arabidopsis* defense genes by virulent and avirulent *Pseudomonas syringae* strain and by a cloned avirulence gene. *Plant Cell* 8, 155–168.

FAOSTAT (2021) FAOSTAT. FAO, Rome. Available at: https://www.fao.org/faostat/en/#home (accessed 19 June 2024).

Franck-Duchenne, M., Wang, Y., Ben Tahar, S., and Beachy, R.N. (1998) *In vitro* stem elongation of sweet pepper in media containing 24-epi-brassinolide. *Plant Cell, Tissue and Organ Culture* 53, 79–84.

Garg, R., Loganathan, M., Saha, S., and Roy, B.K. (2014) Chilli anthracnose: a review of causal organism, resistance source and mapping of gene. In: Kharwar, R.N., Upadhyay, R.S., Dubey, N.K., and Richa Raghuwanshi (eds.) *Microbial Diversity and Biotechnology in Food Security*. Springer, pp. 589–610.

Gopal, G.V., Lakshmi, K.V., Babu, B.S., Varma, P.K. (2018) Seasonal incidence of chilli thrips, *Scirtothrips dorsalis* Hood in relation to weather parameters. *Journal of Entomology and Zoological Studies* 6, 466–471.

Guzmán, F.A., Ayala, H., Azurdia, C., Duque, M.C., and de Vicente, C. (2005) AFLP assessment of genetic diversity of *Capsicum* genetic resources in Guatemala: home gardens as an option for conservation. *Crop Science* 45, 363–370.

Harpster, M.H., Brummell, D.A., and Dunsmuir, P. (2002) Suppression of a ripening-related endo-1,4-β-glucanase in transgenic pepper fruit does not prevent depolymerization of cell wall polysaccharides during ripening. *Plant Molecular Biology* 50, 345–355.

Heidmann, I., de Lange, B., Lambalk, J., Angenent, G.C., and Boutilier, K. (2011) Efficient sweet pepper transformation mediated by the BABY BOOM transcription factor. *Plant Cell Reports* 30, 1107–1115.

Hernandez-Verdugo, S., Lopez, Espana, R.G., Porras, F., Parra-Terraza, S., Villarreal-Romero, M., and Tomas, O. (2010) Variation in germination among populations and plants of wild chili pepper. *Agrociencia* 44, 667–677.

Hofmann, C., Niehl, A., Sambade, A., Steinmetz, A., and Heinlein, M. (2009) Inhibition of *Tobacco mosaic virus* movement by expression of an actin-binding protein. *Plant Physiology* 149, 1810–1823.

Husain, S., Jain, A., and Kothari, S.L. (1999) Phenylacetic acid improves bud elongation and *in vitro* plant regeneration efficiency in *Capsicum annuum* L. *Plant Cell Reports* 19, 64–68.

Hyde, C.L., and Phillips, G.C. (1996) Silver nitrate promotes shoot development and plant regeneration of chile pepper (*Capsicum annuum* L.) via organogenesis. *In Vitro-Plant* 32, 72–80.

Ishibashi, K., Nishikiori, M., Mawatari, N., and Ishikawa, M. (2007) [Constitutive resistance against viruses in plants]. *Protein, Nucleic Acid, Enzyme* 52, 686–691.

Jayewar, N.E., Bhosle, B.B., and Bhede, B.V. (2018) Short term climatic condition influencing thrips infesting chilli. *Journal of Entomology and Zoology Studies* 6, 2459–2463.

Jarret, R.L. and Dang, P. (2004) Revisiting the waxy locus and the *Capsicum annuum* complex. *Georgia Journal of Science* 62, 117–133.

Kim, C.G., Kim, D.I., Kim, H.J., Park, J.E., Lee, B., *et al.* (2009a) Assessment of gene flow from genetically modified anthracnose-resistant chili pepper (*Capsicum annuum* L.) to a conventional crop. *Journal of Plant Biology* 52, 251–258.

Kim, C.G., Park, K.W., Lee, B., Kim, D.I., Park, J.Y., *et al.* (2009b) Gene flow from genetically modified to conventional chili pepper (*Capsicum annuum* L.). *Plant Science* 176, 406–412.

Kim, D.H., Kang, J.G., Kim, S., and Kim, B.D. (2001a) Identification of *coxII* and *atp6* region as associated to CMS in *Capsicum annuum* by using RFLP and long and accurate PCR. *Journal of Korean Society of Horticultural Science* 42, 121–127.

Kim, E., Noh, H.M., Phat, C., Lee, G.P., Kim, J.H., *et al.* (2016) Identification and safety assessment of cucumber mosaic virus coat protein in genetically modified pepper (*Capsicum annuum*). *Horticultural Science and Technology* 34, 924–939.

Kim, S., Kim, S.R., Chung Sun, A.N., Hong, Y.N., and Lee, K.W. (2001b) Constitutive expression of rice MADS box gene using seed explants in hot pepper (*Capsicum annuum* L.). *Molecules and Cells* 12, 221–226.

Kim, S.J., Lee, S.J., Kim, B.D., and Paek, K.H. (1997) Satellite-RNA-mediated resistance to cucumber mosaic virus in transgenic plants of hot pepper (*Capsicum annuum* cv. Golden Tower). *Plant Cell Reports* 16, 825–830.

Ko, M.K., Jeon, W.B., Kim, K.S., Lee, H.H., Seo, H.H., *et al.* (2005) A *Colletotrichum gloeosporioides*-induced esterase gene of nonclimacteric pepper (*Capsicum annuum*) fruit during ripening plays a role in resistance against fungal infection. *Plant Molecular Biology* 58, 529–541.

Ko, M.K., Cho, J.H., Seo, H.H., Lee, H.H., Kang, H.Y., *et al.* (2016) Constitutive expression of a fungus-inducible carboxylesterase improves disease resistance in transgenic pepper plants. *Planta* 244, 379–392.

Koeda, S., Sato, K., Saito, H., Nagano, A.J., Yasugi, M., *et al.* (2019) Mutation in the putative ketoacyl-ACP reductase CaKR1 induces loss of pungency in *Capsicum*. *Theoretical and Applied Genetics* 132, 65–80.

Krishnakumari, K. and Peter, K.V. (1986) Genetic distance and heterosis in interspecific crosses of *Capsicum*. *Agriculture Research Journal of Kerala* 24: 122.

Kumar, A.M., Reddy, K.N., Sreevathsa, R., Ganeshan, G., and Udayakumar, M. (2009) Towards crop improvement in bell pepper (*Capsicum annuum* L.): transgenics (*uidA::hpt* II) by a tissue-culture-independent *Agrobacterium*-mediated in planta approach. *Scientia Horticulturae* 119, 362–370.

Kumar, M., Reddy, K.M., Sriram, S., and Yadav, R.K. (2021) Inheritance of resistance in chilli against root and collar rot caused by *Phytophthora capsici*. *Journal of Plant Pathology* 103, 769–775.

Kumar, S., Singh, V., Singh, M., Rai, S., Kumar, S., *et al.* (2007) Genetics and validation of fertility restoration associated RAPD markers in pepper (*Capsicum annuum* L.). *Scientia Horticulturae* 111, 197–202.

Kurbett, A., Gopali, J.B., Allolli, T.B., Patil, S., Kumar, V., and Kurubetta, K. (2018) Evaluation of different IPM modules against pest complex of Chilli (cv. Byadgi dabbi). *Journal of Entomology and Zoology Studies* 6, 1991–1996.

Lee, S.J., Kim, B.-D., and Paek, K.-H. (1993) *In vitro* plant regeneration and *Agrobacterium*-mediated transformation from cotyledon explants of hot pepper (*Capsicum annuum* cv. Golden Tower). *Korean Journal of Plant Tissue Culture* 20, 289–294.

Lee, J.H., Hong, J.P., Oh, S.K., Lee, S., Choi, D., and Kim, W. (2004a) The ethylene-responsive factor like protein 1 (CaERFLP1) of hot pepper (*Capsicum annuum* L.) interacts *in vitro* with both GCC and DRE/CRT sequences with different binding affinities: possible biological roles of CaERFLP1 in response to pathogen infection and high salinity conditions in transgenic tobacco plants. *Plant Molecular Biology* 55, 61–81.

Lee, Y.H., Kim, H.S., Kim, J.Y., Jung, M., Park, Y.S., *et al.* (2004b) A new selection method for pepper transformation: callus-mediated shoot formation. *Plant Cell Reports* 23, 50–58.

Lee, J.M., Nahm, S.H., Kim, Y.M., and Kim, B.D. (2004c) Characterization and molecular genetic mapping of microsatellite loci in pepper. *Theoretical and Applied Genetics* 108, 619–627.

Lee, W.P., Lee, J.D., Han, J.H., Kang, B.C., and Yoon, J.B. (2012) Validity test for molecular markers associated with resistance to *Phytophthora* root rot in chili pepper (*Capsicum annuum* L.). *Horticultural Science and Technology* 30, 64–72.

Lee, Y.H., Jung, M., Shin, S.H., Lee, J.H., Choi, S.H., *et al.* (2009) Transgenic peppers those are highly tolerant to a new CMV pathotype. *Plant Cell Reports* 28, 223–232.

Li, D., Zhao, K., Xie, B., Zhang, B., and Luo, K. (2003) Establishment of a highly efficient transformation system for pepper (*Capsicum annuum* L.). *Plant Cell Reports* 21, 785–788.

Liu, W., Parrott, W.A., Hildebrand, D.F., Collins, G.B., and Williams, E.G. (1990) *Agrobacterium* induced gall formation in bell pepper (*Capsicum annuum* L.) and formation of shoot-like structures expressing introduced genes. *Plant Cell Reports* 9, 360–364.

Lozano-Duran, R., Rosas-Diaz, T., Gusmaroli, G., Luna, A.P., Taconnat, L., *et al.* (2011) Gemini viruses subvert ubiquitination by altering CSN-mediated derubylation of SCF E3 ligase complexes and inhibit jasmonate signaling in *Arabidopsis thaliana*. *Plant Cell* 23, 1014–1032.

Maga, J.A., and Todd, P.H. (1975) Capsicum. *Critical Reviews in Food Science & Nutrition* 6, 177–199.

Mahto, B.K., Singh, A., Pareek, M., Rajam, M.V., Dhar-Ray, S., and Reddy, P.M. (2020) Host-induced silencing of the *Colletotrichum gloeosporioides* conidial morphology 1 gene (*CgCOM1*) confers resistance against anthracnose disease in chilli and tomato. *Plant Molecular Biology* 104, 381–395.

Maligeppagol, M., Manjula, R., Navale, P.M., Babu, K.P., *et al.* (2016) Genetic transformation of chilli (*Capsicum annuum* L.) with *Dreb*1A transcription factor known to impart drought tolerance. *Indian Journal of Biotechnology* 15, 17–24.

Manoharan, M., Vidya, C.S., and Sita, G.L. (1998) *Agrobacterium*-mediated genetic transformation in hot chilli (*Capsicum annuum* L. var. Pusajwala). *Plant Science* 131, 77–83.

Marino, D., Peeters, N., and Rivas, S. (2012) Ubiquitination during plant immune signaling. *Plant Physiology* 160, 15–27.

Mathew, D. (2020) Analysis of QTL *Bw1* and marker CAMS451 associated with the bacterial wilt resistance in hot pepper (*Capsicum annuum* L.). *Plant Gene* 24: 100260.

Mihálka, V., Balázs, E., and Nagy, I. (2003) Binary transformation systems based on 'shooter' mutants of *Agrobacterium tumefaciens*: a simple, efficient and universal gene transfer technology that permits marker gene elimination. *Plant Cell Reports* 21, 778–784.

Mini, C. (1989) Evaluation of ornamental chillies and selection for dwarf clustered plant types. PhD thesis, College of Horticulture, Vellanikkara, India.

Mishra, R., Mohanty, J.N., Mahanty, B., and Joshi, R.K. (2021) A single transcript CRISPR/Cas9 mediated mutagenesis of CaERF28 confers anthracnose resistance in chilli pepper (*Capsicum annuum* L.). *Planta* 254: 5.

Murthy, N.S.R. and Murthy, B.S. (1962) Natural cross pollination in chilli. *Andhra Agricultural Journal* 9, 161–165.

Mythili, J.B., Rashmi, H.J., Suneetha, C., Saiprasad, G.V.S., Rajeev, P.R., *et al.* (2015) Transgenic chili possessing baculovirus chitinase gene exhibits *in vitro* fungal inhibition. *Journal of Crop Improvement* 29, 159–187.

Naresh, P., Rao, V.K., Lavanya Reddy, B., Anand Reddy, C., Venkatachalapathi, V., and Madhavi Reddy, K. (2016) Genetic analysis for fruit biochemical traits (capsaicinoids and carotenoids) and dry fruit yield in chilli (*Capsicum annuum* L.). *Industrial Crops and Products* 94, 920–931

Ochoa-Alejo, N., and Ramirez-Malagon, R. (2001) *In vitro* chili pepper biotechnology. *In Vitro Cellular & Developmental Biology-Plant* 37, 701–729.

Oelke, L.M., Bosland, P.W., and Steiner, R. (2003) Differentiation of race specific resistance to *Phytophthora* root rot and foliar blight in *Capsicum annuum*. *Journal of the American Society for Horticultural Science* 128, 213–218.

Oh, S.K., Baek, K.H., Park, J.M., Yi, S.Y., Yu, S.H., Kamoun, S., Choi, D. (2008) *Capsicum annuum* WRKY protein CaWRKY1 is a negative regulator of pathogen defense. *New Phytologist* 177, 977–989.

Panda, R.C., Kumar, O.A., and Raja Rao, K.G. (2004) Cytogenetic studies of some $F_1$ hybrids between wild and cultivated taxa of Capsicum L. *Cytologia* 69, 203–208,

Phillips, G.C., Valera-Montero, L.L., Fan, Z., Jayashankar, S., Hubstenberger, J.F., and Watkins, D.D. (2000) Chile improvement through biotechnology: *in vitro* plant regeneration and genetic transformation. In: New Mexico Chile Conference 2000, 8 February 2000, Pepper Institute, Las Cruces, New Mexico (poster).

Pradeepkumar, T., Gopalkrishnan, T.R. and Peter, K.V. (1993) Interspecific hybridization in the genus *Capsicum*. *Vegetable Science* 20, 132–141.

Purkayastha, J., Alam, S.I., Gogoi, H.K., Singh, L., and Veer, V. (2012) Molecular characterization of 'Bhut Jolokia' the hottest chilli. *Journal of Biosciences* 37, 757–768.

Rai, V.P. (2010) Genetic and molecular analyses of pepper leaf curl resistance in chilli (*Capsicum annuum* L.). PhD thesis, Banaras Hindu University, Varanasi, India.

Rai, V.P., Kumar, R., Singh, S.P., Kumar, S., Kumar, S., *et al.* (2014) Monogenic recessive resistance to Pepper leaf curl virus in an interspecific cross of *Capsicum*. Scientia Horticulturae 172, 34–38.

Ramchiary, N., Kehie, M., Brahma, V., Kumaria, S., and Tandon, P. (2013) Application of genetics and genomics towards *Capsicum* translational research. *Plant Biotechnology Reports* 8, 101–123.

Ramírez-Malagón, R. and Ochoa-Alejo, N. (1996) An improved and reliable chili pepper (*Capsicum annuum* L.) plant regeneration method. *Plant Cell Reports* 16, 226–231.

Ravishankar, G.A., Suresh, B., Giridhar, P., Rao, S.R., and Johnson, T.S. (2003) Biotechnological studies on *Capsicum* for metabolite production and plant improvement. In: *Capsicum: The Genus Capsicum*. Harwood Academic Publishers, Reading, UK, pp. 96–128.

Razzaq, A., Zhiying, M., and Haibo, W. (2004) Genetic transformation of wheat (*Triticum aestivum* L): a review. *Molecular Plant Breeding* 2, 457–464.

Roy, A., Zhai, Y., Ortiz, J., Neff, M., Mandal, B., *et al.* (2019) Multiplexed editing of a begomovirus genome restricts escape mutant formation and disease development. *PLOS ONE* 14: e0223765.

Sahu, P.P., Sharma, N., Puranik, S., Muthamilarasan, M., and Prasad, M. (2014a) Involvement of host regulatory pathways during gemini virus infection: a novel platform for generating durable resistance. *Functional & Integrative Genomics* 14, 47–58.

Sahu, P.P., Sharma, N., Puranik, S., and Prasad, M. (2014b) Post-transcriptional and epigenetic arms of RNA silencing: a defense machinery of naturally tolerant tomato plant against *Tomato leaf curl New Delhi virus*. *Plant Molecular Biology Reporter* 32, 1015–1029.

Saxena, A., Raghuwanshi, R., Gupta, V.K., and Singh, H.B. (2016) Chilli anthracnose: the epidemiology and management. *Frontiers in Microbiology* 7: 1527.

Seo, H.H., Park, S., Park, S., Oh, B.J., Back, K., *et al.* (2014) Overexpression of a defensin enhances resistance to a fruit-specific anthracnose fungus in pepper. *PLoS ONE* 9, e97936.

Sharma, A., Kumar, V., Giridhar, P., and Ravishankar, G.A. (2008) Induction of *in vitro* flowering in *Capsicum frutescens* under the influence of silver nitrate and cobalt chloride and pollen transformation. *Electronic Journal of Biotechnology* 11, 84–89.

Sheela, K.B. (1998) Genetic improvement of bird pepper (*Capsicum frutescens* L.) by selection. PhD thesis, Kerala Agricultural University, Thrissur, Kerala.

Shin, R., Park, J.M., An, J.M., and Paek, K.H. (2002) Ectopic expression of *Tsi1* in transgenic hot pepper plants enhances host resistance to viral, bacterial, and oomycete pathogens. *Molecular Plant–Microbe Interactions* 15, 983–989.

Simonne, A.H., Simonne, E.H., Eitenmiller, R.R., Mills, H.A., and Green, N.R. (1997) Ascorbic acid and provitamin A contents in unusually colored bell peppers (*Capsicum annuum* L.). *Journal of Food Composition and Analysis* 10, 299–311.

Singh, D., Mehta, R., and Talati, J.G. (2008) Identification of chilli (*Capsicum annuum* L.) genotypes through RAPD. *Vegetable Science* 35, 10–13.

Sobhakumari, V.P. and Lalithakumari, D. (2005) High frequency shoot regeneration and *Agrobacterium* mediated DNA transfer in red chilli (*Capsicum annuum* L.). *Plant Cell Biotechnology and Molecular Biology* 6, 9–16.

Solís-Ramos, L.Y., González-Estrada, T., Nahuath-Dzib, S., Zapata-Rodriguez, L.C., and Castaño, E. (2009) Overexpression of *WUSCHEL* in *C. chinense* causes ectopic morphogenesis. *Plant Cell, Tissue and Organ Culture* 96, 279–287.

Song, J., Bradeen, J.M., Naess, S.K., Raasch, J.A., Wielgus, S.M., *et al.* (2003) Gene *RB* cloned from *Solanum bulbocastanum* confers broad spectrum resistance to potato late blight. *Proceedings of the National Academy of Sciences USA* 100, 9128–9133.

Steinitz, B., Wolf, D., Matzevitch-Josef, T., and Zelcer, A. (1999) Regeneration *in vitro* and genetic transformation of pepper (*Capsicum* spp.): the current state of art. *Capsicum and Eggplant Newsletter* 18, 9–15.

Subramanyam, K., Sailaja, K.V., Subramanyam, K., Muralidhara Rao, D., and Lakshmidevi, K. (2011) Ectopic expression of an osmotin gene leads to enhanced salt tolerance in transgenic chilli pepper (*Capsicum annum* L.). *Plant Cell, Tissue and Organ Culture* 105, 181–192.

Sunil, K.P. and Rasheed, A. (1998) Cross compatibility in five species of *Capsicum*. *Journal of Spices and Aromatic Crops* 7, 35–38.

Surh, Y.J. (2002) More than spice: capsaicin in hot chili peppers makes tumor cells commit suicide. *Journal of the National Cancer Institute* 94, 1263–1265.

Tatagar, M.H., Prabhu, S.T., and Jagadeesha, R.C. (2001) Screening chilli genotypes for resistance to thrips, *Scirtothrips dorsalis* Hood and mite, *Polyphagotarsonemus latus* (Banks). Pest *Management in Horticultural Ecosystems* 7, 113–116.

Thakur, H., Jindal, S.K., Sharma, A., and Dhaliwal, M.S. (2018) Chilli leaf curl virus disease: a serious threat for chilli cultivation. *Journal of Plant Diseases and Protection* 125, 239–249.

Thakur, H., Jindal, S.K., Sharma, A., and Dhaliwal, M.S. (2020) Molecular mapping of dominant gene responsible for leaf curl virus resistance in chilli pepper (*Capsicum annuum* L.). *3 Biotech* 10: 182.

Than, P., Prihastuti, H., Phoulivong, S., Paul, W., Taylor, J., and Hyde, K.D. (2008) Chilli anthracnose disease caused by *Colletotrichum* species. *Journal of Zhejiang University Science B* 9, 764–778.

Thul, S.T., Darokar, M.P., Shasany, A.K., and Khanuja, S.P.S. (2012) Molecular profiling for genetic variability in *Capsicum* species based on ISSR and RAPD markers. *Molecular Biotechnology* 51, 137–147.

USDA-ARS (US Department of Agriculture Agricultural Research Service) (2011) GRIN species records of *Capsicum*. National Germplasm Resources Laboratory, Beltsville, Maryland.

Venkataiah, P., Christopher, T., and Subhash, K. (2003) Thidiazuron induced high frequency adventitious shoot formation and plant regeneration in *Capsicum annuum* L. *Journal of Plant Biotechnology* 5, 245–250.

Wang, Y., Yang, M., Pan, N., and Chen, Z. (1991) Plant regeneration and transformation of sweet pepper (*Capsicum frutescens*). *Acta Botanica Sinica* 33, 780–786.

Wang, J.E., Li, D.W., Zhang, Y.L., Zhao, Q., He, Y.M., and Gong, Z.H. (2013a) Defence responses of pepper (*Capsicum annuum* L.) infected with incompatible and compatible strains of *Phytophthora capsici*. *European Journal of Plant Pathology* 136, 625–638.

Wang, X., Zhu, X., Tooley, P., and Zhang, X. (2013b) Cloning and functional analysis of three genes encoding polygalacturonase-inhibiting proteins from *Capsicum annuum* and transgenic CaPGIP1 in tobacco in relation to increased resistance to two fungal pathogens. *Plant Molecular Biology* 81, 379–400.

Yang, S., Shi, Y., Zou, L., Huang, J., Shen, L., *et al.* (2020) Pepper CaMLO6 negatively regulates *Ralstonia solanacearum* resistance and positively regulates high temperature and high humidity responses. *Plant and Cell Physiology* 61, 1223–1238.

Zhang, L., Hanada, K., and Palukaitis, P. (1994) Mapping local and systemic symptom determinants of cucumber mosaic cucumovirus in tobacco. *Journal of General Virology* 75, 3185–3191.

Zhang, Y.L., Li, D.W., Gong, Z.H., Wang, J.E., Yin, Y.X., and Ji, J.J. (2013) Genetic determinants of the defence response of resistant and susceptible pepper (*Capsicum annuum*) cultivars infected with *Phytophthora capsici* (Oomycetes; Pythiaceae). *Genetics and Molecular Research* 12, 3605–3621.

Zhu, Y.X., Ou-Yang, W.J., Zhang, Y.F., and Chen, Z.L. (1996) Transgenic sweet pepper plants from *Agrobacterium* mediated transformation. *Plant Cell Reports* 16, 71–75.

Zhu, Z., Xu, X., Cao, B., Chen, C., Chen, Q., *et al.* (2015) Pyramiding of AtEDT1/HDG11 and Cry2Aa2 into pepper (*Capsicum annuum* L.) enhances drought tolerance and insect resistance without yield decrease. *Plant Cell, Tissue and Organ Culture* 120, 919–932.

# 4 Amaranth Genetic Engineering

**K.R.M. Swamy* and B. Varalakshmi**
*Division of Vegetable Crops, ICAR-Indian Institute of Horticultural Research, Hessaraghatta, Bengaluru, Karnataka, India*

**Abstract**
*Amaranthus* is a genus of food plants belonging to the family Amaranthaceae, comprising 70 species. A wide variety of morphological diversity is noted among and even within certain species. The genus *Amaranthus* is unique in having species that are used for grain, vegetable, and ornamental purposes. Amaranth leaves are eaten as a vegetable while the seeds are used in the same way as cereals. Vegetable and grain types are not distinctly different as the leaves of young plants grown for grain can be eaten as both human and animal food. So far, in the genetic improvement of amaranth, conventional selection methods have been primarily involved. However, advances in genomics and biotechnology have enriched the potential to manipulate the amaranth genome, in particular improving the amount and availability of nutrients. In this chapter, we review the origin and distribution of amaranth, its evolution and domestication, its uses, nutritional value and health benefits, its botany; and genetic engineering. The use of biotechnology has much potential for amaranth and it is necessary to further increase research efforts in this area.

**Keywords:** amaranth, biotechnology, botany, genetic engineering, genomics, species

## 4.1 Introduction

*Amaranthus* is a genus of food plants belonging to the family Amaranthaceae, comprising 70 species. A wide variety of morphological diversity is noted among and even within certain species. The genus *Amaranthus* is unique in having species which are used for grain, vegetable and ornamental purposes. Amaranth leaves are eaten as a vegetable,while the seeds are used in the same way as cereals (Pal, 1999). There are two basic types of amaranth:

**1.** Leaf amaranth/green amaranth. Some are very attractive, such as *Coleus* spp. The crop is ready 50 days after sowing. The foliage is very nutritional, being high in vitamins A and C, iron, calcium, and protein. Harvests vary widely depending on variety and growing conditions. Generally speaking, "greens" means vegetable leaves that are eaten cooked (Harrington, 1984).

**2.** Grain/seed amaranth. These closely resemble a cereal-type grain and can be treated as such. Amaranth grain is high in protein and other nutrients and can be harvested and cooked like rice or popped like popcorn. For millenia this ancient grain has been a dietary staple in certain parts of the world, although amaranth has only recently gained popularity as a health food. It

*Email: krmswamy@gmail.com


DOI: 10.1079/9781800627123.0004

has an excellent nutrient profile and has been associated with a number of impressive health benefits. The crop is widely distributed in India. It is cultivated in 17 states of India: Jammu and Kashmir, Himachal Pradesh, Uttaranchal, North Bihar, Sikkim, Assam, Meghalaya, Arunachal Pradesh, Nagaland, Tripura, Gujarat, Madhya Pradesh, Maharashtra, Karnataka, Kerala, Tamil Nadu and Orissa (Shodhganga, 2018). Information on the statistics on area and production of amaranth in India is lacking.

Amaranth is also known as *chaulai* (Uttar Pradesh and Bihar), *rajara* (Gujarat), *chuka* (Bengal), *kala ghasa*, *chumera*, and *ganhar* (Central India), *shravani maath* (Maharashtra), *kolabhaji* (Mumbai), *seol*, *cholai*, and *sil* (Northern Plains), *keerai* and *cheera* (Kerala), *harive* and *keere soppu* (Karnataka), *malan cheerai* and *mulaikkira/mullukurai* (Tamil Nadu), *bathu* (Shimla), *siriara* (Kulu), *tulsi*, *dankar*, and *kalgi* (Kinnaur), *chua* (Kumaun Hills), *marschu* (Garhwal), and *khada saga* (Orissa). There are at least three names for grain amaranth in Sanskrit: *rahadri*, *rajagiri*, and *rajashakini* (Shodhganga, 2018).

Amaranth is eaten as a vegetable while the seeds are used in the same way as cereals. Vegetable and grain types are not distinctly different, as the leaves of young plants grown for grain can be eaten as both human and animal food. When the leaves are harvested in moderation, the grain yield is unaltered. Vegetable amaranth species are utilized for food in different parts of the world. Grain amaranth can be used as seeds or flour to make products such as cookies, cakes, pancakes, bread muffins, crackers, and pasta. The common names used for amaranth in different parts of India are *ramdana* and *chaulai* (John *et al.*, 2008). Both the leaves and the seeds of amaranth are valuable in terms of human health. Whether one chooses to consume it as a leaf vegetable, a cereal grain, or a grain flour depends on what health benefits one is after. The leaves and stems are an interesting vegetable suitable for soups, salads, and other meals. Amaranth is a gluten-free nutritious grain that provides plenty of fiber, protein, and micronutrients. It has a number of health benefits, including reduced inflammation, lowered cholesterol levels, and increased weight loss. Best of all, this grain is easy to prepare and can be added to a variety of dishes, making it an excellent addition to your diet (Rachael, 2018). Nowadays, amaranth sprouts and micro-greens are also being used as nutrition and health foods.

To date, in the genetic improvement of amaranth, conventional selection methods have been used. However, advances in genomics and biotechnology have enriched the potential to manipulate the amaranth genome, in particular to improve the amount and availability of nutrients (Das, 2016).

For a majority of people worldwide, out of thousands of edible plants, only three cereals—corn, wheat, and rice—are the major food. While these crops provide high amounts of calories, they are low in protein and other essential nutrients. The dependence on only a few crops, often with a narrow genetic basis, leads to high vulnerability of modern cropping systems to the predicted climate change and accompanying weather extremes. Broadening our food sources through the integration of so-called orphan crops can help to mitigate the effects of environmental change and improve qualitative food security. Although thousands of traditional crops are known they have received little attention in the last century with limited breeding efforts. Amaranth is one such underutilized pseudocereal and is of particular interest because of its balanced amino acid and micronutrient profiles. Additionally, use of the $C_4$ photosynthetic pathway and the ability to withstand environmental stress make this crop a suitable choice for future agricultural systems. Despite the potential of amaranth, efforts toward genetic improvement lag considerably behind those of major crops. The progress in novel breeding methods and molecular techniques developed in model plants and major crops will allow rapid improvement of underutilized crops. Here, we review the origin and distribution of amaranth; its evolution and domestication, some of its uses, its nutritional value and health benefits, its botany, and use of biotechnology for improvements.

## 4.2 Origin and Distribution of Amaranth

The grain amaranth (*ramdana*, *marcha*, *ganhar*, *lathe*), considered by many as the crop of the future, has been associated with humans since prehistoric times (4800 BC). The genus *Amaranthus* is notable mainly because its species are sometimes deliberately cultivated as grain crops, green vegetables, dye plants, or ornamentals, or more often as unintentionally encouraged weeds. Many amaranth species have spread and diversified with the advance of artificial habitats on every continent (Sauer, 1957). The genus *Amaranthus*, collectively known as amaranth, is a cosmopolitan genus of annual or short-lived perennial plants. Most *Amaranthus* spp. are summer annual weeds and are commonly referred to as pigweeds. Catkin-like cymes of densely packed flowers grow in summer or fall. Approximately 60 species are recognized, with inflorescences and foliage ranging from purple and red to green or gold. Members of this genus share many characteristics and uses with members of the closely related genus *Celosia*. Amaranth is a grain with high nutritional value, comparable to that of corn and rice. Amaranth has been a staple in Mesoamerica for thousands of years, first collected as a wild food, and then domesticated at least as early as 4000 BC. The edible parts are the seeds, which are consumed whole, toasted, or milled into flour. Other uses of amaranth include dye, forage, and ornamental purposes. Amaranth is a member of the family Amaranthaceae. About 60 species are native to the Americas, whereas fewer species originate from Europe, Africa, and Asia. The most widespread species are native to North, Central, and South America; these are *Amaranthus cruentus*, *A. caudatus*, and *A. hypochondriacus*. *A. cruentus* and *A. hypochondriacus* are native to Mexico and Guatemala. *A. cruentus* is used in Mexico to produce sweets called *alegría*, in which the amaranth grains are toasted and mixed with honey or chocolate. *A. caudatus* is a widely distributed staple food in both South America and India.

Amaranth originated as one of the staple foods for the ancient inhabitants of the Andean region (Sauer, 1967; Maestri, 2015). Some species of green amaranth, especially *Amaranthus gangeticus*, *A. mangostanus*, *A. paniculatus*, and *A. angustifolius*, are supposed to have originated from India or the Indo-Chinese region. Other species have originated from various other areas such as North America, Central America, Mexico, South America, and the Mediterranean region (Nath and Swamy, 2015).

The main cultivated species of the vegetable-type amaranth, *Amaranthus tricolor* L., originated in South or South-east Asia, particularly India, and Buddhist monks and Muslim invaders took the crop to neighboring countries. Another vegetable-type amaranth, *Amaranthus dubius* L., shows diversity in Central America, Indonesia, India, and Africa, whlie *Amaranthus lividus* seems to have been a popular vegetable in Southern and Central Europe. As a hot-season vegetable, it is cultivated throughout the year in the tropics (Fig. 4.1) (Varalakshmi, 2015).

The cultivated species of grain-type amaranth and their probable native regions are as follows (Fig. 4.2) (Sauer, 1986):

- *A. hypochondriacus* (= *A. frumentaceus*, *A. leucocarpus*, etc.) of North-western and Central Mexico.
- *A. cruentus* (= *A. paniculatus*, etc.) of Southern Mexico and Central America.
- *A. caudatus* of the Andes. In the Argentine Andes, the typical form is grown together with a conspicuous mutant that produces club-shaped inflorescence branches with determinate growth, a trait unknown in wild amaranths. This mutant has commonly been given a specific rank (as *A. edulis*) but may better be treated as *A. caudatus* ssp. *mantegazzianus*.

The wild species that appear most closely related to the above grain-type species are as follows (Sauer, 1986):

- *A. powellii*, a pioneer of canyons, desert washes, and other open habitats in the Western Cordillera of the Americas. An aberrant form with indehiscent utricles has sometimes been given a specific rank as *A. bouchonii*.
- *A. hybridus* (= *A. chlorostachys*, *A. patulus*, etc.), a riverbank pioneer of moister regions of eastern North America and the mild highlands of Central America.
- *A. quitensis*, a riverbank pioneer of highland and subtropical South America.

Amaranth was probably widely used among hunter-gatherers in both North and South America. The wild seeds, although small in size, are produced in abundance by the plant and are easy to collect. Evidence of domesticated amaranth seeds comes from the Coxcatlan Cave in the Tehuacan valley of Mexico and dates as early as 4000 BC. Coxcatlan Cave is a rock shelter and was occupied by humans for nearly 10,000

**Fig. 4.1.** Species of leaf/green amaranth.

**Fig. 4.2.** Species of grain/seed amaranth.

years. Later evidence, such as caches with charred amaranth seeds, have been found throughout the south-west USA and the Hopewell culture of the mid-west USA. As with other grains, the seeds are collected by rubbing the inflorescences between the hands (Sauer, 1967; Maestri, 2015). It was a major food crop of the Aztecs, and some have estimated that amaranth was domesticated between 6000 and 8000 years ago. During the pre-Columbian period, the Aztecs cultivated amaranth as a staple grain crop. Annual grain tributes of amaranth to the Aztec emperor were roughly equal to corn tributes. The Aztecs not only consumed the grains as food but also used them as part of their religious practices. Many ceremonies would include the creation of a deity's image that had been made from a combination of amaranth grains and honey. Once formed, the images were worshipped before being broken into pieces and distributed for people to eat. However, things changed when the Spanish conquistadors (meaning "one that conquers") arrived. The conquistadors were Spanish explorers and warriors who successfully conquered much of America in the 16th century. When Cortez and his Spaniards landed in the New World in the 16th century, they immediately began fervent and often forceful attempts to convert the Aztecs to Christianity. One of their first moves was to outlaw foods involved in religious ceremonies, such as amaranth. Although severe punishment was handed out to anyone found growing or possessing amaranth, complete eradication of this culturally important, fast-growing, and very prevalent plant proved to be impossible (Anon., 2015b).

## 4.3 Evolution and Domestication of Amaranth

Evolution is a change in heritable traits of biological populations over successive generations. Evolutionary processes give rise to diversity at every level of biological organization, including the level of species and individual organisms, and at the level of molecular evolution. Evolution can be defined as a process of gradual and relatively continuous change from a lower, simpler, or worse to a higher, more complex, or better state/growth. It is the historical development of a biological group (as a race or species (phylogeny)) whereby animals and plants have their origin in other pre-existing types and that the distinguishable differences are due to modifications in successive generations. Whatever their origin, the grain *Amaranthus* spp. arose from a wild progenitor or progenitors by domestication. Grain types are not found in a wild condition anywhere in the world. The principles involved during a long selection history resulted first in the development of rather short and weak bracts, in order to make the inflorescence less prickly when rubbed between the hands while extracting grain. The perfectly dehiscent utricle has been an added advantage in this direction. Second, selection has been made for a large plant body, particularly large compound inflorescences, in order to give an enormous grain yield without an increase in grain size. Third, there has been a decided preference for white seeds with good popping qualities and flavor. During these selection processes, there has been inadvertent selection for the best proportions of protein, carbohydrate, and oil components to make it a balanced food with a higher calorific value. These features also distinguish the grain types from their wild ancestors (Sauer, 1967; Pal and Khoshoo, 1973).

Sauer (1957) studied a group of ten related North American species, sharply distinguished from all other amaranths of the world by their dioecious habit. None of the dioecious amaranths have become domesticated plants or pandemic weeds. However, they have also been increasing in geographic and genetic ranges in response to civilization. Sauer reported that the ten dioecious species of *Amaranthus*, all originally pioneer plants of naturally disturbed habitats, have behaved quite differently during increasing human disturbance of their native continent, North America. Those at home in coastal marshes, beaches, and dunes have remained static, while those from river floodplains and desert washes have been expanding and diversifying as weeds in artificially opened habitats. Direct historical evidence, mainly from a 100-year record of herbarium collections, permits partial reconstruction of the geographic and genetic changes. The amount of geographic advance has been very unequal, even along different borders of a single species. In some cases, expansion was local and completed before the present century; elsewhere, migration has extended across whole states and is still active. However, unlike the common weedy monoecious amaranths, even the weediest dioecious species have remained geographically cohesive. Apparent hybrids were derived from a large number of species pairs, involving not only all the dioecious but also some monoecious species. Hybrid populations are almost invariably concentrated in the newer, artificial habitats, perhaps because of less rigorous selection in such places. In natural habitats, the older types of populations have remained relatively immune to invasion and hybridization.

According to Sauer (1986), wild amaranth seeds were commonly gathered by many prehistoric American-Indian people. The wild seeds are as nutritious and as large as those of the cultivated species. Archaeological proof of domestication comes with the appearance of pale white seeds, contrasting starkly with the dark brown wild type; the mutation producing this change has never been recorded historically. Associated with the change in color are improved popping quality and flavor. A small proportion of dark seeds are generally present in the grain crops. Where selection for pale seed color was relaxed, for example when the plants are grown as ornamentals, the dark seeds became predominant. The earliest record of the pale-seeded crop is from Tehuacan, Puebla, Mexico, where *A. cruentus* appeared about 4000 BC and was joined by *A. hypochondriacus* about AD 500. By the 14th century, pale-seeded *A. hypochondriacus* was also cultivated by Arizona cliff-dwellers. The earliest record of *A. caudatus* is from 2000-year-old tombs in north-western Argentina, where its pale seeds were found mixed with those of *Chenopodium quinoa* and with dark seeds of weed amaranths and chenopods. The three grain-type species may have been independently domesticated, but there is an alternative possibility, namely that there was a single primary domestication of *A. cruentus* from *A. hybridus*, with the other two domesticates evolving secondarily by repeated crossing of *A. cruentus* with weedy *A. powellii* and *A. quitensis* as the crop spread into their respective territories. Evolution of all three domesticates has involved an increased size of the whole plant and particularly of the inflorescence, resulting in greatly increased seed yield. All three domesticates also display the effects of selection for striking anthocyanin pigmentation of leaves, stem, and inflorescences. Presumably, the intense red color had ceremonial meaning. At the time of the Spanish Conquest, grain amaranths were important in rituals of the Aztecs and other Mexican peoples. Judging by later ethnographic evidence, ceremonial use of red amaranths extended from the Pueblo region of the south-western USA to the Andes and was more widespread than use as a grain crop. The ceremonial-dye amaranths are generally extremely deep red forms of *A. cruentus* with dark seeds; in the Andes, some may be *A. cruentus* × *A. quitensis* hybrids (Fig 4.3).

Putnan (1992) stated that crops are truly artifacts of human history. We can ask ourselves this: "What would Africans do without peanuts, the Indians do without chillies, the Irish or Poles do without potatoes, the Italians do without tomatoes or the whole world do without corn?" All were new crops from the Americas (as is amaranth). Each of these species found an important role in the culture of the society to which they were transported, and filled an important need or niche not previously met by an "old" crop. Crops such as amaranth became only curiosities in Europe, largely because those cultures did not discover ways to easily incorporate the crop into their diet; the grains of amaranth were too small to be easily ground by primitive flour mills. Initial mistrust may have also played a role, as it did with the tomato. Such was not the case in the introduction of amaranth to India centuries ago. Not only is amaranth grown by farmers in the highest habitable regions of the Himalayas, it is also grown in arid regions of the plains and throughout the Indian subcontinent. It is eaten as a leafy vegetable or popped as a candy, "chikki". Amaranth is widely adapted and reasonably high yielding. Amaranth is cited in ancient religious texts and has many Indian names. It is hypothesized that amaranth found its way to Surat and South India via Portuguese traders, and eventually made its way to the Himalayas. In the Indian subcontinent, amaranth is still very much an underutilized crop. The tradition here dictates eating habits, and amaranth is not widely utilized in daily diets. In many Indian institutes, research on amaranth has been carried out for decades.

In Spanish America or Hispanic America (the region comprising the Spanish-speaking nations in the Americas) after the 16th century, grain amaranth cultivation was regarded as a symbol of paganism and was repressed; thus, the

**Fig. 4.3.** Different-colored amaranth seeds.

crop nearly disappeared from history. However, by the early 19th century, grain amaranth had appeared as a staple food crop in the Nilgiri Hills of South India and in the Himalayas; it has since been noted over an increasingly wide area of India, as well as across the interior of China to Manchuria and Eastern Siberia. Pale-seeded *A. hypochondriacus* constitutes the bulk of the Asiatic crop, while the dark-seeded *A. hypochondriacus* and pale-seeded *A. caudatus* are minor components. In the 1940s, cultivation of *A. hypochondriacus* began in East Africa to supply grain to the local Indian population. The wide latitudinal spread of these species in the Old World presumably required evolutionary changes, because their flowering is controlled by photoperiod. *A. cruentus* has not become established as a grain crop in the Old World. However, dark-seeded, deep-red forms of this species have been widely planted in tropical Africa and Asia for over a century as ornamentals, dye plants, and potherbs. All three of the domesticated species may have been introduced to the Old World via Europe; they have been grown in European gardens as ornamentals and curiosities for at least 250 years. Only the dark-seeded forms of *A. hypochondriacus* were known to have presence in Europe (Sauer, 1986).

Amaranth is also making a comeback as a popular superfood. The seed, commonly referred to as grain, is gluten free and is a good source of protein. It may also have a political advantage over quinoa, another healthy grain that is growing in popularity. Quinoa has drawn some controversy recently because the price of the grain has increased so dramatically that the indigenous populations in Bolivia, Ecuador, and Peru (where the grain is grown) can no longer afford to eat it. Some argue that the increased demand creates income for otherwise impoverished areas. Amaranth is inexpensive in comparison with quinoa, and is easily cultivated in a wider variety of conditions. It can survive in arid climates and is currently grown everywhere from Long Island and Iowa to India and Fiji. As a bonus, the plant's stems and leaves are also edible and vitamin rich. In Caribbean cultures, certain varieties of amaranth leaves are known as *callaloo* (Wang, 2012). Seeds from the amaranth plant spread around the world and both leaves and grain became important food sources in areas of Africa, India, and Nepal. In the past two decades, amaranth has reached a much larger number of farmers and can now be found in many non-native regions such as China, Russia, Thailand, and Nigeria, as well as Mexico and parts of South America. It prefers high elevation to low but is impressively adaptive and can grow well in moist, loose soil with good drainage at almost any elevation and in just about any temperate climate. Once established, amaranth can continue to thrive in low-water conditions, making it especially valuable in sub-Saharan Africa where water sources are few, especially in the dry season. Amaranth received renewed interested as a food source in the USA back in the 1970s. Today, you can find it growing in small amounts in some pretty surprising locations, including Iowa, Nebraska, Missouri, North Dakota, and even Long Island, New York. At present, amaranth is grown in the USA, South America, India, China, and Russia. The Czech Republic is the most important grower in Europe (Anon., 2015b).

## 4.4 Uses, Nutritional Value and Health Benefits of Amaranth

Amaranth is the common name for more than 60–70 different *Amaranthus* spp., which are usually very tall plants with broad green leaves and impressively bright purple, red, or gold flowers. The name for amaranth comes from the Greek *amarantos*, meaning "one that does not wither," or "the never-fading." True to form, amaranth's bushy flowers retain their vibrancy even after harvesting and drying, and some varieties of ornamental amaranth forego the production of fancy flowers in favor of flashy foliage, sprouting leaves that can range from deep blood-red to light green with purple veining. Although several species can be viewed as little more than annoying weeds, people around the world value amaranths as leaf vegetables, cereals, and ornamental plants (Anon., 2015b).

Amaranth is not a true cereal grain in the sense that oats, wheat, sorghum, and most other grains are. "True cereals" belong to the family Poaceae, while amaranth is often referred to as a "pseudo-cereal," meaning it belongs to a different plant species, but their overall nutrient profile is similar to that of cereals (Anon., 2015b).

Recently, the potential of microcrystalline (1–3 μm) starch granules for possible replacement of talc in the cosmetic industry and edible dyes has emerged. The most common use in the major regions of its cultivation (Peru, Bolivia, Ecuador, Mexico, India, Nepal, and Bhutan) is in the form of cakes or balls (laddoos) prepared by binding the popped seed in jiggery or sugar. The vegetable amaranths are used as pot herbs in most tropical countries of the world, while the related foliage ornamental types add color to the otherwise drab garden surroundings in summer months (Pal, 1999). Vegetable amaranth is widely grown in the tropics and is one of the most important leafy vegetables in the lowlands of Africa and Asia. Amaranth is an annual, fast-growing herb that can easily be cultivated in home gardens and on a commercial scale (Ebert *et al.*, 2011).

Amaranth today is enjoyed many ways. In Mexico and India, the seeds are popped and mixed with sugar to make a confection. In Mexico, they are roasted for the traditional drink "atole." Peruvians use the grain to make a beer. Elsewhere, it is used to treat toothache and fevers, or to color corn and quinoa. In many countries, the leaves are used boiled or fried. In Nepal, the seeds are made into gruel. Amaranth seed and flour can also be found in health-food stores (Green Deane, 2012).

### 4.4.1 Nutritional value of amaranth grains

Amaranth plant is valued for the positive chemical composition of its seed, which do not contain gluten. Amaranth is an interesting crop from the point of its high production potential. It grows intensively, photosynthesizes fast and effectively, does not suffer from major diseases, and is tolerant to various extreme conditions (Anon., 2015a). According to Howard (2013), amaranth is a broad-leafed, bushy plant that grows about 1.8 m tall. It produces a brightly colored flower that can contain up to 60,000 seeds. The seeds are nutritious and can be made into flour. There are around 60–70 different species of amaranth, and a few of them are native to Mesoamerica. It is a whole, healthy food that can be produced locally, and may create the possibility of change in tackling obesity.

Amaranth is gluten free, and its seeds contain about 30% more protein than rice, sorghum, and rye, according to a US Department of Agriculture Forest Service report. It is also relatively high in calcium, iron, potassium, magnesium, phosphorus, and fiber. Amaranth's amino acid profile is as close to perfect as you can get for a protein source. The plant contains eight essential amino acids and is particularly high in the amino acid lysine, which is largely lacking in corn and wheat. Amaranth has been recommended by the World Health Organization as a well-balanced food and recommended by National Aeronautics and Space Administration (NASA) for consumption in space missions. The variety of amaranth consumed in Mexico has 16–18% protein, compared with 14% in wheat and 9–10% in corn. Some studies have shown that amaranth also contains beneficial ω-3s and may help reduce blood pressure (Howard, 2013). It is the only grain documented to contain vitamin C.

Amaranth seeds have a high content of protein, essential amino acids and minerals, usually in the following ranges: 14–19% protein, 5–8% lipids, 62–69% starch, 2–3% total carbohydrates and 4–5% fiber. The seed composition is comparable with that of oat. Amaranth protein bioavailability reaches 78%. The seed does not contain gluten, which causes celiac disease in sensitive individuals. In contrast to cereals, amaranth has a higher content of amino acids, mainly lysine, methionine, threonine, and cysteine. Starch is the major carbohydrate and starch granules are small (1–3 μm) and easily degradable by α-amylases. Amaranth starch is highly stabile during freezing and highly resistant to mechanical stress. The lipid content in amaranth seed in the embryo is linoleic, oleic, and palmitic acid. According to the seed composition, amaranth oil is similar to oils made from cotton or corn but has lower digestibility. Amaranth oil contains about 8% squalene, a sterol precursor, used in medicine and the cosmetic industry. The seeds are a good source of vitamins, mainly ascorbic acid and vitamin B-complex, and antioxidants α-tocopherol and β- and γ-tocotrienols. Amaranth is used as a food ingredient primarily in bread, pasta, baby food, instant drinks, etc. For such purposes, the seeds have to undergo processing including boiling, swelling, flaking, extrusion, puffing, roasting, grinding, and sprouting. The most common product is flour, whole amaranth seeds in breads, muesli bars, breakfast food and porridges, pasta, and biscuits and cookies. The leaves and stems are suitable for soups, salads, and other meals. As a fodder, amaranth is a valuable nutritious feedstuff with high production ability. It also provides starch, protein concentrates, and natural dye, and is a good source of squalene and antioxidants. It is being intensively tested as one of the interesting energy crops (Anon., 2015a).

### 4.4.2 Nutritional value of amaranth leaves

A major attribute of amaranth is the high carotene (pro-vitamin A) content in the leaves (Ebert *et al.*, 2011). Amaranth is a highly nutritious leafy vegetable, both raw and cooked. Its nutritional value is comparable to spinach, but much higher than cabbage and Chinese cabbage. Amaranth is well adapted to high temperatures in the tropics. Amaranth is low in saturated fats and sodium, and cholesterol is absent. It is a good source of calcium, iron, magnesium, phosphorus, potassium, zinc, copper, and manganese. It is a very good source of high-quality protein with well-balanced amino acids. Many vitamins are found in high levels in both raw and cooked amaranth leaves and stems: vitamin A, vitamin C, riboflavin (vitamin $B_2$), vitamin $B_6$, folate (vitamin $B_9$), and niacin (vitamin $B_3$). Amaranth also contains betacyanines (amaranthine and isoamaranthine) responsible for the red-violet colors of the leaves, stems, flowers, and seeds. These pigments dissolve in the cooking water, which is poured off. The leaves and stems contain the antinutrients nitrate (mostly in the stems) and oxalate at a level similar to other leafy vegetables such as spinach and spinach beet. Consumption of 100–200 g of fresh produce day$^{-1}$ is safe. Cooking in water removes the toxic components (Ebert *et al.*, 2011). The fresh, tender leaves and stems are delicious on cooking. It is a cheap vegetable. Besides having the highest protein content among leafy vegetables, it also contains carbohydrates, calcium and iron. The vitamin A content in different species varies from 23,000 to 54,110 IU 100 g$^{-1}$ dry weight. The leaves contain 130–173 mg vitamin C, 100–130 mg vitamin B, 4 g protein, 397 mg calcium, 83 mg phosphorus, 25.5 mg iron, 341 mg potassium, 247 mg magnesium and 9200 IU of vitamin A 100 g$^{-1}$ dry weight.

### 4.4.3 Health benefits of amaranth grain and leaves

Very little research has been conducted on amaranth's beneficial properties, but the studies that have focused on amaranth's role in a healthy diet have revealed three important reasons to add it to our diet.

**1.** It is a protein powerhouse. At about 13–14%, it easily trumps the protein content of most other grains.
**2.** It is good for our heart. Amaranth has shown potential as a cholesterol-lowering whole grain.
**3.** It is naturally gluten free. Gluten is the major protein in many grains and is responsible for the elasticity in dough, allows leavening, and contributes chewiness to baked products. However, more and more people are finding they cannot comfortably—or even safely—eat products containing gluten, often due to celiac disease. Celiac disease is an autoimmune digestive disease that damages the body's ability to absorb nutrients from food. This makes amaranth an important grain for those with this disease (Anon., 2015b).

La Jolla Mom (2013) reported that amaranth helps with hair loss and graying, and lowers cholesterol and risk of cardiovascular disease. Amaranth is full of antioxidants and minerals. It is the only grain to have vitamin C, but it is also high in vitamin E, iron, magnesium, phosphorus, and potassium, which are necessary for overall health. The leaves are high in vitamin C, vitamin A, and folate. Amaranth is traditionally given to patients recovering from illness or people coming off fasts. It is the mix of amino acids that allows very easy digestion.

## 4.5 Botany of Amaranth

Amaranth belongs to the the genus *Amaranthus*, family Amaranthaceae. The genus *Amaranthus* (Magnoliophyta: Caryophyllidae) comprises 70 species grouped into three subgenera (Mosyakin and Robertson, 2003). *Amaranthus* spp. show a wide variety of morphological diversity among and even within certain species. Formerly, Sauer (1955) classified the genus into two subgenera, differentiating only between monoecious and dioecious species: *Acnida* and *Amaranthus* spp. Although this classification was widely accepted, further intrageneric classification was (and still is) needed to differentiate this widely diverse group. Currently, the genus includes three recognized subgenera and 70 species, although species numbers are questionable due to hybridization and species concepts. Intrageneric classification focuses on inflorescence, flower characters, and whether a species is monoecious or dioecious, as in the classification suggested by Sauer (1955). A modified intrageneric classification of *Amaranthus* was published by Mosyakin and Robertson (1996) and includes three subgenera: *Acnida*, *Amaranthus*, and *Albersia*. The taxonomy is further differentiated by sections within each of the subgenera. Sauer (1967) recognized two sections in *Amaranthus*: *Amaranthus* and *Blitopsis*, with equal number of species in each. The section *Amaranthus* consists of species having terminal flower clusters and includes grain types, whereas the section *Blitopsis* consists of species having flower clusters in axils and includes the green or vegetable types.

The most economically important is the subgenus *Amaranthus* proper, which includes the three species domesticated for grain production, *A. hypochondriacus*, *A. cruentus*, and *A. caudatus*.

Other species of amaranths have been domesticated as leaf vegetables, for fodder, as potherbs, or as ornamentals; among these species, *A. tricolor* from South Asia is probably the most important (Sauer, 1967). According to Ebert *et al.* (2011), of the approximately 70 species in the genus *Amaranthus*, 40 are native to the Americas. Within the genus, 17 species are vegetable amaranths with edible leaves, and three are grain amaranths with edible seeds (*A. caudatus*, Inca wheat, 'Love Lies Bleeding'; *A. cruentus*, purple amaranth; *A. hypochondriacus*, Prince's feather).

There are several species of *Amaranthus* used for leaves, grains, or both. There is taxonomic confusion because species quickly adapt in any environment, differences among various species are small, and many specific and common names have been used throughout the world, almost interchangeably, and also due to quick intergradation of a species in the region itself where it thrives well when cultivated and appears adapted. In spite of this confusion, some species are sufficiently recognized to merit universal acceptance.

The best of the species for grains is *A. hypochondriacus* L. and for edible leaves is *A. gangeticus* L., *A. cruentus* L., and *A. dubius* Mart ex. Thell. *A. hypochondriacus* was used as a grain in India and Sri Lanka in the 18th century. It became prevalent in the foothills of the Himalayas during the 19th century, where it became a staple food. It is important now in Nepal, China, Manchuria, and Uganda. In India, *A. hypochondriacus* and *A. caudatus* (grains) and *A. gangeticus* (leaves) are of major importance. *A. mangostanus* (syn. *A. tricolor* var *mangostanus*), *A. lividus* and *A. dubius* (a recognized tetraploid) are grown on a limited scale in Orissa and other Indian states (Nath and Swamy, 2015).

The important species of leafy amaranth, *A. tricolor* L. occupies a predominant position in India with different morphological forms in terms of colour and shape of leaves. *A. dubius*, *A. lividus*, *A. blitum*, *A tristis* L., *A. spinosus* L., and *A. viridis* are other amaranth species under cultivation. Amaranth is an annual herb, erect or trailing, with scarce to profuse branches, shallow to deep tap roots, a green to purple stem, and simple, alternate or opposite leaves, green to purple in colour. The inflorescence is terminal with axillary, branched spikes, a small, regular flower, mostly unisexual, and monoecious. In general, the cultivated species are monoecious (Nath and Swamy, 2015).

Amaranth is an annual plant and carries out $C_4$ type photosynthesis. Depending on species, amaranth leaves vary in shape, size and color (green, red, purple). This plant can grow up to 3 m. Its stem, sometimes branched, is terminated by a branched inflorescence (panicle). The inflorescence is usually indeterminate and reaches different lengths. The basic unit of the inflorescence is called a glomerule containing female or male florets, or both. The seed has a lenticular shape (1–2 mm) (Fig. 4.4). Amaranth has a high coefficient of propagation (Anon., 2015a).

### 4.5.1 Cytogenetic studies on *Amaranthus* spp.

Madhusoodanan and Pal (1981) studied the cytology of five species of *Amaranthus* (section *Blitopsis*): *A. tricolor*, *A. lividus*, *A. graecizans*, *A. viridis*, and *A. albus*. They reported that all were diploid with X = 17 or 16, the former being more common. Mallika (1987) conducted cytogenetic studies on eight species of *Amaranthus*: *A. tricolor*, *A. lividus*, *A. viridis*, *A. spinosus*, *A. dubius*, *A. hypochondriacus*, *A. cruentus*, and *A. caudatus* and their hybrids. The first three belong to the section *Blitopsis* and the remaining five belong to the section *Amaranthus*. Meiosis in both the sections was normal, with regular formation of bivalents. Out of eight species, seven were diploid with n=16 or 17, while one species, *A. dubius*, was polyploid with $n = 32$. Seven intraspecic hybrids were obtained, which showed normal growth and flowering. The chromosome counts indicated *A. tricolor* to be $2n = 2X = 34$, *A. cruentus* to be $2n = 32$ and *A. dubius* to be $2n = 64$.

Queiros (1989) reported the following chromosome numbers in eight *Amaranthus* spp. in Portugal: $2n = 32$ for *A. hybridus*, *A. blitoides*, *A. albus*, and *A. graecizans*; and $2n = 34$ for *A. paniculatus* (*A. cruentus*), *A. retroflexus*, *A. deflexus*, and *A. lividus*. Bao-Hua *et al.* (2002) from China reported chromosome numbers of 14 species: *A. retroflexus*, $2n = 34$; *A. caudatus* $2n = 32$; *A. hybridus* $2n = 32$; *A. spinosus* $2n = 34$; *A. cruentus* $2n = 34$; *A. hypochondriacus* $2n = 32$; *A. paniculdatus* $2n = 32$; *A. roxburghianus* $2n = 34$; *A. blitoides* $2n = 32$; *A. polygonoides* $2n = 34$; *A. albus* $2n = 32$; A. *viridis*

**Fig. 4.4.** Inflorescence and seed heads of amaranth.

$2n = 34$; *A. lividus* $2n = 34$, and *A. tricolor* $2n = 34$. The number in *A. roxburghianus* is reported here for the first time.

Thus, the basic chromosome number in this genus is X = 16 and 17 and both numbers are found in section *Amaranthus* and section *Blitopsis*. The chromosomes in this genus are very small in size, hampering a detailed karyotype analysis.

## 4.6 Genetic Engineering of Amaranth

So far, in the genetic improvement of amaranth, conventional selection methods have been primarily involved. However, advances in genomics and biotechnology have dramatically enriched the potential to manipulate the amaranth genome, in particular to improve the amount and availability of nutrients. Besides conventional breeding methods, application of non-conventional breeding strategies such as induced mutation breeding techniques, genetic engineering, and biotechnological approaches are gaining importance (Das, 2016).

There is potential for the use of biotechnology of amaranth and it is important to build on the research that has been carried out so far. This section summarizes and discusses the research already done regarding tissue culture, molecular biology, and genetic transformation of amaranth. It should be noted that genetic engineering is a modern and frontline application of biotechnology. The main streams that biotechnology touches are cell and tissue culture, genetic engineering, microbiology, embryology, molecular biology, and many others. Gulping beer, tasting wine, favourite chocolate, ever-loving ice cream, and many other products are proud results of biotechnology. The cultivation of food plants producing high-yielding crops, antibiotics, enzymes, and hundreds of other products are often the result of biotechnology. Pharmacology and medicine are some other areas that have been helped by biotechnology. Genetic materials of organisms can be altered using genetic-engineering techniques or recombinant DNA technology. Recombinant DNA technology is the process used to create a recombinant DNA molecule that carries the DNA of interest and vector DNA, while genetic engineering is a broad term used to describe the processes involved in manipulation of the genetic structure of an organism. This is the key difference between genetic engineering and recombinant DNA technology (Rogers, 2020).

Genetically modified organisms (GMOs) are any living entity constructed by altering their genetic composition, without affecting their normal function, in other words organisms whose genetic material has been altered using recombinant DNA technology or gene therapy. Significant improvements in plant hybridization are seen after the introduction of various sophisticated genetics and genomics techniques (Chauhan, 2020).

The aim of constructing GMOs is to either change organisms' characteristics or produce a new variation in nature. Those changes are actually beneficial to humans. GMOs have tremendous application in agriculture, medicine, human science and animal research. Starting from primitive prokaryotes to higher eukaryotes, any living organism's genome can be altered using artificial means. Genetically modified plant species or bacterial strains are created for various applications (Chauhan, 2020).

Genes from two organisms of the same species, different species, or different genera are linked or joined to construct the recombinant DNA. Using artificial vectors or plasmids, the newly formed DNA is inserted into the host organism. Polymerase chain reaction (PCR), DNA sequencing, and microarrays are some of the techniques used to validate the insert (Chauhan, 2020).

The process of GMO construction (Chauhan, 2020) consists of the following steps:

- Select the donor and recipient species. Usually, the donor is a microorganism whose gene is being inserted into the recipient.
- The gene of interest that we wish to express is isolated from the donor.
- Next the foreign DNA is inserted into the recipient organism, for example a plant species. For this, a special type of gene transfer vehicle is used known as vector. This is usually a plasmid (a small, circular bacterial DNA). The gene of interest along with a selectable marker sequence is inserted into the plasmid.
- Additional methods of inserting foreign DNA include plasmid gene guns, particle bombardment, and electroporation.
- The newly formed DNA is now known as recombinant DNA.
- This is then inserted into the host bacterium to infect the plant species. *Agrobacterium tumefaciens* is the first choice for transfection in plants.
- The plasmid is injected into *A. tumefaciens* and the insertion is validated by growing the bacteria in growth medium in which only bacteria carrying the selectable marker can grow.
- These transformed bacteria are then used to infect the plant cells, which are grown in tissue culture.
- A whole new plant is generated from a few cells during tissue culture.

The red-colored vegetative tissue produces high levels of betacyanin pigments that can be used as natural food colorants. Thus, amaranth has been attracting worldwide attention as a high-potential new crop with multiple uses; a spate of recent scientific publications on novel applications of amaranth grain and its fractions suggest that greater demand may be on the horizon (Cai *et al.*, 2004).

Technological innovation in plant biotechnology is an important catalyst in crop improvement. The technique of *Agrobacterium*-mediated transformation has been standardized in amaranth, which has opened up a new avenue to transfer useful genes such as oxalate decarboxylase to reduce the level of oxalic acid, a known antinutrient, in vegetable amaranth. The storage seed albumin protein AmA1 has been characterized from *A. hypochondriacus* and the protein gene has been successfully cloned and transferred into potato showing a tuber-specific expression (Das, 2016).

Marker-assisted selection and molecular breeding have helped to generate knowledge regarding the transfer of simply inherited quantitative traits from genetic resources into elite cultivars and also to understand the molecular control of target traits of interest. Molecular markers such as simple sequence repeats, restriction fragment length polymorphisms (RFLPs), and inter-simple sequence repeats have been found effective in distinguishing genetically similar accessions and in determining individual degrees of heterozygosity and heterozygosity within plant genetic resource populations. Although studies on *in vitro* growth and morphogenesis of amaranth are scarce, such studies are gaining momentum (Das, 2016).

### 4.6.1 Tissue culture of amaranth

Plant tissue culture for biotechnological applications and research is the path of choice for achieving many controlled manipulations on plants resulting in the selection of new and useful traits, such as resistance to chemicals/pathogens, micro-propagation of "hardy" plants, and studies about the physiology, biochemistry, and genetics of specific cultivars, among others. In particular, tissue culture of cereals was a challenge for some time until the discovery of the right conditions for the establishment of embryogenic calli of different Gramineae (Vasil and Vasil, 1984).

The first reported studies about tissue culture of amaranth (Flores and Galston, 1992) were done using *A. cruentus*, *A. hypochondriacus*, and *A. tricolor*. The authors reported that cell cultures from different explants of these species were easily established in several media (B5, Murashige and Skoog, and White's) using combinations of various plant growth regulators, such as 2,4-dichlorophenoxyacetic acid (2,4-D), 1-naphthaleneacetic acid (NAA), kinetin, 6-benzylaminopurine (BAP), and zeatin. Appropriate calli for suspension cultures were obtained from leaf discs, and in some cases embryo-like structures were observed. Using high concentrations of cytokinins, it was possible to get shoot regeneration from hypocotyl-derived calluses.

Bagga *et al.* (1987), using a different species of amaranth (*A. paniculatus*), reported full regeneration of complete plants. This was done with calli from hypocotyl segments, which were induced to form shoots in kinetin and NAA; these shoots were then rooted and transferred to soil.

Bennici *et al.* (1992), using several varieties of four amaranth species (*A. caudatus*, three varieties; *A. cruentus*, four varieties; *A. hybridus*, two varieties; and *A. hypochondriacus*, eight varieties), reported successful (>95%) callus formation from hypocotyl and stem tissue in MS medium. In general, the NAA/BAP combination was more effective for callus growth in *A. cruentus*, while 2,4-D/kinetin was best for *A. hybridus*. Only some of the calli of *A. hypochondriacus* and *A. caudatus* were more efficient at regenerating different organs such as shoots or roots. All regenerated shoots were able to form roots and, after transfer to pots, plantlets developed.

Arya *et al.* (1993) reported on the use of inflorescences of *A. paniculatus* to produce secondary inflorescences and plantlets *in vitro*. The average percentage of floral buds responding per inflorescence was up to 10% independently of the kinetin concentration. If the buds forming leaves and shoots were kept in the same medium, they formed calli at the base, with subsequent elongation but without roots. After transfer of these tissues to medium with 12 mg kinetin $l^{-1}$ and 15% coconut milk, the authors observed the formation of roots and shoots. These plantlets flowered after 20 days *in vitro* without a special photoperiod.

All of these reports show that, although it is easy to get tissue cultures of amaranth (mainly calli), they are still recalcitrant species for plant regeneration, and, so far, there is no reproducible procedure for this goal. The genotype of the different varieties of one species plus the endogenous auxin/cytokinin balance may play an important role. After the results obtained in cereals, plant regeneration of amaranth does not seen impossible and some of the information obtained from similar work already done on Gramineae could be useful (Hibberd, 1984).

### 4.6.2 Molecular biology/genomic resources of amaranth

Molecular biology techniques have been the most powerful tool for the transfer of specific and new genetic information into living organisms. Current techniques for plant genetic transformation range from the initial method based on the natural genetic transformation mediated by the Ti plasmid of *Agrobacterium tumefaciens* to biolistic or gene gun procedures.

According to Herrera-Estrella and Simpson (1988), using the *Agrobacterium* system, it is possible to do several variations by using co-integrative or binary vectors. Shillito and Saul (1988) stated that other procedures for genetic transformation use plant protoplasts. Because of the nature of these cells, which do not have cell walls, it is possible to achieve transformation through different techniques. These techniques include direct gene transfer using polyethylene glycol-mediated transformation, electroporation (i.e. introduction of DNA vectors by electric shock), and transformation via co-cultivation with *A. tumefaciens*. Fromm *et al.* (1986) used an electroporation procedure to transform corn cells, while a "biolistic" procedure uses velocity micro-projectiles to carry DNA past cell walls and membranes (Sanford, 1990). The advantages of the latter technique are that it has been effective regardless of species or tissue type; it is rapid and simple and should facilitate the direct transformation of totipotent tissues such as pollen, embryos, meristems, and cell cultures, and it also seems suitable for organelle transformation. Although it requires special equipment, currently it is commercially available. However, a necessary procedure for the genetic transformation of plants is a good regeneration system from tissue cultures of the target plant. This is required for the appropriate selection of individuals that not only possess but also express the introduced genetic information. Although there are some procedures where this requisite could be overcome, most plant genetic transformation procedures are based on selection for antibiotic resistance conferred by genes contained in the introduced molecular vector (e.g. plasmid, cosmid, phage).

In the case of amaranth, plant regeneration from different tissue cultures is not yet established in an efficient and reproducible way. Even *Agrobacterium* infection, which allows a viral infection mediated by *Agrobacterium* and does not require the tissue culture system (Grimsley *et al.*, 1988), is still beyond the current knowledge regarding amaranth processing.

Once this "bottleneck" is solved in the amaranth model, genetic transformation could be a reality, mainly because the diverse techniques for DNA introduction into plant cells could be assayed until the right one for amaranth is found. This situation does not mean that amaranth is less interesting for molecular biology of plants because, independent of the problem of regeneration, it could be used right now as a source of interesting genes.

Albumins are among the most important seed proteins of amaranth. There are studies about their localization and some of their biochemical and physicochemical properties. They have been isolated and purified, and their amino acid composition is known. Among the albumin fractions of amaranth is a small group of methionine-rich proteins (MRPs) (Segura-Nieto *et al.*, 1992). These proteins have a high methionine content (16–18 mol%) as well as a relatively high lysine content (6.1–6.4 mol%). Constituent polypeptides have been isolated, and the N-terminal amino acid sequences (first 25 amino acids) of two MRPs were determined.

Other interesting information has to do with antimicrobial peptides found in the seeds of *A. caudatus*, where two of these peptides were identified and the complete sequence of 30 amino acids was also determined. These peptides are homologous to the cysteine/glycine-rich domain of chitin-binding proteins. These amaranth peptides inhibit the growth of fungi to a much higher extent than other antifungal chitin-binding proteins (Broekaert *et al.*, 1992). This property suggests that they may play a role in the defense of seeds against fungal attack.

There is a proteinase inhibitor peptide isolated from *A. hypochondriacus* that is a trypsin inhibitor. The sequence of the first 26 N-terminal amino acids was determined (Valdes *et al.*, 1992). The purified inhibitor showed a typical trypsin-like inhibitor activity against proteases from fungi, storage insects, and germinated amaranth grains.

Considering the state of the art regarding these amaranth proteins, there are two possible strategies to isolate the respective genes. One is a classical approach including the raising of specific antibodies against each of these purified proteins. The antibodies can then be used to screen the protein or peptide synthesized in *Escherichia coli* by using expression vectors in a cDNA library. This strategy was successfully used to isolate a gene encoding an albumin with a well-balanced amino acid composition and high level of essential amino acids from *A. hypochondriacus* (Raina and Datta, 1992).

Another currently available strategy is the design and construction of degenerate oligonucleotides based on the flanking and already determined amino acid sequences of amaranth proteins and peptides. These site-directed oligonucleotides can be used to amplify the entire nucleotide sequence in between with PCR (Saiki *et al.*, 1988). PCR is based on the enzymatic amplification of a DNA fragment using two oligonucleotides as primers, which flank the defined target DNA sequence. A typical cycle of the reaction starts with a heating step to denature the double-stranded target DNA, followed by cooling and hybridization of the PCR primers to the single-stranded target DNA and then primer extension by DNA polymerase to produce the desired sequence. Cycle repetition results in the exponential reproduction of a specific DNA sequence. The use of a thermostable DNA polymerase from the bacterium *Thermus aquaticus* allows its use in the reaction mix without having to add new polymerase for each cycle. The goal would be to use the amplified DNA fragment as a homologous probe to screen particular DNA sequences and ultimately to identify and isolate the corresponding gene from a DNA library (Dehmer *et al.*, 1991).

Because the N-terminal or full amino acid sequence of MRPs, antimicrobial peptides, and proteinase inhibitors of amaranth are known, there are good prospects for this strategy. Once the respective genes are isolated, they can be used to transform other plants to increase the nutritional properties or as a plant defense mechanism. Unfortunately, achievement of these objectives is still some way off and there are other problems to be solved first.

Another interesting trait of amaranth is the natural herbicide resistance of the weed *A. hybridus*. Some biotypes of this species are resistant to the herbicide atrazine; a single amino acid mutation in the Q protein is responsible for this resistance. The Q protein is encoded by the chloroplast *psbA* gene. After isolation and cloning of this gene, some modifications were made in order to express it as a nuclear gene (Cheung *et al.*, 1988). Using *Agrobacterium*-mediated transformation, it was possible to obtain transgenic atrazine-tolerant tobacco plants.

The characterization of genetic variability of amaranth species regarding specific traits such as early maturity, plant height, resistance to lodging, pest resistance, male sterility, seed size, drought resistance, and others could, in principle, be addressed at the molecular level by using RFLP and/or random amplified polymorphic DNA (RAPD) methods (Perez de la Vega, 1993).

With RFLPs, differences in the length of DNA fragments and number of sites for different restriction enzymes are determined. In this technique, it is necessary to isolate genomic DNA from leaves; after digestion of samples of the DNA with different restriction enzymes, the resulting fragments are separated electrophoretically. The separated fragments are then denatured and blotted on to a nylon or nitrocellulose member. Then, radioactive or non-radioactively labeled single-strand DNA probes are hybridized to the denatured genomic DNA fragments. After washing and autoradiography or antibody detection of hybrids, the result is a specific pattern of discrete bands for each individual DNA fragment. The development of RFLP molecular markers on amaranth chromosomes that are linked to traits of interest is currently in progress (Komen, 1992).

For the RAPD technique, the genomic DNA is first used as a substrate in PCR to amplify one or several sequences. The resulting fragments are then separated by electrophoresis and the amplified sequences are observed directly on the gel after nucleic acid staining. The distances between the PCR primers may be specific for each DNA assayed, giving a particular pattern of bands that reflects a genetic polymorphism, analogous to RFLP. PCR primers could be either targeted to known sequences or used to amplify unknown sequences.

Despite the economic importance of amaranth, there is very little information available about the extent and nature of genetic diversity present in the genus *Amaranthus* at the molecular level. Ranade *et al.* (1997) reported the RAPD profiles of different species of *Amaranthus* as well as different accessions of the species. The putative hybrid origin of *A. dubius* from *A. hybridus* and *A. spinosus* was also ruled out by the RAPD data. The trends of species relationships among the amaranths determined by RAPDs are consistent with the cytogenetic and evolutionary relationships that have already been determined.

In the last few years, genomics resources have been developed for *Amaranthus* spp. Two research groups independently developed microsatellite markers from *A. hypochondriacus* (Mallory *et al.*, 2007; Lee *et al.*, 2008). In both cases, the markers were demonstrated to be transferable to other cultivated as well as weedy *Amaranthus* spp. A preliminary phylogenetic analysis using some of these markers placed *A. hybridus* within multiple grain amaranth clades, suggesting multiple domestication events from *A. hybridus* (Mallory *et al.*, 2007). Additional *Amaranthus* microsatellite markers were obtained recently from *A. tuberculatus* (Lee *et al.*, 2009). Collectively, these microsatellite markers will be valuable for more detailed phylogenetic studies, as well as for various genetic studies and breeding efforts (e.g. population genetics, construction of genetic maps, map-based cloning, and marker-assisted selection) (Trucco and Tranel, 2011).

Genetic diversity and relationships among six *Amaranthus* spp. from eight phytogeographic regions of the Indo-Gangetic plains were analyzed using an RAPD marker by Ray and Roy (2009). A unique 1371 bp RAPD band specific for *A. tricolor* of a particular phytogeographic region was converted to a sequence characterized amplified region (SCAR) marker. The translated marker sequence showed homology with hemagglutinin protein.

A second genomics resource for *Amaranthus* spp. is a bacterial artificial chromosome (BAC) library from *A. hypochondriacus* (Maughan *et al.*, 2008). This library contains over 35,000 clones averaging 147 kb, or about a tenfold coverage of the genome. The utility of the library was demonstrated by using it to obtain the full-length sequences of the *ALS* and *PPO* genes, both of which encode important herbicide target sites (Maughan *et al.*, 2008). In addition to serving as a source for candidate gene isolation and sequencing, this BAC library could be used further to develop a physical map of the *Amaranthus* genome and could serve as a scaffold for whole-genome sequencing. A third *Amaranthus* genomics resource recently obtained is shotgun sequence data from *A. tuberculatus* (Lee *et al.*, 2009). Using next-generation sequencing technology, over 40 Mbp of sequence was obtained from *A. tuberculatus*. Included in the dataset was a nearly complete sequence of the chloroplast genome and partial sequences of most of the currently known herbicide target-site genes. The dataset also provided leads for microsatellite markers, mentioned above. Although the dataset contains only partial sequences for nuclear genes, such information serves as a starting point for candidate gene isolation in so far as designing primers for PCR-based approaches. Both the genomic and transcriptomic datasets are being made publicly available via the National Center for Biotechnology Information (www.ncbi.nlm.nih.gov/, accessed 21 June 2024). Montgomery *et al.* (2020) published the most contiguous draft assemblies of these three species, *A. tuberculatus*, *A. hybridus*, and *A. palmeri*, to date. These multiple reference genomes will accelerate genomic studies in these species to further our understanding of weedy species evolution within the genus *Amaranthus*. Amaranth landraces belonging to three species, namely *A. tricolor*, *A. dubius* and *A. hypochondriacus*, were studied to assess genetic diversity and interspecific relationships using RAPD markers by Faseela and Joseph (2007).

### 4.6.3 Genetic transformation of amaranth

Plant transformation is a method to insert DNA from another plant into the genome of a plant of interest. Transgenic plants are those plants having genomes modified through genetic-engineering techniques either by the addition of foreign genes or removal of detrimental genes.

A current need is the development of a facile genetic transformation system for *Amaranthus* spp. Only modest success has been reported in regenerating plants from *Amaranthus* callus tissue (Brenner *et al.*, 2000). There is, however, one report of successful transformation of *A. hypochondriacus* by inoculation of mature embryo explants with *Agrobacterium* (Jofre-Garfias *et al.*, 1997). There are not many reports of *Amaranthus* transformation, and it is not known whether this is due to lack of effort or technical challenges. Possibly, the weedy species are more amenable to genetic transformation and could serve as model systems for developing and optimizing protocols that could then be adapted to the cultivated species.

Mature embryos of *A. hypochondriacus* were used by Jofre-Garfias *et al.* (1997) to develop an *in vitro* culture system for plant regeneration and genetic transformation. Transgenic plants were obtained by inoculation of mature embryo explants with a disarmed *Agrobacterium* strain containing the plasmid pGV2260 (pEsc4), which carried genes encoding neomycin phosphotransferase type II (*nptII*) and β-glucuronidase (*gus*). The presence of transgenes in the genome of transformed amaranth plants and their progeny was demonstrated by Southern blot hybridization. Tissue-specific and light-inducible expression directed by a pea chlorophyll *a*/*b*-binding protein promoter was observed in transgenic amaranth plants and their progeny (Jofre-Garfias *et al.*, 1997).

Jofre-Garfias *et al.* (2001) reported that amaranth (*A. hypochondriacus*) is susceptible to transformation by *Agrobacterium tumefaciens* strains. Pal *et al.* (2013) optimized a procedure for genetic transformation of the major leafy vegetable crop *A. tricolor* using epicotyl explant co-cultivation with *Agrobacterium tumefaciens*. The *ARGOS-LIKE* gene of *Arabidopsis thaliana* along with the dahlia mosaic virus promoter can be used to increase primarily the green weight of the shoot and leaf size of amaranth (Kuluev *et al.*, 2017).

Transgenic *A. hypochondriacus* and *A. hybridus* roots were generated and a distinct plant regeneration program via somatic embryos produced from these hairy roots was established by Castellanos-Arévalo *et al.* (2020). Sunil *et al.* (2014) sequenced the genome and transcriptome of *A. hypochondriacus*, a diploid ($2n = 32$) belonging to the order Caryophyllales with an estimated genome size of 466 Mb. Of the 411 linkage single-nucleotide polymorphisms (SNPs) reported for grain amaranths, 355 SNPs (86%) were represented in the scaffolds and 74% of the 8.6 billion bases of the sequenced transcriptome mapped to the genomic scaffolds. The genome of *A. hypochondriacus* encodes at least 24,829 proteins, shares the paleo hexaploidy event with species under the superorders Rosids and Asterids, harbors one SNP in 1000 bases, and contains 13.76% repeat elements. Annotation of all the genes in the lysine biosynthetic pathway using comparative genomics and expression analysis offers insights into the high-lysine phenotype. As the first grain species under Caryophyllales and the first $C_4$ dicot genome reported, these results will be beneficial in improving crops and in expanding our understanding of angiosperm evolution.

## 4.7 References


Anon. (2015a) Amaranth (*Amaranthus* sp. L.). Available at: http://www.vurv.cz/altercrop/amaranth.htm (accessed 20 June 2024).

Anon. (2015b) Amaranth—May Grain of the Month. Whole Grains Council. Boston, Massachusetts. Available at: http://wholegrainscouncil.org/whole-grains-101/amaranth-may-grain-of-the-month-0 (accessed 20 June 2024).

Arya, I.D., Chakravarty, T.N., and Sopori, S.K. (1993) Development secondary inflorescences and *in vitro* plantlets from inflorescence cultures of *Amaranthus paniculatus*. *Plant Cell Reports* 12: 286–188.

Bagga, S., Venkateswarlu, K., and Sopori, S.K. (1987) *In vitro* regeneration of plants from hypocotyl segments of *Amaranthus paniculatus*. *Plant Cell Reports* 6: 183–184.

Bao-Hua, S., Xue-Jie, Z., Fa-Zeng, L. and Peng, W. (2002) Chromosome numbers of 14 species in *Amaranthus* from China. *Acta Phytotaxonomica Sinica* 40, 428–432.

Bennici, A., Schiff, S., and Bovelli, R. (1992) *In vitro* culture of species and varieties of four *Amaranthus* L. species. *Euphytica* 62: 181–186.

Brenner, D.M., Baltensperger, D.D., Kulakow, P.A., Lehmann, J.W., Myers, R.L., *et al.* (2000) Genetic resources and breeding of *Amaranthus*. *Plant Breeding Reviews* 19, 227–283.

Broekaert, W.F., Marien, W., Terras, F.R.G., de Bolle, M.F.C., Proost, P., *et al.* (1992) Antimicrobial peptides from *Amaranthus caudatus* seeds with sequence homology to the cysteine/glycine-rich domain of chitin-binding proteins. *Biochemistry* 31, 4308–4314.

Cai, Y.Z., Corke, H., and Wu, H.X. (2004) Amaranth. In: Wrigley, C. (ed.) *Encyclopedia of Grain Science*, Volume 3. Elsevier Academic Press, pp. 1–10.

Castellanos-Arévalo, A.P., Estrada-Luna, A.A., Cabrera-Ponce, J.L., Valencia-Lozano, E., Herrera-Ubaldo, H., *et al.* (2020) *Agrobacterium rhizogenes*-mediated transformation of grain (*Amaranthus hypochondriacus*) and leafy (*A. hybridus*) amaranths. *Plant Cell Reports* 39, 1143–1160.

Chauhan, T. (2020) Genetically Modified Organisms (GMO)—Definition, Process and Examples. Available at: https://geneticeducation.co.in/genetically-modified-organisms-gmo-definition-process-and-examples/ (accessed 21 June 2024).

Cheung, A.Y., Bogorad, L., Montagu, M.V. and Schell, J. (1988) Relocating a gene for herbicide tolerance: a chloroplast gene is converted into a nuclear gene. *Proceedings of the National Academy of Sciences USA* 85, 391–395.

Das, S. (2016) Breeding of amaranths. In: *Amaranthus: A Promising Crop of Future*,. Springer, pp. 107–145.

Dehmer, A.Y., Graner, A., and Wenzel, G. (1991) Screening for defined DNA sequences in minimal amounts of barley tissue by PCR. *Plant Breeding* 107, 70 –72.

Ebert, A.W., Wu, T., and Wang, S. (2011) *Vegetable Amaranth (*Amaranthus *L.)*. International Cooperator's Guide. AVRDC Publication No. 11-754.

Faseela, K.V. and Joseph, S. (2007) Molecular characterization of amaranth landraces and assessment of interspecific relationships among *Amaranthus* spp. (L.) using RAPD markers. *Indian Journal of Genetics and Plant Breeding* 67, 12–17.

Flores, H.E. and Galston, A.W. (1992) *In vitro* culture of grain and vegetable amaranth (*Amaranthus* spp.). *American Journal of Botany* 69, 1049–1054.

Fromm, M.E., Taylor, L.P., and Walbot, V. (1986) Stable transformation of maize after gene transfer by electroporation. *Nature* 319, 791–793.

Green Deane (2012) Amaranth: Grain, Vegetable, Icon. Eat the Weeds and Other Things Too. Available at: www.eattheweeds.com/amaranth-grain-vegetable-icon/ (accessed 21 June 2024)

Grimsley, N.H., Ramos, C., Hein, T., and Hohn, B. (1988) Meristematic tissues of maize plants are most susceptible to agroinfection with maize streak virus. *Biotechnology* 6, 185–189.

Harrington, G. (1984) Growing Chinese Vegetables in Your Own Backyard. Storey Publishing, North Adams, Massachusetts.

Herrera-Estrella, L. and Simpson, J. (1988) Foreign gene expression in plants. In: Shaw, C.H. (ed.) *Plant Molecular Biology. A Practical Approach*. IRL Press, Oxford, UK, p. 131.

Hibberd, K.A. (1984) Induction, selection and characterization of mutants in maize cell cultures. In: Vasil, I.K. (ed.) *Cell Culture and Somatic Cell Genetics Cell Plants, Vol.* 1. Academic Press, New York, p. 571–576.

Howard, B.C. (2013) Amaranth: Another Ancient Wonder Food, But Who Will Eat It? National Geographic. Available at: www.nationalgeographic.com/culture/article/130812-amaranth-oaxaca-mexico-obesity-puente-food (accessed 21 June 2024).

Jofre-Garfias, A.E., Cabrera-Ponce, J.L., Adame-Alvarez, R.M., Herrera-Estrella L., and Simpson, J. (2001) Transgenic amaranth (*Amaranthus hypochondriacus*). In: Bajaj Y.P.S (ed.) *Genetic Transformation of Amaranth (*Amaranthus hypochondriacus*)*. Springer, pp. 13–25.

Jofre-Garfias, A.E., Villegas-Sepúlveda, N., Cabrera-Ponce, J.L., Adame-Alvarez, R.M., Herrera-Estrella, L., and Simpson, J. (1997). *Agrobacterium*-mediated transformation of *Amaranthus hypochondriacus*: light- and tissue-specific expression of a pea chlorophyll *a*/*b*-binding protein promoter. *Plant Cell Reports* 16, 847–852.

John, H.M., Dorothy, N., Dorothy, N.N., and Dorothy, M. (2008). Efforts to promote amaranth production and consumption in Uganda to fight malnutrition. In: Robertson, G.L. and Lupien, J.R. (eds.) *Using Food Science and Technology to Improve Nutrition and Promote National Development*. International Union of Food Science & Technology, Canada.

Komen, J. (1992) Grains for the tropical regions, biotechnology and development. *Biotechnology and Development Monitor* 10, 3–5.

Kuluev, B.R., Mikhaylova, E.V., Taipova, R.M., and Chemeris, A.V. (2017) Changes in phenotype of transgenic amaranth *Amaranthus retroflexus* L., overexpressing *ARGOS-LIKE* gene. *Russian Journal of Genetics* 53, 67–75.

La Jolla Mom (2013) 10 health benefits of amaranth grain and leaves. In: *Food and Drink*. Available at: http://lajollamom.com/amaranth-grain-leaves-health-benefits/ (accessed 21 June 2024).

Lee, J.R., Hong, G.Y., Dixit, A., Chung, J.W., Ma, K.H., *et al.* (2008) Characterization of microsatellite loci developed for *Amaranthus hypochondriacus* and their cross–amplification in wild species. *Conservation Genetics* 9, 243–246.

Lee, R.M., Thimmapuram, J., Kate, A., Thinglum, G.G., Hernandez, A.G., *et al.* (2009) Sampling the waterhemp (*Amaranthus tuberculatus*) genome using pyrosequencing technology. *Weed Science* 57, 463–469.

Madhusoodanan, K.J. and Pal, M. (1981) Cytology of vegetable amaranths. *Botanical Journal of the Linnean Society* 82, 61–68.

Maestri, N. (2015) Amaranth—The Origin and Use of Amaranth in Ancient Mesoamerica. Available at: http://archaeology.about.com/od/amthroughanterms/a/Amaranth.htm (accessed 21 June 2024).

Mallika, V.K. (1987) Genome analysis in the genus *Amaranthus*. PhD thesis, Kerala Agricultural University, Trichur, Kerala.

Mallory, M.A., Rozaura, V., Hall, A.R., McNabb, D.B., Pratt, E., *et al.* (2007) Development and characterization of microsatellite markers for the grain amaranths. *Crop Science* 48, 1098–1106.

Maughan, P.J., Sisneros, N., Luo, M., Kudrna, D., Ammiraju, J.S., and Wing, R.A. (2008) Construction of an amaranth (*Amaranthus hypochondriacus*) bacterial artificial chromosome library and genomic sequencing of herbicide target genes. *Plant Genome* 48, 85–94.

Montgomery, J.S., Giacomini, D., Waithaka, B., Lanz, C., Murphy, B.P., *et al.* (2020) Draft genomes of *Amaranthus tuberculatus*, *Amaranthus hybridus*, and *Amaranthus palmeri*. *Genome Biology and Evolution* 12, 1988–1993.

Mosyakin, S.L. and Robertson, K.R. (1996) New infrageneric taxa and combinations in *Amaranthus* (Amaranthaceae). *Annales Botanici Fennici* 33, 275–281.

Mosyakin, S.L. and Robertson, K.R. (2003) *Amaranthus*. In: *Flora of North America North of Mexico*. Oxford University Press, New York.

Nath, P. and Swamy, K.R.M. (2015) *Vegetables for the Tropical Region*. Indian Council of Agricultural Research (ICAR), New Delhi.

Pal, A., Swain, S.S., Das, A.B., Mukherjee, A.K., and Chand, P.K. (2013) Stable germ line transformation of a leafy vegetable crop amaranth (*Amaranthus tricolor* L.) mediated by *Agrobacterium tumefaciens*. *In Vitro Cellular & Developmental Biology-Plant* 49, 114–128.

Pal, M. (1999) *Amaranthus*: evolution, genetic resources and utilization. *EnviroNews* 5.

Pal, M. and Khoshoo, T.N. (1973) Evolution and improvement of cultivated amaranths. VI. Cytogenetic relationships in grain types. *Theoretical and Applied Genetics* 43, 242–251.

Perez de la Vega, M. (1993) Biochemical characterization of populations. In: Hayward, M.D., Bosemark, N.O., and Romagosa, L. (eds.) *Plant Breeding: Principles and Prospects*. Chapman & Hall, London, p. 184–200.

Putnan, D. (1992) Amaranth and the currents of history. *Legacy* 5, 1–4.

Queiros, M. (1989) Cytotaxonomic studies of *Amaranthus* in Portugal. *Lazaroa* 11, 9–17.

Rachael, L. (2018) Amaranth: An Ancient Grain with Impressive Health Benefits. *Healthline Wellness Newsletter*, June 2018. Available at: www.healthline.com/nutrition/amaranth-health-benefits (accessed 21 June 2024).

Raina, A. and Datta, A. (1992) Molecular cloning of a gene encoding a seed-specific protein with nutritionally balanced amino acid composition from *Amaranthus*. *Proceedings of the National Academy of Sciences USA* 89, 11774–11778.

Ranade, S.A., Kumar, A., Goswami, M., Farooqui, N., and Sane, Ñ.V. (1997) Genome analysis of amaranths: determination of inter- and intra-species variations. *Journal of Biosciences* 22, 457–464.

Ray, T. and Roy, S.C. (2009) Genetic diversity of *Amaranthus* species from the Indo-Gangetic plains revealed by RAPD analysis leading to the development of ecotype-specific SCAR marker. *Journal of Heredity* 100, 338–347.

Rogers, K. (2020) Genetic engineering. *Encyclopaedia Britannica*. Available at: www.britannica.com/science/genetic-engineering (accessed 21 June 2024).

Saiki, R.K., Gelfand, H.D., Stoffel, S., Scharf, S.J., Higuchi, R., *et al.* (1988) Primer-directed enzymatic amplification of DNA with a thermostable DNA polymerase. *Science* 239, 487–491.

Sanford, J.C. (1990) Biolistic plant transformation. *Physiologia Plantarum* 79, 206–209.

Sauer, J. D. (1967) The grain amaranths and their relatives: a revised taxonomic and geographic. *Annals of the Missouri Botanical Garden* 54, 103–137.

Sauer, J.D. (1955) Revision of dioecious amaranths. *Madrono* 13, 5–46.

Sauer, J.D. (1957) Recent migration and evolution of the dioecious amaranths. *Evolution* 2, 11–31.

Sauer, J.D. (1986) Grain amaranths (*Amaranthus* spp.—Amaranthaceae). In: Simmonds, N.W. (ed.) *Evolution of Crop Plants*. Longman Scientific & Technical, UK, pp. 4–7.

Segura-Nieto, M., Rodrigues-Nester, C., and Olguin-Martinez, L. (1992) Identification of methionine rich proteins from *Amaranthus hypochondriacus* seed. Presented at: *International Symposium on Biotechnology Crop Improvement in Latin America (CQI 10), 1–7 November.*

Shillito, R.D. and Saul, M.W. (1988) Protoplast isolation and transformation. In: Shaw, C.H. (ed.) *Plant Molecular Biology, A Practical Approach*. IRL Press, Oxford, UK, p. 161–168.

Shodhganga *(*2018) General Introduction of Plant. Chapter 1. Available at: http://shodhganga.inflibnet.ac.in/jspui/bitstream/10603/172291/1/chapter%201.pdf (accessed 21 June 2024).

Sunil, M., Hariharan, A.K., Nayak, S., Gupta, S., Nambisan, S.R., *et al.* (2014) The draft genome and transcriptome of *Amaranthus hypochondriacus*: a $C_4$ dicot producing high-lysine edible pseudo-cereal. *DNA Research* 21, 585–602.

Trucco, F. and Tranel, P.J. (2011) Amaranthus. In: Kole, C. (ed.) *Wild Crop Relatives: Genomic and Breeding Resources*. Springer, Berlin/Heidelberg, pp. 11–21.

Valdes, S., Chagolla, A., Martinez, N., Mendiola, E., Segura, M., and Blanco, A. (1992) Study of a protease inhibitor from amaranth (PO- 171). Presented at: *VII PAABS Congress and XIX Congress, 27 September–2 October.* Mexican Biochemistry Society, Mexico.

Varalakshmi, B. (2015) Vegetable amaranth. In: Peter, K.V. and Hazra, P. (eds) *Handbook of Vegetables, Vol.* 3. Studium Press, Texas, pp. 487–520.

Vasil, V. and Vasil, I.K. (1984) Induction and maintenance of embryogenic callus cultures of Gramineae. In: Vasil, I.K. (ed.) *Cell Culture and Somatic Cell Genetics of Plants*, Vol. I. Academic Press, New York, p. 36–42.

Wang, J.Y. (2012) Last Chance Foods: Amaranth's Ancient History. Available at: www.wnyc.org/story/225970-last-chance-foods-amaranths-ancient-history/ (accessed 21 June 2024).

# 5 Genetic Engineering of Cauliflower using Traditional and Innovative Techniques

**Simone Treccarichi, Donata Arena, Giulio Flavio Rizzo and Ferdinando Branca***
*Dipartimento di Agricoltura, Alimentazione e Ambiente (Di3A), Università di Catania, Catania, Italy*

**Abstract**
Cauliflower (*Brassica oleracea* var. *botrytis*) is an important target for breeding programs aimed at transferring nutraceutical or agronomic traits and could be also used as a source of genes for improving other *Brassica* spp. crops. Consumption of brassicas such as cauliflower and broccoli is increasing, supported by several improved $F_1$ hybrids and landraces available on the market. Their diversification represents a source of biodiversity that can be exploited in breeding programs. Genetic engineering has important goals in terms of genetic improvement leading to the development of biotic and abiotic resistance, and could be used to massively improve crop production and nutritional values in a short time compared with traditional breeding. Over the last few decades, new genome editing techniques have been developed following on from *Rhizobium*-mediated transformation exploiting microorganism-mediated protocols, such as clustered regularly interspaced palindromic repeats (CRISPR)-CRISPR-associated protein 9 (Cas9). Genetic engineering also allows us to overcome the interspecific barriers that limit the transfer of interesting genes. *In vitro* techniques developed via organogenesis, such as protoplast fusion and somatic hybridization, also represent an important means of breeding self-incompatible and outcrossing genotypes, including incompatible species. In this chapter, the methods and the different objectives related to the transformation techniques for cauliflower are described.

**Keywords:** biotechnology, *Brassica oleracea*, Brassicaceae, breeding, vegetable

## 5.1 Introduction

Cauliflower (*Brassica oleracea* var. *botrytis* L.) is one of the most important vegetable crops belonging to the family Brassicaceae; it is widespread and cultivated in many countries including China, India, Spain, and Italy. The edible part is represented by its highly suppressed "fleshy prefloral apical meristem" branches called "curd" (Sharma *et al.*, 2004). *B. oleracea* var. *botrytis* is an important crop in the Mediterranean area, and was domesticated and differentiated mostly in southern Italy (Maggioni *et al.*, 2014). Cauliflower and broccoli show wide diversity in terms of their reproductive organs but can easily be crossed (Scalzo *et al.*, 2024). The genus *Brassica* has been the object of several polyploidy and hybridization events followed by genome downsizing and loss of redundant gene copies (Schiessl and Mason, 2020). Cauliflower has

*Email: fbranca@unict.it

 *Genetic Engineering of Vegetable Crops* (Pritam Kalia ed.)
DOI: 10.1079/9781800627123.0005

18 diploid and somatic chromosomes ($n = 9$) and represents the genome CC in the model of the U's triangle proposed by Nagaharu in 1935 (Singh *et al.*, 2018), in contrast to the genomes AA and BB, which are represented by *Brassica rapa* and *Brassica nigra*, respectively; the model is also represented by the amphidiploid species arising from the union of the genomes AA and BB (*Brassica juncea*, $n = 18$, genome AABB), the genomes BB and CC (*Brassica carinata*, $n = 17$, BBCC), and the genomes AA and CC (*Brassica napus*, $n = 19$, AACC) (Arias *et al.*, 2014). A genotyping by sequencing approach using next-generation sequencing techniques performed by Sun *et al.* (2019) revealed that chromosome 3 could be the most ancient one in the cauliflower genome as it has the most syntenic blocks shared with the A, B, and C genomes of *Brassica* spp. and the remaining chromosomes. The development of tools such as molecular markers represents an important perspective to investigate genetic traits and assist in breeding oriented toward the improvement of plant production, resistance to abiotic and biotic stresses, and more recently the plant–microbiome interaction (Makukha and Dubina, 2021; Treccarichi *et al.*, 2021, 2022). Next-generation sequencing technologies and the publishing of the draft genome sequence of cauliflower have allowed us to search, discover, map, and clone new quantitative trait loci such as the downy mildew resistance loci (Shaw *et al.*, 2021). Cauliflower's genetic resources may represent an important source of biochemical traits positively related to nutraceutical properties (Picchi *et al.*, 2020), as well as being a functional food (Di Bella *et al.*, 2021): in this context, the orthologous gene of cauliflower *Orange* (*OR*), involved in β-carotene accumulation, was cloned in rice (*Oryza sativa*) (Endo *et al.*, 2019). The *OR* gene plays an important role in plant development, enhancing petiole elongation and inducing the formation of chromoplasts, which provide a metabolic sink to sequester and deposit carotenoids; it also regulates carotenoid homeostasis, increasing plant tolerance to environmental stress (Kim *et al.*, 2018). It has been cloned (Zhou *et al.*, 2011) and its mutants characterized (Welsch *et al.*, 2020). Genetic engineering can be used as a tool to insert specific added-value traits in existing cultivars to improve them by exploiting different transformation and regeneration techniques (Kumar and Srivastava, 2016). In this chapter, various transformation methods will be discussed, focusing on the best results achieved by a number of authors, summarized in Table 5.1.

## 5.2 Methods and Techniques

### 5.2.1 *Rhizobium*-mediated transformation

For a number of years, *Rhizobium radiobacter* (syn. *Agrobacterium tumefaciens*)) was used as a tool to obtain transformed plants through tissue infection, which develops by transfer of plasmid T-DNA into a host cell (Krenek *et al.*, 2015); genetic-engineering systems can also exploit *Rhizobium*-mediated transformation in new techniques such as clustered regularly interspaced palindromic repeats (CRISPR)-CRISPR-associated protein 9 (Cas9) (Char *et al.*, 2017). Cauliflower is a recalcitrant species for *in vitro* culture for the application of transformation protocols, and a number of factors influence the efficiency of transformation such as explant age, bacterium concentration, and duration of infection; the obtained mutant plants also show a significant loss of sterility (Bhalla and Smith, 1998). The development of an improved protocol for tissue regeneration and organogenesis is key for *in vitro* transformation (Ravanfar *et al.*, 2017).

*Rhizobium*-mediated transformation has been used to transfer selected traits, and one of the first applications in *B. oleracea* was transfer of the *bar* gene encoding phosphinothricin acetyltransferase, involved in herbicide (glufosinate) resistance (de Block *et al.*, 1989).

Several transformation strategies have been performed using binary vectors, such as plasmid vector pART278 carrying the neomycin phosphotransferase II (*nptII*) gene for kanamycin resistance and a β-glucuronidase gene (*uidA*) in the T-DNA sequence of *Rhizobium rhizogenes* (formerly *Agrobacterium rhizogenes*) strain A4T to screen the transformed individuals (Henzi *et al.*, 2000). *A. rhizogenes* represents an alternative to the use of *R. radiobacter* and causes hairy root symptom, which is useful in *in vitro* regeneration (David and Tempé, 1988). *R. radiobacter*-mediated transformation was tested in *B. oleracea* var. *italica* to induce a retarding effect on postharvest yellowing by transformation of the cytokinin-synthesizing isopentenyltransferase (*ipt*) gene, regulating senescence-associated gene promoters from *Arabidopsis* (Chen *et al.*, 2001).

Transformation mediated by *Rhizobium* could modify several biochemical pathways related to abiotic stress resistance; an example is the genetic improvement of heavy-metal tolerance that helps to overcome the weak growth of stressed plants growing in a high metal concentration. The development of transgenic lines showing heavy-metal tolerance was successfully achieved in *B. juncea* (Bhuiyan *et al.*, 2011; Wang *et al.*, 2013) and *B. napus* (Sheng and Xia, 2006; Shen *et al.*, 2011).

Transgenic mutants of cauliflower were obtained by introgression of the yeast metallothionein gene (*CUP1*) regulated by the cauliflower mosaic virus (CaMV) 35S promoter, resulting in cadmium-tolerant transgenic cauliflower, which showed optimal growth in the presence of 400 pM or less of cadmium (Hasegawa *et al.*, 1997).

Gene pyramiding is defined as the join of different genes or specific alleles in the genome of the same organism and can be carried out by conventional techniques (e.g. crossing) or facilitated by bacterial-mediated transformation (Chen *et al.*, 2016). Gene pyramiding was applied successfully in *B. oleracea* var. *botrytis* by *R. radiobacter*-mediated transformation to achieve pest resistance, mostly regarding several lepidopteran species (e.g. *Plutella xylostella*, *Crocidolomia pavonana*, *Hellula undalis*, and *Pieris* spp.). Genes such as *cry1a*, *cry1b*, and *cry1c* from *Bacillus thuringiensis* (*Bt*) encode specific insecticidal crystal proteins that can confer resistance against Lepidoptera (Kalia *et al.*, 2018). Bacterial transformation was also applied successfully using binary plasmid vectors mobilized into *R. radiobacter*, inserting the *cryIA*(*b*) and *cryIC* genes to obtain mutants carrying the *Bt* toxins (Fig. 5.1) (Chakrabarty *et al.*, 2002; Cao *et al.*, 2003).

Genetically modified cauliflowers of the cultivar 'Early Snowball' were obtained by *R. radiobacter*-mediated transformation leading to transgenic plants expressing the 79 kDa viral spike protein of severe acute respiratory syndrome coronavirus (SARS-CoV). Cauliflower plant tissues

**Table 5.1.** Transformation mediated by different techniques using different target genes.

| Technique | Target gene(s) | Reference |
|---|---|---|
| *Rhizobium*-mediated transformation | Organogenesis regeneration protocol | Ravanfar *et al.* (2017) |
| | Metallothionein gene (*CUP1*) for cadmium-tolerant transgenic cauliflower | Hasegawa *et al.* (1997) |
| | Pyramiding of *Bacillus thuringiensis* (*Bt*) genes (*cry1a*, *cry1b*, and *cry1c*) conferring resistance against Lepidoptera species | Chakrabarty *et al.* (2000); Cao *et al.* (2003); Kalia *et al.* (2018) |
| | Transgenic plants expressing the 79 kDa viral spike protein from severe acute respiratory syndrome coronavirus (SARS-CoV) | Pogrebnyak *et al.* (2006) |
| | Elastin-like polypeptides | Kowalczyk *et al.* (2018) |
| CRISPR Cas9-mediated transformation | Acetolactate synthase (ALS) for herbicide resistance | Chen *et al.* (2017); Wang *et al.* (2022) |
| RNA interference | Suppress the activity of acyl CoA oxidase (ACO) | Cheng *et al.* (2005) |
| Protoplast fusion and somatic hybridization | Mesophyll protoplast transformation and selection by hygromycin and kanamycin antibiotics | Radchuk *et al.* (2002) |
| | Insertion of plastome-targeting regions (*accD–rbcL*) | Nugent *et al.* (2006) |
| | Interspecific somatic hybrids resistant to black rot | Wang *et al.* (2011, 2016); Sharma *et al.* (2017) |
| | Transfer of cytoplasmic male sterility from *Brassica napus* and *B. juncea* to cauliflower | Chamola *et al.* (2013) |

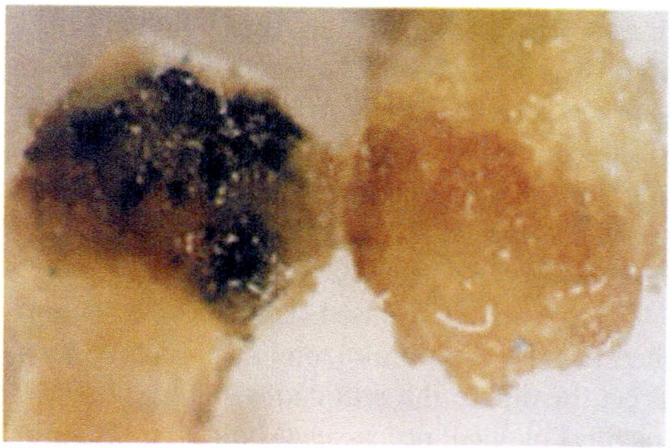

**Fig. 5.1.** Growth of hypocotyls expressing β-glucuronidase (*uidA*) reporter gene from cauliflower explants. Adapted from Chakrabarty *et al.* (2002). Used with permission from Springer Nature.

are sensitive to *Rhizobium* infection, showing massive necrosis at the infection site, so conditions needed to be optimized for successful transformation (Pogrebnyak *et al.*, 2006).

*R. rhizogenes* strains A4 and ATCC 18534 were tested in cauliflower and strain A4 showed a high efficiency of transgenic regenerants using the binary plasmid vector pCAMBIA 1305.2 carrying elastin-like polypeptides. The mutant plants were analyzed and selected by genomic and proteomic analysis (Kowalczyk *et al.*, 2018).

A comparative study of three vector systems and a design for co-transformation of allotetraploid *B. napus* was performed by Liu *et al.* (2020) investigating different transformation frequencies. The three co-transformation methods included four vectors tested in a mixed-strain system: the commercial plasmids pCAMBIA1300 and pCAMBIA3300, which were tested individually and together, and pBID RT. All three co-transformation methods were successful.

### 5.2.2 CRISPR/Cas9-mediated transformation

CRISPR/Cas9 is a genetic-engineering technique based on the enzymatic cleavage of a target region preceded by a protospacer adjacent motif (PAM) flanking the target region in which a guide RNA (gRNA) leads the endonuclease Cas9 to the target site (Doudna and Charpentier, 2014; Jiang and Doudna, 2017). *R. radiobacter*-mediated transformation is helpful in this new gene-editing technique for gene delivery of the Ti plasmid encoding the gRNA (Xu *et al.*, 2014; Zhang *et al.*, 2018).

Liu *et al.* (2018) tested the applicability of the CRISPR-Cas9 system to introduce resistance to CaMV in *Arabidopsis thaliana* plants by designing six different single guide RNAs (sgRNAs) that targeted different sites in the virus coat protein coding sequence; the results obtained suggested that the CRISPR/Cas9 system can be used for CaMV resistance. CRISPR/Cas 9 has also been used to transform the CaMV 35S promoter (driven by the Cas9 nuclease), allowing gene expression in plants (Gao *et al.*, 2016; Jiang *et al.*, 2017; Zhang *et al.*, 2017).

An improved CRISPR/Cas9 system was developed using the *Arabidopsis* promoter *UBIQUITIN10*, which shows high mutation efficiency in *Arabidopsis* (Wolabu *et al.*, 2020) and in tobacco, and when added to complexes of tRNA and gRNA (Kumar *et al.*, 2022).

Improving plant architecture is a major challenge in Brassicaceae cultivation, in particular plant height and the number of branches, which are components directly related to rapeseed (*B. napus* subsp. *napus*) yield (Liu *et al.*, 2022). Zheng *et al.* (2020) verified that the two rapeseed gene *BnaMAX1* homologs had redundant functions similar to those of *Arabidopsis MAX1*, which regulates plant height and axillary bud growth, and two sgRNAs were designed to modify *BnaMAX1* through CRISPR/Cas9. $T_0$ plants showed homozygous, heterozygous, biallelic, and chimeric mutations at the *BnaMAX1* target sites, and the mutations were transmitted to $T_1$ and $T_2$ progeny. Zheng *et al.* (2020) also obtained transgene-free lines created by CRISPR/Cas9 editing where no mutations in potential off-target sites were detected. Simultaneous knockout of all four *BnaMAX1* alleles consequently showed semi-dwarf and branched phenotypes with more silique, contributing to a higher yield per plant than the wild type (Zheng *et al.*, 2020).

Gene-knockout approaches have the potential to be an excellent tool for improving research on biological processes critical for improving crop performance (Jaganathan *et al.*, 2018). Neequaye *et al.* (2021) investigated the role of MYB28 in regulating aliphatic glucosinolate (A-GSL) biosynthesis and associated sulfur metabolism of *B. oleracea* plants grown in the field, using CRISPR-Cas9 gene-editing technology. The authors then described the first *myb28* knockout mutant in *B. oleracea* and thus the first field trial of CRISPR-Cas9-modified plants in the UK. It was observed that the elimination of *myb28* caused downregulation of the biosynthesis genes for A-GSL and a reduction in the accumulation of the glucosinolates derived from methionine, such as glucoraphanin, in the leaves and flowers of *myb28* mutant broccoli plants grown in the field, while leaving unchanged the accumulation of sulfate, *S*-methyl cysteine sulfoxide, and indole glucosinolate in the tissues of leaves and flowers of the mutant plant (Neequaye *et al.*, 2021).

The CRISPR/Cas9 system represents an efficient tool to induce targeted point mutations in selected genes (Li *et al.*, 2018; Tian *et al.*, 2018). In cauliflower hypocotyl explants, the induction of desired mutations can be exploited for herbicide resistance, as reported by Wang *et al.* (2022), who selected acetolactate synthase (*ALS*) and centromere-specific histone H3 variant (*CENH3*) as base editing genes. The mutants obtained by *R. radiobacter*-mediated transformation through the binary vector pHSE901 developed by Chen *et al.* (2017), and the obtained shoots, especially the *ALS* mutant, showed an important resistance to the herbicide tribenuron (Wang *et al.*, 2022). Real-time polymerase chain reaction (PCR) screening methods have been developed to investigate the presence of the CaMV promoter, detecting gene-modifying constructs in food (Becker and Ulrich, 2018; Bak and Emerson, 2020).

### 5.2.3 RNA interference

In recent decades, post-transcriptional silencing has been widely performed in plants, mostly regarding plant protection aspects, modulating or silencing expression of the transgenic plant by expressing double-stranded RNAs (dsRNAs) against selected targets (Dalakouras *et al.*, 2020).

RNA interference (RNAi) mediates post-transcriptional gene silencing (PTGS) directed by small RNAs (sRNAs). PTGS is activated by dsRNA molecules, which are cleaved by Dicer-like proteins (DCLs) into small interfering RNAs (siRNAs) of 21–24 nt. These siRNAs bind to Argonaute proteins (AGO) with the consequent formation of RNA-induced silencing complexes, which target complementary RNAs for degradation or translational inhibition (Agrawal *et al.*, 2003; Parent *et al.*, 2012; Akbar *et al.*, 2022). Plants have developed RNAi as an effective primary defense system against pathogens and parasites, particularly viruses (Zhang *et al.*, 2017). Application of RNAi via host-produced dsRNAs is called host-induced gene silencing (HIGS) (Nowara *et al.*, 2010). The effectiveness of HIGS can be improved by introducing transgenic constructs that transcribe dsRNA (typically hairpin RNA) into the host genome (Koch and Wassenegger, 2021).

Gene silencing-based strategies can be applied in pest crop management, selecting as a target the different insect genes involved in various physiological processes (Mamta and Rajam, 2017). As an example, this approach was carried out against the aphid *Myzus persicae* by Russell *et al.* (2017): RNAi sequence constructs of >20 nt against genes involved in critical functions were designed and inserted in cabbage.

The CaMV 35S promoter has been used to drive the selective marker genes or reporter genes controlling their expression (Yoo *et al.*, 2005), and can modify the activities and patterns of adjacent tissue- and organ-specific gene promoters (Zheng *et al.*, 2007). Several authors have demonstrated that loss of function of the BTB-TAZ domain protein 2 (BT2) is linked to the 35S promoter (Irigoyen *et al.*, 2022). CaMV expression can transcribe a large amount of RNAs (~600 nt) from non-coding regions (Blevins *et al.*, 2011), which overwhelm the host's RNAi system (Miller *et al.*, 2016). Transcript knockout of the viral CaMV, which expresses its genome through a polycistronic 35S RNA, requires the coordinated action of four *A. thaliana* Dicer-like proteins (DCL-1–4) (Moissiard and Voinnet, 2006). Recently, transgene PTGS that hinders the ubiquitin promoter-driven Cas9 system has been evaluated (Mao *et al.*, 2018).

RNAi has been used in cauliflower to suppress acyl CoA oxidase (*ACO*) gene expression, obtaining transgenic plants with downregulation of activity of the ACO enzyme, demonstrated by northern blot analysis (Chen *et al.*, 2005).

### 5.2.4 Protoplast fusion and somatic hybridization

Several *in vitro* techniques have been used to generate transgenic plants by protoplast fusion, obtaining hybrid and cybrid combinations and allowing gene transfer in sexually incompatible species (Ovcharenko *et al.*, 2023). The development of an efficient protoplast isolation protocol could improve the protoplast transformation technique and the growth of double-haploid lines. The choice of explant (hypocotyl, leaf fragment) and the concentration of the medium represent important parameters for regeneration via organogenesis of cauliflower explants (Veera *et al.*, 2009; Sheng *et al.*, 2011). Cultivar reactivity, donor plant age, and explant type also play an important role in *in vitro* regeneration (Ovesna *et al.*, 1993). The formation of haploids and doubled haploids using microspores allows the generation of homozygous genotypes in *Brassica* spp., and this aspect plays an important role in breeding self-incompatible and outcrossing genotypes (El-Esawi, 2016). Adventitious shoot regeneration followed by protoplast isolation and culture were examined from leaf explants of *in vitro* shoot cultures of several cauliflower cultivars, focusing on the development of improved nuclear and plastid transformation protocols. Mesophyll protoplasts from leaves gave the highest yield and division frequencies (Chikkala *et al.*, 2009). Mesophyll protoplast transformation has been developed using polyethylene glycol (PEG), and three plasmid vectors with different lengths were used to obtain stable transformants of *B. oleracea* var. *botrytis*; the seedlings of the individuals that developed were selected using hygromycin and kanamycin antibiotics (Radchuk *et al.*, 2002). Protoplast transformation was performed using PEG-mediated uptake, by plasmid nuclear and plastid vectors having *accD–rbcL* plastome-targeting regions cloned from *B. napus*, in addition to the marker gene to select the explants with the recombinant vector. The transgenic plants were regenerated from protoplast calluses (Fig. 5.2) (Nugent *et al.*, 2006).

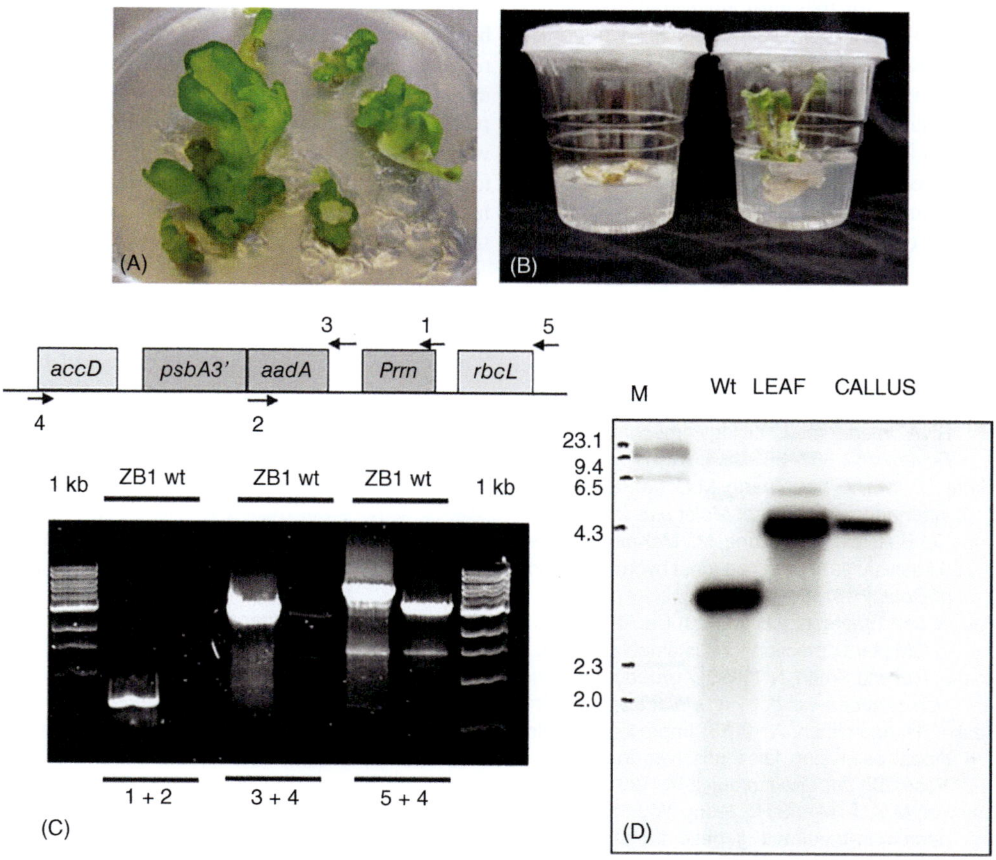


**Fig. 5.2.** Cauliflower nuclear transformants generated from calluses (A), followed by leaf tissue culture for the mutant and wild-type lines (B). (C,D) Polymerase chain reaction (PCR)-based selection of mutants (C) and Southern blot analysis performed using *Eco*RI and *Eco*RV for enzymatic cleavages (D). Adapted from Nugent *et al.* (2006). Used with permission from Elsevier.

In cauliflower, *in vitro* embryo rescue has been performed to generate an interspecific hybrid and backcross progeny between the cultivar 'Pusa Sharad' and the Ethiopian mustard NPC-9 to transfer the black rot disease resistance trait (Sharma *et al.*, 2017). Protoplast fusion allows the setting up of hybrid and cybrid combinations of species that are sexually incompatible, developing somatic hybrids between incompatible species. This technology has allowed not only intragenic hybridizations but also the production of intergenic hybrids and cybrids (Cardoza and Stewart, 2004); however, it is now considered a "dormant" biotechnology (Holmes, 2018). The asymmetric somatic hybridization performed between cauliflower 'Korso' and *B. nigra* showed genetic and epigenetic alteration of the introgression lines exhibiting sequence alterations, deletions, and variation in patterns of cytosine methylation; the hybridization also showed transference of black rot resistance genes (Wang *et al.*, 2011, 2016). *B. nigra* and cauliflower somatic hybrids were identified by morphological observation, and using random amplified polymorphic DNA and sequence-related amplified polymorphism molecular markers (Zhang *et al.*, 2008).

Somatic hybridization has been used to transfer cytoplasmic male sterility from *B. juncea* and *B. napus* (having the mitochondrial genome of *Moricandia arvensis* and *Erucastrum canariense*, respectively) to cauliflower (Chamola *et al.*, 2013).

## 5.3 Conclusion

Among the techniques discussed in this chapter, *R. radiobacter*-mediated transformation methods are the most widely used for the genetic engineering of cauliflower. Construction of plasmid vectors to carried out gene transfer and organogenesis methods represent the basis of other transformation techniques such as CRISPR/Cas 9 and RNAi, while methods such as protoplast fusion were often used in the past but are currently less favored, mostly because of their low efficiency and difficult replicability. Genetic engineering could be exploited in breeding programs to induce targeted mutation to develop improved cultivars for sustainable agriculture, but these methods are not always popular with consumers who may not agree with the use of genetically modified organisms for consumption. However, they are of great interest for identifying new genes and studying their functions.

## 5.4 References

Agrawal, N., Dasaradhi, P.V., Mohmmed, A., Malhotra, P., Bhatnagar, R.K., and Mukherjee, S.K. (2003) RNA interference: biology, mechanism, and applications. *Microbiology and Molecular Biology Reviews* 67, 657–685.

Akbar, S., Wei, Y., and Zhang, M.Q. (2022) RNA interference: Promising approach to combat plant viruses. *International Journal of Molecular Sciences* 23: 5312.

Arias, T., Beilstein, M.A., Tang, M., McKain, M.R., and Pires, C. (2014) Diversification times among *Brassica* (Brassicaceae) crops suggest hybrid formation after 20 million years of divergence. *American Journal of Botany* 101, 86–91.

Bak, A. and Emerson, J.B., (2020) Cauliflower mosaic virus (CaMV) biology, management, and relevance to GM plant detection for sustainable organic agriculture. *Frontiers in Sustainable Food Systems* 4: 21.

Bhalla, P.L. and Smith, N. (1998) *Agrobacterium tumefaciens*-mediated transformation of cauliflower, *Brassica oleracea var. botrytis*. *Molecular Breeding* 4, 531–541.

Becker, R. and Ulrich, A. (2018) Improved detection and quantification of cauliflower mosaic virus in food crops: assessing false positives in GMO screening based on the 35S promoter. *European Food Research and Technology* 244, 1861–1871.

Bhuiyan, M.S.U., Min, S.R., Jeong, W.J., Sultana, S., Choi, K.S., *et al.* (2011) An improved method for *Agrobacterium*-mediated genetic transformation from cotyledon explants of *Brassica juncea*. *Plant Biotechnology* 28, 17–23.

Blevins, T., Rajeswaran, R., Aregger, M., Borah, B.K., Schepetilnikov, M., *et al.* (2011) Massive production of small RNAs from a non-coding region of *Cauliflower mosaic virus* in plant defense and viral counter-defense. *Nucleic Acids Research* 39, 5003–5014.

Cao, J., Brants, A. and Earle, E.D. (2003) Cauliflower plants expressing a *cry1C* transgene control larvae of diamondback moths resistant or susceptible to Cry1A, and cabbage loopers. *Journal of New Seeds* 5(2–3), 193–207.

Cardoza, V. and Stewart, C.N. Jr. (2004) *Brassica* biotechnology: progress in cellular and molecular biology. *In Vitro Cellular and Developmental Biology-Plant* 40, 542–551.

Chakrabarty, R., Viswakarma, N., Bhat, S.R., Kirti, P.B., Singh, B.D., and Chopra, V.L. (2002) *Agrobacterium*-mediated transformation of cauliflower: optimization of protocol and development of Bt-transgenic cauliflower. *Journal of Bioscience* 27, 495–502.

Chamola, R., Balyan, H.S., and Bhat, S.R. (2013) Transfer of cytoplasmic male sterility from alloplasmic *Brassica juncea* and *B. napus* to cauliflower (*B. oleracea* var. *botrytis*) through interspecific hybridization and embryo culture. *Indian Journal of Genetics* 73, 203–210.

Char, S.N., Neelakandan, A.K., Nahampun, H., Frame, B., Main, M., *et al.* (2017) An *Agrobacterium*-delivered CRISPR/Cas9 system for high-frequency targeted mutagenesis in maize. *Plant Biotechnology Journal* 15, 257–268.

Chen, C., Zhao, X., Lei, J., Cao, B., and Chen, G. (2016) Pyramiding of Bt gene into *Brassica alboglabra* LH Bailey by biolistic and *Agrobacterium*-mediated transformation methods. *China Vegetables* 1, 21–28.

Chen, L.F.O., Hwang, J.Y., Charng, Y.Y., Sun, C.W., and Yang, S.F. (2001) Transformation of broccoli (*Brassica oleracea* var. *italica*) with isopentenyltransferase gene via *Agrobacterium tumefaciens* for post-harvest yellowing retardation. *Molecular Breeding* 7, 243–257.

Chen, Y., Wang, Z., Ni, H., Xu, Y., Chen, Q., and Jiang, L. (2017) CRISPR/Cas9-mediated base-editing system efficiently generates gain-of-function mutations in *Arabidopsis*. *Science China Life Sciences* 60, 520–523.

Chen, Y.H., Zhang, J.H., Ouyang, B., Li, H.X., and Ye, Z.B. (2005) Cloning of ACC oxidase gene and inhibition of endogenous gene expression with RNAi in cauliflower. *Acta Genetica Sinica* 32, 764–769.

Chikkala, V.R., Nugent, G.D., Dix, P.J. and Stevenson, T.W. (2009) Regeneration from leaf explants and protoplasts of *Brassica oleracea* var. *botrytis* (cauliflower). *Scientia Horticulturae* 119, 330–334.

Dalakouras, A., Wassenegger, M., Dadami, E., Ganopoulos, I., Pappas, M.L., and Papadopoulou, K. (2020) Genetically modified organism-free RNA interference: exogenous application of RNA molecules in plants. *Plant Physiology* 182, 38–50.

David, C. and Tempé, J. (1988) Genetic transformation of cauliflower (*Brassica oleracea* L. var. *botrytis*) by *Agrobacterium rhizogenes*. *Plant Cell Reports* 7, 88–91.

de Block, M., de Brouwer, D., and Tenning, P. (1989) Transformation of *Brassica napus* and *Brassica oleracea* using *Agrobacterium tumefaciens* and the expression of the *bar* and *neo* genes in the transgenic plants. *Plant Physiology* 91, 694–701.

Di Bella, M.C., Treccarichi, S., Arena, D., Nicotra, R., Mazzaglia, A., *et al.* (2021) Evaluation of Sicilian landraces of broccoli (*B. oleracea* var. *italica* Plenck) for quality traits. *Acta Horticulturae* 1354, 343–350.

Doudna, J.A. and Charpentier, E. (2014) The new frontier of genome engineering with CRISPR-Cas9. *Science* 346: 1258096.

El-Esawi, M.A. (2016) Somatic hybridization and microspore culture in *Brassica* improvement. In: Anis, M. and Ahmad, N. (eds.) *Plant Tissue Culture: Propagation, Conservation and Crop Improvement*. Springer, Singapore, pp. 599–609.

Endo, A., Saika, H., Takemura, M., Misawa, N., and Toki, S. (2019) A novel approach to carotenoid accumulation in rice callus by mimicking the cauliflower *Orange* mutation via genome editing. *Rice* 12: 81.

Gao, X., Chen, J., Dai, X., Zhang, D., and Zhao, Y. (2016) An effective strategy for reliably isolating heritable and Cas9-free *Arabidopsis* mutants generated by CRISPR/Cas9-mediated genome editing. *Plant Physiology* 171, 1794–1800.

Hasegawa, I., Terada, E., Sunairi, M., Wakita, H., Shinmachi, F., *et al.* (1997) Genetic improvement of heavy metal tolerance in plants by transfer of the yeast metallothionein gene (*CUP1*). In: Ando, T., Fujita, K., Mae, T., Matsumoto, H., Mori, S., and Sekiya, J. (eds.) *Plant Nutrition for Sustainable Food Production and Environment. Developments in Plant and Soil Sciences, Vol.* 78. Springer, Dordrecht, pp. 391–395.

Henzi, M.X., Christey, M.C., and McNeil, D.L. (2000) Factors that influence *Agrobacterium* rhizogenes-mediated transformation of broccoli (*Brassica oleracea* L. var. *italica*). *Plant Cell Reports* 19, 994–999.

Holmes, M. (2018) Somatic hybridization: the rise and fall of a mid-twentieth-century biotechnology. *Historical Studies in the Natural Sciences* 48, 1–23.

Irigoyen, S., Ramasamy, M., Misra, A., McKnight, T.D., and Mandadi, K.K. (2022) A BTB-TAZ protein is required for gene activation by *Cauliflower mosaic virus* 35S multimerized enhancers. *Plant Physiology* 188, 397–410.

Jaganathan, D., Ramasamy, K., Sellamuthu, G., Jayabalan, S., and Venkataraman, G. (2018) CRISPR for crop improvement: An update review. *Frontiers in Plant Science* 9: 985.

Jiang, F. and Doudna, J.A. (2017) CRISPR-Cas9 structures and mechanisms. *Annual Review of Biophysics* 46, 505–529.

Jiang, W.Z., Henry, I.M., Lynagh, P.G., Comai, L., Cahoon, E.B., and Weeks, D.P. (2017) Significant enhancement of fatty acid composition in seeds of the allohexaploid, *Camelina sativa*, using CRISPR/Cas9 gene editing. *Plant Biotechnology Journal* 15, 648–657.

Kalia, P., Aminedi, R., Golz, J., Russell, D., Choudhary, P., *et al.* (2018). Development of diamondback moth resistant transgenic cabbage and cauliflower by stacking *Cry1B* and *Cry1C* Bt genes. *Acta Horticulturae* 1282, 237–246.

Kim, H.S., Ji, C.Y., Lee, C.J., Kim, S.E., Park, S.C., and Kwak, S.S. (2018) *Orange*: a target gene for regulating carotenoid homeostasis and increasing plant tolerance to environmental stress in marginal lands. *Journal of Experimental Botany* 69, 3393–3400.

Koch, A. and Wassenegger, M. (2021) Host-induced gene silencing—mechanisms and applications. *New Phytologist* 231, 54–59.

Kowalczyk, T., Gerszberg, A., Durańska, P., Biłas, R., and Hnatuszko-Konka, K. (2018) High efficiency transformation of *Brassica oleracea* var. *botrytis* plants by *Rhizobium rhizogenes*. *AMB Express* 8, 125.

Krenek, P., Samajova, O., Luptovciak, I., Doskocilova, A., Komis, G., and Samaj, J. (2015) Transient plant transformation mediated by *Agrobacterium tumefaciens*: principles, methods and applications. *Biotechnology Advances* 33, 1024–1042.

Kumar, M., Ayzenshtat, D., Marko, A., and Bocobza, S. (2022) Optimization of T-DNA configuration with UBIQUITIN10 promoters and tRNA–sgRNA complexes promotes highly efficient genome editing in allotetraploid tobacco. *Plant Cell Reports* 41, 175–194.

Kumar, P. and Srivastava, D.K. (2016) Biotechnological advancement in genetic improvement of broccoli (*Brassica oleracea* L. var. *italica*), an important vegetable crop. *Biotechnology Letters* 38, 1049–1063.

Liu, F., Wang, P., Xiong, X., Fu, P., Gao, H., Ding, X., and Wu, G. (2020) Comparison of three *Agrobacterium*-mediated co-transformation methods for generating marker-free transgenic *Brassica napus* plants. *Plant Methods* 16: 81.

Liu, H., Soyars, C.L., Li, J., Fei, Q., He, G., *et al.* (2018) CRISPR/Cas9-mediated resistance to cauliflower mosaic virus. *Plant Direct* 2: e00047.

Liu, S., Raman, H., Xiang, Y., Zhao, C., Huang, J., and Zhang, Y. (2022) *De novo* design of future rapeseed crops: challenges and opportunities. *Crop Journal* 10, 587–596.

Maggioni, L., von Bothmer, R., Poulsen, G., Branca, F., and Bagger Jørgensen, R. (2014). Genetic diversity and population structure of leafy kale and *Brassica rupestris* Raf. in south Italy. *Hereditas* 151, 145–158.

Makukha, Y. and Dubina, E. (2021) Molecular marking in breeding *Brassica oleracea* L. for resistance to *Xanthomonas campestris* pv. *campestris*. *E3S Web of Conferences* 285: 03009.

Mamta, B. and Rajam, M.V. (2017) RNAi technology: a new platform for crop pest control. *Physiology and Molecular Biology of Plants* 23, 487–501.

Mao, Y., Yang, X., Zhou, Y., Zhang, Z., Botella, J.R., and Zhu, J.K. (2018). Manipulating plant RNA-silencing pathways to improve the gene editing efficiency of CRISPR/Cas9 systems. *Genome Biology* 19: 149.

Miller, W.A., Shen, R., Staplin, W., and Kanodia, P. (2016) Noncoding RNAs of plant viruses and viroids: sponges of host translation and RNA interference machinery. *Molecular Plant–Microbe Interactions* 29, 156–164.

Moissiard, G. and Voinnet, O. (2006) RNA silencing of host transcripts by cauliflower mosaic virus requires coordinated action of the four *Arabidopsis* Dicer-like proteins. *Proceedings of the National Academy of Sciences USA* 103, 19593–19598.

Nagaharu, U. (1935) Genome analysis in *Brassica* with special reference to the experimental formation of *B. napus* and peculiar mode of fertilization. *Japan Journal of Botany* 7, 389–452.

Neequaye, M., Stavnstrup, S., Harwood, W., Lawrenson, T., Hundleby, P., *et al.* (2021) CRISPR-Cas9-mediated gene editing of *MYB28* genes impair glucoraphanin accumulation of *Brassica oleracea* in the field. *CRISPR Journal* 4, 416–426.

Nowara, D., Gay, A., Lacomme, C., Shaw, J., Ridout, C., *et al.* (2010) HIGS: host-induced gene silencing in the obligate biotrophic fungal pathogen *Blumeria graminis*. *Plant Cell* 22, 3130–3141.

Nugent, G.D., Coyne, S., Nguyen, T.T., Kavanagh, T.A., and Dix, P.J. (2006) Nuclear and plastid transformation of *Brassica oleracea* var. *botrytis* (cauliflower) using PEG-mediated uptake of DNA into protoplasts. *Plant Science* 170, 135–142.

Ovcharenko, O.O., Rudas, V.A., and Kuchuk, M.V. (2023) Protoplast fusion for cellular engineering of the Brassicaceae. *Cytology and Genetics* 57, 432–450.

Ovesna, J., Ptáček, L., and Opatrný, Z. (1993) Factors influencing the regeneration capacity of oilseed rape and cauliflower in transformation experiments. *Biologia Plantarum* 35, 107–112.

Parent, J.S., Martinez de Alba, A.E., and Vaucheret, H. (2012) The origin and effect of small RNA signaling in plants. *Frontiers in Plant Science* 3: 179.

Picchi, V., Lo Scalzo, R., Tava, A., Doria, F., Argento, S., *et al.* (2020) Phytochemical characterization and *in vitro* antioxidant properties of four brassica wild species from Italy. *Molecules* 25: 3495.

Pogrebnyak, N., Markley, K., Smirnov, Y., Brodzik, R., Bandurska, K., *et al.* (2006) Collard and cauliflower as a base for production of recombinant antigens. *Plant Science* 171, 677–685.

Radchuk, V.V., Ryschka, U., Schumann, G., and Klocke, E. (2002) Genetic transformation of cauliflower (*Brassica oleracea* var. *botrytis*) by direct DNA uptake into mesophyll protoplasts. *Physiologia Plantarum* 114, 429–438.

Ravanfar, S.A., Orbovic, V., Moradpour, M., Abdul Aziz, M., Karan, R., *et al.* (2017). Improvement of tissue culture, genetic transformation, and applications of biotechnology to *Brassica*. *Biotechnology and Genetic Engineering Reviews* 33, 1–25.

Russell, D.A., Huang, D., Bhalla, P., Singh, M., Robin, C., *et al.* (2017) Progress in the development of transgenic cabbage, cauliflower and canola expressing stacked bts for caterpillar control and RNAi for aphid suppression. *Mysore Journal of Agricultural Sciences (Special Issue)* A 51, 159–167.

Scalzo, R.L., Bianchi, G., Picchi, V., Campanelli, G., Ficcadenti, N., *et al.* (2024) Compositional traits of hybrid populations of *Brassica oleracea* L. var. *italica* (broccoli) and *Brassica oleracea* L. var. *botrytis* (cauliflower) during four organic breeding cycles. *Journal of Food Composition and Analysis* 131: 106209.

Schiessl, S.V. and Mason, A.S. (2020). Ancient and recent polyploid evolution in *Brassica*. In: Wani, S., Thakur, A., and Jeshima Khan, Y. (eds.) Brassica *Improvement*. Springer, Cham, pp. 49–66.

Sharma, B.B., Kalia, P., Singh, D., and Sharma, T.R. (2017) Introgression of black rot resistance from *Brassica carinata* to cauliflower (*Brassica oleracea botrytis* group) through embryo rescue. *Frontiers in Plant Science* 8: 1255.

Sharma, S.R., Singh, P.K., Chable, V., and Tripathi, S.K. (2004) A review of hybrid cauliflower development. *Journal of New Seeds* 6, 151–193.

Shaw, R.K., Shen, Y., Zhao, Z., Sheng, X., Wang, J., *et al.* (2021) Molecular breeding strategy and challenges towards improvement of downy mildew resistance in cauliflower (*Brassica oleracea* var. botrytis L.). *Frontiers in Plant Science* 12: 1262.

Shen, Q.I., Jiang, M., Li, H.U.A., Che, L.L., and Yang, Z.M. (2011) Expression of a *Brassica napus* heme oxygenase confers plant tolerance to mercury toxicity. *Plant, Cell & Environment* 34, 752–763.

Sheng, X., Zhao, Z., Yu, H., Wang, J., Xiaohui, Z. and Gu, H. (2011) Protoplast isolation and plant regeneration of different doubled haploid lines of cauliflower (*Brassica oleracea* var. *botrytis*). *Plant Cell, Tissue and Organ Culture* 107, 513–520.

Sheng, X.F. and Xia, J.J. (2006) Improvement of rape (*Brassica napus*) plant growth and cadmium uptake by cadmium-resistant bacteria. *Chemosphere* 64, 1036–1042.

Singh, B.K., Singh, B., and Singh, P.M. (2018) Breeding cauliflower: a review. *International Journal of Vegetable Science* 24, 58–84.

Sun, D., Wang, C., Zhang, X., Zhang, W., Jiang, H., *et al.* (2019) Draft genome sequence of cauliflower (*Brassica oleracea* L. var. *botrytis*) provides new insights into the C genome in *Brassica* species. *Horticulture Research* 6: 82.

Tian, S., Jiang, L., Cui, X., Zhang, J., Guo, S., *et al.* (2018) Engineering herbicide-resistant watermelon variety through CRISPR/Cas9-mediated base-editing. *Plant Cell Reports* 37, 1353–1356.

Treccarichi, S., Di Gaetano, C., Di Stefano, F., Gasparini, M., and Branca, F. (2021) Using simple sequence repeats in 9 *Brassica* complex species to assess hypertrophic curd induction. *Agriculture* 11: 622.

Treccarichi, S., Cali, R., Amari, M., Mortada, A., Felis, M.D., *et al.* (2022) New tools for organic plant nutrition: microbial and amino acid treatments for organic seed production of broccoli (*Brassica oleracea* var. *italica* Plenk.). *Acta Horticulturae* 1365, 81–90.

Veera, R.N., Gregory, D.N., Philip, J.D., and Trevor, W.S. (2009) Regeneration from leaf explants and protoplasts of *Brassica oleracea* var. *botrytis* (cauliflower). *Science Horticulture* 119, 330–334.

Wang, G.-X., Yan, H., Zeng, X., Sheng, X., Tang, Y., *et al.* (2011) New alien addition lines resistance to black rot generated by somatic hybridization between cauliflower and black mustard. *Acta Horticulturae Sinica* 38, 1901–1910.

Wang, G.X., Lv, J., Zhang, J., Han, S., Zong, M., *et al.* (2016) Genetic and epigenetic alterations of *Brassica nigra* introgression lines from somatic hybridization: a resource for cauliflower improvement. *Frontiers in Plant Science* 7: 1258.

Wang, G., Zong, M., Liu, D., Wu, Y., Tian, S., *et al.* (2022) Efficient generation of targeted point mutations in the *Brassica oleracea* var. *botrytis* genome via a modified CRISPR/Cas9 system. *Horticultural Plant Journal* 8, 527–530.

Wang, J.W., Li, Y., Zhang, Y.X., and Chai, T.Y. (2013) Molecular cloning and characterization of a *Brassica juncea* yellow stripe-like gene, *BjYSL7*, whose overexpression increases heavy metal tolerance of tobacco. *Plant Cell Reports* 32, 651–662.

Welsch, R., Zhou, X., Koschmieder, J., Schlossarek, T., Yuan, H., *et al.* (2020) Characterization of cauliflower OR mutant variants. *Frontiers in Plant Science* 10: 1716.

Wolabu, T.W., Park, J.J., Chen, M., Cong, L., Ge, Y., *et al.* (2020) Improving the genome editing efficiency of CRISPR/Cas9 in *Arabidopsis* and *Medicago truncatula*. *Planta* 252: 15.

Xu, R., Li, H., Qin, R., Wang, L., Li, L., *et al.* (2014) Gene targeting using the *Agrobacterium tumefaciens*-mediated CRISPR-Cas system in rice. *Rice* 7: 5.

Yoo, S.Y., Bomblies, K., Yoo, S.K., Yang, J.W., Choi, M.S., *et al.* (2005). The 35S promoter used in a selectable marker gene of a plant transformation vector affects the expression of the transgene. *Planta* 221, 523–530.

Zhang, F., LeBlanc, C., Irish, V.F., and Jacob, Y. (2017a) Rapid and efficient CRISPR/Cas9 gene editing in *Citrus* using the *YAO* promoter. *Plant Cell Reports* 36, 1883–1887.

Zhang, J., Khan, S.A., Heckel, D.G., and Bock, R. (2017b) Next-generation insect-resistant plants: RNAi-mediated crop protection. *Trends in Biotechnology* 35, 871–882.

Zhang, L., Zhao, H., Chen, B., and Liu, F. (2008) Development and identification of interspecific somatic hybrids between cauliflower and black mustard. *Chinese Bulletin of Botany* 25, 176–184.

Zhang, S., Zhang, R., Song, G., Gao, J., Li, W., *et al.* (2018) Targeted mutagenesis using the *Agrobacterium tumefaciens*-mediated CRISPR-Cas9 system in common wheat. *BMC Plant Biology* 18: 302.

Zheng, M., Zhang, L., Tang, M., Liu, J., Liu, H., *et al.* (2020) Knockout of two *BnaMAX1* homologs by CRISPR/Cas9-targeted mutagenesis improves plant architecture and increases yield in rapeseed (*Brassica napus* L.). *Plant Biotechnology Journal* 18, 644–654.

Zheng, X., Deng, W., Luo, K., Duan, H., Chen, Y., *et al.* (2007) The cauliflower mosaic virus (CaMV) 35S promoter sequence alters the level and patterns of activity of adjacent tissue- and organ-specific gene promoters. *Plant Cell Reports* 26, 1195–1203.

Zhou, X., Sun, T.H., Wang, N., Ling, H.Q., Lu, S., and Li, L. (2011) The cauliflower *Orange* gene enhances petiole elongation by suppressing expression of eukaryotic release factor 1. *New Phytologist* 190, 89–100.

# 6 Carrot Genetic Engineering

**Magdalena Klimek-Chodacka*, Aneta Lukasiewicz, Kamil Szymonik and Rafal Baranski**
*Department of Plant Biology and Biotechnology, Faculty of Biotechnology and Horticulture, University of Agriculture in Krakow, Krakow, Poland*

**Abstract**
Carrot is one of the ten economically most important vegetables in the world. Currently, it is cultivated over an area of 1.1 million ha in various climatic zones, due to the development of varieties adapted to cultivation both in summer and winter seasons, and in some areas even all year round. It is a very good source of antioxidants and provitamin A, and therefore has beneficial effects on human health. Carrot is also a model species used in research with protoplast, cell, and tissue culture *in vitro*. Genetic transformation has been employed in a wide range of basic and applied research. A well-developed transgene reporter system and cell-labeling methods in conjunction with *in vitro* techniques provide opportunities for establishing carrot as a biotech agricultural model plant. Modifications introduced into significant genes of the carrot carotenoid and anthocyanin biosynthetic pathways have allowed detailed studies on these important nutritional components of the human diet and their biosynthesis in plant cells. *Agrobacterium*-mediated transformation has led to the introduction of new biotic and abiotic stress resistance-related traits in carrot. Genetically engineered carrot has been modified toward better resistance to some leaf and root fungal pathogens. The introduction of osmolyte-related genes such as Δ1-pyrroline-5-carboxylate synthase and betaine aldehyde dehydrogenase has improved the salt tolerance of carrot plants. Efficient protocols have also been developed for the precise editing of the carrot genome using the clustered regularly interspaced palindromic repeats (CRISPR)-CRISPR-associated protein 9 (Cas9) system. However, despite such advanced research, there are still no commercially available genetically modified carrot varieties.

**Keywords**: *Agrobacterium*, biotic and abiotic stress, carrot, *Daucus carota*, genetic transformation

## 6.1 Introduction

The carrot (*Daucus carota* L. subsp. *sativus* Hoffm.) is a root vegetable that was only domesticated around 1000 years ago (Iorizzo *et al*., 2013). The first domesticated carrots had purple or yellow roots, and their occurrence was recorded in Central Asia and Asia Minor, from where they were later transferred to western Europe and England between the 11th and 15th centuries. Mentions of carrot roots with the orange color we know today appeared in the 15th and 16th centuries in Europe, suggesting that this color is due to the second domestication event (Stolarczyk and Janick, 2011). The first large cultivations were located mainly in the temperate regions of Europe and Asia; however, later many varieties were adapted for cultivation in tropical and subtropical climates. This has contributed to the rapid increase in the cultivation area during the last 50 years.

*Email: m.chodacka@urk.edu.pl


DOI: 10.1079/9781800627123.0006

Carrot is a source of important phytonutrients, making this vegetable a functional food with significant pro-health properties. The storage root is especially rich in provitamin A carotenoids such as α- and β-carotenes (Block, 1994), but also contains large amounts of ascorbic acid and tocopherol and is classified as vitaminized food (Hashimoto and Nagayama, 2004). It also contains bioactive polyacetylenes (Hansen *et al.*, 2003; Kidmose *et al.*, 2004) and phenolics (black carrot) (Babic *et al.*, 1993).

Carrots have also been one of the model organisms for plant tissue and cell-culture research, since pioneering work from late 1950s in which the totipotency of plant cells was demonstrated using carrot cell cultures (Steward *et al.*, 1958). Due to its high ability for cell and tissue proliferation and high embryogenic potential (Hardegger and Shakya, 2004), carrot has also been used successfully in research on horizontal gene transfer, focused on studying gene function and regulation, and plant metabolism. Applied research includes the development of plants resistant to both abiotic and biotic stresses, and modifications of biosynthetic pathways. Overexpression of several heterologous genes in transgenic carrots have enhanced plant tolerance to different pathogens causing major leaf and root diseases.

## 6.2 Genetic Resources

The cultivated carrot is the most important member of the Apiaceae family, both economically and nutritionally. It is the second most popular vegetable after the potato. The Apiaceae family contains 466 genera and over 3800 species that occur in many geographic regions (Plunkett *et al.*, 2018). Central Asia is assumed to be the origin of carrot as a root plant, with cultivation sites in Afghanistan and Persia. Molecular analyses have also indicated Central Asia as the center of origin, from where carrot spread to western Asia, North Africa and later to Europe in the 1300s and North America (Iorizzo *et al.*, 2013; Bolton *et al.*, 2020).

The improvement and development of new carrot varieties in breeding programs relies on available genetic diversity. It is estimated that *ex situ* carrot germplasm collections are very numerous around the world and include over 13,400 accessions gathered in 62 institutions (Allender, 2019). Carrot genetic resources are also preserved *in situ* and using on-farm conservation methods, which are particularly valuable to preserve carrot wild relatives and landraces. There is no gene bank in the world that is responsible for preserving the global diversity of carrots. Carrot germplasms are located in many gene banks, mainly in Europe and the USA. According to the Genesys database (www.genesys-pgr.org/, accessed 25 June 2024), 7155 accessions of the genus *Daucus* are deposited in 58 institutions, of which 1972 are *D. carota* (Table 6.1). The largest number of *Daucus* spp. (1574 accessions) is deposited at the North Central Regional Plant Introduction Station, US Department of Agriculture-Agricultural Research Service (USDA-ARS). Significant collections of genetic resources are also located in gene banks in the Warwick Genetic Resources Unit at the University of Warwick in the UK, the Plant Breeding and Acclimatization Institute in Poland, the Leibniz Institute of Plant Genetics, and the Crop Plant Research in Germany.

The European Plant Genetic Resource Search Directory (EURISCO) is also available (https://www.ecpgr.org/resources/germplasm-databases/eurisco-catalogue, accessed 25 June 2024). EURISCO contains information on over 2 million records of crops and their wild relatives, *ex situ* preserved by nearly 400 institutes. Based on a network of national inventories from 43 member countries, it represents a major effort to preserve world agrobiological diversity by providing information on the high genetic diversity maintained by collaborating institutions. It contains information on 4658 carrot accessions collected in 30 countries but does not contain information about collections in the USA. The EURISCO and Genesys databases are similar in terms of information quantity, although EURISCO is a bit on the low side. However, these two databases do not contain all relevant data. Large and noteworthy collections are in the Vavilov Institute, Russia, the Institute of Vegetables and Flowers at the Chinese Academy of Agricultural Sciences, and in India maintained by the National Bureau of Plant Genetic Resources. It is estimated that over 3600 accessions are deposited only in these three institutions (Allender, 2019).

The *ex situ* conservation of carrot genetic resources generally involves the long-term storage

**Table 6.1.** Collections of carrot genetic resources. Data obtained from the Genesys database (www.genesys-pgr.org, accessed 13 November 2024).

| Country | FAO code | Name of institution | No. of accessions |
|---|---|---|---|
| USA | USA020 | North Central Regional Plant Introduction Station (NCRPIS), US Department of Agriculture-Agricultural Research Service (USDA-ARS) | 1574 |
| UK | GBR006 | Warwick Genetic Resources Unit, UK Vegetable Genebank | 1509 |
| Poland | POL003 | Plant Breeding and Acclimatization Institute | 660 |
| Germany | DEU146 | Gene Bank, Leibniz Institute of Plant Genetics and Crop Plant Research | 507 |
| Czech Republic | CZE122 | Gene Bank, Crop Research Institute | 404 |
| Ukraine | UKR021 | Institute of Vegetable and Melon Growing | 165 |
| USA | USA974 | Seed Savers Exchange | 239 |
| UK | GBR004 | Millennium Seed Bank Project, Seed Conservation Department, Royal Botanic Gardens, Kew | 229 |
| Hungary | HUN003 | Institute for Agrobotany | 208 |
| Sweden | SWE054 | Nordic Genetic Resource Center | 196 |
| Other | | | 1464 |
| **Total** | | | **7155** |

of seed samples. Carrots have seeds referred to as orthodox, meaning that the viability is preserved when the seeds are stored in low-humidity and low-temperature conditions.

Varieties created in the course of breeding programs are the largest part of the global collection and account for about 40%. The second most numerous group comprises wild varieties and constitutes about 23%. There are also local accessions (12%).

Despite such large genetic resources of *Daucus* spp., few are characterized in terms of morphological or functional features. This will require a huge investment of time and finances for the gathered resources to be useful. Currently, the Carrot Diversity Set created by the UK Vegetable Genebank is available, which includes 77 properties from 35 countries, including wild carrot, landraces, and advanced varieties. These all are well described in terms of root and leaf characters such as shape or color, and their geographic origin has been validated. Small parts of the collected genetic resources have also been utilized for research purposes, such as phylogenetic studies, domestication, genetics of selected traits, and resistance to biotic and abiotic stresses (Ellis *et al.*, 1993; Grzebelus *et al.*, 2013a; Spooner *et al.*, 2013; Nothnagel *et al.*, 2017; Bolton *et al.*, 2019).

## 6.3 Chromosome Numbers and Ploidy

According to recent studies, about 40 species and two clades are currently distinguished in the genus *Daucus* (Banasiak *et al.*, 2016). Most of these species are diploid with a basic chromosome number ranging from $n = 8$ to $n = 11$. However, several polyploid species have been identified so far, including the tetraploid *Daucus glochidiatus*, *D. incognitus*, *D. melananthus*, and *D. pedunculatus* ($2n = 44$), and the hexaploid *D. montanus* ($2n = 66$) (Banasiak *et al.*, 2016). Cultivated carrots have nine pairs of chromosomes, and their haploid genome is not large, estimated at 473 Mb (Iorizzo *et al.*, 2016).

An important parameter is also the C-value, representing the amount of DNA in the unreplicated gametic nuclear genome. Nowicka *et al.* (2016b) conducted a comprehensive study of a large number of *Daucus* spp., including advanced breeding varieties and landraces with different root colors. The average C-value was 0.96 pg and did not differ much between the examined accessions. In constrast, large variation in the size of the genome was noted among wild varieties, with the highest C-value of 3.02 pg recorded in *D. littoralis* ($2n = 20$). Variations in genome size within a taxon were usually very small, with the exception of *D. guttatus* where the difference in C-values was almost twofold (1.49–2.83 pg).

Carrot chromosomes are small and uniform in both length and shape, which significantly hampers cytological studies or the construction of cytogenetic maps (Iovene *et al.*, 2011). Giemsa C-bands or fluorescence *in situ* hybridization staining using different probes have been used, but it was not easy to distinguish all chromosomes. Effective identification of carrot chromosomes was possible only as a result of using a mixture of several fluorescently labeled probes (Nowicka *et al.*, 2016a). With this approach, precise chromosome discrimination was achieved. The first and longest pair of carrot chromosomes is subtelocentric and has a total length of 4.7 μm. The next five pairs of chromosomes are 4.0–3.39 μm in length and are subtelocentric or metacentric. The smallest pair of metacentric type is 2.75 μm.

## 6.4 Production and Productivity

Carrots are currently grown over an area exceeding 1.1 million ha worldwide (Table 6.2) (FAOSTAT, 2019). Most carrots are grown in Asia, where they are planted on nearly 640,000 ha, but this is only 1.2% of all vegetable crops in this region. The highest percentage of carrot cultivation in relation to other vegetables was recorded in Europe, reaching 7.2%, representing over 234,000 ha. The total cultivated area has remained constant for the last 15 years, ranging from 1 million to 1.2 million ha (Fig. 6.1). In contrast, the highest dynamics of the growth of carrot cultivation area was recorded in 1990–2005, which is an increase of over 3.5 times in under 60 years. The increase in crops in Asia had a significant impact, while in other regions the cultivation area has remained at a constant level since the 1960s.

The average yield in 2019 was 39,659 kg ha$^{-1}$, but this varies greatly depending on the region (Table 6.2). The highest was recorded in the Oceania region, amounting to 57,359.5 kg ha$^{-1}$, while in African countries it was over three times lower and amounted to only 17,555.9 kg ha$^{-1}$. The average global dynamics of the yield increase in the last 60 years is not high and amounts to 2.5-fold, while the increase in yield in the Asian region reached 3.7-fold (Fig. 6.2).

## 6.5 Major Biotic Stresses

### 6.5.1 Major root diseases

*Cavity spot (*Pythium *spp.)*

Cavity spot is caused by several *Pythium* spp., the most common being *Pythium violae* and *P. sulcatum*, and this disease has been documented in nearly all carrot production regions (McDonald, 2002). The disease rarely reduces the root yield but has a significant economic impact as shallow lesions on the surface of the roots render them unsuitable for the fresh and processing market. In addition, lesions on the root surface can be infected with other microorganisms, which often leads to discoloration, and this is especially visible after heat treatment. So far, it has not been possible to obtain

**Table 6.2.** Global harvested area of primary vegetables and carrots with turnips, and mean yield of carrots and turnips in world regions according to the FAOSTAT (2019) database.

| | Harvested area (ha) | | | | |
|---|---|---|---|---|---|
| Region | Primary vegetables (ha) | Carrots and turnips (ha) | Percentage of vegetables[a] | Production (t) | Yield (kg ha$^{-1}$) |
| Africa | 10,372,651 | 134,599 | 1.30 | 2,363,011 | 17,555.9 |
| Americas | 3,473,345 | 113,138 | 3.26 | 4,453,599 | 39,364.3 |
| Asia | 42,436,835 | 639,754 | 1.51 | 29,027,885 | 45,373.5 |
| Europe | 3,235,332 | 234,395 | 7.24 | 8,527,803 | 36,382.2 |
| Oceania | 171,011 | 6,809 | 3.98 | 390,561 | 57,359.5 |
| **World** | **59,689,174** | **1,128,695** | **1.89** | **44,762,859** | **39,659.0** |

[a]Calculated as the percentage of harvested area of carrot and turnip in relation to the harvested area of primary vegetables.

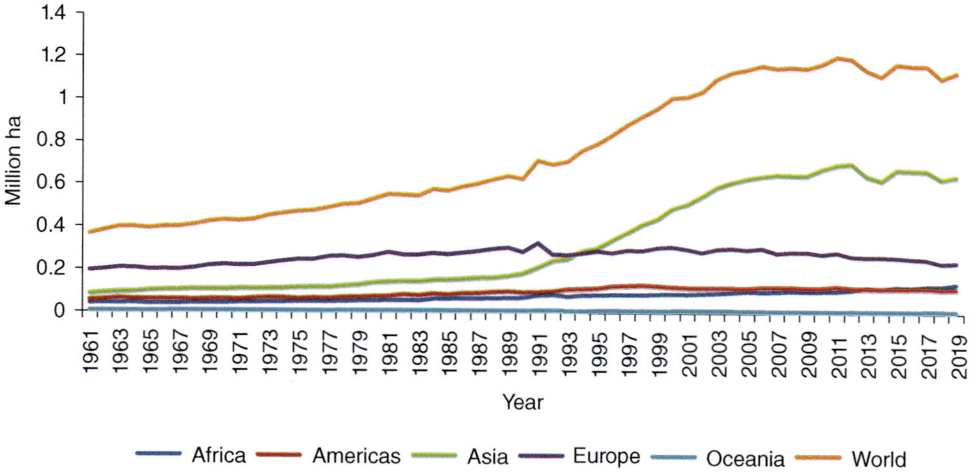


**Fig. 6.1.** Global and regional total harvested area of carrot with turnip from 1961 to 2019.

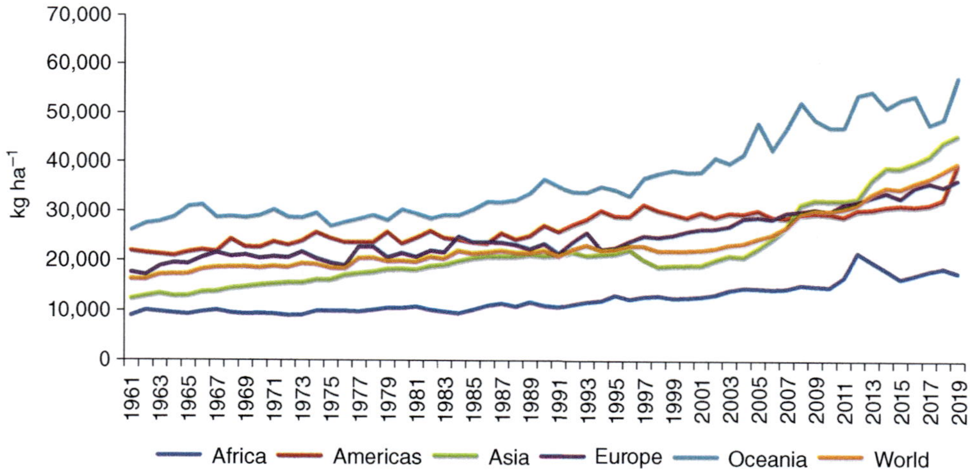


**Fig. 6.2.** Global and regional yield of carrot in years 1961–2019.

varieties of carrots completely resistant to cavity spot. As a result of numerous studies, some cultivars were indicated that are less susceptible to this pathogen (Soroker *et al.*, 1984; Groom and Perry, 1985; Sweet *et al.*, 1986; White, 1988; Vivoda *et al.*, 1991; McDonald, 1994a; McDonald, 2002).

### *Diseases caused by* Rhizoctonia *spp.*

Carrots can be infected by three *Rhizoctonia* spp., *Rhizoctonia carotae*, *R. crocorum*, and *R. solani*. However, the greatest damage is caused by *R. solani*, which causes damping-off of carrot seedlings and root rot in the crown of the root (Nuñez and Westphal, 2002). Development of the disease after germination significantly reduces the number of plants that die at the seedling stage as a result of rotting or root dieback, and some dwarf plants of low vigor may also develop (Nuñez and Westphal, 2002). Disease symptoms on the roots appear mainly in muddy soils rich in humus with high temperature and humidity (Punja, 2002). The petioles and leaves rot, and dark brown, sunken lesions develop near the crown and sometimes further down the root. The symptoms may be similar to those accompanying cavity spot (Davis and Raid, 2002).

### *Black rot (*Alternaria radicina*)*

Black rot is a disease caused by the fungus *Alternaria radicina* (formerly *Stemphylium radicinum*). This pathogen was first recorded in carrots in 1988 in Denmark and other northern European countries, but now occurs in all world regions with carrot production. It can survive in soil for up to 8 years and infect other related species (Farrar *et al.*, 2004). The fungus has also been found to cause parsley leaf blight and stem and celery root rot, and is pathogenic to caraway, fennel, and parsnips (Farrar *et al.*, 2004). Black rot is characterized by dry, black lesions mainly on carrot roots. The greatest losses are recorded at the stage of cold storage, where it can decompose the entire root and spread to other roots. *A. radicina* can infect most carrot tissues, and because it is seed-borne, the disease cycle begins with the sowing of infected seeds (Pryor *et al.*, 1998).

Control of black rot is relatively difficult because the symptoms are mostly visible after harvest, and the fungus is either transmitted by the seeds or survives permanently in the soil. Therefore, disease control is usually done through an integrated approach and mainly involves the use of fungicides at the sowing stage (du Toit *et al.*, 2019). The positive effect of biological control agents as well as yeast *Candida* spp. on the reduction of disease development has also been demonstrated (Chen and Wu, 1999; Kordowska-Wiater *et al.*, 2012). Several screening tests for resistance to the pathogen have been carried out so far (Pryor *et al.*, 2000; Karkleliene *et al.*, 2012). Significant differences in the intensity of disease development between the selected cultivars were demonstrated. The cultivars ‘Panther’ and ‘Caropak’ were indicated as relatively tolerant, while ‘Magi’ and ‘Nogales’ were highly susceptible. Tolerance to *Alternaria* can be tested by wounding the root disks with toothpicks incubated for several days in a fungus suspension, selecting *in vitro* at the protoplast level, or by inoculating leaf petioles or seedlings with agar disks (Pryor *et al.*, 2000; Grzebelus *et al.*, 2013b; Cwalina-Ambroziak *et al.*, 2014).

### *Rubbery brown rot or* Phytophthora *root rot (*Phytophthora *spp.)*

This is a disease that can cause particularly great damage in wetlands, where it occurs particularly often after a wave of heavy rainfall. Its negative effects have been reported in Canada, France, Norway, Australia, and the USA. It can be caused by several *Phytophthora* spp., including *Phytophthora cactorum*, *P. cryptogea*, *P. megasperma*, and *P. porri* (Browne, 2002). The disease results in the appearance of dark brown and even black roots, which often turn rubbery, especially in the middle and crown of the root (Saude *et al.*, 2007). Symptoms appear most often in the preharvest period but most often during storage. Additionally, the pathogen can produce white mycelium on diseased roots, which can lead to secondary invasion of bacterial and fungal lesions and result in soft rot.

To date, no screening for tolerance to rubbery brown rot has been performed. Such tests can be performed using inoculated agar blocks. The symptoms are visible after 7 days at 20°C and after a few weeks at 0°C (McDonald, 1994b). Humidity plays an important role and must exceed 95% for optimal disease development. As has been shown so far, it is not necessary to wound the root to assess the susceptibility of the variety to this pathogen (Saude *et al.*, 2007).

## 6.5.2 Major foliar diseases

### *Carrot leaf blights (*Alternaria dauci, Cercospora carotae, Xanthomonas hortorum *pv.* carotae*)*

Carrot leaf diseases are mainly caused by fungal (*Alternaria dauci* and *Cercospora carotae*) or bacterial (*Xanthomonas hortorum* pv. *carotae*) pathogens. The greatest losses are caused by *Alternaria* leaf blight, which was first described in the 1855 in Germany and later in 1890 in the USA (Kuhn, 1855; Ellis and Holliday, 1972). *A. dauci* causes strong defoliation of carrot plants, especially in conditions of high humidity and temperature (Farrar *et al.*, 2004). Symptoms appear on leaves up to 10 days after infection as small, green-brown spots, which increase with time until the tissue becomes black. Symptoms may also appear on inflorescences and on seeds in developing umbels (Farrar *et al.*, 2004). *Cercospora* leaf spot (caused by *C. carotae*) can often be confounded with disease caused by *Alternaria*. However, this fungus infects only the aerial parts of carrot plants, and the spots are rounder and additionally have a distinct dark margin with a lighter brown center (Gugino *et al.*, 2007; Milosavljević *et al.*, 2014). Symptoms of bacterial leaf blight caused by *X. hortorum* pv. *carotae* can also be difficult to distinguish from those caused by the two mentioned fungi, although sometimes the distinction can be easier, especially when gummy bacterial exudate appears on the petioles, umbels, and seeds (du Toit *et al.*, 2005). Phenotypic studies of carrot cultivars and their response to contamination have focused mainly on *A. dauci*. So far, numerous screening tests for resistance to carrot leaf blights have been carried out, and many varieties have been selected that were highly tolerant, but none have showed complete resistance (Strandberg *et al.*, 1972; Boiteux *et al.*, 1993; Silva *et al.*, 2009; Pereira *et al.*, 2012; Carvalho *et al.*, 2015).

Simon and Strandberg (1998) suggested that the increased tolerance of the studied inbreds may have a genetic basis. Le Clerc *et al.* (2009) identified three regions of quantitative trait loci that correlated with a lower sensitivity to the pathogen, thus postulating that this trait is polygenic. A positive contribution of falcarindiol and 6-methoxymellein was also demonstrated in cultivars tolerating *A. dauci*. These phytoalexins reduced the activity of the fungus and, in addition, falcarindiol was able to induce permeabilization of the *A. dauci* plasma membrane (Lecomte *et al.*, 2012).

#### *Powdery mildew (*Erysiphe heraclei*)*

Powdery mildew in carrots is most often caused by the fungus *Erysiphe heraclei* (Oidium mildew type), although the *Oidiopsis* type has also been reported, which in turn is caused by *Leveillula lanuginosa and L. taurica* (Aegerter, 2002). The occurrence of powdery mildew has been confirmed in all regions where carrots are grown, although the greatest intensity occurs in warm and semi-arid regions, which often also have drip or furrow irrigation (du Toit and Derie, 2008). The disease is particularly troublesome when carrots are grown in greenhouses (Geary and Wall, 1965; Watson, 2016). The intensive development of the disease may strongly impede mechanical harvesting (du Toit *et al.*, 2019). The *Oidiopsis* type of powdery mildew identified in carrots is characteristic of a very warm and dry climate, and is generally limited to the Middle East, Armenia, Mediterranean regions of Europe, India, Kazakhstan, and Africa (Davis, 2004).

*E. heraclei* grows on the outer surface of the plant, showing an ectopic effect, and produces haustoria that penetrate the epidermal cells of the host. During asexual development, the fungus produces white mycelium and spores that cover all aboveground parts of the plant (Davis and Nunez, 2007). As a result, the leaves can become chlorotic and senesce prematurely, resulting in significant yield losses. Conidia produced by powdery mildew can be transmitted over long distances by air (Aegerter, 2002). The spores only require humidity and moderate temperatures to germinate and then infect plants. In addition, this disease is particularly severe on shaded parts of plants (light damages the conidia and mycelium). Hence, most often, powdery mildew affects older leaves, and from them spreads to younger leaves, and even to umbels. Plants also become more susceptible to this disease with age (du Toit *et al.*, 2019). Powdery mildew can also cause large losses in carrot seed crops, because a tall and relatively dense canopy causes favorable conditions (du Toit and Derie, 2008; du Toit *et al.*, 2009). While there are no reports of powdery mildew infecting seeds, they may be contaminated with sexual fruiting bodies (Boerema *et al.*, 1963; Vasudeva, 1963).

Carrot varieties resistant to this pathogen have been identified (Umiel *et al.*, 1975; Bonnet 1977). Additionally, a single *Eh* gene was selected, and its usefulness was demonstrated in backcross breeding with a susceptible variety (Bonnet, 1983). As a result of these research studies, a number of varieties that could be a source of resistance have been selected (Bonnet, 1983; Lebeda and Coufal, 1987).

### 6.5.3 Carrot nematode and insect pests

#### *Carrot fly (*Psila rosae*)*

The most significant pest of carrots and other Apiaceae plants is carrot fly (*Psila rosae*) (Collier and Finch, 2009). This pest is a problem mainly in the cooler regions of northern Europe and Canada (du Toit *et al.*, 2019). Females of carrot fly lay eggs in the area of the root crown and the larvae descend down the root to feed. This causes huge losses due to deterioration of the commercial quality of the roots rather than to the yield itself (Ellis, 1999). At the same time, these insects can transmit several carrot diseases that can result in additional losses (Groves *et al.*, 2020). Partial resistance has been identified in *D. carota* 'Sytan' (Ellis and Hardman, 1981). The conducted breeding programs made it possible to obtain $F_3$ and $F_4$ lines derived from crosses between the cultivars of carrot and *D. capillifolius*. Later breeding programs used wild-type objects to search for the source of resistance. This resulted in the release of the partially resistant cultivar 'Flyaway' and others that were much more resistant than 'Sytan' (Ellis, 1999; Simlat *et al.*, 2013). Based on the conducted research, it has been suggested that the preferences of carrot fly may be related to both olfactory and contact chemostimulators, although there was also a correlation between leaf color and morphology (Guerin and Stadler, 1984). Additionally, an increased level of *PAL1* and *PAL2* gene expression was found in the most resistant carrot lines (Simlat *et al.*, 2013).

### *Root-Knot Nematodes (*Meloidogyne *spp.)*

Many species of root-knot nematode (RKN) attack carrots. The most important RKNs with yield losses of up to 100% are nematodes of the genus *Meloidogyne*, including *Meloidogyne hapla*, *M. javanica*, and *M. incognita* (Roberts and Mullens, 2002). Carrots growing in temperate areas are mainly infected by *M. hapla*, while crops in warmer areas are infected by *M. javanica* and *M. incognita* (Bridge and Starr, 2007; Parsons *et al.*, 2015). The main techniques for minimizing nematode damage are the use of nematicides or crop rotation. Nevertheless, the development of genetic determinants allows the most effective defense. In field and greenhouse studies, Nantes Long Chantenay-type roots were found to be more tolerant of RKNs than Danvers type (Yarger and Baker, 1981). Additionally, it has been found that some varieties are more tolerant because their roots are less prone to deformation or, despite being parasitized, do not allow nematode propagation (Wang and Goldman, 1996). It is also likely that the response mechanism to RKNs from different climatic zones is different (Huang *et al.*, 1986). Screening tests were carried out in which 170 varieties of Korean origin were included. From these, more than 60 cultivars were selected that are a potential source of resistance. Additionally, in RKN-resistant cultivars, no giant cell development around infecting roots was found (Seo *et al.*, 2014).

It was possible to identify simple inherited dominant resistance (locus *Mj-1*) to *M. javanica* in the cultivar 'Brasilia', which also exhibited resistance to *M. incognita* (Simon *et al.*, 2000). The RAPD marker linked to the *Mj-1* locus was also developed, which significantly facilitated breeding. At the same time, researchers suggested that resistance is affected by locus dosage (Boiteux *et al.*, 2004). In subsequent years, five quantitative trait loci markers were developed, which correlated with all three species of carrot nematodes and co-localized with the *Mj-1* locus (Parsons *et al.*, 2015). Subsequent studies provided information on the source of resistance to both *M. javanica* and *M. incognita*. A single major gene, *Mj-2*, was identified and located on the same chromosome as the previous gene (Ali *et al.*, 2014).

Several commercial varieties with proven RKN resistance are available on the market. They were obtained from germplasm of *Daucus* spp. or from crosses with 'Brasilia', which is still considered the best source of comprehensive resistance to RKN (Simon *et al.*, 2000; Vieira *et al.*, 2003). In Brazil, there is also a BRS Planalto variety that exhibits high resistance to RKNs (Pinheiro *et al.*, 2011).

### *Aphids*

Several aphid species can feed on carrots, which causes mechanical damage to the plants and also contaminates them with toxic saliva (Rubatzky *et al.*, 1999). Saliva remains on the leaves as sticky and sugary waste (honeydew) and is an ideal medium for the growth of numerous fungi that can cover the entire leaf surface. Many species of aphids are also vectors of viral diseases. Numerous studies have been conducted to assess the sensitivity of different cultivars to aphids, but the results did not provide conclusive results (Lamb, 1953; Dunn, 1970). A hypothesis has also been proposed according to which aphid resistance is given as a dominant trait or as a polygenic trait (Smith and Chuang, 2014).

## 6.6 Major Abiotic Stresses

Carrot cultivation is also limited by numerous factors resulting from abiotic stresses. These are mainly heat, cold, salinity, and drought (Bolton *et al.*, 2020).

Carrot is a cool-season crop that is rather sensitive to heat, which particularly restricts plant growth at early stages but can also limit root development or negatively affect seed production. The increase of temperatures to 35°C during seed germination reduced the growth of seedlings by 50% (Nascimento *et al.*, 2008; Bolton *et al.*, 2019). From among the examined accessions, 'Brasilia' showed exceptional features under high heat stress. The germination efficiency at 37.5°C was reduced only by 10% compared with optimal conditions. Additionally, Broussard *et al.* (2017) studied the effect of temperature during flowering and found that the increase in temperature at this stage of development caused a decrease in the production of volatile terpenoids and in nectar quality, which directly translates into attractiveness of insect pollination. Several genes have also been identified that may be potentially associated with the thermotolerance mechanism, such as alternative oxidase (AOX) (Amirsadeghi *et al.*, 2007) and DcHsp17.7 and other heat-shock proteins (Malik *et al.*, 1999; Park *et al.*, 2013; Huang *et al.*, 2015).

Climate changes observed in recent years have caused the rise of mean temperatures above average, but in many regions temperatures are below average. Carrots are considered to be a plant capable of growing at temperatures below the optimum. Low temperatures can damage the leaves and cause taproot cracking (Palta and Simon, 2004). The selection of different carrot varieties in terms of the response of their leaves and roots to low temperature allowed the development of two frost-resistant hybrid varieties.

Soil salinity is an important abiotic stress affecting carrot yield. Carrots belong to the group of glycophytic plants that are very sensitive to saline soil (Bolton and Simon, 2019). The root yield drops by 14% with an increase in salinity above the threshold of only 1.0 dS $m^{-1}$ (Maas and Hoffman, 1977). The ability to germinate and the rate of germination are significantly limited in saline soil. Additionally, it has been proved that salt stress causes a decrease in photosynthesis and stomatal conductance in carrots (Schmidhalter and Oertli, 1991; Basma *et al.*, 2014). So far, not much research has been conducted on salt stress in carrots. It was found that the level of numerous antioxidant compounds may play a key role in the salt stress tolerance mechanism (Bano *et al.*, 2014). Presumably, the synthesis of phytoene (DcPSY2) may also be involved in the reaction of carrots to salinity (Fuentes *et al.*, 2012; Wang *et al.*, 2014). Subsequent studies have shown that both salt stress and abscisic acid (ABA) can regulate DcPSY2 (Simpson *et al.*, 2018). A local variety tolerating high soil salinity was also selected, and the decrease in the weight of roots and leaves was significantly lower than that of a second salt-sensitive variety (Smoleń *et al.*, 2020). Additionally, a correlation was demonstrated between the salt content in the leaves of the tolerant cultivar and expression of the *DcNHX4* gene, which may also be related to the salt stress tolerance mechanism. Various types of carrot were screened for their salinity tolerance. Basma *et al.* (2014) and Bolton and Simon (2019) tested 10 and 294 cultivars and wild carrots, respectively. These studies confirmed the wide variability of salinity tolerance during germination, which could potentially be a good source of salt tolerance introduction.

Salinity stress is often associated with drought stress, and the initial mechanisms in plants are particularly similar (Munns and Tester, 2008). Salinity stress causes an increase in oxidative stress in cells, which is manifested by a reduced water potential in the leaves and slower cell growth, which in turn adversely affects plant growth and numerous physiological and biochemical processes (Hussain *et al.*, 2018). One of the potential elements determining the adaptation of carrots to drought is a thaumatin-like protein encoded in carrots by the *DcTLP* gene. In carrots, overexpression of this gene was found as a result of dehydration in embryogenic calli, seedlings, and mature plants, and additionally, this factor was not regulated by jasmonic, salicylic or ABA (Jung *et al.*, 2005). There was also a significant accumulation of DcHsp17.7, a small heat-shock protein, increased presence of which was also noted during high temperature stress (Ahn and Song, 2012). Several other experiments have been carried out analyzing the content of sugars and other nutrients in carrot tissues during drought stress. Undoubtedly, research devoted to this topic will be valuable in the coming years.

## 6.7 Transformation Methods

Development of the transformation process in carrot started in 1979 when the liposome-encapsulated [$^3$H]RNA was incorporated into protoplasts (Matthews *et al.*, 1979). In 1985, the first stable protoplast transformation using electroporation was reported (Langridge *et al.*, 1985), along with the method of transformation using *Rhizobium rhizogenes* (formerly *Agrobacterium rhizogenes*) (Tepfer, 1985). Two years later, a successful experiment carried out with *Agrobacterium tumefaciens* (Scott and Draper, 1987) paved the way for easier, more versatile, and efficient transformation, and transformation with *A. tumefaciens* is still the most commonly used method in carrot. Recent years have brought new possibilities of genome modification by utilizing novel tools based on the clustered regularly interspaced short palindromic repeats (CRISPR) CRISPR-associated protein 9 (Cas9) system. This system has been used with success in numerous plant species, including carrot, and can be used not only for basic research but also to benefit agriculture.

### 6.7.1 Vector methods

#### *Agrobacterium-mediated transformation*

The first transformation process with *A. tumefaciens* in carrot was carried out over 30 years ago by Scott and Draper (1987). A suspension culture of highly pro-embryogenic cells of *D. carota* 'Early Nantes' was incubated with the non-oncogenic *A. tumefaciens* strain C58C1 harboring the plasmid pGV3850::1103, which contained a chimeric nopaline synthase (*nos*)– neomycin phosphotransferase gene (*nptII*, conferring kanamycin resistance). The authors obtained transformation frequencies of up to 73% assessed on kanamycin-supplemented medium and confirmed the presence of *nos* and *nptII* sequences by Southern blot analysis. They were also able to obtain plants of normal phenotype from transgenic callus via somatic embryogenesis. The high success rate with this pioneering work gave hope for an easy and efficient transformation process in carrot. Subsequent works of other authors focused on protocol optimization, and, although none can be considered universal, it allowed estimation of major factors influencing the transformation process. One of the main factors that could limit transformation efficiency is the susceptibility of carrot genotypes to transformation with *A. tumefaciens*. Depending on the genotype used, highly different transformation efficiencies have been reported, although due to various methods applied for scoring transformants, unequivocal comparison of the results presented in numerous works is not possible. One of the first studies reporting this matter in carrot was carried out by Thomas *et al.* (1989), who excised hypocotyls from four carrot cultivars, 'Chaptenary Red Cored', 'Danvers 126', 'Imperator 58,' and 'Scarlet Nantes', and transformed them with the *A. tumefaciens* LBA4404 strain containing the β-glucuronidase (GUS) gene (*uidA*) and the nopaline synthase polyadenylation sequence. Kanamycin-resistant calli were obtained with 5.8%, 3.8%, 2.1%, and 0.9% frequencies for 'Danvers 126', 'Imperator 58', 'Chaptenary Red Cored,' and 'Scarlet Nantes,' respectively. In other reports, the interaction of different factors, such as genotype, bacterial strain and plasmid was more pronounced, as well as the type and age of explants used for transformation.

Pawlicki and others (1992) carried out an experiment where they used petiole segments of five varieties/cultivars and transformed them with *A. tumefaciens* C58C1 strain harboring either the pGSTRN943 plasmid, which contained *nptII*, or the pGSGlucl plasmid with *nptII* and *uidA* sequences. Most prone to transformation was the cultivar 'Nanco', for which they obtained 46.7% and 35.8% transformation rates for pGSTRN943 and pGSGlucl, respectively. 'De Chantenay' and 2027H, exhibited a similar reaction, about 30% and 15%, respectively, while the 840217 line rates were much lower at 6.7% and 5.8%, respectively. 'Gold Pack' was recalcitrant to *A. tumefaciens* and no transformants were obtained.

In an experiment where epicotyls of 'Danver Half Long', 'Nanco,' and 'Golden State' were transformed with the *A. tumefaciens* EHA105 strain, transformation efficiency varied from

1.8% for 'Danver Half Long' up to 12.1% for 'Nanco,' but when 'Nanco' and 'Danver Half Long' were transformed with the MOG101 strain, there was a tenfold higher transformation frequency (5.1% compared with 0.5% for 'Nanco'; Gilbert *et al.*, 1996). 'Nantaise' transformed with *A. tumefaciens* GV3101 strain had a 20-fold higher transformation efficiency over the LBA4404 strain (Hardegger and Sturm, 1998). In other reports where the cultivars 'Nantes Scarlet' and 'Kurodagosan' were transformed with two different bacterial strains, transformation with C58C1 strain was twice as efficient as with LBA4404 strain in 'Nantes Scarlet,' while for 'Kurodagosun' the transformation efficiency was sixfold higher with LBA4404 (Takaichi and Oeda, 2000).

The bacterial genetic background also can have an impact on transformation efficiency depending on the type and age of explants. Pawlicki *et al.* (1992) observed that for petioles excised from 2-week-old seedlings, the transformation frequency was two times lower when the *A. tumefaciens* C58C1 strain contained the pGSGluc1 rather than the pGSTRN943 vector, while in older seedlings this difference did not occur. In callus transformation, twofold increases in transformation rate were observed when plasmid pGV3850 rather than pTiA6 was used (Wurtele and Bulka, 1989). Taproots and roots excised from seedlings enabled higher transformation efficiency when transformed with the GV3101 strain in comparison with LBA4404 (95% compared with 5% for taproot and 41% compared with 2% for seedling root). When cell cultures were used for transformation, the success rate did not exceed 0.1% (Hardegger and Sturm, 1998). In addition, a tenfold lower GUS activity was observed when carrot cells were transformed with LBA4404 containing pRGUSII in comparison with BI121.1, but when the same experiment was simultaneously carried out using tobacco, this difference was not observed which suggested genotype dependent interaction (Thomas *et al.* 1989). The most common starting material is hypocotyls as they are easy to maintain in *in vitro* conditions. Some reports suggest that germination conditions can influence further transformation, as better results were obtained when seedlings were grown in the dark than in the light (Pawlicki *et al.*, 1992; Hardegger and Sturm, 1998). Regardless of the plant material, the transformation process is carried out in a similar manner. First, explants are immersed in or covered with inoculum (Wurtele and Bulka, 1989; Pawlicki *et al.*, 1992; Hardegger and Sturm, 1998; Yau *et al.*, 2008; Anon *et al.*, 2014). *Agrobacterium* is able to transfer T-DNA into cells by wounding the plant tissues. Wounded plants produce phenolic compounds such as acetosyringone, which plays a role in bacterium–host recognition (Hooykaas and Schilperoort, 1992). Tissue disruption or addition of exogenous acetosyringone during the transformation process has been proposed to stimulate *Agrobacterium* virulence, although the results obtained in carrot have been not conclusive: researchers who evaluated the influence of such factors reported little or no effect (Wurtele and Bulka, 1989; Pawlicki *et al.*, 1992; Hardegger and Sturm, 1998). It is estimated that the use of additional acetosyringone is advantageous only in plants deficient in endogenous phenolic compounds or if the co-cultivation period is too short (van Wordragen and Dons, 1992). The optimal co-cultivation period is 2–3 days; an extended period has rarely been used (Marquet-Blouin *et al.*, 2003) and 7-day co-cultivation caused difficulties with elimination of bacteria (Pawlicki *et al.*, 1992). After co-cultivation, the explants are placed on medium supplemented with antibiotics to inhibit the growth of bacteria. Selection of transformants can be made at the same time as antibiotic treatment (Kim *et al.*, 2009; Noh *et al.*, 2012), or can be delayed for about 2 weeks (Hardegger and Sturm, 1998; Brodzik *et al.*, 2009; Arango *et al.*, 2010). Several subsequent subcultures allow the development and selection of transgenic callus and further formation of somatic embryos.

### *Rhizobium*-mediated transformation

*Rhizobium*-mediated transformation relies on the use of a soil-borne bacterial vector, *R. rhizogenes*, which causes the hairy root phenotype. Hairy roots are characterized by loss of gravitropism, a branched phenotype, and intensive growth even on hormone-free medium (Sarkar *et al.*, 2018). In carrot, hairy roots do not show the characteristic "hairy" phenotype; hence, they are described as "adventitious root" (Baranski, 2008). Transformation with *Rhizobium* is less common than with *Agrobacterium*, although due to growing demand for production of plant-derived food additives, pharmaceuticals, and other bioactive compounds, the *in vitro* culture of hairy roots can become advantageous and more frequently used for large-scale production. Susceptibility to transformation by *R. rhizogenes* in carrot, similar to *A. tumefaciens*, is genotype specific. Baranski *et al.* (2006) transformed 12 carrot genotypes with four *Rhizobium* strains and observed transformation efficiencies ranging between 25% and 93% for the cultivars 'Danvers', 'Dolanka', and 'Karotan', with a commercially available mix being most prone to transformation.

As mentioned earlier, carrot contains high amounts of endogenous phenolics and the addition of acetosyringone has little or no effect on the transformation process with *A. tumefaciens*. In contrast, the addition of acetosyringone during transformation with *R. rhizogenes* had a positive impact on the number of adventitious roots, especially on explants excised from less susceptible genotypes (Baranski *et al.*, 2006). Carrot varieties of the western type seem to be more susceptible to the transformation process, but recently successful transformation of black carrot (eastern type) with *R. rhizogenes* has been reported (Barba-Espín *et al.*, 2020). The plant material × bacterial strain interaction is less pronounced when using *R. rhizogenes* than *A. tumefaciens* vectors (Baranski *et al.*, 2006).

The transformation protocol first requires preparation of taproots, which are surface sterilized and placed on moist filter paper (Epstein *et al.*, 1991), agar (Cardarelli *et al.*, 1985; Fründt *et al.*, 1998) or the surface of full-, half-, or quarter-strength Murashige and Skoog (MS) medium (Cardarelli *et al.*, 1987; Barba-Espín *et al.*, 2020). Root disks are placed preferably with the apical side facing up. Reports indicate that transformation of the apical side is more effective than the basal side, although it is possible to transform taproots on the basal side when an appropriate bacterial basal-positive strain is used or when the disks are exposed to auxins in either the medium or inoculum (Ryder *et al.*, 1985; Bercetche *et al.*, 1987). Strains A4, TR105, 1855, and 15834 are basal positive and are characterized by higher activity of genes involved in auxin biosynthesis, hence their ability to transform the basal side of the taproot (Baranski, 2008). Root disks are covered with inoculum especially over the cambium area with high potential for neoplasia (Boulanger *et al.*, 1986; Bercetche *et al.*, 1987; Baranski, 2008). Transformation with *R. rhizogenes* does not require immediate elimination of the bacteria as the hairy root phenotype is observed about 10–14 days after inoculation on the root disk surface, having no contact with the medium. Furthermore, roots can be excised from the explant and continuously cultivated in solid or liquid mineral medium without growth regulators, supplemented with antibiotics. Higher hairy root biomass was obtained when half- or full-strength MS medium was used for cultivation rather than quarter-strength MS medium (Barba-Espín *et al.*, 2020).

### 6.7.2 Non-vector transformation

#### *Microprojectile bombardment*

The development of the microprojectile bombardment method by Sanford *et al.* (1987) allowed utilization of another transformation approach, especially for monocot plants for which vector transformation is problematic, as monocots are not host plants for *Agrobacterium*. In this method, cells are bombarded at high velocity with particles made from tungsten or gold covered with DNA. As the carrot is a dicot plant susceptible to transfection by *Agrobacterium*, microprojectile bombardment has not been used extensively. Similar to the vector transformation methods for carrot, it is hard to determine the main factors influencing transformation efficiency; more important is an interaction of several factors. In this method, the effect of genotype on the transformation process is less important, although in the experiments conducted by Deroles *et al.* (2002), the transformation efficiency of two carrot lines was different but not dependent on other factors such as type of microcarrier or gas pressure at which particles were shot. The most frequently used target material for microprojectile bombardment is callus derived either from cell suspension (Deroles *et al.*, 2002), or from petioles or stems (Kumar *et al.*, 2004; Rojas-Anaya *et al.*, 2009), but direct bombardment of root disk has also been used with success (Hibberd *et al.*, 1998). A pre-culture period of about 6 days may have a positive impact on transformation (Deroles *et al.*, 2002).

In microprojectile bombardment, there seem to be two main factors influencing transformation, the type and size of particles. Generally, for carrot, but also for other plant species, particles made of gold 0.6–1.0 μm in diameter resulted in higher efficiency than when tungsten carriers were used (Deroles *et al.*, 2002; Kikkert *et al.*, 2005). The volume of particles per shot (up to 5 μl) also had a positive impact on transformation efficiency (Deroles *et al.*, 2002), while results obtained for different DNA concentrations were variable: Deroles *et al.* (2002) did not observe any influence of higher DNA concentration on transformation rate, whereas Fromm *et al.* (1985) noticed a proportional increase in transformation efficiency, measured as chloramphenicol acetyltransferase (CAT) activity related to DNA concentration. Helium pressure differed significantly among studies, ranging from 500 up to 8963 kPa (Fromm *et al.*, 1985; Deroles *et al.*, 2002; Kumar *et al.*, 2004; Rojas-Anaya *et al.*, 2009). All of the parameters used were sufficient for DNA transfer into protoplasts and transformants were obtained. Deroles *et al.* (2002) observed the highest expression at 600 kPa (the maximum pressure used), Rojas-Anaya *et al.* (2009) used 870 kPa, while Kumar *et al.* (2004) obtained the best results when applying 7500 kPa. The shooting distance used varied from 9 to 16 cm but an optimum of 12–13 cm was consistently found in all studies.

### *Nucleic acid uptake by protoplasts*

Due to the absence of cell walls, protoplasts are amenable to gene manipulation. Direct (non-vector) delivery of nucleic acid into protoplasts can be done by electroporation or by treatment with chemical agents such as polyethylene glycol (PEG). Both approaches lead to a reversible destabilization in the cell outer membrane allowing entry of DNA into the cell (Bower and Birch, 1993). Protocols developed for both methods are relatively similar, but a few main steps can be distinguished. The most common starting materials for protoplast isolation are cell suspensions and petioles for the PEG method (Dröge *et al.*, 1992; Rasmussen and Rasmussen, 1993; Gallie, 1993; Dirks *et al.*, 1996; Aviv *et al.*, 2002), and cell suspensions for electroporation (Langridge *et al.*, 1985; Boston *et al.*, 1987; Bates *et al.*, 1988, 1990). The optimal protoplast density was found to be $10^6$ $ml^{-1}$; increasing or decreasing this value resulted in a lower transformation efficiency (Rasmussen and Rasmussen, 1993). Some sources for other plant species suggest that in PEG, molecular mass is an important parameter that could affect the transformation process, as it is assumed that highly polymerized PEG creates a larger number of pores of larger area in the cell membrane, although in carrot transformation the effect of PEG molecular mass was not observed even when the PEG mass was doubled from 4000 to 8000 (Ballas *et al.*, 1987). The PEG concentration highly influences the efficiency of DNA uptake. Rasmussen and Rasmussen (1993) obtained the highest transformation efficiency (measured as GUS activity) when a 25% concentration of PEG was used; however, Baranski *et al.* (2007b) reported that a similar 24% PEG concentration led to degradation of transformed cells only few days after transformation and suggested lowering the concentration to 20%. Another important component of the reaction mixture is $Ca^{2+}$ and $Mg^{2+}$ divalent cations, as their removal from the solution resulted in a decrease of transformation efficiency by approximately 90% and 20%, respectively (Gallie, 1993). When mRNA was used for protoplast transformation, the efficiency correlated with the concentration of mRNA up to 5 μg; for DNA, the concentrations used ranged between 20 and 100 μg in different studies (Rasmussen and Rasmussen, 1993; Dirks *et al.*, 1996; Aviv *et al.*, 2002), although the optimal concentration has not been assessed.

In the case of electroporation, similar to PEG treatment, the equilibrium between transformation efficiency and protoplast viability must be established. The composition of the medium also plays an important role in the transformation process. It was postulated that in low-salt or salt-free medium or medium lacking $Ca^{2+}$ ions, the transformation process could be ineffective when electroporation was done at low voltage, and the protoplasts could lose their viability (Bates *et al.*, 1988). There are two approaches concerning the time of the pulses and voltage: the general rule is that when increasing the time of pulses up to 50 ms, the voltage should be lowered and adjusted to 200–800 V $cm^{-1}$. When the pulses are shorter with a maximum of 200 μs, the voltage should be increased up to 10 kV $cm^{-1}$. Bates *et al.* (1988) observed that increased voltage favored DNA uptake, assessed by measuring higher CAT expression. However, higher voltage reduced the viability of protoplasts, although the voltage and time of pulses could be reduced when they were applied in short series (Langridge *et al.*, 1985). The process of electroporation could be enhanced by heat shock (Langridge *et al.*, 1985), but another report suggested that best results were obtained at room temperature (Bates *et al.*, 1988). The amount of DNA used for transformation varied from 10 to 40 μg, although Bates *et al.* (1988) observed that doubling the DNA concentration resulted in CAT expression that was twice as high. The efficiency of both electroporation and PEG-mediated transformation increased after the addition of carrier DNA such as salmon sperm (Bates *et al.*, 1988), and some suggest calf thymus DNA, although the results are inconsistent. The efficiency of transformation was also increased when both electroporation and PEG treatments were applied simultaneously (Boston *et al.*, 1987).

#### *Genome editing*

New technologies allow directed changes in the genome. The CRISPR/Cas9 genome editing system relies on two components, a Cas9 protein and guide RNA (gRNA), which directs the Cas9 nuclease to the target DNA where a double-strand break is generated. The break is then repaired by the cell either through homology-directed repair or by non-homologous end joining, which often leads to site mutations. This new tool can be used not only for basic research but also for the development of plants with new or improved agronomic traits. In carrot, the CRISPR/Cas system relies on transformation using vector methods, mostly *A. tumefaciens* (Klimek-Chodacka *et al.*, 2018; Xu *et al.*, 2019a, 2019b), but also *R. rhizogenes* (Dunemann *et al.*, 2019). As CRISPR/Cas9 is a relatively new method, protocols differ among experiments conducted by different scientific groups. The starting material used has included purple callus (Klimek-Chodacka *et al.*, 2018), hypocotyls from purple carrot 'Deep Purple', hypocotyls from orange carrot 'Kurodagosun' (Xu *et al.*, 2019a), and KRD and SHLC carrot lines (Xu *et al.*, 2019b), or taproot for transformation with *R. rhizogenes* from the different colored carrot cultivars 'Blanche', 'Deep Purple', 'Rotin' or 'Yellowstone' (Dunemann *et al.*, 2019). The plant material is immersed in an inoculum of *A. tumefaciens* with the addition of filter paper and co-cultivated for 2–3 days in the dark. Elimination of *Agrobacterium* is done by rinsing the explants with antibiotics and transferring them to solid medium, also supplemented with antibiotics. After 4 weeks, the callus is transferred to medium with a selection agent. As the target genes were responsible for pigmentation, the successful genome editing was directly confirmed by visual detection of mutants showing discoloration, although additional sequencing of polymerase chain reaction (PCR) products was also conducted (Klimek-Chodacka *et al.*, 2018; Xu *et al.*, 2019b).

## 6.8 Basic Research Transformation Objectives

### 6.8.1 Promoter studies

The GUS-based reporter system for detection and evaluation of transformation events is a widely used approach. Unfortunately, it is destructive for generated genetically modified plants because prolonged incubation with the X-Gluc substrate at a relatively high temperature is required.

Cauliflower mosaic virus (CaMV) 35S, maize zein gene, nopaline synthase (*nos*), *A. tumefaciens*-derived and mannopine synthase (*mas*) gene promoters have been successfully introduced into carrot cells and used to activate gene expression. The 35S promoter activity was three times lower than that of the *nos* promoter, as determined by GUS expression (Boston *et al.*, 1987; Rathus *et al.*, 1993). However, fusion of the 35S promoter with the *Agrobacterium* octopine synthase enhancer sequence boosted GUS expression (Rathus *et al.*, 1993). Other 35S promoter modifications, such as GC-rich G-box insertion (Ishige *et al.*, 1999) and CaMV 35S promoter doubling (Wally *et al.*, 2008), improved the 35S promoter activity. The GUS gene (*uidA*) controlled by SRD1 root-specific promoter (derived from sweet potato) was ten times more active than when controlled by the 35S promoter (Noh *et al.*, 2012). Two other root-specific promoters, the discorin 3 protein subunit (pDJ3S) from *Dioscorea japonica* and the cassava pMe1 promoter were tested in carrot root tissue. In comparison with the CaMV 35S promoter, the activity of the pMe1 promoter was at a similar level, while pDJ3S was two times higher (Arango *et al.*, 2010).

The GUS reporter system has been used in promoter activity research. Guan *et al.* (2009) conducted studies on the role of basic leucine zipper domain (bZIP)-type transcription factors during carrot somatic embryogenesis. Transformed carrot cells contained plasmid with a reporter system based on the cDNA fragment of the *CAREB1* or *CAREB2* and the *uidA* gene. Carrot cell lines with overexpression of *CAREB1* or *CAREB2* were also generated. The investigation revealed a network between ABA signaling, sucrose concentration, and expression of the *Dc3* gene encoding LEA (late embryogenesis abundant)-type protein. The increase in ABA concentration was observed during carrot embryo development and was induced by a high sucrose content in the medium. In response to medium with high sucrose content or ABA treatment, upregulation of *CAREB1* gene was seen. In a complementary experiment with application of ABA inhibitor (fluridone), the opposite reaction was seen in carrot embryos and *CAREB1* gene was downregulated. Expression of *Dc3* gene also decreased. Overexpression of *CAREB1* in genetically modified carrot lines increased *Dc3* gene expression and resulted in abortion of somatic embryogenesis.

### 6.8.2 Cell organelle labeling

Carrot male sterility is determined by the cytoplasm and mitochondrial genome characteristics (Szklarczyk *et al.*, 2014). Repetitive backcrossing is the most common approach to generating male sterile lines, but it has additional difficulties in carrot, as there is a problem with severe inbreeding depression. Furthermore, the allogamous and biennial nature of carrot prolongs the backcrossing process (Simon, 2000). An alternative is a somatic hybridization, an essential tool in genetic engineering and modern plant breeding. Detection of heterofusants is a crucial step of the experiment.

The green fluorescent protein (*gfp*) gene is a frequently used reporter for monitoring transient gene expression. The first reports about GFP expression in carrot were in 1998. The microprojectile bombardment technique led to *gfp* delivery and GFP expression in carrot chromoplasts, but no detailed results were presented (Hibberd *et al.*, 1998). A method enabling easy monitoring of transformation success that was non-destructive for plant tissue was proposed by Baranski *et al.* (2006). *R. rhizogenes*-mediated transformation of carrot root disks resulted in visible green fluorescence in hairy roots when they were illuminated by a hand-held ultraviolet lamp. The efficiency of transformation was estimated at 20–25%, but only 10% of hairy roots exhibited intense green fluorescence, allowing easy selection of carrot transformants without the need for antibiotics or herbicide usage. The mGFP5-ER protein has also been used as a visual marker of the carrot transformation process. Stable fluorescence occurred at various developmental stages, starting from the protoplast a few hours after transformation then in different carrot plant parts such as petioles, stem, and flowers (Baranski *et al.*, 2007b).

The approach proposed by Gieniec *et al.* (2020) provided a comprehensive assessment of the carrot protoplast response to an applied electric field as well as identification of the direct current parameters suitable for hybrid formation. Additionally, real-time observations of protoplast fluorescence allowed estimation of electrofusion performance. The non-toxic fluorescent protein (FP) tagging of carrot protoplasts was described. One of four FPs, cyan (eCFP), green (sGFP), yellow (eYFP), or the mCherry variant of red FP (RFP), with a fused mitochondrial targeting sequence, was introduced to carrot cell lines using *Agrobacterium*-mediated transformation. The carrot callus lines with GFP-, YFP-, or RFP-labeled mitochondria showed stable fluorescence and served as protoplast sources. The authors showed that fluorescent markers can be useful for immediate discrimination of parental components during the protoplast fusion stimulated by the application of electric current (Gieniec *et al.*, 2020).

### 6.8.3 Tolerance to herbicides

The introduction of tolerance to herbicides in plants not only enables screening and detection of transgenic events and estimation of transformation success rate but is also an important way to improve crop protection. Despite much controversy and serious concern for probable environmental and social difficulties of this type of improvement in crop production, studies aimed at the development of herbicide-tolerant plants, including carrot, have been conducted.

Dröge *et al.* (1992) proposed the introduction of phosphinothricin *N*-acetyltransferase (*pat*) gene under control of the 35S promoter to carrot protoplasts via PEG-mediated transformation. The *pat* gene was isolated from *Streptomyces viridochromogenes* and fragments allowing expression in plants were added. In the transgenic carrot plants phosphinothricin (Pt), an active compound of many herbicides, underwent acetylation, in contrast to the control plants where it was differently metabolized and led to inhibition of glutamine synthase, the relevant enzyme of plant nitrogen metabolism. The Pt-resistant plants were selected using 20 mg $l^{-1}$ of Pt in the medium (Dröge *et al.*, 1992). Another approach that was used to obtain Pt-resistant carrot was introduction of the *Streptomyces hygroscopicus bar* gene via *Agrobacterium*-mediated petiole transformation. The generated genetically modified carrot callus was selected using 10 mg $l^{-1}$ of Pt and used for plant regeneration. The transgenic carrot lines showed various sensitivities to the herbicide, which were associated with the transgene copy number (Jayaraj and Punja, 2007).

### 6.8.4 Drought tolerance

Several reports show a protective role of osmotically active compounds in the plant response to biotic stress. Carrot transformed with the tobacco osmotin gene under the control of the CaMV 35S promoter, using *Agrobacterium*-mediated transformation, showed better viability during and after drought stress treatment. Slower rates of leaf wilting are associated with drought tolerance. The transgenic carrot lines were able to retain more water in their tissues and exhibited faster and full recovery after irrigation restoration. Drought-stressed transgenic carrot lines also showed a significantly lower level of malondialdehyde accumulation compared with the wild-type carrot. Malondialdehyde is an important marker of oxidative stress and cell membrane stability. The authors suggested involvement of the osmotin gene in the drought response by mitigating cell membrane damage by regulating the cellular redox state. Accumulation of amino acids and proteins is a strategy to increase plant growth during stresses including water deficiency (Annon *et al.*, 2014).

The drought-inducible carrot *TLP* gene (thaumatin-like protein; *dcTLP*) expression pattern was examined under drought stress conditions (Jung *et al.*, 2005). *Agrobacterium*-mediated transformation using the binary vector pBI121 containing a *dcTLP* promoter::*uidA* chimeric gene and *nptII* gene as a plant-selectable marker, under control of the *nos* promoter was used for carrot callus transformation. During the study, the *dcTLP* gene level increased in response to drought stress in embryogenic calli, seedlings, and mature carrot plants. The GUS histochemical staining of genetically modified callus revealed that the *dcTLP* promoter actively responded to drought stress and the activity increased gradually over time. Moreover, the study demonstrated that *dcTLP* promoter activity was not ABA dependent.

### 6.8.5 Salt tolerance

Soil salinity constitutes a real problem for contemporary agriculture and for providing global food security. According to recent estimates, approximately 1 billion ha of land is salt affected (FAO, 2015), and this area is increasing by 2 million ha year$^{-1}$ (Lhissou *et al.*, 2014). It is mainly the use of poor-quality water for crop irrigation that causes arable land salinization (Daliakopoulos *et al.*, 2016). Global climate changes and increasingly frequent drought events elevate the problem (Corwin, 2020). Carrot is considered a highly salt-sensitive plant (Bernstein and Ayers, 1953; Maas and Hoffman, 1977; Bolton and Simon, 2019). Soil salinity strongly reduces the growth and development of carrot and contributes significantly to the decrease in carrot yield and root quality (Maas, 1986).

Investigations concerning improvement, via genetic transformation, of the carrot growth rate during salt stress have been conducted. Genes identified as salinity related in some plants were used. Plasmid pGA748 containing coding sequence of *Vigna aconitifolia* Δ1-pyrroline-5-carboxylate synthetase (P5CS) under the control of the CaMV 35S promoter and *nos* terminator was used for *A. tumefaciens*-mediated transformation of carrot hypocotyls (Han and Hwang, 2003). Genetically modified carrot callus with the *P5CS* gene accumulated a five times higher proline content and was characterized by a higher growth rate during salt treatment than non-transgenic callus lines. In addition, transgenic carrot plants derived from genetically modified calli had a higher proline content than non-transgenic regenerants, but this was unrelated to the initial proline content in the transgenic callus lines. Interestingly, some transgenic carrot plants showed growth inhibition under normal conditions in comparison with the control plants (Han and Hwang, 2003). Proline plays a wide range of roles during different stresses including salinity. It is one of the most common osmolytes accumulating in plant cells, preventing maintenance of water homeostasis under osmotic stress. In addition, proline has been reported to play a role in the non-enzymatic antioxidant system, helping plants scavenge reactive oxygen species (Signorelli, 2016). Accumulation of proline is often associated with salt-tolerant plant genotypes (Ami *et al.*, 2020).

Manipulations of the glycine betaine biosynthesis pathway using genetic engineering are a popular way to enhance salt tolerance (Flowers, 2004). The betaine aldehyde dehydrogenase (*badh*) gene is involved in the biosynthesis of glycine betaine. Glycine betaine plays a role in providing osmotic balance, protecting enzymes, and maintaining stability of membrane proteins (Rezaei *et al.*, 2020). A carrot cell suspension induced from stem segments, placed on Whatman paper filter, was bombarded with the carrot chloroplast transformation vector pDD-Dc-aadA/*badh*, as described by Daniell *et al.* (2002), with a 13.3% transformation efficiency. Transgenic carrot cultures exposed to 100 mM NaCl accumulated up to a three times higher amount of betaine (93.1 mmol $g^{-1}$ dry weight) than non-NaCl-treated culture (26.5 mmol $g^{-1}$ dry weight), and up to 50-fold more betaine than the control, non-transformed carrot cells. The transformed cell cultures grown in the presence of 100 mM NaCl demonstrated a visible distinct cell biomass growth rate, producing 8.75 g of cells (genetically modified carrot line) in comparison with the control (1.29 g). The *badh*-overexpressing carrot plants grew in the presence of high (up to 400 mM) concentrations of NaCl in water used for irrigation. In contrast, the non-transgenic carrot plants exhibited comparable limitation of growth and vitality in the presence of 100 mM NaCl (Kumar *et al.*, 2004).

### 6.8.6 Anthocyanin pathway

Anthocyanins are ubiquitous secondary metabolites occurring in different organs, cells, and subcellular organelles of plants. In many plant species, anthocyanins are responsible for the pigmentation of red, purple, and blue flowers, and of seeds and fruits, and also leaves and roots. Anthocyanins play a role in the plant response to many biotic and abiotic stressors. A wide range of anthocyanin involvement during stress reactions has been proposed (Landi *et al.*, 2015).

In 2013, genes associated with anthocyanin biosynthesis in carrot root were quantified by Yildiz *et al.* (2013). They provided gene expression and mapping data to establish a framework for further studying the anthocyanin pathway in carrot. Publication of the carrot genome in 2016 contributed to further studies (Iorizzo *et al.*, 2016). The first efficient CRISPR/Cas9-based genome editing in carrot cells was another breakthrough in the field of carrot anthocyanin studies (Klimek-Chodacka *et al.*, 2018).

A carrot callus line characterized by an intensive purple color due to anthocyanin accumulation was used for *Agrobacterium*-mediated transformation. The vector T-DNA contained the Cas9 gene driven by a double CaMV 35S promoter followed by two single gRNAs, each controlled by the *AtU3* promoter. It contained also the hygromycin resistance gene under the control of the CaMV 35S promoter. The gRNAs were designed to target the second exon of the flavanone 3-hydroxylase (*F3H*) gene at sites adjacent to the CGG-protospacer adjacent motif (PAM) and GGG-PAM. The gRNA:Cas9 complexes induced loss-of-function mutations in the target carrot *F3H* gene. The mutants had visible suppression of anthocyanin accumulation, which was confirmed spectrophotometrically. Transgenic knockout lines provided evidence that the *F3H* gene is critical for anthocyanin biosynthesis in carrot.

The genes encoding enzymes involved in anthocyanin biosynthesis are controlled by transcription factors (TFs) (Kodama *et al.*, 2018) or long non-coding RNAs (Chialva *et al.*, 2021). One TF gene, *DcMYB113-like*, was successfully edited in the purple carrot cultivar 'Deep Purple' using the CRISPR/Cas9 system, which led to depigmentation of the purple carrot plants. Two target sites located in the first exon and one in the second and third exon of the *DcMYB113-like* gene were selected. *A. tumefaciens*-mediated transformation of hypocotyls resulted in generation of mutations in three of the four target sites. Transgenic callus tissue and callus-derived plants were obtained. The efficiency of mutagenesis was different between the target sites. AtU6-29-driven gRNA induced double-stranded breaks at the target site more frequently than the other gRNAs. The heterozygous mutations occurred at two target sites (targeted by AtU6-1- and AtU3d-driven gRNA). Mutagenesis with AtU3b-driven gRNA was inefficient (Xu *et al.*, 2019a).

Xu *et al.* (2020) identified the R2R3-MYB gene (*DcMYB113*) and confirmed that it is responsible for root-specific anthocyanin pigmentation. They used four carrot cultivars: 'Purple Haze' (purple root periderm and phloem tissues but orange root xylem tissue, non-purple petiole), 'Deep Purple' (purple root and petiole), 'Cosmic Purple' (purple root periderm and petiole but orange root phloem and xylem), and 'Kurodagosun' (orange root and non-purple petiole). The *DcMYB113* gene showed high expression levels in the 'Purple Haze' root but was undetectable in the roots of the remaining cultivars. It is likely that different *DcMYB113* gene expression patterns were related to variations in the gene promoter region. The construct used for transformation contained the *DcMYB113* coding sequence from 'Purple Haze' introduced into the pCAMBIA1301 vector under the control of the CaMV 35S promoter and was used to obtain transgenic 'Kurodagosun' and 'Cosmic Purple' lines. The roots of transgenic 'Kurodagosun' plants and callus of 'Cosmic Purple' showed anthocyanin presence, unlike their non-transformed counterparts. Additionally, the authors showed that the *DcMYB113* gene plays a role in anthocyanin glycosylation and acylation by upregulating two genes encoding a glycosyltransferase and an acyltransferase (Xu *et al.*, 2020).

Two other TFs, AmRosea1 (R2R3-MYB) and AmDelila (bHLH), characterized in snapdragon (*Antirrhinum majus*) (Schwinn *et al.*, 2006) can activate the anthocyanin biosynthesis pathway in orange carrot plants, leading to the biosynthesis and accumulation of anthocyanins in the taproots. Sharma *et al.* (2020) revealed the activating role of these TFs using *Agrobacterium*-transformed orange carrot plants. They used three different constructs, one with the *AmRosea1* gene, the second with the *AmDelila* gene, and the third with both genes. Only simultaneous expression of the AmRosea1 and AmDelila TFs in transformed plants resulted in significant upregulation of the major genes involved in anthocyanin biosynthesis. Expression of the *CHS1*, *CHI1*, *F3H1*, *F3'H1*, *DFR1*, *LDOX1*, and *UCGalT1* genes was 7–14 times higher in transformed carrot compared with the non-transformed plants. Genes associated with the phenylpropanoid pathway, *PAL3*, *C4H1*, and *4CL1*, were also significantly upregulated by approximately four- to fivefold. Therefore, anthocyanin accumulation in transformed plants was visible, and the total anthocyanin content of the transformed orange carrot was comparable to that of the reference 'Deep Purple'. However, significant differences were found in anthocyanin composition as determined by high performance liquid chromatography (Sharma *et al.*, 2020).

Hairy root cultures provide a useful method for large-scale production of plant-derived bioactive compounds. High concentrations of anthocyanins are obtained in black carrot hairy roots *in vitro*. Two types of explant (taproot disks and hypocotyl segments) from four different black carrots, two cultivars ('Deep Purple' $F_1$ and 'Night Bird' $F_1$) and two inbred lines, were used for hairy root induction using the *R. rhizogenes* A4 strain harboring the agropine-type pRiA4 plasmid. The authors showed that transformation efficiencies were higher for taproots than for hypocotyl explants. Moreover, anthocyanin production in hairy roots was stimulated by light and by the ethylene-releasing chemical ethephon (at 200 mg $l^{-1}$ concentration) (Barba-Espín *et al.*, 2020).

### 6.8.7 Carotenoid pathway

Carotenoids are ubiquitous tetraterpene pigments that exhibit a yellow, orange, or red color.

In plants, they act as photoprotectors, antioxidants, color attractants, and hormone precursors. In photosynthetic organs, carotenoids also harvest light energy and transfer it to chlorophylls. They also absorb excess energy from chlorophylls, which protect plants from light-induced damage. Moreover, free radicals generated during photosynthesis are destroyed due to the double bonds in the carotenoid molecule structure (Sathasivam *et al.*, 2020). Orange carrot provides a welcome source of β-carotene, acting as provitamin A in the human diet.

The phytoene desaturase (*PDS*) gene was identified as being associated with total carotenoid, phytoene, phytofluene, α-carotene, and lutein content in non-photosynthetic plant organs (Jourdan *et al.*, 2015). Hypocotyl fragments of orange carrot 'Kurodagosun' were transformed using *A. tumefaciens* with 53% efficiency. The results indicated a different type of mutation in *PDS* gene, generated in two of the four designed target sites. The callus tissue developing on genetically modified explants was white or a mosaic of white and yellow. Moreover, part of callus showed embryogenic properties, and mostly albino transgenic carrot plants were generated (Xu *et al.*, 2019a).

The *A. tumefaciens* GV3101 strain containing binary constructs with either of the ζ-carotene desaturase (*ZDS1* or *ZDS2*) gene fragments in an antisense orientation was used to transform carrot hypocotyls. The transgenic carrot plants with a 98% reduction in *ZDS2* transcript level developed normally and accumulated similar carotenoid levels to the wild-type plants. Nevertheless, suppression by RNA interference of *ZDS1* expression altered the initial plant development and triggered a significant reduction in the expression of other carotenoid-pathway genes, such as *ZDS2* and *PSY2*, resulting in a 50–60% reduction in carotenoid content (Flores-Ortiz *et al.*, 2020). Other genes of the carotenoid biosynthesis pathway were also targeted using CRISPR-Cas9 technology. The system was used to knock out the gene encoding lycopene ε-cyclase and geranylgeranyl diphosphate reductase, which redirects carotenoid precursors to the phytyl chain pathway, and different gene sequence rearrangements were identified, presumably affecting carotenoid metabolism (Klimek-Chodacka *et al.*, 2019).

### 6.8.8 Taste modifications

Miraculin is a homodimeric glycoprotein found in and isolated from *Richardella dulcifica* berries. The unique taste-modifying activity of this protein, which converts sour into a sweet taste, makes miraculin a potential alternative sweetener. Through genetic-engineering techniques, plants such as tomato, lettuce, strawberry, and carrot were modified to produce miraculin with different levels of effectiveness (Ezura and Hiwasa-Tanase, 2018). A binary vector containing the miraculin gene under the control of the *SWPA2* promoter and the kanamycin resistance gene was introduced into carrot callus via *Agrobacterium*-mediated (strain LBA4404) transformation. Four carrot transgenic lines were selected based on miraculin expression and other cell analyses. The average yield of miraculin protein varied between 0.1 and 5.9 ng $\mu l^{-1}$ of cell lysate. The correlation between the cell division cycle and miraculin expression was also investigated. The authors indicated differences in miraculin expression according to the cell-cycle phase. A negative correlation with miraculin expression and the $G_1$ phase was shown, whereas the S and $G_2$/M phases showed a positive correlation. These results suggested that the transgenic carrot callus lines obtained could be a potential source of miraculin protein. The developed *in vitro* system has been postulated as a safe and effective way to produce miraculin for the food industry (Park *et al.*, 2021).

### 6.8.9 Resistance to diseases

Several studies have been started to introduce disease resistance to carrot. They concern both leaf and root pathogens.

Pathogenesis-related (PR) proteins are a heterogeneous group of proteins involved in the plant response during pathogen infection. Chitinase and glucanase genes derived from plants and microorganisms are often considered when engineering plants for enhanced resistance. In general, disease-resistant plants are characterized by a faster accumulation of glucanases than sensitive plants. β-1,3-Glucanases are members of the PR proteins and participate in the defense response in plants (Kim *et al.*, 2020).

The β-1,3-glucanase genes derived from tobacco (*Glu-1*) and wheat (*Glu-638*) were introduced via genetic transformation to carrot, independently or combined with chitinase genes. Resistance to *Botrytis cinerea* and *Sclerotinia sclerotiorum* did not increase in the transgenic carrot lines expressing the *Glu-638* gene. Expression of the wheat chitinase gene *Ch383* partly limited infestation by both pathogens. The limitation was observed only in genetically modified carrot lines with a high chitinase activity (eightfold higher in comparison with the control plants). Co-expression of the *Glu-638* and *Ch383* genes did not generate a synergistic effect (a clearly visible and stronger reduction of infection due to expression of both genes) compared with individual expression (Wally *et al.*, 2009b). The opposite reaction, but with different pathogen species, was seen for the synergistic effect of co-expression of tobacco *Glu-1* and chitinase-1 (*Ch1*) genes (Melchers and Stuiver, 2000). Transgenic carrot plants containing the *Glu-1* and *Ch1* genes were characterized by elevated resistance to pathogens including *A. dauci*, *A. radicina*, *C. carotae* and *E. heraclei*.

Two other genes sourced from microorganisms, bacterial *Pseudomonas fluorescens* microbial factor 3 (*MF3*) (Baranski *et al.*, 2007a) and fungal *Trichoderma harzianum* chitinase gene (*chit36*) (Baranski *et al.*, 2008) were chosen to enhance carrot resistance to leaf diseases. The laboratory-based assays on detached leaflets and petioles inoculated with fungal pathogens (*A. radicina*, *A. dauci*, and *B. cinerea*) showed a slower rate of infestation for transgenic MF3 carrot explants. The tolerance increase estimated at 20–40% was regarded as moderate (Baranski *et al.*, 2007a). The transgenic *chit36* carrots were characterized by higher chitinolytic activity and less severe symptoms of fungal disease. Significant, 50% slower disease progress in the case of *A. radicina* and *B. cinerea* was reported compared with the non-transgenic carrot explants. Enhanced resistance to *A. dauci* was also observed, but only in selected genetically modified carrot lines (Baranski *et al.*, 2008).

*NPR1* (non-expressor of pathogenesis-related) and *NPR1*-like genes are attractive candidates for engineering plants for disease resistance. The *Arabidopsis NPR1* gene is involved in plant responses as a regulator in salicylic acid-induced defense. During infection, salicylic acid levels increase and bind to the NPR1 protein, which induces the expression of PR genes (Li *et al.*, 2018). Carrot petioles were transformed by *A. tumefaciens* with a plasmid containing the *A. thaliana* NPR1 gene in cDNA under the CaMV 35S promoter (Wally *et al.*, 2009a). The assessment of disease symptoms and expression of PR genes in transgenic carrot lines confirmed the activation of systemic acquired resistance induced by the *AtNPR1* gene. Genetically modified carrot cell cultures showed enhanced susceptibility to pathogen disease-related elicitors (*S. sclerotiorum* cell walls, 2,6-dichloroisonicotinic acid, and salicylic acid or methyl jasmonic acid) measured as the transcription activation time and expression level of PR genes. Transgenic material also showed a very high resistance level to invasion by *B. cinerea*, *E. heraclei* and *X. hortorum* in inoculation tests (50%, 90%, and 80%, respectively), assessed 7 days after inoculation, in comparison with the control. The resistance to *A. radicina* and *S. sclerotiorum* was weaker, but also noticeable in NPR1 lines.

Subsequent studies concerned pathogen resistance mechanisms involving regulation of apoptosis. BAX proteins are elements that trigger apoptosis mechanisms in animals and plants. These proteins interact with the external mitochondrial membrane to form pores, which leads to cytochrome *c* release, followed by activation of caspases. In plants, this process is strictly regulated, including modulation of $Ca^{2+}$ signaling and sphingolipid metabolism. BAX inhibitor 1 (BI-1) is a highly conserved protein acting as a cell death suppressor in this signaling cascade (Ren *et al.*, 2021). Disease symptoms on leaves infested by *B. cinerea* were effectively reduced in transgenic carrot expressing the barley *BI-1* gene (Imani *et al.*, 2010). Additionally, resistance to *Chalara elegans* in the BI-1 carrot roots was higher than in the control. The results indicated that the mechanisms of programmed cell death are components of carrot resistance to these pathogens (Imani *et al.*, 2010).

## 6.9 References


Aegerter, B.J. (2002) Powdery mildew. In: Davis R.M. and Raid R.N. (eds.) *Compendium of Umbelliferous Crop Diseases*. American Phytopathological Society, St. Paul, Minnesota, pp. 23–24.

Ahn, Y.J. and Song, N.H. (2012) A cytosolic heat shock protein expressed in carrot (*Daucus carota* L.) enhances cell viability under oxidative and osmotic stress conditions. *HortScience* 47, 143–148.

Ali, A., Matthews, W.C., Cavagnaro, P.F., Iorizzo, M., Roberts, P.A., *et al.* (2014) Inheritance and mapping of *mj-2*, a new source of root-knot nematode (*Meloidogyne javanica*) resistance in carrot. *Journal of Heredity* 105, 288–291.

Allender, C. (2019) Genetic resources for carrot improvement. In: Simon, P., Iorizzo, M., Grzebelus, D., and Baranski, R. (eds) *The Carrot Genome. Compendium of Plant Genomes*. Springer, Cham, Switzerland, pp. 93–100.

Ami, K., Planchais, S., Cabassa, C., Guivarc'h, A., Very, A.A., *et al.* (2020) Different proline responses of two Algerian durum wheat cultivars to *in vitro* salt stress. *Acta Physiologiae Plantarum* 42: 21.

Amirsadeghi, S., McDonald, A.E., and Vanlerberghe, G.C. (2007) A glucocorticoid-inducible gene expression system can cause growth defects in tobacco. *Planta* 226, 453–463.

Annon, A., Rathore, K., and Crosby, K. (2014) Overexpression of a tobacco osmotin gene in carrot (*Daucus carota* L.) enhances drought tolerance. *In Vitro Cellular & Developmental Biology—Plant* 50, 299–306.

Arango, J., Salazar, B., Welsch, R., Sarmiento, F., Beyer, P., *et al.* (2010) Putative storage root specific promoters from cassava and yam: cloning and evaluation in transgenic carrots as a model system. *Plant Cell Reports* 29, 651–659.

Aviv, D., Amsellem, Z., and Gressel, J. (2002) Transformation of carrots with mutant acetolactate synthase for *Orobanche* (broomrape) control. *Pest Management Science* 58, 1187–1193.

Babic, I., Amiot, M.J., Nguyen-The, C., and Aubert, S. (1993) Changes in phenolic content in fresh ready-to-use shredded carrots during storage. *Journal of Food Science* 58, 351–356.

Ballas, N., Zakai, N., and Loyter, A. (1987) Transient expression of the plasmid pCaMVCAT in plant protoplasts following transformation with polyethyleneglycol. *Experimental Cell Research* 170, 228–234.

Banasiak, Ł.A., Wojewódzka, J., Baczyński, J.P., Reduron, M., Piwczyński, R., *et al.* (2016) Phylogeny of Apiaceae subtribe Daucinae and the taxonomic delineation of its genera. *Taxon* 65, 563–585.

Bano, S., Ashraf, M., and Akram, N.A. (2014) Salt stress regulates enzymatic and nonenzymatic antioxidative defense system in the edible part of carrot (*Daucus carota* L.). *Journal of Plant Interactions* 9, 324–329.

Baranski, R. (2008) Genetic transformation of carrot (*Daucus carota*) and other Apiaceae species. *Trans Plant Journal* 2, 18–38.

Baranski, R., Klocke, E., and Schumann, G. (2006) Green fluorescent protein as an efficient selection marker for *Agrobacterium rhizogenes* mediated carrot transformation. *Plant Cell Reports* 25, 190–197.

Baranski, R., Klocke, E., and Nothnagel, T. (2007a) Enhancing resistance of transgenic carrot to fungal pathogens by the expression of *Pseudomonas fluorescence* microbial factor 3 (MF3) gene. *Physiological and Molecular Plant Pathology* 71, 88–95.

Baranski, R., Klocke, E., and Ryschka, U. (2007b) Monitoring the expression of green fluorescent protein in carrot. *ActaPhysiol Plant* 29, 239–246.

Baranski, R., Klocke, E., and Nothnagel, T. (2008) Chitinase CHIT36 from *Trichoderma harzianum* enhances resistance of transgenic carrot to fungal pathogens. *Journal of Phytopathology* 156, 513–521.

Barba-Espín, G., Chen, S.-T., Agnolet, S., Hegelund, J.N., Stanstrup, J., *et al.* (2020) Ethephon-induced changes in antioxidants and phenolic compounds in anthocyanin-producing black carrot hairy root cultures. *Journal of Experimental Botany* 71, 7030–7045.

Basma, K., Zied, B., and Chérif, H. (2014) Effect of sodium chloride on the germination of the seeds of a collection of carrot accessions (*Daucus carota* L.) cultivated in the region of Sidi Bouzid. *Journal of Stress Physiology & Biochemistry* 10, 28–36.

Bates, G.W., Piastuch, W., Riggs, C.D., and Rabussay, D. (1988) Electroporation for DNA delivery to plant protoplast. *Plant Cell, Tissue and Organ Culture* 12, 213–218.

Bates, G.W., Carle, S.A., and Piastuch, W.C. (1990) Linear DNA introduction into carrot protoplasts by electroporation undergoes ligation and recircularization. *Plant Molecular Biology* 14, 899–908.

Bercetche, J., Chriqui, D., Adam, S., and David, C. (1987) Morphogenetic and cellular reorientations induced by *Agrobacterium rhizogenes* (strain 1855, 2659 and 8196) on carrot, pea and tobacco. *Plant Science* 52, 195–210.

Bernstein, L. and Ayers, A.D. (1953) Salt tolerances of five varieties of carrots. *Proceedings of the American Society for Horticultural Science* 61, 360–366.

Block, G. (1994) Nutrient sources of provitamin a carotenoids in the American diet. *American Journal of Epidemiology* 139, 290–293.

Boerema, G.H., Dorenbosch, M.M.J., and van Kesteren, H.A. (1963) Some notable fungus infections. II. Jaarboek (1962). *Review of Applied Mycology* 43: 2185c.

Boiteux, L.S., Della Vecchia, P.T., and Reifschneider, F.J.B. (1993) Heritability estimate for resistance to *Alternaria dauci* in carrot. *Plant Breed* 110, 165–167.

Boiteux, L.S., Hyman, J.R., Bach, I.C., Fonseca, M.E.N., Matthews, W.C., *et al.* (2004) Employment of flanking codominant STS markers to estimate allelic substitution effects of a nematode resistance locus in carrot. *Euphytica* 136, 37–44.

Bolton, A. and Simon, P. (2019) Variation for salinity tolerance during seed germination in diverse carrot (*Daucus carota* (L.)) germplasm. *HortScience* 54, 38–44.

Bolton, A., Nijabat, A., Mahmood-ur-Rehman, M., Naveed, N.H., Mannan, A.T.M.M., *et al.* (2019) Variation for heat tolerance during seed germination in diverse carrot (*Daucus carota* (L.)) germplasm. *HortScience* 54, 1470–1476.

Bolton, A., Klimek-Chodacka, M., Martin-Millar, E., Grzebelus, D., and Simon, P.W. (2020) Genome-assisted improvement strategies for climate-resilient carrots. In: Kole, C. (ed.) *Genomic Designing of Climate-Smart Vegetable Crops*. Springer, Cham, Switzerland, pp. 309–343.

Bonnet, A. (1977) Resistance a l'Oidium. INRA Stat. d'Amelior. *Plantes Maraicheres* 33–35.

Bonnet, A. (1983) *Daucus carota* L. subsp. *dentatus* Bertol., géniteur de résistance à l'oïdium pour l'amélioration de la carotte cultivée. *Agronomie* 3, 33–38.

Boston, R.S., Becwar, M.R., Ryan, R.D., Goldsbrough, P.B., Larkins, B.A., *et al.* (1987) Expression from heterologous promoters in electroporated carrot protoplasts. *Plant Physiology* 83, 742–746.

Boulanger, F., Berkaloff, A., and Richaud, F. (1986) Identification of hairy root loci in the T-regions of *Agrobacterium rhizogenes* Ri plasmids. *Plant Molecular Biology* 6, 271–279.

Bower, R. and Birch, R.G. (1993) Transformation in *Daucus carota* L. (carrot). In: Bajaj, Y.P.S. (ed.) *Plant Protoplasts and Genetic Egnineering III. Biotechnology in Agriculture and Forestry*. Springer Verlag, Berlin/Heidelberg, pp. 172–181.

Bridge, J. and Starr, J.L. (2007) *Plant Nematodes of Agricultural Importance: A Colour Handbook*. CRC Press, London.

Brodzik, R., Spistin, S., Pogrebnyak, N., Bandurska, K., Portocarrero, C., *et al.* (2009) Generation of plant-derived recombinant DTP subunit vaccine. *Vaccine* 27, 3730–3734.

Broussard, M.A., Mas, F., Howlett, B., Pattemore, D., and Tylianakis, J.M. (2017) Possible mechanisms of pollination failure in hybrid carrot seed and implications for industry in a changing climate. *PLOS ONE* 12: e0180215.

Browne, G.T. (2002) *Phytophthora* root rot. In: Davis, R.M. and Raid, R.N. (eds) *Compendium of Umbelliferous Crop Diseases*. American Phytopathological Society, St. Paul, Minnesota, pp. 37–38.

Cardarelli, M., Spanò, L., de Paolis, A., Mauro, M.L., Vitali, G., *et al.* (1985) Identification of the genetic locus responsible for non-polar root induction by *Agrobacterium rhizogenes* 1855. *Plant Molecular Biology* 5, 385–391.

Cardarelli, M., Mariotti, D., Pomponi, M., Spanò, L., Capone, I., *et al.* (1987) *Agrobacterium rhizogenes* T-DNA genes capable of inducing hairy root phenotype. *Molecular Genetics and Genomics* 207, 475–480.

Carvalho, A.D.F., Silva, G.O., Pereira, R.B., and Pinheiro, J.B. (2015) Productivity and tolerance to the leaf blight disease of hybrid and open-pollinated carrot cultivars. *Horticultura Brasileira* 33, 299–304.

Chen, T.W. and Wu, W.S. (1999) Biological control of carrot black rot. *Journal of Phytopathology* 147, 99–104.

Chialva, C., Blein, T., Crespi, M., and Lijavetzky, D. (2021) Insights into long non-coding RNA regulation of anthocyanin carrot root pigmentation. *Scientific Reports* 11: 4093.

Collier, R. and Finch, S. (2009) A review of research to address carrot fly (*Psila rosae*) control in the UK. *EPPO Bulletin* 39, 121–127.

Corwin, D.L. (2020) Climate change impacts on soil salinity in agricultural areas. *European Journal of Soil Science* 72, 842–862.

Cwalina-Ambroziak, B., Amarowicz, R., Glosek, M., and Janiak, M. (2014) Changes in the concentrations of phenolic acids in carrot plants inoculated with *Alternaria radicina* Meier, Drechsler & Eddy. *Acta Scientiarum Polonorum Hortorum Cultus* 13, 97–108.

Daliakopoulos, I.N., Tsanis, I.K., Koutroulis, A., Kourgialas, N.N., Varouchakis, A.E., *et al.* (2016) The threat of soil salinity: a European scale review. *Science of the Total Environment* 573, 727–739.

Daniell, H., Dhingra, A., and Allison, L. (2002) Chloroplast transformation: from basic molecular biology to biotechnology. *Reviews in Plant Physiology and Biochemistry* 1, 1–20.
Davis, R.M. (2004) Carrot diseases and their management. In: Naqvi, S.A.M.H. *Diseases of Fruits and Vegetables. Vol. I.* Springer, Dordrecht, pp. 397–439.
Davis, R.M. and Nunez, J. (2007) Integrated approaches for carrot pests and diseases management. In: Ciancio, A. and Mukerji, K.G. (eds.) *General Concepts in Integrated Pest and Disease Management.* Springer, Dordrecht, pp. 149–88.
Davis, R.M. and Raid, R.N. (eds.) (2002) *Compendium of Umbelliferous Crop Diseases*. American Phytopathological Society, St. Paul, Minnesota.
Deroles, S., Smith, M.A.L., and Lee, C. (2002) Factors affecting transformation of cell cultures from three dicotyledonous pigment-producing species using microprojectile bombardment. *Plant Cell, Tissue and Organ Culture* 70, 69–76.
Dirks, R., Sidorov, V., and Tulmans, C. (1996) A new protoplast culture system in *Daucus carota* L. and its application for mutant selection and transformation. *Theoretical and Applied Genetics* 93, 809–815.
Dröge, W., Broer, I., and Pühler, A. (1992) Transgenic plants containing the phosphinothricin-*N*-acetyltransferase gene metabolize the herbicide l-phosphinothricin (glufosinate) differently from untransformed plants. *Planta* 187, 142–151.
du Toit, L.J. and Derie, M.L. (2008) Effect of powdery mildew on seed yield and quality in a carrot seed crop, 2006–2007. *Plant Disease Management Reports* 2: V007.
du Toit, L.J., Crowe, F.J., Derie, M.L., Simmons, R.B., and Pelter, G.Q. (2005) Bacterial blight in carrot seed crops in the Pacific Northwest. *Plant Disease* 89, 896–907.
du Toit, L.J., Derie, M.L., and Wohleb, C.H. (2009) Effect of powdery mildew on seed yield and quality in a carrot seed crop, 2007–2008. *Plant Disease Management Reports* 3: V136.
du Toit, L., Le Clerc, V., and Briard, M. (2019) Genetics and genomics of carrot biotic stress. In: Simon, P.W., Iorizzo, M., Grzebelus, D., and Baranski, R. (eds.) *The Carrot Genome*. Springer, Cham, Switzerland, pp. 317–62.
Dunemann, F., Unkiel, K., and Sprin, T. (2019) Using CRISPR/Cas9 to produce haploid inducers of carrot through targeted mutations of centromeric histone H3 (CENH3). *Acta Horticulturae* 1264, 211–220.
Dunn, J.A. (1970) The susceptibility of varieties of carrot to attack by the aphid, *Cavariella aegopodii* (Scop.). *Annals of Applied Biology* 66, 301–312.
Ellis, M.B. and Holliday, P. (1972) *Alternaria radicina*. No. 346. In: *C.M.I. Descriptions of Pathogenic Fungi and Bacteria*. Commonwealth Mycological Institute, Kew, UK.
Ellis, P.R. (1999) The identification and exploitation of resistance in carrots and wild umbelliferae to the carrot fly, *Psila rosae* (F.). *Integrated Pest Management Reviews* 4, 259–268.
Ellis, P.R. and Hardman, J.A. (1981) The consistency of the resistance of eight carrot cultivars to carrot fly attack at several centres in Europe. *Annals of Applied Biology* 98, 491–497.
Ellis, P.R., Hardman, J.A., Crowther, T.C., and Saw, P.L. (1993) Exploitation of the resistance to carrot fly in the wild carrot species *Daucus capillifolius*. *Annals of Applied Biology* 122, 79–91.
Epstein, E., Nissen, S.J., and Sutter, E.G. (1991) Indole-3-acetic acid and indole-3-butyric acid in tissues of carrot inoculated with *Agrobacterium rhizogenes*. *Journal of Plant Growth Regulation* 10, 97–100.
Ezura, H. and Hiwasa-Tanase, K. (2018) Mass production of the taste-modifying protein miraculin in transgenic plants. In: Merillon, J.M. and Ramawat, K. (eds.) *Sweeteners. Reference Series in Phytochemistry*. Springer, Cham, Switzerland, pp. 1–18.
FAO (Food and Agriculture Organization of the United Nations) (2015) *Status of the World's Soil Resources: Main Report*. FAO, Rome.
FAOSTAT (2019) FAOSTAT database. Food and Agriculture Organization of the United Nations, Rome. Available at: www.fao.org/faostat/en/#home (accessed 25 June 2024).
Farrar, J.J., Pryor, B.M., and Davis, R.M. (2004) *Alternaria* diseases of carrot. *Plant Disease* 88, 776–784.
Flores-Ortiz, C., Alvarez, L.M., Undurraga, A., Arias, D., Durán, F., *et al.* (2020) Differential role of the two ζ-carotene desaturaseparalogs in carrot (*Daucus carota*): *ZDS1* is a functional gene essential for plant development and carotenoid synthesis. *Plant Science* 291: 110327.
Flowers, T.J. (2004) Improving crop salt tolerance. *Journal of Experimental Botany* 55, 307–319.
Fromm, M., Taylor, L.P., and Albot, V.M. (1985) Expression of genes electroporated into monocot and dicot plant cells. *Proceedings of the National Academy of Sciences USA* 82, 5824–58284.
Fründt, C., Meyer, A.D., Ichikawa, T., and Meins, F. (1998) A tobacco homologue of the Ri-plasmid *orf13* gene causes cell proliferation in carrot root discs. *Molecular Genetics and Genomics* 259, 559–568.

Fuentes, P., Pizarro, L., Moreno, J.C., Handford, M., Rodriguez-Concepcion, M.A., *et al.* (2012) Light-dependent changes in plastid differentiation influence carotenoid gene expression and accumulation in carrot roots. *Plant Molecular Biology* 79, 47–59.

Gallie, D.R. (1993) Introduction of mRNA to protoplasts using polyethylene glycol. *Plant Cell Reports* 13, 119–122.

Geary, J.R. and Wall, C.J. (1965) New or uncommon plant diseases and pests. *Plant Pathology* 14, 93–94.

Gieniec, M., Siwek, J., Oleszkiewicz, T., Maćkowska, K., Klimek-Chodacka, M., *et al.* (2020) Real-time detection of somatic hybrid cells during electrofusion of carrot protoplasts with stably labelled mitochondria. *Scientific Reports* 10: 18811.

Gilbert, M.O., Zhang, Y.Y., and Punja, Z.K. (1996) Introduction and expression of chitinase encoding genes in carrot following *Agrobacterium*-mediated transformation. *In Vitro Cellular & Developmental Biology—Plant* 32, 171–178.

Groom, M.R. and Perry, D.A. (1985) Induction of 'cavity spot-like' lesions on roots of *Daucus carota* by *Pythium violae*. *Transactions of the British Mycological Society* 84, 755–757.

Groves, R.L., Clements, J.R., and Bradford, B.Z. (2020) Carrot diseases resulting from phytoplasmas and viruses. In: Geoffriau, E. and Simon, P.W. (eds.) *Carrot and Other Cultivated Apiaceae*. CAB International, Wallingford, UK, pp. 148–155.

Grzebelus, D., Iorizzo, M., Senalik, D., Ellison, S., Cavagnaro, P., *et al.* (2013a) Diversity, genetic mapping, and signatures of domestication in the carrot (*Daucus carota* L.) genome, as revealed by diversity arrays technology (DArT) markers. *Molecular Breeding* 33, 625–637.

Grzebelus, E., Kruk, M., Macko-Podgórni, A., and Grzebelus, D. (2013b) Response of carrot protoplasts and protoplast-derived aggregates to selection using a fungal culture filtrate of *Alternaria radicina*. *Plant Cell, Tissue and Organ Culture* 115, 209–222.

Guan, Y., Ren, H., Xie, H., Ma, Z., and Chen, F. (2009) Identification and characterization of bZIP-type transcription factors involved in carrot (*Daucus carota* L.) somatic embryogenesis. *Plant Journal* 60, 207–217.

Guerin, P.M. and Stadler, E. (1984) Carrot fly cultivar preferences: some influencing factors. *Ecological Entomology* 9, 413–420.

Gugino, B.K., Carroll, J.E., Widmer, T.L., Chen, P., and Abawi, G.S. (2007) Field evaluation of carrot cultivars for susceptibility to fungal leaf blight diseases in New York. *Crop Protection* 26, 709–714.

Han, K.H. and Hwang, C.H. (2003) Salt tolerance enhanced by transformation of a *P5CS* gene in carrot. *Journal of Plant Biotechnology* 5, 157–161.

Hansen, S.L., Purup, S., and Christensen, L.P. (2003) Bioactivity of falcarinol and the influence of processing and storage on its content in carrots (*Daucus carota* L). *Journal of the Science of Food and Agriculture* 83, 1010–1017.

Hardegger, M. and Shakya, R. (2004) Transformation of carrot. In: Curtis I.S. (ed.) *Transgenic Crops of the World*. Springer, Dordrecht, Netherlands pp. 291–300.

Hardegger, M. and Sturm, A. (1998) Transformation and regeneration of carrot (*Daucus carota* L.). *Molecular Breeding* 4, 119–127.

Hashimoto, T. and Nagayama, T. (2004) Chemical composition of ready-to-eat fresh carrot. *Journal of the Food Hygienic Society of Japan* 39, 324–328.

Hibberd, J.M., Linley, P.J., Khan, M.S., and Gray, J.C. (1998) Transient expression of green fluorescent protein in various plastid types following microprojectile bombardment. *Plant Journal* 16, 627–632.

Hooykaas, P.J.J., and Schilperoort, R.A. (1992) *Agrobacterium* and plant genetic engineering. *Plant Molecular Biology* 19, 15–38.

Huang, S.P., Vecchia, P.T.D., and Ferreira, P.E. (1986) Varietal response and estimates of heritability of resistance to *Meloidogyne javanica* in carrots. *Journal of Nematology* 18, 496–501.

Huang, Y., Li, M.Y., Wang, F., Xu, Z.S., Huang, W., *et al.* (2015) Heat shock factors in carrot: genome-wide identification, classification, and expression profiles response to abiotic stress. *Molecular Biology Reports* 42, 893–905.

Hussain, H.A., Hussain, S., Khaliq, A., Ashraf, U., Anjum, S.A., *et al.* (2018) Chilling and drought stresses in crop plants: Implications, cross talk, and potential management opportunities. *Frontiers in Plant Science* 9: 393.

Imani, J., Baltruschat, H., Stein, E., Jia, G., Vogelsberg, J., *et al.* (2010) Expression of barley BAX Inhibitor-1 in carrots confers resistance to *Botrytis cinerea*. *Molecular Plant Pathology* 7, 279–284.

Iorizzo, M., Senalik, D.A., Ellison, S.L., Grzebelus, D., Cavagnaro, P.F., *et al.* (2013) Genetic structure and domestication of carrot (*Daucus carota* subsp. *sativus*) (*Apiaceae*). *American Journal of Botany* 100, 930–938.

Iorizzo, M., Ellison, S., Senalik, D., Zeng, P., Satapoomin, P., *et al.* (2016) A high-quality carrot genome assembly provides new insights into carotenoid accumulation and asterid genome evolution. *Nature Genetics* 48, 657–666.

Iovene, M., Cavagnaro, P.F., Senalik, D., Buell, C.R., Jiang, J., *et al.* (2011) Comparative FISH mapping of *Daucus* species (*Apiaceae* family). *Chromosome Research* 19, 493–506.

Ishige, F., Takaichi, M., Foster, R., Chua, N.H., and Oeda, K. (1999) AG-box motif (GCCACGTGCC) tetramer confers high-level constitutive expression in dicot and monocot plants. *Plant Journal* 18, 443–448.

Jayaraj, J. and Punja, Z.K. (2007) Combined expression of chitinase and lipid transfer protein genes in transgenic carrot plants enhances resistance to foliar fungal pathogens. *Plant Cell Reports* 26, 1539–1546.

Jourdan, M., Gagné, S., Dubois-Laurent, C., Maghraoui, M., Huet, S., *et al.* (2015) Carotenoid content and root color of cultivated carrot: a candidate-gene association study using an original broad unstructured population. *PLOS ONE* 10: e0116674.

Jung, Y.C., Lee, H.J., Yum, S.S., Soh, W.Y., Cho, D.Y., *et al.* (2005) Drought-inducible – but ABA-independent – thaumatin-like protein from carrot (*Daucus carota* L.). *Plant Cell Reports* 24, 366–373.

Karkleliene, R., Radzevicius, A., Dambrauskiene, E., Surviliene, E., Bobinas, C., *et al.* (2012) Root yield, quality and disease resistance of organically grown carrot (*Daucus sativus* Röhl.) hybrids and cultivars. *Zemdirbyste-Agriculture* 99, 393–398.

Kidmose, U., Hansen, S.L., Christensen, L.P., Edelenbos, M., Larsen, E., *et al.* (2004) Effects of genotype, root size, storage, and processing on bioactive compounds in organically grown carrots (*Daucus carota* L.). *Journal of Food Science* 69, S388–S394.

Kikkert, J.R., Vidal, J.R., and Reisch, B.I. (2005) Stable transformation of plant cells by particle bombardment/biolistics. *Methods in Molecular Biology* 286, 61–78.

Kim, S.Y., Bengtsson, T., Olsson, N., Hot, V., Zhu, L.H., *et al.* (2020) Mutations in two aphid-regulated β-1, 3-glucanase genes by CRISPR/Cas9 do not increase barley resistance to *Rhopalosiphum padi* L. *Frontiers in Plant Science* 11: 1043.

Kim Y.-S., Kim, M.-Y., Kim T.-G., and Yang M.-S. (2009) Expression and assembly of cholera toxin B subunit (CTB) in transgenic carrot (*Daucus carota* L.). *Molecular Biotechnology* 41, 8–14.

Klimek-Chodacka, M., Oleszkiewicz, T., Lowder, L.G., Qi, Y., and Baranski, R. (2018) Efficient CRISPR/Cas9-based genome editing in carrot cells. *Plant Cell Reports* 37, 575–586.

Klimek-Chodacka, M., Oleszkiewicz, T., Qi, Y., and Baranski, R. (2019) Carrot genome editing using CRISPR-based systems. *Acta Horticulturae* 1264, 53–66.

Kodama, M., Brinch-Pedersen, H., Sharma, S., Holme, B.I., Joernsgaard, B., *et al.* (2018) Identification of transcription factor genes involved in anthocyanin biosynthesis in carrot (*Daucus carota* L.) using RNA-Seq. *BMC Genomics* 19: 811.

Kordowska-Wiater, M., Wagner, A., and Hetman, B. (2012) Efficacy of *Candida melibiosica* for control of post-harvest fungal diseases of carrot (*Daucus carota* L.). *Acta Scientiarum Polonorum Hortorum Cultus* 11, 55–65.

Kuhn, J. (1855) Uber das befallen des rapses und die krankheit der mohrenblatter. *Hedwigia* 1, 86–92.

Kumar, S., Dhingra, A., and Daniell, H. (2004) Plastid-expressed betaine aldehyde dehydrogenase gene in carrot cultured cells, roots, and leaves confers enhanced salt tolerance. *Plant Physiology* 136, 2843–2854.

Lamb, K.P. (1953) Observations on yield and varietal susceptibility of some carrot varieties to insect attack in the field. *New Zealand Journal of Science and Technology A* 34, 351–357.

Landi, M., Tattini, M., and Gould, K.S. (2015) Multiple functional roles of anthocyanins in plant-environment interactions. *Environmental and Experimental Botany* 119, 4–17.

Langridge, W.H.R., Li, B.J., and Szalay, A.A. (1985) Electric field mediated stable transformation of carrot protoplasts with naked DNA. *Plant Cell Reports* 4, 355–359.

Le Clerc, V., Pawelec, A., Birolleau-Touchard, C., Suel, A., and Briard, M. (2009) Genetic architecture of factors underlying partial resistance to *Alternaria* leaf blight in carrot. *Theoretical and Applied Genetics* 118, 1251–1259.

Lebeda, A., and Coufal, J. (1987) Evaluation of susceptibility of *Daucus carota* varieties to natural infection with *Erysiphe heraclei*. *Arch Züchtungsforsch Berlin* 17, 73–76.

Lecomte, M., Berruyer, R., Hamama, L., Boedo, C., Hudhomme, P., *et al.* (2012) Inhibitory effects of the carrot metabolites 6-methoxymellein and falcarindiol on development of the fungal leaf blight pathogen *Alternaria dauci*. *Physiological and Molecular Plant Pathology* 80, 58–67.

Lhissou, R., El Harti, A., and Chokmani, K. (2014) Mapping soil salinity in irrigated land using optical remote sensing data. *Eurasian Journal of Soil Science* 3, 82–88.

Li, M., Chen, H., Chen, J., Chang, M., Palmer, I.A., *et al.* (2018) TCP transcription factors interact with NPR1 and contribute redundantly to systemic acquired resistance. *Frontiers in Plant Science* 9: 1153.

Maas, E.V. (1986) Salt tolerance of plants. *Journal of Applied Agricultural Research* 1, 12–25.

Maas, E. and Hoffman, G. (1977) Crop salt tolerance–current assessment. *Journal of the Irrigation and Drainage Division* 103, 115–134.

Malik, M.K., Slovin, J.P., Hwang, C.H., and Zimmerman, J.L. (1999) Modified expression of a carrot small heat shock protein gene, *hsp17.7*, results in increased or decreased thermotolerance double dagger. *Plant Journal* 20, 89–99.

Marquet-Blouin, E., Bouche, F.B., Steinmetz, A., and Muller, C.P. (2003) Neutralizing immunogenicity of transgenic carrot (*Daucus carota* L.)-derived measles virus hemagglutinin. *Plant Molecular Biology* 51, 459–469.

Matthews, B., Dray, S., Wildholm, J., and Ostro, M. (1979) Liposome-mediated transfer of bacterial RNA into carrot protoplasts. *Planta* 145, 37–44.

McDonald, M.R. (1994a) Cavity spot of carrot (*Pythium* spp.): etiology, epidemiology and control. PhD dissertation, University of Guelph, Guelph, Ontario, Canada.

McDonald, M.R. (1994b) Rubbery brown rot. In: Howard, R.J., Garland, J.A., and Seaman, W.L. (eds.) *Diseases and Pests of Vegetable Crops in Canada*. Canadian Phytopathological Society, Guelph Ontario, Canada, pp. 71–72.

McDonald, M.R. (2002) Cavity spot. In: Davis, R.M. and Raid, R.N. (eds.) *Compendium of Umbelliferous Crop Diseases*. American Phytopathological Society, St. Paul, Minnesota, pp. 27–29.

Melchers, L.S. and Stuiver, M.H. (2000) Novel genes for disease-resistance breeding. *Current Opinion in Plant Biology* 3, 147–152.

Milosavljević, A., Pfaf-Dolovac, E., Mitrović, M., Jović, J., Toševski, I., *et al.* (2014) First report of *Cercospora apii*, causal agent of *Cercospora* early blight of celery, in Serbia. *Plant Disease* 98: 1157.

Munns, R. and Tester, M. (2008) Mechanisms of salinity tolerance. *Annual Review of Plant Biology* 59, 651–681.

Nascimento, W.M., Vieira, J.V., Silva, G.O., Reitsma, K.R., and Cantliffe, D.J. (2008) Carrot seed germination at high temperature: effect of genotype and association with ethylene production. *HortScience* 43, 1538–1543.

Noh, S.A., Lee, H.-S., Huh, G.H., Oh, M.-J., Paek, K.-H., *et al.* (2012) A sweetpotato *SRD1* promoter confers strong root-, taproot-, and tuber-specific expression in *Arabidopsis*, carrot, and potato. *Transgenic Research* 21, 265–278.

Nothnagel, T., Budahn, H., and Kramer, R. (2017) Characterization of resistance to *Alternaria* spp. in wild relatives of carrot (*Daucus carota* L. ssp. *carota*). *Acta Horticulturae* 1153, 251–257.

Nowicka, A., Grzebelus, E., and Grzebelus, D. (2016a) Precise karyotyping of carrot mitotic chromosomes using multicolour-FISH with repetitive DNA. *Plant Biology* 60, 25–36.

Nowicka, A., Sliwinska, E., Grzebelus, D., Baranski, R., Simon, P.W., *et al.* (2016b) Nuclear DNA content variation within the genus *Daucus* (*Apiaceae*) determined by flow cytometry. *Scientia Horticulturae* 209, 132–138.

Nuñez, J.J. and Westphal, A. (2002) Damping-off. In: Davis, R.M. and Raid, R.N. (eds.) *Compendium of Umbelliferous Crop Diseases*. American Phytopathological Society, St. Paul, Minnesota.

Palta, J.P. and Simon, G. (2004) Developing and successfully implementing a strategy for breeding frost-hardy carrots. *HortScience* 39: 880.

Park, H., Ko, E., Jang, E., Park, S., Lee, J., *et al.* (2013) Expression of *DcHsp17.7*, a small heat shock protein gene in carrot (*Daucus carota* L.). *Horticulture, Environment, and Biotechnology* 54, 121–27.

Park, Y.J., Han, J.E., Lee, H., Lee, J.Y., Ho, T.T., *et al.* (2021) Production of recombinant miraculin protein in carrot callus via *Agrobacterium*-mediated transformation. *Plant Cell, Tissue and Organ Culture* 145, 615–623.

Parsons, J., Matthews, W., Iorizzo, M., Roberts, P., and Simon, P. (2015) *Meloidogyne incognita* nematode resistance QTL in carrot. *Molecular Breeding* 35: 114.

Pawlicki, N., Sangwan, R.S., and Sangwan-Norreel, B.S. (1992) Factors influencing the *Agrobacterium tumefaciens*-mediated transformation of carrot (*Daucus carota* L.). *Plant Cell, Tissue and Organ Culture* 31, 129–139.

Pereira, R.B., de Carvalho, A.D.F., Pinheiro, J.B., da Silva, G.O., and Vieira, J.V. (2012) Resistência de populações de cenoura à queima-das-folhas com diferentes níveis de germoplasma tropical [Carrot populations resistance to leaf blight with different levels of tropical germplasm]. *Horticultura Brasileira* 30, 489–493.

Pinheiro, J.B., de Mendonça, J.L., and de Santana, J.P. (2011) Reaction of wild *Solanaceae* to *Meloidogyne incognita* race 1 and *M. javanica*. *Acta Horticulturae* 917, 237–241.

Plunkett, G.M., Pimenov, M.G., Reduron, J.P., Kljuykov, E.V., van Wyk, B.E., *et al.* (2018) Apiaceae. In: Kadereit, J. and Bittrich, V. (eds.) *Flowering Plants. Eudicots. The Families and Genera of Vascular Plants. Vol. 15*. Springer, Cham. Switzerland, pp. 9–206.

Pryor, B.M., Davis, R.M., and Gilbertson, R.L. (1998) Detection of soilborne *Alternaria radicina* and its occurrence in California carrot fields. *Plant Disease* 82, 891–895.

Pryor, B.M., Davis, R.M., and Gilbertson, R.L. (2000) A toothpick inoculation method for evaluating carrot cultivars for resistance to *Alternaria radicina*. *HortScience* 35, 1099–1102.

Punja, Z.K. (2002) Crown rot of carrot. In: Davis, R.M. and Raid, R.N. (eds.) *Compendium of Umbelliferous Crop Diseases*. American Phytopathological Society, St. Paul, Minnesota, p. 31.

Rasmussen, J.O. and Rasmussen, O.S. (1993) PEG mediated DNA uptake and transient GUS expression in carrot, rapeseed and soybean protoplasts. *Plant Science* 89, 199–207.

Rathus, C., Bower, R., and Birch, R.G. (1993) Effects of promoter, intron and enhancer elements on transient gene expression in sugar-cane and carrot protoplasts. *Plant Molecular Biology* 23, 613–618.

Ren, K., Feng, L., Sun, S., and Zhuang, X. (2021) Plant mitophagy in comparison to mammals: what is still missing? *International Journal of Molecular Sciences* 22: 1236.

Rezaei, F., Bolagh, Q., Solouki, A., Tohidfar, M., Mehrjerdi, M.Z., *et al.* (2020) *Agrobacterium*-mediated transformation of Persian walnut using *BADH* gene for salt and drought tolerance. *Journal of Horticultural Science and Biotechnology* 96, 162–171.

Roberts, P.A. and Mullens, T.R. (2002) Diseases caused by nematodes. In: Davis, R.M. and Raid, R.N. (eds.) *Compendium of Umbelliferous Crop Diseases*. American Phytopathological Society, St. Paul, Minnesota, pp. 45–46.

Rojas-Anaya, E., Loza-Rubio, E., Olivera-Flores, M.T., and Gomez-Lim, M. (2009) Expression of rabies G protein in carrots (*Daucus carrota*). *Transgenic Research* 18, 911–919.

Rubatzky, V.E., Quiros, C.F., and Simon, P.W. (1999) Diseases, disorders, insects and other pests. In: *Carrots and Related Vegetables Umbelliferae*. CAB International, New York, pp. 173–220.

Ryder, M.H., Tate, M.E., and Kerr, A. (1985) Virulence properties of strains of *Agrobacterium* on the apical and basal surfaces of carrot root discs. *Plant Physiology* 77, 215–221.

Sanford, J.C., Klein, T.M., Wolf, E.D., and Allen, N. (1987) Delivery of substances into cells and tissues using a particle bombardment process. *Journal of Particle Science and Technology* 5, 27–37.

Sarkar, S., Ghosh, I., Roychowdhury, D., and Jha, S. (2018) The effects of *rol* genes of *Agrobacterium rhizogenes* on morphogenesis and secondary metabolite accumulation in medicinal plants. In: Kumar, N. (ed.) *Biotechnological Approaches for Medicinal and Aromatic Plants*. Springer, Singapore, pp. 27–51.

Sathasivam, R., Radhakrishnan, R., Kim, J.K., and Park, S.U. (2020) An update on biosynthesis and regulation of carotenoids in plants. *South African Journal of Botany* 140, 290–302.

Saude, C., Hausbeck, M.K., Hurtado-Gonzales, O., Rippetoe, C., and Lamour, K.H. (2007) First report of *Phytophthora cactorum* causing root rot of processing carrots (*Daucus carota*) in Michigan. *Plant Disease* 91: 459.

Schmidhalter, U. and Oertli, J.J. (1991) Transpiration/biomass ratio for carrots as affected by salinity, nutrient supply and soil aeration. *Plant and Soil* 135, 125–132.

Schwinn, K., Venail, J., Shang, Y., Mackay, S., Alm, V., *et al.* (2006) A small family of MYB-regulatory genes controls floral pigmentation intensity and patterning in the genus antirrhinum. *Plant Cell* 18, 831–851.

Scott, R.J. and Draper, J. (1987) Transformation of carrot tissues derived from proembryogenic suspension cells: A useful model sytem for gene expression studies in plants. *Plant Molecular Biology* 8, 265–274.

Seo, Y., Park, J., Kim, Y.S., Park, Y., and Kim, Y.H. (2014) Screening and histopathological characterization of Korean carrot lines for resistance to the root-knot nematode *Meloidogyne incognita*. *Plant Pathology Journal* 30, 75–81.

Sharma, S., Holme, I.B., Dionisio, G., Miyako Kodama, M., Dzhanfezova, T., *et al.* (2020) Cyanidin based anthocyanin biosynthesis in orange carrot is restored by expression of *AmRosea1* and *AmDelila*, MYB and bHLH transcription factors. *Plant Molecular Biology* 103, 443–456.

Signorelli, S. (2016) The fermentation analogy: a point of view for understanding the intriguing role of proline accumulation in stressed plants. *Frontiers in Plant Science* 7: 1339.

Silva, G.O., Vieira, J.V., Vilela, M.S., Reis, A., and Boiteux, L.S. (2009) Parâmetros genéticos da resistência ao complexo da queima-das-folhas em populações de cenoura [Genetic parameters of the resistance to the leaf blight disease complex in carrot populations]. *Horticultura Brasileira* 27, 354–356.

Simlat, M., Stobiecki, M., and Szklarczyk, M. (2013) Accumulation of selected phenolics and expression of PAL genes in carrots differing in their susceptibility to carrot fly (*Psila rosae* F.). *Euphytica* 190, 253–266.

Simon, P.W. (2000) Domestication, historical development, and modern breeding of carrot. *Plant Breeding Reviews* 19, 157–190.

Simon, P.W. and Strandberg, J.O. (1998) Diallel analysis of resistance in carrot to *Alternaria* leaf blight. *Journal of the American Society for Horticultural Science* 123, 412–415.

Simon, P.W., Matthews, W.C., and Roberts, P.A. (2000) Evidence for simply inherited dominant resistance to *Meloidogyne javanica* in carrot. *Theoretical and Applied Genetics* 100, 735–742.

Simpson, K., Fuentes, P., Quiroz-Iturra, L.F., Flores-Ortiz, C., Contreras, R., *et al.* (2018) Unraveling the induction of phytoene synthase 2 expression by salt stress and abscisic acid in *Daucus carota*. *Journal of Experimental Botany* 69, 4113–4126.

Smith, C.M. and Chuang, W.P. (2014) Plant resistance to aphid feeding: behavioral, physiological, genetic and molecular cues regulate aphid host selection and feeding. *Pest Management Science* 70, 528–540.

Smoleń, S., Lukasiewicz, A., Klimek-Chodacka, M., and Baranski, R. (2020) Effect of soil salinity and foliar application of jasmonic acid on mineral balance of carrot plants tolerant and sensitive to salt stress. *Agronomy* 10: 659.

Soroker, E., Bashan, Y., and Okon, Y. (1984) Reproducible induction of cavity spot in carrots and physiological and microbial changes occurring during cavity formation. *Soil Biology and Biochemistry* 16, 541–548.

Spooner, D., Rojas, P., Bonierbale, M., Mueller, L.A., Srivastav, M., *et al.* (2013) Molecular phylogeny of *Daucus* (*Apiaceae*). *Systematic Botany* 38, 850–857.

Steward, F.C., Mapes, M.O., and Mears, K. (1958) Growth and organized development of cultured cells. II. organization in cultures grown from freely suspended cells. *American Journal of Botany* 45, 705–708.

Stolarczyk, J. and Janick, J. (2011) Carrot: history and iconography. *Chronica Horticulturae* 51, 13–18.

Strandberg, J.O., Bassett, M.J., Peterson, C.E., and Berger, R.D. (1972) Sources of resistance to *Alternaria dauci*. *HortScience* 7: 345.

Sweet, J.B., Lake, S.E., Wright, I.R., and Priestley, R.H. (1986) Resistance of carrot varieties to cavity spot disease. *Aspects of Applied Biology* 12, 235–245.

Szklarczyk, M., Szymański, M., Wójcik-Jagła, M., Simon, P.W., Weihe, A., *et al.* (2014) Mitochondrial *atp9* genes from petaloid male-sterile and male-fertile carrots differ in their status of heteroplasmy, recombination involvement, post-transcriptional processing as well as accumulation of RNA and protein product. *Theoretical and Applied Genetics* 127, 1689–1701.

Takaichi, M. and Oeda, K. (2000) Transgenic carrots with enhanced resistance against two major pathogens, *Erysiphe heraclei* and *Alternaria dauci*. *Plant Science* 153, 135–144.

Tepfer, D. (1985) Transformation of several species of higher plants by *Agrobacterium rhizogenes*: sexual transmission of the transformed genotype and phenotype. *Cell* 37, 359–367.

Thomas, J.C., Guiltinan, M.J., Bustos, S., Thomas, T., and Nessler, C. (1989) Carrot (*Daucus carota*) hypocotyls transformation using *Agrobacterium tumefaciens*. *Plant Cell Reports* 8, 354–357.

Umiel, N., Jacobson, R., and Globerson, D. (1975) Pollination of the cultivated carrot (*Daucus carota* L.) by the wild carrot (*D. carota* var. *maximus*) and its implication on commercial seed production. *Hassadeh* 56, 478–480.

van Wordragen, M.F. and Dons, H.J.M. (1992) *Agrobacterium tumefaciens*-mediated transformation of recalcitrant crops. *Plant Molecular Biology Reporter* 10, 12–36.

Vasudeva, R.S. (1963) Report of Division of Mycology and Plant Pathology. In: *Scientific Reports of the Agricultural Research Institute*, New Delhi, pp. 87–100.

Vieira, J.V., Charchar, J.M., Aragão, F.A.S., and Boiteux, L.S. (2003) Heritability and gain from selection for field resistance against multiple root-knot nematode species (*Meloidogyne incognita* Race 1 and *M. javanica*) in carrot. *Euphytica* 130, 11–16.

Vivoda, E., Davis, R.M., Nunez, J.J., and Guerard, J.P. (1991) Factors affecting the development of cavity spot of carrot. *Plant Disease* 75, 519–522.

Wally, O., Jayaraj, J., and Punja, Z.K. (2008) Comparative expression of β-glucuronidase with five different promoters in transgenic carrot (*Daucus carota* L.) root and leaf tissues. *Plant Cell Reports* 27, 279–287.

Wally, O., Jayaraj, J., and Punja, Z.K. (2009a) Broad-spectrum disease resistance to necrotrophic and biotrophic pathogens in transgenic carrots (*Daucus carota* L.) expressing an *Arabidopsis NPR1* gene. *Planta* 231, 131–141.

Wally, O., Jayaraj, J., and Punja, Z.K. (2009b) Comparative resistance to foliar fungal pathogens in transgenic carrot plants expressing genes encoding for chitinase, β-1,3-glucanase and peroxidise. *European Journal of Plant Pathology* 123, 331–342.

Wang, H., Ou, C.G., Zhuang, F.Y., and Ma, Z.G. (2014) The dual role of phytoene synthase genes in carotenogenesis in carrot roots and leaves. *Molecular Breeding* 34, 2065–2079.
Wang, M. and Goldman, I.L. (1996) Resistance to root knot nematode (*Meloidogyne hapla* Chitwood) in carrot is controlled by two recessive genes. *Journal of Heredity* 87, 119–123.
Watson, A. (2016) Managing carrot powdery mildew. *Australasian Plant Pathology* 45, 29–35.
White, J.G. (1988) Studies on the biology and control of cavity spot of carrots. *Annals of Applied Biology* 113, 259–268.
Wurtele, E. and Bulka, K. (1989) A simple, efficient method for the *Agrobacterium*-mediated transformation of carrot callus cells. *Plant Science* 61, 253–262.
Xu, Z.S., Feng, K., and Xiong, A.S. (2019a) CRISPR/Cas9-mediated multiply targeted mutagenesis in orange and purple carrot plants. *Molecular Biotechnology* 61, 191–199.
Xu, Z.S., Yang, Q.Q., Feng, K., and Xiong, A.S. (2019b) Changing carrot color: insertions in *DcMYB7* alert the regulation of anthocyanin biosynthesis and modification. *Plant Physiology* 181, 195–207.
Xu, Z.S., Yang, Q.Q., Feng, K., Yu, X., and Xiong, A.S. (2020) DcMYB113, a root-specific R2R3-MYB, conditions anthocyanin biosynthesis and modification in carrot. *Plant Biotechnology Journal* 18, 1585–1597.
Yarger, L.W. and Baker, L.R. (1981) Tolerance of carrot to *Meloidogyne hapla*. *Plant Disease* 65, 337–339.
Yau, Y.Y., Davis, S.J., Ipek, A., and Simon, P.W. (2008) Early identification of stable transformation events by combined use of antibiotic selection and vital detection of green fluorescent protein (GFP) in carrot (*Daucus carota* L.) callus. *Agricultural Sciences in China* 7, 664–671.
Yildiz, M., Willis, D.K., Cavagnaro, P.F., Iorizzo, M., Abak, K., *et al.* (2013) Expression and mapping of anthocyanin biosynthesis genes in carrot. *Theoretical and Applied Genetics* 126, 1689–1702.

# 7 Genetic Engineering of Cucurbits: Abiotic and Biotic Stress Resistance

**T.R. Usharani*, K.N. Poornima, Megha Ganachari and H.K. Shreyas**
*Division of Basic Sciences, ICAR-Indian Institute of Horticultural Research, Hesaraghatta, Bengaluru, Karnataka, India*

**Abstract**

Cucurbits largely comprise nutritive vegetables that serve as cheaper sources of various nutrients and were some of the earliest domesticated crops for human consumption worldwide in different climatic and geographical patterns. They are rich in digestive fibers, which are important for intestinal health. Squashes, melons and gourds of various types, cultivars and subspecies are widely cultivated. Their production is greatly affected by biotic and abiotic stresses, the main ones being cucumber mosaic virus, fruit flies, wilt complexes, nematodes, salinity, low and high temperatures, and drought. Biotic stresses affect not only the yield but also the quality of the produce to a large extent. As a monoecious species, the ratio of male and female flowers is greatly affected by abiotic stresses. Genetic engineering offers a wider scope for expressing genes from other plant species proven to enhance tolerance to biotic and abiotic stresses. The utilization of wild species to impart biotic and abiotic stress resistance has been limited due to linkage dragging, which affects the taste and quality of cucurbits. In this context, genetic engineering approaches play a major role by expressing the genes of interest isolated from wild curcurbit and other plant species. In the absence of resistance sources, genetic engineering approaches such as transgenic development, genome editing, and targeted mutagenesis can be successfully utilized to develop cucurbit varieties with higher productivity levels. This chapter discusses the tools, genes, methods of gene transfer, genome-editing approaches and scope of genetic engineering.

**Keywords:** abiotic stress, biotic stress, cucurbits, transgenics

## 7.1 Introduction

The genus *Cucurbita* belongs to the family Cucurbitaceae, which includes cucumber, melon, squash, pumpkin, and gourds exclusively native to the Americas. The most important species of Cucurbitaceae in India are *Trichosanthes* (22 species), *Cucumis* (11) *Momordica* (eight), and *Zehneria* (five) (Renner and Pandey, 2013). *Cucurbita* spp. are grown primarily for their seeds, fruits, and gourds. Mature and immature *Cucurbita* fruits are edible and are eaten as a vegetable (OECD, 2016). Several *Cucurbita* spp. are used in traditional medicine as an anthelmintic, as a diuretic, to heal burns and inflammation, and as an antihyperglycemic agent (Chiej, 1984; Argueta, 1994; Andrade-Cetto and Heinrich, 2005; Xia and Wang, 2006). Cucurbits are widely cultivated as a warm season, annual vegetable crop. However, cucurbit cultivation has suffered from various abiotic and biotic stresses, which have led to huge losses in production and quality (Choudhuri *et al.*, 2017). The major abiotic stresses include drought, high/low

*Email: usharani.tr@icar.gov.in


DOI: 10.1079/9781800627123.0007

temperatures, and salinity, which adversely affect cucurbits with a vast impact on overall productivity, while the major biotic stresses are pests, diseases, and nematodes (Fig. 7.1) (Lira-Saade and Montes-Hernandez, 1994). Development of stress-resistant plants is needed and sources of resistance are being researched. Although there are many sources of resistance to biotic stresses in cucurbits, breeding for resistant varieties and hybrids has been difficult due to cross-incompatibility (Dempewolf *et al.*, 2017). This problem can be overcome by alternative methods such as genetic engineering and genome editing.

Genetic-engineering approaches have advanced to the extent that modification of the genome is possible with genes for improving adaptability and leading to resilience against different stressors (Gaba *et al.*, 2004). Cisgenics is another widely adopted technique where crops are transformed using genes from sexually compatible plants including gene introns and regulative regions such as promoters and terminators (Schouten *et al.*, 2006a,b). One of the most important advantages of cisgenesis is the avoidance of linkage drag (transmission of undesirable associated genes), which occurs during gene introgression using classical crosses (Giudice *et al.*, 2021). Gene silencing using small-interfering RNA (siRNA) has been used to develop resistance against many plant viruses in cucurbits, as all plant viruses encode silencing suppressors to overcome the host silencing pathway (Csorba *et al.*, 2015). In recent years, the clustered regularly interspaced palindromic repeats (CRISPR)-CRISPR-associated protein 9 (Cas9) system has been a powerful tool widely applied to crop improvement and gene function studies (Zhang *et al.*, 2021). CRISPR/Cas genome editing has been used to manipulate cucurbit genomes to develop novel traits for high-yielding and stress-resistant crop varieties (Chandrasekaran *et al.*, 2016; Hu *et al.*, 2017). The watermelon acetolactate synthase (*ALS*) gene was used as target for base editing to produce herbicide-resistant cucumber (Tian *et al.*, 2018). However, applications of genome editing depend on validation of the CRISPR/Cas construct using efficient transformation procedures (Nguyen *et al.*, 2022).

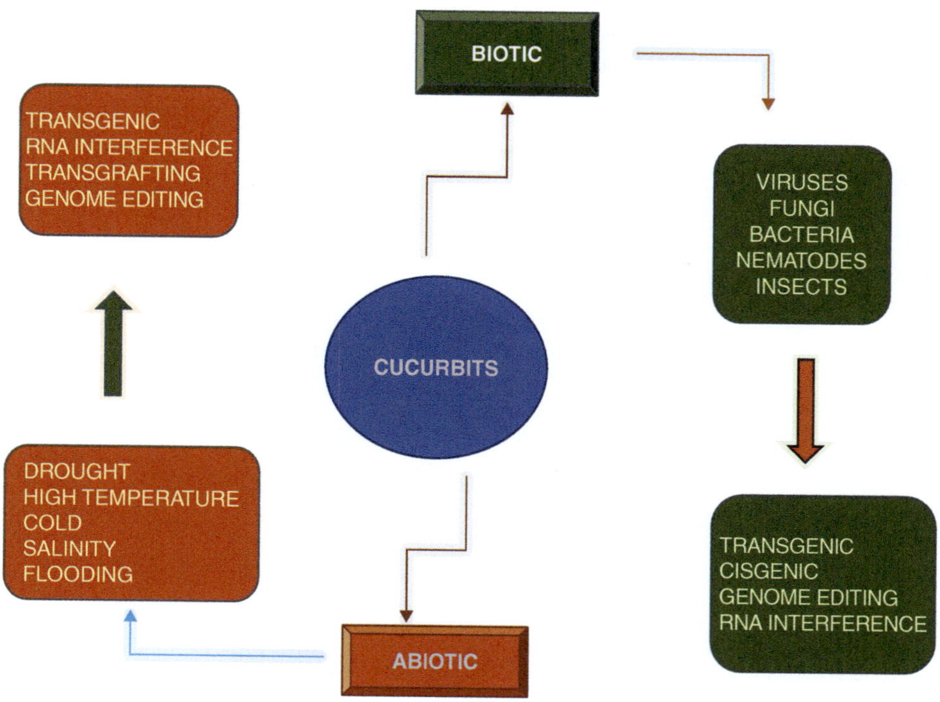


**Fig 7.1.** Classification of biotic and abiotic stresses among cucurbits.

A reproducible tissue-culture and transformation protocol is a prerequisite for successful genetic engineering (Vasudevan *et al.*, 2007). Transformation with marker and reporter genes either through an *Agrobacterium*-mediated system (Trulson *et al.*, 1986; Sarmento *et al.*, 1989) or direct gene transfer (Chee and Slightom, 1992) has been reported. Various types of stress tolerance genes have been introduced to overcome abiotic stresses in cucurbits. These transgenes can be categorized as: (i) genes that encode stress-tolerant proteins; (ii) genes responsible for signaling; and (iii) genes expressing transcription factors. However, the success of transgenic plants depends on the stable integration and expression of the heterologous genes. Therefore, development of suitable procedures and supporting tools are necessary to accelerate genetic engineering in cucurbits.

## 7.2 Abiotic Stress Management using a Transgenic Approach

Cucurbits are often threatened by abiotic stressors, namely cold and high temperatures, salinity, and drought, which affect plant growth as well as the quality and ripening of fruits. The management of abiotic stress is therefore an important target in commercial crop production programs.

### 7.2.1 Emphasis on genes involved in abiotic stress

Plants respond to stress by triggering cell signaling, transcriptional regulation, and production of reactive oxygen species (ROS), scavengers, compatible solutes, and phytohormones (Rodríguez *et al.*, 2005). Routine exposure to stress makes it essential to produce plants with stress tolerance for sustainable productivity. It is well known that all abiotic stresses are interconnected leading to a series of morphological, physiological, biochemical, and molecular changes (Baker *et al.*, 1994). Under low temperature, cellular water crystallizes and dehydrates tissues (Pearce, 2001), inhibiting the closure of stomata and resulting in water stress (Sarad *et al.*, 2004). Cold stress stimulates the formation of free radicals resulting in oxidative stress (Beck *et al.*, 2004). Similarly, high salt concentration is attributed to stomata closure and reduced transpiration, as well as preventing the roots from taking up water and nutrients (Estaji *et al.*, 2019). Some of the genes that have been targeted are shown in Fig. 7.2.

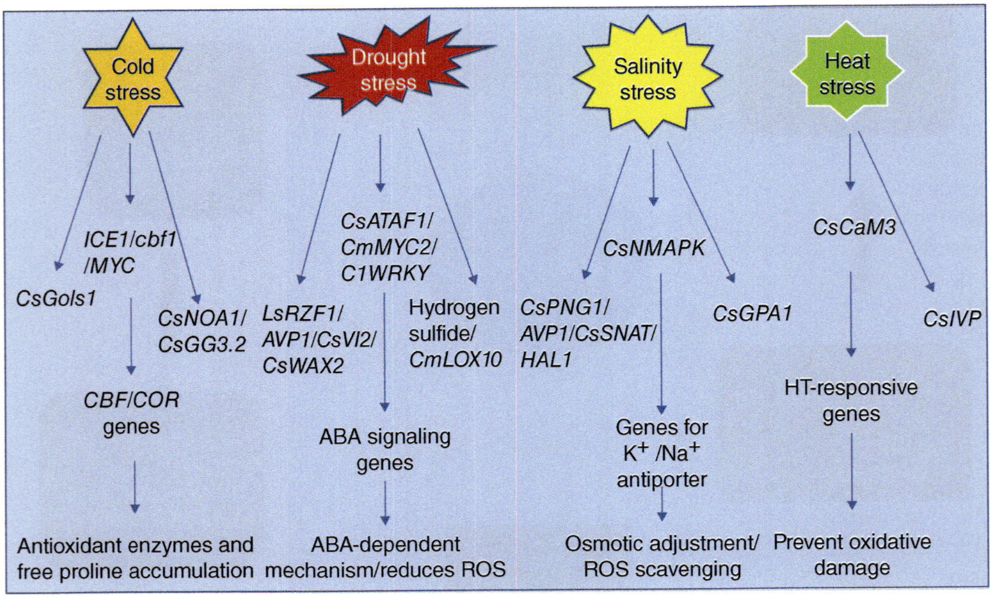


**Fig 7.2.** Genes targeted for transgenic development to combat abiotic stress in cucurbits.

### 7.2.2 Cold tolerance

Cold temperatures affect plant growth as well as fruit ripening. Therefore, developing cold tolerance in cucurbit crops would be advantageous. During cold stress, transcription factors regulate the expression of a number of genes involved in imparting tolerance in cucurbits (Bhatnagar-Mathur *et al.*, 2008). These transcription factors include CBF1, CBF2, and CBF3, which bind to the C-repeat/dehydration-responsive region in the promoter of the cold-responsive (*COR*) gene (Stockinger *et al.*, 1997; Liu *et al.*, 1998). *COR* genes encode a diverse array of proteins such as the enzymes involved in respiration, and metabolism of carbohydrates, lipid, antioxidants, and antifreeze proteins (Guy, 1990; Thomashow, 1999), and are efficient targets for genetic transformation. ICE1 is a positive regulator that specifically binds to the MYC recognition sequences in the *CBF3* promoter (Chinnusamy *et al.*, 2003). Overexpression of *ICE1* in cucumbers induced the expression of *COR* genes, further activating the accumulation of soluble sugars and inhibiting malondialdehyde accumulation, regulating osmotic potential under cold stress (Liu *et al.*, 2010). The expression of stress-inhibiting dehydrin protein (DHN24) from wild potato in cucumber resulted in improved frost tolerance (Yin *et al.*, 2006). Marker-free transgenic cucumber 'Poinsett 76 SR' plants containing the *Arabidopsis cbf1* gene were less sensitive to chilling stress (4°C) and showed a significant increase in antioxidant enzymes and free proline content in the leaves (Gupta *et al.*, 2012). The *Arabidopsis Nit* gene comprises *cis* elements conferring cold resistance and was overexpressed in cucumber (Jang *et al.*, 2013). Constitutive overexpression of nitric oxide synthase associated gene 1 (*CsNOA1*), a gaseous signaling molecule, led to a greater accumulation of soluble sugars and starch and upregulation of cold-regulatory C-repeat binding factor 3 (*CBF3*) expression in transgenic cucumber plants leading to chilling tolerance (X. Liu *et al.*, 2016). A cucumber group II WRKY gene, *CsWRKY46*, was expressed in the nucleus and upregulated in response to cold stress and exogenous abscisic acid (ABA) treatment (Zhang *et al.*, 2016). The *Arabidopsis* transgenic lines overexpressing *CsWRKY46* had higher seedling survival rates on freezing treatment compared with the wild type due to higher proline accumulation and lower malondialdehyde levels. Rubisco activase-overexpressing transgenic cucumber plants exhibited low-temperature and low-light tolerance (Bi *et al.*, 2017). The rootstock of *CBF3*-overexpressing transgenic bottle gourd promoted watermelon fruit production in cold weather and at a lower soil temperature (Cho *et al.*, 2017). Transgenic cucumber plants constitutively overexpressing cold signal transducer G-protein γsubunit (*CsGG3.2*) exhibited tolerance to chilling conditions and increased expression of *CBF* genes and their regulon (Bai *et al.*, 2018). The overexpression of galactinol synthase (*CsGolS1*) catalyzing raffinose family oligosaccharide biosynthesis enhanced the assimilate translocation efficiency and accelerated the growth rates of leaves, fruits, and whole plants under cold stress in cucumber (Dai *et al.*, 2022).

### 7.2.3 Salinity tolerance

Salinity is an important limiting factor in cucurbit production as it induces oxidative injury damaging cellular structures of crops. The tolerance to salt stress is a complex phenomenon being attempted by transferring genes involved in the synthesis of osmolytes and/or osmoprotectants (Tarczynski *et al.*, 1993; Zhang *et al.*, 2000). The osmolytes and stress-related proteins detoxify plants by scavenging ROS (Zhu, 2001). Genes coding ROS-scavenging enzymes such as glutathione *S*-transferase and peroxidase (Roxas *et al.*, 1997, 2000) or superoxide dismutases (van Camp *et al.*, 1996) enhance salt-stress tolerance in crop species. The *Saccharomyces cerevisiae HAL1* gene helps in salt tolerance of plants by increasing intracellular $K^+$ and decreasing $Na^+$ levels (Serrano 1996; Rios *et al.*, 1997). *Agrobacterium*-mediated transformation of watermelon plants with *HAL1* conferred better performance under salt-stress conditions (Ellul *et al.*, 2003). Mitogen-activated protein kinase (MAPK) cascades play a key role in biotic and abiotic stress response. The cucumber CsNMAPK is involved in positive regulation of ROS scavenging and osmotic adjustment under salt stress. Suppression of *CsNMAPK* in transgenic cucumber plants led to increased sensitivity to salt stress compared with the wild-type plants (Xu *et al.*, 2011).

Transgenic bottle gourd lines expressing the *Arabidopsis* $H^+$-pyrophosphatase *AVP1* gene maintained a higher relative water content under a salt-stress regime. Watermelon plants grafted on to the transgenic bottle gourd rootstock expressing AVP1 exhibited greater salt tolerance (Han *et al.*, 2015). Previous studies have shown that exogenous melatonin (*N*-acetyl-5-methoxytryptamine) promotes plant defense against various abiotic stresses in a concentration-dependent manner. The overexpression of serotonin *N*-acetyltransferase (*CsSNAT*) in cucumber exhibited stronger salt tolerance than wild type by promoting melatonin synthesis (Qi *et al.*, 2020). The cucumber G-protein α-subunit (CsGPA1) interacts with CsTIP1.1 (aquaporin family member) and positively regulates salt-stress tolerance. However, the RNA interference (RNAi) lines showing suppression of *CsGPA1* resulted in the reverse expression of *CsAQPs* in leaves and roots, decreasing the water content of cucumber seedlings under salt stress (Yan *et al.*, 2020). Similarly, Hou *et al.* (2020) demonstrated that peptide:*N*-glycanase (CsPNG1) also plays a positive role in improving salt tolerance in cucumber by deglycosylating misfolded glycoproteins in the endoplasmic reticulum. It was found that overexpression of *CsPNG1* in tobacco increased the salt-stress tolerance of transgenic plants.

### 7.2.4 Drought tolerance

Engineering of drought stress tolerance in cucurbits allows them to be grown in water-scarce environments as drought is one of the main factors causing preharvest losses. Drought enhances ABA accumulation and induces the expression of ABA signaling genes, leading to reduced transpiration by closure of the stomatal apertures (S. Liu *et al.*, 2016). Genetic manipulation of the vacuolar proton ($H^+$) pumps is a promising approach for generating salt- and drought-resistant crops by increasing vacuolar solute accumulation (Gaxiola *et al.*, 2001). Transgenic bottle gourd lines overexpressing the *Arabidopsis* $H^+$-pyrophosphatase *AVP1* gene showed increased auxin polar transport from shoots to roots in seedlings, suggesting that the robust root system is a prerequisite for drought tolerance (Li *et al.*, 2005; Park *et al.*, 2014). Furthermore, the *AVP1* transcripts were not found in wild-type watermelon scions grafted on to transgenic bottle gourd rootstocks (Kim *et al.*, 2015). Transgenic cucumber with abnormal expression of *CsWAX2* exhibited different sensitivities to biotic and abiotic stresses affecting the synthesis of alkanes and the function of pollen (Wang *et al.*, 2015a). Similarly, abnormal expression of *CsCER1* in transgenic cucumber plants influenced very-long-chain alkane biosynthesis and imparted drought resistance (Wang *et al.*, 2015b). Cucumber NAC transcription factor CsATAF1 functioned as a positive regulator in response to drought stress by an ABA-dependent pathway and decreasing ROS accumulation in cucumber. CsATAF1-suppressed RNAi cucumber plants showed a higher water loss rate in leaves and superoxide radicals (Wang *et al.*, 2018). *Lagenaria siceraria* RING zinc finger 1 (LsRZF1) is involved in the drought response through the proline metabolism-mediated pathway. Transgenic watermelon lines expressing antisense *LsRZF1* showed reduced expression of *GdRZF1*, a watermelon homolog of *LsRZF1*, and less sensitivity to drought stress than the wild type (Chung *et al.*, 2017). Overexpression of pumpkin phloem protein CmPP16 in pumpkin plants in the cotyledonary stage by direct inoculation of *Agrobacterium tumefaciens* and *Rhizobium rhizogenes* (formerly *Agrobacterium rhizogenes*) displayed higher photosynthetic capacity during drought than wild-type pumpkin or pumpkin transformed with another construct (Ramírez-Ortega *et al.*, 2014). Drought stress induced the transcription of the lipoxygenase gene (*CmLOX10*), a master regulator for lipid peroxidation in leaves of oriental melon seedlings. The ectopic overexpression of *CmLOX10* in *Arabidopsis* enhanced drought tolerance and decreased the stomatal aperture (Xing *et al.*, 2020). Vacuolar invertase (*CsVI2*)-overexpressing transgenic lines showed enhanced drought tolerance compared with the wild type, as invertase regulates sucrose metabolism during drought stress in cucumber seedlings (Chen *et al.*, 2021).

### 7.2.5 Heat stress tolerance

High-temperature stress limits the growth and production of cucurbits, but germplasms with high heat tolerance are very scarce in this crop. Calmodulin (CaM) has been confirmed to be related to the regulation of high-temperature stress resistance in plants. Overexpression of *CsCaM3* in cucumber plants has the potential to improve their heat tolerance by regulating the expression of high-temperature-responsive genes and protecting against oxidative damage during high-temperature stress (Yu *et al.*, 2018). *Cucumis sativus* Irregular Vasculature Patterning (*CsIVP*)-RNAi-expressing cucumber plants can improve resistance to high-temperature stress. The high survival rate of *CsIVP*-RNAi plants under high-temperature condition has been attributed to the significant expression of *Hsfs*, *Hsps*, *DREB2C*, *MBF1b*, and *WRKY33* under heat stress (Yan *et al.*, 2022).

### 7.2.6 Hypoxia and multiple-stress tolerance

The overexpression of the sucrose synthase gene (*CsSUS3*) in cucumber under the control of a stress-inducible promoter imparted higher resistance to hypoxia due to the increased supply of sugars for glycolysis as energy source for maintaining metabolism in plant under hypoxia conditions (Wang *et al.*, 2014). The WRKY proteins constitute a large family of transcription factors that regulate plant tolerance to various abiotic stresses. The expression of *WRKY* genes (*ClWRKYs*) in different tissues of watermelon revealed diverse responses to drought, salt, and cold stress (Yang *et al.*, 2018). Overexpression of multifunctional enzyme 16 putative *O*-methyltransferase (ClOMT) genes in watermelon significantly increased melatonin, a signaling molecule during abiotic stress. *ClCOMT1* knockout using the CRISPR/Cas-9 system decreased melatonin content in watermelon calluses, indicating the positive role of ClCOMT1 in abiotic stress tolerance (Chang *et al.*, 2021). Trehalose is a protective molecule under a variety of abiotic stress conditions in many organisms because of its highly hydrophilic nature (Eleutherio *et al.*, 1993). A cotyledonary node explant of cucumber co-cultivated with the gene encoding a bifunctional fusion of the trehalose 6-phosphate synthase and trehalose-6-phosphate phosphatase of *Escherichia coli* showed increased trehalose accumulation. However, all of the transgenic plants showed abnormal morphology (Kim *et al.*, 2010). Activation of the *PR-2d* promoter was investigated in transgenic cucumber using the β-galactosidase gene (*uidA*) construct in response to biotic and abiotic stress (Yin *et al.*, 2004). Hydrogen sulfide ($H_2S$) is a significant gasotransmitter involved in abiotic stress resistance. The coding DNA sequence of $H_2S$-producing *CsaLCD* or *CsaDES2* under the control of 35S promoter was transformed into a *lcd/des* mutant of *Arabidopsis*, and their heterologous expression restored the $H_2S$ production rate, as well as enhancing the resistance to osmosis stress of the transgenic seedlings (Liu *et al.*, 2021).

## 7.3 Biotic Stress Management Using a Transgenic Approach

### 7.3.1 Emphasis on genes involved in biotic stress

There has been tremendous advancement by researchers in the development of transgenic plants that are resistant to certain important biotic stresses in cucurbits. Viruses are a very important biotic stress, especially in cucurbits, so many studies have focused on the development of virus resistance in these plants (Fig. 7.3).

### 7.3.2 Transgenics for virus resistance

Transgenic plants of the inbred squash lines ZW-20 and CZW-3 carrying coat protein (CP) genes for three potyviruses, cucumber mosaic virus (CMV), zucchini yellow mosaic virus (ZYMV), and watermelon mosaic virus (WMV), were produced to have resistance to multiple viruses (Tricoli *et al.*, 1995). In 2018, virus-resistant squash was grown on a commercial scale representing 6% of total squash production in the USA. The genetically engineered squash showed higher yields than conventional squash, resulting in a net gain to farmers of US$10.1 million (Brookes and Barfoot, 2020). The genetically engineered squash line CZW3 has been permitted for import as a food by Health Canada (Health Canada, 1999). Ribozyme genes were employed to protect transgenic melon plants against two potyviruses, WMV2 and ZYMV (Huttner *et al.*, 2001). Transgenic plants resistant to all three aphid-borne viruses of squash were developed

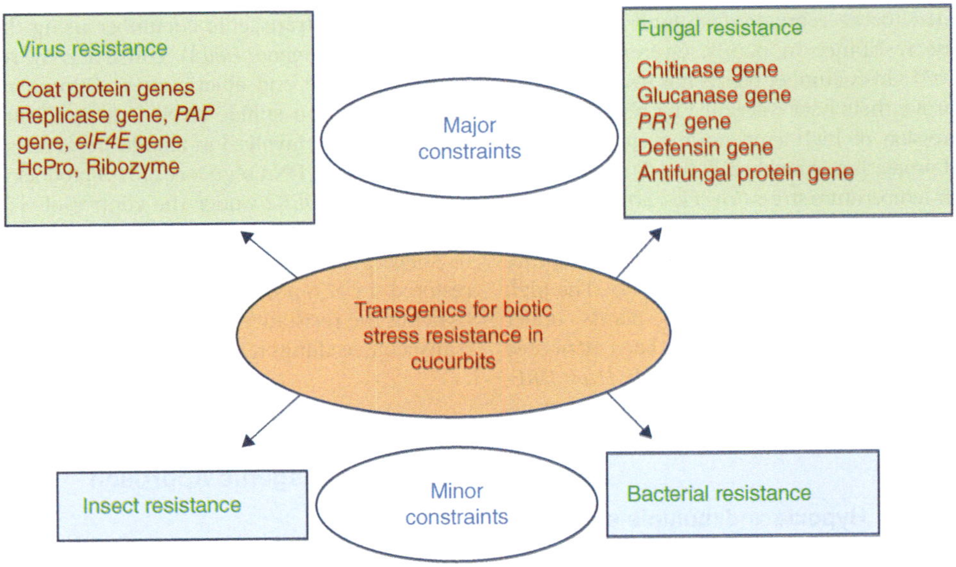


**Fig 7.3.** Genes targeted for transgenic development to combat biotic stress in various cucurbits.

and compared in terms of the fitness cost (Fuchs, *et al.*, 2004). Sheng-Niao *et al.* (2005) developed transgenic watermelon using a trivalent vector containing WMV CP genes and replicase genes of ZYMV and CMV. The watermelon lines showed resistance to these three virus diseases. A plant-expressible papaya ring spot virus PRSV replicase gene (*NIb*) was transferred into muskmelon and gene integration was demonstrated by polymerase chain reaction (PCR) amplification. However, the resistance of transgenic muskmelon plants carrying the replicase gene to PRSV was not determined (Kottearachchi *et al.*, 2000). Transformation of pokeweed antiviral protein (*PAP*) gene into plants was shown to improve plant resistance to several viral or fungal pathogens with little negative effect on plant growth. The non-virulent defective *PAP* gene (*PacPAP*) was introduced into cucumber successfully by *Agrobacterium*-mediated transformation. Compared with the non-transgenic susceptible plants, all transgenic plants with *PacPAP* showed resistance in different degrees to CMV (Cao *et al.*, 2011). Melon lines were engineered for resistance to ZYMV using a construct that contained the translatable CP sequence coupled to the 3′-non-translatable region of the virus. The elite cultivar of oriental melon ('Silver Light') was improved by transformation of the this using *Agrobacterium*-mediated transformation of cotyledon cuttings (Wu *et al.*, 2009). An effort has been made to develop a multiple virus infection-resistant transgenic watermelon ('Feeling') using a construct with a single chimeric transgene comprising a silencer DNA from the partial N gene of watermelon silver mottle virus fused to the partial CP gene sequences of CMV, cucumber green mottle mosaic virus (CGMMV), and WMV (Lin *et al.*, 2012). CGMMV in melon was effectively controlled using the transgenic approach wherein the direct repeat of the movement protein gene of CGMMV was used for transforming melon plants using *A. tumefaciens*. Compared with the non-transgenic control, these plants were shown to have high virus resistance (Emran *et al.*, 2012). Single gene-based resistance to control plant viruses may be achieved using genetic engineering, but identification of pathogen-specific genes has been very tedious. Hence, broad-spectrum and durable resistance is highly desirable. The eukaryotic translation initiation factors (eIFs) of the 4E family required for the multiplication of many different viruses have been a target. *Cucumis melo eIF4E* (*Cm-eIF4E*) knockdown melon plants were generated by transformation of explants with a construct that was designed to induce silencing of this gene. On challenge of these transgenic melon plants with eight agronomically important melon-infecting viruses, Rodríguez-Hernández *et al.* (2012) showed that they were resistant to cucumber vein yellowing virus, Moroccan watermelon mosaic virus, melon necrotic spot virus, and ZYMV, demonstrating that *Cm*-eIF4E controls susceptibility of melons to all four viruses. Therefore, *Cm-eIF4E* is an efficient target for the identification of new resistance alleles able to confer broad-spectrum virus resistance in melon. Transgenic watermelon expressing artificial microRNAs that target the CMV *2a*/*2b* genes displayed resistance to CMV infection (L. Liu *et al.*, 2016). Transgenic cucumber and melon lines harboring a hairpin construct of the zucchini yellow mosaic potyvirus (ZYMV) *HC-Pro* gene accumulated different levels of transgenic siRNA (6–44% of total siRNA) and exhibited resistance to systemic ZYMV infection (Leibman *et al.*, 2011). Cas9/subgenomic RNA technology was also employed in the development of virus resistance in cucumber by disrupting the function of the recessive eukaryotic translation initiation factor 4E (*eIF4E*) gene (Chandrasekaran *et al.*, 2016).

### 7.3.3 Transgenics for nematode resistance

RNAi constructs for root knot nematode parasitism genes have been transformed successfully into cucumber and melon. However, none of the transgenic lines was resistant to the nematode *Meloidogyne javanica* (Davis *et al.*, 2013).

### 7.3.4 Transgenics for fungal resistance

Fungal diseases also cause a lot of yield reduction in cucurbits. Genetic engineering approaches have been useful to incorporate genes from other plant species successfully and have been found to be very effective. Efforts have been made to develop downy mildew resistance using the cisgenesis approach. Wild species provide a wide gene pool for development of resistance. Downy mildew resistance has been reported in snap melon (*Cucumis melo* var. *momordica*), and fruit fly resistance has been reported in *Cucumis callosus*. Benjamin *et al.* (2009) overexpressed glyoxylate aminotransferase-encoding cisgenes *At1* and At2, identified for resistance in wild melon, in susceptible genotypes. The class I chitinase cDNA (*RCC2*) of rice was introduced into cucumber by *Agrobacterium*-mediated transformation, and the transgenic cucumbers showed varying levels of disease resistance to gray mold (Kishimoto *et al.*, 2002). In another study, Kishimoto *et al.* (2003) overexpressed the rice class I chitinase gene in cucumber. The transgenic cucumber line (CR32) exhibited resistance to *Phytophthora* rot (*Phytophthora nicotianae* var. *parasitica*) but not to *Fusarium* wilt (*Fusarium oxysporum* f. sp. *cucumerinum*). A class III chitinase gene (*CHI2*) has been induced in cucumber plants in response to infection by pathogenic microorganisms such as *Botrytis cinerea*, the causal agent of gray mold disease on cucumber (Kishimoto *et al.*, 2004). Egusi melon (*Colocynthis citrullus* L.) genotypes 'Ejagham' and NHC1-130,were transformed with *A. tumefaciens* strain EHA101 harboring the wasabi defensin gene (isolated from *Wasabia japonica* L.) in the binary vector pEKH1 (Ntui *et al.*, 2010). Transformation of oriental melon was performed with *A. tumefaciens* strain LBA4404 containing antifungal protein (*AFP*) and chitinase (*CHI*) fusion genes driven by cauliflower mosaic virus 35S promoter. The *AFP–CHI* fusion gene was effective in protecting the transgenic melon plants against fungal disease caused by *Rhizoctonia solani* and *F. oxysporum* (Bezirganoglu *et al.*, 2013). An attempt has been made to produce polygenic resistance to *Fusarium* wilt and powdery mildew with a transgene cassette of three antifungal pathogenesis-related (PR) protein genes (chitinase, glucanase, and *PR1*) into two Iranian cultivars, 'Khatooni' and 'Samsoori,' of melon (Raji *et al.*, 2022).

## 7.4 The Path Ahead for Improvement of Cucurbit Species

Genetic engineering approaches have been shown to be stable and versatile for genetic improvement of various traits including biotic and abiotic stress tolerance and quality parameters, while maintaining quality in major crops of cucurbits. Genetically engineered squash has been under commercial cultivation for more than 25 years and has achieved a remarkably positive socio-economic impact by reducing pests and diseases and increasing the quality of the end product. However genetic improvement of cucurbits is a continuous process and often changes with climatic conditions, geographical patterns, and consumer preferences. The use of new molecular breeding technologies such as transgrafting and genome editing serve as a promising approach for commercial production of cucurbits with resistance to biotic and abiotic stresses and will also help to address some of the regulatory constraints that limit the cultivation of transgenic crops. In the near future, development of varieties combining biotic and abiotic stress tolerances with high quality and yield will be highly preferred to increase production and productivity and in turn increase the income of the farmers by reducing the cost of cultivation. Transgene-free genome-editing methods can be used to develop smart hybrids/varieties/populations in cucurbits that may require less input and are widely adapted for changes in climatic conditions. These novel tools and techniques are highly useful for the accelerated development of cucurbit crops in a sustainable and environmentally friendly manner.

## 7.5 References

Andrade-Cetto, A. and Heinrich, M. (2005) Mexican plants with hypoglycaemic effect used in the treatment of diabetes. *Journal of Ethnopharmacology* 99, 325–348.

Argueta, A. (1994) Two decades of Mexican ethnobotany and research on plant derived drugs. *Atlas de las Plantas de la Medicina Tradicional Mexicana* 3: 172.

Bai, L., Liu, Y., Mu, Y., Anwar, A., He, C., *et al.* (2018) Heterotrimeric G-protein γ subunit CsGG3.2 positively regulates the expression of *CBF* genes and chilling tolerance in cucumber. *Frontiers in Plant Science* 9: 488.

Baker, A.J.M., Reeves, R.D., and Hajar, A.S.M. (1994) Heavy metal accumulation and tolerance in British populations of the metallophyte *Thlaspi caerulescens* J. & C. Presl (Brassicaceae). *New Phytologist* 127, 61–68.

Beck, E.H., Heim, R., and Hansen, J. (2004) Plant resistance to cold stress: mechanisms and environmental signals triggering frost hardening and dehardening. *Journal of Biosciences* 29, 449–459.

Benjamin, I., Kenigsbuch, D., Galperin, M., Abrameto, J.A., and Cohen, Y. (2009) Cisgenic melons over expressing glyoxylate-aminotransferase are resistant to downy mildew. *European Journal of Plant Pathology* 125, 355–365.

Bezirganoglu, I., Hwang, S.Y., Fang, T.J., and Shaw, J.F. (2013) Transgenic lines of melon (*Cucumis melo* L. var. *makuwa* cv. 'Silver Light') expressing antifungal protein and chitinase genes exhibit enhanced resistance to fungal pathogens. *Plant Cell, Tissue and Organ Culture* 112, 227–237.

Bhatnagar-Mathur, P., Vadez, V., and Sharma, K.K. (2008) Transgenic approaches for abiotic stress tolerance in plants: retrospect and prospects. *Plant Cell Reports* 27, 411–424.

Bi, H., Liu, P., Jiang, Z., and Ai, X. (2017) Overexpression of the rubisco activase gene improves growth and low temperature and weak light tolerance in *Cucumis sativus*. *Physiologia Plantarum* 161, 224–234.

Brookes, G. and Barfoot, P. (2020) *GM Crops: Global Socio-Economic and Environmental Impacts 1996–2018*. PG Economics Ltd, Dorchester, UK.

Cao, B., Lei, J., Chen, G., Cao, P., Liu, X., *et al.* (2011) Testing of disease-resistance of pokeweed antiviral protein gene (*PacPAP*) in transgenic cucumber (*Cucumis sativus*). *African Journal of Biotechnology* 10, 6883–6890.

Chandrasekaran, J., Brumin, M., Wolf, D., Leibman, D., Klap, C., *et al.* (2016) Development of broad virus resistance in non-transgenic cucumber using CRISPR/Cas9 technology. *Molecular Plant Pathology* 17, 1140–1153.

Chang, J., Guo, Y., Yan, J., Zhang, Z., Yuan, L., *et al.* (2021) The role of watermelon caffeic acid *O*-methyltransferase (*ClCOMT1*) in melatonin biosynthesis and abiotic stress tolerance. *Horticulture Research* 8: 210.

Chee, P.P. and Slightom, J.L. (1992) Transformation of cucumber tissues by microprojectile bombardment: identification of plants containing functional and non-functional transferred genes. *Gene* 118, 255–260.

Chen, L., Zheng, F., Feng, Z., Li, Y., Ma, M., *et al.* (2021) A vacuolar invertase CsVI2 regulates sucrose metabolism and increases drought tolerance in *Cucumis sativus* L. *International Journal of Molecular Sciences* 23: 176.

Chiej, R. (1984) *Encyclopaedia of Medicinal Plants (MacDonald Encyclopedias)*. Little, Brown and Company, London.

Chinnusamy, V., Ohta, M., Kanrar, S., Lee, B.H., Hong, X., *et al.* (2003) ICE1: a regulator of cold-induced transcriptome and freezing tolerance in *Arabidopsis*. *Genes & Development* 17, 1043–1054.

Cho, S.H., Joung, Y.H., Karna, S., Lee, H.E., Kim, J.H., *et al.* (2017) The development of cold resistance rootstock using *Agrobacterium*-mediated transformation of *Arabidopsis* CBF3/DREB1A in bottle gourd (*Lageneraria siceraria* Standl.). *Scientia Horticulturae* 214, 141–146.

Choudhuri, P., Mondal, B., and Laskar, N. (2017) *Cucurbits: Biotic and Abiotic Stresses*. New India Publishing Agency, New Delhi.

Chung, J.S., Park, S.H., Min, J.H., Min, K.H., Lee, S., *et al.* (2017) Reduced expression of gongdae ring zinc finger 1 (GdRZF1) enhances drought stress tolerance in Watermelon (*Citrullus lanatus*). *Horticultural Science and Technology* 35, 637–646.

Csorba, T., Kontra, L., and Burgyán, J. (2015) Viral silencing suppressors: tools forged to fine-tune host-pathogen coexistence. *Virology* 479–480, 85–103.

Dai, H., Zhu, Z., Wang, Z., Zhang, Z., Kong, W., and Miao, M. (2022) Galactinol synthase 1 improves cucumber performance under cold stress by enhancing assimilate translocation. *Horticulture Research* 9: uhab063.

Davis, E.L., Oka, Y., Gal-On, A., Wehner, T., and Zelcer, A. (2013) *Broad-spectrum Resistance to Root-knot Nematodes in Transgenic Cucurbits*. US Department of Agriculture National Agricultural Library. Available at: https://search.nal.usda.gov/discovery/fulldisplay?context=L&vid=01NAL_INST:MAIN&search_scope=MyInstitution&tab=LibraryCatalog&docid=alma9916346487507426 (accessed 26 June 2024).

Dempewolf, H., Baute, G., Anderson, J., Kilian, B., Smith, C., and Guarino, L. (2017) Past and future use of wild relatives in crop breeding. *Crop Science* 57, 1070–1082.

Eleutherio, E.C., Araujo, P.S., and Panek, A.D. (1993) Protective role of trehalose during heat stress in *Saccharomyces cerevisiae*. *Cryobiology* 30, 591–596.

Ellul, P., Rios, G., Atares, A., Roig, L.A., Serrano, R., and Moreno, V. (2003) The expression of the *Saccharomyces cerevisiae* HAL1 gene increases salt tolerance in transgenic watermelon [*Citrullus lanatus* (Thunb.) Matsun. & Nakai.]. *Theoretical and Applied Genetics* 107, 462–469.

Emran, A.M., Tabei, Y., Kobayashi, K., Yamaoka, N., and Nishiguchi, M. (2012) Molecular analysis of transgenic melon plants showing virus resistance conferred by direct repeat of movement gene of *Cucumber green mottle mosaic virus*. *Plant Cell Reports* 31, 1371–1377.

Estaji, A., Kalaji, H.M., Karimi, H.R., Roosta, H.R., and Moosavi-Nezhad, S.M. (2019) How glycine betaine induces tolerance of cucumber plants to salinity stress. *Photosynthetica* 57, 753–761.

Fuchs, M., Chirco, E.M., McFerson, J.R., and Gonsalves, D. (2004) Comparative fitness of a wild squash species and three generations of hybrids between wild × virus-resistant transgenic squash. *Environmental Biosafety Research* 3, 17–28.

Gaba, V., Zelcer, A., and Gal-On, A. (2004) Cucurbit biotechnology—the importance of virus resistance. *In Vitro Cellular & Developmental Biology—Plant* 40, 346–358.

Gaxiola, R.A., Li, J., Undurraga, S., Dang, L.M., Allen, G.J., *et al.* (2001) Drought-and salt-tolerant plants result from overexpression of the AVP1 $H^+$-pump. *Proceedings of the National Academy of Sciences USA* 98, 11444–11449.

Giudice, G., Moffa, L., Varotto, S., Cardone, M.F., Bergamini, C., *et al.* (2021) Novel and emerging biotechnological crop protection approaches. *Plant Biotechnology Journal* 19, 1495–1510.

Gupta, N., Rathore, M., Goyary, D., Khare, N., Anandhan, S., *et al.* (2012) Marker-free transgenic cucumber expressing *Arabidopsis cbf1* gene confers chilling stress tolerance. *Biologia Plantarum* 56, 57–63.

Guy, C.L. (1990) Cold acclimation and freezing stress tolerance: role of protein metabolism. *Annual Review of Plant Biology* 41, 187–223.

Han, J.S., Park, K.I., Jeon, S.M., Park, S., Naing, A.H., and Kim, C.K. (2015) Assessments of salt tolerance in a bottle gourd line expressing the *Arabidopsis* $H^+$-pyrophosphatase *AVP1* gene and in a watermelon, plant grafted onto a transgenic bottle gourd rootstock. *Plant Breeding* 134, 233–238.

Health Canada (1999) Novel Food Information—Virus Resistant Squash Line CZW-3. Available at: www.canada.ca/en/health-canada/services/foodnutrition/genetically-modified-foods-other-novel-foods/approved-products/virus-resistant-squash-line-czw-3.html.

Hou, K., Wang, Y., Tao, M.Q., Jahan, M.S., Shu, S., *et al.* (2020) Characterization of the *CsPNG1* gene from cucumber and its function in response to salinity stress. *Plant Physiology and Biochemistry* 150, 140–150.

Hu, B., Li, D., Liu, X., Qi, J., Gao, D., *et al.* (2017) Engineering non-transgenic gynoecious cucumber using an improved transformation protocol and optimized CRISPR/Cas9 system. *Molecular Plant* 10, 1575–1578.

Huttner, E., Tucker, W., Vermeulen, A., Ignart, F., Sawyer, B., and Birch, R. (2001) Ribozyme genes protecting transgenic melon plants against potyviruses. *Current Issues in Molecular Biology* 3, 27–34.

Jang, H.A., Lim, K.M., Kim, H.A., Park, Y.I., Kwon, S.Y., and Choi, P.S. (2013) Development of transgenic cucumbers expressing *Arabidopsis Nit* gene. *Journal of Plant Biotechnology* 40, 198–202.

Kim, B.O., Han, J.S., Park, K.I., Jeon, S.M., and Kim, C.K. (2015) Absence of AVP1 transcripts in wild type watermelon scions grafted onto transgenic bottle gourd rootstocks. *Journal of Plant Biotechnology* 42, 13–18.

Kim, H.A., Min, S.R., Choi, D.W., Choi, P.S., and Hong, S.G. (2010) Development of transgenic cucumber expressing TPSP gene and morphological alterations. *Journal of Plant Biotechnology* 37, 72–76.

Kishimoto, K., Nishizawa, Y., Tabei, Y., Hibi, T., Nakajima, M., and Akutsu, K. (2002) Detailed analysis of rice chitinase gene expression in transgenic cucumber plants showing different levels of disease resistance to graymold (*Botrytis cinerea*). *Plant Science* 162, 655–662.

Kishimoto, K., Nakajima, M., Nishizawa, Y., Tabei, Y., Hibi, T., and Akutsu, K. (2003) Response of transgenic cucumber expressing a rice class I chitinase gene to two fungal pathogens with different infectivities. *Journal of General Plant Pathology* 69, 358–363.

Kishimoto, K., Nakajima, M., Nishizawa, Y., Tabei, Y., Hibi, T., and Akutsu, K. (2004) Transgenic cucumber expressing an endogenous class III chitinase gene has reduced symptoms from *Botrytis cinerea*. *Journal of General Plant Pathology* 70, 314–320.

Kottearachchi, N.S., Kertbundit, S., and Juricek, M. (2000) *Agrobacterium* mediated transformation of *Cucumis melo* with replicase gene from papaya ringspot virus and regeneration of transformed plants. *Tropical Agriculture Research and Extension* 3, 94–97.

Leibman, D., Wolf, D., Saharan, V., Zelcer, A., Arazi, T., *et al.* (2011) A high level of transgenic viral small RNA is associated with broad potyvirus resistance in cucurbits. *Molecular Plant–Microbe Interactions* 24, 1220–1238.

Li, J., Yang, H., Peer, W.A., Richter, G., Blakeslee, J., *et al.* (2005) *Arabidopsis* $H^+$-PPase AVP1 regulates auxin-mediated organ development. *Science* 310, 121–125.

Lin, C.Y., Ku, H.M., and Chiang, Y.H. (2012) Development of transgenic watermelon resistant to *Cucumber mosaic virus* and *Watermelon mosaic virus* by using a single chimeric transgene construct. *Transgenic Research* 21, 983–993.

Lira-Saade, R., and Montes-Hernandez, M. (1994) Cucurbits (*Cucurbita* spp.). In: Hernando Bermejo, J., and Leon, J. (eds.) *Neglected Crops: 1492 from a Different Perspective.* Plant Production and Protection Series No. 26. Food and Agriculture Organization of the United Nations, Rome, pp. 63–77.

Liu, L., Duan, L., Zhang, J., Zhang, Z., Mi, G., and Ren, H. (2010) Cucumber (*Cucumis sativus* L.) overexpressing cold-induced transcriptome regulator *ICE1* exhibits changed morphological characters and enhances chilling tolerance. *Scientia Horticulturae* 124, 29–33.

Liu, L., Gu, Q., Ijaz, R., Zhang, J., and Ye, Z. (2016) Generation of transgenic watermelon resistance to *Cucumber mosaic virus* facilitated by an effective *Agrobacterium*-mediated transformation method. *Scientia Horticulturae* 205, 32–38.

Liu, Q., Kasuga, M., Sakuma, Y., Abe, H., Miura, S., *et al.* (1998) Two transcription factors, DREB1 and DREB2, with an EREBP/AP2 DNA binding domain separate two cellular signal transduction pathways in drought-and low-temperature-responsive gene expression, respectively, in *Arabidopsis*. *Plant Cell* 10, 1391–1406.

Liu, S., Li, H., Lv, X., Ahammed, G.J., Xia, X., *et al.* (2016) Grafting cucumber onto luffa improves drought tolerance by increasing ABA biosynthesis and sensitivity. *Scientific Reports* 6: 20212.

Liu, X., Liu, B., Xue, S., Cai, Y., Qi, W., *et al.* (2016) Cucumber (*Cucumis sativus* L.) nitric oxide synthase associated gene1 (*CsNOA1*) plays a role in chilling stress. *Frontiers in Plant Science* 7: 1652.

Liu, Z., Qu, J., Zhang, L., Liu, X., Yang, G., and Pei, Y. (2021) Cloning of cucumber LCD and DES gene and their response to abiotic stress. *Scientia Horticulturae* 278: 109802.

Nguyen, D.V., Hoang, T.T., Le, N.T., Tran, H.T., Nguyen, C.X., *et al.* (2022) An efficient hairy root system for validation of plant transformation vector and CRISPR/Cas construct activities in cucumber (*Cucumis sativus* L.). *Frontiers in Plant Science* 12: 770062.

Ntui, V.O., Thirukkumaran, G., Azadi, P., Khan, R.S., Nakamura, I., and Mii, M. (2010) Stable integration and expression of wasabi defensin gene in "Egusi" melon (*Colocynthis citrullus* L.) confers resistance to *Fusarium* wilt and *Alternaria* leaf spot. *Plant Cell Reports* 29, 943–954.

OECD (2016) Squashes, pumkins, zucchinis and gourds (*Cucurbita* species). In: *Safety Assessment of Transgenic Organisms in the Environment, Vol. 5: OECD Consensus Documents.* OECD Publishing, Paris.

Park, M., Han, J., Ahn, Y., Kim, J., Lee, H., *et al.* (2014) Ectopic expression of *Arabidopsis* $H^+$-pyrophosphatase AVP1 enhances drought resistance in bottle gourd (*Lagenaria siceraria* Standl.). *Plant Cell, Tissue and Organ Culture* 118, 383–389.

Pearce, R.S. (2001) Plant freezing and damage. *Annals of Botany* 87, 417–424.

Qi, C., Zhang, H., Liu, Y., Wang, X., Dong, D., *et al.* (2020) CsSNAT positively regulates salt tolerance and growth of cucumber by promoting melatonin biosynthesis. *Environmental and Experimental Botany* 175: 104036.

Raji, M.R., Lotfi, M., Tohidfar, M., Ramshini, H., Sahebani, N., *et al.* (2022) Multiple fungal diseases resistance induction in *Cucumis melo* through co-transformation of different pathogenesis related (PR) protein genes. *Scientia Horticulturae* 297: 110924.

Ramírez-Ortega, F.A., Herrera-Pola, P.S., Toscano-Morales, R., Xoconostle-Cázares, B., and Ruiz-Medrano, R. (2014) Overexpression of the pumpkin (*Cucurbita maxima*) 16 kDa phloem protein CmPP16 increases tolerance to water deficit. *Plant Signaling & Behavior* 9: e973823.

Renner, S.S., and Pandey, A.K. (2013) The Cucurbitaceae of India: accepted names, synonyms, geographic distribution, and information on images and DNA sequences. *PhytoKeys* 20, 53–118.

Rios, G., Ferrando, A., and Serrano, R. (1997) Mechanism of salt tolerance conferred by overexpression of the *HAL1* gene in *Saccharomyces cerevisiae*. *Yeast* 13, 515–528.

Rodríguez, M., Canales, E., and Borrás-Hidalgo, O. (2005) Molecular aspects of abiotic stress in plants. *Biotecnología Aplicada* 22, 1–10.

Rodríguez-Hernández, A.M., Gosalvez, B., Sempere, R.N., Burgos, L., Aranda, M.A., and Truniger, V. (2012) Melon RNA interference (RNAi) lines silenced for Cm-eIF4E show broad virus resistance. *Molecular Plant Pathology* 13, 755–763.

Roxas, V.P., Smith, R.K., Allen, E.R., and Allen, R.D. (1997) Overexpression of glutathione *S*-transferase/glutathione peroxidase enhances the growth of transgenic tobacco seedlings during stress. *Nature Biotechnology* 15, 988–991.

Roxas, V.P., Lodhi, S.A., Garrett, D.K., Mahan, J.R., and Allen, R.D. (2000) Stress tolerance in transgenic tobacco seedlings that overexpress glutathione *S*-transferase/glutathione peroxidase. *Plant and Cell Physiology* 41, 1229–1234.

Sarad, N., Rathore, M., Singh, N.K., and Kumar, N. (2004) Genetically engineered tomatoes: new vista for sustainable agriculture in high altitude regions. In: *Proceedings of the Fourth International Crop Science Congress*. The Regional Institute Ltd., Brisbane, Australia.

Sarmento, G.G., Alpert, K.B., Punja, Z.K., and Tang, F.A. (1989) Transformation of pickling cucumber (*Cucumis sativus* L.) by *Agrobacterium tumefaciens* and expression of kanamycin resistance in regenerated transgenic plants. *Journal of Cellular Biochemistry* 13: 268.

Schouten, H.J., Krens, F.A., and Jacobsen, E. (2006a) Cisgenic plants are similar to traditionally bred plants: international regulations for genetically modified organisms should be altered to exempt cisgenesis. *EMBO Reports* 7, 750–753.

Schouten, H.J., Krens, F.A., and Jacobsen, E. (2006b) Do cisgenic plants warrant less stringent oversight? *Nature Biotechnology* 24, 753–753.

Serrano, R. (1996) Salt tolerance in plants and microorganisms: toxicity targets and defense responses. *International Review of Cytology* 165, 1–52.

Sheng-Niao, N., Xue-Sen, H., Sek-Man, W., Jia-Lin, Y., Fu-Xing, Z., *et al.* (2005) Creation of trivalent transgenic watermelon resistant to virus infection. *Chinese Journal of Agricultural Biotechnology* 2, 179–185.

Stockinger, E.J., Gilmour, S.J., and Thomashow, M.F. (1997) *Arabidopsis thaliana CBF1* encodes an AP2 domain-containing transcriptional activator that binds to the C-repeat/DRE, a *cis*-acting DNA regulatory element that stimulates transcription in response to low temperature and water deficit. *Proceedings of the National Academy of Sciences USA* 94, 1035–1040.

Tarczynski, M.C., Jensen, R.G., and Bohnert, H.J. (1993) Stress protection of transgenic tobacco by production of the osmolyte mannitol. *Science* 259, 508–510.

Thomashow, M.F. (1999) Plant cold acclimation: freezing tolerance genes and regulatory mechanisms. *Annual Review of Plant Biology* 50, 571–599.

Tian, S., Jiang, L., Cui, X., Zhang, J., Guo, S., *et al.* (2018) Engineering herbicide-resistant watermelon variety through CRISPR/Cas9-mediated base-editing. *Plant Cell Reports* 37, 1353–1356.

Tricoli, D.M., Carney, K.J., Russell, P.F., McMaster, J.R., Groff, D.W., *et al.* (1995) Field evaluation of transgenic squash containing single and multiple protein gene constructs for resistance to *Cucumber mosaic virus*, *Watermelon mosaic virus 2*, and *Zucchini yellow mosaic virus*. *Biotechnology* 13, 1458–1465.

Trulson, A.J., Simpson, R.B., and Shahin, E.A. (1986) Transformation of cucumber (*Cucumis sativus* L.) plants with *Agrobacterium rhizogenes*. *Theoretical and Applied Genetics* 73, 11–15.

van Camp, W., Capiau, K., van Motagu, M., Inze, D., and Slooten, L. (1996) Enhancement of oxidative stress tolerance in transgenic tobacco plants overproducing Fe-superoxide dismutase in chloroplasts. *Plant Physiology* 112, 1703–1714.

Vasudevan, A., Selvaraj, N., Ganapathi, A., and Choi, C.W. (2007) *Agrobacterium*-mediated genetic transformation in cucumber (*Cucumis sativus* L.). *American Journal of Biochemistry and Biotechnology* 3, 24–32.

Wang, H., Sui, X., Guo, J., Wang, Z., Cheng, J., *et al.* (2014) Antisense suppression of cucumber (*Cucumis sativus* L.) sucrose synthase 3 (*CsSUS3*) reduces hypoxic stress tolerance. *Plant, Cell & Environment* 37, 795–810.

Wang, J., Zhang, L., Cao, Y., Qi, C., Li, S., *et al.* (2018) CsATAF1 positively regulates drought stress tolerance by an ABA-dependent pathway and by promoting ROS scavenging in cucumber. *Plant and Cell Physiology* 59, 930–945.

Wang, W., Liu, X., Gai, X., Ren, J., Liu, X., *et al.* (2015a) *Cucumis sativus* L. WAX2 plays a pivotal role in wax biosynthesis, influencing pollen fertility and plant biotic and abiotic stress responses. *Plant and Cell Physiology* 56, 1339–1354.

Wang, W., Zhang, Y., Xu, C., Ren, J., Liu, X., *et al.* (2015b) Cucumber *ECERIFERUM1* (*CsCER1*), which influences the cuticle properties and drought tolerance of cucumber, plays a key role in VLC alkanes biosynthesis. *Plant Molecular Biology* 87, 219–233.

Wu, H.W., Yu, T.A., Raja, J.A., Wang, H.C., and Yeh, S.D. (2009) Generation of transgenic oriental melon resistant to *Zucchini yellow mosaic virus* by an improved cotyledon-cutting method. *Plant Cell Reports* 28, 1053–1064.

Xia, T., and Wang, Q. (2006) D-*chiro*-Inositol found in *Cucurbita ficifolia* (Cucurbitaceae) fruit extracts plays the hypoglycaemic role in streptozocin-diabetic rats. *Journal of Pharmacy and Pharmacology* 58, 1527–1532.

Xing, Q., Liao, J., Cao, S., Li, M., Lv, T., and Qi, H. (2020) *CmLOX10* positively regulates drought tolerance through jasmonic acid-mediated stomatal closure in oriental melon (*Cucumis melo* var. *makuwa* Makino). *Scientific Reports* 10: 17452.

Xu, H., Sun, X., Wang, X., Shi, Q., Yang, X., and Yang, F. (2011) Involvement of a cucumber MAPK gene (*CsNMAPK*) in positive regulation of ROS scavengence and osmotic adjustment under salt stress. *Scientia Horticulturae* 127, 488–493.

Yan, S., Yu, B., Ming, F., Liang, Y., Zhong, Y., *et al.* (2022) CsIVP modulates low nitrogen and high-temperature resistance in cucumber. *Plant and Cell Physiology* 63, 605–617.

Yan, Y., Sun, M., Li, Y., Wang, J., He, C., and Yu, X. (2020) The *CsGPA1-CsAQPs* module is essential for salt tolerance of cucumber seedlings. *Plant Cell Reports* 39, 1301–1316.

Yang, X., Li, H., Yang, Y., Wang, Y., Mo, Y., *et al.* (2018) Identification and expression analyses of *WRKY* genes reveal their involvement in growth and abiotic stress response in watermelon (*Citrullus lanatus*). *PLOS ONE* 13: e0191308.

Yin, Z., Hennig, J., Szwacka, M., and Malepszy, S. (2004) Tobacco PR-2d promoter is induced in transgenic cucumber in response to biotic and abiotic stimuli. *Journal of Plant Physiology* 161, 621–629.

Yin, Z., Rorat, T., Szabala, B.M., Ziółkowska, A., and Malepszy, S. (2006) Expression of a *Solanum sogarandinum* SK3-type dehydrin enhances cold tolerance in transgenic cucumber seedlings. *Plant Science* 170, 1164–1172.

Yu, B., Yan, S., Zhou, H., Dong, R., Lei, J., *et al.* (2018) Overexpression of *CsCaM3* improves high temperature tolerance in cucumber. *Frontiers in Plant Science* 9: 797.

Zhang, D., Zhang, Z., Unver, T., and Zhang, B. (2021) CRISPR/Cas: a powerful tool for gene function study and crop improvement. *Journal of Advanced Research* 29, 207–221.

Zhang, J., Klueva, N.Y., Wang, Z., Wu, R., Ho, T.H.D., and Nguyen, H.T. (2000) Genetic engineering for abiotic stress resistance in crop plants. *In Vitro Cellular & Developmental Biology-Plant* 36, 108–114.

Zhang, Y., Yu, H., Yang, X., Li, Q., Ling, J., *et al.* (2016) CsWRKY46, a WRKY transcription factor from cucumber, confers cold resistance in transgenic-plant by regulating a set of cold-stress responsive genes in an ABA-dependent manner. *Plant Physiology and Biochemistry* 108, 478–487.

Zhu, J.K. (2001) Plant salt tolerance. *Trends in Plant Science* 6, 66–71.

# 8 Genetic Engineering in Potato: Status and Future Perspectives

S. Sundaresha, Jagesh K. Tiwari*, Vinay Bhardwaj, Sanjeev Sharma, A. Jeevalatha, Ravinder Kumar, Vinay Sagar, B.P. Singh, S.K. Chakrabarti and Manoj Kumar

*ICAR-Central Potato Research Institute, Shimla, Himachal Pradesh, India*

**Abstract**

Potato is the world's most important vegetable crop. Important traits related to quality production can be detected easily at the phenotypic level in wild/cultivated potatoes. The development of varieties through conventional breeding can take 10–15 years due to sexual incompatibility barriers between diploid and tetraploid species. Hence, genetic engineering is necessary for fast development of cultivars. Potato is easily grown using tissue culture and it is relatively easy to introduce new traits through genetic transformation. The advantages of genetically engineered potatoes are pest and disease resistance, increased production/productivity, better processing traits, improved nutritional quality, and reductions in production costs, but there are environmental risks. Here, we highlight the progress made in traits utilized in potato improvement using genetic-engineering methods.

**Keywords:** abiotic stress, biotic stress, potato, quality trait, transgenics

## 8.1 Introduction

Potato is the primary major non-grain food crop in the world. Molecular cloning or recombinant DNA technology is an important practice in research laboratories that is used to create copies of a particular gene for downstream applications, such as sequencing, mutagenesis, genotyping, or heterologous expression of a protein. Heterologous expression is used to produce large amounts of a protein of interest for functional and biochemical analyses. It is a traditional technique frequently used to understand biochemical and functional aspects, and involves the transfer of a DNA fragment of interest from one organism to other via a self-replicating genetic material called a plasmid. Cloned genes have been integrated randomly into the host genome for the production of transgenic organisms which are then said to be genetically modified. Foreign DNA is introduced into the organism and then transmitted through the germ line so that every cell, including germ cells, of the organism contains the same modified genetic material. However, in the case of vegetatively propagated crops, the stability of the transgene needs to be checked in the first two generations for inheritance of the gene and traits. This has been demonstrated in transgenic potato events from various research groups including in India (Sanju *et al.*, 2015; Shandil *et al.*, 2017; Tomar *et al.*, 2018).

Gene cloning has become much easier with the availability of huge bioinformatics databases. For example, the National Institutes of Health's GenBank database (www.ncbi.nlm.nih.gov/genbank/,

*Email: jageshtiwari@gmail.com


DOI: 10.1079/9781800627123.0008

accessed 12 July 2024) maintains more than 71 million sequences, out of which around 90,781 are from potato. Many of these nucleotide sequences represent identified genes. For potato, about 39,800 gene sequences are available in the Potato Genome Sequencing Consortium database (http://solanaceae.plantbiology.msu.edu/pgsc_download.shtml, accessed 12 July 2024). This information can be utilized for functional genomic research studies including the development of transgenic plants. The availability of a suitable regeneration protocol is a pre-requisite for undertaking genetic transformation work in any crop. A rapid and efficient *Agrobacterium tumefaciens*-mediated transformation protocol based on direct organogenesis from internodal stem explants of *in vitro* potato plants has been well standardized. It ensures the development of large numbers of transgenic plants in a short period of 4–6 weeks without the formation of an intermediary callus phase. The protocol gives about 70% regeneration efficiency and is being used routinely for genetic transformation of potato cultivars (Shandil *et al.*, 2017).

Before a cloning experiment is undertaken for such genes, scientists must understand how many genes are involved, how they are regulated, what other effects they might have on the plant, and how they interact with other genes active in the same biochemical pathway. Public and private research programs are investing heavily into newer technologies of gene expression to investigate these issues and to rapidly sequence and determine functions of complex genes. These efforts should result in identification of a large number of genes potentially useful for producing transgenic varieties and for functional validation.

## 8.2 Possible Targeted Traits for Genetically Engineered Potato

Conventional breeding of potato is a complex and time-consuming exercise. Mutation breeding is also not feasible in a crop like potato that is autotetraploid, heterozygous, and vegetative propagated. Transformation of potato with transgenes or cisgenes is a new possibility to improve existing varieties without causing any major phenotypic side effects. The Indian Council of Agricultural Research (ICAR)—Central Potato Research Institute (CPRI), Shimla, Himachal Pradesh, India, has the policy of using genetic transformation techniques for improving those traits that cannot be manipulated by conventional breeding. The institute has adequate facilities and expertise for cloning and characterization of genes, for designing gene constructs for plant expression, and also for gene silencing through RNA interference (RNAi) technology. Many of the gene sequences available from the Potato Genome Sequence Consortium have been used for cloning and genetic transformation in potato.

Needs-based potato research and development has prioritized a number of traits for improvement using the transgenic approach and is working on the cloning of novel genes that could be exploited in potato transgenic plants. Some examples are as follows:

- *Transgenic potato for durable resistance to late blight*. As none of the varieties have durable resistance, lines have been developed expressing the *RB* gene in popular Indian potato cultivars using the cisgene approach followed by backcross breeding.
- *Virus-resistant transgenic potato*. Due to unavailability of host resistance sources, transgenics have been developed in potato against tomato leaf curl New Delhi virus (ToLCNDV) using the replicase gene of RNAi constructs and against potato virus Y using the coat protein gene.
- *Transgenic potato for reduced cold-induced sweetening*. Indian potato varieties lack resistance against sweetening during cold storage. ICAR-CPRI has developed potatoes with reduced cold-induced sweetening using RNAi technology.
- *Dwarf transgenic potato*. Dwarf potato lines have been developed for a better harvest index, particularly in hill conditions.
- *Bacterial wilt resistance potato using RNAi technology*. None of the germplasms have a resistance source, so RNAi technology has been used to suppress the susceptible factor to achieve resistance against bacterial wilt.
- *Transgenic potato for heat and drought stress, and early maturing*. Transgenic potato has been developed for tolerance to heat and drought stress, as well as early-maturing potato genotypes.

### 8.2.1 Transgenic potato for late blight resistance using overexpression and RNAi technology

Since the Irish famine in 1845–1849 and continuing today, potato production across the world has been severely affected by the notorious disease late blight, caused by *Phytophthora infestans*. Various management strategies including host resistance, disease forecasting, use of chemicals, and cultural practices are being employed to manage this disease, but *P. infestans* is a notorious oomycete that is able to rapidly evolve and adapt to the host background due to the predominance of transposable elements in the pathogen genome.

The *RB/Rpi-blb1* gene was isolated from the wild potato species *Solanum bulbocastanum*, cloned in pCLD04541 and transferred into the potato cultivar 'Katahdin' using *Agrobacterium tumefaciens* (Song *et al.*, 2003; Vleeshouwers *et al.*, 2008). This gene imparts a broad spectrum of resistance against all known races of *P. infestans* to RB-Katahdin lines, both in the greenhouse and in field experiments (Kuhl *et al.*, 2007; Halterman *et al.*, 2008). The wild diploid species *S. bulbocastanum* originated from Mexico and Guatemala, and is highly resistant to late blight. The event SP951 (potato 'Katahdin') contains the coiled coil– nucleotide-binding site–leucine-rich repeat class of resistance (*R*) gene (*RB*) isolated from the sexually incompatible wild species *S. bulbocastanum*. This event was transferred to ICAR from the University of Wisconsin through the Agricultural Biotechnology Support Project II. At ICAR-CPRI, Shimla, India, a confined field trial was carried out with permission from the Review Committee on Genetic Manipulation and validated late blight field resistance in this event under Indian conditions (Fig. 8.1). The *RB*

**Fig. 8.1.** (A) Resistance response of SP51 event in potato 'Kufri Jyoti' lines. (B) Control plants shown with over 90% infection.

gene was introgressed in $F_1$ hybrids by crossing the popular cultivar 'Kufri Jyoti' with SP951 and identifying five promising genotypes with very high late blight resistance and other desirable traits. The site of integration and copy number of the inserted *RB* gene in the event SP951 has been determined through molecular analysis (Shandil *et al.*, 2017). The *RB* transgene can be used effectively to reduce foliar late blight infection in cultivated potatoes. This enhanced resistance over the generations may be ascribed to the pooling of *R* gene activation or reshuffling of *R* genes in the background of 'Kufri Jyoti' that imparts resistance to *P. infestans*.

The expression of RB protein, like other R proteins in plants, is extremely low due to its intractable nature. Because of its intractable nature, RB protein could not be purified from either the transgenic event SP951 or from the host species *S. bulbocastanum*. As a consequence, no antibody could be produced for this protein. Similarly, no detectable level of recombinant RB protein could be expressed in either bacterial or yeast hosts. It posed a serious challenge to estimate RB protein in the transgenic events through Western blotting. Therefore, the allergenicity of this protein was determined using a bioinformatics tool (AllergenOnline: https://farrp.unl.edu/resources/allergenonline, accessed 12 July 2024). The tool did not identify any significant alignment with known allergens. PubMed searches also failed to show any published evidence indicating an allergenic or toxic nature of the RB protein (Goodman, 2015). It is therefore argued that a regulatory food safety study on pure RB protein cannot be undertaken due to the intractable nature of this common plant protein, which is suggested to be non-toxic and non-allergenic. However, it would be possible to generate biosafety and food safety data for event SP951 or its $F_1$ progeny at the time of Biosafety Research Level 1 and 2 trials.

### 8.2.2 RNAi transgenic potato for *P. infestans* resistance

When potato and *P. infestans* come into contact, a complex and dynamic communication occurs between the two organisms. This activates the defense mechanism of potato, and many specific genes may be induced. Several studies have demonstrated that the challenge with *P. infestans* leads to transcriptional activation of specific genes in potato. Furthermore, the activation of potato defense responses by *P. infestans* causes the differential expression of numerous genes. A comprehensive analysis of differentially expressed genes could contribute to a better understanding of the molecular processes involved in plant–pathogen interactions. Therefore, powerful methods such as microarrays and RNAi could be used to anticipate these constraints. Potato cultivars possessing *R* genes are not resistant to all races of the pathogen and provide only short-lived resistance in the field, and this has been well demonstrated globally by various groups. Host-induced gene silencing of pathogen genes has provided a gene-knockout strategy for a wide range of biotechnological applications. The RXLR effector *Avr3a* gene is largely responsible for virulence of the oomycete plant-pathogen *P. infestans*. We attempted to silence the *Avr3a* gene of *P. infestans* through RNAi technology (Thakur *et al.*, 2015). *P. infestans* inoculation resulted in lower disease progression and a reduction in pathogen load, as demonstrated by disease scoring and quantification of pathogen biomass in terms of Pi08 repetitive elements, respectively. The RXLR effector *Avr3a* has been demonstrated to be secreted from haustoria and translocated from the extrahaustorial matrix to inside the plant cell (Whisson *et al.*, 2007). This would explain why the higher accumulation of short interfering RNA (siRNA) silencing molecules in the host cell led to elevated endogenous silencing of the target *Avr3a* gene. These observations suggest that the nature of silencing molecules involved in the host-induced gene silencing of *P. infestans* seems to be siRNAs generated in the host.

Transgenic plants induced moderate silencing of *Avr3a*, and the presence and/or expression of siRNAs, as determined by Northern blot hybridization, indicated siRNA targeted against *Avr3a* conferred moderate resistance to *P. infestans*. The single effector gene did not provide complete resistance against *P. infestans*. Although the *Avr3a* effector gene could confer moderate resistance, for complete resistance the cumulative effect of effector genes in addition to *Avr3a* needs to be considered. We have demonstrated that host-induced RNAi is an effective strategy for functional genomics in oomycetes.

Whisson *et al.* (2012) suggested that RXLR effector genes are located preferentially in genomic regions heavily populated with transposons, and small RNAs involved in silencing can extend from the genome region spanning the effector and transposons. In future, further mining of the siRNA inhibitor and stacking of other effector genes on single or multiple constructs is required; the influence of nearby transposons on the evolution and expression of effectors that provide high levels and durable resistance against late blight also needs to be studied.

### 8.2.3 Genetically engineered potato expressing ToLCNDV replicase small RNAs for resistance to ToLCNDV disease

Viruses are important ubiquitous pathogens and cause 80% losses in potato yield. Potato is infected by more than 40 viruses and two viroids (Jeffries *et al.*, 2005), of which so far only nine viruses and two viroids are of economic significance for potato. There are numerous variable factors, such as plant genetic diversity, biology, life cycle of the host plant/pathogen, vector species, biotype, and environmental conditions, that affect the incidence and severity of viral diseases (Yi and Gray, 2020). In India, potato virus Y, potato leaf roll virus, potato virus X (PVX), potato virus S (PVS), potato virus A (PVA) and potato virus M (PVM) cause yield reductions of 10–30% depending on the intensity and vector pressure (Fox *et al.*, 2017; Kumar *et al.*, 2017). The incidence of ToLCNDV has increased immensely year on year and it has been the most important potato virus in India for the last two decades. In the Indo-Gangetic plains, disease incidence is 40–100%, which incurs heavy yield losses in potato (Lakra, 2002; Venkatasalam *et al.*, 2011). Numerous data on the incidence of potato apical leaf curl virus have been documented and are available in several published annual reports and research literature (Usharani *et al.*, 2003; Jeevalatha *et al.*, 2013, 2018).

### 8.2.4 RNAi transgenic tool for inducing bacterial wilt disease resistance in potato

Bacterial wilt caused by *Ralstonia solanacearum* is one of the most serious diseases in potatoes, with a wide host range of over 200 plant species. Worldwide losses in potato yield due to bacterial wilt range from 33% to 90%, and it can be as high as 70% in India (Sagar *et al.*, 2014). The few available resistant germplasms (wild diploid sources) in potato have failed to provide durable resistance against Indian *R. solanacearum* races, as the $2n$ endosperm balance number of diploid potato is difficult to introgress with the commercial cultivated tetraploid potatoes with a $4n$ endosperm balance number, and restricts their use in resistance breeding programs.

Earlier studies identified the *PAP2* gene controlling pathogenesis in various plant species. PAP2 acts as a negative regulator in response to *R. solanacerum* infection (Nakano *et al.*, 2015). PAP2 inhibits the accumulation of phosphatidic acid, which further interferes with the hypersensitive response in the host. Hence, PAP2 can be silenced to help control *R. solanacearum* in potato. We developed RNAi lines that showed an immune response to wilt symptoms, and a few of the lines showed delayed infection compared with the control plants. Eighty 'Kufri Jyoti' RNAi lines were developed and subjected to an efficient, quick, *in vitro* hydroponic method of bioassay to avoid escape of the pathogen, as well as whole plant bioassays using the root inoculation method (Thakur *et al.*, 2020). There is still no clear mechanism to understand the conversion or competition of virulent to avirulent types in resistant genotypes. This might involve a complex host-signaling cascade and pathogen effector interactions inside the xylem vessel of the host system. Thus, investigations on imparting resistance in potato through the plant's immune system provides a novel genetic source in potato for bacterial wilt resistance, as none of the available germplasms and cultivars have resistance loci for durable bacterial wilt resistance.

### 8.2.5 Genetically engineered protein-rich potatoes

Proteins, one of the principal constituents of a balanced diet, impart nutritional value to food due to their structural constituents, amino acids. However, it is very rare in nature to find all of the essential amino acids in a single food crop. The United Nations declared 2008 as the "International Year of the Potato," affirming the need to focus on the role that the potato can play in providing food security (United Nations General Assembly Resolution, 2005). Although in developing economies the majority of potatoes grown are used for direct consumption, a shift toward the use of potato in convenience foods, for example in potato chips and fries, has dramatically increased in developed countries. Unfortunately, the nutritional quality of potato tubers is greatly compromised because they contain less protein and are deficient in lysine, tyrosine, and the sulfur-containing amino acids (Jaynes *et al.*, 1986).

The seed protein amaranth albumin 1 (AmA1) has great agricultural importance because it is well balanced in terms of amino acid composition. More importantly, because it is a non-allergenic protein that originated from an edible crop, transgenic crops expressing *AmA1* would have greater acceptability. Chakraborty *et al.* (2010) reported the development of transgenic potatoes with enhanced nutritive value as a result of tuber-specific expression of AmA1 in seven genotypic backgrounds suitable for cultivation in different agro-climatic regions. Analyses of the transgenic tubers revealed up to a 60% increase in total protein content. In addition, the concentrations of several essential amino acids were increased significantly in the transgenic tubers that are otherwise limited in potato. Moreover, the transgenic potatoes also exhibited enhanced photosynthetic activity with a concomitant increase in total biomass. These results are striking because this genetic manipulation also resulted in a moderate increase in tuber yield. The comparative protein profiling suggested that the proteome rebalancing might cause increased protein content in transgenic tubers. Furthermore, the data on field performance and safety evaluation indicated that the transgenic potatoes are suitable for commercial cultivation. *In vitro* and *in vivo* studies on experimental animals demonstrated that the transgenic tubers are also safe for human consumption. Altogether, these results suggest that the expression of AmA1 is a potential strategy for nutritional improvement of food crops.

## 8.3 Conclusion

The International Food Policy Research Institute launched its 2020 Vision, a call for a new "green revolution." A total of 1.02 billion people, residing mostly in the developing world, suffer from a lack of availability of food causing undernutrition. Therefore, to meet the demand of poverty populations, advanced biotechnology tools are necessary to boost food-based crop-improvement research programs, including those for rice and wheat. The biotechnology-based strategies to modify genomes, especially in seed crops, are major challenges where homozygosity is of concern, whereas in cases of vegetatively propagated potatoes, maintenance of heritability of genes via tubers is easy and does not require backcrossing to maintain homozygosity. An efficient, rapid, and genotype-independent transformation method has now been developed for potato. To the best of our knowledge, the increased and durable resistance to late blight developed by our group at ICAR-CPRI is one of first observed in Indian potato cultivars, comparable with US-developed transgenic event SP951. It is hoped that these promising late blight resistance events will get the green light for conducting biosafety research trials for environmental release.

## 8.4 Acknowledgements

The authors are thankful to the Competent Authority, ICAR-CPRI, Shimla, India, for providing the necessary facilities to carry out the research work.

## 8.5 References

Chakraborty, S., Chakraborty, N., Agrawal, L., Ghosh, S., Narula, K., *et al.* (2010) Next-generation protein-rich potato expressing the seed protein gene *AmA1* is a result of proteome rebalancing in transgenic tuber. *Proceedings of the National Academy of Sciences USA* 107, 17533–17538.

Fox, A., Collins, L.E., Macarthur, R., Blackburn, L.F., and Northing, P. (2017) New aphid vectors and efficiency of transmission of *Potato virus A* and strains of *Potato virus Y* in the UK. *Plant Pathology* 66, 325–335.

Goodman, R.E. (2015) Bioinformatic analysis and literature search for potential allergenicity and toxicity test of late blight resistance potato event SP951, with *RB* and *Npt*II proteins. In: *Food Allergy Research and Resource Program*. Study No. RFG-LBR-2015. Department of Food Science and Technology, University of Nebraska, Lincoln, Nebraska, pp. 1–26.

Halterman, D.A., Kramer, L.C., Wielgus, S., and Jiang, J.M. (2008) Performance of transgenic potato containing the late blight resistance gene *RB*. *Plant Diseases* 92, 339–343.

Jaynes, J.M., Yang, M.S., Espinoza, N., and Dodds, J.H. (1986) Plant protein improvement by genetic engineering: use of synthetic genes. *Trends in Biotechnology* 4, 314–320.

Jeevalatha, A., Kaundal, P., Venkatasalam, E.P., Chakrabarti, S.K., and Singh, B.P. (2013) Uniplex and duplex PCR detection of geminivirus associated with potato apical leaf curl disease in India. *Journal of Virological Methods* 193, 62–67.

Jeevalatha, A., Kaundal, P., Kumar, R., Raigond, B., Kumar, R., *et al.* (2018) Optimized loop-mediated isothermal amplification assay for *Tomato leaf curl New Delhi virus-[potato]* detection in potato leaves and tubers. *European Journal of Plant Pathology* 150, 565–573.

Jeffries, C., Barker, H., and Khurana, S.M.P. (2005) Potato viruses and viroids. In: Gopal, J. and Khurana, S.M.P. (eds.) *Handbook of Potato Production, Improvement and Post-harvest Management*. The Haworth's Food Products Press, New York.

Kuhl, J.C., Zarka, K., Coombs, J., Kirk, W.W., and Douches, D.S. (2007) Late blight resistance of *RB* transgenic potato lines. *Journal of the American Society for Horticultural Science* 132, 783–789.

Kumar, R., Jeevalatha, A., Raigond, B., Kumar, R., Sharma, S., and Nagesh, M. (2017). A multiplex RT PCR assay for simultaneous detection of five viruses in potato. *Journal of Plant Pathology* 99, 37–45.

Lakra, B.S. (2002) Leaf curl: a threat to potato crop in Haryana. *Journal of Mycology and Plant Pathology* 32: 367.

Nakano, M., Yoshioka, H., Ohnishic, K., Hikichia, Y., and Kiba, A. (2015) Cell death-inducing stresses are required for defense activation in DS1-phosphatidic acid phosphatase-silenced *Nicotiana benthamiana*. *Journal of Plant Physiology* 184, 15–19.

Sagar, V., Gurjar, M.S., Arjunan, J., Bakade, R.R., Chakrabarti, S.K., Arora, R.K., and Sharma, S. (2014) Phylotype analysis of *Ralstonia solanacearum* strains causing potato bacterial wilt in Karnataka in India. *African Journal of Microbiology Research* 8, 1277–1281.

Sanju, S., Sundaresha, S., Thakur, A., Shukla, P.K., Srivastava, N., *et al.* (2015) Host induced gene silencing of *Phytophthora infestans* effector gene *Avr3a* in potato. *Functional and Integrative Genomics* 15, 697–706.

Shandil, R.K., Chakrabarti, S.K., Singh, B.P., Sharma, S., Sundaresha, S., *et al.* (2017) Genotypic background of the recipient plant is crucial for conferring *RB* gene mediated late blight resistance in potato. *BMC Genetics* 18: 22.

Song, J.Q., Bradeen, J.M., Naess, S.K., Raasch, J.A., Wielgus, S.M., *et al.* (2003) Gene *RB* cloned from *Solanum bulbocastanum* confers broad spectrum resistance to potato late blight. *Proceedings of the National Academy of Sciences USA* 100, 9128–9133.

Thakur, A., Sanju, S., Sundaresha, S., Sharma, S., and Singh, B.P. (2015) Artificial microRNA mediated gene silencing of *Phytophthora infestans AVr3a* gene for imparting late blight resistance in potato. *Journal of Plant Pathology* 14, 1–12.

Thakur, K., Sagar, V., Sundaresha, S., Salaria, N., Goutam, U., and Sharma, S. (2020) An efficient and quick *in-vitro* method of evaluation of potato genotypes against *Ralstonia solanacearum*. *Plant Cell Biotechnology and Molecular Biology* 21, 46–52.

Tomar, G., Chakrabarti, S.K., Sharma, N.N., Jeevalatha, A., Sundaresha, S., *et al.* (2018) RNAi based transgene conferred extreme resistance to the geminivirus causing apical leaf curl disease in potato. *Plant Biotechnology Reports* 12, 195–205.

United Nations General Assembly Resolution (2005) Resolution No. A/RES/60/191. United Nations, New York. Available at: https://daccess-ods.un.org/access.nsf/Get?OpenAgent&DS=A/RES/60/191&Lang=E (accessed 12 July 2024).

Usharani, K.S., Surendranath, B., Khurana, S.M.P., Garg, I.D., and Malathi, V.G. (2003) Potato leaf curl—a new disease of potato in Northern India caused by a strain of *Tomato leaf curl New Delhi virus*. *New Diseases Reports* 8: 2.

Venkatasalam, E.P., Singh, S., Sivalingam, P.N., Malathi, V.G., and Garg, I.D. (2011). Polymerase chain reaction and nucleic acid spot hybridisation detection of begomovirus(es) associated with apical leaf curl disease of potato. *Archives of Phytopathology and Plant Protection* 44, 987–992.

Vleeshouwers, V.G.A.A., Rietman, H., Krenek, P., Champouret, N., Young, C., *et al.* (2008) Effector genomics accelerates discovery and functional profiling of potato disease resistance and *Phytophthora infestans* avirulence genes. *PLOS ONE* 3: e2875.

Whisson, S.C., Boevink, P.C., Moleleki, L., Avrova, A.O., Morales, J.G., *et al.* (2007). A translocation signal for delivery of oomycete effector proteins into host plant cells. *Nature* 450, 115–118.

Whisson, S., Vetukuri, R., Avrova, A., and Dixelius, C. (2012) Can silencing of transposons contribute to variation in effector gene expression in *Phytophthora infestans*? *Mobile Genetic Elements* 2, 110–114.

Yi, X.U. and Gray, S.W. (2020) Aphids and their transmitted potato viruses: a continuous challenges in potato crops. *Journal of Integrative Agriculture* 19, 367–375.

# 9 Status of Tropical Tuber Crop Improvement Through Genetic Engineering

**K.M. Senthilkumar, K. Divya, B.G. Sangeetha, M.N. Sheela and T. Makeshkumar***
*ICAR-Central Tuber Crops Research Institute, Thiruvananthapuram, Kerala, India*

**Abstract**
Tropical tuber crops including cassava, sweet potato, yams, and aroids serve as an important energy source for humans, especially for marginal communities in tropical regions. These crops remain an important raw material for many industries including the food, animal, and pharma industries. Tropical tuber crops are considered to be climate resilient, with huge potential to meet the growing food, energy, and feed demands of the future. Hence, development of high-yielding, climate-resilient, nutritionally rich elite cultivars would be helpful in meeting the future demand for these crops. Wild relatives and germplasms of these crops serve as excellent resources for genetic improvement; however, the non-flowering nature of some germplasms, asynchronous flowering, heterozygosity, inbreeding depression, ploidy levels, and other problems remain an important genetic bottleneck for introgression of agronomically important traits through conventional breeding strategies. The availability of standardized transformation and regeneration protocols for cassava, sweet potato, yam, and aroids would be helpful in genetic improvement of desirable traits through genetic-engineering strategies.

**Keywords:** gene cloning, genomics, transformation, transgenics, tuber crops

## 9.1 Introduction

The main tropical tuber crops are cassava (*Manihot esculenta*), sweet potato (*Ipomoea batatas*), taro (*Colocasia esculenta*, yam, especially greater yam (*Dioscorea alata*) and elephant foot yam (*Amorphophallus paeoniifolius*), remain an important staple food among many communities in tropical regions of the world (Prakash *et al.*, 2020). Cultivation and production of these crops are under constant pressure from many different abiotic (e.g. drought, temperature) and biotic (e.g. insect, viruses, bacteria, and fungi) stresses, which impair their growth and development and result in intense crop loss (Jeeva *et al.*, 2004; Babu *et al.*, 2011; Legg *et al.*, 2015, 2022). Tropical tuber crops have been widely used in many industries including the animal, food, and bioenergy industries. For instance, processed tubers as well as starch extracted from cassava and sweet potato remain a major feed stock in many food industries (Ray and Sivakumar, 2009; Sajeev *et al.*, 2021). Moreover, most of the tuber crops are rich in vitamins, minerals, and other bioactive compounds (El Sheikha and Ray, 2017; Anish *et al.*, 2021). Hence, these crops are considered a climate-smart future crop with huge potential to address the growing demands of energy and nutritional security in the future (Githunguri and

*Email: makeshctcri@gmail.com or makeshkumar.t@icar.gov.in


DOI: 10.1079/9781800627123.0009

Njiru, 2021; Feyisa, 2022). These crops have a profound impact on food security and economic stability of a notably massive group of people, especially in tropical regions of Africa and a few regions in Asia. Although conventional breeding programs have helped in the development of high-yielding elite cultivars in many tuber crops, breeding for agronomically important traits such as disease/pest resistance and reducing antinutritional factors, and enhancing quality traits has been limited in tuber crops, mainly due to lack of genetic variation of the desirable traits in germplasm (e.g. waxy starch in cassava, sweet potato weevil resistance in sweet potato), heterozygosity barriers, their non-flowering nature, inbreeding depression, and ploidy levels. Hence, incorporation of genomics and genetic engineering assisted-breeding approaches would be helpful in improving the various agronomically important traits such as resistance against pathogens, reducing antinutritional factors, and enhancement of bioactive compounds (Kayondo *et al.*, 2018; Abah *et al.*, 2022; Okwuonu *et al.*, 2022). Some of the problems in breeding, including time consumption, labor intensiveness, expenditure, scarcity of resistance genes in wild germplasm, loss of agronomically and economically important traits during crosses, technical difficulty in transferring resistance genes from diploid wild varieties to polyploid cultivars, and adaptability/evolutionary potential of pathogens, have pushed researchers toward the use of genetic-engineering tools for improvement of tuber crops. In this chapter, we discuss the breeding challenges and genetic-engineering strategies used for improvement of tropical tuber crops. Details of some of the transgenic studies undertaken in these crops are given in Table 9.1.

## 9.2 Important Tropical Tuber Crops, Breeding Challenges and Biotechnological Interventions

### 9.2.1 Cassava

Cassava (*Manihot esculenta* Crantz) belongs to family Euphorbiaceae, and the major cultivated species is native to the Amazon in South America (Olsen and Schaal, 1999; Nassar and Ortiz, 2009). It is one of the major root crops of Africa, the Indian subcontinent, Latin America, and several South-east Asian countries (Karthikeyan *et al.*, 2016). It is the sixth most important crop in the world after wheat, rice, maize, potato, and barley, and was cultivated by American Indians more than 5000 years. This crop was introduced to Africa in the 16th century by Portuguese sailors and spread to tropical regions in Asia during colonization. As a monoecious perennial shrub, the roots of cassava enlarge to form starchy storage roots, and these roots are the major source of carbohydrates for approximately 800 million people in developing regions of the tropics. In parts of Asia and Latin America, the crop is used as an animal feed and raw material for starch products and commercial fuel (bioethanol) production (Wangpor *et al.*, 2017). It is a high-yielding crop and is well adapted to drought, poor soils, and high temperature. Moreover, the crop serves as a dietary carbohydrate source for nearly 200 million people in sub-Saharan Africa (Liu *et al.*, 2011).

#### *Breeding challenges in cassava*

Tuber production in cassava is influenced by various environmental and biotic factors. Although cassava withstands drought, a good amount of irrigation is required during the critical growth period, which is the initial 3 months of establishment (de Oliveira *et al.*, 2017). Physiological alterations in response to drought and survival mechanisms compromise tuber yield in cassava (Adjebeng-Danquah *et al.*, 2016; de Almeida Silva *et al.*, 2021). During severe drought conditions, cassava reduces transpiration by shedding its leaves, and the stored dry matter is also mobilized to the root tissues for root elongation in search of the water (Adjebeng-Danquah *et al.*, 2016). Hence, identification of drought-tolerant genotypes and development of abiotic stress-tolerant genotypes remains one of the important breeding objectives in global crop-breeding programs (Okogbenin *et al.*, 2013; de Oliveira *et al.*, 2017; Sivan *et al.*, 2023; Koundinya *et al.*, 2024). Biotic stress factors including diseases and insect pests; mites, whiteflies, white grubs, scales, and termites are reported as serious threats for cassava production in major cassava-growing areas (Garcia-Oliveira *et al.*, 2020; Perez-Fons *et al.*, 2020; Rauwane and Ntushelo, 2020). About 200 species of arthropod pests are known to associate with cassava and

**Table 9.1.** Details of transgenic studies on the improvement of agronomically important traits in tuber crops.

| Crop | Gene | Gene description | Transgenic host | Trait | Reference |
|---|---|---|---|---|---|
| Cassava | *MePMP3-2* | Plasma membrane proteolipid 3 (*PMP3*) | Rice | Salt and drought stress tolerance | Yu *et al.* (2016) |
| | *MeFT1* | *FLOWERING LOCUS T* | Cassava | Early flowering | Odipio *et al.* (2020) |
| | Zeolin | Storage protein | Cassava | Protein content | Abhary *et al.* (2011) |
| | *ASP1* | Artificial storage protein | Cassava | Protein content | Zhang *et al.* (2003) |
| | *MeCBF1* | C-repeat binding factor 1 | *Arabidopsis* and cassava | Cold stress tolerance | An *et al.* (2017) |
| | *MeSCL30* | SR protein | *Arabidopsis* | Drought stress tolerance | Weng *et al.* (2021) |
| | *AtCBF3* | C-repeat binding factor 3 | Cassava | Cold and drought stress tolerance | An *et al.* (2016) |
| | *MeTCP4* | Transcription factor | Cassava | Cold stress tolerance | Cheng *et al.* (2019) |
| | HNL | Hydroxynitrile lyase | Cassava | Reduced acetone cyanohydrin and cyanide levels in roots | Narayanan *et al.* (2011) |
| | Bacterial *glgC* | ADP-glucose pyrophosphorylase | Cassava | Starch self-digestibility and yield increase | Ligaba-Osena *et al.* (2018) |
| | *MeMinE* | Small protein involved in plastid cell division | Cassava and *Escherichia coli* | Plastid division | Wang *et al.* (2019) |
| | *nptII* | Neomycin phosphotransferase gene | Cassava | Aminoglycoside resistance | Niklaus *et al.* (2011) |
| | *AtVIT1* | Vacuolar sequestration of iron | Cassava | Increased iron content | Narayanan *et al.* (2015) |
| | EPSPS | 5-Enolpyruvylshikimate-3-phosphate synthase | Cassava | Glyphosate tolerance | Hummel *et al.* (2018) |
| Sweet potato | *ASP1* | Artificial storage protein | Sweet potato | Protein content | Zhang *et al.* (2003) |
| | *IbMYB1* | Transcription factor | Sweet potato | Transgenic lines with high levels of anthocyanins | Park *et al.* (2015a) |
| | *IbOr* | Sweet potato orange (IbOr) protein | Sweet potato | Transgenic lines with high levels of carotenoids | Park *et al.* (2015b) |
| | *IbBBX24* | Transcription factor | Sweet potato | *Fusarium* wilt resistance | Zhang *et al.* (2020) |

*Continued*

**Table 9.1.** Continued.

| Crop | Gene | Gene description | Transgenic host | Trait | Reference |
|---|---|---|---|---|---|
| | *IbpreproHypSys* | Small signaling peptides, hydroxyproline-rich glycopeptides (*IbpreproHypSys*) | Sweet potato | Suppresses the growth of *Spodoptera litura* | Li *et al.* (2016) |
| Yam | *DrPDS* | Phytoene desaturase gene | *Dioscorea rotundata* | Due to disruption of the phytoene desaturase gene the transgenic lines displayed albinism | Syombua *et al.* (2021) |
| Taro | *gf2.8* | Oxalate oxidase gene from wheat | Taro | Increased resistance to *Phytophthora colocasiae* | He *et al.* (2013) |
| | *RicChi11* | Rice chitinase gene | Taro | Improved resistance to the fungal pathogen *Sclerotium rolfsii* | He *et al.* (2010) |

can cause moderate to severe losses of the crop. The viral disease cassava mosaic disease is transmitted primarily through whitefly and remains a major threat in all cassava-growing regions of the world (Legg *et al.*, 2015). Among the whitefly species of greatest economic importance are *Aleurotrachelus socialis* Bondar, *Trialeurodes variabilis* (Quaintance), *Bemisia tuberculata* Bondar, *Bemisia tabaci* (Gennadius), and *Aleurothrixus aepim* (Goeldi) (Bellotti *et al.*, 1999). Other major insect pests include cassava mealybug (*Phenacoccus manihoti*), African root and tuber scale (*Stictococcus vayssierei*), and cassava green mite (*Mononychellus tanajoa*) (Alabi *et al.*, 2011; Abaca *et al.*, 2014; Chikoti *et al.*, 2015). Mealybugs are one of the main pests that directly damage the plants by sucking the sap and contaminating the plant with their toxic saliva, which indirectly favors the development of sooty molds (Barilli *et al.*, 2014). Hence, host-plant resistance is a valuable tool for all integrated pest/disease management programs. Host-plant resistance studies for whitefly resistance in cassava were initiated at the International Center for Tropical Agriculture (CIAT) more than 15 years ago (Bellotti and Arias, 2001). Traditionally, hybridization was used as one of the methods for developing new varieties in cassava. Although host-plant resistance is not frequent, alternative but promising sources of resistance were identified and hybrids were developed (Bellotti, 2002). The wild parents of *M. esculenta* are known sources of genes resistant to insect pests. Moderate to high levels of resistance to the cassava green mite *Mononychellus tanajoa* and the whitefly *A. socialis* were reported from *Manihot peruviana* and *Manihot flabellifolia*, respectively (Burbano *et al.*, 2003; Carabalí *et al.*, 2010). Cassava also contains the cyanogenic glycoside toxins linamarin, lotaustralin, and amygdalin (Bolarinwa *et al.*, 2016). These toxins are found in the leaves, stem, and tuber tissues of cassava. Linamarin is the major cyanogenic compound found in cassava, while linamarin and linamarase are localized mainly to the vacuole and cell-wall organelles, respectively (Zidenga *et al.*, 2017). Although these toxic cyanogens can easily be removed through processing techniques, consumption of poorly processed tubers and raw cyanogen-rich tubers causes various disorders in humans. Hence, development of cassava with reduced levels of cyanogenic glycosides remains one of the major breeding objectives in cassava improvement programs. Postharvest physiological deterioration remains one of the important constraints that limits the shelf life of harvested cassava tubers. This physiological process is triggered mainly through enzymatic responses to wounding during harvest and postharvest handling of the tubers, involving changes in gene expression and accumulation of secondary metabolites resulting in deterioration of the tubers. Several environmental conditions also influence the onset of postharvest physiological deterioration, which remains a challenge for screening of genotypes (Zainuddin *et al.*, 2018). Research into identification of the genomic regions responsible for important agronomic traits such as abiotic, biotic, and quality traits in cassava is in progress, and the molecular genetic markers closely linked to these genes can be used effectively for the identification of superior germplasm in breeding programs and to improve the efficiency of cassava breeding (Muthusamy *et al.*, 2023). However, a high degree of genetic heterozygosity, genetic overloading, serious separation of progeny, few flowers, low pollen fertility, self-incompatibility, and low fruit set remain limitations for cassava breeding (Ceballos *et al.*, 2010). Therefore, genetic engineering shows great potential in cassava genetic improvement for important agronomic traits including postharvest physiological deterioration tolerance, low cyanogen levels, disease resistance, insect resistance, and high yield and quality to overcome the bottlenecks of conventional cassava breeding (Taylor *et al.*, 2004; Zhang *et al.*, 2017). Thus, the genetic diversity in existing germplasms can be transferred to high-yielding cultivars through marker-assisted breeding and genetic-engineering strategies for the development of superior cultivars (Taylor *et al.*, 2004; Zhang *et al.*, 2017; Sivan *et al.*, 2023).

### *Genetic engineering in cassava*

The first reports of genetic transformation systems for cassava were in 1996 by two research groups: one from the Potrykus laboratory at Eidgenossische Technische Hochschule Zurich (Li *et al.*, 1996) and another from the Fauquet Laboratory at Scripps Research Institute (Schöpke *et al.*, 1996). Friable embryogenic callus (FEC) of the West African cultivar 60444 (also known as TMS60444) transformed using *Agrobacterium tumefaciens* strain ABI was reported by Gonzalez *et al.* (1998). Biolistic transformation using cassava TMS60444 suspensions was conducted by Schöpke *et al.* (1996), while particle bombardment to transfer the plasmid pHMG into TMS60444 embryogenic suspensions using a selectable marker (hygromycin-mannose) was carried out by Zhang and Puonti-Kaerlas (2000). The transgenic nature of the plants was confirmed by a β-glucuronidase assay and Southern blot analysis. FECs and embryogenic suspensions using *A. tumefaciens* or particle bombardment have subsequently become the usual methods for cassava transformation. Zhang *et al.* (2000) transformed embryogenic suspensions of TMS60444 with *A. tumefaciens* LBA440. An efficient and reproducible transformation protocol for cassava embryogenic suspension culture with *A. tumefaciens* was reported by Schreuder *et al.* (2001). *Agrobacterium*-mediated transformation protocols for TMS60444 FECs were developed from model cultivars to farmer-preferred varieties (Liu *et al.*, 2011). The successful transformation of a synthetic artificial storage protein 1 (*ASP1*) gene encoding a storage protein rich in essential amino acids (80%) into embryogenic suspensions of cassava was done by Zhang *et al.* (2003a) using the same system as González *et al.* (1998). Further cassava transformation for the verification of a tissue-specific promoter (Zhang *et al.*, 2003b; Beltran *et al.*, 2010), for resistance to African cassava mosaic virus (Chellappan *et al.*, 2004; Zhang *et al.*, 2005; Vanderschuren *et al.*, 2007, 2009), for increased protein content (Abhary *et al.*, 2011), and for improved cassava brown streak virus resistance (Yadav *et al.*, 2011) have been reported. The entire protocol from induction of FECs to regeneration of plants was optimized and reported by (Zhang and Gruissem, 2004; Bull *et al.*, 2009).

Transgenic approaches for engineering resistance include the pathogen-derived resistance (PDR) mechanism, expression of antimicrobial compounds, silencing of susceptible genes and mining of resistance (*R*) genes, silencing of pathogen genes, upregulation of defense pathways, and transgenic expression of detoxifying pathogenic toxins. PDR and expression of antimicrobial compounds are the mostly widely used mechanisms in tuber crops. PDR is the resistance conferred by pathogenic genes in the host plants to subsequent infection by the same pathogen through RNA interference (RNAi). The proposed operational mechanisms of PDR include competition of the transgene with the virus, annealing of transgene-derived mRNA with viral mRNA, and prevention of uncoating of the virus. Initially, it was thought to be protein mediated, as evidenced from different studies in which expression of full-length viral coat protein, movement protein, and replicase protein conferred resistance. Further studies showed the possibility of PDR driven by RNA, while transgenic plants harboring truncated coding regions or non-coding regions also developed resistance against respective viruses. Later, post-transcriptional gene silencing (PTGS), a component of the plant's adaptive immune system, which functions based on the small interfering RNA (siRNA) silencing pathway, was identified as the mechanism of resistance (Hamilton and Baulcombe, 1999; Waterhouse *et al.*, 2001). Hairpin or inverted repeat constructs harboring viral genomic regions were shown to be more efficient than the use of viral genomic segments. siRNAs derived from hairpin or inverted repeat constructs are directed to and bind with the target site based on sequence complementarities. This subsequently leads to degradation or translational repression of target mRNA at the post-transcriptional level or epigenetic modification in the form of RNA-directed DNA methylation. As target identification and silencing in siRNA-mediated PTGS is based on complementary base pairing between the target and transgenic construct, it operates in a sequence-specific manner. PDR can be engineered by RNAi using sense and antisense RNAs, inverted repeat sequences/hairpin RNA, artificial microRNAs, and *trans*-activating siRNA. These PDR strategies are advantageous in the case of biosafety concerns as the transgene is degraded as siRNA and is not stably expressed in the host. Various compounds such as peptides, enzymes, and enzyme inhibitors with antimicrobial activity are present in many plant species. Transgenic expression of such compounds in target plants of interest can confer resistance against specific pathogens. PTGS through RNAi has been used as a pest-management strategy against *B. tabaci* infesting various crops. Exogenous dsRNA-mediated gene silencing of glutathione *S*-transferase of *B. tabaci* was reported by Ashokan *et al.*

(2015). The *Dnmt1* gene sequence has been identified in *B. tabaci* and successfully knocked down using RNAi (Dai *et al.*, 2017). Transgenic cassava plants overexpressing Cry1Aa (*Bt* cassava) were developed with high resistance to *Helicoverpa armigera* (Duan *et al.*, 2013). The draft genome of the cassava whitefly *B. tabaci* sub-Saharan Africa—East and Central Africa, associated with the rapid spread of viruses causing the pandemics of cassava mosaic disease and cassava brown streak disease, was reported by Chen *et al.* (2019). Recently, the first genome-wide characterization of pathogenesis-related (*PR*) genes in cassava's transcriptome response to phloem-feeding whiteflies (*A. socialis*), salicylic acid, jasmonic acid, and other biotic stresses was reported from Africa (Irigoyen *et al.*, 2020). This study revealed a positive correlation of cassava *PR* genes to whiteflies, the bacterium *Xanthomonas axonopodis* and cassava brown streak virus, but a negative correlation to cassava mosaic virus.

Silencing of cytochrome P450 genes (*CYP79D1*/*CYP79D2*) resulted in inhibition of linamarin biosynthesis in transgenic cassava lines (Siritunga and Sayre, 2004). However, the transgenic lines failed to grow in the absence of ammonia and produced a lower yield due to a reduction in tuber size compared with the wild cassava plants. Thus, suppression of cyanogen synthesis affected the nitrogen content in the transgenic lines (Siritunga and Sayre, 2004).

In conclusion, the well-established genetic transformation system of cassava can be utilized for genetic improvement of cassava for various agronomically important traits combining modern genomics tools such as genetic engineering with functional genomics, proteomics, and marker-assisted selection, along with traditional hybridization.

### 9.2.2 Sweet potato

Sweet potato *Ipomoea batatas* (L.) Lam was domesticated at least 8000–10,000 years ago in tropical America (Woolfe, 1992). It belongs to the family Convolvulaceae, and the number of chromosomes is $2n = 6X = 90$. Sweet potato as a highly heterozygous, generally self-incompatible crop with a polyploid genome poses numerous challenges for conventional breeding (Dhir *et al.*, 1998; Cervantes-Flores *et al.*, 2011). The crop originated in north-western parts of South America and was introduced to India by the Portuguese in the 16th century. It is the seventh most important food crop in the world (FAO, 2014). Asia is the largest sweet potato-producing region, with approximately 80% of the world's production and more than 50% of the world's area. China accounts for approximately 68% of world production, with more than 42% of the global area (FAO, 2014). The crop is used mainly as an important industrial raw material for animal feed, alcohol production, manufacturing starch and its products, and antioxidant pigment synthesis including anthocyanins and carotenoids (Oki *et al.*, 2002; Teow *et al.*, 2007; Liu, 2017). The leaves and roots of sweet potato are rich in biologically active anthocyanin, which is used for the production of natural food colorants (Islam, 2006), while the tubers are rich in carbohydrates, vitamin C, fiber, and minerals such as iron, zinc, and potassium (Woolfe, 1992). It is a short-duration crop and can be grown in different agro-ecological conditions (Mukherjee and Naskar, 2012). Tubers of this crop are rich in carbohydrates and minerals, and the orange- and purple-fleshed tubers are a cheap source of β-carotene and anthocyanin (Pradhan *et al.*, 2015). The salt-tolerant, β-carotene-rich ST-14, anthocyanin-rich ST-13, and high-starch ST-10 varieties were developed through breeding and progressive evaluation (Moorthy *et al.*, 2010; Mukherjee and Naskar, 2012).

#### Breeding challenges in sweet potato

Sweet potato production is affected by both abiotic and biotic stresses (Korada *et al.*, 2000; Senthilkumar *et al.*, 2023). Sweet potato weevil is a noxious pest, and globally four species of weevils (*Euscepes postfasciatus* (Fairmaire), *Cylas formicarius* (F.), *Cylas puncticollis* Boheman, and *Cylas brunneus* (F.)) have been reported as major pests of sweet potato (Chalfant *et al.*, 1990; Korada *et al.*, 2000). *C. formicarius* is an Asian species but is usually found throughout tropical regions worldwide, including North America, the Caribbean, Europe, Africa, Asia, and Oceania, while *C. brunneus* and *C. puncticollis* are restricted to Africa. *C. formicarius* is the single most devastating insect pest of sweet potato in the world (Sutherland, 1986; Kuriwada *et al.*, 2010; Prentice *et al.*, 2015; Reddy and Chi, 2015).

In India, sweet potato weevil incidence was first recorded from Tranquebar near Pondicherry (Waterhouse and Norris, 1987). Genetic engineering has been used to improve various traits of sweet potato, but crop-breeding programs for developing weevil-resistant cultivars are slow because the heritability of the trait is extremely low (Amritha *et al.*, 2024). Therefore, weevil resistance through genetic transformation is gaining more importance in sweet potato.

### *Genetic engineering in sweet potato*

Due to limitations in the conventional breeding of sweet potato, genetic engineering is a potential approach for improving important traits, as well as for incorporating novel traits in sweet potato (Kreuze *et al.*, 2008). Several groups have used transgenic techniques to improve various traits in sweet potato such as weevil resistance, drought tolerance, virus resistance, and improved nutritional quality (Kreuze *et al.*, 2008). Transgenic overexpression of the R2R3-type *IbMYB1* gene displayed enhanced synthesis of anthocyanin pigments in orange-fleshed sweet potato transgenic lines (Park *et al.*, 2015a). Similarly, overexpression of the *IbOr* gene, encoding a sweet potato orange protein, resulted in enhanced synthesis of carotenoid pigments in anthocyanin-rich, purple-fleshed sweet potato transgenic lines (Park *et al.*, 2015b). Transgenic lines overexpressing *IbBBX24* displayed enhanced biosynthesis of jasmonic acid, and when challenged with the fungal pathogen *Fusarium*, these lines displayed enhanced resistance (Zhang *et al.*, 2020). Newell *et al.* (1995) developed transgenic sweet potato 'Jewel' with a cowpea (*Vigna unguiculata*) trypsin inhibitor and the mannose-binding snowdrop lectin that showed resistance to sweet potato weevil. Initial results in no-choice feeding tests of these two transgenic plants showed a moderate increase in weevil resistance against the West Indian sweet potato weevil (*E. postfasciatus*). However, later this technique was abandoned due to the development of resistance by the weevil and issues related to the nutritional impact of the proteins on the human diet. Transgenic sweet potato plants with a soybean (*Glycine max*) Kunitz-type trypsin inhibitor (SKTI) were developed by Cipriani *et al.* (1999) using genetic transformation. Later, the toxins from *Bacillus thuringiensis* (*Bt*) were tested against two African weevil species, *C. puncticollis* and *C. brunneus*, and against the American and Asian species *C. formicarius*. The toxicity of eight *Bt* proteins was measured by the diet incorporation method and proved toxic to weevils at less than 1 ppm (Maingi *et al.*, 2002; Moar *et al.*, 2007). Several gene constructs have been developed using chemically synthesized sequences to produce transgenic sweet potato. Sweet potato clones of transgenic plants of the cultivar 'Jewel', expressing the *cry-IIIA* gene from *B. thuringiensis* subsp. *tenebrionis*, were developed by Morán *et al.* (1998). García *et al.* (2000) transformed the stems and leaves of sweet potato with *A. tumefaciens* carrying *nptII* and the *Bt* endotoxin gene.

Using artificial diet bioassays, three Cry protoxins (Cry7Aa1, Cry3Ca1, and CryET33-34) were found to be active against *C. puncticollis* and *C. brunneus* with $LC_{50}$ (50% lethal concentration) values of less than 1 ppm in Africa (Ekobu *et al.*, 2010). Recent studies have shown that *A. tumefaciens* strain EHA105 and embryogenic suspension cultures can be used for sweet potato transformation experiments to develop other agronomically important traits (Gao *et al.*, 2011, 2012; Liu, 2011; Fan *et al.*, 2012, 2015; Liu *et al.*, 2014a,b; Wang *et al.*, 2016). About ten transgenic sweet potato events expressing the Cry7Aa1, Cry3Ca1, and ET33-34 proteins were evaluated for resistance against *C. puncticollis*, but none of the transgenic events tested provided weevil control, probably because of low Cry protein expression in the storage roots (Rukarwa *et al.*, 2013a). The weevil-resistant gene *cry7Aa1* from the transgenic event CIP410008.7 of the cultivar 'Jewel' was introduced to African cultivars 'New Kawogo,' 'Tanzania' and 'NASPOT1,' and segregation of genes in the $F_1$ progeny was studied (Rukarwa *et al.*, 2013b). The three *Cry* protein (Cry7Aa1, CryET33/CryET34, and Cry3Ca1) gene constructs with sweet potato sporamin and amylase regulatory sequences were selected based on these artificial diet bioassay studies, and genetic transformation was carried out using sweet potato cultivars. However, low toxicity was observed based on Cry protein content in the storage roots and further studies are required to develop sweet potato weevil-resistant events (Ghislain *et al.*, 2013). The insect-resistant genes *cry7Aa1* and *cry3Ca1* were assembled into a plasmid vector for genetic transformation in African sweet potato cultivars against *C. puncticollis and C. brunneus* (Sefasi *et al.*, 2014). This genetic transformation protocol could be used for further weevil-resistant transgenic studies. Transgenic sweet potato plants overexpressing the Cry1Aa toxin were developed against *Spodoptera litura*, and analysis revealed that the transgenic lines were able to prevent insect infestation (Zhong *et al.*, 2019). The insecticidal activity of the proteins Cry23Aa and Cry37Aa from *Bt* was tested individually against larvae of *C. punctifollis*. The study revealed the binding specificity of Cry23Aa proteins for the brush border membrane vesicles of larvae independently of Cry37Aa (Hernández-Martínez *et al.*, 2020). This Cry protein combination could be used effectively for weevil management in future transgenic studies. Transgenic sweet potato lines expressing the hydroxyproline-rich glycopeptide *IbpreproHypSys* gene, a small signaling peptide, successfully suppressed growth of *S. litura* compared with the non-transgenic lines (Li *et al.*, 2016). Despite of the absence of weevil-resistant sources in the research field, management of these sweet potato weevils remains a challenge. In this context, developing transgenic sweet potato plants with insect resistance genes from different sources is one of the promising strategies for future molecular breeding programs.

### 9.2.3 Yams

Yams (*Dioscorea* spp.) remain an important staple food crop among communities in many tropical and subtropical regions. They constitute around 10% of global tuber production and are considered the fourth most important tuber crop after potato, cassava, and sweet potato (Viruel *et al.*, 2016). There are many species of yam, but the most exploited species for human consumption are greater yam (*D. alata*), lesser yam (*D. esculenta*), and white yam (*D. rotundata*). They are monocotyledonous, dioecious plants in nature, cultivated mainly for their starchy tubers. Most species are polyploid and are mainly propagated vegetatively from tubers. Yams are rich in diversity, and worldwide there are more than 600 different species in the genus *Dioscorea* (Amanze *et al.*, 2011). Yams have good medicinal value and are widely preferred in many regions for their organoleptic properties. They have several health values including hypocholesterolemic, hypoglycemic, immunomodulatory, and antimicrobial properties that can be exploited to treat many lifestyle diseases (Padhan and Panda, 2020).

#### *Breeding challenges in yam*

Yams are dioecious in nature, and the non-flowering and asynchronous flowering characteristics of the wild species of yams remain a bottleneck in breeding programs. Yam dieback anthracnose disease caused by the fungus *Colletotrichum* sp. remains a serious threat for cultivation of yam (Ntui *et al.*, 2021), and among the cultivated yam species, *D. alata* is the most sensitive. In addition, development of climate-resilient varieties that have the ability to perform better in low soil fertility, drought and heat, as well as the development of dwarf, early-maturing varieties, remain important breeding challenges in yams. Hence, development of a robust transformation protocol would be helpful in genetic improvement of yams.

#### *Genetic engineering in yam*

Manoharan *et al.* (2016) studied the embryogenic potential of six explants (axillary bud, immature leaf, node, stem internode, root segment, and petiole) using growth hormones in *D. rotundata*. Among these, axillary buds cultured on Murashige and Skoog (MS) medium supplemented with picloram (0.5–12 mg $l^{-1}$) were effective in the production of calluses. A successful transformation protocol was standardized in *D. rotundata* using axillary buds as explants using *Agrobacterium*-mediated transformation (Nyaboga *et al.*, 2014). Several factors including the choice of cultivar, *Agrobacterium* strain, and marker genes influenced the transformation efficiency in yam. Yam mosaic virus resistance, anthracnose resistance, herbicide tolerance, and nematode resistance are some of the potential traits for genetic-engineering studies in yam (Darkwa *et al.*, 2020). A clustered regularly interspaced palindromic repeats (CRISPR)-CRISPR-associated protein 9 (Cas9)-based genome-editing technique was standardized through successful silencing of the phytoene desaturase (*DrPDS*) gene in yam (Syombua *et al.*, 2021).

### 9.2.4 Taro

Taro (*Colocasia esculenta* (L.) Schott), belonging to the family Araceae (subfamily Aroideae), remains a staple food in many tribal communities. It is grown mainly for its succulent leaves, which are consumed as a leafy vegetable, and its starchy corm. Taro is mainly propagated using vegetative organs (i.e. corms). Most of the cultivars flower; however, poor germination vigor of the seeds and a slow growth rate reduce the chance of development and maturity of taro (Ivancic and Lebot, 2000; Nelson *et al.*, 2011).

#### *Breeding challenges in taro*

Taro is a vegetatively propagated crop. Its poor flowering nature, asynchronous flowering between male and female lines, poor seed germination, a slow growth rate of seedlings, and long vegetative period remain the major bottlenecks in taro breeding. The disease taro leaf blight (TLB) caused by the pathogen *Phytophthora colocasiae* is a serious threat in taro cultivation, and studies have shown that it can reduce leaf yield by 95% and corm yield by up to 50% in susceptible genotypes (Singh *et al.*, 2012; Alexandra *et al.*, 2020). During epidemics, TLB destroys the leaves in less than 10 days, with a subsequent reduction in corm yield (Nelson *et al.*, 2011). Hence, development of resistance cultivars to TLB is an important objective in taro breeding programs.

#### *Genetic engineering in taro*

Fukino *et al.* (2000) transformed the β-glucuronidase (*uidA*) gene into taro using particle bombardment; however, the efficiency of transformation was very low. Transgenic taro lines expressing the wheat oxalate oxidase gene (*gf2.8*) displayed enhanced tolerance to the pathogen *P. colocasiae* (He *et al.*, 2013). Transgenic expression of rice chitinase gene (*RicChi11*) in taro resulted in enhanced resistance of transgenic lines to the pathogen *Sclerotium rolfsii* (He *et al.*, 2010). These studies showed that the utilization of genetic-engineering tools would be helpful in genetic improvement of taro.

## 9.3 References

Abaca, A., Kiryowa, M., Awori, E., Andema, A., Dradiku, F., *et al.* (2014) Cassava pests and diseases prevalence and performance as revealed by adaptive trial sites in North Western agroecological zone of Uganda. *Journal of Agricultural Science* 6, 116–122.

Abah, S.P., Ezenwaka, L., Njoku, D.N., Okwuonu, I.C., Onyeka, T.J., and Egesi, C.N. (2022) Current and future advances in genomics-assisted Cassava breeding. In: *Agricultural Biotechnology, Biodiversity and Bioresources Conservation and Utilization*. CRC Press, Boca Raton, Florida, pp. 29–47.

Abhary, M., Siritunga, D., Stevens, G., Taylor, N.J., and Fauquet, C.M. (2011) Transgenic biofortification of the starchy staple cassava (*Manihot esculenta*) generates a novel sink for protein. *PLOS ONE* 6: e16256.

Adjebeng-Danquah, J., Gracen, V.E., Offei, S.K., Asante, I.K., and Manu-Aduening, J. (2016) Genetic variability in storage root bulking of cassava genotypes under irrigation and no irrigation. *Agriculture and Food Security* 5: 9.

Alabi, O.J., Kumar, P.L., and Naidu, R.A. (2011) *Cassava Mosaic Disease: A Curse to Food Security in Sub-Saharan Africa.* American Phytopathological Society, St. Paul, Minnesota. Available at: www.apsnet.org/edcenter/apsnetfeatures/Pages/cassava.aspx (accessed 27 June 2024).

Alexandra, S., Jamora, N., Smale, M., and Ghanem, M.E. (2020) The tale of taro leaf blight: a global effort to safeguard the genetic diversity of taro in the Pacific. *Food Security* 12, 1005–1016.

de Almeida Silva, M., Zanetti, S., Luiz Santos, H., and Melo Pereira, L.F. (2021) Nutritional status, yield components, and yield of cassava as influenced by phenological stages and water regimes. *Journal of Plant Nutrition* 44, 2912–2927.

Amanze, N.J., Agbo, N.J., Eke-Okoro, O.N., and Njoku, D.N. (2011) Selection of yam seeds from open pollination for adoption in yam (*Dioscorea rotundata* Poir) production zones in Nigeria. *Journal of Plant Breeding and Crop Science* 3, 68–73.

Amritha, M.S., Dhanya, O.G., Nair, A.G., Vidya, P., Muthusamy, S.K., and Mohan, C. (2024) Genetic diversity studies between *Ipomoea* wild species and cultivated sweet potato (*Ipomoea* batatas (L.) Lam.) using SSR markers. *South African Journal of Botany* 169, 452–463.

An, D., Ma, Q., Yan, W., Zhou, W., Liu, G., and Zhang, P. (2016) Divergent regulation of CBF regulon on cold tolerance and plant phenotype in cassava overexpressing *Arabidopsis CBF3* gene. *Frontiers in Plant Science* 7: 1866.

An, D., Ma, Q., Wang, H., Yang, J., Zhou, W., and Zhang, P. (2017) Cassava *C-repeat binding factor 1* gene responds to low temperature and enhances cold tolerance when overexpressed in *Arabidopsis* and cassava. *Plant Molecular Biology* 94, 109–124.

Anish, R.J., Ambrose, R.P.K., Sajeev, M.S., and Moorthy, S.N. (2021). *Tropical Tuber Starches: Structural and Functional Characteristics*. CAB International, Wallingford, UK.

Asokan, R., Rebijith, K.B., Roopa, H.K., and Kumar, N.K. (2015) Non-invasive delivery of dsGST is lethal to the sweet potato whitefly, Bemisia *tabaci* (G.) (Hemiptera: Aleyrodidae). *Applied Biochemistry and Biotechnology* 175, 2288–2299.

Babu, B., Hegde, V., Makeshkumar, T., and Jeeva, M.L. (2011) Molecular detection and identification of Dasheen mosaic virus infecting *Amorphophallus paeoniifolius*. *Archives of Phytopathology and Plant Protection* 44, 1248–1260.

Barilli, D.R., Pietrowski, V., da S. Wengrat, A.P., Gazola, D., and Ringenberg, R. (2014). Biological characteristics of the cassava mealybug *Phenacoccus manihoti* (Hemiptera: Pseudococcidae). *Revista Colombiana de Entomología* 40, 21–24.

Bellotti, A.C. (2002) Arthropod pests. In: Hillocks, J.M., Thresh, J.M., and Bellotti, A.C. (eds.) *Cassava: Biology, Production and Utilization*. CAB International, Wallingford, UK, pp. 209–235.

Bellotti, A.C. and Arias, B. (2001) Host plant resistance to whiteflies with emphasis on cassava as a case study. *Crop Protection* 20, 813–823

Bellotti, A.C., Smith, L., and Lapointe, S.L. (1999) Recent advances in cassava pest management. *Annual Review of Entomology* 44, 343–370.

Beltrán, J., Prías, M., Al-Babili, S., Ladino, Y., López, D., *et al.* (2010) Expression pattern conferred by a glutamic acid-rich protein gene promoter in field-grown transgenic cassava (*Manihot esculenta* Crantz). *Planta* 231, 1413–1424.

Bolarinwa, I.F., Oke, M.O., Olaniyan, S.A., and Ajala, A.S. (2016) A review of cyanogenic glycosides in edible plants. In: Soloneski, S. and Larramendy, M.L. (eds.) *Toxicology—New Aspects to This Scientific Conundrum*. InTech Open, pp. 179–191.

Bull, S.E., Owiti, J.A., Niklaus, M., Beeching, J.R., Gruissem, W., and Vanderschuren, H. (2009) *Agrobacterium*-mediated transformation of friable embryogenic calli and regeneration of transgenic cassava. *Nature Protocols* 4, 1845–1854.

Burbano, M.M., Carabalí, M.A., Montoya, L.J., and Bellotti, A.C. (2003) Resistance of *Manihot* species to *Mononychellus tanajoa* (Acariformes), *Aleurotrachelus socialis*, and *Phenacoccus herreni* (Hemiptera). *Revista Colombiana de Entomologia* 33, 110–115.

Carabalí, A., Bellotti, A.C., Montoya-Lerma, J., and Fregene, M. (2010) Resistance to the whitefly, *Aleurotrachelus socialis*, in wild populations of cassava, *Manihot tristis*. *Journal of Insect Science* 10: 170.

Ceballos, H., Okogbenin, E., Pérez, J.C., López-Valle, L.A.B., and Debouck, D. (2010) Cassava. In: Bradshaw, J.E. (ed.) *Root and Tuber Crops*. Springer, New York, pp. 53–96.

Cervantes-Flores, J.C., Sosinski, B., Pecota, K.V., Mwanga, R.O.M., Catignani, G.L., *et al.* (2011) Identification of quantitative trait loci for dry-matter, starch, and β-carotene content in sweetpotato. *Molecular Breeding* 28, 201–216.

Chalfant, R.B., Jansson, R.K., Seal, D.R., and Schalk, J.M. (1990) Ecology and management of sweet potato insects. *Annual Review of Entomology* 35, 157–180.

Chellappan, P., Masona, M.V., Vanitharani, R., Taylor, N.J., and Fauquet, C.M. (2004) Broad spectrum resistance to ssDNA viruses associated with transgene-induced gene silencing in cassava. *Plant Molecular Biology* 56, 601–611.

Chen, W., Wosula, E.N., Hasegawa, D.K., Casinga, C., Shirima, R.R., *et al.* Genome of the African cassava whitefly *Bemisia tabaci* and distribution and genetic diversity of cassava-colonizing whiteflies in Africa. *Insect Biochemistry and Molecular Biology* 110, 112–120.

Cheng, Z., Lei, N., Li, S., Liao, W., Shen, J., and Peng, M. (2019) The regulatory effects of *MeTCP4* on cold stress tolerance in *Arabidopsis thaliana*: a transcriptome analysis. *Plant Physiology and Biochemistry* 138, 9–16.

Chikoti, P.C., Tembo, M., Chisola, M., Ntawuruhungu, P., and Ndunguru, J. (2015) Status of cassava mosaic disease and whitefly population in Zambia. *African Journal of Biotechnology* 14, 2539–2546.

Cipriani, G., Michaud, D., Brunelle, F., Golmirzaie, A., and Zhang, D.P. (1999) Expression of soybean proteinase inhibitor in sweet potato. In: *CIP Program Report 1997–1998*. International Potato Center, Lima, pp. 271–277.

Dai, T.M., Lü, Z.C., Liu, W.X., Wan, F.H., and Hong, X.Y. (2017) The homology gene *BtDnmt1* is essential for temperature tolerance in invasive *Bemisia tabaci* Mediterranean cryptic species. *Science Reports* 7: 3040.

de Oliveira, E.J., Morgante, C.V., de Tarso Aidar, S., de Melo Chaves, A.R., Antonio, R.P., *et al.* (2017) Evaluation of cassava germplasm for drought tolerance under field conditions. *Euphytica* 213: 188.

Dhir, S.K., Oglesby, J. and Bhagsari, A.S. (1998) Plant regeneration via somatic embryogenesis, and transient gene expression in sweet potato protoplasts. *Plant Cell Reports* 17, 665–669.

Duan, X., Xu, J., Ling, E., and Zhang, P. (2013) Expression of Cry1Aa in cassava improves its insect resistance against *Helicoverpa armigera*. *Plant Molecular Biology* 83, 131–141.

Darkwa, K., Olasanmi, B., Asiedu, R., and Asfaw, A. (2020) Review of empirical and emerging methods and tools for yam (Dioscorea spp.) improvement: Status and prospects. *Plant Breeding* 139, 474–497.

Ekobu, M., Solera, M., Kyamanywa, S., Mwanga, R.O.M., Odongo, B., *et al.* (2010) Toxicity of seven *Bacillus thuringiensis cry* proteins against *Cylas puncticollis* and *Cylas brunneus* (Coleoptera: Brentidae) using a novel artificial diet. *Journal of Economic Entomology* 103, 1493–1502.

El Sheikha, A.F., and Ray, R.C. (2017) Potential impacts of bioprocessing of sweet potato. *Critical Reviews in Food Science and Nutrition* 57, 455–471.

Fan, W.J., Zhang, M., Zhang, H.X. and Zhang, P. (2012) Improved tolerance to various abiotic stresses in transgenic sweet potato (*Ipomoea batatas*) expressing spinach betaine aldehyde dehydrogenase. *PLOS ONE* 7: e37344.

Fan, W.J., Deng, G.F., Wang, H.X., Zhang, H.X., and Zhang, P. (2015) Elevated compartmentalization of $Na^+$ into vacuoles improves salt and cold stress tolerance in sweet potato (*Ipomoea batatas*). *Physiologia Plantarum* 154, 560–571.

FAO (Food and Agriculture Organization of the United Nations: Home) (2014) GIEWS—Global Information and Early Warning System on Food and Agriculture. FAO, Rome. Available at: www.fao.org/giews/english/fo/index.htm (accessed 27 June 2024).

Feyisa, A.S. (2022) Review of the role of orphan crops in food security. *Research on World Agricultural Economy* 3: 501.

Fukino, N., Hanada, K., Ajisaka, H., Sakai, J., Hirochika, H., *et al.* (2000) Transformation of taro (*Colocasia esculenta* Schott) using particle bombardment. *Japan Agricultural Research Quarterly* 34, 159–165.

Gao, S., Yu, B., Zhai, H., He, S.Z., and Liu, Q.C. (2011) Enhanced stem nematode resistance of transgenic sweet potato plants expressing *Oryzacystatin*-I gene. *Agricultural Sciences in China* 10, 519–525.

Gao, S., Yuan, L., Zhai, H., Liu, C.L., He, S.Z., and Liu, Q.C. (2012) Over expression of *SOS* genes enhanced salt tolerance in sweet potato. *Journal of Integrative Agriculture* 11, 378–386.

García, R., Morán, R., Mena, J., Somontes, D., Zaldúa, Z., *et al.* (2000) Sweet potato (*Ipomoea batatas* L.) regeneration and transformation technology to provide weevil (*Cylas formicarius*) resistance. *Field Trial Results Developments in Plant Genetics and Breeding* 5, 112–117.

Garcia-Oliveira, A.L., Kimata, B., Kasele, S., Kapinga, F., Masumba, E., *et al.* (2020) Genetic analysis and QTL mapping for multiple biotic stress resistance in cassava. *PLOS ONE* 15: e0236674.

Ghislain, M., Tovar, J., Prentice, K., Ormachea, M., Rivera, C., *et al.* (2013) Weevil resistant sweet potato through biotechnology. *Acta Horticulturae* 974, 91–98.

Githunguri, C.M. and Njiru, E.N. (2021) Role of cassava and sweetpotato in mitigating drought in semi-arid Makueni County in Kenya. In: *African Handbook of Climate Change Adaptation*. Springer, Cham. pp. 241–259.

González, A.E., Schöpke, C., Taylor, N.J., Beachy, R.N., and Fauquet, C.M. (1998) Regeneration of transgenic cassava plants (*Manihot esculenta* Crantz) through *Agrobacterium*-mediated transformation of embryogenic suspension cultures. *Plant Cell Reports* 17, 827–831.

Hamilton, A.J., and Baulcombe, D.C. (1999) A species of small antisense RNA in posttranscriptional gene silencing in plants. *Science* 286, 950–952.

He, X., Miyasaka, S.C., Zou, Y., Fitch, M.M., and Zhu, Y.J. (2010) Regeneration and transformation of taro (*Colocasia esculenta*) with a rice chitinase gene enhances resistance to *Sclerotium rolfsii*. *HortScience* 45, 1014–1020.

He, X., Miyasaka, S.C., Fitch, M.M., Khuri, S., and Zhu, Y.J. (2013) Taro (*Colocasia esculenta*) transformed with a wheat oxalate oxidase gene for improved resistance to taro pathogen *Phytophthora colocasiae*. *HortScience* 48, 22–27.

Hernández-Martínez, P., Khorramnejad, A., Prentice, K., Andrés-Garrido, A., Vera-Velasco, N.M., *et al.* (2020) The independent biological activity of *Bacillus thuringiensis* Cry23Aa protein against *Cylas puncticollis*. *Frontiers in Microbiology* 11: 1734.

Hummel, A.W., Chauhan, R.D., Cermak, T., Mutka, A.M., Vijayaraghavan, A., *et al.* (2018) Allele exchange at the EPSPS locus confers glyphosate tolerance in cassava. *Plant Biotechnology Journal* 16, 1275–1282.

Irigoyen, M.L., Garceau, D.C., Bohorquez-Chaux, A., Lopez-Lavalle, L.A.B., Perez-Fons, L., *et al.* (2020) Genome-wide analyses of cassava *Pathogenesis-related* (*PR*) gene families reveal core transcriptome responses to whitefly infestation, salicylic acid and jasmonic acid. *BMC Genomics* 21: 93.

Islam, S. (2006) Sweetpotato (*Ipomoea batatas* L.) leaf: its potential effect on human health and nutrition. *Journal of Food Science* 71, R13–R121.

Ivancic, A. and Lebot, V. (2000) *The Genetics and Breeding of Taro*. Centre de Coopération Internationale en Recherche Agronomique pour le Développement (CIRAD), France.

Jeeva, M.L., Hegde, V., Makeshkumar, T., Sriram, S., Nair, R.R., *et al.* (2004) Rust of queensland arrowroot (*Canna edulis*) caused by *Puccinia thaliae*: a new record for India. *Plant Pathology* 53: 261.

Kayondo, S.I., Pino Del Carpio, D., Lozano, R., Ozimati, A., Wolfe, M., *et al.* (2018) Genome-wide association mapping and genomic prediction for CBSD resistance in *Manihot esculenta*. *Scientific Reports* 8: 1549.

Karthikeyan, C., Patil, B.L., Borah, B.K., Resmi, T.R., Turco, S., *et al.* (2016) Emergence of a latent Indian cassava mosaic virus from cassava which recovered from infection by a non-persistent Sri Lankan cassava mosaic virus. *Viruses* 8: 264.

Korada, R.R., Naskar, S.K., Palaniswami, M.S., and Ray, R.C. (2000) Management of sweet potato weevil (*Cylas formicarius* (Fab.)): an overview. *Journal of Root Crops* 36, 14–26.

Koundinya, A.V.V., Muthusamy, S.K., Ajeesh, B.R., Mohan, C., Sreekumar, J., *et al.* (2024) Elucidation of novel drought-responsive genes from tuber transcriptome of cassava under water deficit stress. *South African Journal of Botany* 169, 255–267.

Kreuze, J.F., Klein, I.S., Lazaro, M.U., Chuquiyuri, W.J.C., Morgan, G.L., *et al.* (2008) RNA silencing-mediated resistance to a crinivirus (Closteroviridae) in cultivated sweetpotato (*Ipomoea batatas* L.) and development of sweetpotato virus disease following co-infection with a potyvirus. *Molecular Plant Pathology* 9, 589–598.

Kuriwada, T., Hosokawa, T., Kumano, N., Shiromoto, K., Haraguchi, D., and Fukatsu, T. (2010) Biological role of *Nardonella* endosymbiont in its weevil host. *PLOS ONE* 5: e13101.

Legg, J.P., Kumar, P.L., Makeshkumar, T., Tripathi, L., Ferguson, M., *et al.* (2015) Cassava virus diseases: biology, epidemiology, and management. *Advances in Virus Research* 91, 85–142.

Legg, J.P., Diebiru-Ojo, E., Eagle, D., Friedmann, M., Kanju, E., *et al.* (2022) Commercially sustainable cassava seed systems in Africa. In: Thiele, G., Friedmann, M., Campos, H., Polar, V., and Bentley, J.E. (eds.) *Root, Tuber and Banana Food System Innovations*. Springer, Cham, pp. 453–482.

Li, H.Q., Sautter, C., Potrykus, I., and Puonti-Kaerlas, J. (1996) Genetic transformation of cassava (*Manihot esculenta* Crantz). *Nature Biotechnology* 14, 736–740.

Li, Y.C., Wan, W.L., Lin, J.S., Kuo, Y.W., King, Y.C., *et al.* (2016) Signal transduction and regulation of *Ibpre-proHypSys* in sweet potato. *Plant, Cell and Environment* 39, 1576–1587.

Ligaba-Osena, A., Jones, J., Donkor, E., Chandrayan, S., Pole, F., *et al.* (2018) Novel bioengineered cassava expressing an archaeal starch degradation system and a bacterial ADP-glucose pyrophosphorylase for starch self-digestibility and yield increase. *Frontiers in Plant Science* 9: 192.

Liu, Q. (2011) Sweet potato omics and biotechnology in China. *Plant Omics* 4, 295–301.

Liu, Q. (2017) Improvement for agronomically important traits by gene engineering in sweetpotato. *Breeding Science* 67, 15–26.

Liu, D.G., He, S.Z., Zhai, H., Wang, L.J., Zhao, Y., *et al.* (2014a) Overexpression of IbP5CR enhances salt tolerance in transgenic sweet potato. *Plant Cell, Tissue and Organ Culture* 117, 1–16.

Liu, D.G., Wang, L.J., Zhai, H., Song, X.J., He, S.Z., and Liu, Q.C. (2014b) A novel α/β-hydrolase gene *IbMas* enhances salt tolerance in transgenic sweet potato. *PLOS ONE* 9: e115128.

Liu, J., Zheng, Q., Ma, Q., Gadidasu, K.K., and Zhang, P. (2011) Cassava genetic transformation and its application in breeding. *Journal of Integrative Plant Biology* 53, 552–569.

Maingi, D., Sivasupramaniam, S., Brown, G., Wagner, R., and Folk, W.R. (2002) *Bacillus thuringiensis* and cholesterol oxidase activity against the sweet potato weevil (Coleoptera: Curculionidae) in semi artificial diet assays. In: *Proceedings of the 5th Annual Meeting of the Entomological Society of America*, 17–20 November, Fort Lauderdale, Florida.

Manoharan, R., Tripathi, J.N., and Tripathi, L. (2016) Plant regeneration from axillary bud derived callus in white yam (*Dioscorea rotundata*). *Plant Cell, Tissue and Organ Culture* 126, 481–497.

Moar, W.J., Mwanga, R.O.M., Odongo, B., Ekobu, M., Solera, M., and Ghislain, M. (2007) Progress towards engineering resistance to weevil in sweet potato using *Bt* gene technology. In: *Biotechnology, Breeding and Seed Systems for African Crops*. The Rockefeller Foundation, New York/Maputo, Mozambique, p. 162.

Moorthy, S.S., Naskar, S.K., Shanavas, S., Radhika, G.S., and Mukherjee, A. (2010) Physicochemical characterization of selected sweet potato cultivars and their starches. *International Journal of Food Properties* 13, 1280–1289.

Morán, R., García, R., López, A., Zaldúa, Z., Mena, J., *et al.* (1998) Transgenic sweet potato plants carrying the delta-endotoxin gene from *Bacillus thuringiensis* var. *tenebrionis*. *Plant Science* 139, 175–184.

Mukherjee, A. and Naskar, S.K. (2012) Performance of orange and purple fleshed sweet potato genotypes in coastal locations of Odisha. *Journal of Root Crops* 38, 26–31.

Muthusamy, S.K., Pushpitha, P., Makeshkumar, T., and Sheela, M.N. (2023) Genome-wide identification and expression analysis of Hsp70 family genes in Cassava (*Manihot esculenta Crantz*). 3 Biotech, 13: 341.

Narayanan, N.N., Ihemere, U., Ellery, C., and Sayre, R.T. (2011) Overexpression of hydroxynitrile lyase in cassava roots elevates protein and free amino acids while reducing residual cyanogen levels. *PLOS ONE* 6: e21996.

Narayanan, N., Beyene, G., Chauhan, R.D., Gaitán-Solis, E., Grusak, M.A., *et al.* (2015) Overexpression of *Arabidopsis VIT1* increases accumulation of iron in cassava roots and stems. *Plant Science* 240, 170–181.

Nassar, N.M.A., and Ortiz, R. (2009) Cassava genetic resources: manipulation for crop improvement. *Plant Breeding Reviews* 31, 247–275.

Nelson, S., Brooks, F., and Teves, G. (2011) *Taro Leaf Blight in Hawai'i*. PD-71. College of Tropical Agriculture and Human Resources University of Hawai'i at Manoa. Available at: http://www.ctahr.hawaii.edu/oc/freepubs/pdf/PD-71.pdf (accessed 27 June 2024).

Newell, C.A., Lowe, J.M., Merryweather, A., Rooke, L.M., and Hamilton, W.D.O. (1995) Transformation of sweet potato (*Ipomoea batatas* (L.) Lam.) with *Agrobacterium tumefaciens* and regeneration of plants expressing cowpea trypsin inhibitor and snowdrop lectin. *Plant Science* 107, 215–227.

Niklaus, M., Gruissem, W., and Vanderschuren, H. (2011) Efficient transformation and regeneration of transgenic cassava using the neomycin phosphotransferase gene as aminoglycoside resistance marker gene. *GM Crops* 2, 193–200.

Ntui, V.O., Uyoh, E.A., Ita, E.E., Markson, A.A.A., Tripathi, J.N., *et al.* (2021) Strategies to combat the problem of yam anthracnose disease: status and prospects. *Molecular Plant Pathology* 22, 1302–1314.

Nyaboga, E., Tripathi, J.N., Manoharan, R., and Tripathi, L. (2014) *Agrobacterium*-mediated genetic transformation of yam (*Dioscorea rotundata*): an important tool for functional study of genes and crop improvement. *Frontiers in Plant Science* 5: 463.

Odipio, J., Getu, B., Chauhan, R.D., Alicai, T., Bart, R., *et al.* (2020) Transgenic overexpression of endogenous FLOWERING LOCUS T-like gene *MeFT1* produces early flowering in cassava. *PLOS ONE* 15: e0227199.

Oki, T., Masuda, M., Furuta, S., Nichiba, Y., Terahara, N., and Suda, I. (2002) Involvement of anthocyanins and other phenolic compounds in radical-scavenging activity of purple-fleshed sweet potato cultivars. *Journal of Food Science* 67, 1752–1756.

Okogbenin, E., Setter, T.L., Ferguson, M., Mutegi, R., Ceballos, H., *et al.* (2013) Phenotypic approaches to drought in cassava. *Frontiers in Physiology* 4: 93.

Okwuonu, I.C., Obidiegwu, J.E., Onyeka, T.J., and Egesi, C.N. (2022) Genome engineering: a new frontier for improving root and tuber crops. In: Obembe, O.O., Ekundayo, M.O., Okoli, A.S., Gidado, A., Adentunji, C.O., *et al.* (eds.) *Agricultural Biotechnology, Biodiversity and Bioresources Conservation and Utilization*. CRC Press, Boca Raton, Florida, pp. 49–68.

Olsen, K., and Schaal, B.A. (1999) Evidence on the origin of cassava: phylogeography of *Manihot esculenta*. *Proceedings of the National Academy of Sciences USA* 96, 5586–5591.

Padhan, B. and Panda, D. (2020) Potential of neglected and underutilized yams (*Dioscorea* spp.) for improving nutritional security and health benefits. *Frontiers in Pharmacology* 11: 496.

Park, S.C., Kim, Y.H., Kim, S.H., Jeong, Y.J., Kim, C.Y., *et al.* (2015a) Overexpression of the *IbMYB1* gene in an orange-fleshed sweet potato cultivar produces a dual-pigmented transgenic sweet potato with improved antioxidant activity. *Physiologia Plantarum* 153, 525–537.

Park, S.C., Kim, S.H., Park, S., Lee, H.U., Lee, J.S., *et al.* (2015b) Enhanced accumulation of carotenoids in sweetpotato plants overexpressing *IbOr-Ins* gene in purple-fleshed sweetpotato cultivar. *Plant Physiology and Biochemistry* 86, 82–90.

Perez-Fons, L., Ovalle, T.M., Maruthi, M.N., Colvin, J., Lopez-Lavalle, L.A.B., and Fraser, P.D. (2020) The metabotyping of an East African cassava diversity panel: a core collection for developing biotic stress tolerance in cassava. *PLOS ONE* 15: e0242245.

Pradhan, D.M.P., Mukherjee, A., George, J., Chakrabarti, S.K., Vimala, B., *et al.* (2015) High starch, beta carotene and anthocyanin rich sweet potato: ascent to future food and nutrition security in coastal and backward areas. *International Journal of Tropical Agriculture* 10, 9–22.

Prakash, P., Jaganathan, D., Immanuel, S., and Sivakumar, S.P. (2020) Analysis of global and national scenario of tuber crops production: trends and prospects. *Indian Journal of Economics and Development* 16, 500–510.

Prentice, K., Pertry, I., Christiaens, O., Bauters, L., Bailey, A., *et al.* (2015) Transcriptome analysis and systemic RNAi response in the African sweet potato weevil (*Cylas puncticollis*, Coleoptera, Brentidae) *PLOS ONE* 10: e0115336.

Rauwane, M. and Ntushelo, K. (2020) Understanding biotic stress and hormone signalling in cassava (*Manihot esculenta*): potential for using hyphenated analytical techniques. *Applied Sciences* 10: 8152.

Ray, R.C. and Sivakumar, P.S. (2009) Traditional and novel fermented foods and beverages from tropical root and tuber crops. *International Journal of Food Science and Technology* 44, 1073–1087.

Reddy, G.V. and Chi, H. (2015) Demographic comparison of sweetpotato weevil reared on a major host, *Ipomoea batatas*, and an alternative host, *I. triloba*. *Science Report* 5: 11871.

Rukarwa, R.J., Mukasa, S.B., Sefasi, A., Ssemakula, G., Mwanga, R.O.M., and Ghislain, M. (2013a) Segregation analysis of *cry7Aa1* gene in $F_1$ progenies of transgenic and non-transgenic sweet potato crosses. *Journal of Plant Breeding and Crop Science* 5, 209–213.

Rukarwa, R.J., Prentice, K., Ormachea, M., Kreuze, J.F., Tovar, J., *et al.* (2013b) Evaluation of bioassays for testing Bt sweet potato events against sweet potato weevils. *African Crop Science Journal* 21, 235–244.

Sajeev, M.S., Padmaja, G., Sheriff, J.T., and Jyothi, A.N. (2021) Entrepreneurial opportunities in tuber crops processing. In: Sudheer, K.P. and Indira, V. (eds.) *Entrepreneurship and Skill Development in Horticultural Processing*. CRC Press, Boca Raton, Florida, pp. 295–322.

Schöpke, C., Taylor, N., Cárcamo, R., Konan, N.K., Marmey, P., *et al.* (1996) Regeneration of transgenic cassava plants (*Manihot esculenta* Crantz) from microbombarded embryogenic suspension cultures. *Nature Biotechnology* 14, 731–735.

Schreuder, M.M., Raemakers, C.J.J.M., Jacobsen, E., and Visser, R.G.F. (2001) Efficient production of transgenic plants by *Agrobacterium*-mediated transformation of cassava (*Manihot esculenta* Crantz). *Euphytica* 120, 35–42.

Sefasi, A., Ssemakula, G., Ghislain, M., Prentice, K., Kiggundu, A., Mwanga, R., and Mukasa, S.B. (2014) Transient expression of β-glucuronidase in recalcitrant Ugandan sweet potato and putative transformation with two *cry* genes. *African Crop Science* 22, 215–227.

Senthilkumar, K.M., Raju, S., Velumani, R., and Gutam, S. (2023) Transcriptome analysis and identification of leaf, tuberous root and fibrous root tissue-specific high temperature stress-responsive genes in sweet potato. *Journal of Horticultural Sciences*, 18, 53–59.

Singh, D., Jackson, G., Hunter, D., Fullerton, R., Lebot, V., Taylor, M., *et al.* (2012) Taro leaf blight—a threat to food security. *Agriculture* 2, 182–203.

Siritunga, D. and Sayre, R. (2004) Engineering cyanogen synthesis and turnover in cassava (*Manihot esculenta*). *Plant Molecular Biology* 56, 661–669.

Sivan, S., Arya, K., Sheela, M.N., Revathi, B.S., Krishnan, P., and Muthusamy, S.K. (2023) Genetic diversity analysis of Indian Cassava (*Manihot esculenta* Crantz) accessions using morphological and molecular markers. *South African Journal of Botany* 161, 347–357.

Sutherland, J.A. (1986) A review of the biology and control of the sweet potato weevil, *Cylas formicarius elegantulus* (F). *Tropical Pest Management* 32, 304–315.

Syombua, E.D., Zhang, Z., Tripathi, J.N., Ntui, V.O., Kang, M., *et al.* (2021) A CRISPR/Cas9-based genome-editing system for yam (*Dioscorea* spp.). *Plant Biotechnology Journal* 19: 645.

Taylor, N., Chavarriaga, P., Raemakers, K., Siritunga, D., and Zhang, P. (2004) Development and application of transgenic technologies in cassava. *Plant Molecular Biology* 56, 671–688.

Teow, C.C., Truong, V.D., McFeeters, R.F., Thompson, R.L., Pecota, K.V., and Yencho, G.C. (2007) Antioxidant activities, phenolic and β-carotene contents of sweet potato genotypes with varying flesh colors. *Food Chemistry* 103, 829–838.

Vanderschuren, H., Akbergenov, R., Pooggin, M.M., Hohn, T., Gruissem, W., and Zhang, P. (2007) Transgenic cassava resistance to African cassava mosaic virus is enhanced by viral DNA-A bidirectional promoter-derived siRNAs. *Plant Molecular Biology* 64, 549–557.

Vanderschuren, H., Alder, A., Zhang, P., and Gruissem, W. (2009) Dose-dependent RNAi-mediated geminivirus resistance in the tropical root crop cassava. *Plant Molecular Biology* 70, 265–272.

Viruel, J., Segarra-Moragues, J.G., Raz, L., Forest, F., Wilkin, P., *et al.* (2016) Late Cretaceous–Early Eocene origin of yams (Dioscorea, Dioscoreaceae) in the Laurasian Palaearctic and their subsequent Oligocene–Miocene diversification. *Journal of Biogeography* 43, 750–762.

Wang, B., Zhai, H., He, S.Z., Zhang, H., Ren, *et al.* (2016) A vacuolar $Na^+/H^+$ antiporter gene, *IbNHX2*, enhances salt and drought tolerance in transgenic sweet potato. *Scientia Horticulturae* 201, 153–166.

Wang, C.C., Ke, L., Cao, L.J., Yao, Y., Geng, M.T., *et al.* (2019) Overexpression of *MinE* gene affects the plastid division in cassava. *Bioscience, Biotechnology, and Biochemistry*, 83, 95–105.

Wangpor, J., Prayoonyong, P., Sakdaronnarong, C., Sungpet, A., and Jonglertjunya, W. (2017) Bioethanol production from cassava starch by enzymatic hydrolysis, fermentation and *ex-situ* nanofiltration. *Energy Procedia* 138, 883–888.

Waterhouse, D.F. and Norris, K.R. (1987) *Biological Control Pacific Prospects*. Inkata Press, Melbourne, Australia.

Waterhouse, P.M., Wang, M.B., and Lough, T. (2001) Gene silencing as an adaptive defence against viruses. *Nature* 411, 834–842.

Weng, X., Zhou, X., Xie, S., Gu, J., and Wang, Z.Y. (2021) Identification of cassava alternative splicing-related genes and functional characterization of MeSCL30 involvement in drought stress. *Plant Physiology and Biochemistry* 160, 130–142.

Woolfe, J.A. (1992) Sweet potato revisited. *Nutrition Bulletin* 17, 180–189.

Yadav, J.S., Ogwok, E., Wagaba, H., Patil, B.L., Bagewadi, B., *et al.* (2011) RNAi-mediated resistance to *Cassava brown streak Uganda virus* in transgenic cassava. *Molecular Plant Pathology* 12, 677–687.

Yu, Y., Cui, Y.C., Ren, C., Rocha, P.S.C.F., Peng, M., *et al.* (2016) Transgenic rice expressing a cassava (*Manihot esculenta* Crantz) plasma membrane gene *MePMP3*-2 exhibits enhanced tolerance to salt and drought stresses. *Genetics and Molecular Research* 15: 15017336.

Zainuddin, I.M., Fathoni, A., Sudarmonowati, E., Beeching, J.R., Gruissem, W., and Vanderschuren, H. (2018) Cassava post-harvest physiological deterioration: from triggers to symptoms. *Postharvest Biology and Technology* 142, 115–123.

Zhang, H., Zhang, Q., Zhai, H., Gao, S., Yang, L., *et al.* (2020) IbBBX24 promotes the jasmonic acid pathway and enhances *Fusarium* wilt resistance in sweet potato. *Plant Cell* 32, 1102–1123.

Zhang, P. and Gruissem, W. (2004) Production of transgenic cassava (*Manihot esculenta* Crantz). In: Curtis, I.A. *Transgenic Crops of the World—Essential Protocols*. Kluwer Academic Publishers, Dordrecht, Netherlands, pp. 301–319.

Zhang, P., and Puonti-Kaerlas, J. (2000) PIG-mediated cassava transformation using positive and negative selection. *Plant Cell Reports* 19, 1041–1048.

Zhang, P., Jaynes, J., Potrykus, I., Gruissem, W., and Pounti-Kaerlas, J. (2003a) Transfer and expression of an artificial storage protein (*ASP1*) gene in cassava (*Manihot esculenta* Crantz). *Transgenic Research* 12, 243–250.

Zhang, P., Bohl-Zenger, S., Pounti-Kaerlas, J., Potrykus, I., and Gruissem, W. (2003b) Two cassava promoters related to vascular expression and storage root formation. *Planta* 218, 192–203.

Zhang, P., Vanderschuren, H., Fütterer, J., and Gruissem, W. (2005) Resistance to cassava mosaic disease in transgenic cassava expressing antisense RNAs targeting virus replication genes. *Plant Biotechnology Journal* 3, 385–397.

Zhang, P., Ma, Q., Naconsie, M., Wu, X., Zhou, W., and Yang, J. (2017) Advances in genetic modification of cassava. *Achieving Sustainable Cultivation of Cassava* 2, 117–136.

Zhong, Y., Ahmed, S., Deng, G., Fan, W., Zhang, P., and Wang, H. (2019) Improved insect resistance against *Spodoptera litura* in transgenic sweet potato by overexpressing Cry1Aa toxin. *Plant Cell Reports* 38, 1439–1448.

Zidenga, T., Siritunga, D., and Sayre, R.T. (2017) Cyanogen metabolism in cassava roots: impact on protein synthesis and root development. *Frontiers in Plant Science* 8: 220.

# 10 Genetic Engineering of Melon

**N. Nagesha*, Suresh H. Antre, Upasana Mohapatra, T.M. Ningaraju and P.H. Ramanjini Gowda**

*Department of Plant Biotechnology, College of Agriculture, University of Agricultural Sciences, Bengaluru, Karnataka, India*

**Abstract**

Melons are sweet, edible, and highly nutritious fleshy fruit-bearing plants belonging to the family Cucurbitaceae. Globally, there are about 13 different types of melon available with unique characteristics. Melon production is greatly affected by biotic and abiotic stresses leading to poor quality and quantity of melon production. The conventional methods used for controlling these stresses have various limitations. Genetic engineering is one of the best alternatives to conventional methods to address these biotic and abiotic stresses along with qualitative improvement of the melon fruits. In this chapter, we present in detail the importance and different types of melon along with its global production status, and genetic-engineering approaches used to address the biotic and abiotic stresses of melon, with suitable examples as well as a discussion of the advantages and limitations. Melon is considered as one of the best options for expression of recombinant proteins of pharmaceutical importance, as melon fruits are highly nutritious, and their seeds contain high amounts of protein. We also present the use of melon fruits for the expression of recombinant proteins/vaccines with suitable examples.

**Keywords:** abiotic stress, biotic stress, edible vaccines, genetic engineering, melon, recombinant proteins

## 10.1 Introduction

Melons comprise a variety of plants in the family Cucurbitaceae with sweet, edible, fleshy fruit. The word "melon" can refer to either the plant or specifically to the fruit. Botanically, a melon is a type of berry, specifically a "pepo." The word "melon" derives from the Latin *melopepo*.

Melons are usually consumed as desserts, snacks, breakfast or picnic foods. Processed melon products (e.g. pre-cut product displays, in-store salad bars) have been developed to appeal to the single-serving market and to smaller households. Seedless varieties have also helped spur consumption. Improved harvesting and handling techniques, as well as the development of sweeter hybrid varieties, have improved quality and consumer acceptance.

Melons are produced globally, and in 2019, melon production in China amounted to nearly 13.5 million t. In comparison, there were approximately 3760 t of melons produced in New Zealand in 2019 (APAC, 2019). Since 2014, China melon production has increased by 3.1% year$^{-1}$, and in 2019, the country was ranked number one in melon production, with Turkey, Iran, and India, respectively, ranked second, third and fourth (Table 10.1).

Cantaloupe melons have several useful properties and contain 85–92% water, 8–15% dry matter, 0.8% protein, 1.8% cellulose and 6.2% other carbohydrates, 0.9% fat, 0.6% ash, 20–30 mg% vitamin C, and 0.03–0.07 mg% other vitamins and

*Email: nagesha.n@uasbangalore.edu.in


DOI: 10.1079/9781800627123.0010

microelements, such as zinc, iron, calcium, magnesium, potassium, and phosphorus. The sugar content in Central Asian melons can reach up to 14–16% (Ahrolovich *et al.*, 2020).

There is a decrease in total carotenoids and vitamin C after harvest, even under the best conditions. A serving of cantaloupe is about 180 g and contains 60 calories including a little fat and only 1 g of protein per serving. The carbohydrates are composed of 14 g of sugar and 2 g of dietary fiber. The yellow- and orange-fleshed melons contain β-carotene (provitamin A), and all melons are a good source of vitamin C. A serving of cantaloupe supplies 120% and 110% of the daily requirements of vitamins A and C, respectively. In addition to this, cantaloupe is also a good source of niacin, vitamin $B_6$, folate, and potassium (Saltveit, 2011).

**Table 10.1.** Top melon-producing countries in 2019. Source: www.nationmaster.com/nmx/ranking/melons-production, accessed 28 June 2024.

| Country | Production (t) |
|---|---|
| China | 12,961,399.93 |
| Turkey | 1,779,369 |
| Iran | 1,680,252 |
| India | 1,218,697 |
| Kazakhstan | 929,150 |
| USA | 846,397 |
| Egypt | 703,889 |
| Guatemala | 642,270 |
| Spain | 630,825 |
| Italy | 613,080 |
| Mexico | 601,622 |
| Brazil | 589,900 |
| Morocco | 518,926 |
| Afghanistan | 353,785 |
| Honduras | 287,216 |
| France | 260,639 |
| Australia | 248,174 |
| Bangladesh | 241,209 |
| Venezuela | 195,290 |
| South Korea | 161,818 |

## 10.2 Types of Melon and its Global Production Status

Melons belong to the family Cucurbitaceae, with members of the genus *Cucumis* widely cultivated across the world for the delicious fruits they produce. Most melon varieties belong to the species *Cucumis melo* L. Melon fruits have been grown for over 4000 years as an important food source in Egypt and Africa. There are 13 types of melon, which include watermelon, muskmelon (cantaloupe), cucumber, Galia melon, canary melon, winter melon, snap melon, honeydew melon, Santa Claus melon, casaba melon, honey globe melon, gac melon, and Crenshaw melon (Fig. 10.1). The different requirements for their growth including the countries growing these melons are briefly described in Table 10.2.

**Fig. 10.1.** 13 different types of melons. (Source: https://www.trees.com/melon-plants#Types-of-Melon-Plants, accessed 28 June 2024).

**Table 10.2.** Different types of melons. Source: https://www.conserve-energy-future.com/types-of-melons.php, accessed 28 June 2024).

| Melon type | Mature size | Hardiness zone | Light | Water | Soil | Major growing countries |
|---|---|---|---|---|---|---|
| Watermelon (*Citrullus lanatus*) | 1.5 feet tall, 15 feet long | 8–11 | Full sun | Medium moisture needs | Fertile, well draining | China accounts for 70% of watermelon production, followed by Brazil, Turkey, Iran, USA |
| Cantaloupe (*Cucumis melo* var. *cantalupensis)* | 1 foot tall, 4 feet long | 9–11 | Full sun | Medium moisture needs | Well draining, loamy | North America (Mexico, Canada, USA), |
| Armenian cucumber (*Cucumis melo* var. *flexuosus*) | Up to 9 feet long and 3 feet wide | 7–9 | Full sun | Medium moisture | Rich, well draining | Western Asia, Egypt, and Armenia |
| Galia melon (*Cucumis melo* var. *reticulatus* 'Galia') | Up to 1 foot tall, 8 feet long | 5–11 | Full sun | Medium moisture needs | Rich and well draining | South America, Europe and Southern USA |
| Canary melon (*Cucumis melo* var. *inodorus* 'Canary') | Up to 2 feet tall, and 10 feet long | 9–12 | Full sun | Medium moisture needs | Well draining | Asia, South America |
| Winter melon (*Benincasa hispida*) | Up to 20 feet long | 9–11 | Full sun | Medium moisture, somewhat drought tolerant | Well draining | Asia, China |
| Snap melon (*Cucumis melo momordica*) | Up to 5 feet long | 9–11 | Full sun | Medium moisture | Well draining | Asia |
| Honeydew melon (*Cucumis melo* var. *inodorus* H. Jacq.) | Up to 5 feet long | 9–11 | Full sun | Medium moisture | Well draining, moisture retentive | China and Turkey |
| Santa Claus melon (*Cucumis melo* var. *inodorus* 'Sancho') | Up to 10 feet long | 9–11 | Full sun | Medium moisture | Well draining | USA, Spain, South America |
| Casaba melon (*Cucumis melo* var. *inodorus* 'Golden Casaba') | Up to 4 feet long | 9–11 | Full sun | Medium moisture | Well draining | Turkey (Casaba) and USA |
| Honey globe melon (*Cucumis melo* var. *inodorus* 'Honey Globe') | Up to 6 feet long | 9–11 | Full sun | Medium moisture | Well draining | South-east Asia |
| Gac melon (*Momordica cochinchinensis*) | Up to 66 feet long | 9–11 | Full sun | Medium moisture | Well draining | Vietnam, Australia, China and South Asian countries |
| Crenshaw melon (*Cucumis melo* var. *inodorus* 'Crenshaw') | Up to 18 feet long | 8–11 | Full sun | Medium moisture | Well draining, fertile | North and South America, Middle East and countries in the Mediterranean region |

## 10.3 Constraints on Melon Production and Control Measures

Biotic and abiotic stresses continue to be a major limitation for crop productivity in melons and their severity may increase with the changing climate scenario. Cucurbits, mainly melons, are infested with various insects from the primordial stages of the crop to harvest of the products. As well as causing direct damage, many pests act as vectors for viruses (Muthusamy *et al.*, 2017). Depending on the environmental conditions and susceptibility of the crop species, the extent of losses varies between 30% and 100%. The major insect and pests that are the most devastating toward yield loss are melon fruit fly (*Bactrocera cucurbitae*), melon aphid (*Aphis gossypii*), melon thrips (*Thrips palmi*), and whitefly (*Bemisia tabaci*) (Haldhar *et al.*, 2015; Bhowmik and Saha, 2017). Major abiotic stresses such as drought, high temperature, and salinity have been reported to have adverse effects on melon by reducing yield significantly and the quality markedly. Under drought stress, up to 50% yield reduction has been recorded in muskmelon (Ansari *et al.*, 2019). Plant pathogens are a significant challenge in agriculture, despite the best efforts to combat them (van Esse *et al.*, 2020). The biotic stresses of melons include viral, fungal, oomycete, phytoplasma, and bacterial diseases causing great economic loss to the melon farmers. Some of the biotic factors affecting melons are depicted in Table 10.3.

**Table 10.3.** Diseases of cucurbits. (Source: www.apsnet.org/edcenter/resources/commonnames/Pages/Cucurbits.aspx, accessed 28 June 2024).

| Disease type | Disease and causative organisms |
|---|---|
| Bacterial diseases | Angular leaf spot: *Pseudomonas syringae* |
| | Bacterial brown spot: *Pantoea ananatis* |
| | Bacterial fruit blotch: *Acidovorax citrulli* |
| | Bacterial rind necrosis: *Erwinia* spp. |
| | Bacterial soft rot: *Pectobacterium carotovorum* (Jones) Waldee (syn. *Erwinia carotovora* subsp. *carotovora* (Jones) Dye) |
| | Bacterial wilt: *Erwinia tracheiphila* |
| | Cichorii seedling blight: *Pseudomonas cichorii* (Swingle)[a] |
| | Cucurbit yellow vine disease: *Serratia marcescens* |
| | Syringae blight and leaf spot: *Pseudomonas syringae* pv. *syringae* |
| | Viridiflava white spot: *Pseudomonas viridiflava* (Burkholder)[a] |
| | *Xanthomonas* leaf spot: *Xanthomonas cucurbitae* |
| Phytoplasma disease | Aster yellows: 'Candidatus Phytoplasma asteris' |
| Fungal diseases | Acrocalymma vine decline: *Acrocalymma vagum* |
| | *Alternaria* leaf blight: *Alternaria citrullicola* |
| | *Alternaria* rot: *Alternaria alternata* |
| | Anthracnose: *Colletotrichum magnum* |
| | Belly rot: *Rhizoctonia solani* |
| | Black root rot: *Thielaviopsis basicola* |
| | Blue mold rot: *Penicillium* spp. |
| | *Cercospora* leaf spot: *Cercospora citrullina* |
| | Charcoal rot (vine decline and fruit rot): *Macrophomina phaseolina* |
| | Choanephora fruit rot: *Choanephora cucurbitarum* |
| | Crater rot (fruit), myrothecium canker: *Myrothecium roridum* |
| | Damping-off: *Acremonium* spp., *Fusarium equiseti* |
| | *Fusarium* crown and foot rot: *Fusarium solani* |
| | *Fusarium* foot rot: *Fusarium acuminatum* |
| | *Fusarium* root and stem rot: *Fusarium oxysporum*[a] |
| | *Fusarium* wilt: *Fusarium oxysporum* Schltdl. |
| | Gray mold: *Botrytis cinerea* |
| | Gummy stem blight, black rot (fruit): *Stagonosporopsis caricae* |
| | Phoma blight: *Boeremia exigua* |
| | Phomopsis black root rot: *Diaporthe sclerotioides* |

*Continued*

**Table 10.3.** Continued.

| Disease type | Disease and causative organisms |
|---|---|
| | Phomopsis fruit rot: *Diaporthe cucurbitae* |
| | Pink mold rot: *Trichothecium roseum* |
| | Pink root: *Setophoma terrestris* |
| | Root and hypocotyl rot: *Plectosphaerella melonis* |
| | Plectosporium blight: *Plectosphaerella cucumerina* |
| | Powdery mildew: *Golovinomyces cichoracearum* |
| | Purple stem vine decline: *Diaporthe melonis* |
| | Red rot: *Epicoccum nigrum* |
| | *Rhizopus* soft rot: *Rhizopus stolonifer* |
| | Scab, gummosis: *Cladosporium cucumerinum* |
| | *Sclerotinia* stem rot, *Sclerotinia* rot (fruit): *Sclerotinia sclerotiorum* |
| | *Septoria* leaf spot: *Septoria cucurbitacearum* |
| | Southern blight: *Athelia rolfsii* |
| | Target spot (*Corynespora* blight): *Corynespora cassiicola* |
| | *Ulocladium* leaf spot: *Ulocladium cucurbitae* |
| | *Verticillium* wilt: *Verticillium albo-atrum* |
| | Web blight: *Rhizoctonia solani* |
| Oomycete diseases | Damping-off: *Phytophthora capsici* |
| | Downy mildew: *Pseudoperonospora cubensis* |
| | *Phytophthora* crown and root rot, *Phytophthora* fruit rot: *Phytophthora capsici* |
| | *Pythium* cottony leak (*Pythium* fruit rot): *Pythium aphanidermatum*, *P. oligandrum* |
| | Root rot: *Phytophthora drechsleri*, *Pythium aphanidermatum* |
| | Sudden wilt: *Pythium aphanidermatum* |
| Virus diseases | Cucumber green mottle mosaic: cucumber green mottle mosaic virus |
| | Cucumber leaf spot: cucumber leaf spot virus |
| | Cucumber mosaic: cucumber mosaic virus |
| | Cucumber pale fruit: hop stunt viroid |
| | Cucumber vein yellowing: cucumber vein yellowing virus[a] |
| | Cucurbit aphid-borne yellows: cucurbit aphid-borne yellows virus |
| | Cucurbit chlorotic yellows: cucurbit chlorotic yellows virus[a] |
| | Cucurbit leaf crumple: cucurbit leaf crumple virus |
| | Cucurbit yellow stunting disorder: cucurbit yellow stunting disorder virus |
| | Curly top: beet curly top virus |
| | Lettuce infectious yellows: lettuce infectious yellows virus |
| | Melon chlorotic leaf curl: melon chlorotic leaf curl virus[a] |
| | Melon necrotic spot: melon necrotic spot virus |
| | Melon yellow spot: melon yellow spot virus[a] |
| | Melon yellowing: melon yellowing-associated virus[a] |
| | Moroccan watermelon mosaic: Moroccan watermelon mosaic virus[a] |
| | Papaya ring spot: papaya ring spot virus Watermelon strain |
| | Pseudo yellows: beet pseudo yellows virus |
| | Squash leaf curl: squash leaf curl virus |
| | Squash mosaic: squash mosaic virus |
| | Squash vein yellowing: squash vein yellowing virus |
| | Squash yellow mild mottle: squash yellow mild mottle virus[a] |
| | Tobacco ring spot: tobacco ring spot virus |
| | Tomato leaf curl New Delhi: tomato leaf curl New Delhi virus[a] |
| | Tomato ring spot: tomato ringspot virus |
| | Watermelon chlorotic stunt: watermelon chlorotic stunt virus[a] |
| | Watermelon mosaic: watermelon mosaic virus |
| | Watermelon silver mottle: watermelon silver mottle virus[a] |
| | Yellow vein mosaic: squash leaf curl China virus, squash leaf curl Philippines virus[a] |
| | Zucchini yellow mosaic: zucchini yellow mosaic virus |

[a]Not known to occur naturally on cucurbits in the USA.

## 10.4 Genetic-Engineering Approaches for the Control of Biotic and Abiotic Stresses of Melon Fruits

Conventional methods of controlling biotic and abiotic stresses are not reaching 100% satisfaction. These include the use of chemicals, which increases the cost of cultivation as well as polluting the environment. Conventional breeding methods play an important role in crop improvement and are available to develop disease resistance lines of melons. However, the development of resistance lines is a time-consuming process, which usually entails growing and examining large populations of crops over multiple generations, a lengthy and laborious task, which also needs to retain the many good characteristics of melons while transferring traits from different plant sources.

One of the most effective and sustainable ways of managing plant pathogens is to use genetic modification (GM) and genome editing, expanding the breeder's toolkit. For use in the field, these solutions must be efficacious, with no negative effect on plant agronomy, they must be deployed thoughtfully, and they must not introduce a potential allergen or toxin. Expensive regulation of biotech crops is prohibitive for local solutions. However, with 11–30% average global yield losses and greater local impacts, tackling plant pathogens is an ethical imperative.

World food production needs to increase by at least 60%, using the same amount of land, by 2050. Hence, there is a need to adopt the new solutions that GM provides to manage plant pathogens (van Esse *et al.*, 2020).

Advances in the field of plant biotechnology such as genetic engineering, RNA interference, and, most recently, clustered regularly interspaced palindromic repeats (CRISPR)-CRISPR-associated protein 9 (Cas9)-mediated technologies which are mainly involved in the direct alteration of an organism's genetic material (Christou, 2013; Dong and Ronald, 2019), are playing a vital role in the development crop plants resistant or tolerant to biotic and abiotic stresses. The major advantages of these technologies include the following:

1. These methods enable the introduction, removal, modification, or fine-tuning of specific genes of interest with minimal undesired changes to the rest of the crop genome, leading to the exhibition of desired agronomical traits by the crop within fewer generations compared with conventional plant-breeding programs.
2. Genetic engineering allows the interchange of genetic material across species, thereby avoiding restriction of genes to particular crop species.
3. Plant transformation by genetic engineering allows the introduction of new genes into a wide variety of crop plants including vegetatively propagated crops such as banana (*Musa* sp.), cassava (*Manihot esculenta*), and potato (*Solanum tuberosum*). These features make genetic engineering a powerful tool for enhancing resistance against plant pathogens (Dong and Ronald, 2019).

### 10.4.1 Genetic engineering for fungal disease resistance

Application of fungicides is the principal practice for managing many cucurbit diseases. Fungicides that are systemic or have translaminar activity are especially valuable. Unfortunately, these fungicides generally have a high risk of developing resistance because they have specific modes of action. Dealing with resistance continues to be challenging, as it is not always possible to predict the resistance risk of a fungicide or the potential for a pathogen to develop resistance (McGrath, 2004).

Fungal diseases cause reduced production levels of oriental melon (*C. melo* L. var. *makuwa* cv. 'Silver Light') in tropical and subtropical regions. Bezirganoglu *et al.* (2013) developed a transgenic oriental melon by transferring the antifungal protein (*AFP*) and chitinase (*CHI*) fusion genes to control fungal diseases caused by *Rhizoctonia solani* and *Fusarium oxysporum* through *Agrobacterium tumefaciens*-mediated transformation. The gene construct consisted of *AFP* and *CHI* fusion genes under the control of the cauliflower mosaic virus (CaMV) 35S promoter and the neomycin phosphotransferase (*nptII*) gene (conferring kanamycin resistance) as a selectable marker. Putative transgenics were confirmed by polymerase chain reaction (PCR). Southern blot analysis was carried out to confirm the stable integration of the *AFP–CHI* fusion gene in the genomic DNA of the PCR-positive lines. Reverse transcriptase PCR analysis showed that the *AFP–CHI* fusion gene was expressed in individual transgenic lines, and Western blot analysis revealed accumulation of CHI protein in the leaves. A segregation analysis of the $T_1$ generation confirmed inheritance of the transgene and demonstrated that the *AFP–CHI* fusion gene was effective in protecting the transgenic melon plants against fungal disease caused by *R. solani* and *F. oxysporum*.

Akbari *et al.* (2013) reported the development of Iranian melon (*C. melo*) resistant to fungal disease by transforming two antifungal proteins, chitinase and β-glucanase, which are known to increase plant resistance to fungal pathogens. The 'Khatooni' cultivar of Iranian melon transformed with gene constructs consisting of the chitinase and β-1,3-glucanase genes, a selector gene (*ntpII*), and the CaMV 35S promoter were transferred by *Agrobacterium*-mediated transformation. Transgenic plants were regenerated from medium containing kanamycin (25 mg $l^{-1}$), cefotaxime (300 mg $l^{-1}$) and 6-benzylaminopurine (1 mg $l^{-1}$). The presence of the gene constructs was demonstrated by PCR using specific primers with amplification of fragments of 870 bp for the chitinase gene and 680 bp for the glucanase gene. The authors concluded that production of melon resistant to fungal disease provides a reliable source of resistance genes in the melon gene pool.

### 10.4.2 Genetic engineering for virus disease resistance

Diseases caused by viruses are increasing in importance, with recent reports of new viruses detected in cucurbits and reports of previously detected viruses occurring in new areas. This adds to the already long list of different viruses affecting cucurbits (Zitter *et al.*, 1996). They can cause substantial economic losses and limit production. For example, in Lebanon, virus diseases are considered the major cause of economic losses in cucurbits, which are among the most important vegetable crops grown there (Abou Jawdah *et al.*, 2000; McGrath, 2004).

Lin *et al.* (2012) reported the development of transgenic watermelon resistant to multiple virus infections by transforming a single chimeric transgene comprising a silencer DNA from the partial N gene of watermelon silver mottle virus fused to the partial coat protein (CP) gene sequences of cucumber mosaic virus (CMV), cucumber green mottle mosaic virus (CGMMV), and watermelon mosaic virus (WMV) gene constructs using *Agrobacterium*-mediated transformation. Single or multiple transgene copies in various locations in the genome were confirmed by Southern blot analysis. The transgenic watermelon $R_0$ plants were individually challenged with CMV, CGMMV, or WMV, or with a mixture of these three viruses for validation. Two lines of watermelon were identified to exhibit resistance to CMV, CGMMV, and WMV individually, and to a mixed inoculation of the three viruses. The $R_1$ progeny of the two resistant $R_0$ lines showed resistance to CMV and WMV but not to CGMMV. These studies demonstrated that fusion of different viral CP gene fragments in transgenic watermelon can contribute to multiple virus resistance via post-transcriptional gene silencing.

Papaya ring spot virus Watermelon strain (PRSV-W) and zucchini yellow mosaic virus (ZYMV) are two potyviruses that limit the production of oriental melon (*C. melo* var. *makuwa*) in Asia. To combat this, Wu *et al.* (2010) developed an untranslatable chimeric DNA comprising partial CP sequences of ZYMV and PRSV-W and used it to transform an elite cultivar of oriental melon, 'Silver Light,' by *Agrobacterium*-mediated transformation. Greenhouse evaluation by mechanical challenges with ZYMV and PRSV-W, alone or together, identified transgenic lines exhibiting different levels of resistance or complete immunity to ZYMV and PRSV-W. Molecular analyses of transgenic lines revealed random insertion of the transgene into the host genome, with insert numbers differing among transformants. Based on these results, the authors concluded that there was no correlation between transgene insert number and the degree of resistance expressed by transgenic lines. The levels of accumulation of transgene transcripts varied among the transgenic lines, and an inverse correlation was observed between the level of accumulation of transgene transcripts and the degree of virus resistance. These transgenic melon lines with high degrees of resistance to ZYMV and PRSV-W have great potential for the control of ZYMV and PRSV-W in *C. melo* in Asia.

### 10.4.3 Genetic engineering for drought tolerance

Drought is one of the major abiotic stresses responsible for loss of crops including muskmelon (*C. melo*). Ansari *et al.* (2019) studied muskmelon genotype SC-15, which exhibits high drought resistance, and exposed it to deficient water conditions and looked for alterations in physiological, molecular, and proteomic profile changes in the leaves. Drought stress resulted in reduced net photosynthetic rate, stomatal conductance, and transpiration rate. By expanding the severity of drought, a decline was recorded in the content of total chlorophyll and carotenoid, while there was an increase in phenol content indicating generation of oxidative stress. The activities of catalase, superoxide dismutase, ascorbate peroxidase, and guaiacol were increased under drought stress. Peptide mass fingerprinting showed that drought increased the relative abundance of 38 protein spots while decreasing ten protein spots. The identified proteins were involved in protein synthesis, photosynthesis, nucleotide biosynthesis, stress response, transcription regulation, metabolism, energy, and DNA binding. A drought-induced MADSbox transcription factor was identified. Their results indicated that, under drought, muskmelon elevates the abundance of defense proteins and suppresses catabolic proteins. The data obtained suggest possible mechanisms adopted by muskmelon to counter the impacts of drought-induced stress.

Liu *et al.* (2020) studied the roles of *CmCAD1*, *-2*, and *-3* in lignin synthesis and their role in drought tolerance in oriental melon. Good lignification led to efficient axial water transport and a barrier to lateral water flow in oriental melon tolerating drought. Cinnamyl alcohol dehydrogenase was found to be an important enzyme in the monolignol synthesis pathway; five cinnamyl alcohol dehydrogenase (*CAD*) genes were generally induced in melon seedlings by drought stress. Overexpressing CmCAD2 or -3 caused strong recovery of CAD activities and lignin synthesis and composition in *Arabidopsis cadc cadd* plants, whose lignin synthesis is disrupted, while CmCAD1 functioned modestly (Anderson *et al.*, 2015). In melon seedlings, silenced *CmCAD2* and *-3* individually or collectively decreased CAD activities and lignin depositions drastically, resulting in dwarfed phenotypes. Reduced lignin, mainly composed of guaiacyl units catalyzed by CmCAD3, was mainly due to the limited lignification in tracheary elements and development of Casparian strips, while CmCAD1 and -2 exhibited catalysis of *p*-coumaraldehyde and sinapaldehyde, respectively. Compared with *CmCAD1*, drought treatments revealed higher sensitivity of *CmCAD2*- and/or *-3*-silenced melon seedlings, accompanied by lower relative water content and water potential, and relatively higher total soluble sugar content. Slightly upregulated expression of aquaporin genes together with limited lignification might imply higher lateral water loss in the stems of the silenced lines. Based on these results, the authors concluded that lignin synthesized by CmCAD2 and -3 has an important function for drought tolerance in melon.

## 10.5 Genetic Engineering for Qualitative Improvements of Melons and Expression of Recombinant Proteins

Melons are one of the most economically important and widely cultivated vegetable crops in the world, and there is a wide genetic diversity in the cultivated and wild species. The process of fruit development and ripening is very complex and involves many biochemical and physiological changes including the breakdown of chlorophyll, degradation of the cell wall, increase in sugar content, alterations in pigment biosynthesis, and the accumulation of flavor and aromatic compounds. The application of genetic engineering and molecular biology of melon is very limited, but with the use of these technologies a large number of quality/ripening-regulated genes involved in pigmentation, vitamins, soluble carbohydrate metabolism, cell-wall metabolism, and ethylene biosynthesis have been identified. Some genetic manipulations of melons have proved to be useful to improve their quality, sensory attributes, shelf life and other agronomic traits (Li *et al.*, 2006). Some of the applications of genetic engineering of melons for qualitative improvements are discussed.

Chatzopoulou *et al.* (2020) reported the suppression of ascorbate oxidase (AO, EC 1.10.3.3) expression in melon fruit to understand the role of AO in plant metabolism, fruit growth, and physiology. AO is a copper-containing enzyme localized at the apoplast, where it catalyzes the oxidation of ascorbic acid to dehydroascorbic acid via a monodehydroascorbic acid intermediate. They reported that reduction of AO activity increased ascorbic acid content in melon fruit, which was the result of repression of ascorbic acid oxidation and simultaneous induction of certain biosynthetic and recycling genes, leading to an altered state of ascorbate redox in the apoplast. Early ripening of transgenic melon fruit was observed due to the increased ethylene production rate, which coincided with elevated levels of 1-aminocyclopropane-1-carboxylic acid oxidase (ACO, EC 1.14.17.4) activity and gene expression.

Tian *et al.* (2010) studied the activity of sucrose phosphate synthase (SPS) activity in transgenic muskmelon (*C. melo*) plants by silencing the SPS using an antisense approach. An 830 bp cDNA fragment of the muskmelon *SPS* gene was expressed in the antisense orientation under the control of the CaMV 35S promoter. They reported clear phenotypic differences between the control and transgenic muskmelon plants. The leaves of the transgenic plant were markedly smaller, and the plant height and stem diameter were shorter and thinner compared with the control plants. Transmission electron microscope observation showed that membrane degradation of the chloroplast occurred in the transgenic leaves and the numbers of grana and grana lamella in the chloroplast were significantly lower, suggesting that the slow growth and weaker phenotype of the transgenic plants might be due to damage to the chloroplast ultrastructure resulting in a decrease in the net photosynthetic rate. The sucrose concentration and levels of SPS were decreased in transgenic mature fruit, and the fruit size was smaller than the control fruit. Based on these studies, the authors concluded that SPS may play an important role in regulating muskmelon plant growth and fruit development.

Lucchetta *et al.* (2014) studied the physiological and biochemical changes in postharvest ripening of melons. They reported on expression of the apple 1-aminocyclopropane-1-carboxylic acid oxidase (ACC) oxidase gene following ethylene treatment in transgenic melons. Furthermore, they evaluated the cell-wall expression (MPG1, MPG2), ethylene synthesis by ACC oxidase (ACCO1), flavor synthesis by alcohol acyltransferase (AAT), and physiochemical parameters such as pulp firmness, titratable acidity, soluble solid content, volatile esters, ethylene production, antioxidant assay capacity, and ascorbic acid content. In cell-wall expression analysis, MPG1 increased when the fruit of transgenic melons was exposed to ethylene, indicating that they are ethylene dependent. MPG2 decreased gradually when the fruit were subjected to ethylene application. Fruit firmness was modified in transgenic fruits when ethylene was applied, and there was a large reduction similar to that of non-transgenic fruit. Titratable acidity in the transgenic fruit was lower than in the non-transgenic fruit. Ethylene applied to the transgenic fruit made the titratable acid increase over 48 h and after that there was a reduction in titratable acid. In relation to soluble solid content, transgenic fruit treated with or without ethylene did not reduce gradually compared with wild-type melons in all periods. Ethylene production in transgenic fruit was reestablished when ethylene was applied, exhibiting the same behavior as transgenic fruit. Antioxidant assay levels were more active in transgenic fruit when ethylene was applied than in control fruit. Based on these studies, the authors concluded that application of ethylene in transgenic ACC oxidase melons is able to change the metabolism of the cell wall, flavors, and antioxidant capacity levels in the fruit during the ripening process.

In melon, substantial amounts of pectin depolymerization and solubilization occur during ripening, suggesting a role for polygalacturonase (PG) in ripening-associated cell-wall disassembly in melons (McCollum *et al.*, 1989; Hadfield *et al.*, 1998; Rose *et al.*, 1998). In 'Charentais' melon, the expression of six PG genes was detected in a variety of tissues; three of the PG genes, *MPG1–MPG3*, were expressed in ripening fruit but differed in the relative abundance of mRNA accumulation (Hadfield *et al.*, 1998). Among these genes, *MPG1* may be important in the ripening process of melon fruit. Suppression of *MPG1* gene expression in transgenic melons will provide a critical assessment of this gene (Li *et al.*, 2006).

The molecular basis of ethylene sensitivity and fruit enlargement in melon has been analyzed. Ezura (2001) reported genetic engineering of melon for studying the shelf life and fruit size. Two ethylene-receptor genes related to shelf life, *Cm-ERS1* and *Cm-ETR1*, were isolated and methods were developed for altering the ethylene sensitivity of the plants. The fruit enlargement-responsive gene, 3-hydroxy-3-methylglutaryl coenzyme A reductase (*Cm-HMGR*) was also isolated and was expected to be involved in altering fruit size. Genetic technologies were also developed to alter ethylene sensitivity and cell division in the pericarp for further improving postharvest performance and fruit size in melon.

Isolation and subsequent functional characterization of the genes encoding invertase, sucrose synthase, and SPS enzymes in melon may offer an insight for understanding its sugar metabolic mechanism and the possibility of improving fruit quality using biotechnology approaches. To improve the storage and handling characteristics of 'Charentais' melon, a transgenic line with the ACO antisense gene has been reported (Ayub *et al.*, 1996). The antisense fruit can stay on the vine longer and accumulate higher amounts of soluble sugars than non-transformed fruit (Ayub *et al.*, 1996; Guis *et al.*, 1997; Flores *et al.*, 2001). Thus, for commercial purposes, this result suggests that antisense fruit can accumulate more sugars on the vine and be harvested later without the risk of over-ripening (Li *et al.*, 2006).

Flavor, a complex of aroma and taste, is one of the most important quality criteria of fruits and vegetables. Fruit flavor is a complex mixture of numerous compounds, with the volatile compounds being biosynthesized through isoprenoid, shikimate, and β-oxidation pathways. Many are formed naturally by enzymes within intact tissue, and others by enzymes in damaged tissue through a series of chemical reactions (Li *et al.*, 2006).

Volatile esters containing a branched alkyl chain originating from valine, isoleucine, and other amino acids (Yabumoto *et al.*, 1977; Wyllie and Leach, 1990), while aliphatic esters and alcohols are produced from free fatty acids such as linoleic and linolenic acids. Fatty acids can generate compounds with shorter chains by β-oxidation. Both pathways converge in the formation of aldehydes, which are then reduced to alcohols in a reaction catalyzed by alcohol dehydrogenase (Speirs *et al.*, 1998). The last step in the production, esterification, is the result of transacylation from acyl-CoA to an alcohol. The enzyme catalyzing the reaction, AAT, is a key enzyme in aroma biochemistry (Aharoni *et al.*, 2000; Li *et al.*, 2006).

Shalit *et al.* (2001) demonstrated that acetyl-CoA:AAT activity was high in climacteric melons of the reticulatus group but was negligible in non-climacteric melons of the indorus group. Two AAT genes, *CmAAT1* and *CmAAT2*, involved in the formation of aroma volatile esters have recently been isolated from Charentais melon fruit (Yahyaoui *et al.*, 2002; El-Sharkawy *et al.*, 2005). Expression studies indicated that both genes were specifically expressed in fruit at increasing rates in the early and mid-stages of ripening. Expression was sharply reduced in ethylene-suppressed antisense-ACO fruit and in wild-type fruit treated with the ethylene antagonist 1-methylcyclopropene. Cloning of the two genes in yeast revealed that the CmAAT1 protein exhibited AAT activity, while no such activity could be detected for CmAAT2. CmAAT1 was capable of producing esters from a wide range of compounds including alcohols and acyl-CoAs (Li *et al.*, 2006).

The antisense-ACO transgenic melon resulted in a considerable loss of production of aroma volatiles, mainly esters. The total volatiles of antisense fruit can be 60–85% lower than that of non-transgenic fruit (Bauchot *et al.*, 1998). This indicates a regulatory role of ethylene in the biogenesis of aroma volatiles, but the steps on which ethylene acts remain unknown. Recent studies have shown that the production of aliphatic esters such as hexyl and butyl acetate is blocked in antisense-ACO fruit and can be reversed by ethylene treatment (Flores *et al.*, 2002). The reduction of fatty acids and aldehydes appears essentially to be ethylene dependent, while the last step of alcohol acetylation has ethylene-dependent and -independent components, probably corresponding to differentially regulated alcohol acetyltransferases (Flores *et al.*, 2002). Isolating other fruit flavor biosynthesis genes will allow us to understand the pathway for natural flavor production. Thus, use of the biotechnological approach in flavor studies can provide insights into improving flavor characteristics of commercial fruits through plant breeding and selection (Li *et al.*, 2006).

### 10.5.1 Genetic engineering for the expression of recombinant proteins/vaccines

Vaccination is the only means to control some of the deadly neuroviral diseases such as rabies. The use of genetically altered plants to produce vaccines is gaining importance as it is cost-effective and guarantees freedom from human or animal pathogens. The use of transgenic plants as a source of oral vaccines is a tantalizing possibility that has been pursued with varying degrees of success (Haq *et al.*, 1995; Mason and Arntzen, 1995; Thanavala *et al.*, 1995; Modelska *et al.*, 1998; Sandhu *et al.*, 2000; Huang *et al.*, 2001).

Nagesha *et al.* (2007) reported *Agrobacterium*-mediated transformation of *C. melo* with a rabies virus (strain ERA) glycoprotein gene (*PRGSpRgp*). The presence of the rabies glycoprotein gene in the DNA of transformed melon leaves was confirmed by PCR amplification. Synthesis of the rabies glycoprotein was detected in transformed cantaloupe fruit by sodium dodecyl sulfate–polyacrylamide gel electrophoresis (SDS-PAGE) and western immunoblotting. Partially purified, recombinant rabies glycoprotein from transformed cantaloupe melon fruit was injected intramuscularly and intraperitoneally into mice, and resulted in the induction of anti-PRGSpRgp antibodies. The mice were then challenged with rabies virus strain ERA and survival of mice was observed due to the accumulation of sufficient quantities of anti-rabies glycoprotein antibodies (0.6–1.6 IU $ml^{-1}$). Based on these results, they concluded that the accumulation of rabies glycoprotein in edible transgenic cantaloupe melon fruit offers the promise of biomedical exploitation.

## 10.6 Summary and Conclusion

Cantaloupes have several useful properties with highly nutritious components including water, dry matter, protein, cellulose, carbohydrates, fat, ash, vitamin C and other vitamins, and microelements, and are also a good source of sugar. The production constraints of melons can effectively be overcome by the application of modern biotechnological tools including recombinant genetic engineering and CRISPR-CAS genome editing compared with conventional methods of managing the biotic and abiotic stresses of melons. The quality and quantity of production can also be increased by these technologies. In addition, melons act as a good source for the expression of recombinant proteins of biopharmaceutical importance. There is need to explore the applications of these genetic-engineering technologies in melon crop improvement programs.

## 10.7 References


Abou Jawdah, Y., Sobh, H., Fayad, A., Lecoq, H., Delecolle, B., and Ferre, J.T. (2000) Cucurbit yellow stunting disorder virus: a new threat to cucurbits in Lebanon. *Journal of Plant Pathology* 82, 55–60.

Aharoni, A., Keizer, L.C.P., Bouwmeester, H.J., Sun, Z., Alvarez-Huerta, M., *et al.* (2000) Identification of the *SAAT* gene involved in strawberry flavor biogenesis by use of DNA microarrays. *Plant Cell* 12, 647–662.

Ahrolovich, R.N., Madiyarovich, S.S. and Urinbaevana, M.H. (2020) Melon and its environmental characteristics. *Journal of Critical Reviews* 7, 480–490.

Akbari, M., Nadaf, M.E., Lotfi, M., and Tohidfar, M. (2013) Transformation of Iranian melon for increasing resistance to fungal diseases. *Research in Plant Sciences* 1, 1–3.

Anderson, N.A., Tobimatsu, Y., Ciesielski, P.N., Ximenes, E., Ralph, J., *et al.* (2015) Manipulation of guaiacyl and syringyl monomer biosynthesis in an *Arabidopsis* cinnamyl alcohol dehydrogenase mutant results in atypical lignin biosynthesis and modified cell wall structure. *Plant Cell* 27, 2195–2209.

Ansari, W.A., Atri, N., Ahmad, J., Qureshi, M.I., Singh, B., *et al.* (2019) Drought mediated physiological and molecular changes in muskmelon (*Cucumis melo* L.) *PLOS ONE* 14: e0222647.

APAC (2019) Watermelon production in the Asia-Pacific region in 2022, by country or territory. Available at: https://www.statista.com/statistics/679419/asia-pacific-melon-production-by-country/ (accessed 29 June 2024).

Ayub, R., Guis, M., Ben Amor, M., Gillot, L., Roustan, J.P., *et al.* (1996) Expression of ACC oxidase antisense gene inhibits ripening of cantaloupe melon fruits. *Nature Biotechnology* 14, 862–866.

Bauchot, A.D., Mottram, D.S., Dodson, A.T. and John, P. (1998) Effect of aminocyclopropane-1-carboxylic acid oxidase antisense gene on the formation of volatile esters in Cantaloupe *Charentais melon* (cv. Védrantais). *Journal of Agricultural and Food Chemistry* 46, 4787–4792.

Bezirganoglu, I., Hwang, S.Y., Fang, T.J. and Shaw, J.F. (2013) Transgenic lines of melon (*Cucumis melo* L. var. *makuwa* cv. 'Silver Light') expressing antifungal protein and chitinase genes exhibit enhanced resistance to fungal pathogens. *Plant Cell, Tissue and Organ Culture* 112, 227–237.

Bhowmik, S. and Saha, S. (2017) Study on the pest complex of bottle gourd in the gangetic plains of West Bengal. *Journal of Entomology and Zoology Studies* 5, 725–727.

Chatzopoulou, F., Sanmartin, M., Mellidou, I., Pateraki, I., Koukounaras, A., *et al.* (2020) Silencing of ascorbate oxidase results in reduced growth, altered ascorbic acid levels and ripening pattern in melon fruit. *Plant Physiology and Biochemistry* 156, 291–303.

Christou, P. (2013) Plant genetic engineering and agricultural biotechnology 1983–2013. *Trends in Biotechnology* 31, 125–127.

Dong, O.X. and Ronald, P.C. (2019) Genetic engineering for disease resistance in plants: recent progress and future perspectives. *Plant Physiology* 180, 26–38.

El-Sharkawy, I., Manriquez, D., Flores, F.B., Regad, F., Bouzayen, M., *et al.* (2005) Functional characterization of a melon alcohol acyltransferase gene family involved in the biosynthesis of ester volatiles. Identification of the crucial role of a threonine residue for enzyme activity. *Plant Molecular Biology* 59, 5–62.

Ezura, H. (2001) Genetic engineering of melon (*Cucumis melo* L.). *Plant Biotechnology* 18, 1–6.

Flores, F., El Yahyaoui, F., de Billerbeck, G., Romojaro, F., Latcé, A., *et al.* (2002) Role of ethylene in the biosynthetic pathway of aliphatic ester aroma volatiles in charentais cantaloupe melons. *Journal of Experimental Botany* 53, 201–206.

Flores, F.B., Martınez-Madrid, M.C., Sanchez-Hidalgo, F.J. and Romojaro, F. (2001) Differential rind and pulp ripening of transgenic antisense ACC oxidase melon. *Plant Physiology and Biochemistry* 39, 37–43.

Guis, M., Botondi, R., Ben-Amor, M., Ayub, R., Bouzayen, M., *et al.* (1997) Ripening-associated biochemical traits of cantaloupe Charentais melons expressing an antisense ACC oxidase transgene. *Journal of the American Society for Horticultural Science* 122, 748–751.

Hadfield, K.A., Rose, J.K.C., Yaver, D.S., Berka, R.M., Bennett, A.B. (1998) Polygalacturonase gene expression in ripe melon fruit supports a role for polygalacturonase in ripening-associated pectin disassembly. *Plant Physiology* 117, 363–373.

Haldhar, S.M., Choudhary, B.R., and Bhargava, R. (2015) Antixenotic resistance traits of muskmelon *Cucumis melo* (L.) against fruit fly (*Bactrocera cucurbitae* (Coquillett)) in arid region of India. *Indian Journal of Applied Entomology* 29, 81–87.

Haq, T.A., Mason, H.S., Clements, J.D. and Arntzen, C.J. (1995) Oral immunization with recombinant bacterial antigen produced in transgenic plants. *Science* 268, 714–716.

Huang, Z., Webster, D., Strugnell, R. and Wesselingh, S. (2001) Plant-derived measles virus hemaggluatinin protein induces neutralizing antibodies in mice. *Vaccine* 19, 2163–2171.

Li, Z., Ya, L., Yang, Y. and Li, A. (2006) Transgenic approach to improve quality traits of melon fruit. *Scientia Horticulturae* 108, 268–277.

Lin, C.-L., Ku, H.-M., Chiang, Y.-H., Ho, H.-Y., Yu, T.-A., and Jan, F.-J. (2012) Development of transgenic watermelon resistant to *Cucumber mosaic virus* and *Watermelon mosaic virus* by using a single chimeric transgene construct. *Transgenic Research* 21, 983–993.

Liu, W., Jiang, Y., Wang, C., Zhao, L., Jin, Y., *et al.* (2020) Lignin synthesized by *CmCAD2* and *CmCAD3* in oriental melon (*Cucumis melo* L.) seedlings contributes to drought tolerance. *Plant Molecular Biology* 103, 689–704.

Lucchetta, L., Rombaldi, C.V., Silva, J.A., Pegoraro, C., Zanuzo, M.R., and Vieira, C.V. (2014) The effect of ethylene on transgenic melon ripening and fruit quality. *African Journal of Biotechnology* 13, 3252–3261.

Mason, H.S. and Arntzen, C.J. (1995) Transgenic plants as vaccine production systems. *Trends in Biotechnology* 13, 388–392.

McCollum, T.G., Huber, D.J., and Cantliffe, D.J. (1989) Modification of polyuronides and hemicelluloses during muskmelon fruit softening. *Plant Physiology* 76, 303–308.

McGrath, M.T. (2004) Diseases of cucurbits and their management. In: Naqvi, S.A.M.H. (ed.) *Diseases of Fruits and Vegetables, Vol. I*. Kluwer Academic Publishers, Dordrecht, Netherlands, pp. 455–510.

Modelska, A., Dietzschold, B.N., Fang, S.Z.F., Steplewski, K.D., Hooper, C., *et al.* (1998) Immunization against rabies with a plant derived antigen. *Proceedings of the National Academy of Sciences USA* 95, 2481–2485.

Muthusamy, S.K., Sivalingam, P.N., Sridhar, J., Singh, D. and Haldhar, S.M. (2017) Biotic stress inducible promoters in crop plants—a review. *Journal of Agriculture and Ecology* 4, 14–24.

Nagesha, N., Ramanjini Gowda, P.H., Madhusudana, S.N., Lokesh, J., Vinay, N.J., *et al.* (2007) Genetic transformation of cantaloupe melon (*Cucumis melo* L.) with the rabies virus glycoprotein gene (*PRG-SpRgp*) and immunization studies in mice. *Journal of Horticultural Science and Biotechnology* 82, 383–386.

Rose, J.K.C., Hadfield, K.A., and Bennett, A.B. (1998) Temporal sequence of cell wall disassembly in rapidly ripening melon fruit. *Plant Physiology* 117, 345–361.

Saltveit, M.E. (2011) Melon (*Cucumis melo* L.). In: *Postharvest Biology and Technology of Tropical and Subtropical Fruits*. Woodhead Publishing, Sawston, UK, pp. 31–45.

Sandhu, J.S., Kranyanski, S.F., Domier, L.L., Korban, S.S., Osadjan, M.D., and Buetow, D.E. (2000) Oral immunization of mice with transgenic tomato fruit expressing respiratory syncytial virus-F protein induces a systemic immune response. *Transgenic Research* 9, 127–135.

Shalit, M., Katzir, N., Tadmor, Y., Larkov, O., Burger, Y., *et al.* (2001) Acetyl-CoA: Alcohol acetyltransferase activity and aroma formation in ripening melon fruits. *Journal of Agricultural and Food Chemistry* 49, 794–799.

Speirs, J., Lee, E., Holt, K., Yong-Duk, K., Scott, N.S., *et al.* (1998) Genetic manipulation of alcohol dehydrogenase levels in ripening tomato fruit affects the balance of some flavor aldehydes and alcohols. *Plant Physiology* 117, 1047–1058.

Thanavala, Y., Yang, Y.F., Lyons, P., Mason, H.S., and Arntzen, C. (1995) Immunogenicity of plant-derived hepatitis B surface antigen. *Proceedings of the National Academy of Sciences USA* 92, 3358–3361.

Tian, H., Ma, L., Zhao, C., Hao, H., Gong, B., *et al.* (2010) Antisense repression of sucrose phosphate synthase in transgenic muskmelon alters plant growth and fruit development. *Biochemical and Biophysical Research Communications* 393, 365–370.

van Esse, H.P., Reuber, T.L., and van der Does, D. (2020) Genetic modification to improve disease resistance in crops. *New Phytologist* 225, 70–86.

Wu, H.W., Yu, T.A., Raja, J.A.J., Christopher, S.J., Wang, S.L. and Yeh, S.D. (2010) Double-virus resistance of transgenic *Oriental melon* conferred by untranslatable chimeric construct carrying partial coat protein genes of two viruses. *Plant Disease* 94, 1341–1347.

Wyllie, S.G. and Leach, D.N. (1990) Aroma volatiles of *Cucumis melo* cv. Golden Crispy. *Journal of Agricultural and Food Chemistry* 38, 2042–2044.

Yabumoto, K., Yamaguchi, M., and Jennings, W.G. (1977) Biosynthesis of some volatiles constituents of muskmelon *Cucumis melo*. *Chemische und Microbiologische Technologische Lebensraum* 5, 53–56.

Yahyaoui, F.E., Wongs-Aree, C., Latche, A., Hackett, R., Grierson, D., and Pech, J.C. (2002) Molecular and biochemical characteristics of a gene encoding an alcohol acyl-transferase involved in the generation of aroma volatile esters during melon ripening. *European Journal of Biochemistry* 269, 2359–2366.

Zitter, T.A., Hopkins, D.L., and Thomas, C.E. (1996) *Compendium of Cucurbit Diseases*. APS Press, St. Paul, Minnesota.

# 11 Transgenic Resistance to Vegetable Viral Diseases

**M. Malavika[1], Sneha Yogindran[1,2] and Supriya Chakraborty[1*]**
[1]*School of Life Sciences, Jawaharlal Nehru University, New Delhi, India;* [2]*Cochin University of Science and Technology, Cochin, Kerala, India*

**Abstract**
Global food and nutritional security depend on the cultivation not only of staple grains but also of various vegetables required for proper nutrition in the growing population. Vegetable production is affected by many stress factors, ultimately leading to reduced yield. Viral infection in plants is considered a serious issue due to the difficulty in containing its spread. Plant viruses are spread by insect vectors and hence the use of chemical pesticides is the major means of disease control. However, excessive use of chemicals has resulted in increased pollution, which eventually affects the ecosystem. Scientists are in search of more sustainable ways to contain viral infections without or with limited use of insecticides. The advent of genetic modification is a promising method to develop virus-resistant crop plants. However, frequent recombination and hijacking of the host defense machinery often enable the virus to escape the control strategies. Hence, effective and long-term approaches such as transgenic crops may pave the way in the future. This chapter focuses on various transgenic control methods based on both proteins and antisense RNA molecules in vegetable crops.

**Keywords:** CPMR, CRISPR, host proteins, plant viruses, RNAi, transgenic resistance, vegetable crops

## 11.1 Introduction

One in ten people living in the world is experiencing food insecurity, and the COVID-19 pandemic was estimated to increase this number to around 132 million (FAO *et al.*, 2020). The United Nations 2030 agenda of sustainable development urges the nations of the world to address and effectively act to stop the malnutrition prevailing within the population (General Assembly, 2015). Vegetables and fruits are packed with vitamins, minerals, and micronutrients, which, when eaten daily, have a profound effect on improving physical and mental health, and numerous research studies support the idea that consuming five servings of fruits and vegetables per day can prevent the onset of chronic health-related disorders, thus improving the quality of life (Głąbska *et al.*, 2020; Wallace *et al.*, 2020). The importance of monitoring the production, accessibility, and consumption of nutritious food, especially in low- and middle-income countries, is a vital step toward achieving the United Nations agenda. If policy makers do not intervene to understand the existing food consumption pattern and the percentage of the undernourished population in their country, then diet-related disorders, especially diabetes, are estimated to affect a

*Email: supriyachakrasls@yahoo.com


DOI: 10.1079/9781800627123.0011

majority of the population (Global Panel, 2020). Cultivation of vegetable crops can promote the income of smallholder farmers and can uphold the economy in developing countries (Schreinemachers *et al.*, 2018).

The occurrence of plant viral pandemics in vegetables and staple crops entails huge economic losses to farmers and in turn affects the availability of healthy food for people worldwide where over 3 billion cannot afford a healthy diet (FAO *et al.*, 2020; Jones 2020). Crop losses due to viral diseases can be distressing if a community is dependent on it as their staple crop. The cassava mosaic disease outbreaks in sub-Saharan Africa brought about devastating yield losses and famine (Jones, 2020; Eni *et al.*, 2021). Approximately 11 different species of cassava geminiviruses are responsible for this disease from various parts of the world (Fondong, 2017).

Tomatoes are an economically important food crop rich in antioxidants and vitamin C, and have recommended benefits to combat depression-related symptoms (Głąbska *et al.*, 2020). Conversely, it is the most taxed food crop among the countries of the world (FAO, 2020). The major viruses afflicting the yield of tomato (*Solanum lycopersicum*) are tomato spotted wilt virus (TSWV, genus *Orthotospovirus*, family Tospoviridae), tomato yellow leaf curl virus (TYLCV, genus *Begomovirus*, family Geminiviridae), pepino mosaic virus (genus *Potexvirus*), and tomato brown rugose fruit virus (genus *Tombusvirus*, family Tombusviridae), which can drastically reduce the yield and fruit quality in infected fields globally (Jones, 2020, 2021). In India, this crop is affected by several viruses including tomato leaf curl New Delhi virus (ToLCNDV) and tomato leaf curl Gujarat virus (both genus *Begomovirus*, family Geminiviridae), cucumber mosaic virus (CMV, genus *Cucumovirus*, family Bromoviridae), potato virus Y (PVY, genus Potyvirus, family Potyviridae), TSWV, and capsicum chlorosis virus (genus *Tospovirus*, family Bunyaviridae) (Mandal *et al.*, 2012; Pratap *et al.*, 2012a; Sharma *et al.*, 2014; Chakraborty and Kumar, 2020). Virus evolution occurs at a high rate and leads to the emergence of resistance-breaking strains in most cases, emphasizing the importance of taking all measures needed to effectively curb disease spread (Oladokun *et al.*, 2019; Nilon *et al.*, 2021). In tuber crops such as potato (*Solanum tuberosum* L.), potato necrotic ringspot disease caused by potato leafroll virus (PLRV, genus *Polerovirus*, family Solemoviridae) and PVY is destructive to tuber quality and has resulted in 80% yield losses; therefore, potato necrotic ringspot disease is reported as a major epidemic affecting potato cultivation (Jones, 2021). In India, chili cultivation is deterred by CMV, TSWV, pepper mild mottle virus (genus *Tobamovirus*, family Virgoviridae), and geminiviruses such as chili leaf curl virus and pepper leaf curl Bangladesh virus, which can cause severe crop losses (Kumar *et al.*, 2015; Sharma *et al.*, 2020). Likewise, bhendi yellow vein mosaic virus, okra enation leaf curl virus, and okra leaf curl virus are geminiviruses causing huge yield losses in okra (*Abelmoschus esculentus* L.) cultivation (Malathi *et al.*, 2017; Mishra *et al.*, 2017). Cucurbit-infecting viruses such as zucchini yellow mosaic virus (ZYMV, genus *Potyvirus*), cucumber green mottle mosaic virus (CGMMV, genus *Tobamovirus*), and geminiviruses such as ToLCNDV, squash leaf curl China virus, and tomato leaf curl Palampur virus are also predominant viruses that affect the yield and quality of the cucurbit crops (Vučurović *et al.*, 2012; Malathi *et al.*, 2017; Jones, 2021).

Transgenic resistance implies the use of a foreign gene derived from the pathogen, other plants or bacteria inserted into plants to achieve the desired trait of interest. Vegetable crops have been engineered to be virus resistant by various transgenic approaches. Genetically modified crops in 2018 contributed to a sum of US$18.95 billion of farm income benefits globally. Currently, 190.4 million ha of land are being used for the cultivation of genetically modified crops (ISAAA, 2016; Brookes, 2020). The pioneering work on viral coat protein (CP)-mediated resistance (CPMR) by Roger Beachy's laboratory has helped scientists to successfully implement it in various crops (Abel *et al.*, 1986). Certainly, in papaya cultivation in Hawaii with plants resistant to papaya ringspot virus (PRSV) using CPMR, the average yield almost tripled (Ferreira *et al.*, 2002). Later, various viruses infecting major agricultural crops were counteracted by different methods of transgenic resistance (Cillo and Palukaitis, 2014). The International Service for the Acquisition of Agri-biotech Applications (ISAAA) has approved CPMR as the major event in engineering transgenic plants. In crops such as potato, tomato, and squash (*Cucurbita pepo*),

CPMR is the approach used in developing virus-resistant cultivars among countries of the world (www.isaaa.org, accessed 1 July 2024). Expression of the viral replicase and helicase domains of PLRV is another commercialized method for the production of resistant potato cultivars (Lawson *et al.*, 2001; www.isaaa.org).

Activating the host gene-silencing machinery to target the cassava mosaic virus genome using antisense RNAs, dsRNAs, and hairpin RNAs (hpRNAs) in transgenic cassava (*Manihot esculenta*) plants has shown promising results (Fondong, 2017). As well as conventional approaches, the use of peptide aptamers, artificial zinc-finger proteins (AZPs), and other non-viral proteins offer effective virus resistance to transgenic plants (Sera, 2017). Similarly, the emerging use of clustered regularly interspaced palindromic repeats (CRISPR)-CRISPR-associated protein 9 (Cas9) system to combat viruses including TYLCV infections can help in tackling evolving virus species (Tashkandi *et al.*, 2018).

This chapter focuses on the different approaches involved in the development of virus resistance in plants, concentrating on the studies in vegetable crops, starting with the methods involved in pathogen-derived resistance and use of viral proteins in conferring resistance to various vegetable crops, and later focusing on the antisense RNA technology followed by small RNA (sRNA)-mediated resistance and the use of non-viral genes in various crops. A brief account of CRISPR-Cas9 nuclease-mediated virus resistance in vegetables is also addressed.

## 11.2 Approaches for Transgenic Resistance

### 11.2.1 Coat protein-mediated resistance

Plant virus genome segments such as CPs have been used effectively to confer virus-resistant traits in plants. Expression of the CP gene results in enhanced tolerance to viral diseases and is also stably inherited. To attain this, the CP gene open reading frame (ORF) is cloned in an appropriate plant expression vector. Another important aspect for efficient gene expression is the selection of a transcriptional promoter. The cauliflower mosaic virus (CaMV) 35S promoter derived was reported as the most effective in both model and vegetable crops (Tumer *et al.*, 1987; Hoekema *et al.*, 1989; Lawson *et al.*, 1990). However, the CaMV 35S promoter is not always a good choice. Tobacco mosaic virus (TMV) CP cDNA when expressed under the control of the 35S promoter produced relatively low amounts of CP and eventually did not confer resistance to TMV infection (Bevan *et al.*, 1985; Bevan and Harrison, 1986). In case several copies of genes are incorporated during transformation, it is advisable to use a weaker promoter such as the CaMV 19S promoter. Expression of a chimeric alfalfa mosaic virus CP gene under the control of the 19S promoter resulted in resistance to alfalfa mosaic virus infection. Resistance was observed when two CP gene loci were expressed but not when only one was expressed (Loesch-Fries *et al.*, 1987). However, in another study, TMV CP expressed under the control of the 19S promoter resulted in a low level of CP and eventually did not confer resistance to TMV infection (Powell, 1988). This suggests the important role of promoter selection in achieving an effective resistant trait. In-depth knowledge of the genome organization and virion structure of TMV helped the research related to CPMR. The first report of transgenic tobacco plants expressing the TMV CP displayed a delay in symptom development and enhanced tolerance to infection (Powell-Abel *et al.*, 1986).

Thereafter, many groups reviewed the use of CP for conferring resistance to plants. CPMR can provide either narrow- or broad-spectrum protection from viral infection (Tepfer, 2002). Transgenic potato plants expressing the CP gene of potato mosaic virus strain N605 provided resistance to the viral strain and also to the related strain 0803 (Malnoë *et al.*, 1994). However, the CP gene of PRSV strain HA did not show broad-range tolerance (Tennant *et al.*, 1994). Apart from its role as a CP, the gene is also required by the virus to enhance the expression of movement proteins (MPs) and form the virus replication cascade (Asurmendi *et al.*, 2004). Hence, expression of the CP gene will interfere with this natural mechanism in transgenic plants and affect the movement of protein production, eventually reducing the spread of virus infection (Bendahmane *et al.*, 2002). Table 11.1 lists the use of CPMR in different vegetables against different plant viruses.

**Table 11.1.** Coat protein (CP)-mediated resistance (CPMR) to plant viruses in vegetables.

| Vegetable | Virus | Reference |
|---|---|---|
| Tomato | Tomato yellow leaf curl virus, tomato leaf curl virus, tomato leaf curl Taiwan virus, tomato mosaic virus, tomato spotted wilt virus | Sanders *et al.* (1992); Kunik *et al.* (1994); Gubba *et al.* (2002); Cillo and Palukaitis (2014) |
| | Alfalfa mosaic virus | Tumer *et al.* (1987) |
| | Cucumber mosaic virus | Gielen *et al.* (1996); Pratap *et al.* (2012b) |
| | Physalis mottle virus | Sree Vidya *et al.* (2000) |
| | Tobacco mosaic virus | Nelson *et al.* (1988); Sanders *et al.* (1992) |
| Squash | Cucumber mosaic virus, zucchini yellow mosaic virus, watermelon mosaic virus 2 | Fuchs and Gonsalves (1995); Fuchs *et al.* (1998); Cillo and Palukaitis (2014) |
| Sugarbeet | Beet necrotic yellow vein virus | Kallerhoff *et al.* (1990); Mannerlöf *et al.* (1996) |
| Sweet potato | Sweet potato feathery mottle virus | Okada *et al.* (2001); Okada and Saito (2008) |
| Potato | Potato virus X, potato virus Y | Kaniewski *et al.* (1990); Lawson *et al.* (1990) |
| | Potato virus X | Fehér *et al.* (1992); Jongedijk *et al.* (1992); Doreste *et al.* (2002) |
| | Potato virus Y | Malnoë *et al.* (1994); Romano *et al.* (2001); Gargouri-Bouzid *et al.* (2005) |
| | Potato leaf roll virus | Kawchuk *et al.* (1990); van der Wilk *et al.* (1991); Barker *et al.* (1992); Presting *et al.* (1995) |
| | Potato leaf roll virus, potato virus Y | Zhang *et al.* (1997) |
| | Potato virus S | MacKenzie *et al.* (1991) |
| | Potato mop-top virus | Barker *et al.* (1998) |
| Pepper | Cucumber mosaic virus, pepper mild mottle virus | Shin *et al.* (2002) |
| | Cucumber mosaic virus | Balázs *et al.* (2008); Lee *et al.* (2009); Pack *et al.* (2012) |
| | Cucumber mosaic virus, tomato mosaic virus | Shin *et al.* (2002) |
| | Cucumber mosaic virus, Tobacco mosaic virus | Cai *et al.* (2003) |
| Lettuce | Lettuce mosaic virus | Dinant *et al.* (1997) |
| Chinese cabbage | Turnip mosaic virus | Han *et al.* (2014) |
| Soybean | Soybean mosaic virus | Wang *et al.* (2001); Steinlage *et al.* (2002); Furutani *et al.* (2006) |
| | Bean pod mottle virus | Di *et al.* (1996); Reddy *et al.* (2001) |
| | Soybean dwarf virus | Tougou *et al.* (2007) |

### 11.2.2 Viral protein-mediated resistance

The successful use of CPMR encouraged scientists to look for other viral proteins that conferred virus resistance. These include the replicase protein involved in virus replication, the MP required for cell-to-cell movement, and viral proteases involved in the processing of protein products.

Replicase-mediated resistance was first discovered when the hypothetical 54 kDa subunit was being studied for its activity during virus replication (Golemboski *et al.*, 1990). To the authors' surprise, the transgenic tobacco plants carrying the cDNA copy of the replicase gene displayed a high level of resistance to TMV infection. Thereafter, this method was tested by various groups to confer virus resistance in plants. Initial reports were in model plants; however, later the method was used in crop plants as well. In potato, replicase-mediated resistance was obtained toward PLRV, PVY$^N$, and tobacco rattle virus (Thomas *et al.*, 2000; Lawson *et al.*, 2001; Vazquez Rovera *et al.*, 2001; Schubert *et al.*, 2004; Melander, 2006). Transgenic tomato plants expressing the replicase gene displayed resistance to CMV infection (Gal-On *et al.*, 1998; Wu *et al.*, 2006). Cereal crops were also developed with replicase-mediated virus-resistant traits. Transgenic wheat, rice, and barley showed resistance toward wheat yellow mosaic virus, rice yellow mottle virus, and barley yellow dwarf virus (PAV strain), respectively, through replicase-mediated resistance (Koev *et al.*, 1998; Pinto *et al.*, 1999; Wu *et al.*, 2006). Arguably, the nature of resistance was also found to be different. Some studies suggested that the resistance is mediated via the protein (Carr *et al.*, 1992), whereas others suggested RNA-mediated resistance (Marano and Baulcombe, 1998). Another approach, however, indicated that both RNA and protein components could be playing roles in conferring resistance to TMV through the replicase gene (Goregaokar *et al.*, 2000).

Movement protein-mediated resistance has also been reported. Transgenic potato expressing PLRV MP showed broad-spectrum resistance to PLRV, PVY, and potato virus X (PVX) (Tacke *et al.*, 1996). Expression of defective mutant MP in transgenic sugarbeet (*Beta vulgaris*) conferred resistance to beet necrotic yellow vein virus (Lauber *et al.*, 2001). Transgenic potato expressing a mutant version of potato mop-top virus MP displayed reduced infection in tubers (Melander *et al.*, 2001). Transgenic melon carrying tandem copies of the CGMMV MP gene resulted in virus resistance (Emran *et al.*, 2012).

The NIa protease-coding sequences of tobacco vein mottling virus (TVMV) provided resistance to TVMV infection in tobacco. Likewise, the use of viral protease has been reported in a model system to confer resistance toward various plant viruses (Cillo and Palukaitis, 2014). These results provide insight into the potential use of various virus proteins in vegetable crops for conferring resistance.

### 11.2.3 Antisense RNA

In parallel with CPMR, many reports have successfully shown the use of antisense RNA to confer viral disease tolerance. Antisense RNA-harboring constructs comprise the cDNA sequence cloned downstream of a promoter, which expresses the RNA complementary to the viral genome. Plant viral genomes also contain untranslated regions essential for translation initiation and genome amplification. The antisense approach of expressing the 3′-end untranslated regions of TMV resulted in effective resistance (Powell *et al.*, 1989). Transgenic potato expressing antisense RNA for the capsid protein of PLRV showed enhanced resistance to the viral disease (Kawchuk *et al.*, 1991; Palucha *et al.*, 1998). Similarly, antisense RNA of the capsid protein of ZYMV resulted in delayed symptom development in transgenic melon (Fang and Grumet, 1993). Reduced PVY accumulation was also observed in transgenic potato plants expressing antisense RNA of the capsid protein gene of PVY (Sokolova *et al.*, 1994). Transgenic tobacco plants expressing antisense RNA to the *AL1* gene (encoding the Rep protein) of tomato golden mosaic virus conferred tolerance to disease (Day *et al.*, 1991). Transgenic beans expressing bean golden mosaic virus *Rep-TrAP-REn* and *BC1* genes in antisense orientation also resulted in delayed symptoms and reduced viral titer (Aragão *et al.*, 1998). Cotton plants expressing antisense RNA to the *Rep* gene of cotton leaf curl virus were also reported to display resistance (Amudha *et al.*, 2010). Antisense RNAs of African cassava mosaic virus (ACMV) *Rep*, *TrAP*, and *REn* genes when expressed separately in cassava, showed reduced symptom development and viral DNA accumulation (Zhang *et al.*, 2005). Transgenic tomato plants expressing the antisense *Rep* gene of tomato leaf curl virus (ToLCV) also resulted in delayed symptoms (Praveen *et al.*, 2005). Similarly, transgenic soybean expressing the antisense *Rep* gene of mungbean yellow mosaic India virus displayed resistance to the disease (Singh *et al.*, 2013).

Antisense RNA against potato spindle tuber viroid (PSTVd) also resulted in a delay in infection in transgenic potato (Matousek *et al.*, 1994). Peanut plants expressing antisense RNAs to the TSWV nucleocapsid gene displayed good protection against the virus (Magbanua *et al.*, 2000). Transgenic lettuce expressing antisense RNA of the capsid protein gene of lettuce big vein-associated virus (LBVaV) showed resistance to both LBVaV and Mirafiori lettuce virus (Kawazu *et al.*, 2006). Chinese cabbage expressing the *NIb* sequence of turnip mosaic virus showed virus resistance (Yu *et al.*, 2007). These results clearly indicate the efficiency of antisense technology in improving disease tolerance in plants against viruses.

### 11.2.4 Post-transcriptional gene silencing (siRNAs and miRNAs)

Post-transcriptional gene silencing (PTGS) regulates development, stability of the genome, and defense against stress conditions. PTGS can be triggered by transgene or exogenous genes and by viral infections, and thus serves as an RNA-based immune system in plants (Tan *et al.*, 2020). PTGS is also referred to as RNA interference (RNAi) or RNA silencing (Mello and Conte, 2004). RNAi is a sequence-specific regulation of gene expression for the proper growth and development of plants and also provides protection from foreign nucleic acids. RNAi is initiated by dsRNA derived from either host-plant hpRNA stretches or viral replicative dsRNA intermediates, which are digested by Dicer-like (DCL) enzymes into small non-coding RNAs of 18–27 nt in length known as small-interfering RNAs (siRNAs) and microRNAs (miRNAs) (Fig. 11.1). These small RNAs are further directed to bind to the cognate mRNA sequence to mediate sequence-specific downregulation or cleavage of the mRNA (Baulcombe, 2005; Ding, 2010; Guo *et al.*, 2019).

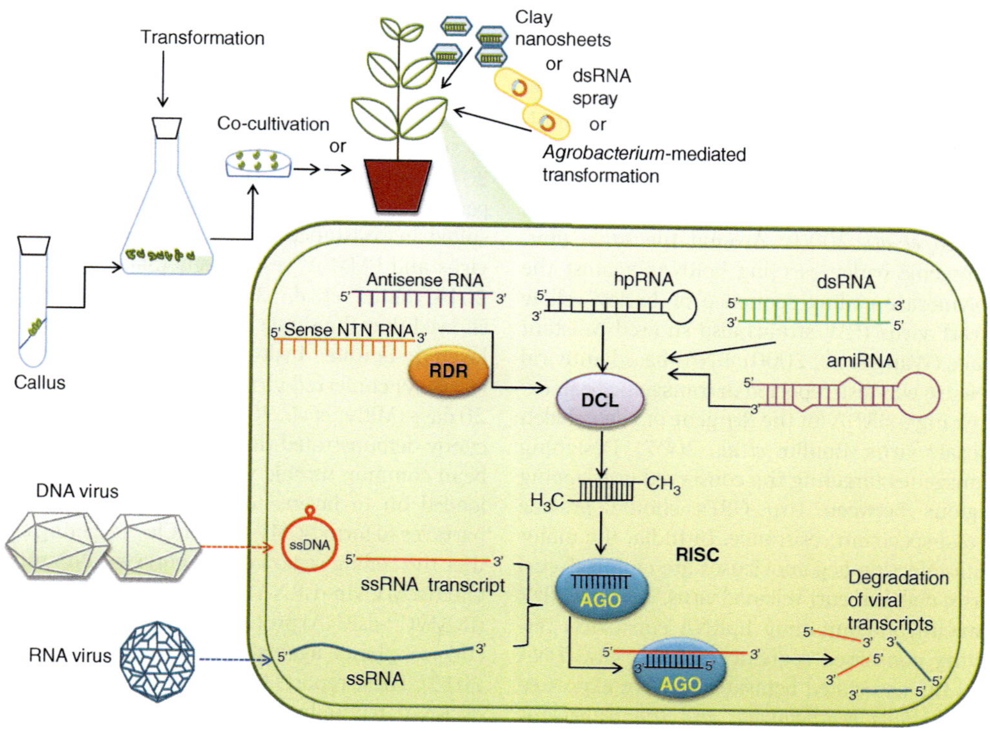


**Fig 11.1.** Approaches to induce RNAi-mediated antiviral resistance in transgenic plants. The transgene can be introduced into the plant using *Agrobacterium*, direct dsRNA spray and dsRNA-coated clay nanosheets. Introduction of the transgene can also occur at early tissue culture stages depending on the plant species. If the introduced transgene forms ssRNA, then it is recognized by RNA-dependent RNA polymerase (RDR) to produce dsRNA, which is acted on by Dicer-like (DCL) enzymes; other dsRNAs are digested directly by DCL enzymes within the plant cell and the plant Argonaute protein (AGO) interacts with the digested ssRNA to form an RNA-induced silencing complex (RISC). Following viral entry and uncoating, the viral transcripts are targeted and cleaved by the RISC complex, which leads to their degradation.

Essentially, siRNAs mediate the defense against invading nucleic acids such as viruses, whereas miRNAs are involved mainly in the regulation of endogenous gene targets involved in development and stress responses.

To express dsRNA or hpRNAs, it is essential to introduce viral sequences in both sense and antisense orientations. Vectors with a multiple cloning site separated by introns are hence used to produce dsRNAs with loops, which ultimately provide stability to the RNA sequences until it is processed by the RNAi machinery (Smith *et al.*, 2000; Wesley *et al.*, 2001). These dsRNAs are then digested by DCL nucleases of plants and produce the siRNA duplexes that will bind to the cognate mRNA sequences of invading viruses (Ding, 2010; Pumplin and Voinnet, 2013). This method can be used to genetically modify crops that are less amenable to transformation and thus reduce the number of lines that need to be screened to get resistant traits.

In addition, expression of chimeric hpRNAs also aids in conferring resistance toward multiple viruses. The first report of the use of hpRNA was demonstrated in transgenic tobacco where the NIa protease-coding gene was targeted (Smith *et al.*, 2000). Around the same time, transgenic barley carrying hpRNAs against the polymerase coding sequence of barley yellow dwarf virus (PAV strain) also showed resistant traits (Wang *et al.*, 2000). RNAi-based antiviral defense was also reported in transgenic bean expressing a siRNA for the *Rep* gene of bean golden mosaic virus (Bonfim *et al.*, 2007). Designing transgenes targeting the conserved overlapping regions between two ORFs should provide broad-spectrum resistance. In India, the major chili-infecting begomoviruses are chili leaf curl virus, chili leaf curl Vellanad virus, and ToLCNDV. Two intron-containing hpRNA constructs targeting conserved sequences of the viral DNA and the associated betasatellite were expressed in *Nicotiana benthamiana*, and the transgenic lines resisted the viral titer up to 90% with decreased symptom development (Sharma *et al.*, 2015). The methodology has also been extrapolated to develop virus-resistant vegetable crops. Table 11.2 lists some of the studies on the successful use of RNAi for generating virus-resistant vegetables.

Although several reports support the use of hpRNA to confer virus resistance in various crops, public acceptance is still limited. To address this problem, exogenous application of hpRNA has been utilized and shown as an effective way of controlling the virus population. Plants are sprayed with dsRNA (or hpRNA) that is complementary to viral sequences (Morozov *et al.*, 2019). Mechanical inoculation or high-pressure spraying has also been used by various groups to confer protection from viruses. The efficiency of exogenous application is greatly dependent on the selected virus target regions. *In vitro*-synthesized dsRNA against pepper mild mottle virus (PMMoV), when applied through mechanical inoculation, conferred virus resistance in pepper (*Capsicum chinense*) (Tenllado and Díaz-Ruíz, 2001). Tomato mechanically inoculated with *in vitro*-synthesized dsRNA resulted in tolerance toward PSTVd, ToLCV, and CMV (Carbonell *et al.*, 2008; Namgial *et al.*, 2019). As well as in vegetable crops, delivery of hpRNA in sugarcane and papaya by means of spray and mechanical inoculation, respectively, improved virus resistance against sugarcane mosaic virus in sugarcane and PRSV in papaya (Gan *et al.*, 2010; Vadlamudi *et al.*, 2019). In peas and cowpea, spraying of *in vitro*-synthesized dsRNA resulted in resistance to pea seed-borne mosaic virus and PMMoV, respectively (Šafářová *et al.*, 2014; Mitter *et al.*, 2017). The latter report showed that the single use of dsRNA loaded on layered double hydroxide clay nanosheets (BioClay) conferred virus protection for at least 20 days (Mitter *et al.*, 2017). The same group recently demonstrated that topical application of bean common mosaic virus (BCMV) CP dsRNA loaded on to layered double hydroxide nanoparticles to form BCMV CP–BioClay targeting either the coding region of the potyviral NIb protein (BCMVNIb-dsRNA) or the CP-coding region (BCMVCP-dsRNA) protected *N. benthamiana* and cowpea plants against BCMV (Worrall *et al.*, 2019). These reports suggest a promising future of RNAi technology in the efficient control of plant viruses either through the transgene method or topical application, or both.

### 11.2.5 Artificial miRNAs (amiRNAs)

Another promising antiviral tool for targeting viral RNAs is the use of artificial miRNAs (amiRNAs). miRNAs are generated endogenously from

**Table 11.2.** RNAi-mediated resistance to plant viruses in vegetables.

| Vegetable | Virus | Gene | Method | Reference |
|---|---|---|---|---|
| Cassava | African cassava mosaic virus | Rep | hpRNAs | Vanderschuren *et al.* (2009) |
| | Cassava brown streak Uganda virus | Capsid protein | hpRNAs | Yadav *et al.* (2011) |
| | Cassava brown streak virus | Capsid protein | hpRNAs | Ogwok *et al.* (2012) |
| | Cassava brown streak virus, Ugandan cassava brown streak virus | Coat protein | Stacked/RNAi tandom | Beyene *et al.* (2017) |
| | Sri Lankan cassava mosaic virus | AV2 and AV1 | RNAi | Ntui *et al.* (2015) |
| Cucurbits | Papaya ringspot virus | Capsid protein | hpRNAs | Krubphachaya *et al.* (2007) |
| | Cucumber green mottle mosaic virus | Capsid protein | hpRNAs | Kamachi *et al.* (2007) |
| | Zucchini yellow mosaic virus, watermelon mosaic virus, papaya ringspot virus | HC-Pro | hpRNAs | Leibman *et al.* (2011) |
| Soybean | Soybean dwarf virus | Capsid protein | hpRNAs | Tougou *et al.* (2006) |
| | Soybean mosaic virus | Capsid protein | hpRNAs | Kim *et al.* (2013); Zhang *et al.* (2013) |
| | Mungbean yellow mosaic India virus | Coat protein | hpRNAs | Kumari *et al.* (2018) |
| | Mungbean yellow mosaic India virus | AC2 | shRNA | Ramesh *et al.* (2019) |
| Potato | Potato virus Y (PVY$^{N}$, PVY$^{NTN}$, PVY$^{O}$) | Capsid protein of PVY$^{N}$ | hpRNAs | Missiou *et al.* (2004) |
| | PVY$^{NTN}$ | Capsid protein | hpRNAs | Bukovinszki *et al.* (2007) |
| | Potato virus X (PVX), PVY$^{O}$, PVY$^{N}$, PVY$^{C}$ | Fused PVX capsid protein and PVY NIb region | Fusion hpRNA | Bai *et al.* (2009) |
| | PVX, PVY, potato leafroll virus | Fused sequence from ORF2 of PVX, HC-Pro of PVY, coat protein of potato leafroll virus | Chimeric hpRNA | Arif *et al.* (2012) |
| | Cucumber mosaic virus | Replicase gene | hpRNAs | Ntui *et al.* (2013) |
| | PVY, PVX, potato virus S | Coat protein | hpRNAs | Hameed *et al.* (2017) |
| | PVY$^{NTN}$ | eIF4E | hpRNAs | Miroshnichenko *et al.* (2020) |
| Sugarbeet | Beet necrotic yellow vein virus | Replicase gene | hpRNAs | Lennefors *et al.* (2006); Pavli *et al.* (2010) |
| Sweet potato | Sweet potato feathery mottle virus | Replicase and capsid protein | hpRNAs | Nyaboga *et al.* (2008) |
| | Sweet potato feathery mottle virus, sweet potato chlorotic stunt virus, sweet potato virus G, sweet potato mild mottle virus | Coat protein | Fusion hpRNA | Sivparsad and Gubba (2014) |

*Continued*

**Table 11.2.** Continued.

| Vegetable | Virus | Gene | Method | Reference |
|---|---|---|---|---|
| Tomato | Tomato yellow leaf curl virus | Rep | hpRNAs | Fuentes *et al.* (2006) |
| | | Capsid gene | hpRNAs | Zrachya et *et al.* (2007) |
| | Potato spindle tuber viroid | IR region | hpRNAs | Schwind *et al.* (2009) |
| | Tomato chlorosis virus, tomato yellow leaf curl virus | Coat protein | dsRNAs | Jin *et al.* (2020) |
| | Pepino mosaic virus (US1, LP, and CH2) | Conserved sequences | hpRNAs | Leibman *et al.* (2021) |
| Chili and tobacco | Chili leaf curl virus, tomato leaf curl New Delhi virus, chili leaf curl Vellanad virus | Rep, TrAP, Beta C1 | hpRNAs | Sharma *et al.* (2015) |

well-defined precursor structures coded by their primary miRNA structures and are ultimately processed into a functional duplex of 20–24 nt in length. Artificial miRNA technology makes use of this natural mechanism wherein the mature miRNA region within the primary miRNA or precursor miRNAs is replaced by a customized 21 nt RNA sequence of interest complementary to viral genes. This replacement will not affect the processing of these precursor miRNAs and AGO loading of the amiRNAs (Khalid *et al.*, 2017). As a result, when amiRNA enters the plant cell, the miRNA machinery will recognize it and process it to finally confer virus resistance by regulating the viral gene expression. This technology has been validated for protection against various plant viruses such as ToLCNDV, CMV, rice black-streaked dwarf virus (RBSDV), rice stripe virus (RSV), wheat streak mosaic virus, and wheat dwarf virus (WDV) (Cisneros and Carbonell, 2020). Most reports discuss the use of amiRNAs in model plants such as *Arabidopsis thaliana* and *N. benthamiana* for studying the efficiency of the technology as these are excellent hosts for various viruses and easy to modify at the genetic level. However, there has also been successful implementation of amiRNA technology in vegetable crops. amiRNA targeting the coding sequence shared by 2a and 2b genes and the conserved region of 3′-untranslated region of CMV in transgenic tomato resulted in effective resistance to the target and non-target viruses such as TMV and TYLCV (Zhang *et al.*, 2011). Similarly, amiRNAs targeting AV1 (CP) and an overlapping region of AV1 + AV2 (pre-CP) of ToLCV in tomato plants displayed resistance only in AV1 + AV2-expressing lines, whereas no amiRNAs were detected when only AV1 was expressed (Vu *et al.*, 2013). This also suggests an important aspect associated with RNA-mediated gene silencing—the selection of potential target genes for obtaining maximum resistance. This technology is also used in cereal crops such as *Zea mays*, *Hordeum vulgare*, *Oryza sativa*, and *Triticum aestivum* where it conferred resistance to RBSDV, WDV, RBSDV + RSV, and WSMV, respectively (Fahim *et al.*, 2012; Xuan *et al.*, 2015; Kis *et al.*, 2016; Sun *et al.*, 2016).

Recently, a comparative analysis of the use of amiRNAs and synthetic *trans*-acting siRNAs (syn-tasiRNAs) in tomato showed that syn-tasiRNA lines were more effective against infection, whereas amiRNA lines were susceptible due to mutation in the target site (Carbonell *et al.*, 2019). This further provides an insight to look for the use of multiple amiRNAs to target more than one viral gene at once to avoid the consequence of mutation, as mutation in all target sites simultaneously would a rare event. The use of syn-tasiRNAs can also be promoted to achieve efficient resistant traits in vegetable crops.

### 11.2.6 Non-viral genes

As well as viral genes and RNAi machinery, the use of heterologous genes to produce transgenic cultivars of antiviral resistance has been carried out. One such gene is *Pac1*, encoding a dsRNA targeting ribonuclease protein produced by yeast (*Saccharomyces pombe*) that provided multi-virus resistance in tobacco (Watanabe *et al.*,

1995). A transgenic potato (*S. tuberosum* L. 'Russet Burbank') plant expressing the *Pac1* gene has been developed. The plants resisted PSTVd accumulation even at 40 days post-infection, with only mild detection of PSTVd observed (Sano *et al.*, 1997). Likewise in *S. tuberosum* 'Desiree', PVX and PVY accumulation were reduced by up to 35% and 59%, respectively, compared with control plants (Colalongo *et al.*, 2005). In transgenic soybean, the *Pac1* gene performed well in conferring multiple resistance to the plant against different strains of soybean mosaic virus, BCMV, watermelon mosaic virus, and bean pod mottle virus (Yang *et al.*, 2019). Similarly, Dicer 1 protein from *Candia albicans* (CaDcr1) possesses an RNase III domain that specifically cleaves dsRNAs, and *in vitro* degradation of CMV dsRNA has been demonstrated. Moreover, CaDcr1 restricts the PVX level in transgenic tobacco and delays symptom development (Alam *et al.*, 2019). The expression of a pancreatic RNase gene from cattle (*Bos taurus*) and S-like RNase from *Zinnia elegans* in transgenic potato (*S. tuberosum* 'Luhivs'ka' and 'Lasnayak') delayed the development of symptoms following PVY inoculation, even though there was no significant difference in viral load (Potrokhov *et al.*, 2021). The ssRNA-specific bovine pancreatic RNase A localizes in the apoplast and its expression in tobacco (*Nicotiana tabacum* 'SR1') provided resistance against unrelated viruses (TMV and CMV), delayed symptom development, and restricted viral load (Trifonova *et al.*, 2007; Sugawara *et al.*, 2016).

The expression of mammalian 2′-5′-oligoadenylate synthetase in potato infected with PVX imparted resistance to plants, but in comparison with plants expressing the CP of PVX, the difference was insignificant (Truve *et al.*, 1993). While observing virus resistance in transgenic tobacco plants expressing RNase L along with 2′-5′-oligoadenylate synthetase, necrotic lesions appeared on tobacco etch virus-, TMV-, and AMV-inoculated resistant transgenic tobacco plants and the resistance remained stable up to the $R_2$ generation (Mitra *et al.*, 1996). However, Ogawa *et al.* (1996) reported virus-induced death of plants co-expressing RNase L and 2′-5′-oligoadenylate synthetase at 20 days post-infection in response to PVY inoculation; interestingly, the plants resisted CMV infection. This resistance is due to the activation of pathogenesis-related proteins such as HIN1 and HSR203J along with defense-related gene upregulation (Honda *et al.*, 2003).

Ribosome-inactivating proteins (RIPs) are produced by plants and can provide resistance to viral infection by depurinating the critical sequence of 28S rRNA and thus inactivating the ribosomes. RIPs do not attack self-ribosomes and can act only on specific heterologous ribosomes (Battelli *et al.*, 1984; Robaglia and Tepfer, 1996). Pokeweed antiviral protein (PAP) is among the well-studied RIPs, and when expressed in transgenic tobacco and potato, it offered multi-virus resistance. The toxicity on self-ribosomes by PAP is prevented by the formation of homodimers in the cytosol (Tourlakis *et al.*, 2010). However, once the level of PAP in recombinant lines increases, phenotypic abnormalities such as mottling were observed (Lodge *et al.*, 1993). The constitutive expression of PAPII in transgenic tobacco exhibited a reduction in lesion numbers following TMV and PVX inoculation (Wang *et al.*, 1998). There are numerous RIPs produced by plants such as *Mirabilis* antiviral protein, α-momorcharin (α-MMC), *Phytolacca insularis* antiviral protein (PIP), *Iris* type-1 RIP, *Sambucus nigra* agglutinin I′, and dianthin (Kataoka *et al.*, 1992; Hong *et al.*, 1996; Robaglia and Tepfer, 1996; Chen *et al.*, 2002; Vandenbussche *et al.*, 2004; Iglesias *et al.*, 2005; Zhu *et al.*, 2013; Citores *et al.*, 2021). α-MMC activates the production of antioxidant enzymes and defense-related genes in addition to ribosomal inactivation. α-MMC-pre-treated plants, when inoculated with viruses (chili veinal mottle virus, CMV, and TMV), restrict the viral titer and extend systemic resistance in tobacco (Zhu *et al.*, 2013, 2020). The constitutive expression of PIP in potatoes can inhibit PVX and PVY well; however, the phloem-associated PLRV can escape inhibition under field conditions because PIP expression is largely confined to the extracellular matrix (Moon *et al.*, 1997). Similarly, aphid-transmitted PLRV and PVY exhibited infectivity and escaped the RIP-induced resistance in transgenic potato plants (Lodge *et al.*, 1993). Expressing RIPs only in virus-affected cells at the time of infection is an advantage. Dianthin, when expressed under the control of a virus-induced promoter in *N. benthamiana*, provided resistance to ACMV (Hong *et al.*, 1996). Co-inoculation of RIP along with virus can avoid establishing resistance by inhibiting virus at an early stage before the infected cell produces reactive oxygen species. Tobacco necrosis virus (TNV) and the RIPs PD-L1 and PD-L4 from *Phytolacca dioica* L. on the leaves of beans (*Phaseolus vulgaris*) almost reduced the lesion number by around 99% (TNV + PD-L4) and prevented virus accumulation (Bulgari *et al.*, 2020).

The exploitation of dominant-negative interactions in nature by engineering peptides targeting conserved sequences of viral proteins could put an end to viral replication and survival within the host plant (Sera, 2017). Transgenic tobacco (*N. benthamiana*) plants expressing peptide aptamers (29 amino acids) that interact with the nucleocapsid protein of tospoviruses were engineered, and the plants did not develop any symptoms and showed resistance to different tospoviruses (Rudolph *et al.*, 2003). The effectiveness of peptide aptamers on DNA viruses was demonstrated by Lopez-Ochoa *et al.* (2006). Transformation of tobacco protoplasts with a 20 amino acid peptide of thioredoxin A, which interacts with the AL1 (Rep) protein of tomato golden mosaic virus and cabbage leaf curl virus, can suppress replication of these viruses. When selected peptide aptamers were expressed in transgenic tomato plants challenged with TYLCV and tomato mottle virus, about 28% of plants presented with no symptoms even at 28 days post-infection, and lower levels of viral DNA accumulation was observed (Reyes *et al.*, 2013).

AZPs are designed in such a way that they can compete with a viral protein for binding sites in the genetic material of the virus, suppressing their replication. Transgenic *A. thaliana* was developed expressing an AZP that binds efficiently to the replicase (Rep) protein-binding site in beet severe curly top virus DNA and prevented systemic spread of the virus (Sera, 2005). The efficiency of AZPs lies with their binding affinity, about a 1000-fold higher than Rep. Experiments on TYLCV infection found that even if the concentration of AZP and Rep differed by a ratio of 1:1000, respectively, in nanomolar levels, the AZP could competently bind to TYLCV DNA. The transgenic plants expressing targeted AZPs exhibited no phenotypic abnormalities and could be more effective in controlling virus proliferation (Takenaka *et al.*, 2007; Koshino-Kimura *et al.*, 2008, 2009; Mori *et al.*, 2013; Sera, 2017).

To develop transgenic plants for virus resistance, several genes and methods have been utilized. The expression of plantibodies or single-chain variable fragment (scFv) against viral proteins is yet another way to confer resistance to viral infections (Paraskevi *et al.*, 1993; Rossmann and Johnson, 1989). The development of a monoclonal antibody (mAbP1-1) against a critical C-terminal RNA-binding domain in the P1 protein of PLRV reduced the concentration of virus in transgenic potato plants constitutively expressing scFvP1-1 and established systemic resistance (Nickel *et al.*, 2008). Moreover, in transgenic tomato plants, resistance has been correlated with the level of scFv expression (Villani *et al.*, 2005). Earlier studies on experimental models such as tobacco revealed that endoplasmic reticulum compartmentalization of scFv may help to achieve stability and higher expression of plantibodies obtained from hybridoma technology or phage display library purification (Vaughan *et al.*, 1996; Fiedler *et al.*, 1997). In sugarbeet, the plantibody targeted to the cytoplasm provided better resistance to beet necrotic yellow vein virus than those targeted to the mitochondria or apoplast (Jafarzade *et al.*, 2019). In some transgenic potato plants expressing scFv targeting the NIa protease of PVY, no detectable virus accumulation was observed and it offered comprehensive resistance irrespective of expression in the cytoplasm or apoplast regions of the cell (Gargouri-Bouzid *et al.*, 2006; Ayadi *et al.*, 2012). A plantibody engineered to target a conserved motif in the RNA-dependent RNA polymerase of the tombusvirus tomato bushy stunt virus provided variable levels of broad-spectrum resistance to the transgenic tobacco plants (Boonrod *et al.*, 2004). A variety of approaches including nanobodies (VHH, a single variable domain on a heavy chain) derived from camels (camelid nanobodies) proved to be an alternative and effective method in transgenic broad beans (*Vicia faba*) to provide resistance against broad bean mottle virus, and the resistance obtained was demonstrated to be highly specific (Ghannam *et al.*, 2015). However, transgenic potato plants expressing only the variable domain of the heavy chain (VH) domain of an antibody targeting the NIa protease of PVY did not effectively suppress virus accumulation (Bouaziz *et al.*, 2009). The introduction of transgenic constructs in a plant cell is under threat of inactivation by the host DNA methylation machinery. It has been observed in transgenic Chinese cabbage (*Brassica rapa*) with overexpression scFv antibody specific to the ssDNA of turnip mosaic virus that susceptibility to virus infection was linked to low expression of the scFv, as well as DNA methylation and inactivation of the transgene by the host (Zhao *et al.*, 2013).

### 11.2.7 CRISPR-Cas9-mediated transformation

Precise genome editing or targeted genome-editing technology exploits the use of bacterial defense machinery of clustered interspaced palindromic repeats (CRISPR) and CRISPR-associated 9 (Cas9) enzymes in plants (Fig. 11.2) (Belhaj *et al.*, 2013). Heritable mutations can be established using the CRISPR-Cas9 system (Brooks *et al.*, 2014). Strategic use of CRISPR/Cas9 can efficiently limit geminivirus infections in plants (Ali *et al.*, 2015). TYLCV resistance induction in tomato plants (*Solanum lycopersicum* 'Moneymaker') has shown promising results. A stable resistance up to the $T_3$ generation was observed in transgenic tomato plants against TYLCV, but a major shortcoming observed in the form of viral escape (Tashkandi *et al.*, 2018). Transgenic plants expressing the CRISPR/Cas9 system have been reported with altered nucleotide sequences in the Cas9 target site of the viral replicase gene sequence (Ali *et al.*, 2019). In the same way, in transgenic cassava that has been developed to resist ACMV, viral escapees have been observed (Mehta *et al.*, 2019). A single nucleotide change occurs as a result of non-homologous

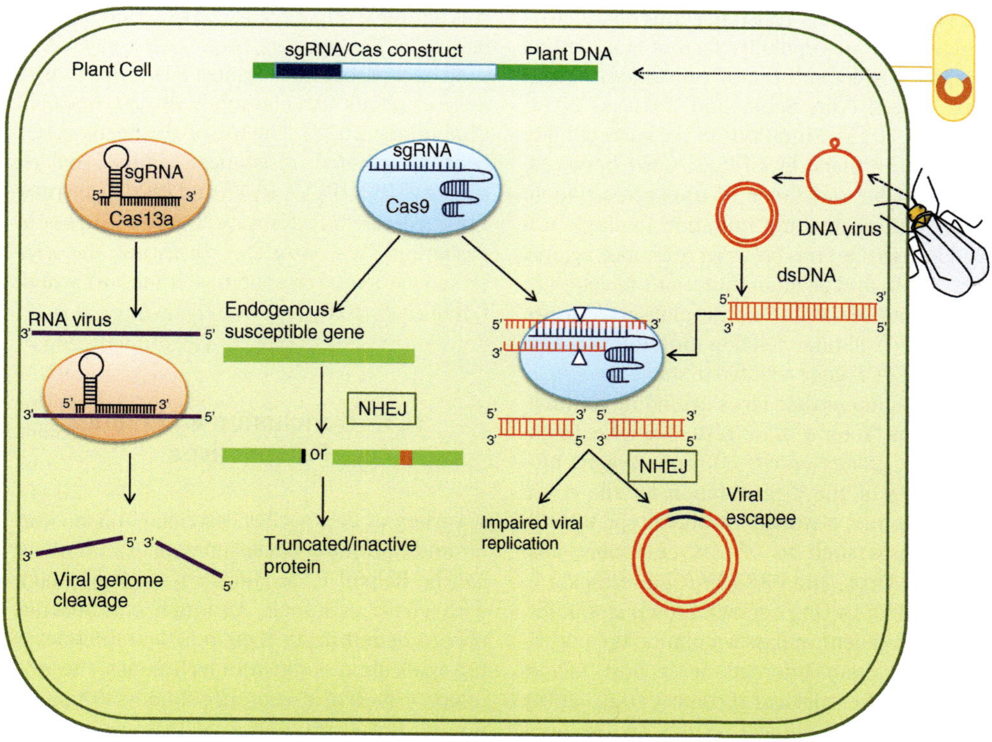


**Fig 11.2.** Genome-editing approaches for plant virus resistance. Upon delivery of the transgene, assembly of single guide RNA (sgRNA) and CRISPR-associated protein 9 (Cas9) protein takes place. Resistance can be achieved by targeting viral DNA or endogenous susceptibility genes of the host plant. However, viral escapees after Cas9-mediated cleavage of viral DNA have been identified in geminiviruses. Using CRISPR/Cas13a, the RNA of the viruses can be targeted for cleavage. NHEJ, non-homologous end joining.

end joining repair after CRISPR/Cas9 cleavage, which provides a loophole for viruses to emerge as cleavage-resistant strains. The reasons for concern with geminivirus escape in cassava is the compromised efficiency of the CRISPR/Cas9 system and that the rate of multiplication of virus outcompeted the cleavage rate in the transgenic plants (Rybicki, 2019).

The construction of single guide RNA (sgRNA) and the time of delivery hold the key to better cultivar development. Introduction of two sgRNAs (CP_sgRNA and IR_sgRNA) targeting the intergenic region and CP suppressed the viral titer appreciably in Cas9-expressing transgenic tomato plants in comparison with a single application during TYLCV infection (Ghorbani Faal *et al.*, 2020). Instead of targeting the viral genes, the use of CRISPR/Cas9 to target a susceptibility gene (*SlPelo*) and the resulting homozygous knockout lines reported high resistance in controlling the viral titer of TYLCV (Pramanik *et al.*, 2021). The recessive resistance thus obtained by targeting the susceptibility factors is a method that has provided substantial resistance in transgenic plants (van Schie and Takken, 2014; Sanfaçon 2015). Mutations in the susceptibility gene *eIF4E1* made by CRISPR/Cas9 have provided variable resistance to transgenic tomato plants. An inactivating mutation (a single nucleotide insertion) has provided resistance against $PVY^N$ strain and deletion mutations (deletion of 9 nt) prevented CMV CP accumulation (Atarashi *et al.*, 2020). Similar deletion mutations created in the *eIF4E1* gene exhibited stable resistance against pepino mosaic virus extending to the $E_2$ generation (Yoon *et al.*, 2020). Likewise, in cucumber (*Cucumis sativus* L.), homozygous mutant plants of the $T_3$ generation for the *eIF4E* gene presented a resistance phenotype to multiple viruses such as ZYMV, cucumber vein yellowing virus, and PRSV-W (Chandrasekaran *et al.*, 2016). During cassava brown streak disease development in cassava plants, the potyviral VPg protein interacts with host factors (eIF4(iso)E) for survival (Léonard *et al.*, 2000; Yeam *et al.*, 2007; Shi *et al.*, 2017). Two isoforms of eIF4E, novel cap-binding proteins (nCBP1 and nCBP2), have been shown to interact with VPg, and using CRISPR/Cas9 knockout technology the homozygous double-mutant transgenic cassava plants had a significantly reduced cassava brown streak virus titer, with a reduction of around 45% in RNA levels in root compared with the wild-type plants (Gomez *et al.*, 2019). Interestingly, editing of the endogenous host gene coilin, required for ribonucleoprotein assembly, in potato (*S. tuberosum* 'Chicago') plants led to resistance to the potyvirus $PVY^O$, as well as conferring abiotic stress tolerance to the transgenic cultivars (Makhotenko *et al.*, 2019). Similar to vegetable crops, in the model plant *N. benthamiana*, broad-spectrum resistance has been observed by targeting the CRISPR RNA toward the non-coding region and overlapping viral ORFs, and in *A. thaliana*, several successful attempts on CRISPR/Cas9-mediated endogenous gene targeting, using a viral gene targeting both RNA and DNA viruses, have been reported (Varanda *et al.*, 2021).

The use of CRISPR/Cas9 system in RNA virus resistance has been developed, and transgenic tobacco and *Arabidopsis* plants showed resistance to CMV and TMV, and the resistance was heritable (Zhang *et al.*, 2018). Moreover, the use of Cas13a nuclease targets RNA and can be used as a precise tool against RNA virus resistance in plants (Abudayyeh *et al.*, 2016, 2017; Chaudhary, 2018). The use of this method successfully created resistance against multiple strains of PVY ($PVY^N$, $PVY^O$, and $PVY^{N:O}$) in transgenic potato (*S. tuberosum* 'Desiree') plants by designing Cas13a/sgRNA targeting the conserved sequences present in all three PVY strains (Zhan *et al.*, 2019).

## 11.3 Conclusion and Future Perspectives

A variety of approaches described in this chapter and upcoming advancements in technology can be helpful in producing transgenic plants with virus resistance. Although scientists are able to demonstrate their practical design and the application of different techniques, there remains a lack of acceptance among the public around the world, and most transgenic plants never reach the farmer's field. Several reports have provided examples of the use of heterologous proteins in plants to confer virus resistance, but a lack of sequence specificity in RNases will be a hindrance for practical application of this approach on a large scale in vegetable crops.

The use of amiRNA-mediated virus resistance has great potential in inducing resistance, but a major drawback is the production of viral escapees. By virtue of the viral RNA-dependent RNA polymerase, which lacks proofreading (exonuclease) activity, and selection pressure, synonymous mutations are likely to occur at amiRNA target sites leading to viral escape. Alternatively, use of syn-tasiRNA has provided better resistance in TSWV infection in tomato plants because syn-tasiRNAs can effectively inhibit viral RNAs, even at low transgene expression levels (Carbonell *et al.*, 2019). Similarly, geminivirus escapees accumulate mutations after CRISPR/Cas9-mediated cleavage and can survive inside the plant (Tashkandi *et al.*, 2018; Mehta *et al.*, 2019). Viral escapes must be considered and judicious transgene design with expression levels higher than the virus and targeting multiple genes may avoid these viral escapes and confer resistance to the plant. Instead of targeting viral genes, editing host-susceptibility genes can effectively curb virus proliferation, providing resistance to the plant. Successful implementation of recessive resistance has been observed in the resistance of various potyviruses in tomato, cucumber, soybean, and other plant species using RNAi and genome-editing techniques (Yeam *et al.*, 2007; Mazier *et al.*, 2011; Sanfaçon, 2015; Chandrasekaran *et al.*, 2016; Atarashi *et al.*, 2020; Gao *et al.*, 2020; Yoon *et al.*, 2020).

Genome editing in RNA viruses using CRISPR/Cas13a has been proven to offer specific resistance to PVY in transgenic potato plants (Zhang *et al.*, 2018; Zhao *et al.*, 2020). Recent studies on the guide-induced gene-silencing approach, where a guide RNA executes gene silencing without Cas13, may widen horizons and provide better solutions for virus resistance in plant development in the future (Sharma *et al.*, 2022). With the advent of new transgenic and genome-editing approaches, producing plant resistance to multiple viruses and vectors will be feasible.

## 11.4 Acknowledgements

This study was supported by a grant from the Department of Biotechnology, Ministry of Science and Technology, Government of India (no. BT/PR31648/GET/119/283/2019).

## 11.5 References


Abel, P.P., Nelson, R.S., De, B., Hoffmann, N., Rogers, S.G., *et al.* (1986) Delay of disease development in transgenic plants that express the tobacco mosaic virus coat protein gene. *Science* 232, 738–743.

Abudayyeh, O.O., Gootenberg, J.S., Konermann, S., Joung, J., Slaymaker, I.M., *et al.* (2016) C2c2 is a single-component programmable RNA-guided RNA-targeting CRISPR effector. *Science* 353: aaf5573.

Abudayyeh, O.O., Gootenber, J.S., Essletzbichler, P., Han, S., Joung, J., *et al.* (2017) RNA targeting with CRISPR-Cas13. *Nature* 550, 280–284.

Alam, C.M., Jain, G., Kausar, A., Singh, A.K., Mandal, B., *et al.* (2019) Dicer 1 of *Candida albicans* cleaves plant viral dsRNA *in vitro* and provides tolerance in plants against virus infection. *Virus Disease* 30, 237–244.

Ali, Z., Abulfaraj, A., Idris, A., Ali, S., Tashkandi, M., and Mahfouz, M.M. (2015) CRISPR/Cas9-mediated viral interference in plants. *Genome Biology* 16: 238.

Ali, Z., Zaidi, S.S., Tashkandi, M., and Mahfouz, M.M.. (2019) A simplified method to engineer CRISPR/Cas9-mediated geminivirus resistance in plants. *Methods in Molecular Biology* 2028, 167–183.

Amudha, J., Balasubramani, G., Malathi, V.G., Monga, D., Bansal, K.C., and Kranthi, K.R. (2010) Cotton transgenics with antisense *AC1* gene for resistance against cotton leaf curl virus. *Electronic Journal of Plant Breeding* 1, 360–369.

Aragão, F.J.L., Ribeiro, S.G., Barros, L.M.G., Brasileiro, A.C.M., Maxwell, D.P., *et al.* (1998) Transgenic beans (*Phaseolus vulgaris* L.) engineered to express viral antisense RNAs show delayed and attenuated symptoms to bean golden mosaic geminivirus. *Molecular Breeding* 4, 491–499.

Arif, M., Azhar, U., Arshad, M., Zafar, Y., Mansoor, S., and Asad, S. (2012) Engineering broad-spectrum resistance against RNA viruses in potato. *Transgenic Research* 21, 303–311.

Asurmendi, S., Berg, R.H., Koo, J.C., and Beachy, R.N. (2004) Coat protein regulates formation of replication complexes during tobacco mosaic virus infection. *Proceedings of the National Academy of Sciences USA* 101, 1415–1420.

Atarashi, H., Wikum, H.J., Joon, K., Hangil, K., Yosuke, T., *et al.* (2020). Artificially edited alleles of the eukaryotic translation initiation factor 4E1 gene differentially reduce susceptibility to cucumber mosaic virus and potato virus Y in tomato. *Frontiers in Microbiology* 11: 564310.

Ayadi, M., Bouaziz, D., Nouri-Ellouz, O., Rouis, S., Drira, N., and Gargouri-Bouzid, R. (2012) Efficient resistance to potato virus Y infection conferred by cytosolic expression of anti-viral protease single-chain variable fragment antibody in transgenic potato plants. *Journal of Plant Pathology* 94, 561–569.

Bai, Y., Guo, Z., Wang, X., Bai, D., and Zhang, W. (2009) Generation of double-virus-resistant marker-free transgenic potato plants. *Progress in Natural Science* 19, 543–548.

Balázs, E., Bukovinszki, Á., Csányi, M., Csillėry, G., Divéki, Z., *et al.* (2008) Evaluation of a wide range of pepper genotypes for regeneration and transformation with an *Agrobacterium tumefaciens* shooter strain. *South African Journal of Botany* 74, 720–725.

Barker, H., Reavy, B., and McGeachy, K.D. (1998) High level of resistance in potato to potato mop-top virus induced by transformation with the coat protein gene. *European Journal of Plant Pathology* 104, 737–740.

Barker, H., Reavy, B., Kumar, A., Webster, K.D., and Mayo, M.A. (1992) Restricted virus multiplication in potatoes transformed with the coat protein gene of potato leafroll luteovirus: similarities with a type of host gene-mediated resistance. *Annals of Applied Biology* 120, 55–64.

Battelli, M.G., Lorenzoni, E., Stirpe, F., Cell, R., and Parsis, B. (1984) Differential effect of ribosome-inactivating proteins on plant ribosome activity and plant cells growth. *Journal of Experimental Botany* 35, 882–889.

Baulcombe, D. (2005) RNA silencing. *Trends Biochemical Sciences* 30, 290–293.

Belhaj, K., Chaparro-Garcia, A., Kamoun, S., and Nekrasov, V. (2013) Plant genome editing made easy: targeted mutagenesis in model and crop plants using the CRISPR/Cas system. *Plant Methods* 9: 39.

Bendahmane, M., Szecsi, J., Chen, I., Berg, R.H., and Beachy, R.N. (2002) Characterization of mutant tobacco mosaic virus coat protein that interferes with virus cell-to-cell movement. *Proceedings of the National Academy of Sciences USA* 99, 3645–3650.

Bevan, M.W. and Harrison, B.D. (1986) Genetic engineering of plants for tobacco mosaic virus resistance using the mechanisms of cross-protection. In: Arntzen, C.J., Alan C., and Ryan, R. (eds.) *Molecular Strategies for Crop Protection*. Liss, New York, pp. 215–220.

Bevan, M.W., Mason, S.E., and Goelet, P. (1985) Expression of tobacco mosaic virus coat protein by a cauliflower mosaic virus promoter in plants transformed by agrobacterium. *EMBO Journal* 4, 1921–1926.

Beyene, G., Chauhan, R.D., Ilyas, M., Wagaba, H., Fauquet, C.M., and Miano, D. (2017) A virus-derived stacked RNAi construct confers robust resistance to cassava brown streak disease. *Frontiers in Plant Science* 7: 2052.

Bonfim, K., Faria, J.C., Nogueira, E.O.P.L., Mendes, E.A., and Aragao, F.J.L. (2007) RNAi-mediated resistance to bean golden mosaic virus in genetically engineered common bean (*Phaseolus vulgaris*). *Molecular Plant–Microbe Interactions* 20, 717–726.

Boonrod, K.J., Galetzka, D., Nagy, P.D., Conrad, U., and Krczal, G. (2004) Single-chain antibodies against a plant viral RNA-dependent RNA polymerase confer virus resistance. *Nature Biotechnology* 22, 856–862.

Bouaziz, D., Ayadi, M., Bidani, A., Rouis, S., Nouri-Ellouz, O., *et al.* (2009) A stable cytosolic expression of VH antibody fragment directed against PVY NIa protein in transgenic potato plant confers partial protection against the virus. *Plant Science* 176, 489–496.

Brookes, G. (2020) *GM Crops: Global Socio-Economic and Environmental Impacts 1996–2018*. PG Economics Ltd, UK.

Brooks, C., Nekrasov, V., Lippman, Z.B., and van Eck, J. (2014) Efficient gene editing in tomato in the first generation using the clustered regularly interspaced short palindromic repeats/CRISPR-associated9 system. *Plant Physiology* 166, 1292–1297.

Bukovinszki, A., Divéki, Z., Csányi, M., Palkovics, L., and Balázs, E. (2007) Engineering resistance to PVY in different potato cultivars in a marker-free transformation system using a 'shooter mutant' *Agrobacterium tumefaciens*. *Plant Cell Reports* 26, 459–465.

Bulgari, D., Landi, N., Ragucci, S., Faoro, F., and di Maro, A. (2020) Antiviral activity of PD-L1 and PD-L4, type 1 ribosome inactivating proteins from leaves of *Phytolacca dioica* L. in the pathosystem *Phaseolus vulgaris*–tobacco necrosis virus (TNV). *Toxins* 12: 524.

Cai, W.Q., Rong, X.F., Hong, S.S., Xu, W., Feng, L.Z., *et al.* (2003) Development of CMV- and TMV-resistant transgenic chili pepper: Field performance and biosafety assessment. *Molecular Breeding* 11, 25–35.

Carbonell, A., de Alba, A.E.M., Flores, R., and Gago, S. (2008) Double-stranded RNA interferes in a sequence-specific manner with the infection of representative members of the two viroid families *Virology* 371, 44–53.

Carbonell, A., Lisó, P., and Daròs, J.A. (2019) Multi-targeting of viral RNAs with synthetic *trans*-acting small interfering RNAs enhances plant antiviral resistance. *Plant Journal* 100, 720–737.

Carr, J.P., Marsh, L.E., Lomonossoff, G.P., Sekiya, M.E., and Zaitlin, M. (1992) Resistance to tobacco mosaic virus induced by the 54-kDa gene sequence requires expression of the 54-kDa protein. *Molecular Plant–Microbe Interactions* 5, 397–404.

Chakraborty, S. and Kumar, M. (2020) Tomato Leaf Curl New Delhi Virus (Geminiviridae). In: Bamford, D.H. and Zuckerman, M. (eds.) *Encyclopedia of Virology*, 4th edn. Elsevier, pp. 749–760.

Chandrasekaran, J., Brumin, M., Wolf, D., Leibman, D., Klap, C., *et al.* (2016) Development of broad virus resistance in non-transgenic cucumber using CRISPR/Cas9 technology. *Molecular Plant Pathology* 17, 1140–1153.

Chaudhary, K. (2018) CRISPR/Cas13a targeting of RNA virus in plants. *Plant Cell Reports* 37, 1707–1712.

Chen, Y., Peumans, W.J., and van Damme, E.J. (2002) The *Sambucus nigra* type-2 ribosome-inactivating protein. SNA-I' exhibits in planta antiviral activity in transgenic tobacco. FEBS Letters 516, 27–30.

Cillo, F. and Palukaitis, P. (2014) Transgenic resistance. *Advances in Virus Research* 90, 35–146.

Cisneros, A.E. and Carbonell, A. (2020) Artificial small RNA-based silencing tools for antiviral resistance in plants. *Plants* 9: 669.

Citores, L., Iglesias, R., and Ferreras, J.M. (2021) Antiviral activity of ribosome-inactivating proteins. *Toxins* 13: 80.

Colalongo, C., Contaldo, N., Facciolo, G., Pagan, P., Gianni, F., and Paolo, P. (2005) The use of *pac1* gene from *Schizosaccaromyces pombe* to protect potato from potato virus Y (PVY) and potato virus X (PVX) infections. *Phytopathologia Mediterranea* 44, 226–232.

Day, A.G., Bejarano, E.R., Buck, K.W., Burrel, M., and Lichtenstein, C.P. (1991) Expression of an antisense viral gene in transgenic tobacco confers resistance to the DNA virus tomato golden mosaic virus. *Proceedings of the National Academy of Sciences USA* 88, 6721–6725.

Di, K., Purcell, V., Collins, G.B., and Ghabrial, S.A. (1996) Production of transgenic soybean lines expressing the bean pod mottle virus coat protein precursor gene. *Plant Cell Reports* 15, 746–750.

Dinant, S., Maisonneuve, B., Albouy, J., Chupeau, Y., Chupeau, M.-C., *et al.* (1997) Coat protein gene-mediated protection in *Lactuca sativa* against lettuce mosaic potyvirus strains. *Molecular Breeding* 3, 75–86.

Ding, S.W. (2010) RNA-based antiviral immunity. *Nature Reviews Immunology* 10, 632–644.

Doreste, V., Ramos, P.L., Enríquez, G.A., Rodríguez, R., Peral, R., and Pujol, M. (2002) Transgenic potato plants expressing the potato virus X (PVX) coat protein gene developed resistance to the viral infection. *Phytoparasitica* 30, 177–185.

Emran, A.M., Tabei, Y., Kobayashi, K., Yamaoka, N., and Nishiguchi, M. (2012) Molecular analysis of transgenic melon plants showing virus resistance conferred by direct repeat of movement gene of cucumber green mottle mosaic virus. *Plant Cell Reports* 31, 1371–1377.

Eni, A.O., Efekemo, O.P., Onile-Ere, O.A., and Pita, J.S. (2021) South West and North Central Nigeria: assessment of cassava mosaic disease and field status of *African cassava mosaic virus* and *East African cassava mosaic virus*. *Annals of Applied Biology* 178, 466–479.

Fahim, M., Millar, A.A., Wood, C.C., and Larkin, P.J. (2012) Resistance to wheat streak mosaic virus generated by expression of an artificial polycistronic microRNA in wheat. *Plant Biotechnology Journal* 10, 150–163.

Fang, G., and Grumet, R. (1993) Genetic engineering of potyvirus resistance using constructs derived from the zucchini yellow mosaic virus coat protein gene. *Molecular Plant–Microbe Interactions* 6, 358–367.

FAO (Food and Agriculture Organization of the United Nations) (2020) *The State of Food Security and Nutrition in the World 2020. Transforming Food Systems for Affordable Healthy Diets*. FAO, Rome. Availabel at: www.fao.org/publications/home/fao-flagship-publications/the-state-of-food-security-and-nutrition-in-the-world/en (accessed 30 June 2024).

Fehér, A., Skryabin, K.G., Balázs, E., Preiszner, J., Shulga, O.A., *et al.* (1992) Expression of PVX coat protein gene under the control of extensin-gene promoter confers virus resistance on transgenic potato plants. *Plant Cell Reports* 11, 48–52.

Ferreira, S.A., Pitz, K.Y., Manshardt, R., Fitch, M., and Gonsalves, D. (2002) Virus coat protein transgenic papaya provides practical control of papaya ringspot virus in Hawaii. *Plant Disease* 86, 101–105.

Fiedler, U., Phillips, J., Artsaenko, O., and Conrad, U. (1997) Optimization of scFv antibody production in transgenic plants. *Immunotechnology* 3, 205–216.

Fondong, V.N. (2017) The search for resistance to cassava mosaic geminiviruses: how much we have accomplished, and what lies ahead. *Frontiers in Plant Science* 8: 408.

Fuchs, M. and Gonsalves, D. (1995) Resistance of transgenic hybrid *Squash Zw-20* expressing the coat protein genes of zucchini yellow mosaic virus and watermelon mosaic virus 2 to mixed infections by both potyviruses. *Bio/Technology* 13, 1466–1473.

Fuchs, M., Tricoli, D.M., Carney, K.J., Schesser, M., McFerson, J.R., and Gonsalves, D. (1998) Comparative virus resistance and fruit yield of transgenic squash with single and multiple coat protein genes. *Plant Disease* 82, 1350–1356.

Fuentes, A., Ramos, P.L., Fiallo, E., Callard, D., Sánchez, Y., *et al.* (2006) Intron-hairpin RNA derived from replication association protein *c1* gene confers immunity to tomato yellow leaf curl virus infection in transgenic tomato plants. *Transgenic Research* 15, 291–304.

Furutani, N., Hidaka, S., Kosaka, Y., Shizukawa, Y., and Kamenatsu, S. (2006) Coat protein gene-mediated resistance to soybean mosaic virus in transgenic soybean. *Breeding Science* 56, 119–124.

Gal-On, A., Wolf, D., Wang, Y., Faure, J.-M., Pilowsky, M., and Zelcer, A. (1998) Transgenic resistance to cucumber mosaic virus in tomato: blocking of long-distance movement of the virus in lines harboring a defective viral replicase gene. *Phytopathology* 88, 1101–1107.

Gan, D., Zhang, J., Jiang, H., Jiang, T., Zhu, S., and Cheng, B. (2010) Bacterially expressed dsRNA protects maize against SCMV infection. *Plant Cell Reports* 29, 1261–1268.

Gao, L., Luo, J., Ding, X., Wang, T., Hu, T., *et al.* (2020) Soybean RNA interference lines silenced for eIF4E show broad potyvirus resistance. *Molecular Plant Pathology* 21, 303–317.

Gargouri-Bouzid, R., Leila, J., Riadh, B.M., Yemna, H., Malika, A., and Radhouane E. (2005) PVY resistant transgenic potato plants (cv Claustar) expressing the viral coat protein. *Journal of Plant Biotechnology* 7, 143–148.

Gargouri-Bouzid, R., Jaoua, L., Mansour, R.B., Hathat, Y., Ayadi, M., and Ellouz, R. (2006) PVY-resistant transgenic potato plants expressing an anti-NIa protein scFv antibody. *Molecular Biotechnology* 33, 133–140.

General Assembly (2015) *Transforming Our World: The 2030 Agenda for Sustainable Development*. United Nations. New York.

Ghannam, A., Kumari, S., Muyldermans, S., and Abbady, A.Q. (2015) Camelid nanobodies with high affinity for broad bean mottle virus: A possible promising tool to immunomodulate plant resistance against viruses. *Plant Molecular Biology* 87, 355–369.

Ghorbani Faal, P., Farsi, M., Seifi, A., and Kakhki, A.M. (2020) Virus-induced CRISPR-Cas9 system improved resistance against tomato yellow leaf curl virus. *Molecular Biology Reports* 47, 3369–3376.

Gielen, J., Ultzen, T., Bontems, S., Loots, W., and van Schepen, A. (1996) Coat protein-mediated protection to cucumber mosaic virus infections in cultivated tomato. *Euphytica* 88, 139–149.

Głąbska, D., Guzek, D., Groele, B., and Gutkowska, K. (2020) Fruit and vegetable intake and mental health in adults: a systematic review. *Nutrients* 12: 115.

Global Panel (2020) *Future Food Systems. Vol. 3*. Apple Academic Press.

Golemboski, D.B., Lomonossoff, G.P., and Zaitlin, M. (1990) Plants transformed with a tobacco mosaic virus nonstructural gene sequence are resistance to the virus. *Proceedings of the National Academy of Sciences USA* 87, 6311–6315.

Gomez, M.A., Lin, Z.D., Moll, T., Chauhan, R.D., Hayden, L., *et al.* (2019) Simultaneous CRISPR/Cas9-mediated editing of cassava eIF4E isoforms nCBP-1 and nCBP-2 reduces cassava brown streak disease symptom severity and incidence. *Plant Biotechnology Journal* 17, 421–434.

Goregaoker, S.P., Eckhardt, L.G., and Culver, J.N. (2000) Tobacco mosaic virus replicase-mediated cross-protection: contributions of RNA and protein-mediated mechanisms. *Virology* 273, 267–275.

Gubba, A., Gonsalves, C., Stevens, M.R., Tricoli, D.M. and Gonsalves, D. (2002) Combining transgenic and natural resistance to obtain broad resistance to tospovirus infection in tomato (*Lycopersicon esculentum* Mill). *Molecular Breeding* 9, 13–23.

Guo, Z.-X., Li, Y., and Ding, S.-W. (2019) Small RNA-based antimicrobial immunity. *Nature Reviews Immunology* 19, 31–44.

Hameed, A., Tahir, M.N., Asad, S., Bilal, R., van Eck, J., *et al.* (2017) RNAi-mediated simultaneous resistance against three RNA viruses in potato. *Molecular Biotechnology* 59, 73–83.

Han, L.W., Ge, Y., Yao, L., Cui, C.S., and Qu, S. (2014) Coat protein-mediated resistance to turnip mosaic virus in Chinese cabbage (*Brassica rapa* ssp. *pekinensis*) by the pollen-tube pathway. *International Journal of Agriculture and Biology* 16, 1232–34.

Hoekema, A., Huisman, M.J., MolenDijk, L., van den Elzen, P.J.M., and CorNelissen, B.J.C. (1989) The genetic engineering of two commercial potato cultivars for resistance to potato virus X. *Bio/Technology* 7, 273–278.

Honda, A., Takahashi, H., Toguri, T., Ogawa, T., Hase, S., *et al.* (2003) Activation of defense-related gene expression and systemic acquired resistance in cucumber mosaic virus-infected tobacco plants expressing the mammalian 2′5′oligoadenylate system. *Archives of Virology* 148, 1017–1026.

Hong, Y., Saunders, K., Hartley, M.R., and Stanley, J. (1996) Resistance to geminivirus infection by virus-induced expression of dianthin in transgenic plants. *Virology* 220, 119–127.

Iglesias, R., Pérez, Y., de Torre, C., Ferreras, J.M., Antolín, P., *et al.* (2005) Molecular characterization and systemic induction of single-chain ribosome-inactivating proteins (RIPs) in sugar beet (*Beta vulgaris*) leaves. *Journal of Experimental Botany* 56, 1675–84.

ISAAA (International Service for the Acquisition of Agri-biotech Applications) (2016) *Pocket K No. 16—Biotech Crop Highlights in 2019*. ISAAA. Available at: www.isaaa.org/resources/publications/pocketk/16/ (accessed 30 June 2024)

Jafarzade, M., Ramezani, M., Hedayati, F., Mokhtarzade, Z., Zare, B., *et al.* (2019) Antibody-mediated resistance to rhizomania disease in sugar beet hairy roots. *Plant Pathology Journal* 35, 692–697.

Jin, F.M., Song, J., Xue, J., Sun, H.B., Zhnag, Y., *et al.* (2020) Successful generation of anti-ToCV and TYLCV transgenic tomato plants by RNAi. *Biologia Plantarum* 64, 490–496.

Jones, R.A.C. (2020) Disease pandemics and major epidemics arising from new encounters between indigenous viruses and introduced crops. *Viruses* 12: 1388.

Jones, R.A.C. (2021) Global plant virus disease pandemics and epidemics. *Plants* 10: 233.

Jongedijk, E., de Schutter, A.A., Stolte, T., van den Elzen, P.J., and Cornelissen, B.J. (1992) Increased resistance to potato virus X and preservation of cultivar properties in transgenic potato under field conditions. *Nature Biotechnology* 10, 422–429.

Kallerhoff, J., Perez, P., Bouzoubaa, S., Tahar, S.B., and Perret, J. (1990) Beet necrotic yellow vein virus coat protein-mediated protection in sugarbeet (*Beta vulgaris* L.) protoplasts. *Plant Cell Reports* 9, 224–228.

Kaniewski, W., Lawson, C., Sammons, B., Haley, L., Hart, J. *et al.* (1990) Field resistance of transgenic Russet Burnak potato to effects of infection by potato virus X and potato virus Y. *Nature Biotechnology* 8, 750–754.

Kataoka, J., Noriyuki, H., Masashi, M., Chikara, M., and Akira, K. (1992) Adenine depurination and inactivation of plant ribosomes by an antiviral protein of *Mirabilis jalapa* (MAP). *Plant Molecular Biology* 20, 1111–1119.

Kawazu, Y., Fujiyama, R., Sugiyama, K., and Sasaya, T. (2006) A transgenic lettuce line with resistance to both lettuce big-vein associated virus and Mirafiori lettuce virus. *Journal of the American Society for Horticultural Science* 131, 760–763.

Kawchuk, L.M., Martin, R.R., and McPherson, J. (1990) Resistance in transgenic potato expressing the potato leafroll virus coat protein gene. *Molecular Plant–Microbe Interactions* 3, 301–307.

Kawchuk, L.M., Martin, R.R., and McPherson, J. (1991) Sense and antisense RNA- mediated resistance to potato leafroll virus in Russet Burbank potato plants. *Molecular Plant–Microbe Interactions* 4, 247–253.

Khalid, A., Zhang, Q., Yasir, M., and Liu, F. (2017) Small RNA based genetic engineering for plant viral resistance: application in crop protection. *Frontiers in Microbiology* 8: 43.

Kim, H.-J., Kim, M.-J., Pak, J.H., Jung, H.W., Choi, H.K., and Lee, Y.-H. (2013) Characterization of SMV resistance of soybean produced by genetic transformation of SMV-CP gene in RNAi. *Plant Biotechnology Reports* 7, 425–433.

Kis, A., Tholt, G., Ivanics, M., Varallyay, E., Jenes, B., and Havelda, Z. (2016) Polycistronic artificial miRNA-mediated resistance to wheat dwarf virus in barley is highly efficient at low temperature. *Molecular Plant Pathology* 17, 427–437.

Koev, G., Mohan, B.R., Dinesh-Kumar, S.P., Torbert, K.A., Somers, D.A., and Miller, W.A. (1998). Extreme reduction of oats transformed with the 5′ half of the barley yellow dwarf virus-PAV genome. *Phytopathology* 88, 1013–1019.

Koshino-Kimura, Y., Kosuke, T., Fumiya, D., Yasuhiro, A., and Takashi, S. (2008) Generation of plants resistant to tomato yellow leaf curl virus by using artificial zinc-finger proteins. *Nucleic Acids Symposium Series* 52, 189–190.

Koshino-Kimura, Y., Kosuke, T., Fumiya, D., Masayoshi, O., Toshihide, M., *et al.* (2009) Construction of plants resistant to TYLCV by using artificial zinc-finger proteins. *Nucleic Acids Symposium Series* 53, 281–282.

Krubphachaya, P., Jurícek, M., and Kertbundit, S. (2007) Induction of RNA-mediated resistance to papaya ringspot virus type W. *Journal of Biochemisty and Molecular Biology* 40, 404–411.

Kumar, R.V., Singh, A.K., Singh, A.K., Yadav, T., Basu, S., *et al.* (2015) Complexity of begomovirus and betasatellite populations associated with chilli leaf curl disease in India. *Journal of General Virology* 96, 3143–3158.

Kumari, A., Hada, A., Subramanyam, K., Theboral, J., Misra, S., *et al.* (2018) RNAi-mediated resistance to yellow mosaic viruses in soybean targeting coat protein gene. *Acta Physiologiae Plantarum* 40: 32.

Kunik, T., Salomon, R., Zamir, D., Navot, N., Zeidan, M., *et al.* (1994). Transgenic tomato plants expressing the tomato yellow leaf curl virus capsid protein are resistant to the virus. *Bio/Technology* 12, 500–504.

Lauber, E., Janssens, L., Weyens, G., Jonard, G., Richards, K.E., and Lefebvre, M. (2001) Rapid screening for dominant negative mutations in the beet necrotic yellow vein virus triple gene block proteins P13 and P15 using a viral replicon. *Transgenic Research* 10, 293–302.

Lawson, C., Kaniewski, W., Haley, L., Rozman, R., and Newell, C. (1990) Engineering resistance to mixed virus infection in a commercial potato cultivar: resistance to potato virus X and potato virus Y in transgenic Russet Burbank. *Nature Biotechnology* 8, 27–34.

Lawson, E.C., Weiss, J.D., Thomas, P.E., and Kaniewski, W.K. (2001) NewLeaf Plus® Russet Burbank potatoes: replicase-mediated resistance to potato leafroll virus. *Molecular Breeding* 7, 1–12.

Lee, Y.H., Min, J., Sun, H.S., Ji, H.L., Soon, H.C., *et al.* (2009). Transgenic peppers that are highly tolerant to a new CMV pathotype. *Plant Cell Reports* 28, 223–232.

Leibman, D., Wolf, D., Saharan, V., Zelcer, A., Arazi, T., and Shiboleth, Y. (2011) A high level of transgenic viral small RNA is associated with broad potyvirus resistance in cucurbits. *Molecular Plant–Microbe Interactions* 24, 1220–1238.

Leibman, D., Ortega-Parra, N., Dalia, W., Meital, S., Inge, H., and Gal-On, A. (2021) A transgenic RNAi approach for developing tomato plants immune to pepinomosaic virus. *Plant Pathology* 70, 1003–1012.

Lennefors, B.-L., Savenkov, E.I., Bensefelt, J., Wremerth-Weich, E., Roggen, P., and Tuvesson, S. (2006) dsRNA-mediated resistance to beet necrotic yellow vein virus infections in sugar beet (*Beta vulgaris* L. spp. *vulgaris*). *Molecular Breeding* 18, 313–325.

Léonard, S., Plante, D., Wittmann, S., Daigneault, N., Fortin, M.G., and Laliberté, J.F. (2000) Complex formation between potyvirus VPg and translation eukaryotic initiation factor 4E correlates with virus infectivity. *Journal of Virology* 74, 7730–7737.

Lodge, J.K., Wojciech, K.K., and Nilgun, E.T. (1993) Broad-spectrum virus resistance in transgenic plants expressing pokeweed antiviral protein. *Proceedings of the National Academy of Sciences USA* 90, 7089–7093.

Loesch-Fries, L.S., Merlo, D., Zinnen, T., Burhop, L., and Hall, K. (1987) Expression of alfalfa mosaic virus RNA 4 in transgenic plants confers virus resistance. *EMBO Journal* 6, 845–851.

Lopez-Ochoa, L., Ramirez-Prado, J., and Hanley-Bowdoin, L. (2006) Peptide aptamers that bind to a geminivirus replication protein interfere with viral replication in plant cells. *Journal of Virology* 80, 5841–5853.

MacKenzie, D.J., Tremaine, J.H., and McPherson, J. (1991) Genetically engineered resistance to potato virus S in potato cultivar russet burbank. *Molecular Plant–Microbe Interactions* 4, 95–102.

Magbanua, Z.V., Wilde, H.D., Roberts, J.K., Chowdhury, K., Abad, J., and Moyer, J.M. (2000) Field resistance to tomato spotted wilt virus in transgenic peanut (*Arachis hypogaea* L.) expressing an antisense nucleocapsid gene sequence. *Molecular Breeding* 6, 227–236.

Makhotenko, A.V., Khromov, A.V., Snigir, E.A., Makarova, S.S., Makarov, V.V., *et al.* (2019) Functional analysis of coilin in virus resistance and stress tolerance of potato *Solanum tuberosum* using CRISPR-Cas9 editing. *Doklady Biochemistry and Biophysics* 484, 88–91.

Malathi, V.G., Renukadevi, P., Chakraborty, S., Biswas, K.K., Anirban, R., *et al.* (2017) Begomoviruses and their satellites occurring in India: distribution, diversity and pathogenesis. In: Mandal, B., Rao, G., Baranwal, V. and Jain, R. (eds.) *A Century of Plant Virology in India*. Springer, Singapore, pp. 75–177.

Malnoë, P., Laurent, F., Collet, G.F., and Werner, R. (1994) Small-scale field tests with transgenic potato, cv. Bintje, to test resistance to primary and secondary infections with potato virus Y. *Plant Molecular Biology* 25, 963–975.

Mandal, B., Jain, R.K., Krishnareddy, M., Krishna Kumar, N.K., Ravi, K.S., and Hanu, R.P. (2012). Emerging problems of tospoviruses (Bunyaviridae) and their management in the Indian subcontinent. *Plant Disease* 96, 468–479.

Mannerlöf, M., Lennerfors, B.L., and Tenning, P. (1996). Reduced titer of BNYVV in transgenic sugar beets expressing the BNYVV coat protein. *Euphytica* 90, 293–299.

Marano, M.R. and Baulcombe, D. (1998) Pathogen-derived resistance targeted against the negative-strand RNA of tobacco mosaic virus: RNA strand-specific gene silencing? *Plant Journal* 13, 537–546.
Matousek, J., Schroder, A.R.M., Trnena, L., Reimers, I., Baumstark, T., and Dedic, P. (1994) Inhibition of viroid infection by antisense RNA expressed in transgenic plants. *Biological Chemistry* 375, 765–777.
Mazier, M., Flamain, F., Nicolaï, M., Sarnette, V., and Caranta, C. (2011) Knock-down of both *eIF4E1* and *eIF4E2* genes confers broad-spectrum resistance against potyviruses in tomato. *PLOS ONE* 6: e29595.
Mehta, D., Alessandra, S., Hirsch-Hoffmann, M., Wilhelm, G., and Hervé, V. (2019) CRISPR-Cas9 interference in cassava linked to the evolution of editing-resistant geminiviruses. *Genome Biology* 20: 80.
Melander, M. (2006) Potato transformed with a 57-kDa readthrough portion of the tobacco rattle virus replicase gene displays reduced tuber symptoms when challenged by viruliferous nematodes. *Euphytica* 150, 123–130.
Melander, M., Lee, M., and Sandgren, M. (2001) Reduction of potato mop-top virus accumulation and incidence in tubers of potato transformed with a modified triple gene block gene of PMTV. *Molecular Breeding* 8, 197–206.
Mello, C.C. and Conte, D. (2004) Revealing the world of RNA interference. *Nature* 431, 338–342.
Miroshnichenko, D., Timerbaev, V., and Okuneva, A. (2020) Enhancement of resistance to PVY in intragenic marker-free potato plants by RNAi-mediated silencing of eIF4E translation initiation factors. *Plant Cell Tissue and Organ Culture* 140, 691–705.
Mishra, G.P., Singh, B., Seth, T., Singh, A.K., Halder, J., *et al.* (2017) Biotechnological advancements and begomovirus management in okra (*Abelmoschus esculentus* L.): status and perspectives. *Frontiers in Plant Science* 8: 360.
Missiou, A., Kalantidis, K., Boutla, A., Tzortzakaki, S., Tabler, M., and Tsagris, M. (2004) Generation of transgenic potato plants highly resistant to potato virus Y (PVY) through RNA silencing. *Molecular Breeding* 14, 185–197.
Mitra, A., Higgins, D.W., Langenberg, W.G., Nie, H., Sengupta, D.N., and Silverman, R.H. (1996) A mammalian 2-5A system functions as an antiviral pathway in transgenic plants. *Proceedings of the National Academy of Sciences USA* 93, 6780–6785.
Mitter, N., Worrall, E.A., Robinson, K.E., Li, P., Jain, R.G., *et al.* (2017) Clay nanosheets for topical delivery of RNAi for sustained protection against plant viruses. *Nature Plants* 3: 16207.
Moon, Y.H., Sang, K.S., Kyu, W.C., and Jong, S.L. (1997) Expression of a cDNA encoding *Phytolacca insularis* antiviral protein confers virus resistance on transgenic potato plants. *Molecules and Cells* 7, 807–815.
Mori, T., Kosuke, T., Fumiya, D., Yasuhiro, A., and Takashi, S. (2013) Inhibition of binding of tomato yellow leaf curl virus rep to its replication origin by artificial zinc-finger protein. *Molecular Biotechnology* 54, 198–203.
Morozov, S.Y., Solovyev, A.G., Kalinina, N.O., and Taliansky, M.E. (2019) Double-stranded RNAs in plant protection against pathogenic organisms and viruses in agriculture. *Acta Naturae* 11, 13–21.
Namgial, T., Kaldis, A., Chakraborty, S., and Voloudakis, A. (2019) Topical application of double-stranded RNA molecules containing sequences of tomato leaf curl virus and cucumber mosaic virus confers protection against the cognate viruses. *Physiological and Molecular Plant Pathology* 108: 101432.
Nelson, R.S., McCormick, S.M., Delannay, X., Dubé, P., Layton, J., *et al.* (1988) Virus tolerance, plant performance of transgenic tomato plants expressing coat protein from toracco mosaic virus. *Bio/Technology* 6, 403–409.
Nickel, H., Kawchuk, L., Twyman, R.M., Zimmermann, S., Junghans, H., *et al.* (2008) Plantibody-mediated inhibition of the *Potato leafroll virus* P1 protein reduces virus accumulation. *Virus Research* 136, 140–145.
Nilon, A., Robinson, K., Pappu, H.R., and Mitter, N. (2021) Current status and potential of RNA interference for the management of tomato spotted wilt virus and thrips vectors. *Pathogens* 10: 320.
Ntui, V.O., Kynet, K., Azadi, P., Khan, R.S., Chin, D.P., and Nakamura, I. (2013) Transgenic accumulation of a defective cucumber mosaic virus (CMV) replicase derived double stranded RNA modulates plant defence against CMV strains O and Y in potato. *Transgenic Research* 22, 1191–1205.
Ntui, V.O., Kynet, K., Raham, S.K., Tomoko, I., Gnanaguru, J.J., *et al.* (2015). Resistance to Sri Lankan cassava mosaic virus (SLCMV) in genetically engineered cassava cv. KU50 through RNA silencing. *PLOS ONE* 10: e0120551.

Nyaboga, E.N., Ateka, E.M., Gichuki, S.T., and Bulimo, W.D. (2008) Reaction of transgenic sweet potato (*Ipomoea batatas* L.) lines to virus challenge in the glasshouse. *Journal of Applied Biosciences* 9, 362–371.

Ogawa, T., Tamaki, H., and Isao, I. (1996) Virus-induced cell death in plants expressing the mammalian 2′,5′ oligoadenylate system. *Nature Biotechnology* 14, 1566–1569.

Ogwok, E., Odipio, J., Halsey, M., Gaitan-Solıs, E., Bua, A., and Taylor, N.J. (2012) Transgenic RNA interference (RNAi)-derived field resistance to cassava brown streak disease. *Molecular Plant Pathology* 13, 1019–1031.

Okada, Y. and Saito, A. (2008) Evaluation of resistance to complex infection of SPFMVs in transgenic sweet potato. *Breeding Science* 58, 243–250.

Okada, Y., Saito, A., Nishiguchi, M., Kimura, T., and Mori, M., *et al.* (2001) Virus resistance in transgenic sweetpotato [*Ipomoea batatas* L. (Lam)] expressing the coat protein gene of sweet potato feathery mottle virus. *Theoretical and Applied Genetics* 103, 743–751.

Oladokun, J.O., Halabi, M.H., Barua, P., and Nath, P.D. (2019) Tomato brown rugose fruit disease: current distribution, knowledge and future prospects. *Plant Pathology* 68, 1579–1586.

Pack, I.S., Kim, Y.J., Youk, E.S., Lee, W.K., Yoon, W.K., *et al.* (2012) A molecular framework for risk assessment of a virus-tolerant transgenic pepper line. *Journal of Crop Science and Biotechnology* 15, 107–115.

Palucha, A., Zagorski, W., Chrzanowska, M., and Hulanicka, D. (1998) An antisense coat protein gene confers immunity to potato leafroll virus in a genetically engineered potato. *European Journal of Plant Pathology* 104, 287–293.

Paraskevi, T., Benvenuto, E., Trinca, S. de Martinis, D., Cattaneo, A., and Galeffi, P. (1993) Transgenic plants expressing a functional single-chain Fv antibody are specifically protected from virus attack. *Nature* 366, 469–472.

Pavli, O.I., Panopoulos, N.J., Goldbach, R., and Skaracis, G.N. (2010) BNYVV-derived dsRNA confers resistance to rhizomania disease of sugar beet as evidenced by a novel transgenic hairy root approach. *Transgenic Research* 19, 915–922.

Pinto, Y.M., Kok, R.A., and Baulcombe, D.C. (1999) Resistance to rice yellow mottle virus (RYMV) in cultivated African rice varieties containing RYMV transgenes. *Nature BioTechnology* 17, 702–707.

Potrokhov, A., Sosnovska, D., Ovcharenko, O., Budzanivska, I., Rudas, V., and Kuchuk, M. (2021) Increased ribonuclease activity in *Solanum tuberosum* L. transformed with heterologous genes of apoplastic ribonucleases as a putative approach for production of virus resistant plants. *Turkish Journal of Biology* 45, 79–87.

Powell, P.A. (1988) Genetically engineered protection against tobacco mosaic virus infection by the expression of viral sequences in transgenic plants. PhD thesis, Washington University, St. Louis, Missouri.

Powell, P.A., Stark, D.M., Sanders, P.R., and Beachy, R.N. (1989) Protection against tobacco mosaic virus in transgenic plants that express tobacco mosaic virus antisense RNA. *Proceedings of the National Academy of Sciences USA* 86, 6949–6952.

Powell-Abel, P.A., Nelson, R.S., De, B., Hoffman, N., and Rogers, S.G. (1986) Delay of disease development in transgenic plants that express the tobacco mosaic virus coat protein gene. *Science* 232, 738–743.

Pramanik, D., Shelake, R.M., Park, J., Kim, M.J., Hwang, I., *et al.* (2021). CRISPR/Cas9-mediated generation of pathogen-resistant tomato against tomato yellow leaf curl virus and powdery mildew. *International Journal of Molecular Sciences* 22: 1878.

Pratap, D., Kumar, S., Snehi, S.K., and Raj, S.K. (2012a) Biological and molecular characterization of cucumber mosaic virus isolate causing shoestring disease of tomato in India which has closer affinity to European or east Asian isolates of CMV. *Indian Journal of Virology* 23, 57–63.

Pratap, D., Raj, S.K., Kumar, S., Snehi, S.K., Gautam, K.K., and Sharma, A.K. (2012b) Coat protein-mediated transgenic resistance in tomato against a IB subgroup cucumber mosaic virus strain. *Phytoparasitica* 40, 375–82.

Praveen, S., Kushwaha, C.M., Mishra, A.K., Singh, V., Jain, R.K., and Varma, A. (2005) Engineering tomato for resistance to tomato leaf curl disease using viral rep gene sequences. *Plant Cell, Tissue and Organ Culture* 83, 311–318.

Presting, G.G., Smith, O.P., and Brown, C.R. (1995) Resistance to potato leafroll virus in potato plants transformed with the coat protein gene or with vector control constructs. *Phytopathology* 85, 436–442.

Pumplin, N. and Voinnet, O. (2013) RNA silencing suppression by plant pathogens: defence, counter-defence, and counter-counter-defence. *Nature Reviews Microbiology* 11, 745–760.

Ramesh, S.V., Shivakumar, M., Praveen, S., Chouhan, B.S., and Chand, S. (2019) Expression of short hairpin RNA (shRNA) targeting *AC2* gene of mungbean yellow mosaic India virus (MYMIV) reduces the viral titre in soybean. *3 Biotech* 9: 334.

Reddy, M., Ghabrial, S.A., Redmond, C.T., Dinkins, R.D., and Collins, G.B. (2001) Resistance to bean pod mottle virus in transgenic soybean lines expressing the capsid polyprotein. *Phytopathology* 91, 831–838.

Reyes, M.I., Nash, T.E., Dallas, M.M., Ascencio-Ibanez, J.T., and Hanley-Bowdoin, L. (2013) Peptide aptamers that bind to geminivirus replication proteins confer a resistance phenotype to *Tomato yellow leaf curl virus* and *Tomato mottle virus* infection in tomato. *Journal of Virology* 87, 9691–9706.

Robaglia, C. and Tepfer, M. (1996) Using nonviral genes to engineer virus resistance in plants. *Biotechnology Annual Review* 2, 185–204.

Romano, E., Ferreira, A.T., Dusi, A.N., Proite, K., Buso, J.A., *et al.* (2001). Extreme resistance to two Brazilian strains of potato virus Y (PVY) in transgenic potato, cv. Achat, expressing the PVYo coat protein. *Horticultura Brasileira* 19, 118–122.

Rossmann, M.G., and Johnson, J.E. (1989) Icosahedral RNA virus structure. *Annual Review of Biochemistry* 58, 533–573.

Rudolph, C., Schreier, P.H., and Uhrig, J.F. (2003) Peptide-mediated broad-spectrum plant resistance to tospoviruses. *Proceedings of the National Academy of Sciences USA* 100, 4429–4434.

Rybicki, E.P. (2019) CRISPR-Cas9 strikes out in cassava. *Nature Biotechnology* 37, 727–728.

Šafářová, D., Brazda, P., and Navrátil, M. (2014) Effect of artificial dsRNA on infection of pea plants by pea seed-borne mosaic virus. *Czech Journal of Genetics and Plant Breeding* 50, 105–108.

Sanders, P.R., Sammons, B., Kaniewski, W., Haley, L., Layton, J., *et al.* (1992) Field resistance of transgenic tomatoes expressing the tobacco mosaic virus or tomato mosaic virus coat protein genes. *Molecular Plant Pathology* 82, 683–690.

Sanfaçon, H. (2015) Plant translation factors and virus resistance. *Viruses* 7, 3392–3419.

Sano, T., Nagayama, A., Ogawa, T., Ishida, I., and Okada, Y. (1997) Transgenic potato expressing a double-stranded RNA-specific ribonuclease is resistant to potato spindle tuber viroid. *Nature Biotechnology* 15, 1290–1294.

Schreinemachers, P., Simmons, E.B., and Wopereis, M.C.S. (2018). Tapping the economic and nutritional power of vegetables. *Global Food Security* 16, 36–45.

Schubert, J., Matousek, J., and Mattern, D. (2004) Pathogen-derived resistance in potato to potato virus Y—aspects of stability and biosafety under field conditions. *Virus Research* 100, 41–50.

Schwind, N., Zwiebel, M., Itaya, A., Ding, B., Wang, M.-B., and Krczal, G. (2009) RNAi-mediated resistance to potato spindle tuber viroid in transgenic tomato expressing a viroid hairpin RNA construct. *Molecular Plant Pathology* 10, 459–469.

Sera, T. (2005) Inhibition of virus DNA replication by artificial zinc finger proteins. *Journal of Virology* 79, 2614–2619.

Sera, T. (2017) Use of peptide aptamers, cationic peptides and artificial zinc finger proteins to generate resistance to plant viruses. *Current Opinion in Virology* 26, 120–124.

Sharma, P., Sahu, A.K., Verma, R.K., Mishra, R., Choudhary, D.K., and Gaur, R. K. (2014) Current status of potyvirus in India. *Archives of Phytopathology and Plant Protection* 47, 906–918.

Sharma, S., Sharma, A., Kaur, S., and Kang, S.S. (2020) Occurrence and incidence of pepper mottle virus infecting chilli in Punjab, India. *Indian Phytopathology* 73, 573–576.

Sharma, V.K., Saumik, B., and Supriya, C. (2015) RNAi mediated broad-spectrum transgenic resistance in *Nicotiana benthamiana* to chilli-infecting begomoviruses. *Plant Cell Reports* 34, 1389–1399.

Sharma, V.K., Marla, S., Zheng, W.G., Mishra, D., Huang, J., *et al.* (2022) RNA silencing by CRISPR in plants does not require Cas13. *Genome Biology* 23: 6.

Shi, S., Zhang, X., Mandel, M.A., Zhang, P., Zhang, Y., *et al.* (2017) Variations of five eIF4E genes across cassava accessions exhibiting tolerant and susceptible responses to cassava brown streak disease. *PLOS ONE* 12: e0181998.

Shin, R., Han, J.H., Lee, G.J., and Peak, K.H. (2002). The potential use of a viral coat protein gene as a transgene screening marker and multiple virus resistance of pepper plants coexpressing coat proteins of cucumber mosaic virus and tomato mosaic virus. *Transgenic Research* 11, 215–219.

Singh, V.B., Haq, Q.M.R., and Malathi, V.G. (2013) Antisense RNA approach targeting Rep gene of *Mungbean yellow mosaic India virus* to develop resistance in soybean. *Archives of Phytopathology and Plant Protection* 46, 2191–2207.

Sivparsad, B.J. and Gubba, A. (2014) Development of transgenic sweet potato with multiple virus resistance in South Africa (SA). *Transgenic Research* 23, 377–388.

Smith, N.A., Singh, S.P., Wang, M.-B., Stoutjesdijk, P.A., Green, A.G., and Waterhouse, P.M. (2000) Total silencing by intron-spliced hairpin RNAs. *Nature* 407, 319–320.

Sokolova, M.A., Pugin, M.M., Shulga, O.A., and Skryabin, K.G. (1994) Construction of transgenic potato plants resistant to potato-virus-Y. *Molecular Biology* 28, 646–649.

Sree Vidya, C.S., Manoharan, M., Ranjit Kumar, C.T., Savithri, H.S., and Lakshmi Sita, G. (2000) *Agrobacterium*-mediated transformation of tomato (*Lycopersicon esculentum* var. Pusa Ruby) with coat-protein gene of *Physalis* mottle tymovirus. *Journal of Plant Physiology* 156, 106–110.

Steinlage, T.A., Hill, J.H., and Nutter, F.W. (2002) Temporal and spatial spread of soybean mosaic virus (SMV) in soybeans transformed with the coat protein gene of SMV. *Phytopathology* 92, 478–486.

Sugawara, T., Trifonova, E.A., Kochetov, A.V., and Kanayama, Y. (2016) Expression of an extracellular ribonuclease gene increases resistance to cucumber mosaic virus in tobacco. *BMC Plant Biology* 16, 147–152.

Sun, L., Lin, C., Du, J., Song, Y., Jiang, M., *et al.* (2016) Dimeric artificial microRNAs mediate high resistance to RSV and RBSDV in transgenic rice plants. *Plant Cell, Tissue and Organ Culture* 126, 127–139.

Tacke, E., Salamini, F., and Rohde, W. (1996) Genetic engineering of potato for broad-spectrum protection against virus infection. *Nature Biotechnology* 14, 1597–1601.

Takenaka, K., Koshino-Kimura, Y., Aoyama, Y., and Sera, T. (2007) Inhibition of tomato yellow leaf curl virus replication by artificial zinc-finger proteins. *Nucleic Acids Symposium Series* 51, 429–430.

Tan, H., Li, B., and Guo, H. (2020) The diversity of post-transcriptional gene silencing mediated by small silencing RNAs in plants. *Essays in Biochemistry* 64, 919–930.

Tashkandi, M., Ali, Z., Aljedaani, F., Shami, A., and Mahfouz, M.M. (2018) Engineering resistance against tomato yellow leaf curl virus via the CRISPR/Cas9 system in tomato. *Plant Signaling and Behavior* 13: e1525996.

Tenllado, F. and Díaz-Ruíz, J.R. (2001) Double-stranded RNA-mediated interference with plant virus infection. *Journal of Virology* 75, 12288–12297.

Tennant, P.F., Gonsalves, C., Ling, K.-S., Fitch, M., Manshardt, R., and Slightom, J.L. (1994) Differential protection against papaya ringspot virus isolates in coat protein gene transgenic papaya and classically cross-protected papaya. *Phytopathology* 84, 1359–1366.

Tepfer, M. (2002) Risk assessment of virus-resistant transgenic plants. *Annual Review of Phytopathology* 40, 467–491.

Thomas, P.E., Lawson, E.C., Zalewski, J.C., Reed, G.L., and Kaniewski, W.K. (2000) Extreme resistance to potato leafroll virus in potato cv. Russet Burbank mediated by the viral replicase gene. *Virus Research* 71, 49–62.

Tougou, M., Furutani, N., Yamagishi, N., Shizukawa, Y., Takahata, Y., and Hidaka, S. (2006) Development of resistant transgenic soybeans with inverted repeat-coat protein genes of soybean dwarf virus. *Plant Cell Reports* 25, 1213–1218.

Tougou, M., Noriko, Y., Noriyuki, F., Yoshiaki, S., Yoshihito, T., and Soh, H. (2007) Soybean dwarf virus-resistant transgenic soybeans with the sense coat protein gene. *Plant Cell Reports* 26, 1967–1975.

Tourlakis, M.E., Karran, R.A., Desouza, L., Siu, K.W., and Hudak, K.A. (2010) Homodimerization of pokeweed antiviral protein as a mechanism to limit depurination of pokeweed ribosomes. *Molecular Plant Pathology* 11, 757–767.

Trifonova, E.A., Sapotsky, M.V., Komarova, M.L., Scherban, A.B., Shumny, V.K., *et al.* (2007) Protection of transgenic tobacco plants expressing bovine pancreatic ribonuclease against tobacco mosaic virus. *Plant Cell Reports* 26, 1121–1126.

Truve, E., Aaspôllu, A., Honkanen, J., Puska, R., Mehto, M., *et al.* (1993) Transgenic potato plants expressing mammalian 2′-5′ oligoadenylate synthetase are protected from potato virus X infection under field conditions. *Nature Biotechnology* 11, 1048–1052.

Tumer, N.E., O'Connell, K.M., Nelson, R.S., Sanders, P.R., Beachy, R.N., *et al.* (1987) Expression of alfalfa mosaic virus coat protein gene confers cross-protection in transgenic tobacco and tomato plants. *EMBO Journal* 6, 1181–1188.

Vadlamudi, T., Patil, B.L., Kaldis, A., Gopal, D.V.R.S., Mishra, R., Berbati, M., and Voloudakis, A. (2019) DsRNA-mediated protection against two isolates of papaya ringspot virus through topical application of dsRNA in papaya. *Journal of Virological Methods* 275: 113750.

Vandenbussche, F., Peumans, W.J., Desmyter, S., Proost, P., Ciani, M., and van Damme, E.J. (2004) The type-1 and type-2 ribosome-inactivating proteins from iris confer transgenic tobacco plants local but not systemic protection against viruses. *Planta* 220, 211–221.

Vanderschuren, H., Alder, A., Zhang, P., and Gruissem, W. (2009) Dose-dependent RNAi- mediated geminivirus resistance in the tropical root crop cassava. *Plant Molecular Biology* 70, 265–272.

van der Wilk, F., Posthumus-Lutke Willink, D., Huisman, M.J., Huttinga, H., and Goldbach, R. (1991) Expression of the potato leafroll luteovirus coat protein gene in transgenic potato plants inhibits viral infection. *Plant Molecular Biology* 17, 431–439.

van Schie, C.C.N. and Takken, F.L.W. (2014) Susceptibility genes 101: how to be a good host. *Annual Review of Phytopathology* 52, 551–81.

Varanda, C.M., Félix, M.D.R., Campos, M.D., Patanita, M., and Materatski, P. (2021) Plant viruses: from targets to tools for CRISPR. *Viruses* 13: 141.

Vaughan, T.J., Williams, A.J., Pritchard, K., Osbourn, J.K., Pope, A.R., *et al.* (1996) Human antibodies with sub-nanomolar affinities isolated from a large non-immunized phage display library. *Nature Biotechnology* 14, 309–314.

Vazquez Rovere, C., Asurmendi, A., and Hopp, H.E. (2001) Transgenic resistance in potato plants expressing potato leaf roll virus (PLRV) replicase gene sequence is RNA-mediated and suggests the involvement of post-transcriptional gene silencing. *Archives of Virology* 146, 1337–1353.

Villani, M.E., Roggero, P., Bitti, O., Benvenuto, E., and Franconi, R. (2005) Immunomodulation of cucumber mosaic virus infection by intrabodies selected in vitro from a stable single-framework phage display library. *Plant Molecular Biology* 58, 305–316.

Vu, T.V., Choudhury, N.R., and Mukherjee, S.K. (2013) Transgenic tomato plants expressing artificial microRNAs for silencing the pre-coat and coat proteins of a begomovirus, *Tomato leaf curl New Delhi virus*, show tolerance to virus infection. *Virus Research* 172, 35–45.

Vučurović, A., Bulajić, A., Stanković, I., Ristić, D., Berenji, J., *et al.* (2012) Non-persistently aphid-borne viruses infecting pumpkin and squash in Serbia and partial characterization of zucchini yellow mosaic virus isolates. *European Journal of Plant Pathology* 133, 935–947.

Wallace, T.C., Bailey, R.L., Blumberg, J.B., Burton-Freeman, B., Chen, C.O., *et al.* (2020) Fruits, vegetables, and health: a comprehensive narrative, umbrella review of the science and recommendations for enhanced public policy to improve intake. *Critical Reviews in Food Science and Nutrition* 60, 2174–2211.

Wang, M.-B., Abbott, D.C., and Waterhouse, P.M. (2000) A single copy of a virus-derived transgene encoding hairpin RNA gives immunity to barley yellow dwarf virus. *Molecular Plant Pathology* 1, 347–356.

Wang, P., Zoubenko, O., and Tumer, N.E. (1998) Reduced toxicity and broad spectrum resistance to viral and fungal infection in transgenic plants expressing pokeweed antiviral protein II. *Plant Molecular Biology* 38, 957–964.

Wang, X., Eggenberger, A.L., Nutter, F.W., and Hill, J.H. (2001) Pathogen-derived transgenic resistance to soybean mosaic virus in soybean. *Molecular Breeding* 8, 119–127.

Watanabe, Y., Ogawa, T., Takahashi, H., Ishida, I., Takeuchi, Y., *et al* (1995) Resistance against multiple plant viruses in plants mediated by a double stranded-RNA specific ribonuclease. *FEBS Letters* 372, 165–168.

Wesley, S.V., Helliwell, C.A., Smith, N.A., Wang, M.-B., Rouse, D.T., and Liu, Q. (2001) Construct design for efficient, effective and high throughput gene silencing in plants. *Plant Journal* 27, 581–590.

Worrall, E.A., Bravo-Cazar, A., Nilon, A.T., Fletcher, S.J., Robinson, K.E., Carr, J.P., and Mitter, N. (2019) Exogenous application of RNAi-inducing double-stranded RNA inhibits aphid-mediated transmission of a plant virus. *Frontiers in Plant Science* 10: 265.

Wu, Y.F., Chen, Y., Liang, X.M., and Wang, X.Z. (2006) An experimental assessment of the factors influencing agrobacterium-mediated transformation in tomato. *Russian Journal of Plant Physiology* 53, 252–256.

Xuan, N., Zhao, C., Peng, Z., Chen, G.F., Bian, M.L., Liu, G., Wang, X., and Bi, Y. (2015) Development of transgenic maize with anti-rough dwarf virus artificial miRNA vector and their disease resistance. *Chinese Journal of Biotechnology* 31, 1375–1386.

Yadav, J.S., Ogwok, E., Wagaba, H., Patil, B.L., Bagewadi, B., and Gaitin-Solis, E. (2011) RNAi-mediated resistance to cassava brown streak Uganda virus in transgenic cassava. *Molecular Plant Pathology* 12, 677–687.

Yang, X., Niu, L., Zhang, W., He, H., Yang, J., *et al.* (2019) Increased multiple virus resistance in transgenic soybean overexpressing the double-strand RNA-specific ribonuclease gene *PAC1*. *Transgenic Research* 28, 129–140.

Yeam, I., Cavatorta, J.R., Ripoll, D.R., Kang, B.C., and Jahn, M.M. (2007) Functional dissection of naturally occurring amino acid substitutions in eIF4E that confers recessive potyvirus resistance in plants. *Plant Cell* 19, 2913–2928.

Yoon, Y.J., Venkatesh, J., Lee, J.H., Kim, J., Lee, H.E., *et al.* (2020) Genome editing of *eIF4E1* in tomato confers resistance to pepper mottle virus. *Frontiers in Plant Science* 11: 1098.

Yu, Z., Zhao, S., and He, Q. (2007) High level resistance to turnip mosaic virus in chinese cabbage (*Brassica campestris* spp. *pekinensis* (Lour) Olsson) transformed with the antisense NIb gene using marker-free *Agrobacterium tumefaciens* infiltration. *Plant Science* 172, 920–929.

Zhan, X., Zhang, F., Zhong, Z., Chen, R., Wang, Y., *et al.* (2019) Generation of virus-resistant potato plants by RNA genome targeting. *Plant Biotechnology Journal* 17, 1814–1822.

Zhang, H.L., Peng, X.-X., Song, Y.R., Li, T.R., Meng, Q., *et al.* (1997) Resistance to mixed PVY and PLRV infection in potato cultivars expressing dual PVY and PLRV coat protein genes. *Acta Botanica Sinica* 39, 236–240.

Zhang, J.-Q., Li, H.-Y., Hu, X.-N., Shan, Z.-H., and Tang, G.-X. (2013) *Agrobacterium tumefaciens* mediated transformation of RNAi CP gene into soybean (*Glycine max* L.) *Acta Agronomica Sinica* 39, 1594–1601.

Zhang, P., Vanderschuren, H., Fütterer, J., and Gruissemm W. (2005) Resistance to cassava mosaic disease in transgenic cassava expressing antisense RNAs targeting virus replication genes. *Plant Biotechnology Journal* 3, 385–397.

Zhang, T., Zheng, Q., Yi, X., An, H., Zhao, Y., *et al.* (2018) Establishing RNA virus resistance in plants by harnessing CRISPR immune system. *Plant Biotechnology Journal* 16, 1415–1423.

Zhang, X., Li, H., Zhang, J., Zhang, C., Gong, P., *et al.* (2011) Expression of artificial microRNAs in tomato confers efficient and stable virus resistance in a cell-autonomous manner. *Transgenic Research* 20, 569–581.

Zhao, M.-A., An, S.-J., Lee, S.-C., Kim, D.-S., and Kang, B.-C. (2013) Overexpression of a single-chain variable fragment (*scFv*) antibody confers unstable resistance to TuMV in Chinese cabbage. *Plant Molecular Biology Reporter* 31, 1203–1211.

Zhao, Y., Yang, X., Zhou, G., and Zhang, T. (2020) Engineering plant virus resistance: from RNA silencing to genome editing strategies. *Plant Biotechnology Journal* 18, 328–336.

Zhu, F., Ping, Z., Yan, F.M., Fei, X., Da, W.Z., Jian, C., Hong, H.L., and De, H.X. (2013) Alpha-momorcharin, a RIP produced by bitter melon, enhances defense response in tobacco plants against diverse plant viruses and shows antifungal activity in vitro. *Planta* 237, 77–88.

Zhu, F., Peng, X.Z., Fei, X., Yan, P.C., Yi, M.M., and Zhao, L.J. (2020) Alpha-momorcharin enhances *Nicotiana benthamiana* resistance to tobacco mosaic virus infection through modulation of reactive oxygen species. *Molecular Plant Pathology* 21, 1212–1226.

Zrachya, A., Kumar, P.P., Ramakrishnan, U., Levy, Y., Loyter, A., and Arazi, T. (2007) Production of siRNAs targeted against TYLCV coat protein transcripts leads to silencing of its expression and resistance to the virus. *Transgenic Research* 16, 385–398.

# 12 Embryogenic Cell Suspension Culture: A Potential Tool for Crop Improvement

**S. Backiyarani*, R. Karthic, S. Kalpana and S. Uma***
*ICAR-National Research Center for Banana, Trichy, Tamil Nadu, India*

**Abstract**
Breeding crops to increase yield, pest and disease resistance, and nutritional quality is a continuous process to meet the demands of the growing population, as well as to tackle the consequences of climate change on crops. Most breeding approaches are based on exploiting existing variability or selecting the best lines by creating the variability through recombination/mutation. However, these approaches are time consuming and laborious. Moreover, many crops are not amenable to conventional breeding approaches. These problems can be overcome by recent techniques such as genetic transformation and gene editing. Somatic embryogenic cells are the basic requirement for genetic modification through genetic engineering. Developing somatic embryogenesis protocols in all crops is essential not only in developing improved varieties but also for conservation of germplasm, producing secondary metabolites, and planting material production. In this chapter, production of embryogenic cell suspension cultures, maturation and development of plantlets from somatic embryos, and application of this technique in many vegetable crops are described. Use of this technique in crop improvement and need to understand the molecular mechanisms of the somatic embryogenic pathway are also covered.

**Keywords:** genetic modification, mutation, somatic embryogenesis, somatic hybrid, synthetic seed

## 12.1 Introduction

The population is growing exponentially placing demand on food production, but the availability of land sources restricts production. Moreover, it is obvious that global warming will also affect sustainable food production. One of the possible ways of increasing agricultural productivity is the cultivation of climate-resilient, high-yielding varieties that can withstand the impact of climate changes. Variability available in a particular crop species is the primary source for crop improvement programs. To improve the trait of interest in desirable varieties, natural variability can be exploited through conventional breeding approaches using appropriate breeding strategies specific to self-pollinated, cross-pollinated, and clonally propagated crops (Acquaah, 2016). In the absence of the variability for specific traits in a crop species, or wherever it is not amenable to incorporate the traits through sexual reproduction, breeding strategies such as mutation breeding or molecular breeding (genetic engineering) are the best approaches. These approaches are highly cost effective as they facilitate the breeder to improve a specific trait or create novel variants without changing their genetic makeup over a short period of time (Jain

*Email: backiyarani@gmail.com or umabinit@yahoo.co.in


DOI: 10.1079/9781800627123.0012

*et al.*, 1998; Krishna *et al.*, 2016). Culturing of cells and tissue to produce somatic embryos is the best technique for the production of true-to-type progenies, as well as for genetic manipulation without the formation of chimeras. Developing somatic embryogenesis protocols in all crops is imperative not only for mass propagation and improvement of traits but also to unravel the function of unknown genes.

## 12.2 Embryogenic Cell Suspension

### 12.2.1 Somatic embryogenesis

Somatic embryogenesis is the process of development of an embryo from somatic cells. During *in vitro* culture, cells of an explant will differentiate and develop into embryos (Fig. 12.1). Based on the process of embryo formation, somatic

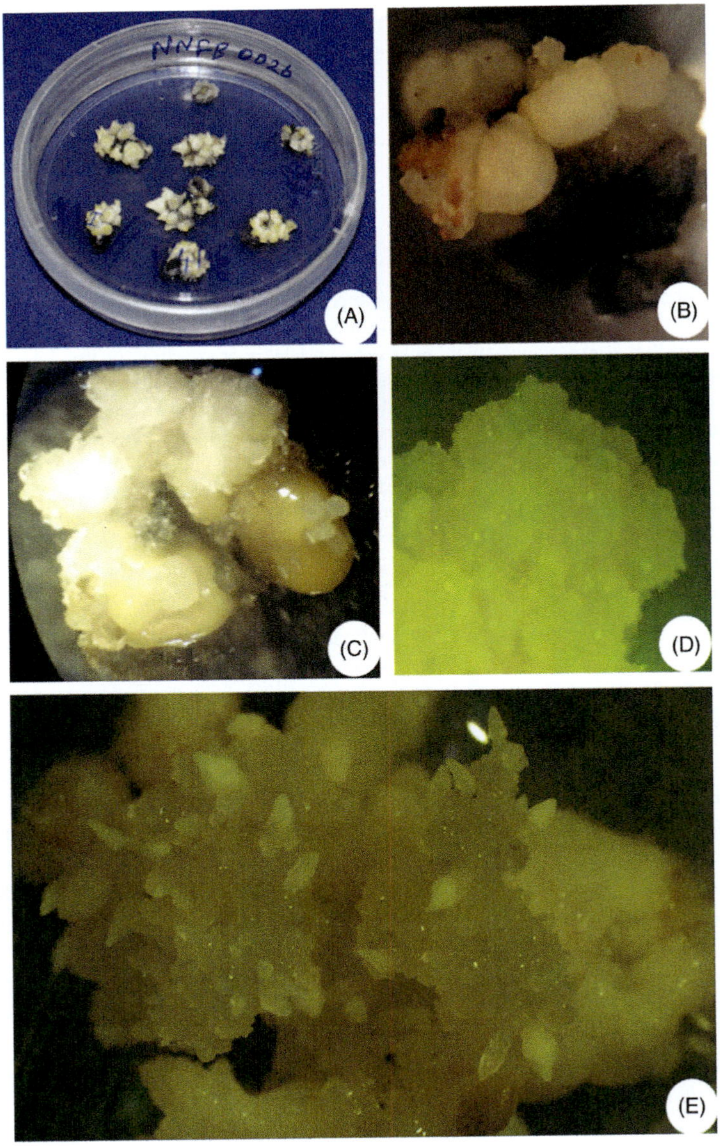


**Fig. 12.1.** Different types of callus during induction of somatic embryogenesis. (a) Explants in somatic embryo induction medium. (b) Non-embryogenic callus. (c) Callus with an embryogenic response. (d) Non-embryogenic friable callus. (e) Embryogenic friable callus.

embryogenesis is classified into two types: (i) formation of somatic embryogenic cells directly from the surface of explants without any dedifferentiation, known as direct somatic embryogenesis; and (ii) formation of somatic embryogenic cells by reiterating numerous cycles of cell divisions, resulting in callus growth, pro-embryogenic mass production, and somatic cell formation, known as indirect somatic embryogenesis. Differentiation of embryogenic callus from non-embryogenic callus is a critical step in the production of embryogenic cells (Fig. 12.1B, C).

### 12.2.2 Production of somatic embryogenic cells

#### *Embryogenic callus initiation*

Callus is dedifferentiated plant tissue that further induces somatic embryogenesis by redifferentiation (Fehér, 2019). The callus has the potential to regenerate somatic embryos known as embryogenic callus. Initially, the explants undergo dedifferentiation (convert to a cell mass) and then redifferentiation (convert to an organ). The pattern of somatic embryogenesis from cultured tissue is defined epigenetically and influenced by the physiological phase of the explant, type of basal medium, auxin concentrations, environmental factor, etc. The specific developmental stage of the explant on a specific growth medium with appropriate environmental conditions results in an embryogenic response. A classic example for indirect somatic embryogenesis is carrot somatic embryogenesis, which showed that embryogenic suspension cells regenerate embryos at a frequency of 85–90% (Fujimura and Komamine, 1979a,b). Later, several reports described somatic embryogenesis in carrot using leaf, stem, seed, and root as explants (Halperin and Jensen, 1967; Nomura and Komamine, 1985; Ammirato, 1986; Kamada *et al.*, 1989; Liu *et al.*, 1994; Shiota *et al.*, 1999; Raghavan, 2006; Ceasar and Ignacimuthu, 2010). The role of auxins and cytokinins in the induction of somatic embryogenesis has been investigated extensively in vegetable crops. Supplementation with the auxins 2,4-dichlorophenoxyacetic acid (2,4-D), 1-naphthaleneacetic acid (NAA) or 4-amino-3,5,6-trichloropicolinic acid (picloram), and with the cytokinins benzyloadenine and thidiazuron have been reported to promote callus proliferation and somatic embryogenesis (Liu *et al.*, 1993; Khan *et al.*, 2006; Kaparakis and Alderson, 2008). In particular, 2,4-D plays an important role in governing the induction of a friable embryogenic callus (Lazzeri *et al.*, 1987; Raghavan, 2004; Thiruvengadam *et al.*, 2006). It has been reported that the auxins (4-chlorophenoxy) acetic acid 2-(dimethylamino)ethyl ester (centrophenoxine) and quinolinecarboxylic acid (quinclorac) induced direct somatic embryogenesis in zygotic embryo explants of sweet pepper (Steinitz *et al.*, 2003).

#### *Establishment of an embryogenic cell suspension*

The establishment of an embryogenic cell suspension (ECS) is a crucial phase requiring friable embryogenic callus with translucent embryos (5–20 mg in 25 ml) and frequent observations and replacement of fresh nutrient medium at regular intervals. The introduction of liquid medium on friable embryogenic callus allows growth as individual cells or tiny clusters. The callus undergoes rapid growth with biochemical changes resulting in either fast multiplication or mortality by browning or blackening due to exudation of phenolics. The success rate of establishing a high-quality ECS depends on the quality and quantity of the selected friable embryogenic callus (Fig. 12.1E). Finer and Nagasawa (1988) reported that initiation and proliferation of an embryogenic suspension culture of soybean were dependent on the inoculum, method of subculture, and composition of the subculture medium. It is most important that replacement of fresh medium occurs at regular intervals and that there is a sufficient aeration by shaking the vessel, so the cultured callus slowly adopts the liquid culture system and grows like microbial cells. ECS cells are larger than those of bacteria or fungi, with diameters of 60–100 μm, and are isodiametric in shape, containing a dense cytoplasm rich in starch content (Fig. 12.2C) and a large nucleus with a single enlarged nucleolus (Carman, 1990). The ECS frequently contains numerous aggregates or clumps made up of a varied number of cells (Fig. 12.2A, B). These aggregates may contain up to 200 cells or more and reach up to 2–3 mm in diameter (Godoy-Hernández and Vázquez-Flota, 2012). The non-embryogenic

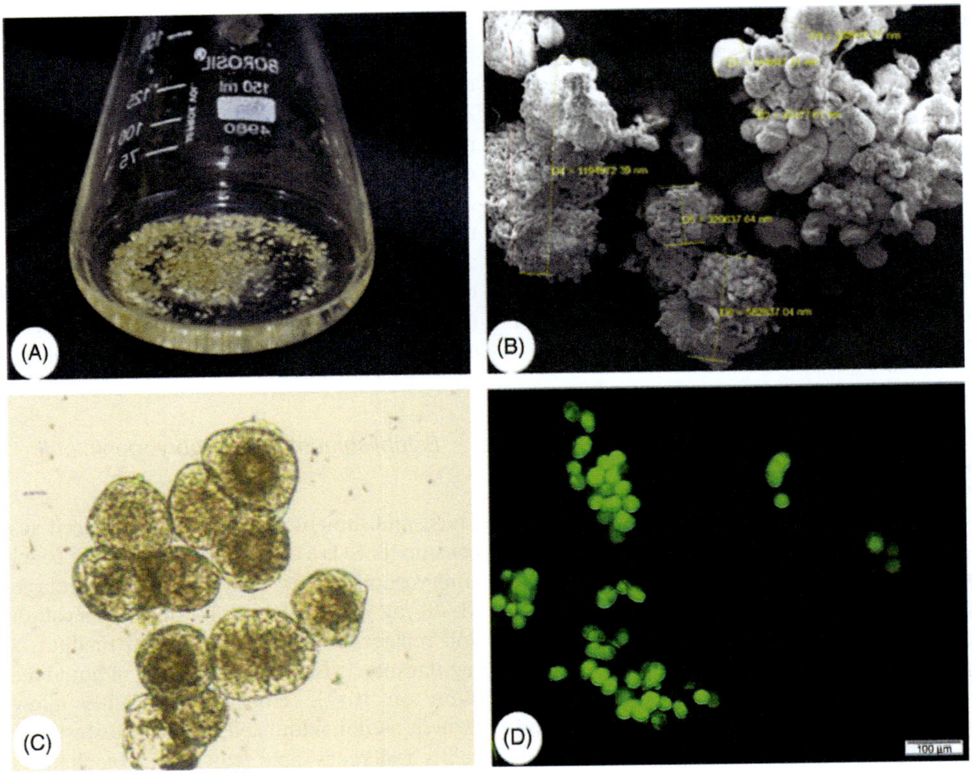


**Fig. 12.2.** Embryogenic cell suspension (ECS) culture. (a) ECS in a flask. (b) Scanning electron microscope image showing aggregated ECS cells. (c) Stereomicroscopic image showing suspension cells rich in starch granules. (d) Fluorescein diacetate-stained suspension cells show live cells.

cells are larger (100–1000 µm) and elongated, with intercellular non-embryogenic cells from the suspension to obtain a highly regenerative, homogenous ECS. Hypocotyl-derived callus of okra introduced to liquid medium formed cell clusters after 3 weeks of culture with a weekly subculture cycle (Ganesan *et al.*, 2007). Bayliss (1977) reported that, in carrot suspension cell cultures, the changes in the time between subcultures is an important regulator of the chromosomal constitution of cells treated with colchicine. Thiruvengadam *et al.* (2006) established a robust protocol for bitter melon ECS where the recovery of plantlets could be completed within 3 months.

### *Growth and quantification*

The ECS grows as clumps or aggregates, sometimes containing large organelles, and assessment of viability of the cells is routinely done by fluorescein diacetete (FDA) staining (Widholm, 1972). The live cells take up the stain and can be observed by fluorescence microscopy (Fig. 12.2D).

Quantitative measurement of growth is an essential component of the ECS culture system. There are several methods to assess the growth kinetics of plant cells such as fresh cell weight, settled cell volume (SCV), packed cell volume, cell optical density, cell size and number, nitrogen content, sugar content, protein content, nucleic acid content, organic matter, mitotic index, electrical conductivity, respiration, and pH measurement (Lazzeri *et al.*, 1987). An ECS is rich in starch content and also settles quickly compared with non-embryogenic cells, and hence measurement of SCV is preferred globally. Assessment of growth index, specific growth rate, and doubling time of the ECS culture system can also be calculated using the following formulas (Godoy-Hernández and Vázquez-Flota, 2012; Chin *et al.*, 2014; Karthic *et al.*, 2017):

$$\text{Growth index}(i) = \frac{\text{Final SCV}}{\text{Initial SCV}}$$

$$\text{Specific growth rate } (\mu_{max})(\text{day}^{-1}) = \frac{\ln(\text{final SCV}) - \ln(\text{initial SCV})}{\text{culture duration in days}}$$

$$\text{Doubling time } (T_D) = \frac{\ln(2)}{\mu_{max}}$$

### *Regeneration of somatic embryos from ECS*

Embryogenic cells are unique, and develop from somatic embryos into plants, a typical example of the totipotency of plant cells (Fig. 12.3). The potential of single cells to develop into somatic embryos is not limited to the appearance of cell morphology. Toonen *et al.* (1994) observed single-suspension cells of carrot developing into somatic embryos using a video-based cell-tracking system. There are a number of factors affecting the development of somatic embryos from a cell suspension such as explant type, growth regulators, nutrient medium, light, temperature, humidity of growth vessel, and oxygen tension (Dantu and Tomar, 2010). A choice of plant material must be made for the explant used to induce somatic embryogenesis; it is generally recommended to use actively growing meristem portions. ECS contains rapidly growing pro-embryogenic cells, which have the potential to regenerate somatic embryos when provided with proper *in vitro* conditions. The regeneration ratio will be influenced by the quality of the ECS, and the pro-embryogenic callus used for establishment of the ECS should be examined carefully before use. Media components and auxins play an important role in the regeneration of somatic embryos. Christianson *et al.* (1983) reported that 2,4-D exposure and withdrawal, and replacement of the nitrogen in a Murashige and Skoog (MS) salts-based medium (Murashige and Skoog, 1962) with 20 mM ammonium citrate produced embryoids that were capable of germinating into plantlets in suspension culture. In some cases, growth regulator-free medium has facilitated the regeneration of somatic embryos. Smith and Krikorian (1989, 1990) reported the release of somatic embryogenic potential from excised zygotic embryos of carrots and maintenance of pro-embryonic cultures in growth regulator-free medium. Suspension cells of *Zea mays* cultured on N6 medium (PCH with modifications) containing 2,4-D formed globular and early scutellar somatic embryos within 2 weeks (Guo and Zhang, 2005). Somatic embryogenesis and subsequent plant regeneration are both significantly influenced by light and dark. As auxins are light sensitive, maintaining auxin concentrations and darkness are necessary for somatic embryo regeneration, but embryo germination and the plant transition to autotropism require light conditions and low auxin levels (Ferreira *et al.*, 2017). Pelkonen and Kauppi (1999) studied the effect of light and auxin source on regeneration of somatic embryos in *Lilium regale* cells; somatic embryogenesis was observed in auxin-free treatments maintained in the dark. Humidity inside the culture vessel and starvation of cultures influences somatic

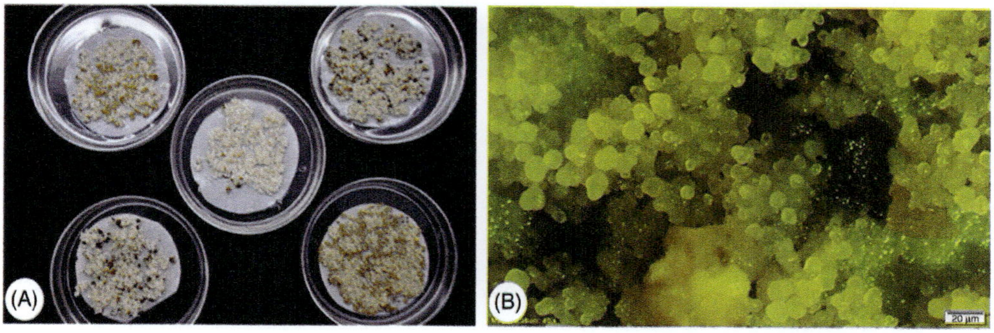


**Fig. 12.3.** Regeneration and maturation of somatic embryos from embryogenic cell suspension. (a) Somatic embryos in plate culture. (b) Matured somatic embryos.

embryogenesis in carrot (Lee *et al.*, 1997). Oxygen tension has been an important factor to promote embryogenic development in cultures. It was reported that by reducing the dissolved oxygen in carrot cultures, somatic embryogenesis could be induced (Shimazu and Kurata, 2003).

### *Germination of plantlets*

Germination is a fundamental process by which plant species grow from a single seed (embryo) into a plant (Fig. 12.4). The process starts with rapid utilization of water to soften the seedcoat by swelling, and optimum humidity and temperature are essential to activate the internal physiological processes and start respiration and synthesis of proteins, and to utilize the stored carbohydrates for further development. Monocots and dicots have different embryo development patterns. Dicots go through three stages, globular, heart-shaped, and torpedo, whereas monocots go through globular, scutellar and coleoptilar stages. The developmental process depends mainly on *in vitro* conditions, namely optimum environmental conditions, nutrients, and growth regulators. Castillo *et al.* (1998) studied germination of plants from encapsulated somatic embryos of papaya (*Carica papaya*) and observed a higher frequency of germination in half-strength MS medium. In another study, MS medium supplemented with 0.4 mg $l^{-1}$ 6-benzylaminopurine (BAP) and 0.02 mg $l^{-1}$ NAA favored germination and well-developed shoots and roots (Anandan *et al.*, 2012). Medium supplemented with 1.0 mg $l^{-1}$ BAP or 6-furfurylaminopurine (kinetin) significantly improved somatic embryo conversion into plantlets in cabbage and cauliflower (Pavlović *et al.*, 2013). In medium enriched with $10^{-6}$ M zeatin, 75% of *Brassica napus* embryos developed into plantlets (Koh and Loh, 2000). Woody plant medium containing 1 μM BAP, 1.5% sucrose, 0.3% activated charcoal, and 0.75% agar favored *Vitis vinifera* transgenic somatic embryos (Scorza

**Fig. 12.4.** Germination of plantlets from somatic embryos. (a) Somatic embryos start germination in plate culture. (b) Development of shoots. (d) Development of shoots and roots. (c) Development of plantlets.

*et al.*, 1995). Fang *et al.* (2022) reported that somatic embryos cultured in DKW medium with 2.0 mg $l^{-1}$ gibberellic acid improved germination percentage. Light is required for shoot regeneration from vegetative explants but does not appear to be required for shoot regeneration from somatic embryos. Darkness improved the somatic embryo germination of various monocots (Nhut *et al.*, 2000; von Aderkas *et al.*, 2015). Somatic embryos of *Curcuma amada* were cultured in darkness on half-strength MS medium containing 1.44 µM gibberellic acid, which was ideal for plantlet development (Chellappan *et al.*, 2014). Half-strength MS medium supplemented with NaCl affected the survival and growth of germinated plantlets under a 16 h photoperiod at a photon flux density of 50–70 µmol $m^{-2}$ $s^{-1}$ from cool white, fluorescent tubes in guava (Rai *et al.*, 2010).

#### *Hardening*

Plantlets developed by *in vitro* techniques must be acclimatized to survive in *ex vitro* conditions. Plantlets developed from somatic embryos transferred to *ex vitro* conditions undergo exposure to abiotic (altered temperature, light intensity, and humidity) and biotic (soil microflora) conditions. The successful *ex vitro* establishment and survival of *in vitro*-grown plantlets depends mainly on the physiological development of functional secondary roots and stomata (Hazarika, 2003). The process depends on environmental factors such as humidity, temperature, and light intensity. Another important factor is the microflora of the potting mixture, as plantlets grown by *in vitro* techniques are sterile, so sudden exposure to soil microflora could infect the plantlets and result in high mortality (Chandra *et al.*, 2010). A high mortality is observed in plantlets that are exposed to *ex vitro* conditions without a proper hardening process, which is due mainly to non-functional stomata, a weak root system, and a poorly developed cuticle of the cultured plants (Mathur *et al.*, 2008). Agnihotri *et al.* (2004) stated that transplanting *Carica papaya* plantlets grown in culture tubes with Soilrite to a potted mixture of garden soil and Soilrite in a 1:1 ratio resulted in 80% transplant success. Khan *et al.* (2006) achieved an 85% survival rate for acclimatization of plants regenerated from somatic embryos of *Capsicum annuum* in Soilrite. Venkataiah *et al.* (2016) acclimatized somatic embryo-derived plantlets of *Capsicum baccatum* with a rise in temperature from 25°C to 28°C and a fall in relative humidity from 80% to 30%, and achieved 68% survival in the field after transplantation. Ritchie *et al.* (1991) reported that a relative humidity of 81% was optimal for acclimatization of sugarbeet plantlets, which showed increased epicuticular wax, improved stomatal functioning, and reduced leaf dehydration.

## 12.3 Application of Somatic Embryogenesis

### 12.3.1 Germplasm conservation

*Ex situ* conservation of genetic resources in field repositories is expensive, as it requires considerable inputs such as land, labor, and the risk of exposure to endemic pests, diseases, and natural disasters. *In vitro* conservation is the best strategy for long-term conservation of germplasm. Conservation of embryogenic cells or somatic embryos of germplasm accessions under ultra-low temperatures (−196± 1°C) in a minimum storage space reduces the cost of maintenance and the number of duplicates to be maintained (Perez-Reyes *et al.*, 1998). These cryo-conserved embryogenic cells/somatic embryos are good sources for genetic manipulation, as they can be immediately retrieved and used as an explant for improving traits through advanced biotechnological tools such as mutation, genetic transformation, and gene editing.

### 12.3.2 Mass propagation of planting material

The production of a large number of somatic embryos from a single explant is easy by scaling up the ECS, which facilitates the mass production of quality planting materials. Mass production of hybrid seed in high-valued vegetable crops (e.g. solanaceous vegetables, cucurbits, root and bulb crops) that exhibit significant heterosis is a highly challenging one (Backiyarani *et al.*, 2021). For commercial hybrid seed production, hybridization has to be done throughout the year, which requires a higher work load for emasculation and pollination, as well as requiring conducive climatic conditions for seed production. Tissue-culture techniques such as the somatic embryogenic system are an efficient alternative for conventional hybrid seed/seedling production (Table 12.1). Schuabb Heringer *et al.* (2013) established a feasible protocol for somatic embryogenesis from immature zygotic embryos for commercial-scale production of plantlets in the papaya hybrid UENF/CALIMAN 01. Mass production of hybrid plants using ECS culture has been demonstrated in cooking banana (Daniels *et al.*, 2002; Gómez Kosky *et al.*, 2002, 2006).

### 12.3.3 Virus-free planting material

Viral diseases are the major constraint in the production of vegetable crops. It has been reported that planting of virus-free propagated material is the best approach to reduce yield loss (Conci *et al.*, 2003). In vegetatively propagated crops in particular, production of virus-free planting material is highly challenging as viral infection is systemic, and its production in an area free from the virus is expensive (Pramesh and Baranwal, 2015). *In vitro* propagation is an alternate way, and requires a larger number of virus-free explants to produce true-to-type virus-free plantlets. Among the *in vitro* techniques, developing plantlets

**Table 12.1.** Examples of somatic embryogenesis from various tissues of vegetable crops.

| Crop | Explant used | Direct/indirect | Reference |
|---|---|---|---|
| Broccoli | Stigma | I | Maciej Zenkteler *et al.* (2006) |
| | Root | D | Yang *et al.* (2010) |
| | Cotyledon and hypocotyl | I | Qin *et al.* (2007) |
| | Stem | D and I | Abdullah *et al.* (2021) |
| *Capsicum* | Immature zygotic embryo | D | Binzel *et al.* (1996) |
| | Cotyledon and leaf explants | D | Venkataiah *et al.* (2016) |
| | Leaf | I | Montcho *et al.* (2021) |
| Carrot | Seed | I | Kiyosue *et al.* (1989) |
| | Leaf | I | Krikorian (1992) |
| | Seed | D | Smith and Krikorian (1988) |
| Garlic | Roots of *in vitro*-sprouted garlic cloves | I | Keles *et al.* (2010) |
| | Basal tissue of garlic clove | D | Sata *et al.* (2000) |
| | Basal plate and receptacle | I | Xue *et al.* (1991) |
| | Shoot tips | I | Novak (1981) |
| | Root tips | | Haque *et al.* (1999) |
| | Shoot tips | | Linsmaier and Skoog (1965) |
| | Root tips | D | Haque *et al.* (1997) |
| | Stem disc | D and I | Ayabe and Sumi (1998) |
| Onion | Zygotic embryo | | Sivanesan *et al.* (2015); Zhang *et al.* (2004) |
| | Flower bud | | Luthar and Bohanec (1999) |
| | Seed | | Saker (1997) |
| | Radicle | | Aswath *et al.* (2006) |
| | Shoot apex | | Ramakrishnan *et al.* (2013) |
| Potato | Nodal cutting | I | Vargas *et al.* (2005); Sharma *et al.* (2007) |
| | Field-grown tuber and microtuber | | Nassar *et al.* (2011) |
| | Unripe zygotic embryos | I | Pretova and Dedicova (1992) |
| | Leaf and stem nodal section | | Iriawati and Aprilianty (2014) |
| | Internode and leaf segments | I | Byarugaba *et al.* (2018) |
| Sweet potato | Bud meristem | | Sonnino and Mini (1992) |
| | Lateral meristems | | Manrique-Trujillo *et al.* (2013) |

through somatic embryogenesis is the best approach, as the somatic embryos are induced from non-vascular tissues and viruses are mostly confined to vascular tissue (Bhat and Rao, 2020). These authors also reported that cyclic somatic embryogenesis can produce large numbers of virus-free plantlets in a short span of time with less expense. Sasi and Bhat (2018) showed that piper yellow mottle virus could be eliminated in India through cyclic somatic embryos, which were pre-treated with ribavirin before induction.

### 12.3.4 Synthetic seed production

Among the types of propagating materials, seed propagation is the best way for cultivation owing to its low cost and convenience for handling and storage. The seeds should be genetically pure, free from pests and diseases and other weed seeds, and physically viable. Production of genetically pure seeds in cross-pollinated crops such as asparagus, spinach, cucumbers, broccoli, cabbage, cauliflower, carrot, beetroot, and okra is highly expensive due to the involvement of a huge work load and inputs (Sudha *et al.*, 2006). Unfavorable climatic conditions also lead to the production of small seeds, seeds without endosperm, and fungus-infected seeds, which results in poor quality seeds (Saiprasad, 2001). As seed production is done in open-field conditions and is specific to season, protecting the seed plots from pests and disease outbreaks is a challenge. Following prolonged storage of the seeds, their germination ability will be reduced, and most vegetable seeds are not used for edible purposes, which are some of the reasons for increased seed costs (Sharma, 2011). Use of synthetic seeds for vegetable cultivation is an alternate approach to overcome the problem of zygotic seed production systems.

Synthetic seed production requires *in vitro* propagules as the explant (somatic embryo, any active meristematic tissue), which mimics the embryo of the seeds. A mixture of nutrients, plant growth regulators, and antifungal compounds act as an endosperm, and the gelling material that encompasses the *in vitro* propagules acts as a seedcoat. The synthetic seeds can be germinated under *in vitro or ex vitro* conditions and can be stored for a longer period without affecting their germination ability (Nongdam, 2016). Production of synthetic seeds using somatic embryos has commercial value as they are genetically true to type and easy to handle, store, and transport (Rai *et al.*, 2009). To reduce the cost of seeds/hybrid seeds of vegetable crops, efforts were made from the 1990s onwards by many researchers to encapsulate embryogenic cells and somatic embryos, and the results of these studies are shown in Table 12.2.

### 12.3.5 Production of therapeutic proteins through cell suspension cultures of vegetable crops

Plant cell suspension culture is the best approach for molecular farming (production of recombinant proteins) (Table 12.3). Therapeutic proteins can be produced through culture of transgene cells, which has the advantages of high efficiency, no requirement for post-translational modifications, and extraction of purified proteins on a large scale in a short timespan.

### 12.3.6 Somatic embryogenesis in crop improvement programs

Using the successful protocols of somatic embryogenesis, breeders have created more variability in vegetable crops through genetic manipulation using different strategies such as somatic hybrid mutation, insertion of genes of interest, knockout of genes, and gene editing.

### 12.3.7 Somatic hybrids

Introgression of desirable genes from related species through sexual hybridization takes a long time as they require extensive backcrossing (Chelelat *et al.*, 1993). *Lycopersicon chmielewskii*, a wild relative of the cultivated tomato (*Solanum lycopersicum*), contains double the concentration of soluble sugar than tomato fruit. Although these two species are sexually compatible, to reduce the breeding cycle Chen *et al.* (2002) tried incorporation of the sucrose accumulator gene into cultivated tomato through somatic hybridization. Cho and Park (2014) also reported that incompatibility in potato breeding programs can

**Table 12.2.** Synthetic seed production in vegetable crops.

| Crop | Cell/embryo | Application | Reference |
|---|---|---|---|
| Asparagus | Somatic embryo | Enhance germination rate | Abbasi *et al.* (2020) |
| Broccoli | Somatic embryo | Production of single and multiple shoots | |
| | Somatic embryo | Enhance germination rate | Al Shamari *et al.* (2015) |
| Carrot | Somatic embryo | Dehydrated capsule, enhance the germination | Liu *et al.* (1992); Patel *et al.* (2000) |
| | Somatic embryo | Storage efficiency | Latif *et al.* (2007); Kitto and Janick (1985) |
| Cauliflower | Somatic embryo | Enhance germination rate | Redenbaugh *et al.* (1987) |
| Celery | Somatic embryo | Enhance germination rate | Onishi *et al.* (1992) |
| Cucumber | Somatic embryo | Enhance germination rate | Tabassum *et al.* (2010); Koochani *et al.* (2020) |
| Eggplant | Somatic embryo | Enhance germination rate | Rao and Singh (1991); Huda and Bari (2007); Huda *et al.* (2007) |
| | Somatic embryo | Enhance the germination percentage | Rao and Singh (1991) |
| Garlic | Somatic embryo | Enhance germination rate | Mehta *et al.* (2013) |
| *Parkia speciosa* | Somatic embryo | Storage and germination | Ummi *et al.* (2011) |
| Pepper | Somatic embryo | | Büyükalaca and Mavituna (1995) |
| Potato | Somatic embryo | Enhance germination rate | Schafer-Menuhr *et al.* (2003) |
| | ECS | Enhance somatic embryo formation | Schafer-Menuhr *et al.* (2003) |
| Sweet pepper | Somatic embryo | | Phonkajornyod *et al.* (2004) |
| Sweet potato | Somatic embryo | Enhance hardiness in somatic embryo | Mukherjee (2002) |
| Sugarbeet | Somatic embryo | Enhance germination rate | Saunders and Tsai (1999) |
| Tomato | Somatic embryo | Enhance germination percentage | Al-Dabagh (2012) |

**Table 12.3.** Production of therapeutic proteins from different crop cell-line cultures.

| Crop (cell-line culture) | Protein | Therapeutic value | Reference |
|---|---|---|---|
| *Solanum lycopersicum* | Human granulocyte–macrophage colony-stimulating factor | Immunosuppressive and immunomodulator | Kwon *et al.* (2003) |
| *Glycine max* | Hepatitis B virus surface antigen | Vaccine against hepatitis B virus | Smith *et al.* (2002) |
| *Ipomoea batatas* | Human lactoferrin | Immunosuppressive and immunomodulator | Min *et al.* (2006) |
| *Daucus carota* | Taliglucerase alfa | Drug for Gaucher disease | Shaaltiel *et al.* (2007) |
| | PEGylated recombinant human acetylcholinesterase (PRX-105) | Biodefense program | Protalix Biotherapeutics (www.protalix.com, accessed 2 July 2024) |
| | $\alpha_1$-Antitrypsin (PRX-107) | Drug for emphysema | Protalix Biotherapeutics (www.protalix.com) |

be overcome through somatic hybridization via protoplast fusion. These research findings showed that somatic embryogenesis protocol is a prerequisite for incorporating the desired traits from wild relatives and incompatible parents in a short period of time.

### 12.3.8 Exploitation of somaclonal variation from somatic embryogenesis

It has been reported that plants regenerated through embryogenesis produce more genetically uniform plants than through organogenesis (Vázquez-Boland *et al.*, 2001), as the level of methylation is less in the initial stages of embryogenesis (Sahijram *et al.*, 2003). However, variation was observed in brinjal plants developed through somatic embryogenesis (Naseer and Mahmood, 2014). Kuznetsova *et al.* (2006) reported that prolonged culture under *in vitro* conditions may lead to an increased percentage of somaclonal variation. It was also reported that the number of subcultures (Sun *et al.*, 2013), inadequate control of cell-cycle *in vitro* conditions (Nwauzoma and Jaja, 2013), or a higher concentration of plant growth regulators or in combination with other growth regulators (d'Amato and Bayliss, 1985) may act as mutagens and induce variation. Hence, indirect organogenesis is also an important means of retrieving genetic variation through somaclones with useful traits for agronomic or industrial use. Stable heritable changes might arise through point mutations, chromosomal aberrations, gene duplications, changes of ploidy level, epigenetic variation, or activation of transposable elements (Seibt *et al.*, 2012).

Somatic cell screening against pathotoxins has enabled the recovery of stable resistant plants. *In vitro* selection can shorten the time for selection of desirable traits with minimal environmental interaction and can complement field selection (Jain, 2001). These positive somaclonal variants of a single trait arise through cell culture of the existing ruling varieties and could be exploited in crop improvement programs (Table 12.4). However, care should be taken to avoid somaclonal

**Table 12.4.** Positive somaclonal variants for improved traits in vegetable crops.

| Crop | Trait | Reference |
|---|---|---|
| Brinjal | Stress tolerance | Ferdausi *et al.* (2009) |
| *Capsicum* | Yellow-fruited var. Bell sweet | Morrison *et al.* (1989) |
| Carrot | Resistance to leaf spot (*Alternaria dauci*) | Dugdale *et al.* (2000) |
| | Resistant to drought | Rabiei *et al.* (2011) |
| Celery | *Fusarium* resistant var. UC-TC | Heath-Pagliuso and Rappaport (1990) |
| | Multiple-resistance against *Spodoptera exigua* and *Fusarium* wilt | Diawara *et al.* (1996) |
| Garlic | Stable higher bulb yield | Vidal *et al.* (1993) |
| | Resistance to *Sclerotium cepivorum* | Zhang *et al.* (2012) |
| Ginger | Tolerant to wilt pathogen (*Fusarium oxysporum* f. sp. *zingiberi* Trujillo) | Bhardwaj *et al.* (2012) |
| Pepper | Early flowering and increase of yield components | Hossain *et al.* (2003) |
| Potato | Resistance to *Alternaria solani* | Matern *et al.* (1978); Veitia-Rodriguez *et al.* (2002) |
| | Resistance to *Verticillium dahlia* | Sebastiani *et al.* (1994) |
| | Resistance to *Phytophthora infestans* | Shepard *et al.* (1980) |
| | Improved tuber morphology, starch content | Thieme and Griess (2005) |
| | Heat tolerance | Das *et al.* (2000) |
| | Resistance to *Streptomyces scabiei* | Veitia-Rodriguez *et al.* (2002) |
| | Superior processing traits | Nassar *et al.* (2011) |
| | High-yielding genotype | Hoque and Morshad (2014) |
| | Increased phytonutrient and antioxidant components | Nassar *et al.* (2014) |
| Sugarbeet | Resistance to Fiji disease virus | Krishnamurthi and Tlaskal (1974) |
| | Improved cane diameter, stalk length, and weight | Lat and Lantin (1976) |
| | Resistance to downy mildew and culmicolous smut diseases | Liu and Chen (1978) |
| | Eyespot disease | Heinz *et al.* (1977) |
| Sweet potato | Tolerant to salinity | Anwar *et al.* (2010) |
| Tomato | Resistance to bacterial canker | van den Bulk and Dons (1993) |

variation while using cell culture for other applications such genetic transformation or gene editing.

### 12.3.9 Induction of mutations in cell-line culture

Several crops showed limited success in pollination and restricted crop improvement against biotic and abiotic stresses due to complications in sexual incompatibility with their wild relatives. Mutation breeding is gaining advantages and allowing us to develop crop varieties that can withstand biotic and abiotic stress, such as diseases and drought, frequently caused by climate change. The Food and Agricultural Organization/iety database now includes over 3402 mutant varieties spanning more than 225 crops and 233 plant species. Among these, there are 94 food vegetables, five fruit vegetables, and three banana and plantain varieties. These mutant varieties offer higher yields and enhanced tolerance to pests, diseases, and environmental stresses (Dixit, 2018; FAO/IAEA, 2022). Mutation breeding is a technique of exposing plant seeds, cuttings, or cell cultures to mutagens (physical, chemical or biological) and growing the mutated material in a sterile medium or open field, which results in the formation of a genetically altered plantlet. After multiple generations of selection for better agronomic qualities, the mutant plants are propagated and tested for agronomic performance (Chaleff, 1983; Nelshoppen and Widholm, 1990). Mutagen treatments have gained international interest because of the advancement of tissue-culture techniques. These techniques have the capacity to manage a high number of totipotent cells and organs, as well as plantlet regeneration, and have increased the possibility of producing desired mutations (Maliga, 1984). To accelerate the breeding process, tissue-culture and molecular techniques are applied (McCallum *et al.*, 2000a,b). The induction of mutations, both *in vivo* and *in vitro*, is viewed with caution as a strategy for producing superior plants, as the alterations cannot be controlled because they occur by chance. The probability of alteration increases when the investigation is conducted *in vitro*, and a large number of individuals can be regenerated in this way. The type of plant material (explant) that is treated with mutagen and the type of mutagen are also significant factors in obtaining successive mutants. In vegetatively propagated crops, actively dividing tissues such as apical or auxiliary buds are used for mutagenic treatment; the different cells in the tissue tend to have different types of mutations, which necessitate a thorough examination for the chimeric nature of mutants. ECS is an outstanding clonal propagation technique to create variability through mutagens/radiation. Because somatic embryos are unicellular in origin, it may help prevent chimeras in the regenerated plants, making ECS an excellent system for mutagenesis compared with other explant methods. This technique enables the treatment of enormous populations of cells, as well as the rapid generation of homozygous plants. Hence, emphasis must be directed to the establishment of ECS culture systems in vegetable crops in order to generate a larger number of mutant variants for vegetable crop improvement.

### 12.3.10 Genetic modification

Molecular breeding using genetic transformation techniques avoids the risk of chimeric plant formation obtained through organogenesis, while using the embryogenic cells for genetic transformation. Therefore, somatic embryogenesis is the best system to develop stable transgenic lines. Ashakiran *et al.* (2011) developed a somatic embryogenesis protocol in tomato from cotyledons and found that the percentage transformation frequency was enhanced by *Agrobacterium*-mediated transformation. Venkataiah *et al.* (2016) demonstrated successful regeneration system from cotyledon and leaf tissues of *Capsicum baccatum* through somatic embryogenesis. Kandasami *et al.* (2016) patented a method for the regeneration of okra (*Abelmoschus esculentus*) through somatic embryogenesis for the development of transgenic plants for virus resistance. *Agrobacterium*-mediated genetic transformation and plant regeneration was carried out using organogenesis and somatic embryogenesis from cotyledon leaves in eggplant (*Solanum melongena* L. cv. 'Kecskeméti Lila') (Fári *et al.*, 1995).

### 12.3.11 Genome editing

Development of transgene-free plants is the only way to reduce the concerns of biosecurity and ethics. Transgene-free mutant lines can be developed using the CRISPR/Cas9 genome-editing technique. Zhang *et al.* (2021) reported that protoplast culture is an efficient explant for CRISPR/Cas9 genome editing as it enhances editing efficiency. Many reports have shown that cell suspension cultures are a good source for the isolation of totipotent protoplasts.

## 12.4 Conclusions

Somatic embryogenesis plays a major role in *in vitro* conservation of germplasm, production of secondary metabolites, large-scale production of planting materials of elite clones, creating new variability for selection of desirable traits, and modification of genes for improving traits of interest. Although somatic embryogenesis systems have been developed in vegetable crops, understanding the pathway responsible for somatic embryogenesis in each vegetable crop is paramount to fine tuning the somatic embryogenesis protocol in recalcitrant varieties. Irrespective of the success of somatic embryogenesis in vegetable crops, enhancing the regeneration efficiency is urgently needed to accelerate research activities in vegetable crops. The use of bioreactors in somatic embryogenesis regeneration is the best approach to hasten the process of somatic embryogenesis. Standardizing protocols for somatic embryogenesis in lesser-known/underutilized vegetable crops will be essential for mass production of planting materials. This will encourage large-scale cultivation of underutilized vegetables that are rich sources of vitamins, minerals, and other health-promoting factors such as high antioxidant activity, and that could play a major role in the diversification of diet leading to more balanced sources of micronutrients. Mass propagation of quality planting materials of vegetable crops with lower prices will indirectly reduce vegetable costs and encourage poorer populations to undertake a balanced diet, which will increase their food security and improve their health. In-depth research on the long-term storage of various stages of somatic embryogenesis is mandatory to reduce the time and cost of planting material, as well as to accelerate molecular breeding to develop resilient varieties to face the future challenges of climate change.

## 12.5 References


Abbasi, F., Majd, A., Farahvash, F., Nejadsattari, T., and Tarinejad, A. (2020) Synthetic seed technology for encapsulation and regrowth of *in vitro*-derived shoot-tips and somatic embryos of *Asparagus officinalis* L. *Nexo Revista Científica* 33, 276–285.

Abdullah, R.M., Salih, S.M., and Al-Nema, Q.S. (2021) Thidiazuron induced plant regeneration via organogenesis and somatic embryogenesis in broccoli *Brassica oleracea* var. *italica*. *Natural Volatiles & Essential Oils Journal* 8, 12658–12673.

Acquaah, G. (2016) Conventional plant breeding principles and techniques. In: Al-Khayri, J.M., Jain, S.M., and Johnson, D.V. (eds.) *Advances in Plant Breeding Strategies: Breeding, Biotechnology and Molecular Tools, Vol. 1*. Springer, Cham Switzerland, pp. 115–158.

Agnihotri, S., Singh, S.K., Jain, M., Sharma, M., Sharma, A.K., and Chaturvedi, H.C. (2004) *In vitro* cloning of female and male carica papaya through tips of shoots and inflorescences. *Indian Journal of Biotechnology* 3, 235–240.

Al-Dabagh, F.M.K. (2012) Artificial seed production by encapsulating somatic embryos and shoot tips of two tomato hybrids. *Iraqi Journal of Agricultural Sciences* 43, 52–64.

Al Shamari, M., Rihan, H.Z., Al Swedi, F., and Fuller, M.P. (2012) The use of somatic embryogenesis in artificial seeds production in cauliflower (*Brassica oleraceae* var. *botrytis*). *Acta Horticulturae* 1099, 399–406.

Ammirato, P.V. (1986) Organizational events during somatic embryogenesis. In: Green, C.E. (ed.) *Plant biology, Vol. 3, Plant Tissue and Cell Culture*. Alan R. Liss, New York, pp. 57–81.

Anandan, R., Sudhakar, D., Balasubramanian, P., and Gutieórrez-Mora, A. (2012) *In vitro* somatic embryogenesis from suspension cultures of *Carica papaya* L. *Scientia Horticulturae* 136, 43–49.

Anwar, A., Kikuchi, A., and Watanabe, K.N. (2010) Assessment of somaclonal variation for salinity tolerance in sweet potato regenerated plants. *African Journal of Biotechnology* 9, 7256–7265.

Ashakiran, K., Chidambareswaran, M., Govindasamy, V., Sivankalyani, V., and Girija, S. (2011) Somatic embryogenesis for *Agrobacterium* mediated transformation of tomato—*Solanum lycopersicum* L. *International Journal of Biotechnology Applications* 3, 72–79.

Aswath, C.R., Mo, S.Y., Kim, D.H., and Park, S.W. (2006) *Agrobacterium* and biolistic transformation of onion using non-antibiotic selection marker phosphomannose isomerase. *Plant Cell Reports* 25, 92–99.

Ayabe, M. and Sumi, S. (1998) Establishment of a novel tissue culture method, stem-disc culture, and its practical application to micropropagation of garlic (*Allium sativum* L.). *Plant Cell Reports* 17, 773–779.

Backiyarani, S., Uma, S., Saranya, S., *et al.* (2021) Multiple shoot induction in zygotic embryos: a strategy for acceleration of banana breeding. *Plant Cell, Tissue and Organ Culture* 147, 339–350.

Bayliss, M.W. (1977) Factors affecting the frequency of tetraploid cells in a predominantly diploid suspension culture of *Daucus carota*. *Protoplasma* 92, 109–115.

Bhardwaj, S.V., Thakur, T., Sharma, R., and Sharma, P. (2012) *In vitro* selection of resistant mutants of ginger (*Zingiber* offici*nale* Rosc.) against wilt pathogen (*Fusarium oxysporum* f. sp. *zingiberi* Trujillo). Plant Disease Research 27, 194–199.

Bhat, A.I. and Rao, G.P. (2020) *Characterization of Plant Viruses*. Humana, New York.

Binzel, M.L., Sankhla, N., Joshi, S., and Sankhla, D. (1996) Induction of direct somatic embryogenesis and plant regeneration in pepper (*Capsicum annuum* L.). *Plant Cell Reports* 15, 536–540.

Büyükalaca, S. and Mavituna, F. (1995) Large scale production of pepper somatic embryos using bioreactors. *Acta Horticulturae* 412, 58–63.

Byarugaba, A.A., Magembe, E., Barekye, A., and Skilton, R. (2018) *In vitro* somatic embryogenesis and regeneration potential of two potato varieties in Uganda. *African Journal of Biotechnology* 17, 515–520.

Carman, J.G. (1990) Embryogenic cells in plant tissue cultures: occurrence and behavior. *In Vitro Cellular & Developmental Biology* 26, 746–753.

Castillo, B., Smith, M.A.L., and Yadava, U.L. (1998) Plant regeneration from encapsulated somatic embryos of *Carica papaya* L. *Plant Cell Reports* 17, 172–176.

Ceasar, S.A. and Ignacimuthu, S. (2010) Effects of cytokinins, carbohydrates and amino acids on induction and maturation of somatic embryos in kodo millet (*Paspalum scorbiculatum* Linn.). *Plant Cell, Tissue and Organ Culture* 102, 153–162.

Chaleff, R.S. (1983) Isolation of agronomically useful mutants from plant cell cultures. *Science* 219, 676–682.

Chandra, S., Bandopadhyay, R., Kumar, V., and Chandra, R. (2010) Acclimatization of tissue cultured plantlets: from laboratory to land. *Biotechnology Letters* 32, 1199–1205.

Chelelat, R., Canady, M., Foolad, M., Francis, D., Heisey, R., *et al.* (1993) *Minutes from Tomato CGC meeting on September 15, 2014 at the Tomato Breeders Roundtable in Asheville, NC*. Available at: www.ars-grin.gov/npgs/cgc_reports/tomato2014min.pdf (accessed 2 July 2024).

Chellappan, S.R., Aslam, A., Kathiravan, K., Palani, P., and Shajahan, A. (2014) Direct somatic embryogenesis and plant regeneration from leaf sheath explants of mango ginger (*Curcuma amada* Roxb.). *In Vitro Cellular and Developmental Biology—Plant* 50, 752–759.

Chen, L., Murai, K., Inoue, M., Kaneko, Y., Sato, Y., and Adachi, T. (2002) Somatic hybrids between *Lycopersicon esculentum* and *Lycopersicon chmielewskii*. *Plant Biotechnology* 19, 389–396.

Chin, W.Y.W., Annuar, M.S.M., Tan, B.C., and Khalid, N. (2014) Evaluation of a laboratory scale conventional shake flask and a bioreactor on cell growth and regeneration of banana cell suspension cultures. *Scientia Horticulturae* 172, 39–46.

Cho, K.S. and Park, T.H. (2014) Potato breeding via protoplast fusion. *Journal of Plant Biotechnology* 41, 65–72.

Christianson, M.L., Warnick, D.A., and Carlson, P.S. (1983) A morphogenetically competent soybean suspension culture. *Science* 222, 632–634.

Conci, V.C., Canavelli, A., Lunello, P., Di Rienzo, J., Nome, S.F., *et al.* (2003) Yield losses associated with virus-infected garlic plants during five successive years. *Plant Disease* 87, 1411–1415.

d'Amato, F. and Bayliss, M.W. (1985) Cytogenetics of plant cell and tissue cultures and their regenerates. *Critical Reviews in Plant Sciences* 3, 73–112.

Daniels, D., Gómez Kosky, R., and Reyes Vega, M. (2002) Plant regeneration system via somatic embryogenesis in the hybrid cultivar FHIA-21 (*Musa* sp. AAAB group). *In Vitro Cellular & Developmental Biology—Plant* 38, 330–333.

Dantu, P.K. and Tomar, U.K. (2010) Somatic embryogenesis. In: Tripathi, G. (ed.) *Cellular and Biochemical Science*, 1st edn. I.K. International House, New Delhi, pp. 892–908.

Das, A., Gosal, S.S., Sidhu, J.S., and Dhaliwal, H.S. (2000) Induction of mutations for heat tolerance in potato by *using in vitro* culture and radiation. *Euphytica* 114, 205–209.

Diawara, M.M., Trumble, J.T., Lacy, M.L., White, K.K., and Carson, W.G. (1996) Potential of somaclonal celeries for use in integrated pest management. *Journal of Economic Entomology* 89, 218–223.

Dixit, A. (2018) *IAEA Fact sheet Food and Agriculture, Mutation Breeding for Crop Improvement*. International Atomic Energy Agency, Vienna.

Dugdale, L.J., Mortimer, A.M., Isaac, S., and Collin, H.A. (2000) Disease response of carrot and carrot somaclones to *Alternaria dauci*. *Plant Pathology* 49, 57–67.

Fang, H., Dong, Y., Zhou, R., Wang, Q., Duan, Q., *et al.* (2022) Optimization of the induction, germination, and plant regeneration system for somatic embryos in apomictic walnut (*Juglans regia* L.). *Plant Cell, Tissue and Organ Culture* 150, 289–297.

FAO/IAEA (Food and Agriculture Organization of the United Nations/International Atomic Energy Agency) (2022) Mutant Variety Database. FAO/IAEA. Available at: https://nucleus.iaea.org/sites/mvd/SitePages/Search.aspx (accessed 2 July 2024).

Fári, M., Nagy, I., Csányi, M., Mitykó, J., and Andrásfalvy, A. (1995) *Agrobacterium*-mediated genetic transformation and plant regeneration via organogenesis and somatic embryogenesis from cotyledon leaves in eggplant (*Solanum melongena* L. cv. 'Kecskeméti lila'). *Plant Cell Reports* 15, 82–66.

Fehér, A. (2019) Callus, dedifferentiation, totipotency, somatic embryogenesis: what these terms mean in the era of molecular plant biology? *Frontiers in Plant Science* 10: 536.

Ferdausi, A., Nath, U.K., Das, B.L., and Alam, M.S. (2009) *In vitro* regeneration system in brinjal (*Solanum melongena* L.) for stress tolerant somaclone selection. *Journal of the Bangladesh Agricultural University* 7, 253–258.

Ferreira, L.T., de Araújo Silva, M.M., Ulisses, C., Camara, T.R., and Willadino, L. (2017) Using LED lighting in somatic embryogenesis and micropropagation of an elite sugarcane variety and its effect on redox metabolism during acclimatization. *Plant Cell, Tissue and Organ Culture* 128, 211–221.

Finer, J.J., and Nagasawa, A. (1988) Development of an embryogenic suspension culture of soybean (*Glycine max* Merrill.). *Plant Cell, Tissue and Organ Culture* 15, 125–136.

Fujimura, T. and Komamine, A. (1979a) Involvement of endogenous auxin in somatic embryogenesis in a carrot cell suspension culture. *Zeitschrift für Pflanzenphysiologie* 95, 13–19.

Fujimura, T. and Komamine, A. (1979b) Synchronization of somatic embryogenesis in a carrot cell suspension culture. *Plant Physiology* 64, 162–164.

Ganesan, M., Chandrasekar, R., Ranjitha Kumari, B.D., and Jayabalan, N. (2007) Somatic embryogenesis and plant regeneration of *Abelmoschus esculentus* through suspension culture. *Biologia Plantarum* 51, 414–420.

Godoy-Hernández, G. and Vázquez-Flota, F.A. (2012) Growth measurements: estimation of cell division and cell expansion. *Methods in Molecular Biology* 877, 41–48.

Gómez Kosky, R., de Feria Silva, M., Posada Pérez, L., Gilliard, T., Martinez, F.B., *et al.* (2002) Somatic embryogenesis of the banana hybrid cultivar FHIA-18 (AAAB) in liquid medium and scaled-up in a bioreactor. *Plant Cell, Tissue and Organ Culture* 68, 21–26.

Gómez Kosky, R., Barranco, L.A., Pérez, B.C., Daniels, D., Vega, M., and de Feria Silva, M. (2006) Trueness-to-type and yield components of the banana hybrid cultivar FHIA-18 plants regenerated via somatic embryogenesis in a bioreactor. *Euphytica* 150, 63–68.

Guo, Y. and Zhang, Z. (2005) Establishment and plant regeneration of somatic embryogenic cell suspension cultures of the *Zingiber officinale* Rosc. *Scientia Horticulturae* 107, 90–96.

Halperin, W. and Jensen, W.A. (1967) Ultrastructural changes during growth and embryogenesis in carrot cell cultures. *Journal of Ultrastructure Research* 18, 428–443.

Haque, M.S., Wada, T., and Hattori, K. (1997) High frequency shoot regeneration and plantlet formation from root tip of garlic. *Plant Cell, Tissue and Organ Culture* 50, 83–89.

Haque, M.S., Wada, T., and Hattori, K. (1999) Anatomical changes during *in vitro* direct formation of shoot bud from root tips in garlic (*Allium sativum* L.). *Plant Production Science* 2, 146–153.

Hazarika, B.N. (2003) Acclimatization of tissue-cultured plants. *Current Science* 85, 2–4.

Heath-Pagliuso, S., and Rappaport, L. (1990) Somaclonal variant UC-T3: the expression of *Fusarium* wilt resistance in progeny arrays of celery, *Apium graveolens* L. *Theoretical and Applied Genetics* 80, 390–394.

Heinz, D.J., Krishnamurthy, M., Nickell, L.G., and Maretzki, A. (1977) Cell, tissue and organ culture in surgarcane improvement. In: Reinert, J. and Bajaj, Y.P.S. (eds.) *Applied and Fundamental Aspects of Plant Cell, Tissue and Organ Culture*. Springer, Berlin/Heidelberg/New York, pp. 3–17.

Hoque, M.E. and Morshad, M.N. (2014) Somaclonal variation in potato (*Solanum tuberosum* L.) using chemical mutagens. *Agriculturists* 12, 15–25.

Hossain, M.A., Konisho, K., Minami, M., and Nemoto, K. (2003) Somaclonal variation of regenerated plants in chili pepper (*Capsicum annuum* L.) *Euphytica* 130, 233–239.

Huda, A.K.M.N. and Bari, M.A. (2007) Production of synthetic seed by encapsulating asexual embryo in eggplant (*Solanum melongena* L.). *International Journal of Agricultural Research* 2, 832–837.

Huda, A.K.M.N., Bari, M.A., Rahman, M., and Nahar, N. (2007) Somatic embryogenesis in two varieties of eggplant (*Solanum melongena* L.). *Research Journal of Botany* 2, 195–201.

Iriawati, I. and Aprilianty, F. (2014) Induction of somatic embryos from leaf and stem nodal section explants of potato (*Solanum tuberosum* L.). *Journal of Mathematical and Fundamental Sciences* 46, 301–312.

Jani, S.M. (2001) Tissue culture-derived variation in crop improvement. *Euphytica* 118: 153166.

Jain, S.M., Ahloowalia, B.S., and Veilleux, R.E. (1998) Somaclonal variation in crop improvement. In: Jain, S.M., Brar, D.S., and Ahloowalia, B.S. (eds.) *Somaclonal Variation and Induced Mutations in Crop Improvement*. Current Plant Science and Biotechnology in Agriculture, Vol. 32. Springer, Dordrecht, Netherlands, pp. 203–218.

Kamada, H., Kobayashi, K., Kiyosue, T., and Harada, H. (1989) Stress induced somatic embryogenesis in carrot and its application to synthetic seed production. *In Vitro Cellular & Developmental Biology* 25, 1163–1166.

Kandasami, P., Ravindran, S., Perumal, M., Elangovan, M., Rajendran, P.M., *et al.* (2016) Regeneration and genetic transformation of okra through somatic embryogenesis. Patent. International Publication No. WO2016189543A1.

Kaparakis, G. and Alderson, P.G. (2008) Role for cytokinins in somatic embryogenesis of pepper (*Capsicum annuum* L.)? *Journal of Plant Growth Regulation* 27, 110–114.

Karthic, R., Vel, K., Uma, S., and Suthanthiram, B. (2017) Growth characteristics of embryogenic cell suspension of banana (*Musa* spp.). In: *International Symposium on Horticulture: Priorities & Emerging Trends, 5–8 September, 2017.* Society for Promotion of Horticulture, Bengaluru, India, p. 202.

Keles, D., Taskın, H., Baktemur, G., Yucel, N.K., and Buyukalaca, S. (2010) Somatic embryogenesis in garlic (*Allium sativum* L.). In: *XXVIII International Horticultural Congress on Science and Horticulture for People (IHC2010): International Symposium on Micro 923*, pp. 39–45.

Khan, H., Siddique, I., and Anis, M. (2006) Thidiazuron induced somatic embryogenesis and plant regeneration in *Capsicum annuum*. *Biologia Plantarum* 50, 789–792.

Kitto, S.L. and Janick, J. (1985) Production of synthetic seeds by encapsulating asexual embryos of carrot. *Journal of the American Society for Horticultural Science* 110, 277–282.

Kiyosue, T., Kamada, H., and Harada, H. (1989) Induction of somatic embryogenesis from carrot seeds by hypochlorite treatment. *Plant Tissue Culture Letters* 6, 138–143.

Koh, W.L. and Loh, C.S. (2000) Direct somatic embryogenesis, plant regeneration and *in vitro* flowering in rapid-cycling *Brassica napus*. *Plant Cell Reports* 19, 1177–1183.

Koochani, M., Majd, A., Arbabian, S., Ghanati, F., and Jafari Marandi, S. (2020). A comparative study on the effects of ultrasound and some growth factors on somatic embryogenesis and artificial seed production in cucumber (*Cucumis sativus* L.). *Notulae Botanicae Horti Agrobotanici Cluj-Napoca* 48, 1915–1928.

Krikorian, A.D., Levine, H.G., Kann, R.P., and O'Connor, S.A. (1992) Effects of spaceflight on growth and cell division in higher plants. *Advances in Space Biology and Medicine* 2, 181–209.

Krishna, H., Alizadeh, M., Singh, D., Singh, U., Chauhan, N., *et al.* (2016) Somaclonal variations and their applications in horticultural crops improvement. *3 Biotech* 6: 54.

Krishnamurthi, M. and Tlaskal, J. (1974) Fiji disease resistant *Saccharum officinarum* var. 'Pindar' sub clones from tissue cultures. *Proceedings of the Society of Sugarcane Technology* 15:130–137.

Kuznetsova, O.I., Ash, O.A., and Gostimsky, S.A. (2006) The effect of the duration of callus culture on the accumulation of genetic alterations in pea *Pisum sativum* L. *Russian Journal of Genetics* 42, 555–562.

Kwon, T.-H., Kim, Y.-S., Lee, J.-H., and Yang, M.-S. (2003) Production and secretion of biologically active human granulocyte-macrophage colony stimulating factor in transgenic tomato suspension cultures. *Biotechnology Letters* 25, 1571–1574.

Lat, J.B. and Lantin, M.M. (1976) Agronomic performance of sugarcane clones derived from callus culture. *Philippine Journal of Crop Science* 1, 117–123.

Latif, Z., Nasir, I.A., and Riazuddin, S. (2007) Indigenous production of synthetic seeds in *Daucus carota*. *Pakistan Journal of Botany* 39, 849–855.

Lazzeri, P.A., Hildebrand, D.F., and Collins, G.B. (1987) Soybean somatic embryogenesis: effects of hormones and culture manipulations. *Plant Cell, Tissue and Organ Culture* 10, 197–208.

Lee, E.K., Cho, D.Y., and Soh, W.Y. (1997) Effect of humidity on somatic embryogenesis in cotyledon explant culture of carrot. *Journal of Plant Biology* 40, 89–94.
Linsmaier, E.M. and Skoog, F. (1965) Organic growth factor requirements of tobacco tissue culture. *Physiologia Plantarum* 18, 100–127.
Liu, C.M., Xu, Z.H., and Chua, N.H. (1993) Auxin polar transport is essential for the establishment of bilateral symmetry during early plant embryogenesis. *Plant Cell* 5, 621–630.
Liu, J.R., Jeon, J.H., Yang, S.G., Lee, H.S., Song, N.H., and Jeong, W.J. (1992) Dry type of carrot (*Daucus carota* L.) artificial seeds. *Scientia Horticulturae* 51, 1–11.
Liu, M.C., and Chen, W.H. (1978) Tissue and cell culture as aids to sugarcane breeding II. Performance and yield potential of callus derived lines. *Euphytica* 27, 273–282.
Liu, Z.B., Ulmasov, T., Shi, X., Hagen, G., and Guilfoyle, T.J. (1994) Soybean GH3 promoter contains multiple auxin-inducible elements. *Plant Cell* 6, 645–657.
Luthar, Z., and Bohanec, B. (1999) Induction of direct somatic organogenesis in onion (*Allium cepa* L.) using a two-step flower or ovary culture. *Plant Cell Reports* 18, 797–802.
Maliga, P. (1984) Isolation and characterization of mutants in plant cell culture. *Annual Review of Plant Physiology* 35, 519–561.
Manrique-Trujillo, S., Díaz, D., Reaño, R., Ghislain, M., and Kreuze, J. (2013) Sweetpotato plant regeneration via an improved somatic embryogenesis protocol. *Scientia Horticulturae* 161, 95–100.
Mathur, A., Mathur, A.K., Verma, P., Yadav, S., Gupta, M.L., and Darokar, M.P. (2008) Biological hardening and genetic fidelity testing of micro-cloned progeny of *Chlorophytum borivilianum* Sant. et Fernand. *African Journal of Biotechnology* 7, 1046–1053.
Matern, U., Strobel, G. and Shepard, J. 1978. Reaction to phytotoxins in a potato population derived from mesophyll protoplasts. *Proceedings of the National Academy of Sciences USA* 75, 4935–4939.
McCallum, C.M., Comai, L., Greene, E.A., and Henikoff, S. (2000a) Targeting induced local lesions IN genomes (TILLING) for plant functional genomics. *Plant Physiology* 123, 439–442.
McCallum, C.M., Comai, L., Greene, E.A., and Henikoff, S. (2000b) Targeted screening for induced mutations. *Nature Biotechnology* 18, 455–457.
Mehta, J., Sharma, A., Sharma, N., Megwal, S., Sharma, G., *et al.* (2013) An improved method for callus culture and *in vitro* propagation of garlic (*Allium sativum* L.) *International Journal of Pure and Applied Bioscience* 1, 1–6.
Min, S.R., Woo, J.W., Jeong, W.J., Han, S.K., Lee, Y.B., and Liu, J.R. (2006) Production of human lactoferrin in transgenic cell suspension cultures of sweet potato. *Biologia Plantarum* 50, 131–134.
Montcho, D., Idohou, R., Hegbe, C., Hodehou, T., Atou, R., *et al.* (2021) Selection of drought tolerant mutant of *Amaranthus cruentus* L. in green house. *International Journal of Plant Breeding and Crop Science* 8, 1009–1015.
Morrison, R.A., Loh, W.H.T., Green, S.K., Griggs, T.D., and McLean, B.T. (1989) Tissue culture of tomato and pepper: new tools for plant breeding. Tomato and pepper production in the tropics. In: *Proceedings of international symposium on integrated management practices, Tainen, Taiwan*. AVRDC, Shanhua, Taiwan, pp 44–50.
Mukherjee, A. (2002) Effect of NaCl on *in vitro* propagation of sweet potato (*Ipomoea batatas* L.). *Applied Biochemisty and Biotechnology* 102, 431–441.
Murashige, T., and Skoog, F. (1962) A revised medium for rapid growth and bio assays with tobacco tissue cultures. *Physiologia Plantarum* 15, 473–497.
Naseer, S. and Mahmood, T. (2014) Tissue culture and genetic analysis of somaclonal variations of *Solanum melongena* L. cv. Nirrala. *Open Life Sciences* 9, 1182–1195.
Nassar, A.M., Abdulnour, J., Leclerc, Y., Li, X.Q., and Donnelly, D.J. (2011) Intraclonal selection for improved processing of NB 'Russet Burbank' potato. *American Journal of Potato Research* 88, 387–397.
Nassar, A.M., Kubow, S., Leclerc, Y.N., and Donnelly, D.J. (2014) Somatic mining for phytonutrient improvement of 'Russet Burbank' potato. *American Journal of Potato Research* 91, 89–100.
Nelshoppen, J.M. and Widholm, J.M. (1990) Mutagenesis techniques in plant tissue cultures. *Plant Cell and Tissue Culture* 6, 413–430.
Nhut, D.T., van Le, B., and Thanh, K.T. (2000) Somatic embryogenesis and direct shoot regeneration of rice (*Oryza sativa* L.) using thin cell layer culture of apical meristematic tissue. *Journal of Plant Physiology* 157, 559–565.
Nomura, K. and Komamine, A. (1985) Identification and isolation of single cells that produce somatic embryos at a high frequency in a carrot suspension culture. *Plant Physiology* 79, 988–991.

Nongdam, P. (2016) Development of synthetic seed technology in plants and its applications: a review. *International Journal of Current Science* 19, E86–E101.
Novak, F.J. (1981) Chromosomal characteristics of long-term callus cultures of *Allium sativum* L. *Cytologia* 46, 371–379.
Nwauzoma, A.B. and Jaja, E.T. (2013) A review of somaclonal variation in plantain (*Musa* spp): mechanisms and applications. *Journal of Applied Biosciences* 67, 5252–5260.
Onishi, N., Mashito, T., and Okamoto, A. (1992) Cultural system producing encapsulatable units of synthetic seeds in celery. *Acta Horticulturae* 319, 113–118.
Patel, A.V., Pusch, I., Mix-Wagner, G., and Vorlop, K.D. (2000) A novel encapsulation technique for the production of artificial seeds. *Plant Cell Reports* 19, 868–874.
Pavlović, S., Vinterhalter, B., Zdravković-Korać, S., Vinterhalter, D., Zdrackovic, J., *et al.* (2013) Recurrent somatic embryogenesis and plant regeneration from immature zygotic embryos of cabbage (*Brassica oleracea* var. *capitata*) and cauliflower (*Brassica oleracea* var. *botrytis*). *Plant Cell, Tissue and Organ Culture* 113, 397–406.
Pelkonen, V.P. and Kauppi, A. (1999) The effect of light and auxins on the regeneration of lily (*Lilium regale* Wil.) cells by somatic embryogenesis and organogenesis. *International Journal of Plant Sciences* 160, 483–490.
Perez-Reyes, E., Cribbs, L.L., Daud, A., Lacerda, A.E., Barclay, J., *et al.* (1998) Molecular characterization of a neuronal low-voltage-activated T-type calcium channel. *Nature* 391, 896–900.
Phonkajornyod, P., Pawelzik, E., and Vearasilp, S. (2004) Dry synthetic seed production and desiccation tolerance induction in somatic embryo of sweet pepper. *Deutscher Tropentag* 5–7 October, Berlin.
Pramesh, D. and Baranwal, V.K. (2015) Production of virus-free garlic (*Allium sativum* L.) through meristem tip culture after solar or hot air treatment of cloves. *Journal of Horticultural Science and Biotechnology* 90, 180–186.
Pretova, A. and Dedicova, B. (1992) Somatic embryogenesis in *Solanum tuberosum* L. cv. Desiree from unripe zygotic embryos. *Journal of Plant Physiology* 139, 539–542.
Qin, Y., Li, H.L., and Guo, Y.D. (2007) High-frequency embryogenesis, regeneration of broccoli (*Brassica oleracea* var. *italica*) and analysis of genetic stability by RAPD. *Scientia Horticulturae* 111, 203–208.
Rabiei, K., Khodambashi, N., and Omidi, M. (2011) Use of somaclonal variation on improvement of drought resistant lines of carrot (*Daucus carota* L.). *Journal of Horticultural Sciences* 25, 156–169.
Raghavan, V. (2004) Role of 2,4-dichlorophenoxyacetic acid (2,4-D) in somatic embryogenesis on cultured zygotic embryos of *Arabidopsis*: cell expansion, cell cycling, and morphogenesis during continuous exposure of embryos to 2,4-D. *American Journal of Botany* 91, 1743–1756.
Raghavan, V. (2006) Can carrot and *Arabidopsis* serve as model systems to study the molecular biology of somatic embryogenesis? *Current Science* 1336–1343.
Rai, M.K., Asthana, P., Singh, S.K., Jaiswal, V.S., and Jaiswal, U. (2009) The encapsulation technology in fruit plants—a review. *Biotechnology Advances* 27, 671–679.
Rai, M.K., Jaiswal, V.S., and Jaiswal, U. (2010) Regeneration of plantlets of guava (*Psidium guajava* L.) from somatic embryos developed under salt-stress condition. *Acta Physiologiae Plantarum* 32, 1055–1062.
Rao, P.V.L. and Singh, B. (1991) Plantlet regeneration from encapsulated somatic embryos of hybrid *Solanum melongena* L. *Plant Cell Reports* 10, 7–11.
Ramakrishnan, M., Ceasar, S.A., Duraipandiyan, V., Daniel, M.A., and Ignacimuthu, S. (2013) Efficacious somatic embryogenesis and fertile plant recovery from shoot apex explants of onion (*Allium cepa* L.). *In Vitro Cellular & Developmental Biology—Plant* 49, 285–293.
Redenbaugh, K., David, S., Viss, P., and Fujii, J.A. (1987) Encapsulation of somatic embryos in synthetic seed coats. *HortScience* 22, 803–809.
Ritchie, G., Short, K.C., and Davey, M.R. (1991) *In vitro* acclimatization of chrysanthemum and sugar beet plantlets by treatment with paclobutrazol and exposure to reduced humidity. *Journal of Experimental Botany* 42, 1557–1563.
Sahijram, L., Soneji, J.R., and Bollamma, K.T. (2003) Analyzing somaclonal variation in micropropagated bananas (*Musa* spp.). *In Vitro Cellular & Developmental Biology—Plant* 39, 551–556.
Saiprasad, G.V.S. (2001) Artificial seeds and their applications. *Resonance* 6, 39–47.
Saker, M.M. (1997) *In vitro* regeneration of onion through repetitive somatic embryogenesis. *Biologia Plantarum* 40, 499–506.
Sasi, S. and Bhat, A.I. (2018) *In vitro* elimination of *Piper yellow mottle virus* from infected black pepper through somatic embryogenesis and meristem-tip culture. *Crop Protection* 103, 39–45.

Sata, S.J., Bagatharia, S.B., and Thaker, V.S. (2000) Induction of direct somatic embryogenesis in garlic (*Allium sativum*). *Methods in Cell Science* 22, 299–304.

Saunders, J.W. and Tsai, C.J. (1999) Production of somatic embryos and shoots from sugarbeet callus: effects of abscisic acid, other growth regulators, nitrogen source, sucrose concentration and genotype. *In Vitro Cellular and Developmental Biology—Plant* 35, 18–24.

Schafer-Menuhr, A., Mix-Wagner, G., and Vorlop, K.D. (2003) Regeneration of plants from cell suspensions cultures and encapsulated cell suspension cultures of *Solanum tuberosum* L. Clarissa. *Landbauforsch Volk* 53, 53–59.

Schuabb Heringer, A., Moura Vale, E., Barroso, T., Santa-Catarina, C., and Silveira, V. (2013) Polyethylene glycol effects on somatic embryogenesis of papaya hybrid UENF/CALIMAN 01 seeds. *Theoretical and Experimental Plant Physiology* 25, 116–124.

Scorza, R., Cordts, J.M., Ramming, D.W., and Emershad, R.L. (1995) Transformation of grape (*Vitis vinifera* L.) zygotic-derived somatic embryos and regeneration of transgenic plants. *Plant Cell Reports* 14, 589–592.

Sebastiani, L., Lenzi, A., Pugliesi, C., and Fambrini, M (1994) Somaclonal variation for resistance to *Verticillium dahliae* in potato (*Solanum tuberosum* L.) plants regenerated from callus. *Euphytica* 80, 5–11.

Seibt, K.M., Wenke, T., Wollrab, C., Junghans, H., Muders, K., *et al.* (2012) Development and application of SINE-based markers for genotyping of potato varieties. *Theoretical and Applied Genetics* 125, 185–196.

Shaaltiel, Y., Bartfeld, D., Hashmueli, S., Baum, G., Brill-Almon, E., *et al.* (2007) Production of glucocerebrosidase with terminal mannose glycans for enzyme replacement therapy of Gaucher's disease using a plant cell system. *Plant Biotechnology Journal* 5, 579–590.

Sharma, J.P. (2011) *Quality Seed Production of Vegetable Crops Technological Interventions, Vol.* II. Kalyani Publishers, New Delhi.

Sharma, S.K., Bryan, G.J., and Millam, S. (2007) Auxin pulse treatment holds the potential to enhance efficiency and practicability of somatic embryogenesis in potato. *Plant Cell Reports* 26, 945–950.

Shepard, J.F., Bidney, D., and Shahin, E. (1980) Potato protoplasts in crop improvement. *Science* 208, 17–24.

Shimazu, T. and Kurata, K. (2003) Dynamic dissolved oxygen concentration control for enhancing the formation rate of torpedo-stage embryos in carrot somatic embryo culture. *Journal of Bioscience and Bioengineering* 95, 384–390.

Shiota, H., Tachibana, K., Watabe, K., and Kamada, H. (1999) Successful long-term preservation of abscisic acid-treated and desiccated carrot somatic embryos. *Plant Cell Reports* 18, 749–753.

Sivanesan, I., Kyoung, K.E., Kyoung, K.M., Young, K.E., and Park, S.W. (2015) Somatic embryogenesis and plant regeneration from zygotic embryo explants of onion. *Horticultura Brasileira* 33, 441–447.

Smith, D.L. and Krikorian, A.D. (1988) Production of somatic embryos from carrot tissues in hormone-free medium. *Plant Science* 58, 103–110.

Smith, D.L. and Krikorian, A.D. (1989) Release of somatic embryogenic potential from excised zygotic embryos of carrot and maintenance of proembryonic cultures in hormone-free medium. *American Journal of Botany* 76, 1832–1843.

Smith, D.L. and Krikorian, A.D. (1990) Somatic embryogenesis of carrot in hormone-free medium: external pH control over morphogenesis. *American Journal of Botany* 77, 1634–1647.

Smith, M.L., Mason, H.S., and Shuler, M.L. (2002) Hepatitis B surface antigen (HBsAg) expression in plant cell culture: kinetics of antigen accumulation in batch culture and its intracellular form. *Biotechnology and Bioengineering* 80, 812–822.

Sonnino, A. and Mini, P. (1992) Somatic embryogenesis in sweet potato *Ipomoea batatas* (L.) Lam. *Acta Horticulturae* 336, 239–244.

Steinitz, B., Küsek, M., Tabib, Y., Paran, I., and Zelcer, A. (2003) Pepper (*Capsicum annuum* L.) regenerants obtained by direct somatic embryogenesis fail to develop a shoot. *In Vitro Cellular & Developmental Biology—Plant* 3, 296–303.

Sudha, M., Gajanana, T.M., and Murthy, D.S. (2006) Economic impact of commercial hybrid seed production in vegetables on farm income, employment and farm welfare—a case of tomato and okra in Karnataka. *Agricultural Economics Research Review* 19, 251–268.

Sun, S., Zhong, J., Li, S., and Wang, X. (2013) Tissue culture-induced somaclonal variation of decreased pollen viability in torenia (*Torenia fournieri* Lind.). *Botanical Studies* 54: 36.

Tabassum, B., Nasir, I.A., Farooq, A.M., Rehman, Z., Latif, Z., and Husnain, T. (2010). Viability assessment of *in vitro* produced synthetic seeds of cucumber. *African Journal of Biotechnology* 9, 7026–7032.

Thieme, R. and Griess, H. (2005) Somaclonal variation in tuber traits of potato. *Potato Research* 48, 153–165.

Thiruvengadam, M., Varisai Mohamed, S., Yang, C.H., and Jayabalan, N. (2006) Development of an embryogenic suspension culture of bitter melon (*Momordica charantia* L.). *Scientia Horticulturae* 109, 123–129.

Toonen, M.A.J., Hendriks, T., Schmidt, E.D.L., *et al.* (1994) Description of somatic-embryo-forming single cells in carrot suspension cultures employing video cell tracking. *Planta* 194, 565–572.

Ummi, K.H., Dayana, A.D., Norahimah, A.R., Natrisya Nur, A.R., and Firdaus, A. (2011) Effects of different storage duration on encapsulated *Parkia speciosa* zygotic embryo germination. *Empowering Science, Technology and Innovation Towards a Better Tomorrow* 120, 758–763.

van den Bulk, R.W. and Dons, J.J.M. (1993) Somaclonal variation as a tool for breeding tomato for resistance to bacterial canker. *Acta Horticulturae* 336, 347–356.

Vargas, T.E., de García, E., and Oropeza, M. (2005) Somatic embryogenesis in *Solanum tuberosum* from cell suspension cultures: histological analysis and extracellular protein patterns. *Journal of Plant Physiology* 162, 449–456.

Vázquez-Boland, J.A., Kuhn, M., Berche, P., Chakraborty, T., Domínguez-Bernal, G., *et al.* (2001) *Listeria* pathogenesis and molecular virulence determinants. *Clinical Microbiology Reviews* 14, 584–640.

Veitia-Rodriguez, N., Francisco-Cardoso, J., Perez, J.N., Garcia-Rodriguez, L., Bermudez-Caraballosos, I., *et al.* (2002) Evaluations in field of somaclones of Irish potatoes (*Solanum tuberosum* Lin.) of the variety Desiree obtained by somaclonal variation and *in vitro* mutagenesis. *Biotecnología Vegetal* 2, 21–26.

Venkataiah, P., Bhanuprakash, P., Suman Kalyan, S., and Subhash, K. (2016) Somatic embryogenesis and plant regeneration of *Capsicum baccatum* L. *Journal of Genetic Engineering and Biotechnology* 14, 55–60.

Vidal, B.C., Mello, M.L.S., and Illg, R.D. (1993) Chromosome number and DNA content in cells of a biotechnologically selected somaclone of garlic (*Allium sativum* L.). *Revista Brasileira de Genética* 16, 347–356.

von Aderkas, P., Teyssier, C., Charpentier, J.P., Gutmann, M., Pâques, L., *et al.* (2015) Effect of light conditions on anatomical and biochemical aspects of somatic and zygotic embryos of hybrid larch (*Larix* × *marschlinsii*). *Annals of Botany* 115, 605–615.

Widholm, J.M. (1972) The use of fluorescein diacetate and phenosafranine for determining viability of cultured plant cells. *Stain Technology* 47, 189–194.

Xue, H.M., Araki, H., Shi, L., and Yakuwa, T. (1991) Somatic embryogenesis and plant regeneration in basal plate and receptacle derived-callus cultures of garlic (*Allium sativum* L.). *Journal of the Japanese Society for Horticultural Science* 60, 627–634.

Yang, J.L., Seong, E.S., Kim, M.J., Ghimire, B.K., Kang, W.H., *et al.* (2010) Direct somatic embryogenesis from pericycle cells of broccoli (*Brassica oleracea* L. var. *italica*) root explants. *Plant Cell, Tissue and Organ Culture* 100, 49–58.

Zenkteler, M., Zenkteler, E.K., and Dostatnia, I. (2006) Somatic embryogenesis from broccoli stigmas in tissue culture. *Acta Biologica Cracoviensia Series Botanica* 48, 121–125.

Zhang, L.Q., Cheng, Z.H., Khan, M.A., and Zhou, Y.L. (2012) *In vitro* selection of resistant mutant garlic lines by using crude pathogen culture filtrate of *Sclerotium cepivorum* Australas. *Plant Pathology* 41, 211–217.

Zhang, S., Zhang, R., Gao, J., Song, G., Li, J., *et al.* (2021) CRISPR/Cas9-mediated genome editing for wheat grain quality improvement. *Plant Biotechnology Journal* 19, 1684–1686.

Zhang, W., Lin, X., Takano, H., Takio, S., and Ono, K. (2004) Efficient plant regeneration from suspension cells of *Allium cepa* L. *Plant Cell Reports* 23, 371–376.

# 13 The Rise of Genome Editing: A Promise for Vegetable Crop Improvement

**Rehna Augustine*, K. Anjala, K. Varsha, Sanjay Sathian and T.M. Brindha**
*Centre for Plant Biotechnology and Molecular Biology, Kerala Agricultural University, Thrissur, Kerala, India*

**Abstract**

As the global population hits 8 billion, one in seven people face chronic hunger. According to reports from the Food and Agriculture Organization of the United Nations, there is a need to double food production by 2050 to evade food scarcity. Conventional production technologies are insufficient to meet this crisis in near future. In this context, technologies that could produce better-quality and higher quantities of food are crucial. Genome-editing technologies are in the frontline to meet these goals. The most recent is the clustered regularly interspaced palindromic repeats (CRISPR)-CRISPR-associated protein 9 (Cas9) technology, which can produce genome-edited plants with higher yield and numerous other economically important characteristics. Genome-edited plants can be made transgene free and hence perceived as genetically harmless to humans. Vegetable crops contribute significantly to agriculture production and hence play an important role in the economy. They also constitute a significant portion of our healthy diet. The demand for production of vegetables is huge and conventional breeding methods are inadequate to attain sustainability. Hence, recent techniques such as CRISPR/Cas9-mediated genome-editing technology is a boon for vegetable crop breeding in terms of the short time requirement, simple and user-friendly design and execution, and high editing efficiency. CRISPR/Cas technology has great potential for all aspects of vegetable crop improvement. This chapter describes the discovery and basics of the technology and its suitability for vegetable-breeding programs, describing some seminal research.

**Keywords:** CRISPR/Cas, crop improvement, genome editing, vegetable crops

## 13.1 The History of Genome Editing

### 13.1.1 Discovery of CRISPR/Cas system

The clustered regularly interspaced palindromic repeats (CRISPR)-CRISPR-associated protein 9 (Cas9) technique is currently in the spotlight for molecular biology research in plants. It was discovered in 1987 in an incident that identified the first CRISPR from *Escherichia coli*. Yoshizumi Ishino, a Japanese scientist, was conducting research on genes related to the isozyme conversion of the enzyme alkaline phosphatase in *E. coli* (Ishino *et al.*, 1987). He and his team sequenced the gene and accidentally discovered a partial DNA sequence, later named CRISPR, in the process. However, he could not determine the exact function of these repeats due to the lack of sequence evidence and technologies. These interspaced and palindromic DNA repeat sequences were found in the intergenic region upstream of the *iap* gene in *E. coli*. Later, another

*Email: rehna.augustine@kau.in


DOI: 10.1079/9781800627123.0013

scientist, Francisco Mojica detected similar repeats from the archaea *Haloferax mediterranei and Haloferax volcanii*, which they called short regularly spaced repeats (Mojica *et al.*, 1995, 2000). In 2002, after discussion with Mojica, Jansen *et al.* (2002) coined the term CRISPR for such repeat sequences. An *in silico* study revealed that a novel family of repetitive DNA sequences was present among both archaea and bacteria but was absent in eukaryotes and viruses. This family of sequences consisted of direct repeats of 21 bp in *Salmonella typhimurium* up to 37 bp in *Streptococcus pyogenes*, interspaced by non-repetitive sequences of similar length. The repeat sequences were found to be conserved within a species. A distinctive gene sequence was also identified that was unique to CRISPR-containing organisms and was located adjacent to the CRISPR locus and hence named the CRISPR-associated gene or *cas* gene (Jansen *et al.*, 2002). The first hint of the biological activity of CRISPR was revealed years later as a defense mechanism against bacteriophages, which involved degrading DNA based on a mechanism similar to RNA interference (Bolotin *et al.*, 2005; Makarova *et al.*, 2006). Barrangou *et al.* (2007) suggested that, after invasion by a virus, bacteria integrate new spacers from phage genomic sequences; this serves as a genetic memory, which is required for subsequent expression to neutralize the reinvading phages. Thus, CRISPR together with the Cas protein provide resistance based on phage spacer-sequence similarity. It was DNA that was targeted instead of RNA for degradation. In 2012, a molecular biologist, Jennifer Doudna, and her group at the University of California, Berkeley, and Emmanuelle Charpentier, a microbiologist then at Umea University, Sweden, reported that they had engineered the CRISPR/Cas9 system to cut isolated DNA at sites of their selection. The discovery of this new gene-editing method that harnesses the CRISPR/Cas9 system was published, and the potential of CRISPR/Cas9 to edit DNA was elucidated (Jinek *et al.*, 2012). Later, in 2013, several groups including that of Feng Zhang of the Broad Institute of MIT and Harvard in Cambridge, Massachusetts, reported that CRISPR/Cas9 could also be utilized for living eukaryotic cells and demonstrated its utility in genome editing of higher organisms (Wang *et al.*, 2013). Doudna and Charpentier received the Nobel prize in Chemistry for this breakthrough technology in 2020. Currently, the CRISPR/Cas9 system has all genome-editing capabilities including knockin, knockout, knockdown, and expression activation (Fiaz *et al.*, 2019).

### 13.1.2 Discovery of the *Cas* genes

Due to the availability of enormous genomic sequence data, scientists were able to identify and evaluate the CRISPR gene content of various organisms. This led to the discovery of four genes that were conserved and found regularly with the adjacent regions of CRISPR regions in genomes. These genes were termed as CRISPR-associated genes 1–4 (*cas1–4*). Cas1 and Cas2 do not show similarity in their functional domains to any other known proteins. Cas2 has been shown to be involved in spacer acquisition (Beloglazova *et al.*, 2008). Cas3 is related to the superfamily 2 helicases and Cas4 is related to RecB DNA exonucleases, which function with the RecBCD complex to facilitate the homologous recombination of double-stranded breaks (DSBs). Cas3 and Cas4 were predicted to be involved in DNA metabolism, repair, recombination, and chromosome segregation. All of the *cas* genes are predicted to be involved in the CRISPR loci genesis (Ishino *et al.*, 2018).

### 13.1.3 Classification of the CRISPR/Cas system

Among the various bacteria and archaea investigated so far, around 30–40% exhibit the existence of CRISPR systems (Koonin *et al.*, 2017). According to the new system of classification, CRISPR/Cas includes two classes, six types, and 33 subtypes (Makarova *et al.*, 2020). The class 1 CRISPR system is a complex system with multi-subunit effectors, whereas the class 2 systems are simpler with single effectors. Even though class 1 accounts for about 90% of the CRISPR/Cas systems, it is less exploited as a genome-editing tool (Li *et al.* 2019). The class 2 CRISPR/Cas system includes three types (types II, V, and VI) and 17 subtypes. The class 2 type II system is widely used for genome editing; however, it constitutes only 5–7% of CRISPR/Cas systems

(Chylinski *et al.*, 2014; Bernheim *et al.*, 2019). It has not been observed in many archaea or other bacterial phyla, such as photosynthetic bacteria (e.g. cyanobacteria) and thermophilic bacteria (e.g. *Thermotogae* and *Chlamydia* spp.)

Three categories of type II systems were found to exist based on the organization of *cas* genes in the operon and phylogenies of Cas proteins, which are named types IIA, IIB, and IIC. Type IIA includes the system in *S. pyogenes*, with a typical operon made of the *cas9*, *cas1*, *cas2*, and *csn2* genes. In type IIB, *cas4* is present in the operon instead of *csn2*. However, in type IIC, both the *csn2* and *cas4* genes are absent. The type II system also depends on a Cas9 effector module and a group of other *cas* genes such as *cas1*, *cas2*, *cas4*, or *csn2*, which are involved in the spacer acquisition process. For the expression of type II systems, transcription of the precursor CRISPR RNA (pre-crRNA), which is later processed by an endogenous non-specific RNase III in the later stage, is essential. The active complex is finally formed when the matured individual crRNA/transactivating crRNA (tracrRNA) duplexes are bound to Cas9 protein, which directs the targeted DNA cleavage. The target specificity of the type II systems comprises recognition of a short protospacer adjacent motif (PAM), which is located immediately downstream of the protospacer on the non-target DNA strand. The PAM sequence is essential for both the spacer acquisition process and for the precise target recognition and cleavage process. Because of their simplicity and great potential as programmable DNA cleavage tools compared with others, type II CRISPR systems have generated a tremendous pool of information on the various components and functional aspects of the system (Nishimasu *et al.*, 2014, 2015).

### 13.1.4 General structure of the CRISPR loci

As name suggests, CRISPR are a family of short repeats of 21–40 bp in length, which are unique among organisms but have a similar length and are interspaced from each other by similarly sized non-repetitive and variable sequence spacers. The repeats are palindromic and the spacers have a role in bacterial immunity, as they store the memory of sequences from previous attacks by invaders such as phages. The CRISPR loci consist of repeats that flank the spacer sequence, and the spacer in turn matches the sequences of virus, plasmid, or any other pathogen genome elements (Bolotin *et al.*, 2005). An AT-rich leader sequence must be present upstream of the CRISPR array, with conserved genes coding for the Cas proteins at the end of the loci (Barman *et al.*, 2019). Within a locus, the repeats are generally similar in length and sequence, whereas the spacers generally have similar lengths but highly variable sequences. The number of spacers may also vary from species to species. This makes the crRNA highly customizable for use in genome-editing experiments.

### 13.1.5 Structure of Cas9 protein

Cas9 endonuclease is a large multi-domain protein of 1368 amino acids (160 kDa) consisting six domains. The primary domain that binds to the guide RNA (gRNA) is called RECI. The function of second domain, RECII, is not clearly understood. The arginine-rich bridge helix domain initiates the cleavage activity on binding to target DNA. The PAM-interacting domain is responsible for binding with the target sequence conferred via the interaction with PAM. The nuclease domains HNH and RuvC cleave the target sequences. The HNH domain breaks the DNA strand that is complementary to the crRNA, whereas the RuvC domain cuts the DNA strand complimentary to the target strand. Once triggered, Cas9 searches the target sequence by binding to it with a sequence that matches the PAM sequence (5′-NGG-3′) from the 3′ end. Cas9 then cuts the dsDNA 3–4 bp upstream of the PAM using its HNH and RuvC domains (Jiang and Doudna, 2017).

### 13.1.6 CRISPR/Cas system as a genome-editing tool

CRISPR/Cas9 system is composed of two essential components. The Cas9 nuclease protein and the single guide RNA (sgRNA). Functional sgRNA is composed of crRNA or spacer (17–24 nt), and tracrRNA or binding scaffold (50–150 nt). The guide sequence of crRNA or spacer is designed by the user based on the target gene sequences and defines the genomic target to be modified. The tracrRNA is the universal sequence from *S. pyogenes* near to which the sgRNA is cloned in the construct. The PAM is present just downstream of the target DNA, and is NGG for *S. pyogenes* (Barman *et al.*, 2019; Tahir *et al.*, 2020). The PAM sequence varies with different Cas proteins. The sgRNA recognizes the PAM sequence and binds to that region in the target. The Cas9 protein cleaves the target at region about 3–4 bp upstream of the PAM sequence. The double-stranded break in the DNA is repaired by natural repair mechanisms such as non-homologous end joining (NHEJ), which is highly error prone, or homology-dependent repair, which is more precise. This generates mutations such as insertions/deletions (InDels) at the target site, which can lead to either activation or knockout/knockdown of the gene (Fig. 13.1).

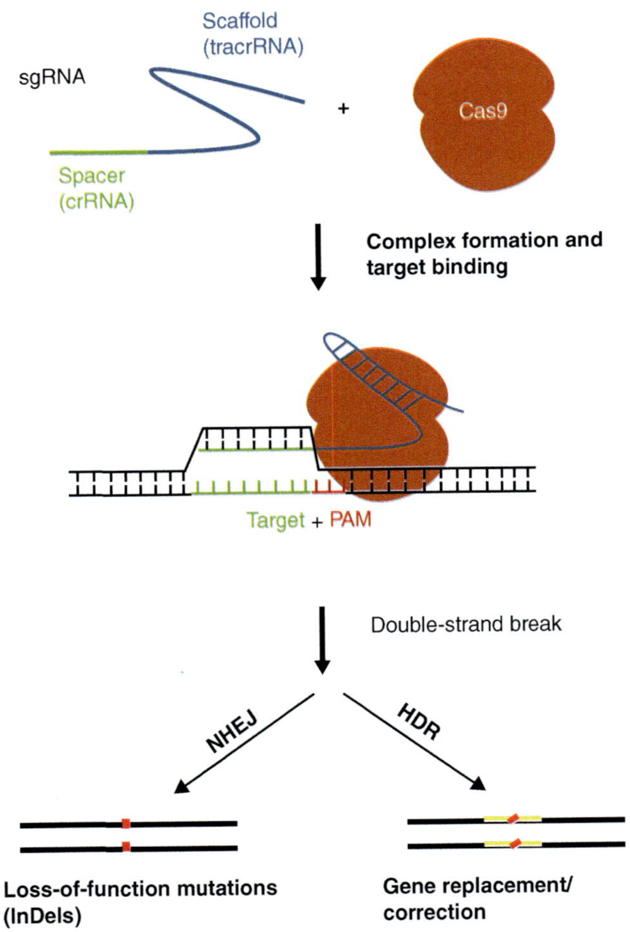


**Fig. 13.1.** Mechanism of CRISPR/Cas9-mediated genome editing. When present together, single guide RNA (sgRNA) and Cas9 form a ribonucleoprotein complex. Cas9 will be guided to the genomic target by the guide sequence and the complex binds to the genomic target based on its sequence complementarity. Cas9, which is an endonuclease, will cleave the DNA and thus introduce double-stranded breaks (DSBs). The DSBs will be repaired by error-prone non-homologous end joining (NHEJ) or homology-dependent repair (HDR) leading to mutations. crRNA, CRISPR RNA; InDels, insertions/deletions; PAM, protospacer adjacent motif; tracrRNA, transactivating crRNA.

## 13.2 Major Advancements in the CRISPR/Cas System

The CRISPR/Cas system has been found to have vast applications in genome. Recently, many variations for Cas proteins have been identified and are widely in use for genome editing in both animals and plant systems. These include the Cpf1 system, SpCas9-VQR, SpCas9-EQR, Cas9-NG, and xCas9 3.7 (Y. Zhang *et al.*, 2019). Depending on the requirements, there are many variants, of which a few are described below.

### 13.2.1 CRISPR/Cpf1 or CRISPR/Cas12a system

The CRISPR/Cpf1 system has been demonstrated to be a more precise and efficient genome editing tool than CRISPR/Cas9 (Bin Moon *et al.*, 2018). The Cpf1 endonuclease represents CRISPR from *Prevotella* and *Francisella* 1, previously known as Cas12a. Compared with Cas9, the Cpf1 endonuclease is smaller in size and utilizes a shorter crRNA for its function (Z.H. Li *et al.*, 2018). The Cpf1 protein is guided by a sgRNA, which binds upstream of the PAM sequence and cleaves the DNA at the proximal end of PAM or far away from the seed region and introduces a 5 bp staggered cut in the DNA (Zetsche *et al.*, 2015). The CRISPR/Cpf1 system does not require a tracrRNA for its target binding activity (Zetsche *et al.*, 2015; Zhang *et al.*, 2017). This system cleaves DNA proximal to a short T-rich PAM, whereas the Cas9 prefers a G-rich PAM. Numerous studies have demonstrated the reliability Cpf1 endonuclease for more precise and targeted genome editing in both prokaryotes and eukaryotes (Zetsche *et al.*, 2015; Jiang *et al.*, 2017; Kim *et al.*, 2017). Various homologs of Cpf1 endonucleases are being isolated and tested from different bacteria for their potential in genome editing; of these, AsCpf1 and LbCpf1 isolated from *Acidaminococcus* sp. BV3L6 and Lachnospiraceae bacterium ND2006, respectively, are extensively used for genome editing (Tóth *et al.*, 2016; Moreno-Mateos *et al.*, 2017; Wang *et al.*, 2017). The Cpf1 database is an online platform providing information on potential genomic targets and helps to design gRNAs to target the desired gene (Park and Bae, 2018).

### 13.2.2 Base editors

Base editors use a catalytically inactivated or "dead" Cas9, dCas9 (D10A and H840A), or a nickase, nCas9, to accurately convert one target DNA nucleotide to another. Base editors producing individual nicks are repaired by an accurate and consistent base excision repair pathway, in contrast to SpCas9-generated DSBs, which are naturally repaired by error-prone NHEJ (Dianov and Hübscher, 2013; Ran *et al.*, 2013). This mechanism of repair reduces the number of unwanted or off-target edits. The conversion of cytosine (C) to thymine (T) is catalyzed by cytosine base editors using a cytosine deaminase (CDA). The system utilizes a rat APOBEC1 or a human activation-induced cytidine deaminase/*Petromyzon marinus* CDA1, attached to nCas9 for its activity (Komor *et al.*, 2016; Nishida *et al.*, 2016). The adenine base editors make use of a DNA processing deoxyadenosine deaminase (TadA) associated with nCas9 to catalyze adenine (A) to guanine (G) conversions (Gaudelli *et al.*, 2017). The PAM-containing DNA strand is shifted to produce an R-loop when the sgRNA binds to the target DNA strand, which generates an RNA–DNA hybrid (Jiang *et al.*, 2016). The single-stranded R-loop exposes and makes the DNA available to CDA or TadA, which facilitates the conversion of the appropriate bases inside the R-loop (transcriptional RNA/DNA hybrid). Base editing has been optimized and used in a variety of plant species (Shimatani *et al.*, 2017; Zong *et al.*, 2017; Shan and Voytas, 2018; Li *et al.*, 2020). The transition mutations of C→T and A→G are created by cytosine and adenine base editors, but glycosylase base editors create transversion mutations such as C→A and C→G, which further expands the scope of base editing (Zhao *et al.*, 2020).

### 13.2.3 Prime editing

Prime editing (PE) is a "search-and-replace" mechanism of genome editing without DSBs or a donor DNA repair template. The PE approach has recently been developed to integrate the desired edit directly within the gRNA (Anzalone *et al.*, 2019). PE is a precise tool that may create targeted InDels, as well as directly introducing both transition and transversion mutations at specific genomic loci with high accuracy. The principle of PE is a target-primed reverse-transcription mechanism similar to retrotransposons. The system is composed of two components: (i) a PE protein, which is a fusion of nickase (nCas9) and an engineered reverse transcriptase (RT) enzyme that creates complementary DNA from an RNA template; and (ii) a PE gRNA (pegRNA) that encodes the primer-binding site and RT template. When nCas9 nicks the PAM-containing DNA strand, it hybridizes with the primer-binding site in the pegRNA, causing the RT to copy the genetic material on the RT template into the target DNA locus. The modified PE2 incorporates five mutations in the Moloney murine leukemia virus RT (D200N, L603W, T330P, T306K, and W313F), whereas PE3 include an additional sgRNA to the non-target strand as well, which is usually 14–116 nt away from pegRNA-induced nicks, which will minimize DSBs. The additional nick facilitates the incorporation of the edit into hetero-duplex DNA (Anzalone *et al.*, 2019). There are diverse online tools available for PE and pegRNA design that are applicable to all species studied. Several PE studies have been reported, which are mainly focused on cereal crops such as rice and wheat (Butt *et al.*, 2020; Lin *et al.*, 2020; Tang *et al.*, 2020; Xu *et al.*, 2020). While PE will likely replace base editors, the technology is still in its infancy, and it is often less efficient and effective than the current base-editing systems, which have better on-target and off-target DNA-editing profiles (Anzalone *et al.*, 2020). Moreover, the construct design is also simpler in conventional CRISPR/Cas systems.

### 13.2.4 Other variants of the system

The other CRISPR systems such as SpCas9-VQR, SpCas9-EQR, Cas9-NG, and xCas9 3.7 are variants of Cas9 with NGA, NGAG, NG, and NG/GAA/GAT as the PAM sequence requirement, respectively. The application of these is well established in plant species such as *Arabidopsis*, rice, potato, tomato, and earth moss (Y. Zhang *et al.*, 2019). Other orthologs of Cas9 sych as SaCas9 from *Staphylococcus aureus* and St1Cas9 from *Streptococcus thermophilus* with the PAM sites NNGRRT and NNGGGT, respectively, have also been successfully demonstrated in plants such as *Arabidopsis*, potato, tobacco, rice, and citrus with higher editing efficiencies (Steinert *et al.*, 2015; Kaya *et al.*, 2016; Jia *et al.*, 2017; Veillet *et al.*, 2020).

Another variant of Cas, denoted Cas12 nuclease, belonging to the class 2, type V CRISPR system, is also being exploited for genome editing. It requires a single crRNA of approximately 42 nt compared with the Cas9 sgRNA of approximately 100 nt. Cas12 effectors do not possess the HNH domain and have only an RuvC-like domain that breaks both strands of the target DNA in the genome leading to the generation of a staggered cut with a 4–5 nt overhang at the 5′ end (Zetsche *et al.*, 2015). LbCas12a is the most widely used Cas12 variant for genetic manipulation in plants with a PAM-containing TTTV sequence (J. Zhang *et al.*, 2019). There are a few more engineered variants of Cas12a with high potential and target ranges that are used for genome-editing experiments (Kleinstiver *et al.*, 2019).

CRISPR-CasΦ is a new member that has been added to the CRISPR toolbox and is a hypercompact type V CRISPR system. It consists of a single CasΦ protein of about 70 kDa, which is about half the size of the Cas9 or Cas12a protein. This system of CRISPR/Cas has a short PAM sequence prerequisite of TBN (where B = G, T, or C). CasΦ does not involve a tracrRNA and produces a staggered cut with 5′ overhangs similar to that of the Cas12a protein. The potential of the system has been demonstrated in *Arabidopsis*, although the editing efficiency was low (Pausch *et al.*, 2020).

It has also been demonstrated that by enabling the expression of Cas9 by cell-type-specific promoters, highly specific somatic gene knockouts in different plant cell types/tissues can be developed, which may lead to a tissue-specific knockout system in *Arabidopsis* (Decaestecker *et al.*, 2019; Ali *et al.*, 2020).

## 13.3 CRISPR/*Cas*-based Genome Editing in Plants

The ultimate goal of plant genetic engineering is to develop desirable plant phenotypes by modifying plant genomes. Technologies such as CRISPR/Cas9 and other sequence-specific nucleases such as zinc-finger nucleases (ZFNs) and transcription activator-like effector nucleases (TALENS) are geared toward revolutionizing basic research and plant breeding. Genome-editing technologies can help to achieve the zero-hunger goal and maintain feeding of the growing human population in the coming decades. Since the invention of this technology, it has been widely used for genome editing in crops such as rice, wheat, tomato, potato, lettuce, cucumber, banana, soybean, and canola to modify various traits. The repertoire of genome-editing resources is expanding day by day and hence it is difficult to include all of them in a chapter.

Vegetables are considered to have a higher share in production and contribute a major portion to a farmer's income. By considering the economic value and availability of efficient plant regeneration or transformation protocols, along with the abundance of genomic resources, genome editing in vegetable crops has mostly progressed in crops such as tomato and potato (Cardi *et al.*, 2017), although there are many reports from other vegetable crops. Most of these studies are in the proof-of-concept stage. Many traits have also been modified in vegetables such as lettuce, broccoli, kale, and cucumber. Some of the seminal work done in vegetable crops is discussed here to emphasize the wide applicability of this technology.

### 13.3.1 Solanaceous vegetables

#### *Tomato (Solanum lycopersicum)*

Tomato is a vegetable that is highly amenable for genome editing for the purpose of biofortification. Brooks *et al.* (2014) reported a pioneer study on genome editing in tomato using the CRISPR/Cas9 technology. A CRISPR/Cas9 construct was developed to the target *ARGONAUTE7*, the knockout of which was expected to show the easily visible phenotype of needle-like leaves. Cas9 nuclease was used with two sgRNAs to generate a large deletion of 90 bp in the *ARGONAUTE7* gene. This led to post-transcriptional silencing of the *AUXIN RESPONSE FACTOR* gene, which determines organ polarity. The plants thus developed wiry or needle-like leaves due to homozygous biallelic or chimeric small InDels.

The *PSY1* and *CrtR-b2* genes of carotenoid biosynthesis were edited by CRISPR/Cas, as their knockdown is readily detectable as a change of fruit or flower color. Neighboring sequences on the first exon of each gene were targeted. The type of mutation was such that 34 new different alleles appeared across the four transformation experiments. Sequence characterization of the CRISPR/Cas9-induced mutations displayed insertions and the deletion of one base. This result showed that the CRISPR/Cas9 system is a potent and rapid method for the generation of useful modifications in tomato to use in breeding experiments (d'Ambrosio *et al.*, 2018).

Enhancing the anthocyanin content in tomato by precise targeted editing of the *ANTHOCYANIN1* (*ANT1*) gene and carotenoid production by targeting carotenoid biosynthesis genes such as phytoene desaturase (*SlPDS*), phytochrome-interacting factor 4 (*SlPIF4*) (Pan *et al.*, 2016), and phytoene synthase (*PSY1*) (Filler Hayut *et al.*, 2017) was also attempted. The CRISPR/Cas9 system was exploited to target the *SGR1* gene to increase lycopene biosynthesis, and other genes such as those catalyzing the cyclization of lycopene, (e.g. *LCY-E*, *LCY-B1*, and *LCY-B2*) were selected. *LCY-E* was selected to inhibit the cyclization of lycopene to α-carotene, and *LCY-B1* and *LCY-B2* were used to hinder the cyclization of lycopene to β-carotene. High-performance liquid chromatography analysis showed that the levels of lycopene and β-carotene in all lycopene mutants were higher than in the wild-type plants (X.D. Li *et al.*, 2018). A summary of some of the genome-editing studies conducted in tomato is given in Table 13.1.

#### *Potato (Solanum tuberosum)*

A powerful genome-editing system is essential for crop improvement in potato because potato has a tetraploid genome. Butler *et al.* (2015) were the first to report the use of the CRISPR/Cas9 system in potato. The *StALS1* gene was modified using diploid and tetraploid variants. Three of the four major events had more than two mutation types at a single *ALS* site in the diploid background, indicating that somatic mutations were more prevalent. This showed the successful application of the technology in potato. Later, CRISPR/Cas9 systems targeting the potato *GRANULE-BOUND STARCH SYNTHASE I* (*GBSSI*) gene were reported. The mutants with targeted mutagenesis in the *GBSSI* gene showed

**Table 13.1.** Summary of CRISPR/Cas-mediated genome editing in tomato for various traits.

| Gene target(s) | Traits | Reference |
|---|---|---|
| *ANT-1* | Anthocyanin fortification | Čermák *et al.* (2015) |
| *RIN* | Delayed fruit ripening | Ito *et al.* (2015) |
| *SlCLV1* (= *FAB*, *SlCLV2*, *SlCLV3*) | Enlarged meristems, altered inflorescence branching with extra flowers and fruits | Xu *et al.* (2015) |
| *SlPDS* | Carotenoid biofortification | Pan *et al.* (2016) |
| *SlBOP1*, *SlBOP2*, *SlBOP3*, *TFAM1*, *TFAM2* | Modified photoperiod response, flowering and control of determinate growth, changes in earliness, harvest index and yield | Xu *et al.* (2016) |
| *SlPiIF4* | Carotenoid fortification | Filler Hayut *et al.* (2017) |
| *PSY*, *CrtR-b2* | Fruit color | d'Ambrosio *et al.* (2018) |
| *SlIAA9* | Parthenocarpy | Ueta *et al.* (2017) |
| *JOINTLESS2* (*J2*), *ENHANCER-OF-JOINTLESS2* (*EJ2*) | Altered branching, increased yield | Soyk *et al.* (2019) |
| *SLAGAMOUS-LIKE 6* (*SlAGL6*) | Vegetative parthenocarpy, fruit color and higher Brix grade under high temperature stress | Klap *et al.* (2017) |
| *GABA-TP1*, *GABA-TP3*, *SSADH*, *CAT9* | Increased γ-aminobutyric acid content | R. Li *et al.* (2018) |
| *SlMYB12* | Pink fruit color | Deng *et al.* (2018) |
| *SGR1*, *LCY-E*, *LCY-B1*, and *LCY-B2* | Enhanced lycopene content | X.D. Li *et al.* (2018) |
| *PL* | Enhanced fruit firmness | Wang *et al.* (2019) |
| *SlJAZ2* | Improved resistance to bacterial *Pseudomonas syringae* pv. *tomato* DC3000 | Ortigosa *et al.* (2019 ) |
| *SLPMR4* | Improved resistance to powdery mildew fungus | Martínez *et al.* (2020) |
| *eIF4E1* | Improved resistance to the potyvirus pepper mottle virus | Yoon *et al.* (2020) |
| *SlLBD40* | Enhanced drought tolerance | Liu *et al.* (2020) |
| *SLMAX1* | Resistance against the parasitic weed *Phelipanche aegyptiaca* | Bari *et al.* (2021) |

characteristics of less amylose starch in their tubers. This result indicates the possibility of using CRISPR/Cas to facilitate genome-editing events in polyploid plants (Kusano *et al.*, 2018).

In another study, transient expression of CRISPR/Cas9-mediated genome editing in protoplasts of tetraploid potato resulted in mutations in all four alleles in a single transfection. Two target regions from the *GBSS* gene exon 8 were chosen, with one target at the site of allelic variation. The third target was taken from exon 9 of the *GBSS* gene. The result emphasized the possibilities for CRISPR/Cas in multiallelic mutagenesis in potato (Andersson *et al.*, 2017). In potato, tuber quality characters such as tuber size, nutrient composition, and resistance to biotic and abiotic stresses such as drought, heat, salinity, and freezing are also being targeted (reviewed by Tiwari *et al.*, 2022).

### *Eggplant/brinjal (Solanum melongena)*

The presence of polyphenols, which cause browning of eggplant after cutting, negatively affect fruit quality for both industrial and domestic consumption. The polyphenol oxidase (*PPO*) genes (named *SmelPPO1–10*) were targeted by CRISPR/Cas9-based mutagenesis. Editing of the *SmelPPO4*, *SmelPPO5*, and *SmelPPO6* loci reduced PPO activity and browning of the fruits after cutting. These results provided the first example of the use of the CRISPR/Cas9 system in eggplant (Maioli *et al.*, 2020).

*Chili (Capsicum spp.)*

Chili is an important crop belonging to the family Solanaceae. Anthracnose, caused by *Colletotrichum* spp., is a primary disease of chili (*Capsicum annuum*) leading to huge yield losses across the world. Transgene-free homozygous mutant lines have been generated in chili using the CRISPR/Cas9 system to target the *CaERF28* locus, which showed enhanced resistance to *Colletotrichum truncatum*. Expression of other defense-responsive genes was also found to be upregulated in mutant lines (Mishra *et al.*, 2021).

### 13.3.2 Cucurbits

*Cucumber (Cucumis sativus)*

Cucumber (*Cucumis sativus* L.) was the first plant in the family Curcubitaceae to successfully be subject to CRISPR-Cas9 gene editing (Chandrasekaran *et al.*, 2016). Virus resistance in cucumber was achieved using CRISPR/Cas9 to interrupt the function of the recessive eukaryotic translation initiation factor 4E (*eIF4E*) gene. Knockout of the gene interfered with virus multiplication and symptom development in the host. Small deletions were noticed in the *eIF4E* gene-targeted sites in the edited cucumber plants, which also exhibited immunity to cucumber vein yellowing virus (genus *Ipomovirus*) infection and resistance to potyviruses zucchini yellow mosaic virus and papaya ring spot mosaic virus-W (Chandrasekaran *et al.*, 2016).

*Watermelon*

As a proof of concept, the *ClPDS* gene in watermelon was edited using CRISPR/Cas9 technology. Insertions or deletions were seen at the expected positions of mutation in transfected watermelon protoplasts. The transgenic watermelon plants showed a clear or mosaic albino phenotype, indicating that the CRISPR/Cas9 system has 100% genome-editing efficiency in watermelon (Tian *et al.*, 2017). Zhang *et al.* (2020) successfully employed the CRISPR/Cas9 system in watermelon to mutate *ClPSK1* gene, encoding the *PSK* precursor, to confer resistance toward the fungal pathogen *Fusarium oxysporum* f. sp. *niveum*. CRISPR/Cas9 has also been used as a functional tool in watermelon to analyze the role of the *ClBG1* gene involved in seed development and germination. Seed size and weight were significantly reduced in the *ClBG1* mutant lines, suggesting their role in seed size determination and germination through modulation of abscisic acid levels (Wang *et al.*, 2021).

### 13.3.3 Cruciferous vegetables

*Lettuce (Lactuca sativa)*

Direct delivery of ribonucleoproteins into protoplast cells of lettuce and the production of targeted genomic changes were reported by Woo *et al.* (2015). They used sgRNA targeting the *BIN2* gene, encoding a negative regulator of the brassinosteroid signaling pathway, to infect the protoplasts. Another study by Bertier *et al.* (2018) showed that knockout of *NCED4* led to a rise in maximum temperature for seed germination, with over 70% of seeds capable of germinating at 37°C, suggesting that the *LsNCED4* gene controls seed germination and thermo-inhibition in lettuce.

*Broccoli (Brassica oleracea var.* italica*)*

The role of the MYB28 transcription factor in the glucosinolate (a secondary metabolite) pathway was deduced by knockout of the *MYB28* gene in broccoli. The mutants were reported to show downregulation of genes involved in aliphatic glucosinolate biosynthesis and this led to the accumulation of the methionine-derived glucosinolate glucoraphanin in leaves and florets (Neequaye *et al.*, 2020). A glucoraphanin-rich broccoli cultivar was developed using CRISPR/Cas9-mediated DNA-free genome editing of the *MYB28* gene involved in the accumulation of glucosinolate levels. Hydrolysis of the endogenous enzyme, myrosinase, converts glucosinolate into sulforaphane, which is involved in defense of the human body against various chronic diseases (Kim *et al.*, 2022). CRISPR/Cas9 was used to edit the *BoGL5* gene; the resulting plants had a glossy green appearance and limited cuticular wax load, showing this as a promising technique for gene functional confirmation and trait development in broccoli (Liu *et al.*, 2023).

#### Cabbage (*Brassica oleracea var.* capitata)

Precise genome-editing using the CRISPR/Cas9 system was achieved in cabbage using multiple sgRNA-expressing cassettes. The *BoPDS* gene was targeted to demonstrate proof of concept of genome editing in cabbage. The albino phenotype was observed with 37.5% editing efficiency, indicating the adaptability of gene knockout by CRISPR/Cas in cabbage (Ma *et al.*, 2019).

#### Chinese kale (*Brassica alboglabra*)

Sun *et al.* (2018) selected a stretch of homologous genes, *BaPDS1* and *BaPDS2*, as the target site for editing. Several stable transgenic lines that had different mutations were created via *Agrobacterium*-mediated transformation, which included *BaPDS1* and *BaPDS2* double mutants and *BaPDS1* or *BaPDS2* single mutants. The total mutation rate reached 76.47%, and these mutations involved nucleotide changes of about 10 bp. It was also observed that Cas9 cleavage in Chinese kale using CRISPR/Cas9 was biased toward AT-rich sequences. CRISPR/Cas9 was used to alter the expression of the carotenoid isomerase gene (*BoaCRTISO*) of Chinese kale. The results showed a high mutation rate of 81.25%. Phenotypic analysis of the homozygous biallelic *crtiso* mutants showed a reduction in the total and individual carotenoid and chlorophyll concentrations with total levels decreased by 11.89–36.33%. The color of the biallelic and homozygous mutants changed from green to yellow, indicating a reduction in the color-masking property of chlorophyll over carotenoids (Sun *et al.*, 2020).

#### Cauliflower (*Brassica oleracea var.* botrytis)

The CRISPR/Cas9 editing system has been used to confer resistance against cauliflower mosaic virus (CaMV), a pararetrovirus with a dsDNA genome. Multiplex targeting of the viral coat protein sequence was carried out by Liu *et al.* (2018). They further showed that these edited coat proteins in most cases led to premature termination of translation and thus produced non-functional coat proteins. They also recovered wild-type coat protein sequence in these infected transgenic plants, suggesting that the edited viral genomes were packaged by wild-type coat proteins.

### 13.3.4 Other vegetable crops

#### Carrot (*Daucus carota*)

Carrot is one of the most beneficial vegetable crops and has been used as a model species in biotechnology for decades. Klimek-Chodacka *et al.* (2018) published their findings on targeted mutagenesis in this species using the CRISPR/Cas9 system. They used multiplexed CRISPR-Cas9 vectors each expressing two sgRNAs that targeted the carrot flavanone 3-hydroxylase (*F3H*) gene. The study aimed to compare the editing efficacy of three codon-optimized *SpCas9* genes (*AteCas9*, *zCas9*, and *Cas9p*) to find the most efficient carrot editing system. The results revealed that the AteCas9 system was the most efficient in creating targeted alterations, with an editing efficiency of up to 90%. Phenotypic evaluation of the mutants showed discoloration of calluses as the result of knockout of the *F3H* gene, confirming the functional relevance of this gene in the anthocyanin-production pathway. The results also demonstrated successful site-directed mutagenesis in carrot using CRISPR/Cas9 technology and the requirement for a model callus culture to validate genome-editing systems.

#### Cowpea (*Vigna unguiculata*)

Due to its agronomically important traits, including tolerance to low rainfall and low fertilization requirements, as well as its high nutrition and health benefits, cowpea is a widely cultivated vegetable crop around the world. A key trait found in cowpea is the ability to fix nitrogen symbiotically. A representative symbiotic nitrogen fixation (*SNF*) gene in cowpea was efficiently disrupted using CRISPR/Cas9-mediated genome-editing technology. The gRNAs targeting the symbiosis receptor-like kinase (*SYMRK*) gene were transformed into cowpea plants using *Agrobacterium rhizogenes*. Hairy root transformed plants showed a mutagenic efficiency of about 67%. The homozygous biallelic mutants showed complete blocking of nodule formation. This preliminary study demonstrated the applicability of the CRISPR/Cas9 system to cowpea for both functional analysis and crop improvement (Ji *et al.*, 2019).

### Spinach (*Spinacia oleracea*)

The CRISPR/Cas9 system has been successfully adopted in spinach. Precise editing of *Spo23361* and *Spo10340*, cellulose synthase-like D (CSLD) genes involved in root hair formation of spinach, was demonstrated by Cao *et al.* (2022). Mutant lines showed short root hairs compared with wild-type plants. Different types of mutation such as replacement, insertion, deletion, and combined mutations were observed, among which deletions accounted for the vast majority (~64.1%). The mutation rate was variable, depending on the target selected. These results indicate that CRISPR/Cas9 is an effective genome-editing tool in spinach.

### Onion (*Allium cepa*)

Mainkar *et al.* (2023) made significant progress in onion research by successfully using CRISPR/Cas9 to edit the phytoene desaturase gene (*AcPDS*) in the Bhima Super onion variety. This achievement paves the way for further exploration of onion biology. They introduced sgRNAs into embryogenic cali through an *Agrobacterium*-mediated transformation method. Among the regenerated shoots, 21 displayed altered phenotypes, indicating successful genetic modifications. The edited shoots showed reductions in chlorophyll content, particularly in albino shoots. Detailed analysis of deep amplicon sequencing showed a considerable variation in the frequency of InDels between the two sgRNAs, ranging from 1.2% to 63.4%. Additionally, there was a substitution frequency of 53.4%. This pioneering study opens up new avenues for genetic enhancement and functional genomics research in onions and potentially other related crops.

### Chicory (*Cichorium intybus*)

CRISPR/Cas9-mediated targeted mutagenesis in chicory was described by Bernard *et al.* (2019). They chose the *CiPDS* gene to obtain an albino phenotype. They were able to recover albino plants at frequencies ranging from 4.5% to 31.3% depending on the approach used such as *Agrobacterium*-mediated transformation and direct administration in protoplasts, demonstrating the potential for genome editing in chicory.

## 13.4 Conclusions and Future Prospects

CRISPR/Cas is a highly efficient method of genome modification in plants for trait improvement. A large number of studies have been carried out in crop plants and this looks set to continue. Genome editing in cereal crops such as rice is carried out extensively. Research is also happening in vegetable crops but is mainly confined to crops such as tomato. In many other vegetable crops, the research is in its infancy. The availability of a reliable transformation system and genomic data are impeding the research. There are a large number of vegetable crops that still need to be explored. The CRISPR/Cas technology has the potential to solve many existing problems of vegetable cultivation including enhancing resistance to biotic and abiotic stresses, nutritional quality improvement, and improving other economically important traits. In the near future, the technology is expected to revolutionize vegetable-breeding programs.

## 13.5 References

Ali, Z., Mahfouz, M.M., and Mansoor, S. (2020) CRISPR-TSKO: a tool for tissue-specific genome editing in plants. *Plant Cell Reports* 25, 123–126.

Andersson, M., Turesson, H., Nicolia, A., Fält, A.S., Samuelsson, M., and Hofvander, P. (2017) Efficient targeted multiallelic mutagenesis in tetraploid potato (*Solanum tuberosum*) by transient CRISPR-Cas9 expression in protoplasts. *Plant Cell Reports* 36, 117–128.

Anzalone, A.V., Randolph, P.B., Davis, J.R., Sousa, A.A., Koblan, L.W., *et al.* (2019) Search-and-replace genome editing without double-strand breaks or donor DNA. *Nature* 576, 149–157.

Anzalone, A.V., Koblan, L.W., and Liu, D.R. (2020) Genome editing with CRISPR-Cas nucleases, base editors, transposases and prime editors. *Nature Biotechnology* 38, 824–844.

Bari, V.K., Nassar, J.A., and Aly, R. (2021) CRISPR/Cas9 mediated mutagenesis of *MORE AXILLARY GROWTH 1* in tomato confers resistance to root parasitic weed *Phelipanche aegyptiaca*. *Scientific Reports* 11: 3905.

Barman, H.N., Sheng, Z., Fiaz, S., Zhong, M., Wu, Y., *et al.* (2019) Generation of a new thermo-sensitive genic male sterile rice line by targeted mutagenesis of TMS5 gene through CRISPR/Cas9 system. *BMC Plant Biology* 19: 109.

Barrangou, R., Fremaux, C., Deveau, H., Richards, M., Boyaval, P., *et al.* (2007) CRISPR provides acquired resistance against viruses in prokaryotes. *Science* 315, 1709–1712.

Beloglazova, N., Brown, G., Zimmerman, M.D., Proudfoot, M., Makarova, K.S., *et al.* (2008) A novel family of sequence-specific endoribonucleases associated with the clustered regularly interspaced short palindromic repeats. *Journal of Biological Chemistry* 283, 20361–20371.

Bernard, G., Gagneul, D., Santos, H.A.D., Etienne, A., Hilbert, J.L., and Rambaud, C. (2019) Efficient genome editing using CRISPR/Cas9 technology in chicory. *International Journal of Molecular Sciences* 20: 1155.

Bernheim, A., Bikard, D., Touchon, M., and Rocha, E.P. (2019) A matter of background: DNA repair pathways as a possible cause for the sparse distribution of CRISPR-Cas systems in bacteria. *Philosophical Transactions of the Royal Society B* 374: 20180088.

Bertier, L.D., Ron, M., Huo, H., Bradford, K.J., Britt, A.B., and Michelmore, R.W. (2018) High-resolution analysis of the efficiency, heritability, and editing outcomes of CRISPR/Cas9-induced modifications of *NCED4* in lettuce (*Lactuca sativa*). *G3* 8, 1513–1521.

Bin Moon, S.U., Lee, J.M., Kang, J.G., Lee, N.E., Ha, D.I., *et al.* (2018) Highly efficient genome editing by CRISPR-Cpf1 using CRISPR RNA with a uridinylate-rich 3′-overhang. *Nature Communications* 9: 3651.

Bolotin, A., Quinquis, B., Sorokin, A., and Ehrlich, S.D. (2005) Clustered regularly interspaced short palindrome repeats (CRISPRs) have spacers of extrachromosomal origin. *Microbiology* 151, 2551–2561.

Brooks, C., Nekrasov, V., Lippman, Z.B., and van Eck, J. (2014) Efficient gene editing in tomato in the first generation using the clustered regularly interspaced short palindromic repeats/CRISPR-associated9 system. *Plant Physiology* 166, 1292–1297.

Butler, N.M., Atkins, P.A., Voytas, D.F., and Douches, D.S. (2015) Generation and inheritance of targeted mutations in potato (*Solanum tuberosum* L.) using the CRISPR/Cas system. *PLOS ONE* 10: e0144591.

Butt, H., Rao, G.S., Sedeek, K., Aman, R., Kamel, R., and Mahfouz, M. (2020) Engineering herbicide resistance via prime editing in rice. *Plant Biotechnology Journal* 18, 2370–2372.

Cao, Y., Xu, Y., Zhang, Y., Zhang, H., Bai, C., *et al.* (2022) Efficient editing of CSLD2 orthologue by CRISPR/Cas9 affects cell morphogenesis of root hair in spinach. *Horticulturae* 8: 735.

Cardi, T., d'Agostino, N., and Tripodi, P. (2017) Genetic transformation and genomic resources for next-generation precise genome engineering in vegetable crops. *Frontiers in Plant Science* 8: 241.

Čermák, T., Baltes, N.J., Čegan, R., Zhang, Y., and Voytas, D.F. (2015) High-frequency, precise modification of the tomato genome. *Genome Biology* 16: 232.

Chandrasekaran, J., Brumin, M., Wolf, D., Leibman, D., Klap, C., *et al.* (2016) Development of broad virus resistance in non-transgenic cucumber using CRISPR/Cas9 technology. *Molecular Plant Pathology* 17, 1140–1153.

Chylinski, K., Makarova, K.S., Charpentier, E., and Koonin, E.V. (2014) Classification and evolution of type II CRISPR-Cas systems. *Nucleic Acids Research* 42, 6091–6105.

d'Ambrosio, C., Stigliani, A.L., and Giorio, G. (2018) CRISPR/Cas9 editing of carotenoid genes in tomato. *Transgenic Research* 27, 367–378.

Decaestecker, W., Buono, R.A., Pfeiffer, M.L., Vangheluwe, N., Jourquin, J., *et al.* (2019) CRISPR-TSKO: a technique for efficient mutagenesis in specific cell types, tissues, or organs in *Arabidopsis*. *Plant Cell* 31, 2868–2887.

Deng, L., Wang, H., Sun, C., Li, Q., Jiang, H., *et al.* (2018) Efficient generation of pink-fruited tomatoes using CRISPR/Cas9 system. *Journal of Genetics and Genomics* 45, 51–54.

Dianov, G.L. and Hübscher, U. (2013) Mammalian base excision repair: the forgotten archangel. *Nucleic Acids Research* 41, 3483–3490.

Fiaz, S., Ahmad, S., Noor, M.A., Wang, X., Younas, A., *et al.* (2019) Applications of the CRISPR/Cas9 system for rice grain quality improvement: perspectives and opportunities. *International Journal of Molecular Sciences* 20: 888.

Filler Hayut, S., Melamed Bessudo, C., and Levy, A.A. (2017) Targeted recombination between homologous chromosomes for precise breeding in tomato. *Nature Communications* 8: 15605.

Gaudelli, N.M., Komor, A.C., Rees, H.A., Packer, M.S., Badran, A.H., *et al.* (2017) Programmable base editing of A·T to G·C in genomic DNA without DNA cleavage. *Nature* 551, 464–471.

Ishino, Y., Shinagawa, H., Makino, K., Amemura, M., and Nakata, A. (1987) Nucleotide sequence of the *iap* gene, responsible for alkaline phosphatase isozyme conversion in *Escherichia coli*, and identification of the gene product. *Journal of Bacteriology* 169, 5429–5433.

Ishino, Y., Krupovic, M., and Forterre, P. (2018) History of CRISPR-Cas from encounter with a mysterious repeated sequence to genome editing technology. *Journal of Bacteriology* 200: e00580-17.

Ito, Y., Nishizawa-Yokoi, A., Endo, M., Mikami, M., and Toki, S. (2015) CRISPR/Cas9-mediated mutagenesis of the *RIN* locus that regulates tomato fruit ripening. *Biochemical and Biophysical Research Communications* 467, 76–82.

Jansen, R., van Emben, J.D.A., Gaastra, W., and Schouls L.M. (2002) Identification of genes that are associated with DNA repeats in prokaryotes. *Molecular Microbiology* 43, 1565–1575.

Ji, J., Zhang, C., Sun, Z., Wang, L., Duanmu, D., and Fan, Q. (2019) Genome editing in cowpea *Vigna unguiculata* using CRISPR-Cas9. *International Journal of Molecular Sciences* 20: 2471.

Jia, H., Xu, J., Orbović, V., Zhang, Y., and Wang, N. (2017) Editing citrus genome via SaCas9/sgRNA system. *Frontiers in Plant Science* 8: 2135.

Jiang, F. and Doudna, J.A. (2017) CRISPR-Cas9 structures and mechanisms. *Annual Review of Biophysics* 46, 505–529.

Jiang, F., Taylor, D.W., Chen, J.S., Kornfeld, J.E., Zhou, K., *et al.* (2016) Structures of a CRISPR-Cas9 R-loop complex primed for DNA cleavage. *Science* 351, 867–871.

Jiang, Y., Qian, F., Yang, J., Liu, Y., Dong, F., *et al.* (2017) CRISPR-Cpf1 assisted genome editing of *Corynebacterium glutamicum*. *Nature Communications* 8: 15179.

Jinek, M., Chylinski, K., Fonfara, I., Hauer, M., Doudna, J.A., and Charpentier, E. (2012) A programmable dual-RNA-guided DNA endonuclease in adaptive bacterial immunity. *Science* 337, 816–21.

Kaya, H., Mikami, M., Endo, A., Endo, M., and Toki, S. (2016) Highly specific targeted mutagenesis in plants using *Staphylococcus aureus* Cas9. *Scientific Reports* 6: 26871.

Kim, H., Kim, S.T., Ryu, J., Kang, B.C., Kim, J.S., and Kim, S.G. (2017) CRISPR/Cpf1-mediated DNA-free plant genome editing. *Nature Communications* 8: 14406.

Kim, Y.C., Ahn, W.S., Cha, A., Jie, E.Y., Kim, S.W., Hwang, B.H., and Lee, S. (2022) Development of glucoraphanin-rich broccoli (*Brassica oleracea* var. *italica*) by CRISPR/Cas9-mediated DNA-free BolMYB28 editing. *Plant Biotechnology Reports* 16, 123–132.

Klap, C., Yeshayahou, E., Bolger, A.M., Arazi, T., Gupta, S.K., *et al.* (2017) Tomato facultative parthenocarpy results from SlAGAMOUS-LIKE 6 loss of function. *Plant Biotechnology Journal* 15, 634–647.

Kleinstiver, B.P., Sousa, A.A., Walton, R.T., Tak, Y.E., Hsu, J.Y., *et al.* (2019) Engineered CRISPR-Cas12a variants with increased activities and improved targeting ranges for gene, epigenetic and base editing. *Nature Biotechnology* 37, 276–282.

Klimek-Chodacka, M., Oleszkiewicz, T., Lowder, L.G., Qi, Y., and Baranski, R. (2018) Efficient CRISPR/Cas9-based genome editing in carrot cells. *Plant Cell Reports* 37, 575–586.

Komor, A.C., Kim, Y.B., Packer, M.S., Zuris, J.A., and Liu, D.R. (2016) Programmable editing of a target base in genomic DNA without double-stranded DNA cleavage. *Nature* 533, 420–424.

Koonin, E.V., Makarova, K.S., and Zhang, F. (2017) Diversity, classification and evolution of CRISPR-Cas systems. *Current Opinion in Microbiology* 37, 67–78.

Kusano, H., Ohnuma, M., Mutsuro-Aoki, H., Asahi, T., Ichinosawa, D., *et al.* (2018) Establishment of a modified CRISPR/Cas9 system with increased mutagenesis frequency using the translational enhancer dMac3 and multiple guide RNAs in potato. *Science Reports* 8: 13753.

Li, D., Zhou, H., and Zeng, X. (2019) Battling CRISPR-Cas9 off-target genome editing. *Cell Biology and Toxicology* 35, 403-406.

Li, R., Li, R., Li, X., Fu, D., Zhu, B., *et al.* (2018) Multiplexed CRISPR/Cas9-mediated metabolic engineering of $\gamma$-aminobutyric acid levels in *Solanum lycopersicum*. *Plant Biotechnology Journal* 16, 415–427.

Li, S., Zhang, Y., Xia, L., and Qi, Y. (2020) CRISPR-Cas12a enables efficient biallelic gene targeting in rice. *Plant Biotechnology Journal* 18, 1351–1353.

Li, X.D., Wang, Y.N., Chen, S., Tian, H., Fu, D., *et al.* (2018) Lycopene is enriched in tomato fruit by CRISPR/Cas9-mediated multiplex genome editing. *Frontiers in Plant Science* 9: 559.

Li, Z.H., Liu, M., Lyu, X.M., Wang, F.Q., and Wei, D.Z. (2018) CRISPR/Cpf1 facilitated large fragment deletion in *Saccharomyces cerevisiae*. *Journal of Basic Microbiology* 58, 1100–1104.

Lin, Q., Zong, Y., Xue, C., Wang, S., Jin, S., *et al.* (2020) Prime genome editing in rice and wheat. *Nature Biotechnology* 38, 582–585.

Liu, H., Soyars, C.L., Li, J., Fei, Q., He, G., *et al.* (2018) CRISPR/Cas9-mediated resistance to cauliflower mosaic virus. *Plant Direct* 2: e00047.

Liu, L., Zhang, J., Xu, J., Li, Y., Guo, L., *et al.* (2020) CRISPR/Cas9 targeted mutagenesis of SlLBD40, a lateral organ boundaries domain transcription factor, enhances drought tolerance in tomato. *Plant Science* 301: 110683.

Liu, Y., Wei, M., Liu, Y., Fang, Z., Zhang, Y., *et al.* (2023) Functional characterization of *BoGL5* by an efficient CRISPR/Cas9 genome editing system in broccoli. *Scientia Horticulturae* 319: 112136.

Ma, C., Liu, M., Li, Q., Si, J., Ren, X., and Song, H. (2019) Efficient *BoPDS* gene editing in cabbage by the CRISPR/Cas9 system. *Horticultural Plant Journal* 5, 164–169.

Mainkar, P., Manape, T.K., Satheesh, V., and Anandhan, S. (2023) CRISPR/Cas9-mediated editing of *PHYTOENE DESATURASE* gene in onion (*Allium cepa* L.). *Frontiers in Plant Science* 14: 1226911.

Maioli, A., Gianoglio, S., Moglia, A., Acquadro, A., Valentino, D., *et al.* (2020) Simultaneous CRISPR/Cas9 editing of three PPO genes reduces fruit flesh browning in *Solanum melongena* L. *Frontiers in Plant Science* 11, 607161.

Makarova, K.S., Grishin, N.V., Shabalina, S.A., Wolf, Y.I., and Koonin, E.V. (2006) A putative RNA-interference-based immune system in prokaryotes: computational analysis of the predicted enzymatic machinery, functional analogies with eukaryotic RNAi, and hypothetical mechanisms of action. *Biology Direct* 1: 7.

Makarova, K.S., Wolf, Y.I., Iranzo, J., Shmakov, S.A., Alkhnbashi, O.S., *et al.* (2020) Evolutionary classification of CRISPR-Cas systems: a burst of class 2 and derived variants. *Nature Reviews Microbiology* 18, 67–83.

Martínez, M.I.S., Bracuto, V., Koseoglou, E., Appiano, M., Jacobsen, E., *et al.* (2020) CRISPR/Cas9-targeted mutagenesis of the tomato susceptibility gene *PMR4* for resistance against powdery mildew. *BMC Plant Biology* 20: 284.

Mishra, R., Mohanty, J.N., Mahanty, B., and Joshi, R.K. (2021) A single transcript CRISPR/Cas9 mediated mutagenesis of *CaERF28* confers anthracnose resistance in chilli pepper (*Capsicum annuum* L.). *Planta* 254: 5.

Mojica, F.J., Ferrer, C., Juez, G., and Rodriguez-Valera, F. (1995) Long stretches of short tandem repeats are present in the largest replicons of the Archaea *Haloferax mediterranei* and *Haloferax volcanii* and could be involved in replicon partitioning. *Molecular Microbiology* 17, 85–93.

Mojica, F.J., Diez-Villasenor, C., Soria, E., and Juez, G. (2000) Biological significance of a family of regularly spaced repeats in the genomes of Archaea, Bacteria and mitochondria. *Molecular Microbiology* 36, 244–246.

Moreno-Mateos, M.A., Fernandez, J.P., Rouet, R., Vejnar, C.E., Lane, M.A., *et al.* (2017) CRISPR-Cpf1 mediates efficient homology-directed repair and temperature-controlled genome editing. *Nature Communications* 8: 2024.

Neequaye, M., Stavnstrup, S., Harwood, W., Lawrenson, T., Hundleby, P., *et al.* (2020) CRISPR-Cas9-mediated editing of *MYB28* genes impair glucoraphanin accumulation of *Brassica oleracea* in the field. *CRISPR Journal* 4, 416–426.

Nishida, K., Arazoe, T., Yachie, N., Banno, S., Kakimoto, M., *et al.* (2016) Targeted nucleotide editing using hybrid prokaryotic and vertebrate adaptive immune systems. *Science* 353: aa8729.

Nishimasu, H., Ran, F.A., Hsu, P.D., Konermann, S., Shehata, S.I., *et al.* (2014) Crystal structure of Cas9 in complex with guide RNA and target DNA. *Cells* 156, 935–949.

Nishimasu, H., Cong, L., Yan, W.X., Ran, F.A., Zetsche, B., *et al.* (2015) Crystal structure of *Staphylococcus aureus* Cas9. *Cells* 162, 1113–1126.

Ortigosa, A., Gimenez-Ibanez, S., Leonhardt, N., and Solano, R. (2019) Design of a bacterial speck resistant tomato by CRISPR/Cas9-mediated editing of Sl JAZ 2. *Plant Biotechnology Journal* 17, 665–673.

Pan, C., Ye, L., Qin, L., Liu, X., He, Y., *et al.* (2016) CRISPR/Cas9-mediated efficient and heritable targeted mutagenesis in tomato plants in the first and later generations. *Science Reports* 6: 24765.

Park, J., and Bae, S. (2018) Cpf1-Database: web-based genome-wide guide RNA library design for gene knockout screens using CRISPR-Cpf1. *Bioinformatics* 34, 1077–1079.

Pausch, P., Al-Shayeb, B., Bisom-Rapp, E., Tsuchida, C.A., Li, Z., *et al.* (2020) CRISPR-CasΦ from huge phages is a hypercompact genome editor. *Science* 369, 333–337.

Ran, F.A., Hsu, P.D., Wright, J., Agarwala, V., Scott, D.A., and Zhang, F. (2013) Genome engineering using the CRISPR-Cas9 system. *Nature Protocols* 8, 2281–2308.

Shan, Q. and Voytas, D.F. (2018) Editing plant genes one base at a time. *Nature Plants* 4, 412–413.

Shimatani, Z., Kashojiya, S., Takayama, M., Terada, R., Arazoe, T., *et al.* (2017) Targeted base editing in rice and tomato using a CRISPR-Cas9 cytidine deaminase fusion. *Nature Biotechnology* 35, 441–443.

Soyk, S., Lemmon, Z.H., Sedlazeck, F.J., Jiménez-Gómez, J.M., Alonge, M., *et al. Duplication* of a domestication locus neutralized a cryptic variant that caused a breeding barrier in tomato. *Nature Plants* 5, 471–479.

Steinert, J., Schiml, S., Fauser, F., and Puchta, H. (2015) Highly efficient heritable plant genome engineering using Cas9 orthologues from *Streptococcus thermophilus* and *Staphylococcus aureus*. *Plant Journal* 84, 1295–1305.

Sun, B., Zheng, A., Jiang, M., Xue, S., Yuan, Q., *et al.* (2018) CRISPR/Cas9-mediated mutagenesis of homologous genes in Chinese kale. *Scientific Reports* 8: 16786.

Sun, B., Jiang, M., Zheng, H., Jian, Y., Huang, W.L., *et al.* (2020) Color-related chlorophyll and carotenoid concentrations of Chinese kale can be altered through CRISPR/Cas9 targeted editing of the carotenoid isomerase gene *BoaCRTISO*. *Horticulture Research* 7: 161.

Tahir, T., Ali, Q., Rashid, M.S., and Malik, A. (2020) The journey of CRISPR-Cas9 from bacterial defense mechanism to a gene editing tool in both animals and plants. *Biological and Clinical Sciences Research* 17, 1–10.

Tang, X., Sretenovic, S., Ren, Q., Jia, X., Li, M., *et al.* (2020) Plant prime editors enable precise gene editing in rice cells. *Molecular Plant* 13, 667–670.

Tian, S., Jiang, L., Gao, Q., Zhang, J., Zong, M., *et al.* (2017) Efficient CRISPR/Cas9-based gene knockout in watermelon. *Plant Cell Reports* 36, 399–406.

Tiwari, J.K., Buckseth, T., Challam, C., Zinta, R., Bhatia, N., *et al.* (2022) CRISPR/Cas genome editing in potato: current status and future perspectives. *Frontiers in Genetics* 13: 827808.

Tóth, E., Weinhardt, N., Bencsura, P., Huszár, K., Kulcsár, P.I., *et al.* (2016) Cpf1 nucleases demonstrate robust activity to induce DNA modification by exploiting homology directed repair pathways in mammalian cells. *Biology Direct* 11: 46.

Ueta, R., Abe, C., Watanabe, T., Sugano, S.S., Ishihara, R., *et al.* Rapid breeding of parthenocarpic tomato plants using CRISPR/Cas9. *Science Reports* 7: 507.

Veillet, F., Durand, M., Kroj, T., Cesari, S., and Gallois, J.L. (2020) Precision breeding made real with CRISPR: illustration through genetic resistance to pathogens. *Plant Communication* 1: 100102.

Wang, D., Samsulrizal, N.H., Yan, C., Allcock, N.S., Craigon, J., *et al.* (2019) Characterization of CRISPR mutants targeting genes modulating pectin degradation in ripening tomato. *Plant Physiology* 179, 544–557.

Wang, H., Yang, H., Shivalila, C.S., Dawlaty, M.M., Cheng, A.W., *et al.* (2013) One-step generation of mice carrying mutations in multiple genes by CRISPR/Cas-mediated genome engineering. *Cells* 153, 910–918.

Wang, M., Mao, Y., Lu, Y., Tao, X., and Zhu, J.K. (2017) Multiplex gene editing in rice using the CRISPR-Cpf1 system. *Molecular Plant* 10, 1011–1013.

Wang, Y., Wang, J., Guo, S., Tian, S., Zhang, J., *et al.* (2021) CRISPR/Cas9-mediated mutagenesis of *ClBG1* decreased seed size and promoted seed germination in watermelon. *Horticulture Research* 8: 70.

Woo, J.W., Kim, J., Kwon, S.I., Corvalán, C., Cho, S.W., *et al.* (2015) DNA-free genome editing in plants with preassembled CRISPR-Cas9 ribonucleoproteins. *Nature Biotechnology* 33, 1162–1164.

Xu, C., Liberatore, K.L., MacAlister, C.A., Huang, Z., Chu, Y.H., *et al.* (2015) A cascade of arabinosyltransferases controls shoot meristem size in tomato. *Nature Genetics* 47, 784–792.

Xu, C., Park, S.J., van Eck, J., and Lippman, Z.B. (2016) Control of inflorescence architecture in tomato by BTB/POZ transcriptional regulators. *Genes & Development* 30, 2048–2061.

Xu, R., Li, J., Liu, X., Shan, T., Qin, R., and Wei, P. (2020) Development of plant prime-editing systems for precise genome editing. *Plant Communication* 1: 100043.

Yoon, Y.J., Venkatesh, J., Lee, J.H., Kim, J., Lee, H.E., *et al.* (2020) Genome editing of eIF4E1 in tomato confers resistance to pepper mottle virus. *Frontiers in Plant Science* 11: 1098.

Zetsche, B., Gootenberg, J.S., Abudayyeh, O.O., Slaymaker, I.M., Makarova, K.S., *et al.* (2015) Cpf1 is a single RNA-guided endonuclease of a class 2 CRISPR-Cas system. *Cells* 163, 759–771.

Zhang, J., Yang, F., Yang, Y., Jiang, Y., and Huo, Y.X. (2019) Optimizing a CRISPR-Cpf1-based genome engineering system for *Corynebacterium glutamicum*. *Microbial Cell Factory* 18: 60.

Zhang, M., Liu, Q., Yang, X., Xu, J., Liu, G., *et al.* (2020) CRISPR/Cas9-mediated mutagenesis of *Clpsk1* in watermelon to confer resistance to *Fusarium oxysporum* f.sp. *niveum*. *Plant Cell Reports* 39, 589–595.

Zhang, Y., Long, C., Li, H., McAnally, J.R., Baskin, K.K., *et al.* (2017) CRISPR-Cpf1 correction of muscular dystrophy mutations in human cardiomyocytes and mice. *Science Advances* 3: e1602814.

Zhang, Y., Malzahn, A.A., Sretenovic, S., and Qi, Y. (2019) The emerging and uncultivated potential of CRISPR technology in plant science. *Nature Plants* 5, 778–794.

Zhao, J., Fang, H., and Zhang, D. (2020) Expanding application of CRISPR-Cas9 system in microorganisms. *Synthetic and Systems Biotechnology* 5, 269–276.

Zong, Y., Wang, Y., Li, C., Zhang, R., Chen, K., *et al.* (2017) Precise base editing in rice, wheat and maize with a Cas9-cytidine deaminase fusion. *Nature Biotechnology* 35, 438–440.

# 14 Biotechnology as a Tool to Maintain Biodiversity of Vegetable Crops

**D.R. Bhardwaj*** **and Suresh Reddy Yerasu**
*ICAR-Indian Institute of Vegetable Research, Varanasi, Uttar Pradesh, India*

**Abstract**
Plant genetic resources have been contributing immensely to food security since the beginning of agriculture. During its 12,000-year history, out of the 30,000 plant species used as food sources on Earth almost 1100 species have been consumed as vegetables including annuals, biennials, perennials, trees, shrubs, aquatic plants, and herbs. Vegetables are protective foods because they are rich in nutraceutical supplements, including dietary fiber, carbohydrates, minerals, vitamins, amino acids, and many secondary metabolites. These metabolites protect against illness and maintain good physical and mental health in humans. Vegetable diversity presents a variety of options to humanity in light of shifting global temperatures and the problems emerging as the climate changes. Some of these changes include species with extraordinarily high nutritional qualities, which can occasionally outperform cultivated counterparts, and others that can adapt well to changing climatic conditions. A startling rate of habitat destruction, land degradation, over-exploitation of water resources, forestry practices, urbanization, shifting social and cultural norms, and displacement by exotics and better kinds is resulting in the loss of genetic resources with potential value. Vegetable genetic diversity loss and conservation are influenced by a variety of variables, including geographic location, plant species growth habits, applications, and current economic significance. With the introduction of new technologies and scientific discoveries, the field of biotechnology is dynamic and always changing. It has promise for biodiversity conservation in all areas, including distribution, characterization, storage, and gathering. Protecting biodiversity has benefited greatly from biotechnology, as it complements conventional conservation techniques. This chapter deals with the importance of vegetable biodiversity, the causes of its loss, different biotechnological approaches such as molecular markers for estimating genetic diversity in vegetable crops, biotechnological approaches in plant genetic resource storage, and conservation such as tissue-culture techniques, slow-growth cultures, cryopreservation, and DNA banks, and their status in different vegetable crops across the world.

**Keywords:** conservation, cryopreservation, DNA banks, gene banks, *in vitro*

## 14.1 Introduction

Plant genetic resources can be described as the total genetic diversity of the cultivated species and their wild relatives (Ford-Lloyd and Jackson, 1991). Biological diversity includes the variability among living organisms from all sources including terrestrial, marine, and other aquatic ecosystems, and the ecological complexes of which they are a part; thus, it includes diversity within species, between species, and within ecosystems (Article 2 of the Convention on Biological Diversity). Biodiversity is considered the foundation of agriculture as it is the source of all crop

*Email: dram_iivr@yahoo.com


DOI: 10.1079/9781800627123.0014

species that have been domesticated and bred since the beginning of agriculture (Cooper *et al.*, 2005). Genetic resources or germplasm include: (i) land races; (ii) obsolete varieties; (iii) cultivated varieties; (iv) breeding lines; and (v) wild forms and wild relatives. The wild relatives of crop plants are of considerable importance in crop-improvement programs as a source of traits such as disease and pest resistance, and stress tolerance. For example, all the resistance genes available in commercial tomato cultivars have been transferred from wild genetic resources (Reem and Toby, 2007). Since the advent of agriculture, plant genetic resources have largely served as the primary components of the food chain. In its history of some 12,000 years, nearly 30,000 edible plant species out of 240,0000 plant species that have existed on Earth have been utilized as a source of food. Worldwide, about 1100 vegetable species are recognized (Meldrum *et al.*, 2018) with a variety of growth forms including annuals, biennials, perennials, trees, shrubs, herbs, and aquatic plants. Vegetables are becoming increasingly popular due to their indispensable role in human nutrition and agricultural diversification. Vegetables are considered protective foods and also play an important role in providing a balanced diet for humans. Vegetables are plant parts, such as leaves, fruits, or immature pods, that are eaten raw or cooked, used in salads and as part of savoury dishes in general (Grubben and Denton, 2004). They are rich in micronutrients and add diversity of forms, tastes, and colours to food. Vegetables along with fruits play an increasingly important role in new global research and development programs that emphasize nutrition and healthy diets alongside climate action, safeguarding biodiversity, ending poverty, and improving livelihoods (Caron *et al.*, 2018; Willett *et al.*, 2019). Under changing global temperatures and the challenges posed by global climate change, vegetable diversity offers different opportunities to humankind, as it includes species with exceptionally high nutritional values, sometimes better than cultivated ones, and some are adapted to harsh environments. Examples include micronutrient-rich African leafy vegetables adapted to rain-fed conditions (Maundu *et al.*, 2009). These also serve a critical role in protecting human and animal lives during famines by providing food and fodder. For example, the bulb of greater galangal (*Alpinia galangal*, family Zingiberaceae) can also be eaten raw. In the Garwhal areas, *Cornus capitata* and *Cornus controversa* (family Cornaceae) are also eaten raw. In Rajasthana, the whole plant of *Gisekia pharnaceoides* (family Molluginaceae) is widely consumed during food shortages, but in the south and west (Deccan Region), leaves are used as greens, such as the leaves of *Glinus trianthemoides*. These are hardy and resilient plants. An amaranth gene, *Am1*, reported to improve the protein quality in terms of lysine content, has been successfully incorporated into rice using biotechnological tools.

## 14.2 Loss of Genetic Diversity

Genetic resources of potential value are being lost at an alarming rate due to habitat destruction, land degradation, over-exploitation of water resources, forestry practices, urban expansion, changing social and cultural norms, and displacement by exotics and improved varieties. This was clearly noted in Agenda 21 at the Earth Summit in 1992. The report of the Food and Agriculture Organization of the United Nations on the state of the World Plant Genetic Resources, released at Leipzing in 1996, listed the main causes of plant genetic erosion in 154 countries. These mainly included over-replacement of local varieties (80 countries), deforestation (over 60 countries), over-exploitation (nearly 55 countries), and population pressure and urbanization (45 countries).

The main causes of genetic degradation can be summarized as follows:

**1.** Replacement of genetically variable land races ("desi" varieties) by improved, genetically uniform pure lines or hybrid varieties. This has resulted in the extinction of several land races, open-pollinated varieties, and other reservoirs of genetic variability.
**2.** Improved crop management practices have virtually eliminated the weedy forms of many crops. They existed as crop–weed complexes in numerous crops, when gene introgression from weed to crop and vice versa happened, and these complexes used to be a source of considerable genetic diversity.
**3.** Increasing human needs have extended farming and grazing into forests, the habitats of

most wild species. As a result, many wild relatives of crops have become extinct.
**4.** Developmental activities such as hydroelectric projects, roads, industrial areas, railways, and buildings have also disturbed the wild habitat, which often leads to loss of wild relatives of crops.
**5.** Sometimes, the introduction of a weedy species may result in invasion of wild habitats by this species and lead to elimination of the native wild relatives of the crop plants. Genetic erosion may also be contributed by the cultivated forms derived from introduced species.

Loss of genetic diversity and conservation of genetic diversity in vegetables depends on many factors including geographical region, growth habit of the plant species, uses, and present economic importance (Table 14.1). Over 90% of plant species used for food and agriculture

**Table 14.1.** Variability, distribution status, and conservation priority of vegetable crops.

| Crop | Genus | Crop status[a] | Distribution status[b] | Germplasm variability status[c] | Genetic erosion status[c] | General crop priorities[c] |
|---|---|---|---|---|---|---|
| Eggplant (brinjal) | *Solanum melongena* | C | W | W | M | H |
| Chili | *Capsicum* | C | W | H | M | H |
| Tomato | *Solanum lycopersicum* | C | W | H | M | H |
| Cluster bean | *Cyamopsis* | C | R | M | M | M |
| Cowpea | *Vigna* | C | W | H | M | H |
| French bean | *Phaseolus* | C | W | M | M | M |
| Lablab bean | *Lablab* | C | W | H | M | M |
| Velvet bean | *Mucuna* | C | R | H | M | M |
| Pea | *Pisum* | C | R | M | M | H |
| Carrot | *Daucus* | C | R | M | M | H |
| Elephant foot yam | *Amorphophallus* | C | R | H | M | M |
| Giant taro | *Alocasia* | W | R | H | M | M |
| Radish | *Raphanus* | C | W | H | M | H |
| Taro | *Colocasia* | W | W | H | M | M |
| Yam | *Dioscoria* | W | R | H | M | M |
| Garlic | *Allium* | C | W | M | M | H |
| Onion | *Allium* | C | R | M | M | H |
| Okra | *Abelmoschus* | C | W | H | M | H |
| Amaranth | *Amaranthus* | C | W | H | M | M |
| Climbing spinach | *Basella* | C | W | M | M | M |
| Fenugreek | *Trigonella* | W | R | M | M | M |
| Palak | *Beata* | C | R | M | M | H |
| Bitter gourd | *Momordica* | C | W | H | M | H |
| Bottle gourd | *Lagenaria* | C | W | H | M | N |
| Cucumber | *Cucumis* | C | W | H | M | M |
| Pointed gourd | *Trichosanthes* | C | L | H | H | M |
| Pumpkin | *Cucurbita* | C | W | H | M | H |
| Luffa | *Luffa* | C | W | H | H | H |
| Tinda | *Praecitrullus* | C | L | M | M | M |
| Watermelon | *Citrullus* | C | R | H | M | M |
| Ash gourd | *Benincasa* | C | W | H | M | M |
| Snake gourd | *Trichosanthes* | C | W | H | M | H |

[a]C, cultivated; W, wild.
[b]W, wide spread distribution; R, regional distribution; L, localized distribution.
[c]H, high; M, medium; L, low.

are located in the economically developing parts of the world, namely the African, Asian, and Latin American continents, and the Far East (Madeley, 1996). In a reversal of the normal economic pattern in the world, the richest nations are poor in plant genetic diversity. Neglected vegetables can be found in every part of the world; however, a strikingly low amount of research and few gene bank accessions have been dedicated to species from Africa and the Asia-Pacific region (Meldrum *et al.*, 2018). Annual vegetables are by far the best conserved, while other growth forms are comparatively neglected (Meldrum *et al.*, 2018).

The major global emphasis over the last three decades has been to conserve germplasm diversity *ex situ* in medium- and long-term gene banks. The different approaches adapted for this include botanical gardens, arboreta, herbal gardens, field gene banks, seed gene banks, cryo-gene banks, *in vitro* gene banks, and DNA banks. There are several economically important species that are sexually propagated and can be conserved in conventional seed banks and cryobanks. Others are asexually propagated, do not have a natural mechanism of seed production, or possess high heterozygosity, and cannot be propagated uniformity through seeds. These crops include potato, sweet potato, yam, cassava, and a number of bulbous and tuberous crops, and spices; hence, these need to be propagated vegetatively in field gene banks or *in vitro* gene banks.

## 14.3 Biodiversity Conservation using Biotechnological Approaches

The word biotechnology was coined early in the 20th century by an agricultural engineer from Hungary, named Karl Freky. According to him, biotechnology is technology that includes all such work by which products are produced from raw material with the aid of living organisms. Biotechnology provides a range of tools and methods for assessment, monitoring, and managing biological resources, such as clarifying taxonomic and evolutionary relationships among groups of organisms and assessing the effects of ecosystem disturbance on components of biological diversity and biological processes. The tools of biotechnology can also be utilized for *in situ* conservation (i.e. assessment of optimal or minimal population size) and *ex situ* conservation (i.e. enhancing the quality of characteristics and efficiency) through compact storage of DNA libraries and sequence databases. Biotechnology has vast applications in characterization/evaluation, conservation, and utilization of plant genetic resources (Chaudhury and Malik, 2008). Plant multiplication is done under optimized conditions by producing adventitious or non-adventitious organs or adventitious embryos. Advances in biotechnology provide new methods for plant genetic resources and evaluation (Paunescu, 2009). Overall, biotechnology can play a greater role in every step of biodiversity conservation with improved methods of collecting and storing the diversity (as seed and tissue culture), improved long-term storage techniques, detection and elimination of duplications and diseases in gene bank collections, identification of useful new economically important alleles/genes, and safer and more efficient distribution of germplasm.

## 14.4 Molecular Markers for Estimating Genetic Diversity in Vegetable Crops

Genetic diversity refers to the inherited differences within and between populations of living organisms, and groups of related plants among which gene flow occurs freely. This diversity acts as the foundation for selection, leading to improvement among plant populations. In any crop, for proper conservation and better utilization of germplasm resources, it is necessary to understand its genetic diversity and distribution (Rao and Hodgkin, 2002). This helps in our understanding of the taxonomy, origin, and evolution of plant species of interest, which in turn helps in major decisions such as what to conserve and where to conserve it (Rao and Hodgkin, 2002). Many reports are available on diversity in a wide variety of vegetable crops based on morphological and biochemical markers. The main limitations with these markers are that they are limited in number and are highly influenced by environmental factors. DNA-based marker systems are preferable due to their abundance, robustness, and the fact that they are not affected by environmental factors. They are used largely because of their speed, productivity, ability to be reproduced, and ability to distinguish large numbers of polymorphisms in a specific population (Reddy *et al.*, 2002; Kuras *et al.*, 2004). Different types of molecular markers have been used successfully in diversity studies in many vegetable species. Some of these studies in different vegetable crops based on DNA markers are presented in Table 14.2.

## 14.5 Biotechnology for Storage and Conservation of Plant Genetic Resources

*Ex situ* conservation of plant germplasm by storing seeds is arguably the most effective and efficient method. Low storage costs, combined with ease of seed distribution and regeneration of whole plants from genetically diverse material, offer distinct advantages. Cryopreservation is one of the *ex situ* conservation methods and works based on the removal of all freezable water from tissues by physical or osmotic dehydration, followed by ultra-rapid freezing. The two main *ex situ* conservation techniques that are used are: (i) tissue-culture methods (*in vitro* conservation); and (ii) cryopreservation (in liquid nitrogen) (Pawar and Kale, 2011).

### 14.5.1 Tissue-culture techniques

Many national and international organizations have used *in vitro* techniques and applied them to more than 1000 species as a conservation method complementary to field gene banks. Clonal propagation is essentially the production of true-to-type plants through tissue culture on a suitable medium. Realization of *in vitro* conservation came about in the early 1970s at a time when the storage of microbial cultures was a routine procedure. Subsequently, *in vitro* conservation has progressed from mere speculation to development and implementation. Following the cell theory put forward by Matthias Schleiden and Theodor Schwann in 1839, the potential for regenerating a whole plant from a single cell was first hypothesized by the German botanist Gottlieb Haberlandt in 1902. The ultimate aim of tissue-culture systems used for the purpose of genetic conservation is the preservation of specific and unique individual genomes.

*In vitro* conservation is an alternative method for conservation of crop species where regenerable cells or cultures are maintained *in vitro* for a long time. It is especially effective in situations where seed banking is not an option, such as vegetatively propagated plants, recalcitrant seed species, and plants with inaccessible or non-viable seeds owing to grazing or disease damage, as well as those with large and fleshy seeds. Some vegetable species conserved under *in vitro* conditions are *Allium* spp., *Solanum* spp., *Musa* spp., *Colocasia esculenta*, *Manihot* spp., and *Ipomoea batatas* (Henshaw, 1975; Withers, 1985; Zapartan and Deliu, 1994; Ashmore, 1997; Withers and Engelmann, 1997; Engelmann and Engels, 2002; González-Benito *et al.*, 2004; Sarasan *et al.*, 2006). It is sometimes called micropropagation, which denotes plant propagation using small explants. Mostly shoot, leaf, flower pieces, immature embryos, hypocotyl fragments, or cotyledons are used as explants (Paunescu, 2009). Generally, younger and more rapidly growing tissues are suitable. Normal, true-to-type donor plants and vigorous and disease-free explants are the criteria for good-grade explants (Fay, 1992). Fragile tissues such as meristems, embryonic embryos, cotyledons, and hypocotyls, in general, require less sterilization than seeds or lignified organs (Paunescu, 2009). Seedlings grown from sterilized seeds can be used to obtain explants.

#### *Advantages of* in vitro *conservation*

True-to-type and disease-free plants can be obtained through *in vitro* culture in large numbers throughout the year from appropriate explants of a given species. *In vitro* conservation technology offers a large number of advantages:

**Table 14.2.** Genetic diversity studies based on DNA markers in different vegetable crops.

| Family | Vegetable crop | Germplasm material | Marker system used | Reference |
|---|---|---|---|---|
| Asteraceae | *Lactuca sativa* L. | Twenty-five lettuce genotypes | 45 random amplified polymorphic DNA (RAPD) markers | Sharma *et al.* (2018) |
| Brassicaceae | *Brassica rapa* L. | 173 *B. rapa* accessions with a worldwide geographic distribution | 51 simple sequence repeat (SSR) markers | Guo *et al.* (2014) |
| | *Ipomoea batatas* (L.) Lam | 137 landraces from different locations in Puerto Rico | 23 SSR markers | Rodriguez-Bonilla *et al.* (2014) |
| | | 69 sweet potato cultivars from different regions of Latin America | Amplified fragment length polymorphism (AFLP) markers | Zhang *et al.* (2000) |
| Cucurbitaceae | *Cucumis sativus* L. | 115 cucumber lines sampled from 3342 accessions worldwide | Deep resequencing genomic variation | Qi *et al.* (2013) |
| | *Momordica charantia* L. | 38 indigenous genotypes of bitter gourd including commercially released varieties from India and two promising gynoecious lines | Random amplified polymorphic DNA (RAPD) markers and agronomic traits | Dey *et al.* (2006) |
| | *Lagenaria siceraria* (Molina) Standl. | 36 bottle gourd landraces of South African origin | 11 SSR markers | Mashilo *et al.* (2017) |
| | *Sechium edule* (Jacq.) Sw. | 26 accessions of chayote across 12 Indian states | Directed amplification of minisatellite DNA (DAMD) markers | Jain *et al.* (2017) |
| | *Cucurbita pepo* L. | 64 pumpkin accessions from different geographical regions in China | 37 SSR markers | Wang *et al.* (2020) |
| | | Seven selfed and unselfed *C. pepo* landraces from three districts in the KwaZulu-Natal Province, South Africa | 36 RAPD and 55 SSR primers | Ntuli *et al.* (2015) |
| | *Cucurbita moschata* | 47 accessions | AFLP and sequence-related amplified polymorphism markers | Ferriol *et al.* (2004) |
| Malvaceae | *Abelmoschus esculentus* L. | 96 accessions of *Abelmoschus* sp. | 37 SSRs | Kumar *et al.* (2017) |
| Solanaceae | *Solanum melongena* L. | 28 accessions of eggplant | 14 RAPD markers | Singh *et al.* (2006) |
| | | 287 worldwide eggplant accessions | 45 SSR markers | Liu *et al.* (2018) |
| | | 96 Indian eggplant landraces | 32 SSR markers | Chourey *et al.* (2017) |

| | | | | |
|---|---|---|---|---|
| | *Solanum tuberosum* L. | 482 potato accessions conserved in the National Agrobiodiversity Center, South Korea | 24 SSR markers | Lee *et al.* (2021) |
| | | 292 genotypes | 30 SSR markers | Wang *et al.* (2019) |
| | *Solanum lycopersicum* L. | Nine determinate and indeterminate tomato inbred lines collected from China, Japan, S. Korea, and USA | 35 SSR markers | Benor *et al.* (2008) |
| | *Capsicum annuum* L. | 15 accessions of *C. annuum* collected from Cross River and Ebonyi States of Nigeria | 10 start codon-targeted and ten DAMD markers | Igwe *et al.* (2019) |
| Umbelliferae | *Daucus carota* L. | 88 cultivars of carrot from eastern (Asian) and western (European) genetic pools | 30 SSR markers | Baranski *et al.* (2012) |

**1.** *In vitro* conservation enables conservation of plants not producing fertile seeds such as male-sterile plants.
**2.** Heterozygosity can be maintained indefinitely. The same is required in conservation of $F_1$ hybrid varieties.
**3.** A wide range of genotypes can be maintained under similar simulated conditions within a limited space.
**4.** Storage is in pathogen-free state (aseptic conditions).
**5.** The collection remains unexposed to natural and climatic changes.
**6.** It is easier to get rid of fungal and bacterial pathogens, but special techniques such as meristem culture coupled with chemo- or thermotherapy are required to eliminate viral or mycoplasma pathogens.
**7.** Maintenance is independent of seasons.
**8.** Long-term storage of crops with short seed viability or plantlets under various stages is possible.
**9.** The multiplication rate is very high.
**10.** Genetic erosion is decreased to zero under optimal conditions.
**11.** There are reduced labor costs.
**12.** There is a ready availability of material for distribution.
**13.** A constant flow of disease-free plantlets/clones can be achieved.
**14.** Easy transportation within the country is possible for research purposes.
**15.** The micro-propagated plants (*in vitro* material) are free from disease and pests, and are multiplied under aseptic conditions so can be exported/exchanged internationally with limited quarantine restrictions.

### *Disadvantages of* in vitro *conservation*

Some species do not respond desirably in tissue culture, which limits the use of these techniques. The problems or disadvantages with *in vitro* conservation include the following:

**1.** Tissue culture is time consuming for initial preparation and maintenance.
**2.** *In vitro* culture conditions have to be determined for each species (and sometimes genotype) to be conserved, which can often be a matter of trial and error.
**3.** In general, there may be loss of morphogenic potential over time.
**4.** Genetic variability in cultures due to chromosomal changes both structural and numerical (i.e. somaclonal variations) can distort the genetic uniformity of *in vitro* cultures (Larkin and Scowcroft, 1981).
**5.** The maximum period for which *in vitro* techniques can be applied is not ascertained.
**6.** The choice of explant for micropropagation differs from species to species.

Tissue-culture techniques have been adopted to conserve germplasm of wild, endangered, and threatened species, as well as selected elite varieties. Many plant species have a tendency to undergo somatic embryogenesis directly from the explant or following a short intervening callus phase under an appropriate auxin and cytokinin ratio. Thidiazuron is very effective in inducing somatic embryogenesis in many plants. In India, different aspects of plant tissue-culture research have been carried out at institutes of government departments. A number of research groups are working in about 100 different universities and research institutions all over the country. Prominent among these are the National Chemical Laboratory (NCL), Pune; Botany Department, Delhi University; Tata Energy Research Institute (TERI), New Delhi; Central Institute for Medical and Aromatic Plants (CIMAP), Lucknow (Uttar Pradesh); Indian Council of Agricultural Research based institutes like ICAR-Sugarcane Breeding Institute (ICAR-SBI), Coimbatore (Tamil Nadu); ICAR-Indian Agricultural Research Institute (ICAR-IARI), New Delhi; ICAR-Indian Institute of Spices Research (ICAR-IISR), Calicut; ICAR-Central Tuber Crops Research Institute (ICAR-CTCRI), Trivandrum; and ICAR-National Bureau of Plant Genetic Resources (ICAR-NBPGR), New Delhi. The *in vitro* repository at NBPGR maintains a large collection of various tropical and temperate crops of more than 52 genera and 158 species (Mandal and Chaudhury, 2004).

### *Techniques of* in vitro *micropropagation*

The term *in vitro* micropropagation covers a wide range of techniques that can be broadly categorized as: (i) organ culture; (ii) callus culture and embryogenesis; (iii) suspension culture; (iv) isolation and culture of single cell and protoplast; and (v) anther and pollen culture (Fig. 14.1).

Tissue-culture techniques owe their success to the inherent feature of totipotency of plant cells (i.e. each cell in a plant has the ability to

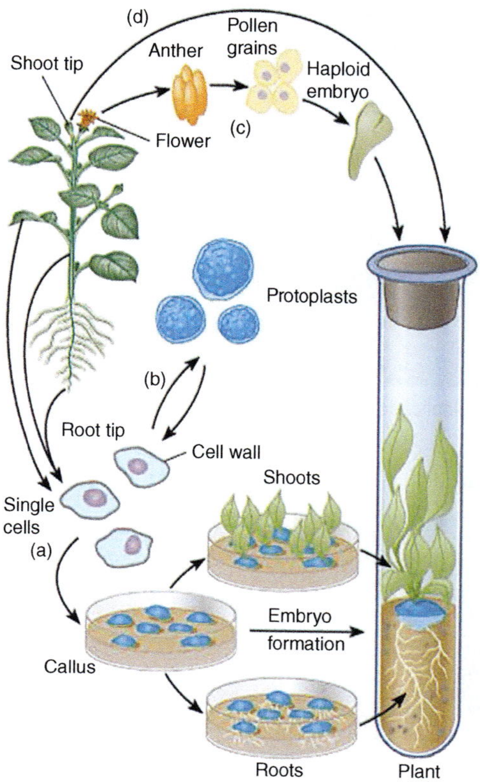


**Fig. 14.1.** Summary of tissue-culture techniques.

produce new progeny). Multiplication in cultures is achieved through subculture of organized tissues (e.g. meristems, somatic/zygotic embryos) or re-differentiated tissues (e.g. cells, protoplasts).

Micropropagation is a prerequisite for *in vitro* conservation. The various steps that are followed for developing a successful micropropagation protocol can be summarized as: (i) selection of stock/mother plant; (ii) establishment of aseptic culture; (iii) multiplication of propagules; (iv) hardening of plantlets; and (v) transfer to the greenhouse or field.

### *Types of* in vitro *culture*

The techniques used for micropropagation are: (i) meristem culture; (ii) proliferation of axillary buds; (iii) growth of nucellar embryos; (iv) growth of somatic embryos; and (v) *in vitro* tuberization and cormlet production.

For an *in vitro* conservation program, organized explants have proved better than unorganized tissue. Suitable explants are processed, surface sterilized, and cultured on solidified or liquid medium under aseptic conditions. The aseptic cultures are then transferred to proliferation medium containing hormones, avoiding any intervening callus phase. The proliferated shoots are then rooted for transplanting from *in vitro* conditions. Currently, this technique is being commercially exploited for large-scale rapid multiplication of elite stock. *In vitro* clonal multiplication aims to rapidly produce larger numbers of disease-free and true-to-type plants. Murashige and Skoog (MS) medium is most commonly used. The stages in plantlet formation include shoot bud initiation, development of buds into shoots, rooting of shoots, and preparation of plantlets for transplantation from culture vessels to soil. Such *in vitro* techniques are very powerful tools, but caution is required in terms of the conformity of progeny plants.

Studies were conducted to standardize the protocol for large-scale *in vitro* multiplication of pointed gourd at ICAR-Indian Institute of Vegetable Research (IIVR), Varanasi (Uttar Pradesh) (Fig. 14.2). Young vines of *Trichosanthes dioica* VRPG-101 (an advanced breeding line) were collected from polyhouse-grown plants. Shoot tips and nodal portions of vine were excised and after surface sterilization were inoculated on to hormone-free, half-strength MS medium. The established shoots served as the mother stock. For further multiplication, shoot tips and nodal portions were excised from the mother stock and cultured on MS medium supplemented with different concentrations of 6-benzylaminopurine (BAP). For root induction, *in vitro*-raised shoots were cultured on MS medium supplemented with different concentrations of indole-3-butyric acid (IBA) or 1-naphthaleneacetic acid. Shoots with primary and secondary roots were transferred to pots containing a soil and sand mixture. The plantlets were kept under controlled conditions and irrigated with quarter-strength MS medium. Finally, they were transferred to the field for evaluation. Shoot multiplication was achieved on medium supplemented with different concentrations of BAP. A significant interaction between growth regulator level and explants was recorded for the number of shoots per explant, number of nodes per shoot, and shoot length. The highest frequency of shoot regeneration (100%) was observed on MS medium supplemented

**Fig. 14.2.** Tissue culture for large-scale multiplication of elite clones of pointed gourd.

with 8.8 µM BAP. Root induction was also achieved with different concentrations of IBA, and the best response was obtained with medium containing 0.49 µM IBA. Plants were acclimatized under controlled conditions and transferred to the field, where they produced normal flowers and fruit.

An efficient protocol for rapid *in vitro* clonal propagation of spine gourd (*Momordica dioica* Roxb.) genotype RSR/DR15 (female) and DR/NKB-28 (male) was developed by enhanced axillary shoot proliferation from nodal segments (Rai *et al.*, 2012). Maximum shoot proliferation of 6.2 shoots per explant with 100% shoot regeneration frequency was obtained from the female genotype on MS medium supplemented with 0.9 µM BAP and 200 mg $l^{-1}$ casein hydrolysate (Table 14.3). From the male genotype, the optimum shoot regeneration frequency was 86.6% and 6.4 shoots per explant were obtained on MS medium supplemented with 2.2 µM BAP. Casein hydrolysate induced vigorous shoots, promoted callus formation, and proved inhibitory for shoot differentiation and shoot length, especially in explants of the male genotype. Rooting was optimum on half-strength MS medium (male 92.8%, female 74.6%) containing 4.9 µM IBA. The plantlets were transferred to plastic cups containing a mixture of cocopeat and perlite (1:1) and then to soil after 2–3 weeks. Overall, 84% of female and 81% of male regenerated plantlets survived and grew vigorously in the field. The genetic stability of the regenerated plants was assessed using random amplified polymorphic DNA (RAPD). The amplification products were monomorphic in the *in vitro*-propagated plants and similar to those of the mother plant. No polymorphisms were detected, revealing the genetic integrity of the *in vitro*-propagated plants. This micropropagation procedure could be useful for raising genetically uniform planting material of known sex for commercial cultivation or for build-up of plant material of a specific sex type.

Rai *et al.*, 2012 also determined the effects of concentrations of IBA and MS medium on root induction in spine gourd (Table 14.4).

**Table 14.3.** Effects of concentrations of 6-benzylaminopurine (BAP) and casein hydrolysate on *in vitro* shoot proliferation from nodal explants of *Momordica dioica* after 30 days. From Rai *et al.* (2012).

| | | Male | | | Female | | |
|---|---|---|---|---|---|---|---|
| BAP (μM) | Casein hydrolysate | Shoot response (%) | No. shoots per explant (mean ± SE) | Average shoot height (cm ± SE) | Shoot response (%) | No. shoots per explant (mean ± SE) | Average shoot height (cm ± SE) |
| – | – | 87.0 | 1.2±0.1 | 6.2±0.6 | 80.4 | 1.2±0.0 | 2.4±0.3 |
| – | + | 85.1 | 1.0±0.0 | 4.0±0.3 | 74.9 | 1.0±0.0 | 3.2±0.4 |
| 0.9 | – | 100.0 | 2.4±0.3 | 1.4±0.1 | 100.0 | 3.1±0.2 | 8.8±0.9 |
| 0.9 | + | 89.0 | 2.4±0.2 | 0.7±0.0 | 100.0 | 6.2±0.4 | 3.4±0.4 |
| 2.2 | – | 86.6 | 6.4±0.2 | 2.0±0.1 | 98.2 | 3.3±0.2 | 0.0±0.2 |
| 2.2 | + | 87.5 | 5.0±0.2 | 2.1±0.1 | 98.2 | 4.5±0.2 | 0.8±0.2 |
| 4.4 | – | 73.6 | 3.6±0.5 | 2.0±0.4 | 72.7 | 3.2±0.5 | 1.6±0.2 |
| 4.4 | + | 69.6 | 3.6±0.4 | 1.5±0.3 | 62.2 | 2.5±0.3 | 1.6±0.3 |
| 8.9 | – | 60.6 | 2.3±0.6 | 1.4±0.2 | 83.7 | 1.5±0.3 | 1.2±0.2 |
| 8.9 | + | 31.1 | 1.2±0.1 | 0.5±0.1 | 70.3 | 1.4±0.2 | 1.1±0.2 |

**Table 14.4.** Effects of concentrations of IBA and MS medium on root induction from *in vitro*-proliferated shoots of *Momordica dioica* after 40 days. From Rai *et al.* (2012).

| | | Male | | | Female | | |
|---|---|---|---|---|---|---|---|
| MS medium strength | IBA (μM) | Root response (%) | Number of roots (mean ± SE) | Average root length (cm ± SE) | Root response (%) | Number of roots (mean ± SE) | Average root length (cm ± SE) |
| Half | – | 38.3 | 2.6±0.3 | 2.5±0.2 | 34.0 | 1.3±0.2 | 1.2±0.2 |
| Full | – | 30.6 | 2.3±0.1 | 2.0±0.1 | 12.9 | 0.9±0.7 | 1.0±0.5 |
| Half | 1.0 | 27.4 | 4.0±0.3 | 2.0±0.2 | 36.1 | 4.0±0.6 | 2.0±0.2 |
| Full | 1.0 | 14.4 | 3.1±0.3 | 0.8±0.1 | 26.4 | 2.6±0.2 | 1.6±0.1 |
| Half | 2.5 | 64.3 | 2.3±0.3 | 2.5±0.1 | 76.2 | 2.6±0.3 | 3.9±0.7 |
| Full | 2.5 | 32.2 | 3.8±0.3 | 1.9±0.1 | 40.1 | 1.6±0.1 | 3.4±0.1 |
| Half | 4.9 | 92.8 | 3.8±0.2 | 4.1±0.4 | 74.6 | 4.6±1.2 | 4.6±0.5 |
| Full | 4.9 | 49.5 | 5.9 ±1.8 | 2.2±0.3 | 69.3 | 2.6±0.6 | 3.2±0.9 |
| Half | 9.9 | 97.1 | 8.6±1.8 | 2.2±0.2 | 76.4 | 4.8±0.6 | 2.2±0.1 |
| Full | 9.9 | 47.4 | 1.4±0.2 | 0.7±0.1 | 25.0 | 1.1±0.2 | 1.4±0.2 |
| Half | 24.6 | 67.5 | 9.1±0.8 | 0.6±0.0 | 73.0 | 7.4±1.0 | 0.5±0.1 |
| Full | 24.6 | 0.0 | 0.0±0.0 | 0.0±0.0 | 0.0 | 0.0±0.0 | 0.0±0.0 |

### *Slow-growth method*

The slow-growth method is an *in vitro* conservation technique where cells, tissues, or organs are conserved in glass or plastic containers under aseptic conditions through slow growth of the cultures. The slow-growth approach is used mainly for short- and medium-term conservation. This method is also known as a growth-limitation approach. Withers (1991) defined the slow-growth method as follows: "Slow-growth methods refer to any tissue-culture procedure aimed at decreasing the growth rate of plant tissues *in vitro*." This involves a reduction in the growth rate of cultures with the aim of increasing the duration of subculturing events. The goal of this method is to obtain the longest period of subculture without detrimental effects on the plant tissues. Cells and/or cultures are stored in non-freezing conditions, but their growth rate is curtailed to a minimum by manipulation of the temperature, media composition, and culture environment. The subculturing period is extended by 100%. Lowering the temperature, use of osmotic retardants such as mannitol and sorbitol, partial desiccation, lowering of atmospheric pressure, and use of chlormequat chloride (Cycocel) are the basis of the slow-growth method. There is no requirement for additional equipment, but this method demands careful imposition of stress so that growth occurs at a slow pace and degenerative processes do not occur. For example, potato meristem, tomato callus, and somatic embryos of carrot have survived storage using this growth-limitation approach. This approach has great commercial application, especially during lean or off-periods. Propagules can be safely stored, and active planting material can be regenerated by removing the growth-limitation factor(s).

Slow growth of cultures can be achieved using one or a combination of the following techniques:

**1.** *Minimal growth at reduced temperature and/or light intensity.* The most commonly used single method for restricting the growth of tissues is by decreasing the temperature in combination with decreased light intensity at which the cultures are maintained. Cultures that are normally kept at a temperature of around 25°C exhibit a reduction in growth when transferred to 10–15°C. A basic storage technique can be developed by decreasing the temperature to prolong the subculture period without significant damage (Withers, 1985). Tropical plant species are generally cold sensitive. The physiological damage induced by low temperature is referred to as chilling injury. Therefore, storage temperature depends on the cold sensitivity of the species.

**2.** *Use of growth retardants in the medium.* Inhibitory levels of growth retardants such as meleic hydrazide, abscisic acid (ABA), N-dimethyl succinic acid (DSA), Cycocel, and phosphon-D have been added to the culture medium in a few cases with fairly good success, although the optimal level of the retardants needs to be determined by experimentation on a particular system. It has generally been found that ABA and DSA (5–50 mg $l^{-1}$) and CCA (2–20 mg $l^{-1}$) can be used safely.

**3.** *Decreasing nutrient content in the culture medium.* Lowering the nutrient content and/or level of sucrose has proved a successful and simple technique to restrict growth. Kartha *et al.* (1981) conserved *Coffee arabica* plantlets for 2 years on a medium devoid of sugar and only half of the mineral solution of the standard medium.

**4.** *Use of osmotic regulators.* Use of osmotic regulators such as mannitol, sorbitol, and sucrose is recommended, as they are relatively metabolic inert, where the intention is to minimize growth by imposing a level of osmotic stress on the culture. The selected osmotic inhibitor is incorporated into fresh culture medium, generally in the range of 3–6% (w/w).

**5.** *Type of enclosure.* It has been observed that the type of enclosures of the culture vessel influences the rate of evaporation of water content of the medium. Therefore, by using propylene caps instead of cotton plugs in the culture vessel, the storage period of the culture can be increased (Balachandran *et al.*, 1990).

**6.** *Size and type of culture vessel.* The size and type of culture vessel play an important role in slow-growth cultures. Roca (1984) indicated that when storing cassava plantlets in 50 × 50 mm bottles instead 25 × 15 mm test tubes, the rate of shoot elongation in the larger vessel was almost doubled; however, leaf fall diminished and culture viability increased. They also observed that the leaves and roots remained healthier in larger vessels than in smaller vessels. Reed (1991) reported the use of sterile, heat-sealable polypropylene bags with the advantage of reduced occupied space in culture rooms.

**7.** *Physiological stage of explants.* The type and physiological stage of explants are very important factors in slow-growth culture. The minimum size of explants should be selected for conservation. The duration between the last transfer and the moment when cultures are placed in storage conditions can be of great importance. It is sometimes better to culture immediately after transfer, which avoids necrosis of the tissues and production of phenolic compounds.

**8.** *Reduction in oxygen concentration.* Minimal growth can be achieved by lowering the oxygen level available to the tissues in culture containers. The simplest method is to cover the tissues with a mineral oil layer. However, some problems associated with decreased oxygen concentration are vitrification and partial or complete necrosis. To simplify the technique, cultures are covered with mineral oil, paraffin oil, or liquid medium.

**9.** *Modification of the gaseous environment.* Some gases such as carbon dioxide and ethylene inside the culture vessel also influence the growth rate of the tissues. However, it is complicated and expensive to control and maintain the gas atmosphere of a large number of individual culture vessels. The development of gas-permeable culture vessels used for strawberry (Reed, 1991) opened up the possibility of restricting growth by controlling the gaseous environment of the entire culture room, which may be more feasible than controlling individual culture vessels. Although this technique is simple with great potential, it is still viewed as an experimental technique by virtue of these limitations.

**10.** *Combined treatments.* Very little systematic work has been done on the judicious use of combinations of the treatments outlined above and this may be more useful for *in vitro* conservation, for example a reduced temperature together with a growth retardant has been used successfully for potato (*Solanum tuberosum*). Investigations on the potential of combined treatments rests with the individual worker, and the safest general approach is to evaluate the effects of each of the treatments before attempting to use them in combination.

### 14.5.2 Cryopreservation

Cryopreservation can be defined as storage of biological samples in viable conditions at ultra-low temperatures such as on solid carbon dioxide (–79°C), in low-temperature deep freezers (–80°C), in the vapor phase of nitrogen (–150°C) or in liquid nitrogen (–150 to –196°C). The science pertaining to cryopreservation is known as cryobiology. Since the discovery of glycerol as a cryoprotectant, it has been used to protect life in the frozen state. Theoretically, cells and tissues can be cryopreserved in a live state indefinitely, and the cells/tissues recovered after thawing should be unchanged and live. This reduces the problems associated with seed regeneration schemes, especially the risk of loss of genes/analyses caused by environmental problems. Cryopreservation is preferred for long-term conservation of germplasm. The basis of cryopreservation is changes in the water present in the cells to a solid state when subjected to subzero temperatures. The cells initially undergo supercooling but regain thermodynamic equilibrium subsequently, either by losing water or by intracellular freezing. The metabolic processes are considerably slowed down and the longevity of the material is enhanced. The seeds of desiccation-tolerant plants can easily be stored at low temperature (–20°C) and low humidity for several years. For desiccation-sensitive and recalcitrant seeds and vegetatively propagated plants, certain tissue-culture methods have to be used for medium- and long-term storage of germplasm without affecting its genetic stability. The utilization of liquid nitrogen as a storage medium is predicated on the ability of seeds to survive liquid nitrogen exposure without significant damage to variability. Long-term storage can be done through cryopreservation of appropriate plant tissue cultures in liquid nitrogen (–196°C) after a slow preliminary cooling to a certain level (Withers and Engelmann, 1997).

#### *Cryopreservation techniques*

Cryopreservation has been applied successfully to calluses, protoplasts, pollen, meristems, zygotic and somatic embryos, and suspension cultures. There are two main cryopreservation technique: classical cryopreservation (slow freezing) and a more recent/new cryopreservation technique.

CLASSICAL CRYOPRESERVATION TECHNIQUE (SLOW FREEZING). This technique was used in all initial studies using plant cell and tissue cultures. It involves chemical cryoprotection and freeze-induced dehydration. Basically two methods have been used:

**1.** *Two-step freezing.* This includes an initail slow controlled cooling of the sample to a defined pre-freezing temperature using a programed freezer followed by rapid immersion in liquid nitrogen. Freezing rates of 0.5–2°C $min^{-1}$ down to a freezing temperature of around –40°C generally give satisfactory results. In order to avoid the use of a sophisticated programmable freezer, domestic or laboratory deep freezers have been used successfully to perform the slow cooling step.

**2.** *Use of cryoprotectants.* In this procedure, samples are submitted to a cryoprotective treatment for a defined period before freezing. Various cryoprotectants including dimethysulphoxide (DMSO), glycerol, sorbitol, mannitol, sucrose, and polyethylene glycol are used either alone or in binary or tertiary mixtures. Mixtures of cryoprotectants have generally proved to be more effective than using a single cryoprotectant.

These techniques have been applied successfully to undifferentiated culture systems such as cell suspensions and calluses (Withers and Engelmann, 1997). In the case of differentiated structures, these techniques have been used for freezing apices of cold-tolerant species (Reed and Chang, 1997).

The cryopreservation process involves a number of successive steps, which have to be defined for each species: (i) selection; (ii) excision of tissues or organs; (iii) culture of the source material; (iv) selection of healthy cultures; (v) application of cryoprotectants; (vi) pregrowth treatments; (vii) cooling/freezing; (viii) storage; (ix) warming and thawing; (x) recovery growth; (xi) viability testing; and (xii) post-thawing.

When selecting a suitable material, rapidly growing meristems that are small and relatively resistant to freezing, possessing fewer or smaller storage cells and dense cytoplasm, are preferred for cryopreservation (Engelmann, 1991). The material can be sampled, preferably from *in vitro* cultures to avoid contamination. The physiological stage of the explant is also important, and the material should be used from fresh active cultures (Harding *et al.*, 1991).

Pregrowth refers to the period in culture before cryopreservation begins. It offers the chance to induce physiological changes in the tissues to increase freeze tolerance. The plant tissues can be cold hardened using effective pregrowth additives such as mannitol, sorbitol, proline, or DMSO in the pregrowth culture medium.

The tissues should undergo cryoprotection just before freezing. Effective cryoprotectants include DMSO, glycerol, proline, sucrose, sorbitol, and polyethylene glycol. These compounds can be added to the medium individually or in combination (Fig. 14.3).

The steps of freezing, storage, and thawing must be followed very carefully to avoid damage during freezing. Slow freezing is carried out using a controlled freezing apparatus and may cause extracellular ice formation. In contrast, during rapid freezing, intracellular ice crystals do not have time to form to a damaging size before all the intracellular water has frozen. Depending on the species, the freezing rate can be very precise as in strawberry, or may fall within a much broader range as in somatic embryos of oil palm (Englemann and Dereuddre, 1988).

It is important to have high viability levels after thawing for recovery growth. Usually, recovery growth is carried out on a defined medium supplemented with activated charcoal and growth substances. Charcoal helps to adsorb toxins released by damaged cells, and growth substances help to stimulate the desired responses. It is sometimes necessary to avoid osmotic shock, which can be done by using an intermediate medium

**Fig. 14.3.** Potato apices embedded in 10% DMSO droplets on aluminium foil.

with low osmotic potential and by successive transfer of material to progressively less concentrated medium (Engelmann *et al.*, 1985). To avoid photo-oxidation, which may be harmful, recovery can be carried out in the dark (Benson *et al.*, 1989).

RECENT CRYOPRESERVATION TECHNIQUE. More recent cryopreservation techniques are based on vitrification where cell desiccation is performed either by exposure of samples to concentrated cryoprotective solutions or by air desiccation. Vitrification-based procedures are superior to classical freezing techniques and are more appropriate for complex organs such as shoot tips and embryos, which contain a variety of cell types. Vitrification procedures are operationally less complex than classical procedures (e.g. they do not require the use of controlled freezers) and have greater applicability (Sakai, 2000). It can be achieved using different procedures, such as desiccation, pregrowth desiccation, encapsulation–dehydration, vitrification, encapsulation–vitrification, and droplet-freezing (Engelmann, 2003).

For desiccation, zygotic embryos are extracted from the seeds and dehydrated under laminar flow or in an air-tight container with silica gel. Freezing is done rapidly by placing the embryos in liquid nitrogen, and regrowth of the embryos after freezing is carried out directly on a suitable medium. Dehydration must be just sufficient and should not be to a high degree, which can induce injury. By using this technique, the excised embryonic axes of tea, jack fruit, trifoliate orange, almond, cocoa, and walnut, and embryos of coconut and coffee have been preserved successfully (Chaudhury and Chandel, 1995a).

Pregrowth desiccation consists of growing/initiating growth of the plant material for different durations (hours to weeks, depending on the material) on medium supplemented with cryoprotectants, and partially desiccating the material before rapid freezing in liquid nitrogen. It has been applied successfully to somatic embryos of oil palm, coffee, pea, and melon, and to zygotic embryos of coconut.

Encapsulation–dehydration is based on the development of "synthetic seeds," which are somatic embryos, shoot/axillary buds, and shoot apices encapsulated in a gel matrix of calcium alginate beads (Fig. 14.4). For cryopreservation, explants are dissected under a microscope and cultured for a defined period on a specific medium. These are subsequently encapsulated in alginate beads and allowed to grow in liquid medium with a high concentration of sucrose (0.5–1.25 M). Encapsulated explants are dehydrated using slow or rapid desiccation techniques under laminar flow to decrease the moisture of the beads to 20%. Freezing is done by placing the beads in liquid nitrogen. Recovery can be made

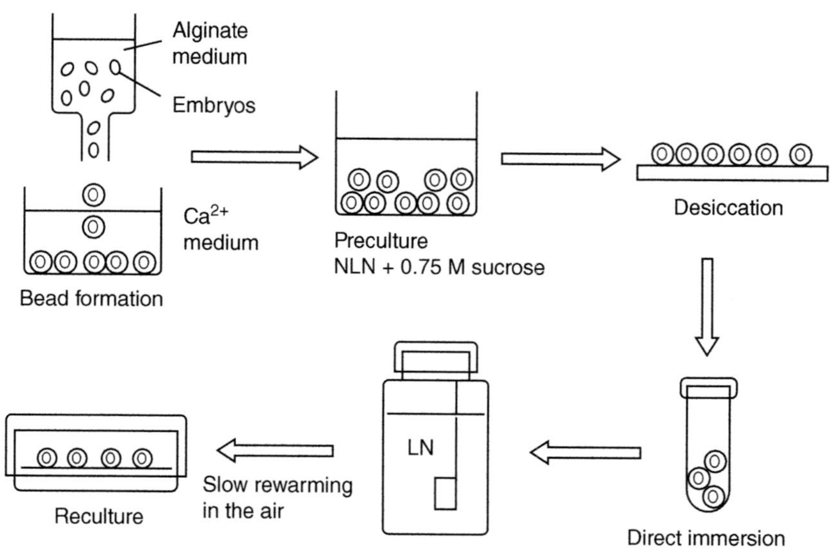


**Fig. 14.4.** The process of encapsulation. PVS2: Plant Vitrification Solution 2.

after slow thawing and using a standard protocol for a particular species. Encapsulation–dehydration has been reported to be successful for potato, sweet potato, and yam, among others.

Vitrification is the phase transition of water from a liquid phase to a non-crystalline or amorphous glass. It consists of treating the plant material with extremely concentrated solutions of cryoprotectant (Plant Vitrification Solution 2 (PVS2), DMSO) and freezing it ultra-rapidly. Consequently, intracellular water vitrifies (i.e. forms an amorphous glassy material). However, it cannot be used for cryoprotectant-sensitive plant material. Vitrification procedures have been developed for more than 20 different species using protoplasts, cell suspensions, apices, and somatic embryos. Vitrification has been successful with shoot tips/meristems of various species such as garlic, yam, cassava, and *Solanum* spp. (Sakai, 2000).

Encapsulation–vitrification involves a combination of the techniques of encapsulation–dehydration and vitrification, whereby vitrification is used to freeze the alginate beads holding the explants. It has been observed that the recovery rate of apices frozen using this technique was 30% higher than with the encapsulation–dehydration technique alone.

DESICCATION TOLERANCE. Optimal survival through freezing was obtained when axes were dried to a moisture level just above that producing desiccation damage. Detailed investigations regarding the nature of water present in recalcitrant seeds using differential scanning calorimetry have revealed that the degree of desiccation tolerance of any seed or axes is determined by the amount of freezable water present at any stage and the ability to withstand the loss of a substantial proportion of non-freezable water. In studies on tea (*Camellia sinensis*) embryonic axes, cooling rates were shown to affect the properties of water in the axes. Increasing the cooling rates and drying the axes to a moisture content between 1.1 and 1.6 g of water $g^{-1}$ could increase the survival of the axes in culture (Wesley-Smith *et al.*, 1992). Thus, in different recalcitrant seed species, various physiological factors play a role in imparting different degrees of desiccation tolerance. Thus, the development of separate freezing protocols is likely to be needed for each of them.

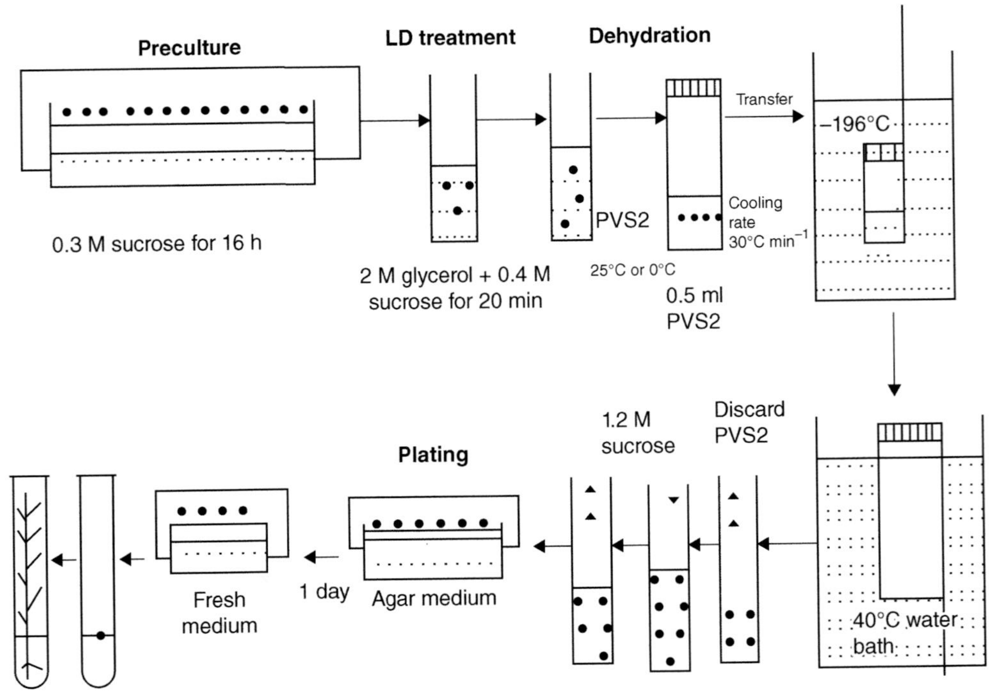


**Fig. 14.5.** Vitrification procedure for embryonic axes.

APPLICATION OF CRYOTECHNIQUES. In most reports of successful cryopreservation, the technique involved has been the aseptic excision of embryonic axes, desiccation to around 11–16% moisture level using air drying or the addition of cryoprotectants, and exposure to liquid nitrogen temperatures either directly or using a controlled rate of freezing. The axes are later thawed rapidly in a water bath maintained at 38°C and cultured on defined media to obtain complete plants. In some cases, plants have also been successfully transferred to the field. More recently, pretreatment of zygotic/somatic embryos by culture on medium with a high sugar content and partial desiccation has proved successful. The results obtained with embryos of coffee, oil palm, and coconut have demonstrated that the critical phase of the various cryopreservation processes is the pretreatment phase, during which desiccation tolerance of embryos is induced using high levels of sugar and/or desiccation (Englemann *et al.*, 1985).

WHAT MATERIALS CAN BE CRYOPRESERVED? Cryopreservation has proved to be extremely useful in the long-term conservation of seeds, vegetative propagules, meristematic tissues and organs, and *in vitro* cultures. Zygotic embryos and embryonic axes of recalcitrant seeds can be used successfully for cryopreservation (Chandel *et al.*, 1995). Embryonic axes are preferred because of their organized small structure, independent identity, and the presence of appreciable protection of meristematic tissues. In fact, cryopreservation is the only technique currently available for long-term storage of vegetatively propagated species and non-orthodox seed species (Bajaj, 1995).

Seeds are suitable for cryostorage. Based on their storage characteristics, seeds have been classified as orthodox or non-orthodox (intermediate and recalcitrant) (Fig. 14.6), and different protocols are used for their cryopreservation (Chaudhury and Malik, 2008). Orthodox seeds are capable of retaining viability even after being dried to less than about 5% moisture content and are tolerant to liquid nitrogen. Seeds of most common agricultural species belong to this category and can be safely cryopreserved, thus providing a means of extending their viability more than is possible through conventional conservation at –20°C. Pre-desiccation followed by direct

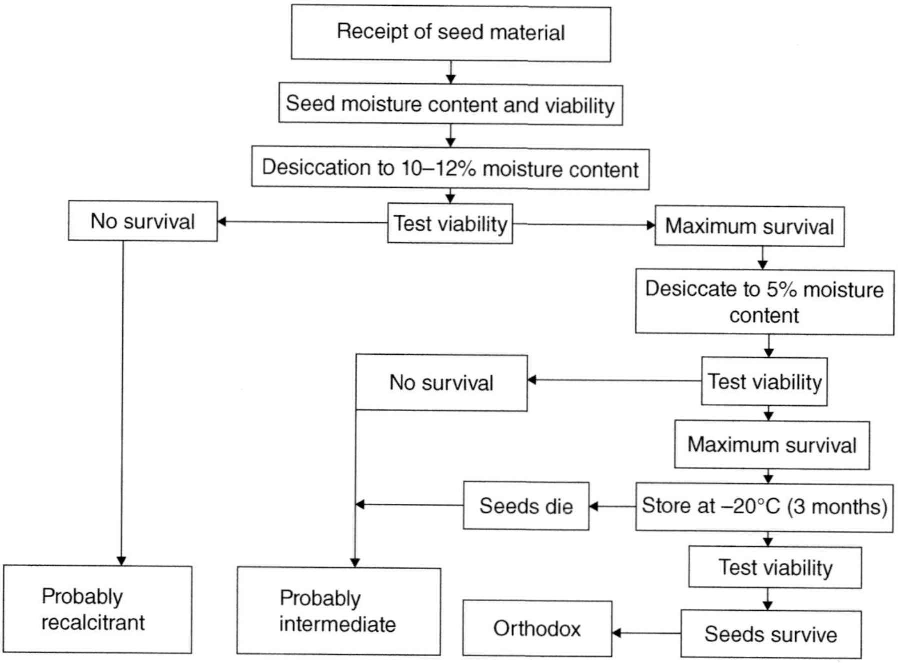


**Fig. 14.6.** Protocol to determine seed storage behavior. Modified from Hong and Ellis (1996).

immersion in liquid nitrogen or after slow cooling is the commonly applied method of cryopreservation for orthodox seeds. At NBPGR, long-term conservation of a large number of orthodox seed collections comprising cereals, pulses, oilseeds, vegetable crops, and other economically important species has been successfully achieved using liquid nitrogen-based cryopreservation (Chaudhury and Malik, 2008). Non-orthodox seeds are shed with high moisture levels and are desiccation and freeze intolerant. They lose viability after being dried below a critical limit, usually between 12% and 30% moisture content. At these moisture levels, the seeds cannot be subjected to subzero temperatures as they undergo freeze injury. In large-seeded species of several tropical and temperate non-orthodox species, the excised embryonic axis is the preferred explant for cryopreservation.

Aseptically excised embryonic axes are either desiccated to around 11–16% moisture level using air drying or treated with cryoprotectants before immersing in liquid nitrogen. The axes are later thawed rapidly in a water bath maintained at 38°C and cultured on defined media to obtain complete plants. Embryonic axes of tea (Chaudhury *et al.*, 1991), jackfruit (Chandel *et al.*, 1995), and neem, black pepper, almond, litchi, and whole seeds of cardamom can be cryopreserved (Chaudhury and Chandel, 1995a). The use of embryonic axes for cryostorage of temperate tree species with short-lived large seeds has been done for *Juglans*, *Aesculus*, *Castanea* and *Quercus* spp. (Pence, 1990).

Preserving pollen of wild species and cultivated crops is important to safeguard the pollen parents in any crop-improvement program and dispenses with the need to grow the same plants frequently. Long-term pollen preservation forms a part of integrated gene resource conservation (Roberts, 1975). Pollen can be made available in abundant quantities and forms an ideal compact material for conservation of recombined genetic information present in pollen at the gametophytic stage. Pollen viability and longevity varies with different crop species, and methods are available for preserving vegetable pollen for extended time periods under cryogenic conditions (Alexander and Ganeshan, 1989). For long-term conservation, potential wild species of vegetable crops can be considered. Wild species of tomato (*Solanum pimpinellifolium* Mill.) and eggplant (*Solanum indicum* L.) and the eggplant (*S. melongena* L.) cultivars 'Arka Kusumakar' and 'Arka Shirish' were successfully conserved through cryostorage (Table 14.5) (Rajasekharan *et al.*, 1998). The results of field pollinations after storage are shown in Table 14.6.

Pollen from *Cucumis sativus*, wild species of *Cucumis hardwickii*, and *Abelmoschus* spp. (*A. callei*, *A. manihot*, and *A. pungens*) have been cryostored for use by breeders in crossing programs (Chaudhury and Malik, 2008). In onion, cryostored pollen gave significant results in fruit and seed set (Ganeshan, 1986). Lotus pollen can be successfully stored at −20±1°C for 14 days with silica gel as a desiccant. Seed formation is possible with stored

**Table 14.5.** Pollen viability in eggplant and tomato under fresh and liquid nitrogen storage. Data from Rajasekharan *et al.* (1998).

| | Eggplant | | | Tomato |
|---|---|---|---|---|
| | 'Arka Kusumakar' (9 years cryostored in liquid nitrogen) | 'Arka Shirish' (3.5 years cryostored in liquid nitrogen) | *Solanum indicum* (9 months cryostored in liquid nitrogen) | *S. pimpinellifolium* (2 years cryostored in liquid nitrogen) |
| Fresh pollen viability (%) | 45.96 (51.71) | 48.32 (55.72) | 50.50 (59.55) | 45.55 (51.00) |
| Stored pollen viability (%) | 20.70 (12.74) | 54.24 (65.88) | 46.73 (53.05) | 44.43 (49.04) |
| SEM | 1.622 | 2.366 | 1.333 | 0.356 |
| LSD ($P$ = 0.05) | 6.369 | NS | NS | NS |
| CV % | 8.430 | 7.992 | 4.749 | 1.373 |

Figures in parentheses indicate original means. SEM, standard error of mean; LSD, least significant difference; CV, coefficient of variation.

**Table 14.6.** Controlled pollinations with cryostored pollen. Data from Rajasekharan *et al.* (1998).

| Crop | Species/cultivar | Parents | | No. flowers pollinated | No. fruits set | No. seeds/fruit recovered |
|---|---|---|---|---|---|---|
| | | Female | Male | | | |
| Eggplant | 'Arka Kusumaker' | 'Arka Kusumaker' | 'Arka Kusumaker' (control) | 46 | 41 | 429.60 |
| | | 'Arka Kusumaker' | 'Arka Kusumaker' (9 years cryostored in liquid nitrogen) | 37 | 34 | 569.50 |
| | 'Arka Shirish' | 'Arka Shirish' | 'Arka Shirish' (control) | 20 | 13 | 237.75 |
| | | 'Arka Shirish' | 'Arka Shirish' (3.5 years cryostored in liquid nitrogen) | 17 | 10 | 405.00 |
| | *Solanum indicum* | *S. indicum* | *S. indicum* (control) | 10 | 6 | 46.33 |
| | | *S. indicum* | *S. indicum* (9 months cryo stored in liquid nitrogen) | 7 | 7 | 25.33 |
| Tomato | *S. pimpinellifolium* | 'Arka Shirish' | *S. pimpinellifolium* (2 years cryostored in liquid nitrogen) | 24 | 23 | 60.00 |

pollen for about 1 year, but the success rate greatly diminishes over time (Kasumi *et al.*, 2000).

CURRENT STATUS OF CRYOPRESERVED MATERIALS. Various desiccation techniques for seeds and embryonic axes are currently being used in the cryolab of NBPGR, New Delhi (Chaudhury and Malik, 1999; Chaudhury *et al.*, 2000a). Cryoproptocols of vitrification and encapsulation have been successfully attempted in embryonic axes of *Artocarpus heterophyllus*, *Litchi chinensis*, *Poncirus trifoliata* and *Citrus* species (Chaudhury *et al.*, 2000; Malik and Chaudhury, 2006) and in *Dioscorea* species using encapsulation–dehydration (Mandal *et al.*, 2009). In 2010, there were around 150 accessions of *Allium* including cultivated and wild species being maintained as *in vitro* cultures and cryopreservation to achieve short-, medium-, and long-term conservation.

Investigations are under way to develop protocols for cryopreservation of more recalcitrant seed species and establish their base collections. As an alternative complementary method, attempts are also being made to cryopreserve pollen in the case of trees and vegetatively propagated species. Attempts at cryopreservation at the NBPGR include successful cryopreservation of embryonic axes of tea (Chaudhury *et al.*, 1991), jackfruit (Chandel *et al.*, 1995), neem, black pepper, almond, litchi, and whole seeds of cardamom (Chaudhury and Chandel, 1991, 1994, 1995a,b).

So far, 34 accessions of legume vegetables have been cryostored. This will complement the conservation in field gene banks where they are difficult to maintain due to various biotic and abiotic stresses. The genetic diversity of highly nutritious and rare underutilized fruits commonly used as vegetables in several parts of India such as *Capparis decidua* (Ker), *Carissa* spp. (Karonda), *Cordia* spp. (Lasora), *Emblica officinalis* (Aonla), and *Moringa oleifera* (drumstick) have also been conserved in the cryogenebank at ICAR-NBPGR (Table 14.7) (Malik *et al.*, 2010).

CRYOGENEBANK CENTRES. Important world cryogenebanks besides NBPGR include those of the

**Table 14.7.** Cryo-conservation status of germplasm of vegetable crops.

| Method of conservation | Location | Crop (no. of accessions) |
|---|---|---|
| Cryobank (–196°C) | ICAR-NBPGR, New Delhi | Okra, eggplant, onion, tomato and cole crops (254) |
| *In vitro* bank | ICAR-NBPGR, New Delhi | Garlic (85), *Allium* spp. (44) |

National Centre for Genetic Resources Preservation (NCGRP), Fort Collins, Colorado; the Association Foret Cellulose (AFOCEL), France; the National Institute of Agrobiological Resources (NIAR), Japan; and the Institute of Plant Genetic Resources and Crop Plant Research (IPK), Germany.

### 14.5.3 *In vitro* gene banks

A cell and organ/shoot-tip gene bank comprises a germplasm collection based on cryopreserved (stored at –196°C in liquid nitrogen) embryogenic cell cultures, shoot tips, and/or somatic/zygotic embryos. The techniques for cryopreservation of plant cells and tissues are being rapidly refined and used for germplasm conservation (e.g. for potato in Germany). In 2006, NBPGR had 7000 germplasm accessions belonging to 710 species cryopreserved in the form of seeds, embryos, embryonic axes, pollen grains, and dormant buds.

Germplasm is conserved as slow-growing cultures of shoot tips and nodal segments in such a gene bank. They are regenerated by subculturing every 6 months to 3 years. This approach offers the following chief merits for the conservation of germplasm of vegetatively propagated crops and tree species: (i) accessions can be preserved indefinitely free of diseases and pests; (ii) they can be used for crops that produce no seed or recalcitrant seeds; (iii) subculture becomes necessary only after relatively long periods (every 6–36 months); and (iv) regeneration (i.e. subculturing) requires a comparatively short time.

In addition, cuttings, bulbs, and tubers can be maintained under controlled humidity and temperature conditions; however, this approach is only practical for short- and medium-term storage, and should be used in conjunction with a field gene bank.

A DNA bank comprises DNA sequences in the genomes of germplasm accessions as the sources of genes required in breeding programs, and is another method of conservation of genetic diversity of plant genetic resources. Cloned genes, genomic libraries, and sequence databases have an important role in germplasm conservation and management.

The term "genetic resources" is defined as "any material of actual or potential value of plant, animal, microbial, or other origin having functional units of heredity" under Article 2 of the Convention on Biodiversity. Using the same logic, genome resources are the whole or parts of the genome (DNA) or its functional units (RNA) of actual or potential value. Unlike genetic resources, which may create entire organisms, genomic resources can only retrieve traits (directly or indirectly) from these organisms.

With advancements in recombinant DNA, molecular markers and other technologies, DNA from the nucleus, mitochondria, and chloroplast is now routinely extracted, purified and used in polymerase chain reaction (PCR)- and non-PCR-based assays, which requires conservation of these genetic resources in the form of nucleic acid (DNA or RNA sequences) popularly referred as "DNA banking."

Current research such as cloning experiments and genome-sequencing projects generates many genome resources. These genomic data are essential for post-genomic research, whether for physiological and morphological characterization of a species, functional studies of genes, comparative genomics, or plant breeding. Therefore, it is essential to maintain an efficient system for conservation and management of these DNA sources.

The availability of techniques that can characterize and utilize DNA sequences (without the requirement for whole organisms), together with the products of gene banks, along with traditional plant breeders, can attract new clients interested in allele mining and cisgenesis, such as molecular biologists and geneticists. Thus, DNA storage should be regarded as an insurance policy rather than a replacement for conventional modes for germplasm storage. To this end, ICAR-NBPGR has established a

National Genomic Resources Repository (NGRR) for conservation of genomic resources (ICAR-NBPGR, 2016).

The discovery of DNA endonucleases, automated synthesis of oligonucleotides, and DNA sequencing and PCR amplification have made available a series of molecular markers for unequivocal classification, cataloguing, clustering, and DNA fingerprinting of diverse germplasm collections. A whole genome in the form of a genomic library or a sequence of DNA within a DNA library can be conserved following an appropriate DNA conservation method. Each technology is selected based on its merits in terms of utility, security, complementary, and the advantages over the others. The ability to store genomic information in the form of DNA libraries has long been established (Peacock, 1989). In DNA banks, DNA segments from the genomes of germplasm accessions are maintained as cosmid clones, phage lysates, or pure DNA (the last only being suitable for relatively short periods). These DNA segments can be analyzed and used to create transgenic plants. This approach can be used for conserved genetic materials from already extinct species, as DNA extracted from well-preserved herbarium specimens can often be cloned. However, this is very expensive and requires highly sophisticated techniques. A major disadvantage of this method is that the existing engineering techniques will not allow recovery of the whole genome and are not able to identify many genes of interest that are to be subsequently cloned and transferred. However, the establishment of these DNA libraries has provided genetic resources that can be used in the future following molecular technological advancements (Ford-Lloyd and Jackson, 1991).

Table 14.8 summarizes the three methods used for *in vitro* germplasm conservation, while Table 14.9 describes some essential observations related to *in vitro* genomics.

In 2006, ICAR-NBPGR had a total of 1904 accessions of germplasm conserved as slow-growth cultures. All the techniques discussed are being utilized at ICAR-NBPGR for conservation of a number of vegetatively propagated species, their wild relative, and elite types for further utilization through optimization of conservation and retrieval protocols.

Preparation and maintenance of core collections is one of the main objectives of all germplasm repositories. Frankel (1984) introduced the term core collection as a concept. A core collection is prepared with an aim of representing the genetic diversity of the entire germplasm collection in a subset of accessions. Core collections greatly facilitate efficient management and utilization of genetic diversity (Brown, 1989; van Heerwaarden *et al.*, 2013). In the initial days of germplasm maintenance, phenotypic data was used to create core collections. Nowadays, molecular markers have become a great help in preparation of core collections due to their abundance, and they are neutral tools for measuring genetic variation. The World Vegetable

**Table 14.8.** Comparison of the three approaches for *in vitro* germplasm conservation.

| Feature | Cryopreservation | Slow-growth culture | DNA bank |
|---|---|---|---|
| Tissue/organ preserved | Shoot tip, zygotic and somatic embryos, cells, protoplast | Shoot tip, slow-growing culture | DNA pieces or phage clones |
| Storage temperature | −196°C | 4–9°C or 15–20°C | 4°C |
| Storage | Liquid nitrogen | Refrigerators | Deep freezers, lyophilized state |
| Metabolic activity | None | Slow | None |
| Monitoring during storage | Replenishing liquid nitrogen | Subculture at every 6–36 months | None |
| Sophistication | Sophisticated | Less sophisticated | Highly sophisticated |
| Applicability | All species amenable to tissue culture | All species amenable to tissue culture | All species |
| Costs involved | Costly equipment required | Relatively cheap | Very costly |
| Risks | Loss of viability with time | Contamination | None |

**Table 14.9.** Essential observations related to *in vitro* genomics. From ICAR-NBPGR (2016).

| Deposition of genomic resources | |
|---|---|
| Types of genomic resources that can be deposited | Cloning vectors, expression vectors, binary vectors, restriction fragment length polymorphism probes.<br>Cloned genes, promoters fused to reporter genes.<br>Subgenomic, cDNA, expressed sequence tags (ESTs), repeat enriched libraries.<br>Bacterial (BAC), yeast (YAC), and P1-derived (PAC) artificial chromosome clone sets from sequencing projects.<br>Genomic, mitochondrial, or chloroplast DNA.<br>Cloned DNA from wild and weedy species produced exclusively for the repository. |
| Who can deposit | Only the principal investigator (PI) of a project can deposit the material (i.e. co-investigator and research associates need the PI's consent; a research student needs the consent of the supervisor guide).<br>The repository shall not be responsible for any disputes once the material is deposited. Disputed material shall be removed from the distribution pipeline/repository as appropriate. |
| Storage methodologies | Genomic resources are conserved for 1–2 years at +4°C; 4–7 years at –20°C and greater than 5 years when stored at –70°C.<br>ESTs, full-length cDNAs, BACs, PACs and YACs should be maintained in 96-well or 384-well microplates at –80°C, whereas cDNA clones as plasmid DNA should be maintained at –20°C.<br>For long-term undisturbed storage, lyophilized DNA should be preserved.<br>All tools and methods should be followed/developed to adopt storage at ambient temperature in order to save energy and the environment. |
| Quality and quantity of material | |
| Genomic DNA | Every sample of genomic DNA meant for deposition should contain a minimum of 100 µg, which is to be used only for the repository and not for distribution. If distribution is expected, replicates should be deposited.<br>The quality of the DNA should be as follows: $A_{260}/A_{280}$ = 1.7–2.0; $A_{262}/A_{230}$ > 1.5.<br>One agarose gel electrophoresis photo with at least one size marker should be submitted. |
| Library | The library should have less than 5% empty vectors and should be free from any contamination.<br>Libraries should preferably be deposited as amplified libraries at $10^{10}$ p.f.u. $ml^{-1}$. No library shall be accepted below $10^6$ p.f.u. $ml^{-1}$.<br>Libraries can be deposited as a 384-well plate (BACs) or a 96-well plate (cDNA, shotgun, ESTs). |
| Accompanying data/voucher | |
| Genomic DNA | Name of the plant, genus, species, common name, family, TaxID, Genbank accession no.<br>Extraction procedure, what the DNA is dissolved in, concentration (ng $\mu l^{-1}$).<br>Importance of the plant.<br>Reasons for choice of variety/landrace/accession.<br>Source of biological material.<br>Clear from Material Transfer Agreement (MTA) for plant material.<br>Information on indigenous technical knowledge (ITK), if any.<br>Publication (copies to be attached).<br>Intellectual property (IP) rights/patents, etc. (copies to be attached). |

*Continued*

**Table 14.9.** Continued.

| Deposition of genomic resources | |
|---|---|
| Cloned DNA/library | Name of the plant, genus, species, common name, family, TaxID, Genbank accession no.<br>Vector, antibiotic marker, cloning sites, fragment size (or size range).<br>Bacterial strain, transformation method, culture conditions, storage conditions.<br>Importance of the plant.<br>Reasons for choice of variety/landrace/accession.<br>Source of biological material.<br>Clear from MTA for plant material.<br>Information on ITK, if any.<br>Publication (copies to be attached).<br>IP rights/patents etc. (copies to be attached).<br>Supplying seeds/other planting material/herbarium specimen along with the genomic resources; if so, ICAR-NBPGR gene bank accession number (if known). |

Center (erstwhile AVRDC) has prepared mini core and core collections of mungbean (*Vigna radiata*) from its *ex situ* germplasm collections, holding the world's second largest mungbean germplasm collection with more than 6700 accessions. A core set of 1481 entries was prepared based on geographical and morphological characterization. A mini core with 289 entries was selected from the core set, which depicted the allele and the genotype diversity of the core collection, prepared with the help of 20 simple sequence repeat markers (Schafleitner *et al.*, 2015).

EcoTILLING is another biotechnological approach that can be used to explore the genetic wealth present in different collections of gene banks leading to breeding of traits and marker development. It is based on the principle of mutation detection used in TILLING (targeting induced local lesions in genomes) (Colbert *et al.*, 2001; Till *et al.*, 2003). EcoTILLING helps in identifying accessions in an entire collection of germplasm with different alleles of a candidate gene. Thus, EcoTILLING greatly reduces the number of samples that need to be phenotyped for any specific character. Ibiza *et al.* (2010) used an EcoTILLING platform on 233 cultivated accessions of the genus *Capsicum* and identified five new eukaryotic translation initiation factors of the 4F family (eIF4E) variants related to potato virus Y-resistance responses.

## 14.6 Conclusions

Biotechnology is a dynamic and ever-evolving area with new technologies being added along with new inventions in science. It has great potential in every aspect of biodiversity conservation including collection, characterization, storage, and distribution. Biotechnology has made a significant contribution to biodiversity protection, working in tandem with traditional conservation methods.

## 14.7 References

Alexander, M.P. and Ganeshan, S. (1989) Preserving viability and fertility of tomato and eggplant pollen in liquid nitrogen. *Indian Journal of Plant Genetic Resources* 2, 140–144.

Ashmore, S.E. (1997) Status Report on the Development and Application of *In Vitro* Techniques for the Conservation and Use of Plant Genetic Resources. International Plant Genetic Resources Institute. Rome.

Bajaj, Y.P.S. (1995) *Cryopreservation of Plant Germplasm I, Vol. 32*. Biotechnology in Agriculture and Forestry. Springer, Berlin.

Balachandran, S.M., Bhat, S.R., and Chandel, K.P.S. (1990) *In vitro* clonal multiplication of turmeric (*Curcuma* species) and ginger (*Zingiber officinale*). *Plant Cell Reports* 8, 521–524.
Baranski, R., Maksylewicz-Kaul, A., Nothnagel, T., Cavagnaro, P.F., Simon, P.W., and Grzebelus, D. (2012) Genetic diversity of carrot (*Daucus carota* L.) cultivars revealed by analysis of SSR loci. *Genetic Resources and Crop Evolution* 59, 163–170.
Benor, S., Zhang, M., Wang, Z., and Zhang, H. (2008) Assessment of genetic variation in tomato (*Solanum lycopersicum* L.) inbred lines using SSR molecular markers. *Journal of Genetics and Genomics* 35, 373–379.
Benson, E.E., Harding, K., and Smith, H. (1989) The effects of pre and post freeze light on the recovery of cryopreserved shoot tips of *Solanum tuberosum*. *Cryo-Letters* 10, 323–344.
Brown, A. (1989) Core collections: a practical approach to genetic resources management. *Genome* 31, 818–824.
Caron, P., Ferrero Y de Loma-Osorio, G., Nabarro, D., Hainzelin, E., Guillou, M., *et al.* (2018) Food systems for sustainable development: proposals for a profound four-part transformation. *Agronomy for Sustainable Development* 38: 41.
Chandel, K.P.S., Chaudhury, R., Radhamani, J., and Malik, S.K. (1995) Desiccation and freezing sensitivity in recalcitrant seeds of tea, cocoa and jackfruit. *Annals of Botany* 76, 443–450.
Chaudhur, R. and Chandel, K.P.S. (1991) Cryopreservation of desiccated seeds of neem (*Azadirachta indica* A. Juss.) for germplasm conservation. *Indian Journal of Plant Genetic Resources* 4, 67–72.
Chaudhur, R. and Chandel, K.P.S. (1994) Germination studies and cryopreservation of seeds of black pepper (*Piper nigrum* L.): a recalcitrant species. *Cryo Letters* 15, 145–150.
Chaudhury, R. and Chandel, K.P.S. (1995a) Cryopreservation of embryonic axes of almond (*Prunus amygdalus*) seeds. *Cryo-Letters* 16, 51–56.
Chaudhury, R. and Chandel, K.P.S. (1995b) Studies on germination and cryopreservation of cardamom (*Elettaria cardamomum* Maton) seeds. *Seed Science* and *Technology* 23, 235–240.
Chaudhury, R. and Malik, S.K. (1999) Cryopreservation and *in vitro* technology for conservation of recalcitrant seeds of some tree species. In: Marzalina, M., Khoo, K.C., Jayanthi, N., Tsan, F.Y., and Krishanapillay, B. (eds.) *Recalcitrant Seeds. Proceedings of the IUFRO Seed Symposium 1998*. FRIM, Kuala Lumpur, pp. 119–131.
Chaudhury, R. and Malik, S.K. (2008) Germplasm storage. In: Keshavachandran, R. and Peter, K.V. (eds.) *Plant Biotechnology: Methods in Tissue Culture and Gene Transfer*. Universities Press, Hyderabad, India, pp. 190–202.
Chaudhury, R., Radhamani, J., and Chandel, K.P.S. (1991) Preliminary observations on the cryopreservation of desiccated embryonic axes of tea (*Camellia sinensis* (L.) O. Kuntze) seeds for genetic conservation. *Cryo-Letters* 12, 31–36.
Chaudhury, R., Mandal, B.B., and Malik, S.K. (2000a) Development of cryopreservation techniques for the long-term conservation of recalcitrant seed and vegetatively propagated plant species in India. IPGRI-NBPGR Project Report (1996-2000). International Plant Genetic Resources Institute (IPGRI), Rome.
Chaudhary, R., Malik, S.K., and Engelmann, F. (2000b) Application of cryopreservation technology for storage of recalcitrant and intermediate seed species. In: *International Conference on Science and Technology for Managing Plant Genetic Diversity in the 21st Century, 12–16 June 2000, Kuala Lumpur*, abstract.
Chourey, S.K., Solanki, S., Gaikward, A.B., Pandaey, C.D., and Archak, S. (2017) SSR marker analysis points to population admixture and continuum of genetic variation among Indian landraces of brinjal (*Solanum melongena* L.). *Scientia Horticulturae* 224, 68–73.
Colbert, T., Till, B.J., Tompa, R., Reynolds, S., Steine, M.N., *et al.* (2001) High-throughput screening for induced point mutations. *Plant Physiology* 126, 480–484.
Cooper, J.C., Lipper, L., and Zilberman, D. (eds.) (2005) *Agricultural Biodiversity and Biotechnology in Economic Development*. Springer, New York.
Dey, S.S., Singh, A.K., Chandel, D., and Behera, T.K. (2006) Genetic diversity of bitter gourd (*Momordica charantia* L.) genotypes revealed by RAPD markers and agronomic traits. *Scientia Horticulturae* 109, 21–28.
Engelmann, F. (1991) *In vitro* conservation of horticultural species. *Acta Horticulturae* 298, 327–334.
Engelmann, F. (2003) Current research status and utilization of plant cryopreservation. In: *Proceedings of the International Workshop on Cryopreservation of Bio-Genetic Resources*. International Technical Cooperation Center, RDA, Suwon, Korea, pp. 19–40.

Engelmann, F. and Dereuddre, J. (1988) Cryopreservation for oil palm somatic embryos: importance of freezing process. *Cryo-Letters* 7, 220–235.

Engelmann, F. and Engels, J.M.M. (2002) Technologies and stratégies for *ex situ* conservation. In: Engels, J.M.M., Rao Ramanatha, V., Brown, A.H.D., and Jackson, M.T. (eds) *Managing Plant Genetic Diversity*. CABI Publishing, Wallingford, UK, pp. 89–103.

Engelmann, F., Duval, Y., and Dereuddre, J. (1985) Survival and proliferation of oil palm (*Elaeis guineesis* Jacq.) somatic embryos after freezing to liquid nitrogen. *Comptes Rendus de l'Académie des Sciences* 301, 111–116.

Fay, M.F. (1992) Conservation of rare and endangered plants using *in vitro* methods. *In Vitro Cellular and Developmental Biology* 28, 1–4.

Ferriol, M., Pico, B., de Cordova, P.F., and Nuez, F. (2004) Molecular diversity of a germplasm collection of squash determined by SRAP and AFLP markers. *Crop Science* 44, 653–664.

Ford-Lloyd, B.V. and Jackson, M.T. (1991) Biotechnology and methods of conservation of plant genetic resources. *Journal of Biotechnology* 17, 247–256.

Frankel, O. (1984) Genetic perspectives of germplasm conservation. In: Arber, W.K., Llinensee, K., and Peacock, W.J. (eds.) *Genetic Manipulation: Impact on Man and Society*. Cambridge University Press, Cambridge, UK, pp. 161–170.

Ganeshan, S. (1986) Viability and fertilizing capacity of onion (*Allium cepa* L.) pollen stored in liquid nitrogen (–196°C). *Tropical Agriculture* 63, 45–48.

González-Benito, M.E., Albert M.J., Iriondo, J.M., Varela, F., and PerezGarcia, F. (2004) Seed germination of four thyme species after short-term storage at low temperatures at several moisture contents. *Seed Science and Technology* 32, 247–254.

Grubben, G. and Denton, O. (2004) *Plant Resources of Tropical Africa, Vol. 2: Vegetables*. PROTA Foundation, Wageningen, Netherlands.

Guo, Y., Cheng, S., Li, Z., and Cowling, W.A. (2014) Centre of origin and centres of diversity in an ancient crop, *Brassica rapa* (turnip rape). *Journal of Heredity* 105, 555–565.

Harding, K., Benson, E.E., and Smith, H. (1991) The effects of *in vitro* culture period on the recovery of cryopreserved shoot tips of *Solanum tuberosum*. *Cryo-letters* 12, 17–22.

Henshaw, G.G. (1975) Technical aspects of storage for genetic conservation. In: Hawkes, J.G., and Frankel, O.H. (eds.) *Crop Genetic Resources for Today and Tomorrow*. Cambridge University Press, Cambridge, UK, pp. 349–358.

Hong, T.D. and Ellis, R.H. (1996) A protocol to determine seed storage behaviour. IPGRI Technical Bulletin No. 1. International Plant Genetic Resources Institute, Rome.

Ibiza, V.P., Cañizares, J., and Nuez, F. (2010) Eco-tilling in *Capsicum* species: searching for new virus resistances. *Genomics* 1:631.

ICAR-NBPGR (2016) *Guidelines for Management of Plant Genetic Resources*. Indian Council of Agricultural Research-National Bureau of Plant Genetic Resources (ICAR-NBPGR), New Delhi.

Igwe, D.O., Afiukwa, C.A., Acquaah, G., and Ude, G.N. (2019) Genetic diversity and structure of *Capsicum annuum* as revealed by start codon targeted and directed amplified minisatellite DNA markers. *Hereditas* 156: 32.

Jain, J.R., Timsina, B., Satyan, K.B., and Manohar, S.H. (2017) A comparative assessment of morphological and molecular diversity among *Sechium edule* (Jacq.) Sw. accessions in India. *3 Biotech* 7: 106.

Kartha, K.K., Mroginski, L.A., Pahl, K., and Leung, N.L. (1981) Germplasm preservation of coffee (*Coffea arabica* L.) by *in vitro* culture of shoot apical meristems. *Plant Science Letters* 22, 301–307.

Kasumi, M., Yashira, K., and Hayashi, M. (2000) Effectiveness of cold stored pollen on seed formation in East Indian lotus (*Nelumbo nucifera* Gaertn.). *Journal of the Japanese Society of Horticultural Science* 65, 732–735.

Kumar, S., Parekh, M.J., Fougat, R.S., Patel, S.K., Patel, C.B., *et al.* (2017) Assessment of genetic diversity among okra genotypes using SSR markers. *Journal of Plant Biochemistry and Biotechnology* 26, 172–178.

Kuras, A., Korbin, M., and Zurawicz, E. (2004) Comparison of suitability of RAPD and ISSR techniques for determination of strawberry (*Fragaria* × *ananassa* Duch.) relationship. *Plant Cell, Tissue and Organ Culture* 79, 189–193.

Larkin, P.J. and Scowcroft, W.R. (1981) Somaclonal variation: a novel source of variability from cell cultures for plant improvement. *Theoretical and Applied Genetics* 60, 197–214.

Lee, K.J., Sebastin, R., Cho, G.T., Yoon, M., Lee, G.A., and Hyun, D.Y. (2021) Genetic diversity and population structure of potato germplasm in RDA-Genebank: utilization for breeding and conservation. *Plants* 10: 752.

Liu, J., Yang, Y., Zhou, X., Bao, S., and Zhuang, Y. (2018) Genetic diversity and population structure of worldwide eggplant (*Solanum melongena* L.) germplasm using SSR markers. *Genetic Resources and Crop Evolution* 65, 1663–1670.

Madeley, J. (1996) *Yours for Food: Plant Genetic Resources and Food Security. Section 1: Plant Genetic Resources*. People's Plan. BUKO Agrar, Hamburg, Germany.

Malik, S.K. and Chaudhury, R. (2006) Successful cryopreservation of embryonic axes of two wild and endangered *Citrus* species of north eastern India for long term conservation. *Plant Genetic Resources: Characterization and Utilization* 4, 204–209.

Malik, S.K., Chaudhury, R., Dhariwal, O.P., and Bhandari, D.C. (2010) *Genetic Resources of Tropical Underutilized Fruits in India*. Indian Council of Agricultural Research-National Bureau of Plant Genetic Resources (ICAR-NBPGR), New Delhi, p. 168.

Mandal, B.B. and Chaudhury, R. (2004) Recent techniques for *ex situ* conservation of plant genetic resources. In: Dhillon, B.S., Tyagi, R.K., Lal, A., and Saxena, S. (eds.) *Plant Genetic Resource Management*. Narosa Publishing House, New Delhi, pp. 194–204.

Mandal, B.B., Dixit-Sharma, S., and Srivastava, P.S. (2009) Cryopreservation of embryogenic cultures of *Dioscorea bulbifera* L. by encapsulation-dehydration. *CryoLetters* 30, 440–448.

Mashilo, J., Shimelis, H.A., Odindo, A.O., and Amelework, B.A. (2017) Genetic differentiation of bottle gourd [*Lagenaria siceraria* (Molina) Standl.] landraces assessed by fruit qualitative traits and simple sequence repeat markers. *Scientia Horticulturae* 216, 1–11.

Maundu, P., Achigan-Dako, E., and Morimoto, Y. (2009) Biodiversity of African vegetables. In: Shackleton, C., Pasquini, M.W., and Drescher, A.W. (eds.) *African Indigenous Vegetables in Urban Agriculture*. Routledge, London, pp. 65–104.

Meldrum, G., Padulosi, S., Lochetti, G., Robitaille, R., and Diulgheroff, S. (2018) Issues and prospects for the sustainable use and conservation of cultivated vegetable diversity for more nutrition-sensitive agriculture. *Agriculture* 8: 112.

Ntuli, N.R., Tongoona, P.B., and Zoboloa, A.M. (2015) Genetic diversity in *Cucurbita pepo* landraces revealed by RAPD and SSR markers. *Scientia Horticulture* 189, 192–200.

Paunescu, A. (2009) Biotechnology for endangered plant conservation: a critical overview. *Romanian Biotechnological Letters* 14, 4095–4103.

Pawar, B.D. and Kale, P.B. (2011) Molecular plant breeding for sustainable agriculture: an overview. *Agrobios Newsletter* 10, 66–68.

Peacock, W.J. (1989) Molecular biology and genetic resources in the use of plant genetic resources. In: Brown, A.H.D., Marshell, D.R., Frankel, O.H., and Willium, J.T. (eds.) Cambridge University Press, Cambridge, UK, pp. 363–376.

Pence, V.V. (1990) Cryostorage of embryo axes of several large-seeded temperate tree species. *Cryobiology* 27, 212–218.

Qi, J., Liu, X., Shen, D., Miao, H., Xie, B., *et al.* (2013) A genomic variation map provides insights into the genetic basis of cucumber domestication and diversity. *Nature Genetics* 45, 1510–1515.

Rai, G.K., Singh, M., Rai, N.P., Bhardwaj, D.R., and Kumar, S. (2012). *In vitro* propagation of spine gourd (*Momordica dioica* Roxb.) and assessment of genetic fidelity of micropropagated plants using RAPD analysis. *Physiology and Molecular Biology of Plants* 18, 273–280.

Rajasekharan, P.E., Alexander, M.P., and Ganeshan, S. (1998) Long term pollen preservation of wild tomato, brinjal species and cultivars in liquid nitrogen. *Indian Journal of Plant Genetic Resources* 11, 117–120.

Rao, V.R. and Hodgkin, T. (2002) Genetic diversity and conservation and utilization of plant genetic resources. *Plant Cell, Tissue and Organ Culture* 68, 1–19.

Reddy, P.M., Sarla, N., and Siddiq, E.A. (2002) Inter simple sequence repeat (ISSR) polymorphism and its application in plant breeding. *Euphytica* 128, 9–17.

Reed, B.M. (1991) Application of gas-permeable bags for *in vitro* cold storage of strawberry germplasm. *Plant Cell Reports* 10, 431–434.

Reed, B.M. and Chang, Y. (1997) Medium and long-term storage of *in vitro* cultures of temperate fruit and nut crops. In: Razdan, M.K. and Cocking, E.C. (eds.) *Conservation of Plant Genetic Resources In Vitro, Vol. 1*. Science Publishers, Enfield, New Hampshire, pp. 67–105.

Reem, H. and Toby, H. (2007) The use of wild relatives in crop improvement: a survey of developments over the last 20 years. *Euphytica* 156, 1–13.

Roberts, E.H. (1975) Problems of long term storage of seeds and pollen for genetic resources conservation. In: Frankel, O.H. and Hawkes, H.G. (eds.) *Crop Genetic Resources for Today and Tomorrow*. Cambridge University Press, Cambridge, UK, pp. 269–296.

Roca, W.M. (1984) Cassava. In: Sharp, W.R., Evans, D.A., Ammirato, P.V., and Yamada Y. (eds.) *Handbook of Plant Cell Culture, Vol. 2*. Macmillan, New York, pp. 269–301.

Rodriguez-Bonilla, L., Cuevas, H.E., Montero-Rojas, M., Bird-Pico, F., Luciano-Rosario, D., and Siritunga, D. (2014) Assessment of genetic diversity of sweet potato in Puerto Rico. *PLOS ONE* 9: e116184.

Sakai, A. (2000) Development of cryopreservation techniques. In: Engelmann, F. and Takagi H. (eds.) *Cryopreservation of Tropical Plant Germplasm*. International Plant Genetic Resources Institute (IPGRI), Rome.

Sarasan, V., Cripps, R., Ramsay, M.M., Atherton, C., McMichen, M., *et al.* (2006) *In vitro* conservation of threatened plants: progress in the past decade. *In Vitro Cellular and Developmental Biology* 42, 206–214.

Schafleitner, R., Nair, R.M., Rathore, A., Wang, Y., Lin, C., *et al.* (2015) The AVRDC: The World Vegetable Center mungbean (*Vigna radiata*) core and mini core collections. *BMC Genomics* 16: 344.

Sharma, S., Kumar, P., Gambhir, G., Kumar, R., and Srivastava, D.K. (2018) Assessment of genetic diversity in lettuce (*Lactuca sativa* L.) germplasm using RAPD markers. *Biotechnology* 8: 9.

Singh, A.K., Singh, R., Kumar, S., and Kalloo, G. (2006) Genetic diversity within the genus *Solanum* (Solanaceae) as revealed by RAPD markers. *Current Science* 90, 711–716.

Till, B.J., Reynolds, S.H., Greene, E.A., Codomo, C.A., Enns, L.C., *et al.* (2003) Large-scale discovery of induced point mutations with high-throughput TILLING. *Genome Research* 13, 524–530.

van Heerwaarden, J., Odong, T., and van Eeuwijk, F. (2013) Maximizing genetic differentiation in core collections by PCA-based clustering of molecular marker data. *Theoretical and Applied Genetics* 126, 763–772.

Wang, Y., Rashid, M.A.R., Li, X., Yao, C., Lu, L., *et al.* (2019) Collection and evaluation of genetic diversity and population structure of potato landraces and varieties in China. *Frontiers in Plant Science* 10: 139.

Wesley-Smith, J., Vertucci, C.W., Berjak, P., Pammenter, N.W., and Crane, J. (1992) Cryopreservation of desiccation-sensitive axes of *Camellia sinensis* in relation to dehydration, freezing rate and the thermal properties of tissue water. *Journal of Plant Physiology* 140, 596–604.

Willett, W., Rockström, J., Loken, B., Springmann, M., Lang, T., *et al.* (2019) Food in the Anthropocene: the EAT–Lancet Commission on healthy diets from sustainable food systems. *Lancet* 393, 447–492.

Withers, L.A. (1985) Cryopreservation and storage of germplasm. In: Dixon, R.A. (ed.) *Plant Cell Culture—A Practical Approach*. IRL Press, Oxford, UK, pp. 169–192.

Withers, L.A. (1991) Tissue culture in the conservation of plant genetic resources. In: Zakri, A.H., Normah, M.T., Senawi, M.T., and Abdul Karim, A.G. (eds.) *Conservation of Plant Genetic Resources Through In Vitro Methods*. FRIM/MNCPGR, Kuala Lumpur, Malaysia, pp. 1–18.

Withers, L.A. and Engelmann, F. (1997) *In vitro* conservation of plant genetic resources. In: Altman, A. (ed.) *Biotechnology in Agriculture*. Marcel Dekker, New York, pp. 57–88.

Wang, Y., Wang, Y., Xu, W., Wang, C., Cui, C., and Qu, S. (2020) Genetic diversity of pumpkin based on morphological and SSR markers. *Pakistan Journal of Botany* 52, 477–487.

Zapartan, M. and Deliu, C. (1994 Conservations of endemic, rare and endangered species in the Romanian flora, *Lilium martagon* Kerner. In: *Proceeding of the 8th National Symposium of Industrial Microbiology and Biotechology*, Bucharest, pp. 423–426.

Zhang, D., Cervantes, J., Huaman, Z., Carey, E. and, Ghislain, M. (2000). Assessing genetic diversity of sweet potato (*Ipomoea batatas* (L.) Lam.) cultivars from tropical America using AFLP. *Genetic Resources and Crop Evolution* 47, 659–666.

# 15 Genomics and Molecular Breeding of Major Vegetable Crops

**Ashish Kaushal[1]*, A.T. Sadashiva[2] and K.V. Ravishankar[3]**
[1]*Graphic Era Hill University, School of Agriculture, Dehradun, Uttarakhand, India;* [2]*Nethra Crop Science, Bengaluru, Karnataka, India;* [3]*ICAR-Indian Institute of Horticultural Research, Bengaluru, Karnataka, India*

**Abstract**

Vegetable crops play a crucial role as a primary source of essential nutrients in the human diet and are cultivated worldwide. The quality attributes of these crops significantly influence consumer acceptance. However, the production of vegetable crops faces challenges from various biotic and abiotic stresses, leading to substantial losses in yield and quality. Breeding vegetables for resistance to these stresses, along with improved quality traits, is a key objective in many research programs. While traditional breeding methods are slow, molecular breeding offers rapid improvement of the trait of interest. Advancements in next-generation sequencing, high-throughput genotyping technologies, availability of reference genomes, gene-editing technologies, marker-assisted selection, and gene pyramiding have revolutionized vegetable breeding. Resequencing approaches applied to extensive germplasm resources have shed light on the domestication and improvement of vegetable crops through human selection. These methods have facilitated genetic mapping and the identification of molecular markers and key genes associated with important traits in vegetables. Genome-based molecular breeding has been implemented successfully in various vegetable crops, enabling molecular breeding at the single-nucleotide level. This chapter provides an overview of current genome-based molecular breeding strategies employed in major vegetable crops, offering insights into the future of molecular breeding to meet the needs of the growing global population.

**Keywords:** genetics, genomics, marker-assisted selection, molecular breeding, vegetables

## 15.1 Introduction

In the present millennium in terms of nutritional security, conservation of plant genetic resources is important for sustainable vegetable production, human health improvement, and enhancement of the economic growth of farmers and countries. Vegetable crops are rich in nutritional sources so can provide essential nutrients for humanity, such as vitamins and minerals, and they have significant effects on economic development. With the increasing world population and the demand for vegetable crops, much attention is required for improvement of vegetables, not only to increase yield but also to improve important agronomic traits, such as disease resistance, fruit flavor, fruit shape, plant architecture, and resistance to biotic and abiotic stresses. Classical breeding approaches are slow and time consuming, and thus advanced approaches such as molecular and genome-based breeding have been used for rapid improvement by shortening the breeding cycle not only to increase the yield of major vegetable crops, but also to improve their

*Email: akiihr04@gmail.com


DOI: 10.1079/9781800627123.0015

quality and to meet consumer requirements and increase convenience to farmers.

With the advent of molecular markers and next-generation sequencing (NGS) technologies, the progress of molecular breeding, genome sequencing, and resequencing of vegetable crops has considerably accelerated the breeding program. The publically available published genome sequences of tomato (Tomato Genome Consortium, 2012), watermelon (Guo *et al.*, 2013), cucumber (Huang *et al.*, 2009), Chinese cabbage (Wang *et al.*, 2011), and other major vegetable crops can help breeders to determine the specific locations of genes of interest on chromosomes. Molecular markers linked to phenotype can be applied on the basis of genotyping technology for marker-assisted selection (MAS) breeding, or can be used to identify genetic relationships among diverse genotypes. During the past decades, progress in molecular mapping has identified chromosomal regions harboring important genes in crop plants using molecular markers. Functional markers (FMs) derived from polymorphic sequences or sites within candidate genes thus more accurately discriminate alleles at one locus and represent ideal markers for MAS in breeding. To improve complex traits, such as yield and tolerance to abiotic stresses in vegetable crops, the modern breeding approach of genomic selection (GS) can be deployed, which aims to improve complex traits by exploiting genome-wide marker data without requiring identification of markers associated with quantitative trait loci (QTLs) for traits of interest (Meuwissen *et al.*, 2001).

Genome-wide association studies (GWAS) are a powerful tool for dissecting the genetic basis of variation for both qualitative and quantitative inherited traits. GWAS is an effective method based on the correlation between genotypes and phenotypes in a group of individuals, to screen the complex trait inheritance variation under the release of reference genome sequences (Saidou *et al.*, 2014). As well as known molecular markers, GWAS tools and reference genomes in vegetable crops can be applied to genome editing. Clustered regulatory interspaced short palindromic repeats (CRISPR)/CRISPR-associated protein 9 (Cas9) is an important and common technique for genome editing in many crops and has also been used for various vegetable crops to improve a number of undesirable agronomic traits efficiently (Khan *et al.*, 2019). This directional breeding is more efficient and time saving than other breeding methods (Chen *et al.*, 2019).

This chapter summarizes the studies on genome-based approaches for gene identification, molecular breeding, genome sequencing of vegetable crops, and molecular breeding, which can provide an insight into modern vegetable crop improvement.

## 15.2 Reference Genomes Available in Major Vegetable Crops

Vegetable genome information has greatly improved the research and progress into vegetable improvement. The recent development of cost-effective, high-throughput NGS technologies such as pyrosequencing (Roche/454), sequencing-by-synthesis (Illumina/Solexa), sequencing-by-ligation Dover (SystemsP), ion semiconductor-based non-optical sequencing (Life Science), single-molecule sequencing (Pacific Biosciences), and nanopore sequencing (Oxford Nanopore) for ultrahigh-throughput platforms has been particularly important for large sequence requirements. This type of genomic research in vegetable crops has become a major target because of their high economic importance and high nutrient content. In the post-genome era, reference genome mapping has been completed successfully in many vegetable crops including wild species and modern cultivars (Table 15.1).

Tomato (*Solanum lycopersicum*), chili/pepper (*Capsicum* spp.), potato (*Solanum tuberosum*), and eggplant (*Solanum melongena*) are important vegetable crops in the family Solanaceae whose genome sequences have been assembled and published (Table 15.1). A high-quality genome sequence size of 900 Mb for *S. lycopersicum* 'Heinz 1706' has been obtained by Sanger and NGS technologies (Tomato Genome Consortium, 2012). The genome sequence was compared with the draft genome (739 Mb) of its closest wild species and revealed existing nucleotide diversity between cultivated and wild tomato genomes (Tomato Genome Consortium, 2012). However, the biotic and abiotic stress tolerance of wild tomato is greater than that of the cultivated tomatoes. Gao *et al.* (2019) constructed a pan-genome using 725 accessions of wild species and revealed

**Table 15.1.** Reference genome sequences available in major vegetable crops.

| Family | Species | Sequenced accession | Sequencing technology used | Genome size (Mb) | Reference |
|---|---|---|---|---|---|
| Solanaceae | *Solanum lycopersicum* | 'Heinz 1706' | Sanger and next-generation technologies | 900 | Tomato Genome Consortium (2012) |
| | *Solanum pimpinellifolium* | LA1589 | Sanger and next-generation technologies | 739 | Tomato Genome Consortium (2012) |
| | *Solanum pennellii* | LA0716 | Illumina | 942 | Bolger *et al.* (2014) |
| | *Solanum sitiens* | LA1974 | Illumina | 1,245 | Molitor *et al.* (2021) |
| | *Solanum pimpinellifolium* | LA0480 | Illumina | 811 | Razali *et al.* (2018) |
| | *S. lycopersicum*, *S. pimpinellifolium*, *S. cheesmaniae*, *S. galapagense* | Pan-genome of 725 accessions belonging to wild species | Illumina | 306,000 | Gao *et al.* (2019) |
| | *S. pimpinellifolium* | LA2093 | Illumina | 923 | Wang *et al.* (2020) |
| | *Solanum melongena* | 'Nakate-Shinkuro' | Illumina | 1,130 | Hirakawa *et al.* (2014) |
| | | HQ-1315 | Illumina, Nanopore and 10X genomics sequencing | 1,170 | Wei *et al.* (2020) |
| | *Solanum tuberosum* | DM1-3 516 R44 | Illumina HiSeq 2000 | 844 | The Potato Genome Sequencing Consortium (2011) |
| | *Capsicum annuum* | CM334 | Illumina | 3,480 | Kim *et al.* (2014) |
| | | Zunla-1 | Illumina | 3,260 | Qin *et al.* (2014) |
| | *Capsicum chinense* | PI159236 | Illumina | 3,200 | Kim *et al.* (2017) |
| | *Capsicum baccatum* | PBC81 | Illumina | 3,900 | Kim *et al.* (2017) |
| Cucurbitaceae | *Cucumis sativus* | 9930 | Sanger and next-generation Illumina | 367 | Huang *et al.* (2009) |
| | *Cucumis melo* | DHL92 | Roche 454 Titanium | 375 | Garcia-Mas *et al.* (2012) |
| | *Citrullus lanatus* | 97103 | Illumina | 425 | Guo *et al.* (2013) |
| | *Momordica charantia* | OHB3-1 | Illumina | 339 | Urasaki *et al.* (2017) |
| | | Dali-11 and wild small-fruited line TR | Illumina | 300 | Cui *et al.* (2020) |
| | *Cucurbita pepo* | MU-CU-16 | Illumina | 263 | Montero-Pau *et al.* (2018) |
| | *Lagenaria siceraria* | USVL1VR-Ls | Illumina | 313.4 | Wu *et al.* (2017) |
| | *Cucurbita maxima* | Rimu | Illumina | 386.6 | Sun *et al.* (2017) |
| | *Cucurbita moschata* | Rifu | Illumina | 372 | Sun *et al.* (2017) |
| | *Cucurbita argyrosperma* | | Illumina HiSeq2000, Illumina MiSeq, and PacBio RS II | 238 | Barrera-Redondo *et al.* (2019) |

| | | | | | |
|---|---|---|---|---|---|
| | *Sechium edule* | Jianshui | Illumina | 606.42 | Fu *et al.* (2021) |
| | *Benincasa hispida* | B227 | Illumina | 1,030 | Xie *et al.* (2019) |
| | *Trichosanthes anguina* | | Illumina | 919.8 | Ma *et al.* (2020) |
| Apiaceace | *Apium graveolens* | Q2-JN11 | Illumina | 22,100 | Li *et al.* (2020) |
| | *Daucus carota* | DH1 | Illumina | 473 | Iorizzo *et al.* (2016) |
| Moringaceae | *Moringa oleifera* | | Illumina | 315 | Tian *et al.* (2015) |
| Compositeae | *Lactuca sativa* | 'Salinas' | Illumina | 2,500 | Reyes-Chin-Wo *et al.* (2017) |
| Fabaceae | *Pisum sativum* | 'Caméor' | Illumina | 3,920 | Kreplak *et al.* (2019) |
| | *Phaseolus vulgaris* | BAT93 | Illumina | 549.6 | Vlasova *et al.* (2016) |
| | *Vigna unguiculata* | | Illumina | 620 | Boukar *et al.* (2019) |
| Chenopodiaceae | *Spinacia oleracea* | Sp75 | Illumina | 1,010 | Xu *et al.* (2017) |
| Brassicaceae | *Brassica rapa* A genome | Chiifu-401-42 | Illumina | 485 | Wang *et al.* (2011) |
| | *Brassica oleracea* C genome | 02-12 | Illumina | 630 | Liu *et al.* (2014) |
| | *Brassica oleracea* L. var. *botrytis* C genome | C-8 | Illumina | 584.6 | D. Sun *et al.* (2019) |
| | *Brassica juncea* AABB genome | T84–66 | Illumina | 922 | Yang *et al.* (2016) |
| | *Brassica oleracea* var. *capitata* | D134 | Single-molecule real-time and 10X-Genomics sequencing | 529.92 | Lv *et al.* (2020) |
| | *Raphanus raphanistrum* | | Illumina | 515 | Moghe *et al.* (2014) |
| | *Raphanus sativus* | 'Aokubi' | Illumina | 529 | Kitashiba *et al.* (2014) |

that 4873 genes were absent from the reference genome. Analyses of the presence/absence of variation reveal substantial gene loss and intense negative selection of genes and promoters during tomato domestication and improvement. Lost or negatively selected genes are enriched for important traits, especially disease resistance.

Wang *et al.* (2020) sequenced a high-quality, chromosome-scale genome of *Solanum pimpinellifolium* accession LA2093 and compared it with the modern *S. lycopersicum* 'Heinz 1706'. Genome comparison identified more than 92,000 structural variants between LA2093 and 'Heinz 1706.' These structural variations including insertions/deletions (indels), inversions, and duplications are the causative genetic variants for many domestication traits of crops and provide rich resources for future research and biodiversity-based breeding. Molitor *et al.* (2021) developed a high-quality *de novo* genome and transcriptome assembly for the wild relative *Solanum sitiens* (accession LA1974) by a hybrid assembly strategy using Illumina short and PacBio long reads, generating a total of approximately 262 Gbp of DNA sequence.

Potato is an important solanaceous vegetable crop and staple food in many countries. The potato genome has been sequenced using a homozygous double haploid, DM1-3 516 R44. The sequence alignment was compared with a heterozygous diploid potato cultivar, which confirmed that some of the genomic variations and high-frequency harmful mutations might be associated with inbreeding decline in potato (Potato Genome Sequencing Consortium, 2011).

In the family Solanaceae, chili/pepper and tomato possess the same number of chromosomes ($2n = 4$), although pepper has a larger genome size because of the long terminal repeat expansion in its genome sequence (Qin *et al.*, 2014). A high-quality *Capsicum* genome has been sequenced by *de novo* assembly of the genomes of two pepper varieties, cultivated *Capsicum annuum* (3.26 Gb) and its progenitor *C. annuum* var. *glabriusculum* (3.07 Gb). As well as *C. annum*, several other pepper species genomes have been sequenced: *C. annuum* (3.48 Gb) (Kim *et al.*, 2014), *C. chinense* (Kim *et al.*, 2017), and *C. baccatum* (Kim *et al.*, 2017) (Table 15.1). The polymorphism level in the whole genome of cultivated pepper is much lower than in its wild species due to continuous selection and breeding (Qin *et al.*, 2014).

Eggplant is another important solanaceous vegetable crop after potato, tomato, and pepper. Genome sequence information (1120 Mb) was generated in cultivated *Solanum melongena* 'Nakate-Shinkuro' using Illumina sequence technology (Hirakawa *et al.*, 2014). Comparative genomic analysis facilitates researcher understanding of molecular breeding in solanaceous plants, and identifies loci that regulate specific flavors (e.g. pungency) or fruit quality (e.g. fruit size and fruit size), as well as resistance to biotic and abiotic stresses.

The Curcurbitaceae is an important vegetable family that plays an important role in the human diet. At present, several reference genomes of the Curcurbitaceae vegetables are available including cucumber (Huang *et al.*, 2009), muskmelon (Garcia-Mas *et al.*, 2012), watermelon (Guo *et al.*, 2013), bitter gourd (Montero-Pau *et al.*, 2018; Cui *et al.*, 2020), pumpkin/squash (Sun *et al.*, 2017; Montero-Pau *et al.*, 2018; Barrera-Redondo *et al.*, 2019), bottle gourd (Wu *et al.*, 2017), and ash gourd (Xie *et al.*, 2019). The Cucurbit Genomics Database (http://cucurbitgenomics.org/, accessed 8 July 2024) provides genomic resources for cucurbit vegetables such as genome and coding sequences, genetic maps, gene annotations, gene ontology, and comparative genomics (Zheng *et al.*, 2019). Cucumber (*Cucumis sativus*) was the first cucurbitaceous vegetable crop to have its genome published, in 2009. The sequencing was done via *de novo* technology, which reduced the cost to a greater level than traditional Sanger technology. The cucumber reference genome of genotype 9930 has been sequenced by a 72.2-fold genomic coverage with a 367 Mb genome. The genome sequence of this cucumber explained the growth and development process during its domestication.

Muskmelon (*Cucumis melo*) and cucumber belong to the same genus but have a different number of chromosomes. The absence of recent whole-genome duplications in the melon lineage since the ancient eudicot triplication explains the increased size of the melon genome (450 Mb) compared with its close relative, cucumber (367 Mb), partially due to transposon amplification; this process may have resulted in phenotypic or quality differences, such as stress-related genes and flavor-related genes, between cucumber and muskmelon (Garcia-Mas *et al.*, 2012).

Following the successful completion of cucumber and melon genome sequencing, other cucurbitaceous crops have also been sequenced with high coverage and efficiency (Table 15.1). The draft genome of watermelon (*Citrullus lanatus*) '97103' has been sequenced, which represents a high-quality genome with 108.6-fold coverage by Illumina sequencing, and *de novo* assembly constructed a genome of 353.5 Mb (Guo *et al.*, 2013). Silver-seed gourd (*Cucurbita argyrosperma*) has the smallest size (238 Mb) genome among the other cucurbitaceous vegetables sequenced.

The draft genome sequence of bitter gourd (*Momordica charantia*) has a size of 339 Mb (Montero-Pau *et al.*, 2018). Genome sequence information for squash (*Cucurbita maxima*), pumpkin (*Cucurbita moschata*), and gourd (*Cucurbita argyrosperma*) is able to explain the principles of their greater tolerance against biotic and abiotic stresses when used as a rootstock (Montero-Pau *et al.*, 2018; Sun *et al.*, 2017). Lettuce (*Lactuca sativa*) is a leafy vegetable belonging to Compositae, and its genome sequence (2500 Mb) provided evidence that a whole-family genome triplication event happened during its evolution (Reyes-Chin-Wo *et al.*, 2017). Spinach (*Spinacia oleracea*) is also a common leafy vegetable that originated from the Mediterranean region. The reported draft genome sequence of spinach (1010 Mb) clarified that its genome is enriched in transposable elements, and the slight difference between the genomes of *S. oleracea* and its progenitor *Spinacia turkestanica* indicated that *S. turkestanica* may be the direct progenitor of cultivated spinach (Xu *et al.*, 2017).

Common bean (*Phaseolus vulgaris*) and pea (*Pisum sativum*) are pod-bearing vegetable crops of the family Leguminosae. The sequencing of the Mesoamerican genotype BAT93 of common bean has identified genes related to disease resistance through comparison with other bean lines (Vlasova *et al.*, 2016). The genome sequence of pea revealed that its genome size is large due to genome expansion and repetitive elements (Kreplak *et al.*, 2019). Carrot (*Daucus carota*) is a root vegetable that contains high levels of carotenoids. A high-quality genome sequence (473 Mb) revealed the evolution of several gene families regulating high carotenoid accumulation; this information may facilitate researchers and breeders to breed cultivars with high-quality nutrient content (Iorizzo *et al.*, 2016).

At present, several reference vegetable genomes of the family Brassicaseae, such as cauliflower (*B. oleracea* L. var. *botrytis*) (D. Sun *et al.*, 2019), cabbage (*B. oleracea*) (Liu *et al.*, 2014), Chinese cabbage (*Brassica rapa*) (Wang *et al.*, 2011), and radish (*Raphanus sativus*) (Kitashiba *et al.*, 2014), have been sequenced. The genome sequencing of these vegetable crops has opened the way for advancing research on genome resequencing, transcriptome analysis, and application of novel gene and marker identification in vegetable crops.

## 15.3 Genome Resequencing of Major Vegetable Crops

Recently, whole-genome resequencing has been widely adopted in gene mapping, candidate gene identification, and marker development. Whole-genome resequencing is a rapid and more efficient in developing linkage markers compared with traditional methods. The availability of different vegetable species genome information enables a comparative phylogenetic analysis study among species, their origins, and evolutionary features (Wendel *et al.*, 2016). In crop plants, genome information allows assessment of the impact of genetic diversity on a gene function, and identification of the genes influencing the most valuable traits in crops.

Genome resequencing has been studied in many vegetable crops and is summarized in Table 15.2. For instance, current large-fruited tomato has been evolved from a small-fruited (cherry) wild species. In order to characterize the gene variations during the domestication process, Gao *et al.*, 2019 studied the pan-genome comprising all genetic information from 725 kinds of cultivated tomatoes and their wild progenitors and found that 4000 genes, including genes regulating fruit weight, flavor, and disease resistance, are absent compared with 95% of the reference genome. Tomato breeders can utilize this genomic information which adds depth and completeness to the reference genome, and is useful for future biological studies and in molecular breeding of tomato (Gao *et al.*, 2019).

Fruit-related traits of cultivated watermelon are the results of human selection and domestication. Guo *et al.* (2019) resequenced

**Table 15.2.** Summary of vegetable species with resequence genome currently available.

| Species | Platform | Resequenced population | Comments | Reference |
|---|---|---|---|---|
| *Citrullus lanatus* | HiSeq 2000 | 414 accessions | Different loci affecting watermelon fruit size have been under selection during speciation, domestication, and improvement | Guo *et al.* (2019) |
| | HiSeq 2500 | 1365 accessions | Identification of genomic variants that underlie phenotypic differences in the reference genome compared with the pan-genome | Wu *et al.* (2019) |
| | HiSeq X-Ten | Six resistant, susceptible, and recombinant inbred lines | Identification of powdery mildew-resistant mapping loci (*ClaPMR2*) | Mandal *et al.* (2020) |
| *Solanum lycopersicum* | GAII Illumina platform | Eight accessions | Variation associated with introgression and breeding events | Causse *et al.* (2013) |
| | Illumina High-Seq 2500 | Two accessions | Mapping of a tomato yellow leaf curl virus resistance gene, *ty-5* | Y. Wang *et al.* (2018) |
| | Illumina NextSeq | 725 accessions | Substantial gene loss and intense negative selection of genes and promoters during tomato domestication and improvement | Gao *et al.* (2019) |
| *Capsicum annum* | Illumina HiSeq X-Ten | Two male-sterile, fertile, and NILs | Identification of candidate gene *Capana02g002096* underlying genic male-sterile *msc-1* | Cheng *et al.* (2018) |
| | HiSeq 4000 | Accessions of *C. annum* and *C. baccatum* | Identification of single-nucleotide polymorphism related to powdery mildew resistance | Ahn *et al.* (2018) |
| *Cucumis melo* | Illumina HiSeq X-Ten | 297 accessions | Two-step selections in two subspecies of melon: ssp. *melo* and ssp. *agrestis* during melon breeding | Liu *et al.* (2020) |
| *Cucumis sativus* | Illumina HiSeq 2500 | Two resistant, susceptible, and chromosome segment substitution lines | Identification of candidate genes governing powdery mildew resistance | Xu *et al.* (2016) |
| | Illumina High-Seq 2000 | 3342 accessions | 112 putative domestication sweeps; one of these regions contains a gene involved in the loss of bitterness in fruits | Qi *et al.* (2013) |
| *Cucurbita pepo* | HiSeq2500 | Eight cultivars | Discovery of genomic variants associated with morphology and horticulturally valuable traits | Xanthopoulou *et al.* (2019) |
| *Lactuca sativa* | BGISEQ-500 | 445 accessions | Identification of genetic variation among the population and genomic region for loss of seed shattering | Wei *et al.* (2021) |
| *Solanum melongena* | Illumina NextSeq | 23 accessions | Substantial gene loss and intense negative selection of genes during eggplant domestication and improvement | Barchi *et al.* (2021) |

| | | | | |
|---|---|---|---|---|
| *Raphanus sativus* | HiSeq 2000 | 11 accessions | Extensive gene flow occurring between wild, weedy, and domesticated radishes | X. Zhang *et al.* (2021) |
| *Brassica oleracea* | HiSeq2000 and HiSeq2500 | Nine accessions | Several genes displaying presence/absence of variation annotated with functions related to major agronomic traits, including disease resistance, flowering time, glucosinolate metabolism, and vitamin biosynthesis | Golicz *et al.* (2016) |
| *Manihot esculenta* | Illumina HiSeq X-Ten | 388 accessions | Allelic variation in heterozygosity associated with key agronomic traits and cassava domestication | Hu *et al.* (2021) |
| *Phaseolus lunatus* | Illumina High-Seq 2000 | 37 accessions | Discovery of single-nucleotide polymorphism markers within previously discovered quantitative trait loci for common bacterial blight and angular leaf spot | Lobaton *et al.* (2018) |

414 accessions representing all extant species in the genus *Citrullus* and their genomic analyses reveal the evolutionary history of *Citrullus*, suggesting independent evolutions in *Citrullus amarus* and the lineage containing *Citrullus lanatus* and *Citrullus mucosospermus*.

Powdery mildew disease is a major constraint of production in watermelon caused by *Podosphaera xanthii*. Mandal *et al.* (2020) conducted whole-genome resequencing analysis of USVL531-PMR, USVL677-PMS, and four introgressed powdery mildew-resistant recombinant inbred lines (USVL531-PMR × USVL677-PMS) and a total of 2566 unique differentially expressed genes were identified between compatible and incompatible interactions with *P. xanthii*. A comparative analysis identified single-nucleotide polymorphisms (SNPs) and indels and based on this, the powdery mildew-resistance gene, identified as *ClaPMR2*.

In pepper, genetic male sterility is an important tool for commercial hybrid seed production, and only a few genes have been cloned and characterized. Cheng *et al.* (2018) resequenced the pepper genome and identified a strong candidate gene, *Capana02g002096*, underlying the *msc-1* locus in pepper which was renamed *Msc-1* based on the evidence at genetic, genomic, transcriptional and post-transcriptional levels.

Liu *et al.* (2020) conducted whole-genome resequencing of 297 wild, landrace, and improved melon accessions. Analysis of the population structure and genetic diversity revealed independent and two-step selections in two subspecies of melon, ssp. *melo* and ssp. *agrestis*, during melon breeding, and detected 233 (~18.35 Mbp) and 159 (~17.71 Mbp) novel potential selective signals during the improvement stage in ssp. *agrestis* and spp. *melo*, respectively. This study provided novel insights into melon crop evolution and new tools that can be exploited in molecular breeding of melon.

Powdery mildew is one of the most destructive diseases affecting cucumber and other cucurbit crops. Xu *et al.* (2016b) performed whole-genome resequencing of resistant and susceptible accessions to identify novel SNPs and indels. On comparison with the reference genome of the inbred cucumber line 9930, a total of 468,616 SNPs and 67,259 indels were identified in SSL508-28, and 537,352 SNPs and 91,698 indels in D8. Two candidate genes resistant to powdery mildew, *Csa2M435460.1* and *Csa5M579560.1*, were identified which can be utilized for breeding for powdery mildew resistance.

Ahn *et al.* (2018) performed whole-genome resequencing of *Capsicum baccatum* (PRH1, a powdery mildew-resistant line) and *Capsicum annuum* (Saengryeg, a powdery mildew-susceptible line) to identify SNPs. In total, 4,887,031 polymorphic SNP loci were identified between the two lines, and 306,871 high-resolution melting marker primer sets were designed. In addition, 6281 SNPs associated with 46 resistance genes were identified and the occurrence of disease resistance gene-related SNPs was predominantly on chromosome 4.

Xanthopoulou *et al.* (2019) conducted whole-genome resequencing of four *Cucurbita pepo* subsp. *pepo* (pumpkin/vegetable marrow, cocozelle, green zucchini, and yellow zucchini) morphotypes and three subsp. *ovifera* (acorn, crookneck, and scallop) morphotypes and a total of 3,823,977 high-confidence SNPs were identified. This study provided new clues and an improved understanding underlying genomic regions, unique SNPs, and candidate genes responsible for traits of high agronomic importance.

Eggplant (*Solanum melongena* L.) is an important vegetable crop that originated from a wild, prickly progenitor with small, round fruit. Barchi *et al.* (2021) resequenced 23 accessions of *S. melongena* representing diverse morphological and genetic diversity of the species including the related species *Solanum insanum* and *Solanum incanum* to compare the variation in genomes with the published reference genome. They constructed a pan-genome containing an additional approximately 51.5 Mb and 816 additional genes compared with the reference genome, and identified 53 selective sweeps related to fruit color, prickliness, and fruit shape in the nuclear genome, highlighting selection leading to the emergence of present-day *S. melongena* cultivars from their wild ancestors. These unexploited genomic variations and candidate genes are an important genomic source for eggplant improvement.

X. Zhang *et al.* (2021) constructed a genus-level pan-genome of radish and revealed genetic variations ranging from SNPs to inversions and translocations of whole ancestral karyotype blocks, providing new insights into the genomic evolution of radish and laying the foundation for molecular breeding and genetic improvement (Table 15.2).

Cassava is a highly heterozygous, clonally propagated vegetable crop. Hu *et al.* (2021) conducted resequencing of 388 cultivated and wild accessions to identify genomic variation underlying important agronomic traits. In their study, they identified that 18 allelic variations in heterozygosity for nine candidate genes are significantly associated with seven key agronomic traits. Eighty-one selective sweeps with decreasing heterozygosity, nucleotide diversity, and harboring 548 genes were enriched in multiple biological processes. Artificial selection for decreased heterozygosity has contributed to the domestication of the large starchy storage root of cassava. Selection for the homozygous GG allele in *MeTIR1* during domestication contributes to increased starch content. This allelic variation in heterozygosity is associated with cassava domestication.

Lettuce is an important salad crop, and genomic tools available for lettuce have greatly facilitated breeding programs; however, its origins and domestication history remain to be elucidated. Wei *et al.* (2021) conducted whole-genome resequencing of 445 *Lactuca* accessions and constructed a comprehensive map of lettuce genome variations, revealing that lettuce was first domesticated near the Caucasus, which was marked by loss of seed shattering. This study also explained the genetic architecture and wild introgressions in major resistance clusters in the lettuce genome.

Common bean (*Phaseolus vulgaris* L.) is the most important grain legume for human consumption and is a major nutritional source in tropical regions. However, production is constrained by abiotic and biotic stresses. To understand the variation in genomic regions, sequence diversity underlying the stress response was unraveled. Lobaton *et al.* (2018) conducted whole-genome sequencing of 37 varieties belonging to species *P. vulgaris*, *Phaseolus acutifolius* (A. Gray), and *Phaseolus coccineus* L., and identified over 40 million genomic variations. A large number of intergene pool introgressions were identified. This genomic dataset will facilitate molecular understanding and provide molecular tools for accelerating bean-breeding programs.

## 15.4 Gene and Marker Identification in Vegetable Crops in the Post-genomic Era

The advent of NGS technology has greatly facilitated the completion of genome sequencing of many vegetable crops, and vegetable crop breeding has entered the post-genomic era. Several new techniques have led to fast gene and marker discovery using genome-based approaches with a known genome sequence. A few important approaches for genome-based gene identification applied to vegetable crops are described below.

### 15.4.1 Bulked segregant RNA sequencing

Bulked segregant RNA sequencing (BSR-seq) combines bulked segregant analysis (BSA-seq; see section 15.4.2) with RNA sequencing (RNA-seq) and requires a reference genome to identify candidate genes. The technique is particularly important for crops that have large and complex genomes such as wheat where resequencing is not cost effective (G. Liu *et al.*, 2012a,b). BSR-seq is suitable for fine mapping for crops that still lack a reference genome sequence. For instance, Lee *et al.* (2014) used BSR-seq to construct a high-density linkage map and determined the candidate gene for the *Rfd1* locus for fertility restoration in radish. Moreover, BSR-seq has been applied simultaneously in pepper to develop gene-regulating chloroplast and immature pepper fruit color (Table 15.3) (Borovsky *et al.*, 2019). In another study, Yan *et al.* (2019) utilized BSR-seq to map the *BoMYB2* gene, the causal gene for the purple trait in kale, kohlrabi, and cabbage, using a map-based cloning approach in cabbage. BSR-seq was also applied to identify the candidate gene *pc1* underlying a QTL region associated with variation in chlorophyll content primarily in *Capsicum chinense*.

### 15.4.2 Bulk segregant analysis sequencing

Bulk segregant analysis sequencing (BSA-seq) is a commonly used approach for whole-genome

**Table 15.3.** Candidate gene discovery using genome-based approaches in major vegetable crops.

| Crop | Traits | Approach | Reference(s) |
|---|---|---|---|
| Tomato | Internode length | BSA-seq | Schrager-Lavelle *et al.* (2019); X. Sun *et al.* (2019) |
| | Early flowering | QTL-seq | Ruangrak *et al.* (2018) |
| | Internode elongation | BSR-seq | Schrager-Lavelle *et al.* (2019) |
| | Leaf mold | QTL-seq | G. Liu *et al.* (2019) |
| | Fruit weight and locule number | QTL-seq | Illa-Berenguer *et al.* (2015) |
| | Yellow-colored fruit | MutMap | Garcia *et al.* (2016) |
| *Capsicum* | Male sterility | BSA-seq | Cheng *et al.* (2018) |
| | Fruit color | BSR-seq | Borovsky *et al.* (2019) |
| | *Phytophthora* root rot | SLAF-seq | Xu *et al.* (2016) |
| | First flower node | SLAF-seq | X. Zhang *et al.* (2018) |
| Watermelon | Partial andromonoecy | BSA-seq | Aguado *et al.* (2020) |
| | Seed coat color | QTL-seq | Paudel *et al.* (2019) |
| | *Fusarium* wilt disease | QTL-seq | Branham *et al.* (2018) |
| | Flesh color | BSA-seq | C.N. Wang *et al.* (2021) |
| | Fruit skin color | BSA-seq | Dou *et al.* (2018a) |
| | Fruit shape | BSA-seq | Dou *et al.* (2018b) |
| Cucumber | Aphid resistance | SLAF-seq | Zhu *et al.* (2016) |
| | Curly leaf | MutMap | Rong *et al.* (2019) |
| | Dwarf plant | MutMap | Lin *et al.* (2016); Xu *et al.* (2018) |
| | Water logging | SLAF-seq | Liang *et al.* (2016) |
| | Organ size | BSR-seq | Yang *et al.* (2018) |
| | Fruit thickness | SLAF-seq | Xu *et al.* (2015) |
| | Light green peel | MutMap | Zhou *et al.* (2015) |
| | Light green coloration | MutMap | Lun *et al.* (2015) |
| | Subgynoecy | QTL-seq | Win *et al.* (2019) |
| | Carpel number | BSA-seq | S. Li *et al.* (2016) |
| | Yellow-green fruit peel color | MutMap | Hao *et al.* (2018) |
| | Tendril-less | BSA-seq | Chen *et al.* (2017) |
| | Leaf variegation | MutMap | Cao *et al.* (2018) |
| | Golden leaf | BSA-seq | Gao *et al.* (2016) |
| | Early flowering | QTL-seq | Lu *et al.* (2014) |
| | Fruit length | BSA-seq | Zhang *et al.* (2019) |
| | Virescent leaf | BSA-seq | K. Zhang *et al.* (2020) |
| | Flesh color | BSA-seq | Wang *et al.* (2021) |
| Melon | Powdery mildew | BSA-seq | Li *et al.* (2017) |
| | Flavor-related | SLAF-seq | Zhang *et al.* (2016) |
| | Striped rind | BSA-seq | G. Liu *et al.* (2019); L. Liu *et al.* (2019) |
| Radish | Restoration of fertility | BSR-seq | Lee *et al.* (2014) |
| Cabbage | Purple traits | BSR-seq | Yan *et al.* (2019) |
| Chinese cabbage | Golden leaf | MutMap | Fu *et al.* (2019) |

BSA-seq, bulked segregant analysis; BSR-seq, bulked segregant RNA sequencing; MutMap, mutation mapping; QTL-seq, quantitative trait locus sequencing; SLAF-seq, specific locus amplified fragment sequencing.

resequencing in crop plants. It relies on sequencing DNA of contrasting bulks for the trait of interest and comparison of the resequencing results with the assembled reference genome. This technique relies mainly on the reference genome to rapidly identify the candidate region. It has been utilized to identify several genes or candidate regions for important traits, such as carpel number and golden leaf in cucumber (Gao *et al.*, 2016; S. Li *et al.*, 2016; Chen *et al.*, 2017), fruit shape and fruit skin color in watermelon (Dou *et al.*, 2018a,b), internode length in tomato (Schrager-Lavelle *et al.*, 2019; Sun *et al.*, 2019), male sterility in pepper (Cheng *et al.*, 2018), striped rind and powdery mildew resistance in melon (Li *et al.*, 2017; G. Liu *et al.*, 2019; L. Liu *et al.*, 2019), and tendril-less and virescent leaf in cucumber (Table 15.3) (Chen *et al.*, 2017; K. Zhang *et al.*, 2020).

### 15.4.3 Quantitative trait locus sequencing

With the availability of reference genomes for various vegetable crops, QTL sequencing (QTL-seq) has been widely used and is a highly efficient method for rapid discovery of candidate genes for the trait of interest in vegetable crops. It has been applied to identify resistance to leaf mold (G. Liu *et al.*, 2019; L. Liu *et al.*, 2019), a QTL linked to fruit weight and locule number (Illa-Berenguer *et al.*, 2015), and early flowering in tomato (Ruangrak *et al.*, 2018). It has been applied to identify the linked marker *Ef2.1* related to flowering time in broccoli and cabbage (Shu *et al.*, 2018). QTL-seq is also an effective method to identify the major genes or QTLs for disease resistance in vegetable crops, such as resistance to *Fusarium wilt* disease for watermelon (Branham *et al.*, 2018) and cucumber mosaic virus for pepper (Guo *et al.*, 2017). A major QTL belonging to *FT* linked to early flowering has been identified in cucumber (Lu *et al.*, 2014). Some genes for other traits, including the genes or loci related to subgynoecy in cucumber (Win *et al.*, 2019) and seed coat color in watermelon, have been identified by QTL-seq (Table 15.3).

### 15.4.4 Specific locus amplified fragment sequencing

Specific locus amplified fragment sequencing (SLAF-seq) is an efficient method of large-scale *de novo* SNP discovery and rapid identification of candidate genes. SLAF-seq has been also used to analyse the genetic diversity of crop species by simplifying genome and bioinformatics, and also to understand the genomic changes in species during evolution (Xu *et al.*, 2014). Genes responsible for aphid resistance, water logging, and fruit thickness have been identified in cucumber by SLAF-seq (Xu *et al.*, 2015; Liang *et al.*, 2016; Zhu *et al.*, 2016). The combination of SLAF-seq and other genome-based sequencing methods can also lead to precise and mean linkage mapping; for instance, SLAF-seq and BSA-seq have been used to characterize the genes related to the first flower node in pepper (X. Zhang *et al.*, 2018) and flavor-related genes in melon (Zhang *et al.*, 2016) (Table 15.3).

### 15.4.5 Mutation mapping techniques for gene identification

The mutation mapping (MutMap) method for candidate gene identification was proposed by Abe *et al.* (2012) in rice to identify genomic regions governing important agronomic traits based on NGS technology. MutMap has also been used to identify several genes in vegetable crops that regulate fruit or leaf color in cucumber (Table 15.3), including yellow-colored fruit (Garcia *et al.*, 2016), curly leaf mutant (Rong *et al.*, 2019), dwarf plant structure (Lin *et al.*, 2016; Xu *et al.*, 2018), two light green peel genes (Lun *et al.*, 2015; Zhou *et al.*, 2015), a yellow-green peel gene (Hao *et al.*, 2018), and a leaf variegation gene (Cao *et al.*, 2018).

## 15.5 Marker-assisted Selection in Vegetable Crops

Production of vegetable crops is constrained by a number of biotic and abiotic stresses. The improvement of vegetable crops through conventional breeding approaches is slow. The efficiency of conventional breeding approaches can be significantly accelerated by using molecular marker technology, and breeders are now using molecular markers in all major crop-breeding programs globally (Gupta *et al.*, 2010). Tomato was the first crop plant for which genetic markers and maps were utilized for breeding purposes (Tanksley, 1983). Since then, a number of traits have been improved through MAS in major vegetable crops; among these, some of the important traits that vegetable breeders are currently focused on are quality of produce (fruit quality and shelf life in tomato), disease resistance, tolerance to insect pest damage, and abiotic stress tolerance (heat, drought, and salinity).

In tomato, markers have been developed for resistant to fungal diseases such as *Fusarium* wilt (Arens *et al.*, 2010; Catanzariti *et al.*, 2015, 2017; Zhang and Dilip, 2021), *Verticillium* wilt

(Diwan *et al.*, 1999; Kawchuk *et al.*, 2001; Arens *et al.*, 2010), late blight (Zhang *et al.*, 2014; Zhi *et al.*, 2021), *Alternaria* stem canker (van der Biezen *et al.*, 1995; Stommel and Zhang, 1998), anthracnose ripe rot (Stommel and Zhang, 1998, 2001), gray leaf spot (Behare *et al.*, 1991; Ji and Scott, 2009), leaf mold (Jones *et al.*, 1993; Balint-Kurti *et al.*, 1994; Laugé *et al.*, 1998; Truong *et al.*, 2011a), and powdery mildew (Chunwongse *et al.*, 1994; van der Beek *et al.*, 1994; Huang *et al.*, 2000; Bai *et al.*, 2003; de Giovanni *et al.*, 2004). Cleaved amplified polymorphic sequences (CAPS), indels, and sequence-characterized amplified region (SCAR) markers were linked to the bacterial disease resistance gene for MAS in tomato (Yu *et al.*, 1995; Yang and Francis, 2005; Pei *et al.*, 2012; Sharlach *et al.*, 2013; Zhang and Dilip, 2021).

The pathogens tomato yellow leaf curl virus (TYLCV), tomato spotted wilt virus (TSWV), and tomato mosaic virus (ToMV) cause major diseases infecting tomato. Several molecular markers have been developed linked to viral disease resistance genes (Table 15.4). For example, a marker linked to the TYLCV resistance genes *Ty-1* (Zamir *et al.*, 1994), *Ty-2* (Yang *et al.*, 2014), *Ty-3* (Dong *et al.*, 2016), *Ty-4* (Ji *et al.*, 2009), *ty-5* (Kadirvel *et al.*, 2013; Lapidot *et al.*, 2015; Wang *et al.*, 2022), and *Ty-6* (Gill *et al.*, 2019) have been developed and utilized for begomovirus resistance breeding in tomato. Kaushal *et al.* (2020b) reported that tomato lines carrying a combination of *Ty-2* and *Ty-3* resistance genes were found to be highly resistant to tomato leaf curl Bangalore virus compared to lines carrying *Ty-2* only. Markers linked to *Sw-5* (Foolad and Panthee, 2012; Zhang and Dilip, 2021), and to *Tm-1* and *Tm-2*/*Tm-22* genes (Dax *et al.*, 1994; Arens *et al.*, 2010; Zhang and Dilip, 2021) resistant to TSWV and ToMV, respectively, have been utilized for MAS in tomato. SCAR markers linked to nematode resistant genes (*Mi-1.2* and *Mi-3*) have been developed for MAS in tomato (Yaghoobi *et al.*, 2005; Seah *et al.*, 2007).

In capsicum, several molecular markers linked to disease resistance have been developed for marker-assisted breeding (MAB). For instance, resistance to bacterial wilt is governed by the *Bw1* gene, and MAS for *Bw1* can be performed with the help of a simple sequence repeat (SSR) marker linked to the gene (Mimura *et al.*, 2009). Markers linked to *Bs2* and *Bs3* bacterial spot resistant genes are available for MAS (Pierre *et al.*, 2000; Truong *et al.*, 2011b; Kim *et al.*, 2017). Thrips are a major insect pest of capsicum and chili pepper, and a marker linked to resistant QTLs is available for MAS (Maharijaya *et al.*, 2015).

Several markers linked to fungal disease resistance have been developed and utilized successfully for MAS in capsicum. Examples include a powdery mildew resistance gene (*PMR1*)-linked marker (Jo *et al.*, 2017), an anthracnose resistance-linked marker (Suwor *et al.*, 2017; Zhao *et al.*, 2020), and a marker linked to a QTL associated with *Phytophthora* root rot resistance (Liu *et al.*, 2014).

### 15.5.1 Functional markers and their application in vegetable breeding

FMs are defined as DNA markers derived from functionally characterized sequence motifs. Following this terminology, molecular markers derived from anonymous regions of the genome are called random DNA markers, which may or may not be developed from a polymorphic site in gene or may not be developed from a gene at all (Varshney *et al.*, 2007). Genetic molecular markers are developed from genomic DNA in coding or non-coding DNA, with or without a known function (Andersen and Lubberstedt, 2003), and with or without phenotypic trait variation (Aggarwal *et al.*, 2007), and therefore they may or may not be FMs (Andersen and Lubberstedt, 2003).

FMs are unaffected by non-functional allelic variation and make it possible to identify an individual gene, which serves as a good "translator" from genomic technologies into improved crop varieties (Andersen and Lubberstedt, 2003). In addition, FMs are useful for screening of alleles in natural as well as in breeding populations, fixation of alleles in the population, construction of linked FM haplotypes, and combination of FM alleles associated with complex traits (Andersen and Lubberstedt, 2003). FMs have been developed for various important traits, such as agronomic traits in a number of vegetables, sex identification in asparagus, male sterility in *Capsicum* spp., and resistance to biotic and abiotic stresses in vegetable crops (Table 15.5).

**Table 15.4.** Molecular markers linked to resistant genes/quantitative trait loci (QTLs) for marker-assisted selection in vegetable crops.

| Crop | Resistance trait | Gene/ QTLs | Marker | Reference(s) |
|---|---|---|---|---|
| Capsicum | Bacterial wilt | *Bw1* | SSR | Mimura *et al.* (2009) |
| | Bacterial spot | *Bs2* | AFLP, SCAR | Truong *et al.* (2011b); Kim *et al.* (2017) |
| | | *Bs3* | SCAR | Pierre *et al.* (2000) |
| | Cucumber mosaic virus | *Cmr1* | HRM | Kim *et al.* (2017) |
| | | *cmr2* | KASP | Choi *et al.* (2018) |
| | Potyviruses resistance | *pvr1* | RFLP | Murphy *et al.* (1998) |
| | | *pvr2* | RAPD, RFLP | Caranta *et al.* (1997); Parrella *et al.* (2002) |
| | | *Pvr4* | AFLP, CAPS | Caranta *et al.* (1999) |
| | | *pvr6* | RFLP | Caranta *et al.* (1996) |
| | Tobacco mosaic virus | *L* | HRM | Yang *et al.* (2009) |
| | Tomato spotted wilt virus | *Tsw* | RAPD, CAPS | Moury *et al.* (2000) |
| | Anthracnose | *AnRGO5* | SCAR | Zhao *et al.* (2020) |
| | Powdery mildew | *PMR1* | HRM, SCAR | Jo *et al.* (2017) |
| | *Phytophthora* root rot | *QTL5* | HRM | Liu *et al.* (2014) |
| | | *Me1* | CAPS, HRM | X. Wang *et al.* (2018) |
| | | *Me3*, *Me4* | AFLP | Djian-Caporalino *et al.* (2001) |
| | | *Me7* | CAPS, HRM | Changkwian *et al.* (2019) |
| Brinjal | *Fusarium* wilt | *Rfo-sa1* | CAPS | Toppino *et al.* (2008) |
| | | *Fm1* | SSR | Miyatake *et al.* (2016) |
| | Bacterial wilt | *ERs1* | SNPs | Salgon *et al.* (2017) |
| Tomato | *Fusarium* wilt | *I* | CAPS | Catanzariti *et al.* (2017) |
| | | *I-2* | SCAR | Arens *et al.* (2010) |
| | | *I-3* | CAPS, SCAR | Catanzariti *et al.* (2015); Zhang and Dilip (2021) |
| | | *I-7* | CAPS | Gonzalez-Cendales *et al.* (2015) |
| | *Verticillium* wilt | *Ve, Ve1, Ve2* | ARMS-PCR | Diwan *et al.* (1999); Arens *et al.* (2010) |
| | | *Ph-2* | CAPS | Zhi *et al.* (2021) |
| | | *Ph-3* | SCAR | Zhang *et al.* (2014) |
| | *Alternaria* stem canker | *Asc* | RFLP | Stommel and Zhang (1998) |
| | Anthracnose ripe rot | *QTLs* | RAPD | Stommel and Zhang (1998) |
| | Gray leafspot | *Sm* | CAPS | Ji and Scott (2009) |
| | Leaf mold | *Cf-1, Cf-2, Cf-4, Cf-5, Cf-9* | RFLP | Jones *et al.* (1993); Balint-Kurti *et al.* (1994); Truong *et al.* (2011a) |
| | Powdery mildew | *Lv, Ol-1, ol-2, Ol-3, Ol-4, Ol-5* | RFLP, PCR, SCAR, CAPS, AFLP | Chunwongse *et al.* (1994); Huang *et al.* (2000); Bai *et al.* (2003); de Giovanni *et al.* (2004) |

*Continued*

**Table 15.4.** Continued.

| Crop | Resistance trait | Gene/ QTLs | Marker | Reference(s) |
|---|---|---|---|---|
| | Bacterial speck | *Pto* | CAPS, SCAR | Yang and Francis (2005); Zhang and Dilip (2021) |
| | | *Rx3* | CAPS | Yang and Francis (2005) |
| | | *Rx4* | InDels | Pei *et al.* (2012) |
| | | *RxopJ4* | CAPS | Sharlach *et al.* (2013) |
| | Tomato yellow leaf curl virus | *Ty-1* | CAPS | Jung *et al.* (2019) |
| | | *Ty-2* | SCAR | Yang *et al.* (2014) |
| | | *Ty-3* | SCAR | Dong *et al.* (2016) |
| | | *Ty-4* | CAPS | Ji *et al.* (2009) |
| | | *ty-5* | SSR, dCAPS, CAPS, indels, SNP, HRM | Kadirvel *et al.* (2013); Lapidot *et al.* (2015); Wang *et al.* (2022) |
| | | *Ty-6* | CAPS | Gill *et al.* (2019) |
| | | *SlATG18f* | CAPS | Prasad *et al.* (2021) |
| | Tomato spotted wilt virus | *Sw-5* | SCAR, CAPS | Foolad and Panthee (2012); Zhang and Dilip (2021) |
| | Tomato mosaic virus | *Tm-1* | SCAR | Arens *et al.* (2010) |
| | | *Tm-2/Tm-22* | SCAR | Zhang and Dilip (2021) |
| | Alfalfa mosaic virus | *Am* | AFLP | Parrella *et al.* (2004) |
| | Cucumber mosaic virus | *Cmr* | RFLP | Stamova and Chetelat (2000) |
| | Root-knot nematode | *Mi-1.2* | SCAR | Seah *et al.* (2007) |
| | | *Mi-3* | SCAR | Yaghoobi *et al.* (2005) |
| Cucumber | Powdery mildew | *pm1.1* | SNP | Xu *et al.* (2016) |
| | | *pm5.2* | CAPS_ CsGy5G015660 | C.Y. Zhang *et al.* (2021) |
| | | *CsLRR-RPK2 and CsaMLO8* | CAPS, indels | Badri *et al.* (2021) |
| | | *pm5.3* | SNP | Xu *et al.* (2016) |
| | Downy mildew | *dm4.1* | *CsDM4-066* | Wang *et al.* (2016) |
| | | *dm5.3* | SNP | K. Zhang *et al.* (2018) |
| | | *dm4.1.1* | Indel | Berg *et al.* (2020) |
| | | *dm4.1.3* | Retrotransposon insertion | Berg *et al.* (2021) |
| | Angular leaf spot | *psl* | SNP | Zhang *et al.* (2019) |
| | Target leaf spot | *cca-3* | SNP | Wen *et al.* (2015) |
| | Anthracnose | *cla* | SNP | Pan *et al.* (2018) |
| | *Fusarium* wilt | *Foc* | SSR | Zhang *et al.* (2014) |
| | Scab | *ccu* | SSR | Kang *et al.* (2011) |
| | Cucumber mosaic virus | *cmv6.1* | SSR | Shi *et al.* (2018) |
| | Papaya ringspot virus Watermelon strain | *PRSV wmv* | SSR | Tian *et al.* (2016) |
| | Zucchini yellow mosaic virus | *zym* | CAPS | Ramırez-Madera and Havey (2017) |

*Continued*

**Table 15.4.** Continued.

| Crop | Resistance trait | Gene/ QTLs | Marker | Reference(s) |
|---|---|---|---|---|
| Cabbage | *Fusarium* wilt | *Foc* | Indel | Lv *et al.* (2013) |
| | | *FOC1* | Indel | Lv *et al.* (2014) |
| | | *Foc-Bo1* | Indel | Shimizu *et al.* (2015) |
| | | *FOC1* | SSR | Liu *et al.* (2017) |
| | Black rot | *BRQTL-C3, BRQTL-C6* | SNP | Lee *et al.* (2015) |
| | *Sclerotinia* rot resistance | QTLs (six for stem, 12 for leaf resistance) | SSR | Mei *et al.* (2013) |
| | Male sterility | *Ms* | SCAR | Wang *et al.* (2000) |
| | | *CDMs399-3* | EST-SSR, SSR | Chen *et al.* (2013) |
| Chinese cabbage | Male sterility | *Ms* | SCAR | Hui *et al.* (2011) |
| | | *CRaki* | SSR | Kato *et al.* (2013) |
| | | *Crr1* | CAPS/indel/SNP/ SSR | Hatakeyama *et al.* (2013); Suwabe *et al.* (2003); Suwabe *et al.* (2006) |
| | | *Crr2* | SNP/SSR | Suwabe *et al.* (2003); Suwabe *et al.* (2006) |
| | | *Crr3* | RAPD | Saito *et al.* (2006) |
| | | *Crr4* | SSR | Suwabe *et al.* (2006) |
| | Downy mildew | *BraDM* | Isozyme/RAPD | Yu *et al.* (2009) |
| | | *K14-1030/ KBrb058m10-1, kbrb006c05-2, KBrb058m10* | SSR | Yu *et al.* (2011) |
| | | *Brh019-SNP137* | SNP | Li *et al.* (2011) |
| | Turnip mosaic virus | *TuRB07* | SSR | Jin *et al.* (2014) |
| | | *trsSCAR* | SCAR | Kim *et al.* (2013) |
| Broccoli | Downy mildew | *Pp523* | OPK17_980, AT.CTA_133/134 | Farinho *et al.* (2004) |
| Cauliflower | Downy mildew resistance | *Ppa207* | SSR | Saha *et al.* (2020) |
| | | *Ppa3* | RAPD, ISSR | Singh *et al.* (2012) |
| | Black rot | *Xca1bo* | ScOPO-04833 and ScPKPS-11635 | Kalia *et al.* (2017) |
| | | *Xca1bo* | RAPD, ISSR | Saha *et al.* (2014) |
| Vegetable pea | Powdery mildew | *er1* | PSMPAD60 | Ek *et al.* (2005) |
| | | *er-2* | ScX17_1400 | Katoch *et al.* (2010) |
| | | *Er3* | SSR | Cobos *et al.* (2018) |
| | *Fusarium* wilt | *Fw* | SCAR | Kwon *et al.* (2013) |
| | Pea seed-borne mosaic virus | *Sbm-1* | STS | Frew *et al.* (2002) |
| Faba bean | *Ascochyta* blight resistance | *Af-1, Af-2* | OPA111045/ OPAB071026, OPE171272/ OPJ18626 | Roman *et al.* (2003) |
| | | *Af-3, Af-4* | OPD161732/ OPG041131, OPJ18655/ OPG111118 | Avila *et al.* (2004) |
| | Rust resistance | *Uvf1* | OPI20900/ OPL181032 | Avila *et al.* (2003) |

*Continued*

**Table 15.4.** Continued.

| Crop | Resistance trait | Gene/ QTLs | Marker | Reference(s) |
|---|---|---|---|---|
| Common bean | Bean common mosaic virus | *bc-1*[2] | SCAR | Miklas *et al.* (2000) |
| | | QTL | STS | Liu *et al.* (2008) |
| | | QTLs | RAPD, SCAR | Yu *et al.* (2000) |
| | White mold | QTLs | RAPD | Park *et al.* (2001) |
| | Anthracnose | *Co* | RAPD | Ragagnin *et al.* (2009) |
| | | *Ur* | RAPD | Ragagnin *et al.* (2009) |
| | Angular leaf spot | *Phg-2* | RAPD primer OPE04 | Namayanja *et al.* (2006) |
| | | *Phg-1* | SCAR | Ragagnin *et al.* (2009) |
| Watermelon | *Fusarium* wilt | *FON-1* | SNP | Fall *et al.* (2018) |
| | | *Fom-2* | CAPS, SCAR | Burger *et al.* (2003) |
| Summer squash | Silver leaf disorder | *sl* | RAPD, SSR | Kabelka and Young (2010) |
| | Powdery mildew | *Pm-0* | CAPS | Holdsworth *et al.* (2016) |
| | Zucchini yellow mosaic virus | *Zym-0*, *Zym-1* and *Zym-2* | SCAR, SSR | Pachner *et al.* (2015) |
| Spinach | Downy mildew | *RPF1*, *RPF2*, *RPF3* | SSR | Bhattarai *et al.* (2021) |
| | Sex-determining locus | X/Y | SNP | She *et al.* (2021) |
| | Monoecious gene | *Xm* | AFLP | Yamamoto *et al.* (2014) |
| Carrot | Nematode resistance | *Mj-1* | STS | Boiteux *et al.* (2004) |
| Sweet potato | Root-knot nematode | *QTLs* | SCAR | Nakayama *et al.* (2012) |
| Lettuce | Die back | *Tvr1* | HRM | Simko *et al.* (2009) |
| | Mosaic virus | *mo-11 and mo-12* | HRM | Nicaise *et al.* (2003) |
| | Corky root resistance | *cor* | HRM | Simko (2013) |
| | Downy mildew | *Dm* | SCI11, OPX11, SCT08, SCW09 | Paran and Michelmore (1993); Maisonneuve *et al.* (1994); Simko (2013); Parra *et al.* (2016) |
| | *Fusarium* wilt | *qFOL1.2* | LG1_ v8_116.506Mbp, WF25-42, SSR, indel, ARMS | Aruga *et al.* (2012); Seki *et al.* (2021) |

AFLP, amplified fragment length polymorphism; ARMS, amplification-refractory mutation system; CAPS, cleaved amplified polymorphic sequence; dCAPS, derived CAPS; EST, expressed sequence tag; HRM, high-resolution melting; indel, insertion/deletion; ISSR, inter-simple sequence repeat; KASP, kompetitive allele specific polymerase chain reaction; PCR, polymerase chain reaction; RAPD, randomly amplified polymorphic DNA; RFLP, restriction fragment length polymorphism; SCAR, sequence-characterized amplified region; SNP, single-nucleotide polymorphism; SSR, single sequence repeat; STS, sequence-tagged site

Vegetable breeders can utilize FMs in precision breeding such as germplasm and genetic diversity analysis, gene pyramiding, MAS, marker-assisted backcross breeding, marker-assisted recurrent selection, and GS for improvement of vegetable crops.

### 15.5.2 Gene pyramiding in vegetable crops

The concept of gene pyramiding was introduced by Watson and Singh in 1952. Gene pyramiding is defined as a method aimed at assembling

**Table 15.5.** Functional markers developed for different traits in major vegetable crops.

| Crop | Trait | Gene | Chromosome | Marker | References |
|---|---|---|---|---|---|
| Tomato | Bacterial wilt | *Bwr-12* | 12 | F-TCAAGGTCCACTACCTTCATCC<br>R-TCGGTATAGAGGGTACGTTG | Kim *et al.* (2018) |
| | Tomato yellow leaf curl virus | *Ty-3* | 6 | F-CCTTATGATGTCTCGTGAAAGG<br>R-GAAGCACAGATTGAAGAAAACC | Nevame *et al.* (2018) |
| | *Fusarium* wilt | *frl* | 9 | F-TACGATGACGTCGGT<br>R-ATGCTACTGCGATGAAAC | Devran *et al.* (2018) |
| | Fruit shape | *OVATE* | 2 | F-GAT CGT CGG TTT CTA CGTCAT CAG ATAG<br>R-GAT AAT GCT TTC CGT TCA ACG ACA GA | Maurya *et al.* (2021) |
| | Fruit size | *fw2.2* | 2 | F-TCTGCTCAGAAGCATGCACA<br>R-TTGTGACCTGTACCCCAGGA | Nesbitt and Tanksley (2002) |
| Capsicum | Male sterility | ms3 | 5 | F-GGTACTTTGA CCCTCATAATTGG<br>R-TTGTTTGT GGTGTACG TGCT | Naresh *et al.* (2018) |
| | Pungency | *Pup1* | 2 | F-CCATGGATTGTTGCTCGGGCCTCC<br>R-CCGTACCGCCCCATTGCGATTCC | Saritnum *et al.* (2008) |
| | Powdery mildew resistance | *InDel1* | 1 | F-AACTTGGTAGCAATTTTATTGGGT<br>R-TGGAGACAATGTGCATAAGTCTCT | Karna and Ahn (2018) |
| Brinjal | Bacterial wilt resistance | *S401* | 6 | F–G ACTGCGTACCAATTCAGTT<br>R-GATGAGTCCTGAGTAACACGATG | Cao *et al.* (2009) |
| Melon | Powdery mildew resistance | *Pm-2 F* | 1 | F-GCCCAACCTTCAACTCGATA<br>R-TTGAATCTCATTTTTCTGTTGCAT | Zhang *et al.* (2013) |
| | Melon necrotic spot virus | *nsv* | 4 | F-GTTTCTGATACGATGTTGTTTCCCTG<br>R-GCCGAGATGCAGCAGGATGCTTTGCAC | Nieto *et al.* (2006) |
| | *Fusarium* wilt | Fom1 | 7 | F-AACGAGAAGGCGGTGGAAAT<br>R-CGATCTCCTCAAGGGAAGGTG | Oumouloud *et al.* (2015) |
| | White flesh | *wf* | 9 | F-TTGGAGGTTCAATGCTTGCC<br>R-CAAAGACCAGAGCACCATCG | Cuevas *et al.* (2009) |
| | Green flesh color | gf | 8 | F-TCTGCAAAATGGTTGCTTTGAA<br>R-AGGTGGATGTGGCACACAAA | Fukino *et al.* (2012) |
| | Short lateral branching | *slb* | 11 | F-CTTGCGCTCCTTGGTATTCC<br>R-CAAGATCGGCAAGAGACAGC | Fukino *et al.* (2012) |
| | Sutures on the rind | *s-2* | 9 | F-GCATCGGAATCTTGTTCGGC<br>R-TCCGGTGGGAGATACCCAAT | Diaz *et al.* (2011) |
| Watermelon | Powdery mildew | *Pm* | 5 | F-ATTTTCTTGCTTCAAATGGA<br>R-ATAAGCAAAAGCATCGAAAG | Gama *et al.* (2015) |
| | Zucchini yellow mosaic virus | *eIF4E* | 3 | F-TGGACITTYTGGTTYGAYAA<br>R-GGRTCYTCCCAYTTIGGYTC | Ling *et al.* (2009) |

*Continued*

**Table 15.5.** Continued.

| Crop | Trait | Gene | Chromosome | Marker | References |
|---|---|---|---|---|---|
| Cucumber | Powdery mildew | *pm-s* | 5 | F-CCCTATGCGTGAAAGCCACT<br>R-CGCCTCAAACCCATACCCAA | Liu *et al.* (2017) |
| | Cauliflower mosaic virus | *cmv6.1* | 6 | F-ACAAAGCTTCTCCGCAAATG<br>R-GGAGGGAAAGGAAGGAGAGA | Shi *et al.* (2018) |
| Cowpea | Drought- and salt-stress tolerance | *CPRD12* | 11 | F-AAAGCATGCCCTAGTGGGAC<br>R-ATGTCGGAAGCTACGGTTTCT | Contour *et al.* (2006) |
| Pea | Powdery mildew resistance | *er1-7* | 6 | F-CGACACCGTATTCAAGCAGG<br>R-TGTTGCCCTGTTTGATCGTT | Sun *et al.* (2016) |
| Faba bean | Tannin and anthocyanin biosynthesis | *VfTTG1* | – | F-TATGAATTCATTTTTAGTTCCCACCTAAC<br>R-GTATCCGGTTGAGGACTCTCATAGATA | Gutierrez and Torres (2019) |
| Common bean | Seed development | *PvMIPSs and PvMIPSv* | 1 | F-TTGCCACGCACCTGCTAATA<br>R-CCTGCAGCTGCGATTTTCAA | Fileppi *et al.* (2010) |
| Lettuce | Bacterial leaf spot resistance | *Xcvr* | 2 | F-TATCAAACGTAAAGTTGGAGCTTGT<br>R-CCAAACACCTTGTGCATTGCT | Wang *et al.* (2016) |
| Turnip | Turnip mosaic virus resistance | *retr02* | 4A | F-TATCAAACGTAAAGTTGGAGCTTGT<br>R-CCAAACACCTTGTGCATTGCT | G.L. Li *et al.* (2016) |
| Cauliflower | Curd architecture | *qCS.C6–1 and qCS.C6–2* | 6C | F-CGGTACTGGAATGTGGACGT<br>R-TGAATTGGTATGAACACGCCTC | Zhao *et al.* (2020) |
| Cabbage | Early and late flowering | *BoFLC1.C9* | – | F-GGAAAGCAACATGGTGATGA<br>R-CATGGTGTGAACCAGAGTCC | Abuyusuf *et al.* (2019) |
| | Male sterility | *CDMs399-3* | 7C | F-TCCCTTTCACATCGTCCACA<br>R-TGCAGCCCAGAACAGTGATA | Chen *et al.* (2013) |
| Sweet potato | β-Carotene & flesh | *QA/QC* | 3 | F-AGTGCGGGACAAGATGATCA<br>R-TCCCGAACATCTGAGCAAGT | Gemenet *et al.* (2020) |
| | Carotenoids | *bCHY-1* | – | F-TCCAGCTTGGGAATTACGTC<br>R-ACAACGAAGCGTGCCATAG | Arizio *et al.* (2014) |
| Carrot | Alternative respiration | *DcAOX1* | 1 | F-AAAATAACAATGATGATGACACG<br>R-CTCCACTTCAGTGATATCCAA | Nogales *et al.* (2016) |
| Celery | Flavonoids | *AgFNSI* | 4 | F-ATGGCTCCATCAAC TATAAC<br>R-CTGCCCTGGCAATCTCCG | Yan *et al.* (2020) |
| Asparagus | Sex identification | *MYB35* | 5 | F-TTGCTTGGCGGATCATATTATG<br>R-TTGCTTGGCGATGTCCCTTTTG | Kim *et al.* (2014) |

F, forward; R, reverse.

multiple desirable genes from multiple parents into a single genotype. The objective of gene pyramiding is to enhance the performance of a trait by combining two or more complementary genes and increasing the durability of the trait. Pyramiding of several genes from multiple parents can be deployed using conventional and molecular MAS. Pyramiding of multiple genes is difficult based on phenotype alone due to epistatics (Mohler and Singrun, 2004). Gene pyramiding has been achieved successfully using molecular MAS in many vegetable crops for a number of fungal, bacterial, and viral diseases (Table 15.6).

### 15.5.3 Marker-assisted selection and marker-assisted back-crossing

As well as for disease resistance, gene pyramiding has also been successfully used for quality-related traits and heat-stress tolerance in pepper (Usman *et al.*, 2018). For instance, marker-assisted gene pyramiding has been used to increase β-carotene content in cauliflower using the *Or* gene through marker-assisted back-crossing (MABC) (Kalia *et al.*, 2017). Jeong *et al.* (2015) used MABC for development of pepper varieties with higher capsinoid content (Table 15.6). The gene pyramiding strategy has greatly improved resistance for several diseases, such as viral diseases in tomato (Robbins *et al.*, 2010; Consuegra *et al.*, 2015; Prasanna *et al.*, 2015a,b; Hanson *et al.*, 2016; Prabhandakavi *et al.*, 2021), begomovirus and late blight (Kaushal *et al.*, 2024), zucchini (Pachner *et al.*, 2015), capsicum (Özkaynak *et al.*, 2014; Jeong *et al.*, 2015; Salamon *et al.*, 2016; Suwor *et al.*, 2017; Moodley *et al.*, 2019), and common bean (Ferreira *et al.*, 2012); for fungal diseases such as late blight, gray leaf spot, and *Fusarium* wilt in tomato (Robbins *et al.*, 2010; Hanson *et al.*, 2016; Kumar *et al.*, 2019), black rot and downy mildew in cauliflower (Saha *et al.*, 2021), late blight in potato (Tan *et al.*, 2010; Stefańczyk *et al.*, 2020), anthracnose and crown root rot in pepper (Suwor *et al.*, 2017; Colak and Yilmaz, 2020), *Fusarium* wilt and gummy stem blight in melon (Zhang *et al.*, 2017; Sousaraei *et al.*, 2018), and anthracnose and angular leaf spot in common beans (Ferreira *et al.*, 2012; Ddamulira *et al.*, 2015; Kiryowa *et al.*, 2015; Lema *et al.*, 2021; Kiryowa *et al.*, 2021; Uwera *et al.*, 2021); and for bacterial disease in tomato (Yang and Francis, 2005).

## 15.6 Genomic Selection and its Application in Vegetable Crop Breeding

To improve complex traits, such as yield and tolerance to abiotic stresses in vegetable crops, GS can be used, which aims to improve complex traits by exploiting genome-wide marker data without requiring identification of markers associated with QTLs for traits of interest (Meuwissen *et al.*, 2001). GS utilizes genome-wide DNA markers and phenotypic data of target traits from a training population to predict genomic estimated breeding values of candidates in a test population, where the latter have been genotyped but not yet phenotyped (Crossa *et al.*, 2017). The advantages of GS over traditional MAS include the fact that GS does not need to detect significant QTLs related to the traits of interest and it greatly increases the efficiency of breeding due to early selection before the phenotypic data are collected. After the hugely successful application of GS in improving genetic gain in animal breeding, GS has been introduced to crop breeding in a number of aspects such as inbred performance prediction, parental selection, and hybrid prediction (Riedelsheimer *et al.*, 2012; Crossa *et al.*, 2014; X. Wang *et al.*, 2017; Millet *et al.*, 2019). GS is an emerging method that bypasses the current limitations of MAS during plant breeding. GS attempts to capture the total additive genetic variance with genome-wide marker coverage and effect estimates (Rutkoski *et al.*, 2011). Therefore, selection of an individual without phenotypic data can be performed based on the individual's predicted breeding value.

### 15.6.1 Genomic selection in vegetable breeding

The usefulness of GS in breeding in rice and other cereals has been proved by several simulations and experimental studies but is still less used in vegetable crops. The use of GS in some vegetable crops is summarized in Table 15.7.

**Table 15.6.** Success stories of marker-assisted selection (MAS)-based gene pyramiding for important traits in major vegetable crops.

| Crop | Trait | Gene(s) | Source | Marker | Strategy | Reference |
|---|---|---|---|---|---|---|
| Cauliflower | Resistance to black rot and downy mildew | *Xca1bo* and *Ppa3* | BR-161 and BR-2 | ScOPO-04833 and BoGMS0624 | MABC | Saha *et al.* (2021) |
| | Increased β-carotene content | *Or* | EC625883 | SA4 | MABC | Kalia *et al.* (2017) |
| Tomato | Resistant to tomato leaf curl New Delhi virus | *Ty-1/Ty-3 and Ty-2, ty-5*, and *ty-6* | CLN3125K and FLA456 | Ty-1/3_K, P6-25, TG105A, SINAC1, SLM 10-46 | MABC | Prabhandakavi *et al.* (2021) |
| | Bacterial wilt, tomato yellow leaf curl virus, late blight, fusarium wilt, gray leaf spot, and tobacco mosaic virus | *Bwr-12, Ty-2, Tm22, Ty-3, Ph-2, Ph-3, I2*, and *Sm* | CLN2777G, G2-6-20-15B, and BR-11 | dTG422, TG328, SLM12-2, SLM12-10, P1-16, P6-25, I2OH, CT55 and NCTm-019 | MAS | Hanson *et al.* (2016) |
| | Resistant to tomato spotted wilt virus and late blight | *Sw-5* and *Ph-3* | NC946 and NC592 | TG328, TG591, SCAR421 | MAS | Robbins *et al.* (2010) |
| | Resistance to tomato yellow leaf curl virus | *Ty-2* and *Ty-3* | FLA478-6-1-11, CLN2498C, and CLN1621E | T0302, P6-25 | MAS | Prasanna *et al.* (2015a) |
| | Resistance to tomato yellow leaf curl virus | *Ty-2* and *Ty-3* | FLA478-6-1-11, CLN2498C and CLN1621E | T0302, P6-25 | MAS | Prasanna *et al.* (2015b) |
| | Resistance to tomato yellow leaf curl virus and tomato spotted wilt virus | *Ty-1/Ty-3* and *Sw-5* | LD3, LA 2779, *Solanum peruvianum* | – | MAS | Consuegra *et al.* (2015) |
| | Resistance to tomato yellow leaf curl virus | – | TY-172 and TY-197, H-902, TY-52 and Fla 595-2, PIM-HIR | – | Phenotyping selection | Vidavski *et al.* (2008) |
| | Resistance to TYLCV and late blight | *Ty-2*, *Ty-3*, *Ph-2* and *Ph-3* | IIHR2834, IIHR2850, and IIHR2895 | TG0302, Ty3-SCAR1, NC-LB-9-66-78, dTG-422 | MAS | Kaushal *et al.* (2024) |
| | Resistance to TYLCV, late blight and root-knot nematode | *Ty-1/Ty-3, Ty-2, Ph-2, Ph-3*, and *Mi-1.2* | EC-654720, EC-654692, and 8-2-1-2-5 | JB-1, FLUW-25, TG105, dTG422, TG328, Mi23 | MAS | Kumar *et al.* (2019) |
| | Bacterial spot and bacterial speck | *Pto* and *Rx3* | Ohio981205 and Ohio9834 | Pto, Rx-3L1 | MAS | Yang and Francis (2005) |
| Pepper | Resistance to potato virus Y, tomato spotted wilt virus, and pepper mild mottle virus | *Pvr4*, *Tsw*, and *L4* | LET-1, ENT-1, RAZ-1 | SCAC568, AP-7/AP-8 | MABC | Özkaynak *et al.* (2014) |
| | Resistance to anthracnose | – | PBC932: PR1 and PBC80: PR2 | SCAR-Indel, SSR-HpmsE032 | MAS | Suwor *et al.* (2017) |

| | | | | | | |
|---|---|---|---|---|---|---|
| | Resistance to tomato spotted wilt virus | *Tsw* | TSR | *Tsw* SCAR | MAS | Salamon *et al.* (2016) |
| | Higher level of capsinoids | Shinhong | Shinhong C | Capsiate_3 rd and Tcc-R3 | MABC | Jeong *et al.* (2015) |
| | Resistance to potato virus Y | *pvr2*[1] and *pvr2*[2] | Benno and Double Up | T200A, G325A and T236G (Tetra primer ARMS-PCR) | MAS | Moodley *et al.* (2019) |
| | Tolerance to heat stress | *Hsp70* and *sHsp* | AVPP0702 | Hsp70-u2 and AGi42 | MABC | Usman *et al.* (2018) |
| | Resistance to *Phytophthora* crown blight | Phyto 5.2 | Criollo de Morelos 334 | OpD04.717 | MAS | Colak and Yilmaz (2020) |
| Potato | Resistance to late blight | *RPi-mcd1* and *RPi-ber* | | TG339 and CT214, TG63, Q133 | MAS | Tan *et al.* (2010) |
| | | *Rpi-rzc1* and *Rpi-phu1* | DG 99-10/36 and DG97-943 | RenSeq1812 and phu1_2069 | MAS | Stefańczyk *et al.* (2020) |
| Melon | Resistance to *Fusarium* wilt | *Fom-2* | Isabelle | Fom2-R409 and Fom2-S253 | MABC | Sousaraei *et al.* (2018) |
| | Resistance to gummy stem blight | *Gsb-6* and *Gsb-4* | 4598 | SGSB1800, CMTA170a | MAS | Zhang *et al.* (2017) |
| Bean | Resistance to bean common mosaic potyvirus | *I, bc-u, bc-1*[2], *bc-22*, and *bc-3* | P90557/G91213, P94207-140, -189, -230, P94207 | SBD51300 | MAS | Miklas *et al.* (2000) |
| | Resistance to anthracnose and *Pythium* root rot | *Co-42*, *Co-5*, *Co-43*, and *Co-9*, *RWR719* | G2333, PI207262, | SAS13, SBB14, SAB3, SB12, PYAA19800 | MAS | Kiryowa *et al.* (2015) |
| | Resistance to anthracnose and potyvirus | *I*, *bc-3*, *Co-2*, *Co-3/9* | A252, A321, A493, Sanilac BC6-Are, BRB130, A252, A321 and A493 | SW13, ROC11, SQ4 SCH20, SI19 | MABC | Ferreira *et al.* (2012) |
| | Resistance to angular leaf spot and anthracnose | Phg-1, Phg-2, Co-14 and CBB QT | AND277, KT-ABC 001 | TGA1.1, CV542014, SAP6 and g796 | MAS | Lema *et al.* (2021) |
| | Resistance to anthracnose | Co-42, Co-43, Co-5, and Co-9 | G 2333 and PI 207262 | SBB14, SAS13, SAB3, SB12 | MAS | Kiryowa *et al.* (2021) |
| | Resistance to anthracnose | *Co*-42 and *Co*-5 | G2333 | SBB14, SAB3 | MABC | Uwera *et al.* (2021) |
| | Resistance to angular leaf spot | *Phg-2*, *Phg*-1, *PhgG5686B* *PhgG5686A*, and *PhgG5686C* | AND277, Mexico 54, G5686 | OPE04, TGA1. 1 and Pv-ctt001, pv-ag004 and Pv-at007 | MAS | Ddamulira *et al.* (2015) |
| Summer squash | Resistant to zucchini yellow mosaic virus | *Zym-0*, *Zym-1*, *Zym-2*, *Zym-3*, and Zym-4 | Nigerian Local and True French | TGM-Zym-0, TGM-Zym-1, TGS-Zym-2 | MABC | Pachner *et al.* (2015) |

MABC, marker-assisted back-crossing.

**Table 15.7.** Summary of genomic selection studies in vegetable crops.

| Crop | Traits | Training population size and type | No. of markers | Statistical model | Correlation values | References |
|---|---|---|---|---|---|---|
| Tomato | Soluble solid content, fruit weight | 96 $F_1$ varieties | 337 SNPs | gBLUP, Bayesian Lasso, wBSR, BayesC, RKHS, RF | 0.56–0.68, 0.22–0.27 | Yamamoto *et al.* (2017) |
| | Metabolic and quality traits | 163 genotypes | 5995 SNPs | RR-BLUP | 0.05–0.81 | Duangjit *et al.* (2016) |
| | Earliness | 172 $F_2$ | 61 SSRs | Bayesian Lasso, BRR BayesA, -B, and -C$\pi$, RKHS, RF | 0.17–0.57 | Hernández-Bautista *et al.* (2020) |
| Capsicum | Fruit length, fruit shape, fruit width, fruit weight, pericarp thickness | 351 genotypes | 8,663 SNPs | gBLUP, Bayesian Lasso, wBSR, BayesB and -C, RKHS, RF | 0.75, 0.73, 0.84, 0.83, and 0.82 | Hong *et al.* (2020) |
| Potato | Late blight and common scab | 1763 and 3885 genotypes | 4,110 SNPs | Bayes B | 0.41–0.71 | Rodriguez *et al.* (2018) |
| | Dry-matter content and chipping quality | 1146 genotypes | 167,637 SNPs | gBLUP | 0.39–0.79 | Sverrisdóttir *et al.* (2018) |
| | Tuber-related traits | 181 genotypes | 183,848 SNPs | BayesA and -B | 0.23–0.81 | Caruana *et al.* (2019) |
| | Tuber starch content, tuber starch yield, and tuber yield | 184 genotypes | 83,000 SNPs | gBLUP, Bayesian Lasso, wBSR, BayesA, -B, and -C$\pi$, RKHS, RF | 0.5–0.8 | Stich and van Inghelandt (2018) |
| | Fry color (off the field and after storage) | 56 genotypes | 46,406 SNPs | RR-BLUP, BayesA, Bayesian Lasso, RF | 0.11–0.77 | Byrne *et al.* (2020) |
| | Tuber-related traits | 192 genotypes | 78,111 silicoDArT | gBLUP, Bayesian Lasso, wBSR, BayesA, -B, and -C$\pi$, RKHS | 0.25–0.76 | Habyarimana *et al.* (2017) |
| Carrot | Root traits | 749 genotypes | 146,816 SNPs | gBLUP | 0.25–0.8 | Brainard *et al.* (2021) |
| Cauliflower | Curd-related traits | 174 accessions | 120,693 SNPs | gBLUP, BayesB | 0.1–0.66 | Thorwarth *et al.* (2018) |

BRR, Bayesian ridge regression; DArT, diversity array technology; gBLUP, genomic best linear unbiased prediction; RKHS, reproducing kernel Hilbert space; RF, random forest; RR-BLUP, ridge regression best linear unbiased prediction; SNP, single-nucleotide polymorphism; SSR, simple sequence repeat; wBSR, weighted Bayesian shrinkage regression.

Hernández-Bautista *et al.* (2020) studied the prediction performance of six statistical models for six quantitative characteristics related to earliness in tomato. The genomic values were estimated by these six models: Bayesian Lasso, Bayesian ridge regression (BRR), BayesA, BayesB, BayesCπ, and reproducing kernel Hilbert spaces (RKHS) regression. The correlation values ranged from 0.17 to 0.57. The highest association values were found in days to flowering of the third inflorescence and 1000-seed weight, which were greater than 0.5. In general, all the models performed in a similar manner because only slight differences were observed among the correlation values. Specifically, BL, BayesB, and RKHS exhibited the highest Pearson correlation values for most traits. Duangjit *et al.* (2016) evaluated the accuracy of GS for 45 metabolomic and quality traits using 163 tomato accessions by cross-validation and estimated the impact of different factors on its accuracy in tomato. Based on the results, it was concluded that the accuracy of GS was significantly influenced by the training population size, marker number/density, and relatedness between individuals. The effect of several parameters on the accuracy of prediction revealed the potential of GS in tomato breeding programs.

Yamamoto *et al.* (2017) assessed the potential of GS for increasing soluble solids content and total fruit weight of tomato. A collection of large-fruited $F_1$ varieties was used to construct the GS models, and the progeny from crosses were used to validate the models. The GS models successfully predicted a better parent even if the phenotypic value did not vary substantially between candidates. The results indicated that GS is a promising strategy for tomato breeding design. Hong *et al.* (2020) demonstrated the potential use of core collections and GS as tools for pepper improvement for fruit-related traits using genotype and phenotype data for a training population to predict phenotypes of a test population with only genotype data, for predicting fruit-related traits such as fruit length, fruit shape, fruit width, fruit weight, and pericarp thickness using 351 accessions. In carrot, Brainard *et al.* (2021) studied the accuracy of GS for root shape and size using genome-wide association analysis and by estimating genomic estimated breeding values for polygenic root shape and size.

In clonally propagated plant species that are highly heterozygous, considering dominance effects in the genomic prediction model has the potential to improve predictions. In potato, Stich and van Inghelandt (2018) examined the prospects of genomic prediction of key performance traits in a diversity panel and investigated the effects of size and makeup of training population, number of test environments, and molecular markers on prediction accuracy. GS in potato has been studied for various traits (Table 15.7) such as dry-matter content and chipping quality (Sverrisdóttir *et al.*, 2018), fry color (Byrne *et al.*, 2020), tuber yield and other tuber-related traits (Habyarimana *et al.*, 2017; Caruana *et al.*, 2019), and tolerance to late blight and common scab (Rodriguez *et al.*, 2018). The application of GS in vegetable-breeding programs has high potential for increasing the gain from selection.

### 15.6.2 Genome-wide association studies in major vegetable crops

GWAS is a powerful tool for dissecting the genetic basis of variation for both qualitative and quantitative inherited traits. GWAS is an effective method based on the correlation between genotypes and phenotypes in a group of individuals, to screen the complex trait inheritance variation under the release of the reference genome sequence (Saidou *et al.*, 2014). It can be conducted by collecting diverse cultivated accessions or inter-crossed lines to sequence and compare genomic diversity among these accessions, especially traits influenced by multiple factors.

The bitterness of cucumber is an undesirable trait, and non-bitter cultivars have been selected during long-term domestication. A significant GWAS on cucumber revealed the molecular mechanism of bitterness biosynthesis using 115 cucumber accessions, and also contributed to identifying bitter flavor accessions by linked molecular markers easily and accurately (Shang *et al.*, 2014). GWAS can also be used to identify the associations related to fruit flavor in tomato, including fruit sugars and organic acid volatiles; the linked markers can be used for flavor-related gene discovery (Zhang *et al.*, 2015; Zhao *et al.*, 2016). Capsaicinoid content is an important factor that affects pepper flavor, and several reports have analyzed capsaicinoid using GWAS; for example, 96 *Capsicum annuum* accessions were used to identify a marker linked to capsaicin and dihydrocapsaicin contents, and several transport factors and an ankyrin-like protein are likely associated with capsaicinoid synthesis (Nimmakayala *et al.*, 2016). In another GWAS population containing 208 *C. annuum* clade accessions, candidate genes or QTLs for regulating capsaicinoid content were identified, and 69 QTL regions and five candidate genes were identified through high-density SNP mapping (Han *et al.*, 2018). GWAS has been employed for genetic analysis of traits such as capsaicinoid content (Nimmakayala *et al.*, 2016; Han *et al.*, 2018), *Phytophthora capsici* root rot resistance (Siddique *et al.*, 2019), potato virus Y resistance (Tamisier *et al.*, 2020), and several other fruit-related traits.

Tomato flavor is an important breeding goal for breeders. Volatile organic compounds are major determinants of tomato flavor. Zhang *et al.* (2015) performed GWAS to identify putative candidate genes and potential chromosome regions associated with flavor volatiles using 174 tomato accessions and 183 SSRs. Although most tomato varieties for the fresh market are $F_1$ hybrids, most studies on the genetic control of flavor-related traits are performed on lines. In another study, Bineau *et al.* (2021) quantified 46 volatile organic compounds in a panel of 121 small-fruited lines, 44 elite cherry lines, and a test-cross panel of 165 hybrids using 7442 SNPs. They identified 205, 183, and 138 associations in the small-fruited lines, cherry lines, and test-cross panel, respectively (Table 15.8). Sauvage *et al.* (2014) identified 44 promising candidate loci that underlie the genetic architecture of traits such as fruit sucrose, ascorbate, malate, and citrate levels, using 163 tomato accessions using GWAS with 5995 SNPs. Zhao *et al.* (2019) studied the genetic control of tomato fruit flavor, using meta-analysis of genome-wide association using 775 tomato accessions and 2,316,117 SNPs from three GWAS panels and revealed 305 significant associations for the contents of sugars, acids, amino acids, and flavor-related volatiles, and found that fruit citrate and malate contents have been impacted by selection during domestication and improvement, while sugar content has undergone less stringent selection.

Campa *et al.* (2020) conducted GWAS for white mold, a devastating fungal disease affecting common bean (*Phaseolus vulgaris* L.), using 294 lines of the Spanish Diversity Panel (SDP) with 9070 SNPs markers. Twenty-two significant associations and 61 potential candidate genes were found for white mold resistance. García-Fernández *et al.* (2021) conducted single- and multi-locus GWAS for pod traits in a panel of 301 bean lines of the SDP. The SDP was genotyped with 32,812 SNPs obtained from genotyping by sequencing. Twenty-one potential candidate genes involving developmental processes were detected, underlying 11 QTLs for pod morphological characteristics, and eight potential candidate genes involved in pigment synthesis were found underlying five QTLs for pod color.

Yam (*Dioscorea* spp.) is predominantly dioecious in nature, with male and female flowers borne on separate plants. Understanding the molecular genetic mechanisms underlying sex determination and cross-compatibility is important for yam breeding programs and for developing genomic tools for water yam breeding programs. Mondo *et al.* (2021) conducted a study using the genome-wide association approach to identify genomic regions linked to sex and cross-compatibility in water yam (*Dioscorea alata* L.). Fifty-four markers associated with flower sex determination and 61 putative gene/protein families with a direct or indirect influence on plant reproduction were detected in chromosomic regions controlling these traits.

Cassava is an important starchy root vegetable crop cultivated in tropical and subtropical countries for fresh consumption and processing purposes. Genetic dissection of traits associated with complex trait inheritance variation under the release of the reference genome sequence have been studied in cassava. Some of the studies are highlighted in Table 15.8. Rabbi *et al.* (2017) identified genetic markers associated with dry matter and total carotenoid content through GWAS, using a panel of 672 clones genotyped at 72,279 SNPs. Two major loci for root yellowness and a single locus for dry-matter content were identified on chromosome 1 at positions 24.1, 30.5, and 24.1 Mbp, respectively.

**Table 15.8.** Identification of candidate genes/QTLs through genome-wide association studies in major vegetable crops.

| Crop | Traits | Panel size | Markers | Association | Candidate genes/QTLs | Validation | References |
|---|---|---|---|---|---|---|---|
| Bottle gourd | *Fusarium* wilt resistant | 89 | 5,330 | – | – | + | Y. Li *et al.* (2021a) |
| Capsicum | Capsaicinoid content | 208 | 109,610 | 69 | 5 | – | Han *et al.* (2018) |
| | Capsaicinoids content and fruit weight | 94 | 7,331 | 30, 56 | 16 | – | Nimmakayala *et al.* (2016) |
| | Resistance to potato virus Y | 256 | 10,307 | 7 | 101 | – | Tamisier *et al.* (2020) |
| | Resistance to *Phytophthora* root rot | 352 | 507,713 | 117 | 3 | – | Siddique *et al.* (2019) |
| | Agronomic traits | 287 | 1,487 | – | | – | Wu *et al.* (2019) |
| Cucumber | Powdery mildew resistant | 231 | 950,382 | – | 13 | – | Jia *et al.* (2021) |
| | Downy mildew resistant | 97 | 74,166 | – | 18 | + | Liu *et al.* (2020) |
| Chinese cabbage | Turnip mosaic virus resistant | 83 | 42,526 | – | 17 | + | R. Zhang *et al.* (2020) |
| Carrot | Root shape | 103 | 81,748 | – | 1 | – | Macko-Podgórni *et al.* (2018) |
| Radish | *Fusarium* wilt resistant | 225 | 188,813 | 42 | – | – | Lee *et al.* (2021) |
| | Skin color | 179 | 588,896 | 14 | – | + | Fan *et al.* (2020) |
| Potato | Potato common scab | 143 | 12,000 | 3 | – | – | Yuan *et al.* (2019) |
| | | 165 | 13,002 | 6 | 1 | – | Koizumi *et al.* (2021) |
| | Resistant to wart disease | 330 | 10,968 | 12 | 1 | – | Prodhomme *et al.* (2020) |
| | Fry color | 219 | 46,406 | 31 | 1 | – | Byrne *et al.* (2020) |
| | Protein content | 277 | 20,000 | 12 | 3 | – | Klaassen *et al.* (2019) |
| | Agronomic and morphological traits | 237 | 12,720 | 36 | 16 | – | Zia *et al.* (2020) |
| | Tuber skin color | 809 | 5,968 | 23 | 3 | – | Berdugo *et al.* (2017) |
| | Anthocyanin color | 96 | 7,520 | 15 | 7 | – | Parra-Galindo *et al.* (2019) |
| Tomato | Flavor | 174 | 182 | 125 | – | – | Zhang *et al.* (2015) |
| | Fruit metabolites | 163 | 5,995 | 44 | – | – | Sauvage *et al.* (2014) |
| | Fruit metabolites | 121, 44, 165 | 7,442 | 205, 183, 138 | – | – | Bineau *et al.* (2021) |
| | Resistance to bacterial wilt | 191 | 38,541 | 8 | 3 | + | Nguyen *et al.* (2021) |
| | Agronomic and morphological traits | 244 | 37,300 | 59 | 6 | – | Tripodi *et al.* (2021) |
| | Fruit metabolites | 163 | 2,316,117 | 305 | 37 | – | Zhao *et al.* (2019) |
| Common bean | Resistance to *Sclerotinia sclerotiorum* | 249 | 9,070 | 22 | 61/15 | – | Campa *et al.* (2020) |
| | Pod-related agronomic traits | 301 | 32,812 | 63 | 25/62 | – | García-Fernández *et al.* (2021) |
| | Drought resistant in seedling stage | 400 | 3,832,340 | 12 | 39 | + | Wu *et al.* (2022) |
| | Resistance to soybean cyst nematode | 317 | 8,796 | 37 | 6 | – | Jain *et al.* (2019) |

*Continued*

**Table 15.8.** Continued.

| Crop | Traits | Panel size | Markers | Association | Candidate genes/QTLs | Validation | References |
|---|---|---|---|---|---|---|---|
| Water yam | Sex determination and cross-compatibility | 74 | 22,140 | 54 | 61 | – | Mondo *et al.* (2021) |
| | Tuber dry matter and oxidative browning | 100 | 22,140 | 3, 2 | 26, 6 | – | Gatarira *et al.* (2020) |
| White guinea yam | Tuber yield and mosaic virus tolerance | 406 | 16,242 | 17, 5 | 25, 4 | – | Agre *et al.* (2021) |
| Cassava | Dry-matter and total carotenoid content | 672 | 72,279 | 1, 2 | 2, 2 | – | Rabbi *et al.* (2017) |
| | Resistance to cassava mosaic disease | 694 | 42,113 | 3 | 14 | – | Wolfe *et al.* (2016) |
| | Yield, starch type, and starch properties | 276 | 89,934 | 31 | 21 | – | Phumichai *et al.* (2021) |
| | Carotenoid content | 591 | 179,310 | 4 | 1 | – | Esuma *et al.* (2016) |
| | Agronomic traits | 158 | 349,827 | 36 | 62 | + | S. Zhang *et al.* (2018) |
| | Resistance to cassava green mite | 845 | 61,307 | 35 | 17 | – | Ezenwaka *et al.* (2018) |
| | Agro-morphological and quality traits | 5130 | 100,000 | 41 | 17 | – | Rabbi *et al.* (2020) |
| | Resistance to brown streak disease | 429, 872 | 173,647 | 83 | 2 | – | Kayondo *et al.* (2018) |
| | Resistance to cassava root rot | 263 | 14,094 | 38 | 31 | – | Brito *et al.* (2017) |
| | Waxing starch quality | 382 | 20,956 | 10 | 2 | – | do Carmo *et al.* (2020) |

+, Validated; –, not validated.

Potato is an important staple food crop for many countries. It has long been the subject of molecular studies owing to its nutritional value and significance for the processing industry. Successful studies have been performed in potato for the detection of diagnostic SNP markers using GWAS related to tuber quality traits, starch content, tuber skin color, disease resistance, and several agronomic and morphological traits. Some of these studies are highlighted in Table 15.8.

### 15.6.3 Genome editing for vegetable improvement

Over the last few decades, there have been a number of major advancements in the technology used in molecular breeding methods in crop plants. The discovery of sequence-specific nucleases and development of the CRISPR/Cas system have facilitated gene editing at the DNA level to generate vegetables with altered and desired traits (Abdallah *et al.*, 2015; Nunez de Caceres Gonzalez and de la Mora Franco, 2020). Over the last few decades, gene/genome-editing technologies have facilitated precise and efficient targeted modification of genomes in various organisms, including vegetable crops to increase yield, resistances, and quality (Chen and Gao, 2014).

Genome editing is defined as a collection of advanced molecular biology techniques that facilitate precise, efficient, and targeted modifications at the genomic loci (Chen and Gao, 2014). There are four main methods of genome editing: knockout or deactivation of a gene, activation or overexpression, repression, and precisely altering the gene including targeted chromosome rearrangements (Choi and Meyerson, 2014). Genome editing results in homozygous, heterozygous, biallelic, and mosaic mutations. Genome editing can be divided in to four classes based on chronological order of their discovery: (i) meganucleases; (ii) zinc-finger nucleases; (iii) transcription activator-like effector nucleases (TALENs); and (iv) CRISPR/Cas (Gupta and Musunuru, 2014). The first three are protein-based genome-editing technologies, while CRISPR/Cas is an RNA–protein hybrid system. The introduction of genome-editing tools into modern breeding programs has facilitated rapid and precise crop improvements.

## 15.7 Application of Genome Editing in Vegetable Crops

The research and application of genome editing using the CRISPR/Cas9 system in the major vegetable crops is summarized in Table 15.9.

### 15.7.1 Potato

Potato is an important vegetable crop worldwide and is considered a staple food crop in Europe and other countries. Potato varieties with higher nutrient content, resistance to diseases and pests, and that can adapt to climate change are the prime breeding goals in potato improvement programs. Gene-editing technology is an excellent tool for potato improvement, and there are a number of traits that have been modified in potato. Butler *et al.* (2015) utilized the CRISPR/Cas9 system for the first time to modify the *StALS1* gene for herbicide tolerance in potato. In the same year, Wang *et al.* (2015) used CRISPR/Cas9 to produce knockouts in the *StIAA2* gene in potato, which is involved in shoot morphogenesis in development of the plant.

Zhao *et al.* (2021) used CRISPR/Cas9 to produce a high amylopectin content in potato by targeting the *Sbe* gene responsible for amylopectin production. Late blight is a major production constraint in potato cultivation. Kieu *et al.* (2021) introduced mutations in the susceptible genes *StDND1*, *StCHL1*, and *StDMR6-1* through CRISPR/Cas9 genome editing, which conferred increased late blight resistance in potato. CRISPR/Cas9 system has been used for a number of other potato trait modification including nutritional content, biotic and abiotic stress tolerance, starch content, postharvest quality, and reduced anti-nutritional factors. Sweet potato virus disease is caused by co-infection of sweet potato chlorotic stunt virus and sweet potato feathery mottle virus; Yu *et al.* (2021) developed germplasm resistant to this disease by targeting the *SPSCV-RNase3* gene in sweet potato using CRISPR/Cas9.

### 15.7.2 Carrot

The CRISPR/Cas9 system has been used successfully for precise genome editing in many plant

**Table 15.9.** Application of CRISPR/Cas9-based editing of genes in vegetables.

| Crop | Target gene | Target trait | Reference |
|---|---|---|---|
| Potato | *StALS1* | Herbicide resistance | Butler *et al.* (2015) |
| | *StIAA2* | Developmental | Wang *et al.* (2015) |
| | *StALS1* | Herbicide resistance | Butler *et al.* (2016) |
| | *StMYB44* | Abiotic stress | Zhou *et al.* (2017) |
| | *Wx1* | High amylopectin content | Andersson *et al.* (2017) |
| | *StDND1*, *StCHL1*, and *StDMR6-1* | Late blight resistance | Kieu *et al.* (2021) |
| | *Sbe* | High amylopectin content | Zhao *et al.* (2021) |
| | *Coilin* | Virus resistance, osmotic stress | Makhotenko *et al.* (2019) |
| | *S-RNase* | Self-incompatibility | Enciso-Rodriguez *et al.* (2019) |
| | *S-RNase* | Self-incompatibility | Ye *et al.* (2018) |
| | *SBE1*, *SBE2* | Starch quality | Tuncel *et al.* (2019) |
| | *GBSS1* | Starch quality | Kusano *et al.* (2018) |
| | *16DOX* | Reduced steroidal glycoalkaloids | Nakayasu *et al.* (2018) |
| | *PPO2* | Reduced enzymatic browning | Gonzalez *et al.* (2020) |
| | *PDS* | Albino phenotype | Tian *et al.* (2017) |
| Sweet potato | *SPSCV-RNase3* | Sweet potato virus disease resistance | Yu *et al.* (2021) |
| Cassava | *SBE2* | High amylose content | Luo *et al.* (2021) |
| Tomato | *SlHyPRP1* | Salt-stress tolerance | Tran *et al.* (2021) |
| | *SlMAPK6* | Growth and development | Y. Li *et al.* (2021b) |
| | *SlPelo* and *SlMlo1* | Tomato yellow leaf curl virus and powdery mildew resistance | Pramanik *et al.* (2021) |
| | *Ms1035* and *GSTAA* | Male sterlity induction | Liu *et al.* (2021) |
| | *SlMIR164a* and *SlMIR164b* | Growth and development | Gupta *et al.* (2021) |
| | *mGFP*, *SlSHR* | Gene functional characterization | Ron *et al.* (2014) |
| | *SLAGO7* | Characterization | Brooks *et al.* (2014) |
| | *RIN* | Slower ripening | Ito *et al.* (2015) |
| | *ANT1* | Characterization | Čermák *et al.* (2015) |
| | *CLV3* | Fruit ripening | Xu *et al.* (2015) |
| | *SlPDS*, *SlPIF4* | Phenotypic | Pan *et al.* (2016) |
| | *SPG5* | Fruit size | Soyk *et al.* (2017) |
| | *PSY1* | Phenotypic | Hayut *et al.* (2017) |
| | *ALC* | Early flowering | Yu *et al.* (2017) |
| | *LRR-XII* family, *PDS* | Phenotypic | Jacobs *et al.* (2017) |
| | *SlAGL6*, *SlIAA9* | Prolonged shelf life | Klap *et al.* (2017) |
| | *DELLA*, *ETR1* | Phenotypic/mutagenized | Shimatani *et al.* (2017) |
| | *SlMlo1* | Populations | Nekrasov *et al.* (2017) |
| | *SlMAPK3* | Parthenocarpy | L. Wang *et al.* (2017) |
| | *SlGAD2*, *SlGAD3* | Enhanced GABA production in fruits | Nonaka *et al.* (2017) |
| | *SlIAA9* | Fungal resistance | Ueta *et al.* (2017) |
| | *GABATP1-3*, *CAT9* | Drought tolerance | R. Li *et al.* (2018a) |
| | *SSADH* | γ-Aminobutyric acid increase | R. Li *et al.* (2018b); X. Li *et al.* (2018) |
| | *SLCBF1* | Parthenocarpy | Dahan *et al.* (2018) |
| | *SGR1*, *LCY-E*, *Blc*, *LCY-B*, *LCY-B2* | Increase in γ-aminobutyric acid | Deng *et al.* (2018) |
| | *CRITISO*, *PSY1* | Abiotic stress | Tomlinson *et al.* (2019) |
| | *SlCLV3* | Phenotypic | Rodríguez-Leal *et al.* (2017) |
| | SlSTR1 | Male sterility induction | Du *et al.* (2020) |

*Continued*

**Table 15.9.** Continued.

| Crop | Target gene | Target trait | Reference |
|---|---|---|---|
| | *SP*, *O*, *FW2.2*, *CycB*, *FAS*, *MULT* | Fruit phenotype | Zsögön *et al.* (2018) |
| Capsicum | *CaERF28* | Anthracnose resistance | Mishra *et al.* (2021) |
| Watermelon | *ClPDS* | Phenotypic | Tian *et al.* (2016) |
| | *ALS* | Herbicide resistance | Tian *et al.* (2018) |
| | *ClPSK1* | Resistance to *Fusarium oxysporum* f. sp. *niveum* | M. Zhang *et al.* (2020) |
| | *ClWIP1* | Knockout/gynoecious watermelon | J. Zhang *et al.* (2020) |
| | *ClBG1* | Seed size reduction, promoted seed germination | J. Wang *et al.* (2021) |
| | *VST1* | Reduce aphid damage | M. Li *et al.* (2021) |
| | *ClNAC68* | Sugar content and seed development | J. Wang *et al.* (2021) |
| | *VST1* | Sugar content, total biomass | Ren *et al.* (2020) |
| | *AGA2*, *SWEET3*, *TST2* | Fruit sugar accumulation | Ren *et al.* (2021) |
| Musk melon | *PDS* | Albino phenotype | Hooghvorst *et al.* (2019) |
| Lettuce | *BIN2* | Development | Woo *et al.* (2015) |
| | *LsNCED4* | Abiotic stress | Bertier *et al.* (2018) |
| | *LsGGP2* | Increased stress tolerance and ascorbate content | H. Zhang *et al.* (2018) |
| Chicory | *ClPDS* | Phenotypic | Bernard *et al.* (2019) |
| Cucumber | *eIF4E* | Virus resistance | Chandrasekaran *et al.* (2016) |
| | *CsWIP1* | Gynoecious induction | Hu *et al.* (2017) |
| Carrot | *F3H* | Albino phenotype | Klimek *et al.* (2018) |
| | DcPDS and DcMYB113 | Albino phenotype | Xu *et al.* (2019) |
| | *psy1* and *psy2* | Cell-wall remodeling | Oleszkiewicz *et al.* (2021) |
| Eggplant | *SmelPPO* | Decreased enzymatic browning | Maioli *et al.* (2020) |
| Chinese kale | *BaPDS1*, *BaPDS2* | Albino phenotype | Sun *et al.* (2018) |
| | *BoaCRTISO* | Yellow color | Sun *et al.* (2020) |
| Cabbage | *BoPDS1*, *BoSRK3*, *BoMS1* | Male sterility induction | Ma *et al.* (2019b) |
| | *BoPDS* | Albino phenotype | Ma *et al.* (2019a) |
| | MYB28 | Glucoraphanin accumulation | Neequaye *et al.* (2021) |
| Chinese cabbage | *BraFLCs* | Avoidance of vernalization | Jeong *et al.* (2019) |

species, including carrot cells. Klimek *et al.* (2018) validated the efficiency of editing in carrot by targeting *F3H* gene knockout in orange- and purple-colored carrot, which generated an albino phenotype in the root. CRISPR/Cas9-mediated knockout of the *DcPDS* gene in orange carrot 'Kurodagosun' resulted in the generation of albino carrot plantlets, with about 35.3% editing efficiency. *DcMYB113-like* was also successfully edited in the purple carrot 'Deep Purple' resulting in purple-depigmented carrot plants, with a rate of mutation of about 36.4% (Xu *et al.*, 2019). Oleszkiewicz *et al.* (2021) generated functional mutations in the *psy1* and *psy2* genes and demonstrated their effect on carotenoid accumulation and mutual regulation of their expression. Inhibition of carotenoid biosynthesis by CRISPR/Cas9 triggered cell-wall remodeling in two carrot *psy* paralogs in a model system and imparted biochemical changes in the cell wall of *psy*-knockout mutants.

### 15.7.3 Cucumber

Cucumber was the first species with a successful report of gene editing by the CRISPR/Cas9 system in the family Curcubitaceae. Chandrasekaran *et al.* (2016) demonstrated the development of virus resistance in cucumber (*Cucumis sativus* L.) using Cas9/subgenomic RNA technology to disrupt the function of the recessive eukaryotic translation initiation factor 4E (*eIF4E*) gene. The gynoecious sex form has been exploited extensively in cucumber breeding. Utilization of a gynoecious line allowed earliness of hybrids, higher yield, and more concentrated fruit set. Hu *et al.* (2017) developed gynoecious lines by utilizing CRISPR/Cas9-mediated targeted mutagenesis in the *CsWIP1* gene.

### 15.7.4 Tomato

Tomato has been used as a model plant and is the most preferred vegetable for research and genome editing using CRISPR/Cas9. This may be because of its economic importance as well as the availability of extensive genomic resources and highly efficient transformation methods, making it an ideal target for gene editing. In fact, tomato was the first vegetable crop to have its genome edited using the CRISPR/Cas9 system (Ron *et al.*, 2014).

Ron *et al.* (2014) were the first to successfully validate the CRISPR/Cas9 system to perform gene knockout for functional characterization in tomato by targeting the *mGFP* promoter and *SlSHR* genes. Later that year, another report was published related to the proof of concept of the system. In the same year, Brooks *et al.* (2014) proved that gene editing can be successful by targeting a gene, *SLARGONAUTE* (*SLAGO7*), which when disrupted produced a very distinctive phenotype. Since Ron *et al.* (2014) and Brooks *et al.* (2014) proved the concept of using CRISPR/Cas9 to edit the genome of tomato, studies have been performed over a broad spectrum of topics. The CRISPR/Cas9 system of genome editing has been studied extensively for the traits such as characterization, growth- and development-related traits, fruit ripening, induction of parthenocarpy, postharvest quality, prolonging fruit shelf life, induction of male sterility, functional gene analysis, and introgression of biotic and abiotic stress tolerance in tomato (Table 15.9). Modifications of several genes involved in developmental pathways have been edited using CRISPR/Cas9. A new male-sterile system in tomato was developed by Du *et al.* (2020) using CRISPR/Cas9-mediated mutagenesis of a stamen-specific gene, *SlSTR1*, and a transgenic maintainer was devised by transforming male-sterile plants with a fertility-restoration gene linked to a seedling-color gene. This study showed that this male-sterility system has great practical potential for hybrid seed production, as it overcomes the problems intrinsic to other male-sterility systems and can easily be adapted for a range of tomato cultivars and other diverse vegetable crops.

### 15.7.5 Watermelon

In 2016, watermelon was first subjected to gene editing using CRISPR/Cas9 by Tian *et al.* (2016). They created knockout mutations in the *ClPDS* gene, which caused an albino phenotype as expected. A point mutation from C to T in the acetolactate synthase (*ALS*) gene in watermelon has generated herbicide-resistant plants in watermelon (Tian *et al.*, 2018). A candidate gene, *ClBG1*, in watermelon was selected for targeted mutagenesis using CRISPR/Cas9, which reduced seed size and weight and promoted seed germination significantly in the *Clbg1* mutant watermelon lines (Y. Wang *et al.*, 2021). Aphids can damage plants through sugar withdrawal by hijacking sugar metabolism and transport genes to increase their sugar sucking ability. M. Li *et al.* (2021) knocked out the sugar transporter *VST1* gene, which reduced aphid damage in watermelon plants. A list of target traits used for editing through the CRISPR/Cas9 system in watermelon is summarized in Table 15.9.

### 15.7.6 Cabbage

Ma *et al.* (2019b) reported a CRISPR/Cas9 gene-editing system based on endogenous tRNA processing to induce high-efficiency and inheritable mutagenesis in cabbage. The target *BoSRK3* gene mutation suppressed self-incompatibility completely, converting the self-incompatible line into a self-compatible line. Neequaye *et al.* (2021) reported that knocking out the *myb28* gene resulted in downregulation of aliphatic glucosinolate biosynthesis genes and a reduction in the accumulation of the methionine-derived glucosinolate glucoraphanin in the leaves and florets of field-grown *myb28*-mutant broccoli plants, whereas accumulation of sulfate, *S*-methyl cysteine sulfoxide, and indole glucosinolate in leaf and floret tissues remained unchanged. Jeong *et al.* (2019) generated early-flowering Chinese cabbage (*Brassica rapa* spp. *pekinensis*) that did not depend on vernalization by target editing in the *FLC* locus.

## 15.8 Conclusion and Future Prospects

The occurrence of biotic and abiotic stresses is considered a major threat to vegetable production and world food security. Sustainable production and improvement of productivity in vegetable crops by developing high-yielding varieties and genetic resistance to biotic and abiotic stresses in vegetable crop is essential. Conventional breeding approaches usually take a long time for the development of new plant varieties. To accelerate the breeding program, molecular breeding approaches and exploitation of genomic tools and genomic information of crop species such as molecular markers linked to traits, whole-genome sequencing-based tools, genome editing, gene pyramiding, and MAS can be used to speed up the breeding cycle.

Exploiting crop genomics, genome sequences, pan-genomes, FMs, and GS tools available for the major vegetable crops discussed in this chapter will provide many more opportunities for practicing GS, MAS, complex trait dissection, and genome editing. Genome sequencing will help in comparisons and resequencing of breeding populations, and in germplasm collections in a rapid and probably cheaper manner.

## 15.9 References


Abdallah, N.A., Prakash, C.S., and McHughen, A.G. (2015) Genome editing for crop improvement: challenges and opportunities. *GM Crops & Food* 6, 183–205.

Abe, A., Kosugi, S., Yoshida, K., Natsume, S., Takagi, H., *et al.* (2012) Genome sequencing reveals agronomically important loci in rice using MutMap. *Nature Biotechnology* 30, 174–178.

Abuyusuf, M., Nath, U.K., Kim, H.T., Islam, M.R., Park, J.I., *et al.* (2019) Molecular markers based on sequence variation in BoFLC1. C9 for characterizing early- and late-flowering cabbage genotypes. *BMC Genetics* 20: 42.

Aggarwal, R.K., Hendre, P.S., Varshney, R.K., Bhat, P.R., Krishnakumar, V., and Singh, L. (2007) Identification, characterization and utilization of EST-derived genic microsatellite markers for genome analyses of coffee and related species. *Theoretical and Applied Genetics* 114, 359–372.

Agre, P.A., Norman, P.E., Asiedu, R., and Asfaw, A. (2021) Identification of quantitative trait nucleotides and candidate genes for tuber yield and mosaic virus tolerance in an elite population of white guinea yam (*Dioscorea rotundata*) using genome-wide association scan. *BMC Plant Biology* 21: 552.

Aguado, E., García, A., Iglesias-Moya, J., Romero, J., Wehner, T.C., *et al.* (2020) Mapping a partial andromonoecy locus in *Citrullus lanatus* using BSA-seq and GWAS approaches. *Frontiers in Plant Science* 11: 1243.

Ahn, Y.K., Manivannan, A., Karna, S., Jun, T.H., Yang, E.Y., *et al.* (2018) Whole genome resequencing of *Capsicum baccatum* and *Capsicum annuum* to discover single nucleotide polymorphism related to powdery mildew resistance. *Scientific Reports* 8: 5188.

Andersen, J.R. and Lubberstedt, T. (2003) Functional markers in plants. *Trends Plant Science* 8, 554–560.

Andersson, M., Turesson, H., Nicolia, A., Fält, A.S., Samuelsson, M., and Hofvander, P. (2017) Efficient targeted multiallelic mutagenesis in tetraploid potato (*Solanum tuberosum*) by transient CRISPR-Cas9 expression in protoplasts. *Plant Cell Reports* 36, 117–128.

Arens, P., Mansilla, C., Deinum, D., Cavellini, L., Moretti, A., *et al.* (2010) Development and evaluation of robust molecular markers linked to disease resistance in tomato for distinctness, uniformity and stability testing. *Theoretical and Applied Genetics* 120, 655–664.

Arizio, C.M., Costa Tártara, S.M., and Manifesto, M.M. (2014) Carotenoids gene markers for sweetpotato (*Ipomoea batatas* L. Lam): applications in genetic mapping, diversity evaluation and cross-species transference. *Molecular Genetics and Genomics* 289, 237–251.

Aruga, D., Tsuchiya, N., Matsumura, H., Matsumoto, E., and Hayashida, N. (2012) Analysis of RAPD and AFLP markers linked to resistance to *Fusarium oxysporum* f. sp. *lactucae* race 2 in lettuce (*Lactuca sativa* L.). *Euphytica* 187, 1–9.

Avila, C.M., Sillero, J.C., Rubiales, D., Moreno, M.T., and Torres, A.M. (2003) Identification of RAPD markers linked to *Uvf-1* gene conferring hypersensitive resistance against rust (*Uromyces viciae-fabae*) in *Vicia faba* L. *Theoretical and Applied Genetics* 107, 353–358.

Avila, C.M., Satovic, Z., Sillero, J.C., Rubiales, D., Moreno, M.T., and Torres, A.M. (2004) Isolate and organ-specific QTLs for ascochyta blight resistance in faba bean. *Theoretical and Applied Genetics* 108, 1071–1078.

Badri Anarjan, M., Bae, I., and Lee, S. (2021) Marker-assisted evaluation of two powdery mildew resistance candidate genes in Korean cucumber inbred lines. *Agronomy* 11: 2191.

Bai, Y., Huang, C.C., van der Hulst, R., Meijer-Dekens, F., Bonnema, G., and Lindhout, P. (2003) QTLs for tomato powdery mildew resistance (*Oidium lycopersici*) in *Lycopersicon parviflorum* G1.1601 co-localize with two qualitative powdery mildew resistance genes. *Molecular Plant–Microbe Interactions*, 16, 169–176.

Balint-Kurti, P.J., Dixon, J.S., Jones, D.A., Norcott, K.A., and Jones, J.D.G. (1994) RFLP linkage analysis of the *Cf-4* and *CF-9* genes for resistance to *Cladosporium fulvum* in tomato. *Theoretical and Applied Genetics* 88, 691–700.

Barchi, L., Rabanus-Wallace, M.T., Prohens, J., Toppino, L., Padmarasu, S., *et al.* (2021) Improved genome assembly and pan-genome provide key insights into eggplant domestication and breeding. *Plant Journal* 107, 579–596.

Barrera-Redondo, J., Ibarra-Laclette, E., Vazquez-Lobo, A., Gutierrez-Guerrero, Y.T., Sánchez de la Vega, G., *et al.* (2019) The genome of *Cucurbita argyrosperma* (silver-seed gourd) reveals faster rates of protein-coding gene and long noncoding RNA turnover and neofunctionalization within *Cucurbita*. *Molecular Plant* 12, 506–520.

Behare, J., Laterrot, H., Sarfatti, M., and Zamir, D. (1991) Restruiction fragment length mapping polymorphism of the *Stemphylium* resistance gene in tomato. *Molecular Plant–Microbe Interactions* 4, 489–492.

Berdugo-Cely, J., Valbuena, R.I., Sánchez-Betancourt, E., Barrero, L.S., and Yockteng, R. (2017) Genetic diversity and association mapping in the Colombian central collection of *Solanum tuberosum* L. andigenum group using SNPs markers. *PLOS ONE* 12: e0173039.

Berg, J.A., Hermans, F.W.K., Beenders, F., Lou, L., Vriezen, W.H., *et al.* (2020) Analysis of QTL DM4.1 for downy mildew resistance in cucumber reveals multiple subQTL: a novel RLK as candidate gene for the most important subQTL. *Frontiers in Plant Science* 11: 569876.

Berg, J.A., Hermans, F.W.K., Beenders, F., Abedinpour, H., Vriezen, W.H., *et al.* (2021) The amino acid permease (AAP) genes *CsAAP2A* and *SlAAP5A/B* are required for oomycete susceptibility in cucumber and tomato. *Molecular Plant Pathology* 22, 658–672.

Bernard, G., Gagneul, D., Alves Dos Santos, H., Audrey, E., Jean-Louis, H., and Caroline, R. (2019) Efficient genome editing using CRISPR/Cas9 technology in chicory. *International Journal of Molecular Sciences* 20: 1155.

Bertier, L.D., Ron, M., Huo, H., Bradford, K.J., Britt, A.B., and Michelmore, R.W. (2018) High-resolution analysis of the efficiency, heritability, and editing outcomes of CRISPR/Cas9-induced modifications of NCED4 in lettuce (*Lactuca sativa*). *G3* 8: 1513.

Bhattarai, G., Yang, W., Shi, A., Feng, C., Dhillon, B., Correll, J.C., and Mou, B. (2021). High resolution mapping and candidate identification of downy mildew race 16 resistance in spinach. *BMC Genomics* 22: 478.

Bineau, E., Rambla, J.L., Priego-Cubero, S., Hereil, A., Bitton, F., *et al.* (2021) Breeding tomato hybrids for flavour: comparison of GWAS results obtained on lines and F1 hybrids. *Genes* 12: 1443.

Boiteux, L.S., Hyman, J.R., Bach, I.C., Fonseca, M.E.N., Matthews, W.C., *et al.* (2004) Employment of flanking codominant STS markers to estimate allelic substitution effects of a nematode resistance locus in carrot. *Euphytica* 136, 37–44.

Bolger, A., Scossa, F., Bolger, M.E., Lanz, C., Maumus, F., Tohge, T., *et al.* (2014) The genome of the stress-tolerant wild tomato species *Solanum pennellii*. *Nature Genetics* 46, 1034–1038.

Borovsky, Y., Monsonego, N., Mohan, V., Shabtai, S., Kamara, I., and Faigenboim, A. (2019) The zinc-finger transcription factor CcLOL1 controls chloroplast development and immature pepper fruit color in *Capsicum chinense* and its function is conserved in tomato. *Plant Journal* 99, 41–45.

Boukar, O., Belko, N., Chamarthi, S., Togola, A., Batieno, J., *et al.* (2019) Cowpea (*Vigna unguiculata*): genetics, genomics and breeding. *Plant Breeding* 138, 415–424.

Brainard, S.H., Ellison, S.L., Simon, P.W., Dawson, J.C., and Goldman, I.L. (2021) Genetic characterization of carrot root shape and size using genome-wide association analysis and genomic-estimated breeding values. *Theoretical and Applied Genetics* 135, 605–622.

Branham, S.E., Wechter, W.P., Lambel, S., Massey, L., Ma, M., *et al.* (2018) QTL-seq and marker development for resistance to *Fusarium oxysporum* f. sp. *niveum* race 1 in cultivated watermelon. *Molecular Breeding* 38: 139.

Brito, A.C., Oliveira, S.A.S., and Oliveira, E.J. (2017) Genome-wide association study for resistance to cassava root rot. *Journal of Agricultural Science* 155, 1424–1441.

Brooks, C., Nekrasov, V., Lippman, Z.B., and Joyce, V.E. (2014) Efficient gene editing in tomato in the first generation using the clustered regularly interspaced short palindromic repeats/CRISPR-associated9 system. *Plant Physiology* 166, 1292–1297.

Burger, Y., Katzir, N., Tzuri, G., Portnoy, V., Saar, U., *et al.* (2003) Variation in the response of melon genotypes to *Fusarium oxysporum* f. sp. *melonis* race 1 determined by inoculation tests and molecular markers. *Plant Pathology* 52, 204–211.

Butler, N.M., Atkins, P.A., Voytas, D.F., and David, S.D. (2015) Generation and inheritance of targeted mutations in potato (*Solanum tuberosum* L.) using the CRISPR/Cas system. *PLOS ONE* 10: e0144591.

Butler, N.M., Baltes, N.J., Voytas, D.F., and David, S.D. (2016) Geminivirus-mediated genome editing in potato (*Solanum tuberosum* L.) using sequence-specific nucleases. *Frontiers in Plant Science* 7: 1045.

Byrne, S., Meade, F., Mesiti, F., Griffin, D., Kennedy, C., and Milbourne, D. (2020) Genome-wide association and genomic prediction for fry color in potato. *Agronomy* 10: 90.

Campa, A., García-Fernández, C., and Ferreira, J.J. (2020) Genome-wide association study (GWAS) for resistance to *Sclerotinia sclerotiorum* in common bean. *Genes* 11: 1496.

Cao, B.H., Lei, J.J., Wang, Y., and Chen, G.J. (2009) Inheritance and identification of SCAR marker linked to bacterial wilt-resistance in eggplant. *African Journal of Biotechnology* 8, 5201–5207.

Cao, W., Du, Y., Wang, C., Xu, L., and Wu, T. (2018) *Cscs* encoding chorismate synthase is a candidate gene for leaf variegation mutation in cucumber. *Breed Science* 68, 571–581.

Caranta, C., Palloix, A., Gebre-Selassie, K., Lefebvre, V., Moury, B., and Daubeze, A.M. (1996) A complementation of two genes originating from susceptible *Capsicum annuum* lines confers a new and complete resistance to pepper veinal mottle virus. *Phytopathology* 86, 739–743.

Caranta, C., Palloix, A., and Lefebvre, V. (1997) QTLs for a component of partial resistance to cucumber mosaic virus in pepper: restriction of virus installation in host-cells. *Theoretical and Applied Genetics* 94, 431–438.

Caranta, C., Thabuis, A., and Palloix, A. (1999) Development of a CAPS marker for the *Pvr4* locus: a tool for pyramiding potyvirus resistance genes in pepper. *Genome* 42, 1111–1116.

Caruana, B.M., Pembleton, L.W., Constable, F., Rodoni, B., Slater, A.T., and Cogan, N.O.I. (2019) Validation of genotyping by sequencing using transcriptomics for diversity and application of genomic selection in tetraploid potato. *Frontiers in Plant Science* 10: 670.

Catanzariti, A.M., Lim, G.T., and Jones, D.A. (2015) The tomato *I-3* gene: a novel gene for resistance to *Fusarium wilt* disease. *New Phytologist* 207, 106–118.

Catanzariti, A.M., Do, H.T.T., Bru, P., de Sain, M., Thatcher, L.F., *et al.* (2017) The tomato *I* gene for *Fusarium* wilt resistance encodes an atypical leucine-rich repeat receptor-like protein whose function is nevertheless dependent on SOBIR1 and SERK3/BAK1. *Plant Journal* 89, 1195–1209.

Causse, M., Desplat, N., Pascual, L., Le Paslier, M.C., Sauvage, C., *et al.* (2013) Whole genome resequencing in tomato reveals variation associated with introgression and breeding events. *BMC Genomics* 14: 791.

Čermák, T., Baltes, N.J., Cegan, R., Zhang, Y., and Voytas, D.F. (2015) High-frequency, precise modification of the tomato genome. *Genome Biology* 16: 232.

Chandrasekaran, J., Brumin, M., Wolf, D., Leibman, D., Klap, C., *et al.* (2016) Development of broad virus resistance in non-transgenic cucumber using CRISPR/Cas9 technology. *Molecular Plant Pathology* 17, 1140–1153.

Changkwian, A., Venkatesh, J., Lee, J.H., Han, J.W., Kwon, J.K., *et al.* (2019) Physical localization of the root-knot nematode (*Meloidogyne incognita*) resistance locus *Me7* in pepper (*Capsicum annuum*). *Frontiers in Plant Science* 10: 886.

Chen, C., Zhuang, M., Fang, Z.Y., Wang, Q.B., Zhang, Y.Y., *et al.* (2013) A co-dominant marker BoE332 applied to marker-assisted selection of homozygous male-sterile plants in cabbage (*Brassica oleracea* var. *capitata* L.). *Journal of Integrative Agriculture* 12, 596–602.

Chen, F., Fu, B., Pan, Y., Zhang, C., Wen, H., *et al.* (2017) Fine mapping identifies *CsGCN5* encoding a histone acetyltransferase as putative candidate gene for *tendril-less1* mutation (*td-1*) in cucumber. *Theoretical and Applied Genetics* 130, 1549–1558.

Chen, K., Wang, Y., Zhang, R., Zhang, H., and Gao, C. (2019) CRISPR/Cas genome editing and precision plant breeding in agriculture. *Annual Review of Plant Biology* 70, 667–697.

Chen, K. and Gao, C. (2014) Targeted genome modification technologies and their applications in crop improvements. *Plant Cell Reports* 33, 575–583.

Cheng, Q., Wang, P., Liu, J., Wu, L., Zhang, Z., and Li, T. (2018) Identification of candidate genes underlying genic male-sterile *msc-1* locus via genome resequencing in *Capsicum annuum* L. *Theoretical and Applied Genetics* 131, 1861–1872.

Choi, P.S. and Meyerson, M. (2014) Targeted genomic rearrangements using CRISPR/Cas technology. *Nature Communications* 5: 3728.

Choi, S., Lee, J.H., Kang, W.H., Kim, J., Huy, H.N., *et al.* (2018) Identification of *Cucumber mosaic resistance 2* (*cmr2*) that confers resistance to a new *Cucumber mosaic virus isolate* P1 (CMV-P1) in pepper (*Capsicum* spp.). *Frontiers in Plant Science* 9: 1106.

Chunwongse, J., Bunn, T.B., Crossman, C., Jinag, J., and Tanksley, S.D. (1994) Chromosomal localization and molecular-marker tagging of the powdery mildew resistance gene (*Lv*) in tomato. *Theoretical and Applied Genetics* 89, 76–79.

Cobos, M.J., Satovic, Z., Rubiales, D., and Fondevilla, S. (2018) *Er3* gene, conferring resistance to powdery mildew in pea, is located in pea LGIV. *Euphytica* 214: 203.

Colak, A.A. and Yilmaz, N. (2020) Molecular marker assisted selection for *Phytophthora capsici* Leon. resistance lines in pepper (*Capsicum annuum* L.). *Acta Scientiarum Polonorum Hortorum Cultus* 19, 179–188.

Consuegra, O.G., Gómez, M.P., and Zubiaur, Y.M. (2015) Pyramiding TYLCV and TSWV resistance genes in tomato genotypes. *Revista de Protección Vegetal* 30: 161.

Contour-Ansel, D., Torres-Franklin, M.L., Cruz de Carvalho, M.H., d'Arcy-Lameta, A., and Zuily-Fodil, Y. (2006) Glutathione reductase in leaves of cowpea: cloning of two cDNAs, expression and enzymatic activity under progressive drought stress, desiccation and abscisic acid treatment. *Annals of Botany* 98, 1279–1287.

Crossa, J.P., Pérez, J., Hickey, J., Burgueño, L., Ornella, J., *et al.* (2014) Genomic prediction in CIMMYT maize and wheat breeding programs. *Heredity* 112, 48–60.

Crossa, J.P., Perez-Rodriguez, J., Cuevas, O., Montesinos-Lopez, D., Jarquin, G., *et al.* (2017) Genomic selection in plant breeding: methods, models, and perspectives. *Trends in Plant Science* 22, 961–975.

Cuevas, H.E., Staub, J.E., Simon, P.W., and Zalapa, J.E. (2009) A consensus linkage map identifies genomic regions controlling fruit maturity and beta-carotene-associated flesh color in melon (*Cucumis melo* L.). *Theoretical and Applied Genetics* 119, 741–756.

Cui, J., Yang, Y., Luo, S., Wang, L., Huang, R., *et al.* (2020) Whole-genome sequencing provides insights into the genetic diversity and domestication of bitter gourd (*Momordica* spp.). *Horticulture Research* 7: 85.

Dahan, T., Filler, S., Melamed, C., Samuel, B., Henryk, C., *et al.* (2018) Efficient *in planta* gene targeting in tomato using geminiviral replicons and the CRISPR/Cas9 system. *Plant Journal* 95, 5–16.

Dax, E., Livneh, O., Edelbaum, O., Kedar, N., Gavish, *et al.* (1994) A random amplified polymorphic DNA (RAPD) molecular marker for the *Tm-2a* gene in tomato. *Euphytica* 74, 159–163.

Ddamulira, G., Mukankusi, C., Ochwo-Ssemakula, M., Edema, R., Sseruwagi, P., and Gepts, P. (2015) Gene pyramiding improved resistance to angular leaf spot in common bean. *Journal of Experimental Agriculture International* 9, 1–12.

de Giovanni, C., Dell'Orco, P., Bruno, A., Ciccarese, F., Lotti, C., and Ricciardi, L. (2004) Identification of PCR-based markers (RAPD, AFLP) linked to a novel powdery mildew resistance gene (*ol-2*) in tomato. *Plant Science* 166, 41–48.

Deng, L., Wang, H., Sun, C., Qian, L., Hongling, J., *et al.* (2018) Efficient generation of pink-fruited tomatoes using CRISPR/Cas9 system. *Journal of Genetics and Genomics* 45, 51–54.

Devran, Z., Kahveci, E., Hong, Y., Studholme, D.J., and Tör, M. (2018) Identifying molecular markers suitable for *Frl* selection in tomato breeding. *Theoretical and Applied Genetics* 131, 2099–2105.

Diaz, A., Fergany, M., Formisano, G., Ziarsolo, P., Blanca, J., *et al.* (2011) A consensus linkage map for molecular markers and quantitative trait loci associated with economically important traits in melon (*Cucumis melo* L.). *BMC Plant Biology* 11: 111.

Diwan, N., Fluhr, R., Eshed, Y., Zamir, D., and Tanksley, S.D. (1999) Mapping of *Ve* in tomato: a gene conferring resistance to the broad-spectrum pathogen, *Verticillium dahliae* race 1. *Theoretical and Applied Genetics* 98, 315–319.

Djian-Caporalino, C., Pijarowski, L., Fazari, A., Samson, M., Gaveau, L., *et al.* (2001) High-resolution genetic mapping of the pepper (*Capsicum annuum* L.) resistance loci *Me3* and *Me4* conferring heat-stable resistance to root-knot nematodes (*Meloidogyne* spp.). *Theoretical and Applied Genetics* 103, 592–600.

do Carmo, C.D., e Sousa, M.B., Brito, A.C., and de Oliveira, E.J. (2020) Genome-wide association studies for waxy starch in cassava. *Euphytica* 216, 1–17.

Dong, P., Han, K., Siddique, M.I., Kwon, J.-K., Zhao, M., *et al.* (2016) Gene-based markers for the tomato yellow leaf curl virus resistance gene *Ty-3*. *Plant Breed Biotechnology* 4, 79–86.

Dou, J., Lu, X., Ali, A., Zhao, S., Zhang, L., *et al.* (2018a) Genetic mapping reveals a marker for yellow skin in watermelon (*Citrullus lanatus* L.). *PLOS ONE* 13: e0200617.

Dou, J., Zhao, S., Lu, X., He, N., Zhang, L., *et al.* (2018b) Genetic mapping reveals a candidate gene (*ClFS1*) for fruit shape in watermelon (*Citrullus lanatus* L.). *Theoretical and Applied Genetics* 131, 947–958.

Du, M., Zhou, K., Liu, Y., Deng, L., Zhang, X., *et al.* (2020) A biotechnology-based male-sterility system for hybrid seed production in tomato. *Plant Journal* 102, 1090–1100.

Duangjit, J., Causse, M., and Sauvage, C. (2016) Efficiency of genomic selection for tomato fruit quality. *Molecular Breeding* 36: 29.

Ek, M., Eklund, M., von Post, R., Dayteg, C., Henriksson, T., *et al.* (2005) Microsatellite markers for powdery mildew resistance in pea (*Pisum sativum* L.). *Hereditas* 142, 86–91.

Enciso-Rodriguez, F., Manrique-Carpintero, N.C., Nadakuduti, S.S., Buell, C.R., Zarka, D., and Douches, D. (2019) Overcoming self-incompatibility in diploid potato using CRISPR-Cas9. *Frontiers in Plant Science* 10: 376.

Esuma, W., Herselman, L., Labuschagne, M.T., Ramu, P., Lu, F., *et al.* (2016) Genome-wide association mapping of provitamin A carotenoid content in cassava. *Euphytica* 212, 97–110.

Ezenwaka, L.C., Del Carpio, D.P., Jannink, J.L., Rabbi, I.Y., Danquah, E.Y., *et al.* (2018) Genome-wide association study of resistance to cassava green mite pest and related traits in cassava. *Crop Science* 58, 1907–1918.

Fall, L.A., Clevenger, J., and McGregor, C. (2018) Assay development and marker validation for marker assisted selection of *Fusarium oxysporum* f. sp. *niveum* race 1 in watermelon. *Molecular Breeding* 38: 130.

Fan, L., Wang, Y., Xu, L., Tang, M., Zhang, X., *et al.* (2020) A genome-wide association study uncovers a critical role of the *RsPAP2* gene in red-skinned *Raphanus sativus* L. *Horticulture Research* 7: 164.

Farinho, M., Coelho, P., Carlier, J., Svetleva, D., Monteiro, A., and Leitao, J. (2004) Mapping of a locus for adult plant resistance to downy mildew in broccoli (*Brassica oleracea* convar. *italica*). *Theoretical and Applied Genetics* 109, 1392–1398.

Ferreira, J.J., Campa, A., Pérez-Vega, E., Rodríguez-Suárez, C., and Giraldez, R. (2012) Introgression and pyramiding into common bean market class fabada of genes conferring resistance to anthracnose and potyvirus. *Theoretical and Applied Genetics* 124, 777–788.

Fileppi, M., Galasso, I., Tagliabue, G., Daminati, M.G., Campion, B., *et al.* (2010) Characterisation of structural genes involved in phytic acid biosynthesis in common bean (*Phaseolus vulgaris* L.). *Molecular Breeding* 25, 453–470.

Foolad, M.R. and Panthee, D.R. (2012) Marker-assisted selection in tomato breeding. *Critical Reviews in Plant Sciences*, 31, 93–123.

Frew, T.J., Russell, A.C., and Timmerman-Vaughan, G.M. (2002) Sequence tagged site markers linked to the *smb1* gene for resistance to pea seed borne mosaic virus in pea. *Plant Breeding* 121, 512–516.

Fu, A., Wang, Q., Mu, J., Ma, L., Wen, C., *et al.* (2021) Combined genomic, transcriptomic, and metabolomic analyses provide insights into chayote (*Sechium edule*) evolution and fruit development. *Horticulture Research* 8: 35.

Fu, W., Ye, X., Ren, J., Li, Q., Du, J.T., *et al.* (2019) Fine mapping of *lcm1*, a gene conferring chlorophyll-deficient golden leaf in Chinese cabbage (*Brassica rapa* ssp. *pekinensis*). *Molecular Breeding* 39: 52.

Fukino, N., Ohara, T., Sugiyama, M., Kubo, N., Hirai, M., *et al.* (2012). Mapping of a gene that confers short lateral branching (*slb*) in melon (*Cucumis melo* L.). *Euphytica* 187, 133–143.

Gama, R.N.C., Santos, C.A.F., Dias, R.C.S., Souza, R.R.C., and Queiróz, M.A. (2015) Microsatellite markers linked to powdery mildew resistance locus in watermelon. *Australian Journal of Crop Science* 9, 92–97.

Gao, L., Gonda, I., Sun, H., Ma, Q., Bao, K., *et al.* (2019) The tomato pan-genome uncovers new genes and a rare allele regulating fruit flavor. *Nature Genetics* 51, 1044–1051.

Gao, M., Hu, L., Li, Y., and Weng, Y. (2016) The chlorophyll-deficient *golden leaf* mutation in cucumber is due to a single nucleotide substitution in *CsChlI* for magnesium chelatase I subunit. *Theoretical and Applied Genetics* 129, 1961–1973.

Garcia, V., Bres, C., Just, D., Fernandez, L., Tai, F.W., *et al.* (2016) Rapid identification of causal mutations in tomato EMS populations via mapping-by-sequencing. *Nature Protocols* 11, 2401–2418.

García-Fernández, C., Campa, A., Garzón, A.S., Miklas, P., and Ferreira, J.J. (2021) GWAS of pod morphological and color characters in common bean. *BMC Plant Biology* 21: 184.

Garcia-Mas, J., Benjak, A., Sanseverino, W., Bourgeois, M., Mir, G., *et al.* (2012) The genome of melon (*Cucumis melo* L.). *Proceedings of the National Academy of Sciences USA* 109, 11872–11877.

Gatarira, C., Agre, P., Matsumoto, R., Edemodu, A., Adetimirin, V., *et al.* (2020) Genome-wide association analysis for tuber dry matter and oxidative browning in water yam (*Dioscorea alata* L.). *Plants* 9: 969.

Gemenet, D.C., Kitavi, M.N., and David, M. (2020) Development of diagnostic SNP markers for quality assurance and control in sweetpotato (*Ipomoea batatas* (L.) Lam.) breeding programs. *PLOS ONE* 15: e0232173.

Gill, U., Scott, J.W., Shekasteband, R., Ogundiwin, E., Schuit, C., *et al.* (2019) *Ty-6*, a major begomovirus resistance gene on chromosome 10, is effective against tomato yellow leaf curl virus and tomato mottle virus. *Theoretical and Applied Genetics* 132, 1543–1554.

Golicz, A.A., Bayer, P.E., Barker, G.C., Edger, P.P., Kim, H., *et al.* (2016) The pangenome of an agronomically important crop plant *Brassica oleracea*. *Nature Communications* 7: 13390.

González, M.N., Massa, G.A., Andersson, M., Turesson, H., Olsson, N., *et al.* (2020). Reduced enzymatic browning in potato tubers by specific editing of a polyphenol oxidase gene via ribonucleoprotein complexes delivery of the CRISPR/Cas9 system. *Frontiers in Plant Science* 10: 1649.

Gonzalez-Cendales, Y., Catanzariti, A.M., Baker, B., McGrath, D.J., and Jones, D.A. (2015) Identification of I-7 expands the repertoire of genes for resistance to *Fusarium* wilt in tomato to three resistance gene classes. *Molecular Plant Pathology* 17, 448463.

Guo, S., Zhang, J., Sun, H., Salse, J., Lucas, W.J., *et al.* (2013) The draft genome of watermelon (*Citrullus lanatus*) and resequencing of 20 diverse accessions. *Nature Genetics* 45, 51–58.

Guo, G., Wang, S., Liu, J., Pan, B., Diao, W., *et al.* (2017) Rapid identification of QTLs underlying resistance to *Cucumber mosaic virus* in pepper (*Capsicum frutescens*). *Theoretical and Applied Genetics* 130, 41–52.

Guo, S., Zhao, S., Sun, H., Wang, X., Wu, S., *et al.* (2019) Resequencing of 414 cultivated and wild watermelon accessions identifies selection for fruit quality traits. *Nature Genetics* 51, 1616–1623.

Gupta, P.K., Kumar, J., Mir, R.R., and Kumar, A. (2010) Marker-assisted selection as a component of conventional plant breeding. *Plant Breeding Reviews* 33, 145–217.

Gupta, R.M. and Musunuru, K. (2014) Expanding the genetic editing tool kit: ZFNs, TALENs, and CRISPR-Cas9. *Journal of Clinical Investigation* 124, 4154–4161.

Gupta, S.K., Vishwakarma, A., Kenea, H.D., Galsurker, O., Cohen, H., *et al.* (2021) CRISPR/Cas9 mutants of tomato *MICRORNA164* genes uncover their functional specialization in development. *Plant Physiology* 187, 1636–1652.

Gutierrez, N. and Torres, A.M. (2019) Characterization and diagnostic marker for TTG1 regulating tannin and anthocyanin biosynthesis in faba bean. *Science Reports* 9: 16174.

Habyarimana, E., Parisi, B., and Mandolino, G., (2017). Genomic prediction for yields, processing and nutritional quality traits in cultivated potato (*Solanum tuberosum* L.). *Plant Breeding* 136, 245–252.

Han, K., Lee, H.Y., Ro, N.Y., Hur, O.S., Lee, J.H., *et al.* (2018) QTL mapping and GWAS reveal candidate genes controlling capsaicinoid content in *Capsicum*. *Plant Biotechnology Journal* 16, 1546–1558.

Hanson, P., Lu, S.F., Wang, J.F., Chen, W., Kenyon, L., *et al.* (2016) Conventional and molecular marker-assisted selection and pyramiding of genes for multiple disease resistance in tomato. *Scientia Horticulturae* 201, 346–354.

Hao, N., Du, Y., Li, H., Wang, C., Wang, C., *et al.* (2018) *CsMYB36* is involved in the formation of yellow green peel in cucumber (*Cucumis sativus* L.). *Theoretical and Applied Genetics* 131, 1659–1669.

Hatakeyama, K., Suwabe, K., Tomita, R.N., Kato, T., Nunome, T., *et al.* (2013) Identification and characterization of *Crr1a*, a gene for resistance to clubroot disease (*Plasmodiophora brassicae* Woronin) in *Brassica rapa* L. *PLOS ONE* 8: e54745.

Hayut, S.F., Bessudo, C.M., and Levy, A.A. (2017) Targeted recombination between homologous chromosomes for precise breeding in tomato. *Nature Communications* 8: 15605.

Hernández-Bautista, A., Lobato-Ortiz, R., García-Zavala, J.J., Cruz-Izquierdo, S., Chávez-Servia, J.L., *et al.* (2020) Prediction accuracy of genomic selection models for earliness in tomato. *Chilean Journal of Agricultural Research* 80, 505–514.

Hirakawa, H., Shirasawa, K., Miyatake, K., Nunome, T., Negoro, S., *et al.* (2014) Draft genome sequence of eggplant (*Solanum melongena* L.): the representative *Solanum* species indigenous to the old world. *DNA Research* 21, 649–660.

Holdsworth, W.L., LaPlant, K.E., Bell, D.C., Jahn, M.M., and Mazourek, M. (2016) Cultivar-based introgression mapping reveals wild species-derived *Pm-0*, the major powdery mildew resistance locus in squash. *PLOS ONE* 11: e0167715.

Hong, J.P., Noh, N., Lee, H.Y., Kim, G.W., Kwon, J.K., *et al.* (2020) Genomic selection for prediction of fruit-related traits in pepper (*Capsicum* spp.). *Frontiers in Plant Science* 11: 1575.

Hooghvorst, I., Lopez-Cristoffanini, C., and Nogues, S. (2019) Efficient knock-out of phytoene desaturase gene using CRISPR/Cas9 in melon. *Scientific Reports* 9: 17077.

Hu, B., Li, D., Liu, X., Qi, J., Gao, D., *et al.* (2017) Engineering non-transgenic gynoecious cucumber using an improved transformation protocol and optimized CRISPR/Cas9 system. *Molecular Plant* 10, 1575–1578.

Hu, W., Ji, C., Liang, Z., Ye, J., Ou, W., *et al.* (2021) Resequencing of 388 cassava accessions identifies valuable loci and selection for variation in heterozygosity. *Genome Biology* 22: 316.

Huang, C.C., Cui, Y.Y., Weng, C.R., Zabel, P., and Lindhout, P. (2000) Development of diagnostic PCR markers closely linked to the tomato powdery mildew resistance gene *Ol-1* on chromosome 6 of tomato. *Theoretical and Applied Genetics* 101, 918–924.

Huang, S., Li, R., Zhang, Z., Li, L.I., Gu, X., *et al.* (2009) The genome of the cucumber, *Cucumis sativus* L. *Nature Genetics* 41, 1275–1281.

Hui, F., Ning, Y., Zhiyong, L., and Hao, W.W. (2011) A genetic male sterile line developed by molecular marker-assisted selection in Chinese cabbage (*Brassica rapa* ssp. *pekinensis*). *African Journal of Biotechnology* 10, 17706–17711.

Illa-Berenguer, E., van Houten, J., Huang, Z., and van der Knaap, E. (2015) Rapid and reliable identification of tomato fruit weight and locule number loci by QTL-seq. *Theoretical and Applied Genetics* 128, 1329–1342.

Iorizzo, M., Ellison, S., Senalik, D., Zeng, P., Satapoomin, P., *et al.* (2016) A high-quality carrot genome assembly provides new insights into carotenoid accumulation and asterid genome evolution. *Nature Genetics* 48, 657–666.

Ito, Y., Nishizawa, A., Endo, M., Masafumi, M., and Seiichi, T. (2015) CRISPR/Cas9-mediated mutagenesis of the *RIN* locus that regulates tomato fruit ripening. *Biochemical and Biophysical Research Communications* 467, 76–82.

Jacobs, T.B., Zhang, N., Patel, D., and Gregory B.M. (2017) Generation of a collection of mutant tomato lines using pooled CRISPR libraries. *Plant Physiology* 174, 2023–2037.

Jain, S., Poromarto, S., Osorno, J.M., McClean, P.E., and Nelson, B.D. (2019) Genome wide association study discovers genomic regions involved in resistance to soybean cyst nematode (*Heterodera glycines*) in common bean. *PLOS ONE* 14: e0212140.

Jeong, H.S., Jang, S., Han, K., Kwon, J.K., and Kang, B.C. (2015) Marker-assisted backcross breeding for development of pepper varieties (*Capsicum annuum*) containing capsinoids. *Molecular Breeding* 35, 1–10.

Jeong, S.Y., Ahn, H., Ryu, J., Oh, Y., Sivanandhan, G., *et al.* (2019) Generation of early-flowering Chinese cabbage (*Brassica rapa* spp. *pekinensis*) through CRISPR/Cas9-mediated genome editing. *Plant Biotechnology Reports* 13, 491–499.

Ji, Y. and Scott, J.W. (2009) A CAPS marker linked to the tomato gray leafspot (*Stemphylium* sp.) resistance gene *Sm*. *Report of the Tomato Genetics Cooperative* 59, 29–31.

Ji, Y., Scott, J.W., Schuster, D.J., and Maxwell, D.P. (2009) Molecular mapping of *ty-4*, a new tomato yellow leaf curl virus resistance locus on chromosome 3 of tomato. *Journal of the American Society for Horticultural Science* 134, 281–288.

Jia, H., Li, X., Song, J., Lin, Y., Zhang, X., *et al.* (2021) Genome-wide association study of powdery mildew resistance of cucumber core germplasms. *Acta Horticulturae Sinica* 48, 1371–1385.

Jin, M., Lee, S.S., Ke, L., Kim, J.S., Seo, M.S., *et al.* (2014) Identification and mapping of a novel dominant resistance gene, *TuRB07* to *Turnip mosaic virus* in *Brassica rapa*. *Theoretical and Applied Genetics* 127, 509–519.

Jo, J., Venkatesh, J., Han, K., Lee, H.Y., Choi, G.J., *et al.* (2017) Molecular mapping of *PMR1*, a novel locus conferring resistance to powdery mildew in pepper (*Capsicum annuum*). *Frontiers in Plant Science* 8: 2090.

Jones, D.A., Dickinson, M.J., Balint-Kurti, P.J., Dixon, M.S., and Jones, J.D.G. (1993) Two complex resistance loci revealed in tomato by classical and RFLP mapping of the *Cf-2*, *Cf-4*, *Cf-5*, and *Cf-9* genes for resistance to *Cladosporium fulvum*. *Molecular Plant–Microbe Interactions* 6, 348–357.

Jung, S., Venkatesh, J., Kang, M.Y., Kwon, J.K., and Kang, B.C. (2019). A non-LTR retrotransposon activates anthocyanin biosynthesis by regulating a MYB transcription factor in *Capsicum annuum*. *Plant Science* 287: 110181.

Kabelka, E.A. and Young, K. (2010) Identification of molecular markers associated with resistance to squash silverleaf disorder in summer squash (*Cucurbita pepo*). *Euphytica* 173, 49–54.

Kadirvel, P., de la Peña, R., Scha Eitner, R., Huang, S., Geethanjali, S., *et al.* (2013) Mapping of QTLs in tomato line FLA456 associated with resistance to a virus causing tomato yellow leaf curl disease. *Euphytica* 190, 297–308.

Kalia, P., Muthukumar, P., and Soi, S. (2017) Marker-assisted introgression of the *Or* gene for enhancing β-carotene content in Indian cauliflower. *Acta Horticuturae* 1203, 121–128.

Kalia, P., Saha, P., and Roy, S. (2017) Development of RAPD and ISSR derived SCAR markers linked to *Xca1Bo* gene conferring resistance to black rot disease in cauliflower (*Brassica oleracea* var. *botrytis* L.). *Euphytica* 213: 232.

Kang, H., Weng, Y., Yang, Y., Zhang, Z., Zhang, S., *et al.* (2011) Fine genetic mapping localizes cucumber scab resistance gene *Ccu* into an *R* gene cluster. *Theoretical and Applied Genetics* 122, 795–803.

Karna, S. and Ahn, Y.K. (2018) Development of InDel markers to identify *Capsicum* disease resistance using whole genome resequencing. *Plant Biotechnology Journal* 45, 228–235.

Kato, T., Hatakeyama, K., Fukino, N., and Matsumoto, S. (2013) Fine mapping of the clubroot resistance gene *CRb* and development of a useful selectable marker in *Brassica rapa*. *Breeding Science* 63, 116–124.

Katoch, V., Sharma, S., Pathania, S., Banayal, D.K., Sharma, S.K., and Rathour, R. (2010) Molecular mapping of pea powdery mildew resistance gene *er2* to pea linkage group III. *Molecular Breeding* 25, 29–237.

Kaushal, A., Sadashiva, A.T., Krishna Reddy, M., Srinivasa Rao, E., Singh, T.H., Sriram, S., Dhananjay, M.V., Venugopalan and Ravishankar, K.V. (2020) Assessment of the effectiveness of Ty genes in tomato against tomato leaf curl Bangalore virus. *Plant Pathology* 69(9), 1777–1786.

Kaushal, A., Sadashiva, A.T., Ravishankar, K.V., Sriram, S. and Reddy, M.K. (2024) Marker-assisted pyramiding of Ty-2, Ty-3, Ph-2, and Ph-3 genes for combined resistance to tomato leaf curl and late blight diseases in tomato (*Solanum lycopersicum* L.). *European Journal of Plant Pathology* 168(3), 557–570.

Kawchuk, L.M., Hachey, J., Lynch, D.R., Kulcsar, F., van Rooijen, G., *et al.* (2001) Tomato *Ve* disease resistance genes encode cell surface-like receptors. *Proceedings of the National Academy of Sciences USA* 98, 6511–6515.

Kayondo, S.I., Del Carpio, D.P., Lozano, R., Ozimati, A., Wolfe, M., *et al.* (2018) Genome-wide association mapping and genomic prediction for CBSD resistance in *Manihot esculenta*. *Scientific Reports* 8: 1549.

Khan, M.Z., Zaidi, S.S., Amin, I., and Mansoor, S. (2019) A CRISPR way for fast-forward crop domestication. *Trends in Plant Science* 24, 293–296.

Kieu, N.P., Lenman, M., Wang, E.S., Petersen, B.L., and Andreasson, E. (2021) Mutations introduced in susceptibility genes through CRISPR/Cas9 genome editing confer increased late blight resistance in potatoes. *Scientific Reports* 11: 4487.

Kim, B., Hwang, I.S., Lee, H.J., Lee, J.M., Seo, E., *et al.* (2018) Identification of a molecular marker tightly linked to bacterial wilt resistance in tomato by genome-wide SNP analysis. *Theoretical and Applied Genetics* 131, 1017–1030.

Kim, H., Yoon, J.B., and Lee, J. (2017) Development of fluidigm SNP type genotyping assays for marker-assisted breeding of chili pepper (*Capsicum annuum* L.). *Horticultural Science and Technology* 35, 465–479.

Kim, J., Kang, W.H., Yang, H.B., Park, S., Jang, C.S., *et al.* (2013) Identification of a broad-spectrum recessive gene in *Brassica rapa* and molecular analysis of the *eIF4E* gene family to develop molecular markers. *Molecular Breeding* 32, 385–398.

Kim, S., Park, M., Yeom, S.I., Kim, Y.M., Lee, J.M., *et al.* (2014) Genome sequence of the hot pepper provides insights into the evolution of pungency in *Capsicum* species. *Nature Genetics* 46, 270–278.

Kim, S., Park, J., Yeom, S.I., Kim, Y.M., Seo, E., *et al.* (2017) New reference genome sequences of hot pepper reveal the massive evolution of plant disease-resistance genes by retroduplication. *Genome Biology* 18: 210.

Kim, S.C., Jung, Y.H., Seong, K.C., Chun, S.J., Kim, C.H., *et al.* (2014) Development of a SCAR marker for sex identification in asparagus. *Korean Journal of Plant Resources* 27, 236–241.

Kiryowa, M., Nkalubo, S., Mukankusi, C., Talwana, H., Gibson, P., and Tukamuhabwa, P. (2015) Effect of marker aided pyramiding of anthracnose and *Pythium* root rot resistance genes on plant agronomic characters among advanced common bean genotypes. *Journal of Agricultural Science* 7, 99–104.

Kiryowa, M., Nkalubo, S.T., Mukankusi, C., Male, A., Gibson, P. *et al.* (2021) Effectiveness of pyramided genes in conferring resistance to anthracnose disease in common bean populations. *Journal of Plant Breeding and Crop Science* 13, 1–13.

Kitashiba, H., Li, F., Hirakawa, H., Kawanabe, T., Zou, Z., *et al.* (2014) Draft sequences of the radish (*Raphanus sativus* L.) genome. *DNA Research* 21, 481–490.

Klaassen, M.T., Willemsen, J.H., Vos, P.G., Visser, R.G., van Eck, H.J., *et al.* (2019) Genome-wide association analysis in tetraploid potato reveals four QTLs for protein content. *Molecular Breeding* 39: 151.

Klap, C., Yeshayahou, E., Bolger, A.M., Tzahi, A., Suresh, K.G., *et al.* (2017) Tomato facultative parthenocarpy results from *SlAGAMOUS-LIKE 6* loss of function. *Plant Biotechnology Journal* 15, 634–647.

Klimek, M., Oleszkiewicz, T., Lowder, L.G., Yiping, Q., and Rafal, B. (2018) Efficient CRISPR/Cas9-based genome editing in carrot cells. *Plant Cell Reports* 37, 575–586.

Koizumi, E., Igarashi, T., Tsuyama, M., Ogawa, K., Asano, K., *et al.* (2021) Association of genome-wide SNP markers with resistance to common scab of potato. *American Journal of Potato Research* 98, 149–156.

Kreplak, J., Madoui, M.A., Capal, P., Novak, P., Labadie, K., *et al.* (2019) A reference genome for pea provides insight into legume genome evolution. *Nature Genetics* 51, 1411–1422.

Kumar, A., Jindal, S.K., Dhaliwal, M.S., Sharma, A., Kaur, S., and Jain, S. (2019) Gene pyramiding for elite tomato genotypes against ToLCV (*Begomovirus* spp.), late blight (*Phytophthora infestans*) and RKN (*Meloidogyne* spp.) for northern India farmers. *Physiology and Molecular Biology of Plants* 25, 1197–1209.

Kusano, H., Ohnuma, M., Mutsuro-Aoki, H., Asahi, T., Ichinosawa, D., *et al.* (2018) Establishment of a modified CRISPR/Cas9 system with increased mutagenesis frequency using the translational enhancer *dMac3* and multiple guide RNAs in potato. *Scientific Reports* 8: 13753.

Kwon, S.J., Smykal, P., Hu, J., Wang, M., Kim, S.J. *et al.* (2013) User-friendly markers linked to *Fusarium* wilt race 1 resistance *Fw* gene for marker assisted selection in pea. *Plant Breed* 132, 642–648.

Lapidot, M., Karniel, U., Gelbart, D., Fogel, D., Evenor, D., *et al.* (2015) A novel route controlling begomovirus resistance by the messenger RNA surveillance factor Pelota. *PLOS Genetics* 11: e1005538.

Laugé, R., Dmitriev, A.P., Joosten, M.H.A.J, and de Wit, P.J.G.M. (1998) Additional resistance gene(s) against *Cladosporium fulvum* present on the *Cf-9* introgression segment are associated with strong PR protein accumulation. *Molecular Plant–Microbe Interactions* 11, 301–308.

Lee, J., Izzah, N.K., Jayakodi, M., Perumal, S., Joh, H.J., *et al.* (2015) Genome-wide SNP identification and QTL mapping for black rot resistance in cabbage. *BMC Plant Biology* 15: 32.

Lee, O., Koo, H., Yu, J.W., and Park, H.Y. (2021) Genotyping-by-sequencing-based genome-wide association studies of *Fusarium* wilt resistance in radishes (*Raphanus sativus* L.). *Genes* 12: 858.

Lee, Y.P., Cho, Y., and Kim, S. (2014) A high-resolution linkage map of the *Rfd1*, a restorer-of-fertility locus for cytoplasmic male sterility in radish (*Raphanus sativus* L.) produced by a combination of bulked segregant analysis and RNA-Seq. *Theoretical and Applied Genetics* 127, 2243–2252.

Lema, M., Demissie, H., and Rezene, Y. (2021) Marker assisted gene introgression in to common bean (*Phaseolus vulgaris* l) against angular leaf spot (*Pseudoserco sporagriseola*) and anthracnose (*Colletotrichum lindemuthianum*). *Journal of Agricultural Science and Practice* 6, 118–129.

Li, B., Zhao, Y., Zhu, Q., Zhang, Z., Fan, C., *et al.* (2017) Mapping of powdery mildew resistance genes in melon (*Cucumis melo* L.) by bulked segregant analysis. *Scientia Horticulturae* 220, 160–167.

Li, G.L., Qian, W., Zhang, S.J., Zhang, S.F., Li, F., *et al.* (2016) Development of gene-based markers for the *Turnip mosaic virus* resistance gene *retr02* in *Brassica rapa*. *Plant Breeding* 135, 466–470.

Li, H., Yu, S.C., Zhang, F.L., Yu, Y.J., Zhao, X.Y., *et al.* (2011) Development of molecular markers linked to the resistant QTL for downy mildew in *Brassica rapa* L. ssp. *pekinensis*. *Yi Chuan* 33, 1271–1278.

Li, M., Guo, S., Zhang, J., Sun, H., Tian, S., *et al.* (2021) Sugar transporter VST1 knockout reduced aphid damage in watermelon. *Plant Cell Reports* 41, 277–279.

Li, M.Y., Feng, K., Hou, X.L., Jiang, Q., Xu, Z.S., *et al.* (2020) The genome sequence of celery (*Apium graveolens* L.), an important leaf vegetable crop rich in apigenin in the Apiaceae family. *Horticulture Research* 7: 9.

Li, R., Li, R., Li, X., Daqi, F., Benzhong, Z., *et al.* (2018a) Multiplexed CRISPR/Cas9-mediated metabolic engineering of $\gamma$-aminobutyric acid levels in *Solanum lycopersicum*. *Plant Biotechnology Journal* 16, 415–427.

Li, R., Zhang, L., Wang, L., Lin, C., Ruirui, Z., *et al.* (2018b) Reduction of tomato-plant chilling tolerance by CRISPR-Cas9-mediated *SlCBF1* mutagenesis. *Journal of Agricultural and Food Chemistry* 66, 9042–9051.

Li, S., Pan, Y., Wen, C., Li, Y., Liu, X., *et al.* (2016) Integrated analysis in bi-parental and natural populations reveals *CsCLAVATA3* (*CsCLV3*) underlying carpel number variations in cucumber. *Theoretical and Applied Genetics* 129, 1007–1022.

Li, X., Wang, Y., Chen, S., Huiqin, T., Daqi, F., *et al.* (2018) Lycopene is enriched in tomato fruit by CRISPR/Cas9-mediated multiplex genome editing. *Frontiers in Plant Science* 9: 559.

Li, Y., Wang, Y., Wu, X., Wang, J., Wu, X., *et al.* (2021a) Novel genomic regions of *Fusarium* wilt resistance in bottle gourd [*Lagenaria siceraria* (Mol.) Standl.] discovered in genome-wide association study. *Frontiers in Plant Science* 12: 650715.

Li, Y., Yue, N., Basit, A., Li, Y., Zhang, D., *et al.* (2021b) Targeted editing of *SlMAPK6* using CRISPR/Cas9 technology to promote the development of axillary buds in tomato plants. *Journal of Agricultural Science* 13, 11–22.

Liang, D., Chen, M., Qi, X., Xu, Q., Zhou, F., and Chen, X. (2016) QTL mapping by SLAF-seq and expression analysis of candidate genes for aphid resistance in cucumber. *Frontiers in Plant Science* 7: 1000.

Lin, T., Wang, S., Zhong, Y., Gao, D., Cui, Q., and Chen, H. (2016) A truncated F-box protein confers the dwarfism in cucumber. *Journal of Genetics and Genomics* 43, 223–226.

Ling, K.S., Harris, K.R., Meyer, J.D.F., Levi, A., Guner, N., *et al.* (2009) Non-synonymous single nucleotide polymorphisms in the watermelon *eIF4E* gene are closely associated with resistance to *Zucchini yellow mosaic virus*. *Theoretical and Applied Genetics* 120, 191–200.

Liu, G., Zhao, T., You, X., Jiang, J., Li, J., and Xu, X. (2019) Molecular mapping of the *Cf-10* gene by combining SNP/InDel-index and linkage analysis in tomato (*Solanum lycopersicum*). *BMC Plant Biology* 19: 15.

Liu, J., Wang, S., Wang, H., Bote, L., Yiyong, C., *et al.* (2021) Rapid generation of tomato male-sterile lines with a marker use for hybrid seed production by CRISPR/Cas9 system. *Molecular Breeding* 41: 25.

Liu, L., Sun, T., Liu, X., Guo, Y., Huang, X., *et al.* (2019) Genetic analysis and mapping of a striped rind gene (*st3*) in melon (*Cucumis melo* L.). *Euphytica* 215: 20.

Liu, P.N., Miao, H., Lu, H.W., Cui, J.Y., Tian, G.L., *et al.* (2017) Molecular mapping and candidate gene analysis for resistance to powdery mildew in *Cucumis sativus* stem. *Genetics and Molecular Research* 16:gmr16039680.

Liu, S., Yu, K., and Park, S.J. (2008) Development of STS markers and QTL validation for common bacterial blight resistance in common bean. *Plant Breeding* 127, 62–68.

Liu, S., Kandoth, P.K., Warren, S.D., Yeckel, G., Heinz, R., *et al.* (2012a) A soybean cyst nematode resistance gene points to a new mechanism of plant resistance to pathogens. *Nature* 492, 256–260.

Liu, S., Yeh, C.T., Tang, H.M., Nettleton, D., and Schnable, P.S. (2012b) Gene mapping via bulked segregant RNA-Seq (BSR-Seq). *PLOS ONE* 7; e36406.

Liu, S., Liu, Y., Yang, X., Tong, C., Edwards, D., Parkin, I.A., *et al.* (2014) The *Brassica oleracea* genome reveals the asymmetrical evolution of polyploid genomes. *Nature Communications* 5: 3930.

Liu, S., Gao, P., Zhu, Q., Zhu, Z., Liu, H., *et al.* (2020) Resequencing of 297 melon accessions reveals the genomic history of improvement and loci related to fruit traits in melon. *Plant Biotechnology Journal* 18, 2545–2558.

Liu, W.Y., Kang, J.H., Jeong, H.S., Choi, H.J., Yang, H.B., *et al.* (2014) Combined use of bulked segregant analysis and microarrays reveals SNP markers pinpointing a major QTL for resistance to *Phytophthora capsici* in pepper. *Theoretical and Applied Genetics* 127, 2503–2513.

Liu, X., Han, F., Kong, C., Fang, Z., Yang, L., *et al.* (2017) Rapid introgression of the *Fusarium* wilt resistance gene into an elite cabbage line through the combined application of a microspore culture, genome background analysis, and disease resistance-specific marker assisted foreground selection. *Frontiers in Plant Science* 8: 354.

Liu, X., Lu, H., Liu, P., Miao, H., Bai, Y., *et al.* (2020) Identification of novel loci and candidate genes for cucumber downy mildew resistance using GWAS. *Plants* 9: 1659.

Lobaton, J.D., Miller, T., Gil, J., Ariza, D., de la Hoz, J.F., *et al.* (2018) Resequencing of common bean identifies regions of intergene pool introgression and provides comprehensive resources for molecular breeding. *Plant Genome* 11: 170068.

Lu, H., Lin, T., Klein, J., Wang, S., Qi, J., *et al.* (2014) QTL-seq identifies an early flowering QTL located near *Flowering Locus T* in cucumber. *Theoretical and Applied Genetics* 127, 1491–1499.

Lun, Y., Wang, X., Zhang, C., Yang, L., Gao, D., and Chen, H. (2015) A *CsYcf54* variant conferring light green coloration in cucumber. *Euphytica* 208, 509–517.

Luo, S., Ma, Q., Zhong, Y., Jianling, J., Zusheng, W., *et al.* (2021) QTL-seq identifies an early flowering QTL located near *Flowering Locus T* in cucumber *Plant Molecular Biology* 108, 429–442.
Lv, H., Yang, L., Kang, J., Wang, Q., Wang, X., *et al.* (2013) Development of InDel markers linked to *Fusarium* wilt resistance in cabbage. *Molecular Breeding* 32, 961–967.
Lv, H., Fang, Z., Yang, L., Zhang, Y., Wang, Q., *et al.* (2014) Mapping and analysis of a novel candidate *Fusarium* wilt resistance gene *FOC1* in *Brassica oleracea*. *BMC Genomics* 15: 1094.
Lv, H., Wang, Y., Han, F., Ji, J., Fang, Z., *et al.* (2020) A high-quality reference genome for cabbage obtained with SMRT reveals novel genomic features and evolutionary characteristics. *Scientific Reports* 10: 12394.
Ma, C., Liu, M., Li, Q., Si, J., Ren, X., and Song, H. (2019a) Efficient *BoPDS* gene editing in cabbage by the CRISPR/Cas9 system. *Horticultural Plant Journal* 5, 164–169.
Ma, C., Zhu, C., Zheng, M., Liu, M., Zhang, D., *et al.* (2019b) CRISPR/ Cas9-mediated multiple gene editing in *Brassica oleracea* var. *capitata* using the endogenous tRNA-processing system. *Horticulture Research* 6: 20.
Ma, L., Wang, Q., Mu, J., Fu, A., Wen, C., *et al.* (2020) The genome and transcriptome analysis of snake gourd provide insights into its evolution and fruit development and ripening. *Horticulture Research* 7: 199.
Macko-Podgórni, A., Stelmach, K., Kwolek, K., Senalik, D., Simon, P.W., and Grzebelus, D. (2018) A genome-wide association study reveals a genomic region associated with the carrot root shape. *Acta Horticulturae* 1264, 129–134.
Maharijaya, A., Vosman, B., Steenhuis-Broers, G., Pelgrom, K., Purwito, A., *et al.* (2015) QTL mapping of thrips resistance in pepper. *Theoretical and Applied Genetics* 128, 1945–1956.
Maioli, A., Gianoglio, S., Moglia, A., Acquadro, A., Valentino, D., *et al.* (2020) Simultaneous CRISPR/Cas9 editing of three PPO genes reduces fruit flesh browning in *Solanum melongena* L. *Frontiers in Plant Science* 11: 607161.
Maisonneuve, B., Bellec, Y., Anderson, P., and Michelmore, R.W. (1994) Rapid mapping of two genes for resistance to downy mildew from *Lactuca serriola* to existing clusters of resistance genes. *Theoretical and Applied Genetics* 89, 96–104.
Makhotenko, A., Khromov, A., Snigir, E., Makarova, S., Makarov, V., *et al.* (2019) Functional analysis of coilin in virus resistance and stress tolerance of potato *Solanum tuberosum* using CRISPR-Cas9 editing. *Doklady Biochemistry and Biophysics* 484, 88–91.
Mandal, M.K., Suren, H., and Kousik, C. (2020) Elucidation of resistance signaling and identification of powdery mildew resistant mapping loci (*ClaPMR2*) during watermelon–*Podosphaera xanthii* interaction using RNA-Seq and whole-genome resequencing approach. *Scientific Reports* 10: 14038.
Maurya, D., Mukherjee, A., Akhtar, S., and Chattopadhyay, T. (2021) Development and validation of the *OVATE* gene-based functional marker to assist fruit shape selection in tomato. *3 Biotech* 11: 474.
Mei, J., Ding, Y., Lu, K., Wei, D., Liu, Y., *et al.* (2013) Identification of genomic regions involved in resistance against *Sclerotinia sclerotiorum* from wild *Brassica oleracea*. *Theoretical and Applied Genetics* 126, 549–556.
Meuwissen, T.H., Hayes, B.J., and Goddard, M.E. (2001) Prediction of total genetic value using genome-wide dense marker maps, *Genetics* 157, 1819–1829.
Miklas, P.N., Larsen, R.C., Riley, R., and Kelly, J.D. (2000) Potential marker-assisted selection for *bc-1*$^2$ resistance to bean common mosaic potyvirus in common bean. *Euphytica* 116, 211–219.
Millet, E.J., Kruijer, W., Coupel-Ledru, A., Alvarez Prado, S., Cabrera-Bosquet, L., *et al.* (2019) Genomic prediction of maize yield across European environmental conditions. *Nature Genetics* 51, 952–956.
Mimura, Y., Kageyama, T., Minamiyama, Y., and Hirai, M. (2009) QTL analysis for resistance to *Ralstonia solanacearum* in *Capsicum* accession 'LS2341'. *Journal of the Japanese Society for Horticultural Science* 78, 307–313.
Mishra, R., Mohanty, J.N., Mahanty, B., and Raj Kumar, J. (2021) A single transcript CRISPR/Cas9 mediated mutagenesis of *CaERF28* confers anthracnose resistance in chilli pepper (*Capsicum annuum* L.). *Planta* 254: 5.
Miyatake, K., Saito, T., Negoro, S., Yamaguchi, H., Nunome, T., *et al.* (2016) Detailed mapping of a resistance locus against *Fusarium* wilt in cultivated eggplant (*Solanum melongena*). *Theoretical and Applied Genetics* 129, 357–367.
Moghe, G.D., Hufnagel, D.E., Tang, H., Xiao, Y., Dworkin, I., *et al.* (2014) Consequences of whole-genome triplication as revealed by comparative genomic analyses of the wild radish *Raphanus raphanistrum* and three other Brassicaceae species. *Plant Cell* 26, 1925–1937.
Mohler, V. and Singrun, C. (2004) General considerations: Marker-assisted selection. In: Lorz, H. and Wenzel, G. (eds.) *Molecular Marker Systems in Plant Breeding and Crop Improvement (Biotechnology in Agriculture and Forestry), Vol. 55*. Springer, Berlin, pp 305–317.

Molitor, C., Kurowski, T.J., Fidalgo de Almeida, P.M., Eerolla, P., Spindlow, D.J., *et al.* (2021) *De novo* genome assembly of *Solanum sitiens* reveals structural variation associated with drought and salinity tolerance. *Bioinformatics* 37, 1941–1945.

Mondo, J.M., Agre, P.A., Asiedu, R., Akoroda, M.O., and Asfaw, A. (2021) Genome-wide association studies for sex determination and cross-compatibility in water yam (*Dioscorea alata* L.). *Plants* 10: 1412.

Montero-Pau, J., Blanca, J., Bombarely, A., Ziarsolo, P., Esteras, C., *et al.* (2018) *De novo* assembly of the zucchini genome reveals a whole-genome duplication associated with the origin of the *Cucurbita* genus. *Plant Biotechnology Journal* 16, 1161–1171.

Moodley, V., Naidoo, R., Gubba, A., and Mafongoya, P.L. (2019) Development of *Potato virus Y* (PVY) resistant pepper (*Capsicum annuum* L.) lines using marker-assisted selection (MAS). *Physiological and Molecular Plant Pathology* 105, 96–101.

Moury, B., Pflieger, S., Blattes, A., Lefebvre, V., and Palloix, A. (2000) A CAPS marker to assist selection of tomato spotted wilt virus (TSWV) resistance in pepper. *Genome* 43, 137–142.

Murphy, J.F., Blauth, J.R., Livingstone, K.D., Lackney, V.K., and Jahn, M. (1998) Genetic mapping of the *pvr1* locus in *Capsicum* spp. and evidence that distinct potyvirus resistance loci control responses that differ at the whole plant and cellular levels. *Molecular Plant–Microbe Interactions* 11, 943–951.

Nakayama, H., Tanaka, M., Takahata, Y., Matsui, K., Iwahori, H., *et al.* (2012) Development of AFLP-derived SCAR markers associated with resistance to two races of southern root-knot nematode in sweetpotato. *Euphytica* 188, 175–185.

Nakayasu, M., Akiyama, R., Lee, H.J., Osakabe, K., Osakabe, Y., *et al.* (2018) Generation of α-solanine-free hairy roots of potato by CRISPR/Cas9 mediated genome editing of the *St16DOX* gene. *Plant Physiology and Biochemistry* 131, 70–77.

Namayanja, A., Buruchara, R., Mahuku, G., Rubaihayo, P., Kimani, P., *et al.* (2006) Inheritance of resistance to angular leaf spot in common bean and validation of the utility of resistance linked markers for marker assisted selection outside the mapping population. *Euphytica* 151, 361–369.

Naresh, P., Lin, S., Lin, C., Wang, Y., Schafleitner, R., *et al.* (2018) Molecular markers associated to two non-allelic genic male sterility genes in peppers (*Capsicum annuum* L.). *Frontiers in Plant Science* 9: 1343.

Neequaye, M., Stavnstrup, S., Harwood, W., Lawrenson, T., Hundleby, P., *et al.* (2021) CRISPR-Cas9-mediated gene editing of *MYB28* genes impair glucoraphanin accumulation of *Brassica oleracea* in the field. *CRISPR Journal* 4, 416–426.

Nekrasov, V., Wang, C., Win, J., Lanz, C., Weigel, D., and Kamoun, S. (2017) Rapid generation of a transgene-free powdery mildew resistant tomato by genome deletion. *Scientific Reports* 7: 482.

Nesbitt, T.C. and Tanksley, S.D. (2002) Comparative sequencing in the genus *Lycopersicon*. Implications for the evolution of fruit size in the domestication of cultivated tomatoes. *Genetics* 162, 365–379.

Nevame, A.Y.M., Xia, L., Nchongboh, C.G., Hasan, M.M., Alam, M.A., *et al.* (2018) Development of a new molecular marker for the resistance to tomato yellow leaf curl virus. *BioMed Research International* 2018: 8120281.

Nguyen, T.T., Le, N.T., and Sim, S.C. (2021) Genome-wide association study and marker development for bacterial wilt resistance in tomato (*Solanum lycopersicum* L.). *Scientia Horticulturae* 289: 110418.

Nicaise, V., German-Retana, S., Sanjuan, R., Dubrana, M.P., Mazier, M., *et al.* (2003) The eukaryotic translation initiation factor 4E controls lettuce susceptibility to the potyvirus *Lettuce mosaic virus*. *Plant Physiology* 132, 1272–1282.

Nieto, C., Morales, M., Orjeda, G., Clepet, C., Monfort, A., *et al.* (2006) An *eIF4E* allele confers resistance to an uncapped and non-polyadenylated RNA virus in melon. *Plant Journal* 48, 452–462.

Nimmakayala, P., Abburi, V.L., Saminathan, T., Alaparthi, S.B., Almeida, A., *et al.* (2016) Genome-wide diversity and association mapping for capsaicinoids and fruit weight in *Capsicum annuum* L. *Scientific Reports* 6: 38081.

Nogales, A., Nobre, T., Cardoso, H.G., Muñoz-Sanhueza, L., Valadas, V., *et al.* (2016) Allelic variation on *DcAOX1* gene in carrot (*Daucus carota* L.): an interesting simple sequence repeat in a highly variable intron. *Plant Gene* 5, 49–55.

Nonaka, S., Arai, C., Takayama, M., Chiaki, M., and Hiroshi, E. (2017) Efficient increase of γ-aminobutyric acid (GABA) content in tomato fruits by targeted mutagenesis. *Scientific Reports* 7: 7057.

Nunez de Caceres Gonzalez, F.F., and de la Mora Franco, D. (2020) Vegetable crop improvement using CRISPR/Cas9. In: Bhattacharya, A., Parkhi, V., and Char, B. (eds.) *CRISPR/Cas Genome Editing: Strategies and Potential for Crop Improvement*. Springer, Cham, Switzerland, pp. 119–129.

Oleszkiewicz, T., Klimek-Chodacka, M., Kruczek, M., Godel-Jędrychowska, K., Sala, K., *et al.* (2021) Inhibition of carotenoid biosynthesis by CRISPR/Cas9 triggers cell wall remodelling in carrot. *International Journal of Molecular Sciences* 22: 6516.

Oumouloud, A., El Otmani, M., and Álvarez, J.M. (2015) Molecular characterization of *Fom-1* gene and development of functional markers for molecular breeding of resistance to *Fusarium* race 2 in melon. *Euphytica* 205, 491–501.

Özkaynak, E., Devran, Z., Kahveci, E., Doğanlar, S., Başköylü, B., *et al.* (2014) Pyramiding multiple genes for resistance to PVY, TSWV and PMMoV in pepper using molecular markers. *European Journal of Horticultural Science* 79, 233–239.

Pachner, M., Paris, H.S., Winkler, J., and Lelley, T. (2015) Phenotypic and marker-assisted pyramiding of genes for resistance to zucchini yellow mosaic virus in oilseed pumpkin (*Cucurbita pepo*). *Plant Breeding* 134, 121–128.

Pan, C., Ye, L., Qin, L., Liu, X., He, Y., *et al.* (2016) CRISPR/Cas9-mediated efficient and heritable targeted mutagenesis in tomato plants in the first and later generations. *Scientific Reports* 6: 24765.

Pan, J.S., Tan, J.Y., Wang, Y.H., Zheng, X.Y., Owens, K., *et al.* (2018) *STAYGREEN* (*CsSGR*) is a candidate for the anthracnose (*Colletotrichum orbiculare*) resistance locus *cla* in Gy14 cucumber. *Theoretical and Applied Genetics* 131, 1577–1587.

Paran, I., and Michelmore, R.W. (1993) Development of reliable PCR-based markers linked to downy mildew resistance genes in lettuce. *Theoretical and Applied Genetics* 85, 985–993.

Park, S.O., Coyne, D.P., Steadman, J.R., and Skroch, P.W. (2001) Mapping of QTL for resistance to white mold disease in common bean. *Crop Science* 41, 1253–1262.

Parra, L., Maisonneuve, B., Lebeda, A., Schut, J., Christopoulou, M., *et al.* (2016) Rationalization of genes for resistance to *Bremia lactucae* in lettuce. *Euphytica* 210, 309–326.

Parra-Galindo, M.A., Piñeros-Niño, C., Soto-Sedano, J.C., and Mosquera-Vasquez, T. (2019) Chromosomes I and X harbor consistent genetic factors associated with the anthocyanin variation in potato. *Agronomy* 9: 366.

Parrella, G., Ruffel, S., Moretti, A., Morel, C., Palloix, A., and Caranta, C. (2002) Recessive resistance genes against potyviruses are localized in colinear genomic regions of the tomato (*Lycopersicon* spp.) and pepper (*Capsicum* spp.) genomes. *Theoretical and Applied Genetics* 105, 855–861.

Parrella, G., Moretti, A., Gognalons, P., Lesage, M.L., Marchoux, G., *et al.* (2004) The *Am* gene controlling resistance to *Alfalfa mosaic virus* in tomato is located in the cluster of dominant resistance genes on chromosome 6. *Phytopathology* 94, 345–350.

Paudel, L., Clevenger, J., and McGregor, C. (2019) Chromosomal locations and interactions of four loci associated with seed coat color in watermelon. *Frontiers in Plant Science* 10: 788.

Pei, C., Wang, H., Zhang, J., Wang, Y., Francis, D.M., and Yang, W. (2012) Fine mapping and analysis of a candidate gene in tomato accession PI128216 conferring hypersensitive resistance to bacterial spot race T3. *Theoretical and Applied Genetics* 124, 533–542.

Phumichai, C., Aiemnaka, P., Nathaisong, P., Hunsawattanakul, S., Fungfoo, P., *et al.* (2021) Genome-wide association mapping and genomic prediction of yield-related traits and starch pasting properties in cassava. *Theoretical and Applied Genetics* 135, 145–171.

Pierre, M., Noel, L., Lahaye, T., Ballvora, A., Veuskens, J., *et al.* (2000) High-resolution genetic mapping of the pepper resistance locus *Bs3* governing recognition of the *Xanthomonas campestris* pv *vesicatora* AvrBs3 protein. *Theoretical and Applied Genetics* 101, 255–263.

Potato Genome Sequencing Consortium (2011) Genome sequence and analysis of the tuber crop potato. *Nature* 475, 189–195.

Prabhandakavi, P., Pogiri, R., Kumar, R., Acharya, S., Esakky, R., *et al.* (2021) Pyramiding *Ty-1*/*Ty-3*, *Ty-2*, *ty-5* and *ty-6* genes into tomato hybrid to develop resistance against tomato leaf curl viruses and recurrent parent genome recovery by ddRAD sequencing method. *Journal of Plant Biochemistry and Biotechnology*, 30, 462–476.

Pramanik, D., Shelake, R.M., Park, J., Kim, M.J., Hwang, I., *et al.* (2021) CRISPR/Cas9-mediated generation of pathogen-resistant tomato against tomato yellow leaf curl virus and powdery mildew. *International Journal of Molecular Sciences* 22: 1878.

Prasad, A., Hari-Gowthem, G., Muthamilarasan, M., Hussain, Z., Yadav, P.K., *et al.* (2021) Molecular characterization of *SlATG18f* in response to *Tomato leaf curl New Delhi* virus infection in tomato and development of a CAPS marker for leaf curl disease tolerance. *Theoretical and Applied Genetics* 134, 1463–1474.

Prasanna, H.C., Kashyap, S.P., Krishna, R., Sinha, D.P., Reddy, S., and Malathi, V.G. (2015b) Marker assisted selection of *Ty-2* and *Ty-3* carrying tomato lines and their implications in breeding tomato leaf curl disease resistant hybrids. *Euphytica* 204, 407–418.

Prasanna, H.C., Sinha, D.P., Rai, G.K., Krishna, R., Kashyap, S.P., *et al.* (2015a) Pyramiding *Ty-2* and *Ty-3* genes for resistance to monopartite and bipartite tomato leaf curl viruses of India. *Plant Pathology* 64, 256–264.

Prodhomme, C., Vos, P.G., Paulo, M.J., Tammes, J.E., Visser, R.G., *et al.* (2020) Distribution of P1(D1) wart disease resistance in potato germplasm and GWAS identification of haplotype-specific SNP markers. *Theoretical and Applied Genetics* 133, 1859–1871.

Qi, J., Liu, X., Shen, D.I., Miao, H., Xie, B., *et al.* (2013) A genomic variation map provides insights into the genetic basis of cucumber domestication and diversity. *Nature Genetics* 45, 1510–1515.

Qin, C., Yu, C., Shen, Y., Fang, X., Chen, L., *et al.* (2014) Whole-genome sequencing of cultivated and wild peppers provides insights into *Capsicum* domestication and specialization. *Proceedings of the National Academy of Sciences USA* 111, 5135–5140.

Rabbi, I.Y., Udoh, L.I., Wolfe, M., Parkes, E.Y., Gedil, M.A., *et al.* (2017) Genome-wide association mapping of correlated traits in cassava: dry matter and total carotenoid content. *Plant Genome* 10: 0094.

Rabbi, I.Y., Kayondo, S.I., Bauchet, G., Yusuf, M., Aghogho, C.I., *et al.* (2020) Genome-wide association analysis reveals new insights into the genetic architecture of defensive, agro-morphological and quality-related traits in cassava. *Plant Molecular Biology* 109, 195–213.

Ragagnin, V.A., de Souza, T.L.P.O., Sanglard, D.A., Arruda, K.M.A., Costa, M.R., *et al.* (2009) Development and agronomic performance of common bean lines simultaneously resistant to anthracnose, angular leaf spot and rust. *Plant Breeding* 128, 156–163.

Ramirez-Madera, A.O. and Havey, M.J. (2017) Different haplotypes encode the same protein for independent sources of zucchini yellow mosaic virus resistance in cucumber. *HortScience* 52, 1040–1042.

Razali, R., Bougouffa, S., Morton, M.J., Lightfoot, D.J., Alam, I., *et al.* (2018) The genome sequence of the wild tomato *Solanum pimpinellifolium* provides insights into salinity tolerance. *Frontiers in Plant Science* 9: 1402.

Ren, Y., Sun, H., Zong, M., Guo, S., Ren, Z., *et al.* (2020) Localization shift of a sugar transporter contributes to phloem unloading in sweet watermelons. *New Phytologist* 227, 1858–1871.

Ren, Y., Li, M., Guo, S., Sun, H., Zhao, J., *et al.* (2021) Evolutionary gain of oligosaccharide hydrolysis and sugar transport enhanced carbohydrate partitioning in sweet water-melon fruits. *Plant Cell* 33, 1554–1573.

Reyes-Chin-Wo, S., Wang, Z., Yang, X., Kozik, A., Arikit, S., *et al.* (2017) Genome assembly with *in vitro* proximity ligation data and whole-genome triplication in lettuce. *Nature Communications* 8: 14953.

Riedelsheimer, C., Czedik-Eysenberg, A., Grieder, C., Lisec, J., Technow, F., *et al.* (2012) Genomic and metabolic prediction of complex heterotic traits in hybrid maize. *Nature Genetics* 44, 217–220.

Robbins, M.D., Masud, M.A., Panthee, D.R., Gardner, R.G., Francis, D.M., and Stevens, M.R. (2010) Marker-assisted selection for coupling phase resistance to tomato spotted wilt virus and *Phytophthora infestans* (late blight) in tomato. *HortScience* 45, 1424–1428.

Rodriguez, E., Douches, F., Lopez-Cruz, D., Coombs, M., and de Los Campos, J.G. (2018) Genomic selection for late blight and common scab resistance in tetraploid potato (*Solanum tuberosum*). G3 8, 2471–2481.

Rodríguez-Leal, D., Lemmon, Z.H., Man, J., Bartlett, M.E., and Lippman, Z.B. (2017) Engineering quantitative trait variation for crop improvement by genome editing. *Cell* 171, 470–480.e8.

Roman, B., Satovic, Z., Avila, C.M., Rubiales, D., Moreno, M.T., and Torres, A.M. (2003) Locating genes associated with *Ascochyta fabae* resistance in *Vicia faba* L. *Australian Journal of Agricultural Research* 54, 85–90.

Ron, M., Kajala, K., Pauluzzi, G., Dongxue, W., Mauricio, A. R., *et al.* (2014) Hairy root transformation using *Agrobacterium rhizogenes* as a tool for exploring cell type-specific gene expression and function using tomato as a model. *Plant Physiology* 166, 455–469.

Rong, F., Chen, F., Huang, L., Zhang, J., Zhang, C., *et al.* (2019) A mutation in class III homeodomain-leucine zipper (HD-ZIP III) transcription factor results in curly leaf (*cul*) in cucumber (*Cucumis sativus* L.). *Theoretical and Applied Genetics* 132, 113–123.

Ruangrak, E., Su, X., Huang, Z., Wang, X., Guo, Y., *et al.* (2018) Fine mapping of a major QTL controlling early flowering in tomato using QTL-seq. *Canadian Journal of Plant Science* 98, 672–682.

Rutkoski, J.E., Heffner, E.L., and Sorrells, M.E. (2011) Genomic selection for durable stem rust resistance in wheat. *Euphytica* 179, 161–173.

Saha, P., Kalia, P., Sonah, H., and Sharma, T.R. (2014) Molecular mapping of black rot resistance locus *Xca1bo* on chromosome 3 in Indian cauliflower (*Brassica oleracea* var. *botrytis* L.). *Plant Breed* 133, 268–274.

Saha, P., Ghoshal, C., Ray, S., Saha, N.D., Srivastava, M., *et al.* (2020) Genetic analysis of downy mildew resistance and identification of molecular markers linked to resistance gene *Ppa207* on chromosome 2 in cauliflower. *Euphytica* 216: 183.

Saha, P., Ghoshal, C., Saha, N.D., Verma, A., Srivastava, M., *et al.* (2021) Marker-assisted pyramiding of downy mildew-resistant gene *Ppa3* and black rot-resistant gene *Xca1bo* in popular early cauliflower variety *Pusa meghna*. *Frontiers in Plant Science* 12: 603600.

Saidou, A.A., Thuillet, A.C., Couderc, M., Mariac, C., and Vigouroux, Y. (2014) Association studies including genotype by environment interactions: prospects and limits. *BMC Genetics* 15: 3.

Saito, M., Kubo, N., Matsumoto, S., Suwabe, K., Tsukada, M., and Hirai, M. (2006) Fine mapping of the clubroot resistance gene, *Crr3*, in *Brassica rapa*. *Theoretical and Applied Genetics* 114, 81–91.

Salamon, P., Mitykó, J., Kalo, P., and Szabo, Z. (2016). Symptoms caused by *Tomato spotted wilt virus* (TSWV) in pepper (*Capsicum* spp.) and marker assisted selection of TSWV resistant pepper lines for hybrid constructions. In: *Proceedings of the XVI Eucarpia Capsicum and Eggplant Meeting, 12–14 September 2016, Kecskemét, Hungary*, pp. 12–14.

Salgon, S., Jourda, C., Sauvage, C., Daunay, M.C., Reynaud, B., *et al.* (2017) Eggplant resistance to the *Ralstonia solanacearum* species complex involves both broad-spectrum and strain-specific quantitative trait loci. *Frontiers in Plant Science* 8: 828.

Saritnum, O., Minami, M., Matsushima, K., Minamiyama, Y., Hirai, M., *et al.* (2008) Inheritance of few-pungent trait in chilli pepper 'S3212' (*Capsicum frutescens*). *Journal of the Japanese Society for Horticultural Science* 77, 265–269.

Sauvage, C., Segura, V., Bauchet, G., Stevens, R., Do, P.T., *et al.* (2014) Genome-wide association in tomato reveals 44 candidate loci for fruit metabolic traits. *Plant Physiology* 165, 1120–1132.

Schrager-Lavelle, A., Gath, N.N., Devisetty, U.K., Carrera, E., Lopez-Diaz, I., *et al.* (2019) The role of a class III gibberellin 2-oxidase in tomato internode elongation. *Plant Journal* 97, 603–615.

Seah, S., Williamson, V.M., Garcia, B.E., Mejıa, L., Salus, M.S., Martin, C.T., and Maxwell, D.P. (2007) Evaluation of a co-dominant SCAR marker for detection of the *Mi-1* locus for resistance to root-knot nematode in tomato germplasm. *Report of the Tomato Genetics Cooperative* 57, 37–40.

Seki, K., Komatsu, K., Hiraga, M., Tanaka, K., Uno, Y., and Matsumura, H. (2021) Development of PCR-based marker for resistance to *Fusarium* wilt race 2 in lettuce (*Lactuca sativa* L.). *Euphytica* 217: 126.

Shang, Y., Ma, Y., Zhou, Y., Zhang, H., Duan, L., *et al.* (2014) Biosynthesis, regulation, and domestication of bitterness in cucumber. *Science* 346, 1084–1088.

Sharlach, M., Dahlbeck, D., Liu, L., Chiu, J., Jiménez-Gómez, J.M., *et al.* (2013) Fine genetic mapping of *RXopJ4*, a bacterial spot disease resistance locus from *Solanum pennellii* LA716. *Theoretical and Applied Genetics* 126, 601–609.

She, H., Xu, Z., Zhang, H., Li, G., Wu, J., *et al.* (2021) Identification of a male-specific region (MSR) in *Spinacia oleracea*. *Horticultural Plant Journal* 7, 341–346.

Shi, L., Yang, Y.H., Xie, Q., Miao, H., Bo, K.L., *et al.* (2018) Inheritance and QTL mapping of cucumber mosaic virus resistance in cucumber (*Cucumis sativus* L.). *PLOS ONE* 13: e0200571.

Shimatani, Z., Kashojiya, S., Takayama, M., Rie, T., and Takayuki, A. (2017) Targeted base editing in rice and tomato using a CRISPR-Cas9 cytidine deaminase fusion. *Nature Biotechnology* 35, 441–443.

Shimizu, M., Pu, Z., Kawanabe, T., Kitashiba, H., Matsumoto, S., *et al.* (2015) Map-based cloning of a candidate gene conferring *Fusarium* yellows resistance in *Brassica oleracea*. *Theoretical and Applied Genetics* 128, 119–130.

Shu, J., Liu, Y., Zhang, L., Li, Z., Fang, Z., *et al.* (2018) QTL-seq for rapid identification of candidate genes for flowering time in broccoli × cabbage. *Theoretical and Applied Genetics* 131, 917–928.

Siddique, M.I., Lee, H.Y., Ro, N.Y., Han, K., Venkatesh, J., *et al.* (2019) Identifying candidate genes for *Phytophthora capsici* resistance in pepper (*Capsicum annuum*) via genotyping-by-sequencing-based QTL mapping and genome-wide association study. *Scientific Reports* 9: 9962.

Simko, I. (2013) Marker-assisted selection for disease resistance in lettuce. In: Varshney, R.K., and Tuberosa, R. (eds.) *Translational Genomics for Crop Breeding, Vol. I: Biotic Stresses*. Wiley-Blackwell, Hoboken, New Jersey, pp. 267–289.

Simko, I., Pechenick, D.A., McHale, L.K., Truco, M.J., Ochoa, O.E., *et al.* (2009) Association mapping and marker-assisted selection of the lettuce dieback resistance gene *Tvr1*. *BMC Plant Biology* 9: 135.

Singh, S., Sharma, S.R., Kalia, P., Deshmukh, R., Vinod, S.P., and Sharma, T.R. (2012) Molecular mapping of the downy mildew resistance gene *Ppa3* in cauliflower (*Brassica oleracea* var. *botrytis* L.). *Journal of Horticultural Science and Biotechnology* 87, 137–143.

Sousaraei, N., Ramshini, H., Lotfi, M., and Sharzei, A. (2018) Marker assisted backcrossing for introgression of *Fusarium* wilt resistance gene into melon. *Euphytica* 214: 7.

Soyk, S., Müller, N.A., Park, S.J., Schmalenbach, I., Jiang, K., (2017) Variation in the flowering gene SELF PRUNING 5G promotes day-neutrality and early yield in tomato. *Nature Genetics* 49, 162–168.

Stamova, B.S. and Chetelat, R.T. (2000) Inheritance and genetic mapping of cucumber mosaic virus resistance introgressed from *Lycopersicon chilense* into tomato. *Theoretical and Applied Genetics* 101, 527–537.

Stefańczyk, E., Plich, J., Janiszewska, M., Smyda-Dajmund, P., Sobkowiak, S., and Śliwka, J. (2020) Marker-assisted pyramiding of potato late blight resistance genes *Rpi-rzc1* and *Rpi-phu1* on di- and tetraploid levels. *Molecular Breeding* 40: 89.

Stich, B. and van Inghelandt, D. (2018) Prospects and potential uses of genomic prediction of key performance traits in tetraploid potato. *Frontiers in Plant Science* 9: 159.

Stommel, J.R. and Zhang, Y. (1998) Molecular markers linked to quantitative trait loci for anthracnose resistance in tomato. *HortScience* 33: 514.

Stommel, J.R. and Zhang, Y. (2001) Inheritance and QTL analysis of anthracnose resistance in the cultivated tomato (*Lycopersicon esculentum*). *Acta Horticulturae* 542, 303–312.

Sun, B., Zheng, A., Jiang, M., Xue, S., Yuan, Q., *et al.* (2018) CRISPR/Cas9-mediated mutagenesis of homologous genes in Chinese kale. *Scientific Reports* 8: 16786.

Sun, B., Jiang, M., Zheng, H., Jian, Y., Huang, W.-L., *et al.* (2020) Color-related chlorophyll and carotenoid concentrations of Chinese kale can be altered through CRISPR/Cas9 targeted editing of the carotenoid isomerase gene *BoaCRTISO*. *Horticulture Research* 7: 161.

Sun, D., Wang, C., Zhang, X., Zhang, W., Jiang, H., *et al.* (2019) Draft genome sequence of cauliflower (*Brassica oleracea* L. var. *botrytis*) provides new insights into the C genome in *Brassica* species. *Horticulture Research* 6: 82.

Sun, H., Wu, S., Zhang, G., Jiao, C., Guo, S., *et al.* (2017) Karyotype stability and unbiased fractionation in the paleo-allotetraploid *Cucurbita* genomes. *Molecular Plant* 10, 1293–1306.

Sun, S., Deng, D., Wang, Z., Duan, C., Wu, X., *et al.* (2016) A novel *er1* allele and the development and validation of its functional marker for breeding pea (*Pisum sativum* L.) resistance to powdery mildew. *Theoretical and Applied Genetics* 129, 909–919.

Sun, X., Shu, J., Ali Mohamed, A.M., Deng, X., Zhi, X., *et al.* (2019) Identification and characterization of *EI* (*Elongated Internode*) gene in tomato (*Solanum lycopersicum*). *International Journal of Molecular Sciences* 20: 2204.

Suwabe, K., Tsukazaki, H., Iketani, H., Hatakeyama, K., Fujimura, M., *et al.* (2003) Identification of two loci for resistance to clubroot (*Plasmodiophora brassicae* Woronin) in *Brassica rapa* L. *Theoretical and Applied Genetics* 107, 997–1002.

Suwabe, K., Tsukazaki, H., Iketani, H., Hatakeyama, K., Kondo, M., *et al.* (2006) Simple sequence repeat-based comparative genomic between *Brassica rapa* and *Arabidopsis thaliana*: the genetic origin of clubroot resistance. *Genetics* 173, 309–319.

Suwor, P., Sanitchon, J., Thummabenjapone, P., Kumar, S., and Techawongstien, S. (2017) Inheritance analysis of anthracnose resistance and marker-assisted selection in introgression populations of chili (*Capsicum annuum* L.). *Scientia Horticulturae* 220, 20–26.

Sverrisdóttir, E., Sundmark, E.H.R., Johnsen, H.Ø., Kirk, H.G., Asp, T., *et al.* (2018) The value of expanding the training population to improve genomic selection models in tetraploid potato. *Frontiers in Plant Science* 9: 1118.

Tamisier, L., Szadkowski, M., Nemouchi, G., Lefebvre, V., Szadkowski, E., *et al.* (2020) Genome-wide association mapping of QTLs implied in potato virus Y population sizes in pepper: evidence for widespread resistance QTL pyramiding. *Molecular Plant Pathology* 21, 3–16.

Tan, M.A., Hutten, R.C., Visser, R.G., and van Eck, H.J. (2010) The effect of pyramiding *Phytophthora infestans* resistance genes $R_{Pi\text{-}mcd1}$ and $R_{Pi\text{-}ber}$ in potato. *Theoretical and Applied Genetics* 121, 117–125.

Tanksley, S.D. (1983) Molecular markers in plant breeding. *Plant Molecular Biology* 1, 3–8.

Thorwarth, P., Yousef, E.A., and Schmid, K.J. (2018) Genomic prediction and association mapping of curd-related traits in gene bank accessions of cauliflower. *G3* 8, 707–718.

Tian, G., Miao, H., Yang, Y.H., Zhou, J., Lu, H.W., *et al.* (2016) Genetic analysis and fine mapping *of Watermelon mosaic virus* resistance gene in cucumber. *Molecular Breeding* 36: 131.

Tian, S., Jiang, L., Gao, Q., Zhang, J., Zong, M., *et al.* (2017) Efficient CRISPR/Cas9-based gene knockout in watermelon. *Plant Cell Reports* 36, 399–406.

Tian, S., Jiang, L., Cui, X., Zhang, J., Guo, S., *et al.* (2018) Engineering herbicide-resistant watermelon variety through CRISPR/Cas9-mediated base-editing. *Plant Cell Reports* 37, 1353–1356.

Tian, Y., Zeng, Y., Zhang, J., Yang, C., Yan, L., *et al.* (2015) High quality reference genome of drumstick tree (*Moringa oleifera* Lam.), a potential perennial crop. *Science China Life Sciences* 58, 627–638.

Tomato Genome Consortium (2012) The tomato genome sequence provides insights into fleshy fruit evolution. *Nature* 485, 635–641.

Tomlinson, L., Yang, Y., Emenecker, R., Smoker, M., Taylor, J., *et al.* (2019) Using CRISPR/Cas9 genome editing in tomato to create a gibberellin-responsive dominant dwarf DELLA allele. *Plant Biotechnology Journal* 17, 132–140.

Toppino, L., Vale, G., and Rotino, G.L. (2008) Inheritance of *Fusarium* wilt resistance introgressed from *Solanum aethiopicum* Gilo and Aculeatum groups into cultivated eggplant (*S. melongena*) and development of associated PCR-based markers. *Molecular Breeding* 22, 237–250.

Tran, M.T., Doan, D.T.H., Kim, J., Song, Y.J., Sung, Y.W., *et al.* (2021) CRISPR/Cas9-based precise excision of *SlHyPRP1* domain(s) to obtain salt stress-tolerant tomato. *Plant Cell Reports* 40, 999–1011.

Tripodi, P., Soler, S., Campanelli, G., Díez, M.J., Esposito, S., *et al.* (2021) Genome wide association mapping for agronomic, fruit quality, and root architectural traits in tomato under organic farming conditions. *BMC Plant Biology* 21: 481.

Truong, H.T.H., Choi, H., Cho, M.C., Lee, H.E., and Kim, J.H. (2011a) Use of Cf-9 gene-based markers in marker-assisted selection to screen tomato cultivars with resistance to *Cladosporium fulvum*. *Horticulture, Environment, and Biotechnology* 52, 20–210.

Truong, H.T.H., Kim, K.T., Kim, S., Cho, M.C., Kim, H.R., and Woo, J.G. (2011b) Development of gene-based markers for the *Bs2* bacterial spot resistance gene for marker-assisted selection in pepper (*Capsicum* spp.). *Horticulture, Environment, and Biotechnology* 52, 65–73.

Tuncel, A., Corbin, K.R., Ahn-Jarvis, J., Harris, S., Hawkins, E., *et al.* (2019) Cas9-mediated mutagenesis of potato starch-branching enzymes generates a range of tuber starch phenotypes. *Plant Biotechnology Journal* 17, 2259–2271.

Ueta, R., Abe, C., Watanabe, T., Shigeo, S.S., Ryosuke, I., *et al.* (2017) Rapid breeding of parthenocarpic tomato plants using CRISPR/Cas9. *Scientific Reports* 7: 507.

Urasaki, N., Takagi, H., Natsume, S., Uemura, A., Taniai, N., *et al.* (2017) Draft genome sequence of bitter gourd (*Momordica charantia*), a vegetable and medicinal plant in tropical and subtropical regions. *DNA Research* 24, 51–58.

Usman, M.G., Rafii, M.Y., Martini, M.Y., Yusuff, O.A., Ismail, M.R., and Miah, G. (2018) Introgression of heat shock protein (Hsp70 and sHsp) genes into the Malaysian elite chilli variety Kulai (*Capsicum annuum* L.) through the application of marker-assisted backcrossing (MAB). *Cell Stress and Chaperones* 23, 223–234.

Uwera, A., Rusagara, J.N., Msolla, S.N., Musoni, A., and Assefa, T. (2021) Molecular marker-assisted backcrossing of anthracnose resistance genes into common beans (*Phaseolus vulgaris* L.) varieties. *American Journal of Plant Sciences* 12, 771–781.

van der Beek, L.G., Pet, G., and Lindhout, P. (1994) Resistance to powdery mildew (*Oidium lycopersicum*) in *Lycopersicon hirsutum* is controlled by an incompletely-dominant gene *Ol-1* on chromosome 6. *Theoretical and Applied Genetics* 89, 467–473.

van der Biezen, E.A., Glagotskaya, T., Overduin, B., Nijkamp, H.J.J., and Hille, J. (1995) Inheritance and genetic mapping of resistance to *Alternaria alternata* f. sp. *lycopersici* in *Lycopersicon pennellii*. *Molecular Genetics and Genomics* 247, 453–461.

Varshney, R.K., Mahendar, T., Aggarwal, R.K., and Börner, A. (2007) Genic molecular markers in plants: development and applications. In; Varshney, R.K. and Tuberosa, R. (eds.) *Genomics-assisted Crop Improvement*. Springer, Dordrecht, Netherlands, pp. 13–29.

Vidavski, F., Czosnek, H., Gazit, S., Levy, D., and Lapidot, M. (2008) Pyramiding of genes conferring resistance to *Tomato yellow leaf curl virus* from different wild tomato species. *Plant Breeding* 127, 625–631.

Vlasova, A., Capella-Gutierrez, S., Rendon-Anaya, M., Hernandez-Onate, M., Minoche, A.E., *et al.* (2016) Genome and transcriptome analysis of the Mesoamerican common bean and the role of gene duplications in establishing tissue and temporal specialization of genes. *Genome Biology* 17: 32.

Wang, C.N., Luan, F.S., Liu, H.Y., Davis, A.R., Zhang, Q.A., *et al.* (2021) Mapping and predicting a candidate gene for flesh color in watermelon. *Journal of Integrative Agriculture* 20, 2100–2111.

Wang, J., Wang, Y., Zhang, J., Ren, Y., Li, M., *et al.* (2021) The NAC transcription factor *ClNAC68* positively regulates sugar content and seed development in watermelon by repressing *ClINV* and *ClGH3.6*. *Horticulture Research* 8: 265.

Wang, L., Chen, L., Li, R., Zhao, R., Yang, M., *et al.* (2017) Reduced drought tolerance by CRISPR/Cas9-mediated *SlMAPK3* mutagenesis in tomato plants. *Journal of Agricultural and Food Chemistry* 65, 8674–8682.

Wang, S., Zhang, S., Wang, W., Xiong, X., Meng, F., *et al.* (2015) Efficient targeted mutagenesis in potato by the CRISPR/Cas9 system. *Plant Cell Reports* 34, 1473–1476.

Wang, X., Fang, Z., Sun, P., Liu, Y., Yang, L., and Zhuang, M. (2000) A SCAR marker applicable in marker assisted selection of a dominant male sterility gene in cabbage. *Acta Horticulturae Sinica* 27, 143–144.

Wang, X., Wang, H., Wang, J., Sun, R., Wu, J., *et al.* (2011) The genome of the mesopolyploid crop species *Brassica rapa*. *Nature Genetics* 43, 1035–1039.

Wang, X., Li, L., Yang, Z., Zheng, X., Yu, S., *et al.* (2017) Predicting rice hybrid performance using univariate and multivariate GBLUP models based on North Carolina mating design II. *Heredity* 118, 302–310.

Wang, X., Fazari, A., Cao, Y., Zhang, Z., Palloix, A., *et al.* (2018) Fine mapping of the root-knot nematode resistance gene *Me1* in pepper (*Capsicum annuum* L.) and development of markers tightly linked to *Me1*. *Molecular Breeding* 38: 39.

Wang, X., Gao, L., Jiao, C., Stravoravdis, S., Hosmani, P.S., *et al.* (2020) Genome of *Solanum pimpinellifolium* provides insights into structural variants during tomato breeding. *Nature Communications* 11: 5817.

Wang, Y., Lu, H., and Hu, J. (2016) Molecular mapping of high resistance to bacterial leaf spot in lettuce PI 358001-1. *Phytopathology* 106, 1319–1325.

Wang, Y., Jiang, J., Zhao, L., Zhou, R., Yu, W., and Zhao, T. (2018) Application of whole genome resequencing in mapping of a tomato yellow leaf curl virus resistance gene. *Scientific Reports* 8: 9592.

Wang, Y., Wang, J., Guo, S., Tian, S., Zhang, J., *et al.* (2021) CRISPR/Cas9-mediated mutagenesis of *ClBG1* decreased seed size and promoted seed germination in watermelon. *Horticulture Research* 8: 70.

Wang, Y., Song, L., Zhao, L., Yu, W., and Zhao, T. (2022) Development of a gene-based high resolution melting (HRM) marker for selecting the gene *ty-5* conferring resistance to tomato yellow leaf curl virus. *Horticulturae* 8: 112.

Wang, Y.H., Vanden Langenberg, K., Wehner, T.C., Kraan, P.A.G., Suelmann, J., *et al.* (2016) QTL mapping for downy mildew resistance in cucumber inbred line WI7120 (PI 330628). *Theoretical and Applied Genetics* 129, 1493–1505.

Watson, I.W. and Singh, D. (1952) The future of rust resistant wheat in Australia. *Journal of the Australian Institute of Agricultural Science* 28, 190–197.

Wei, Q., Wang, J., Wang, W., Hu, T., Hu, H., and Bao, C. (2020) A high-quality chromosome-level genome assembly reveals genetics for important traits in eggplant. *Horticulture Research* 7: 153.

Wei, T., van Treuren, R., Liu, X., Zhang, Z., Chen, J., *et al.* (2021) Whole-genome resequencing of 445 *Lactuca* accessions reveals the domestication history of cultivated lettuce. *Nature Genetics* 53, 752–760.

Wen, C.L., Mao, A.J., Dong, C.J., Liu, H.Y., Yu, S.C., *et al.* (2015) Fine genetic mapping of target leaf spot resistance gene *cca-3* in cucumber, *Cucumis sativus* L. *Theoretical and Applied Genetics* 128, 2495–2506.

Wendel, J.F., Jackson, S.A., Meyers, B.C., and Wing, R.A. (2016) Evolution of plant genome architecture. *Genome Biology* 17: 37.

Win, K.T., Zhang, C., Silva, R.R., Lee, J.H., Kim, Y.C., and Lee, S. (2019) Identificafition of quantitative trait loci governing subgynoecy in cucumber. *Theoretical and Applied Genetics* 132, 1505–1521.

Wolfe, M.D., Rabbi, I.Y., Egesi, C.N., Hamblin, M., Kawuki, R., *et al.* (2016) Genome-wide association and prediction reveals genetic architecture of cassava mosaic disease resistance and prospects for rapid genetic improvement. *Plant Genome* 9: 0118

Woo, J.W., Kim, J., Kwon, S.I., Corvalán, C., Cho, S.W., *et al.* (2015) DNA-free genome editing in plants with preassembled CRISPR-Cas9 ribonucleoproteins. *Nature Biotechnology* 33, 1162–1164.

Wu, L., Wang, P., Wang, Y., Cheng, Q., Lu, Q., *et al.* (2019) Genome-wide correlation of 36 agronomic traits in the 287 pepper (*Capsicum*) accessions obtained from the SLAF-seq-Based GWAS. *International Journal of Molecular Sciences* 20: 5675.

Wu, L., Chang, Y., Wang, L., Wang, S., and Wu, J. (2022) The aquaporin gene *PvXIP1;2* conferring drought resistance identified by GWAS at seedling stage in common bean. *Theoretical and Applied Genetics* 135, 485–500.

Wu, S., Shamimuzzaman, M., Sun, H., Salse, J., Sui, X., *et al.* (2017) The bottle gourd genome provides insights into Cucurbitaceae evolution and facilitates mapping of a *Papaya ring-spot virus* resistance locus. *Plant Journal* 92, 963–975.

Wu, S., Wang, X., Reddy, U., Sun, H., Bao, K., *et al.* (2019) Genome of 'Charleston Gray', the principal American watermelon cultivar, and genetic characterization of 1,365 accessions in the US National Plant Germplasm System watermelon collection. *Plant Biotechnology Journal* 17, 2246–2258.

Xanthopoulou, A., Montero-Pau, J., Mellidou, I., Kissoudis, C., Blanca, J., *et al.* (2019) Whole-genome resequencing of *Cucurbita pepo* morphotypes to discover genomic variants associated with morphology and horticulturally valuable traits. *Horticulture Research* 6: 94.

Xie, D., Xu, Y., Wang, J., Liu, W., Zhou, Q., *et al.* (2019) The wax gourd genomes offer insights into the genetic diversity and ancestral cucurbit karyotype. *Nature Communications* 10: 5158.

Xu, C., Liberatore, K.L., MacAlister, C.A., Huang, Z., Chu, Y.H., *et al.* (2015) A cascade of arabinosyltransferases controls shoot meristem size in tomato. *Nature Genetics* 47, 784–792.

Xu, C., Jiao, C., Sun, H., Cai, X., Wang, X., *et al.* (2017) Draft genome of spinach and transcriptome diversity of 120 *Spinacia* accessions. *Nature Communications* 8: 15275.

Xu, L., Wang, C., Cao, W., Zhou, S., and Wu, T. (2018) CLAVATA1-type receptor-like kinase *CsCLAVATA1* is a putative candidate gene for dwarf mutation in cucumber. *Molecular Genetics and Genomics* 293, 1393–1405.

Xu, Q., Shi, Y., Yu, T., Xu, X., Yan, Y., *et al.* (2016) Whole-genome resequencing of a cucumber chromosome segment substitution line and its recurrent parent to identify candidate genes governing powdery mildew resistance. *PLOS ONE* 11: e0164469.

Xu, X., Xu, R., Zhu, B., Yu, T., Qu, W., *et al.* (2014) A high-density genetic map of cucumber derived from specific length amplified fragment sequencing (SLAF-seq). *Frontiers in Plant Science* 5: 768.

Xu, X., Lu, L., Zhu, B., Xu, Q., Qi, X., and Chen, X. (2015) QTL mapping of cucumber fruit flesh thickness by SLAF-seq. *Scientific Reports* 5: 15829.

Xu, X., Chao, J., Cheng, X., Wang, R., Sun, B., *et al.* (2016a) Mapping of a novel race specific resistance gene to *Phytophthora* root rot of pepper (*Capsicum annuum*) using bulked segregant analysis combined with specific length amplified fragment sequencing strategy. *PLOS ONE* 11: e0151401.

Xu, X., Yu, T., Xu, R., Shi, Y., Lin, X., *et al.* (2016b). Fine mapping of a dominantly inherited powdery mildew resistance major-effect QTL, *Pm1.1*, in cucumber identifies a 41.1 kb region containing two tandemly arrayed cysteine-rich receptor-like protein kinase genes. *Theoretical and Applied Genetics* 129, 507–516.

Xu, Z.S., Feng, K., and Xiong, A.S. (2019) CRISPR/Cas9-mediated multiply targeted mutagenesis in orange and purple carrot plants. *Molecular Biotechnology* 61, 191–199.

Yaghoobi, J., Yates, J.L., and Williamson, V.M. (2005) Fine mapping of the nematode resistance gene *Mi-3* in *Solanum peruvianum* and construction of a *S. lycopersicum* DNA contig spanning the locus. *Molecular Genetics and Genomics* 274, 60–69.

Yamamoto, E., Matsunaga, H., Onogi, A., Ohyama, A., Miyatake, K., *et al.* (2017) Efficiency of genomic selection for breeding population design and phenotype prediction in tomato. *Heredity* 118, 202–209.

Yamamoto, K., Oda, Y., Haseda, A., Fujito, S., Mikami, T., and Onodera, Y. (2014) Molecular evidence that the genes for dioecism and monoecism in *Spinacia oleracea* L. are located at different loci in a chromosomal region. *Heredity* 112, 317–324.

Yan, C., An, G., Zhu, T., Zhang, W., Zhang, L., *et al.* (2019) Independent activation of the *BoMYB2* gene leading to purple traits in *Brassica oleracea*. *Theoretical and Applied Genetics* 132, 895–906.

Yan, J., He, L., Xu, S., Wan, Y., Wang, H., *et al.* (2020) Expression analysis, functional marker development and verification of *AgFNSI* in celery. *Scientific Reports* 10: 531.

Yang, H.B., Liu, W.Y., Kang, W.H., Jahn, M., and Kang, B.C. (2009) Development of SNP markers linked to the *L* locus in *Capsicum* spp. by a comparative genetic analysis. *Molecular Breeding* 24, 433–446.

Yang, J., Liu, D., Wang, X., Ji, C., Cheng, F., *et al.* (2016) The genome sequence of allopolyploid *Brassica juncea* and analysis of differential homoeolog gene expression influencing selection. *Nature Genetics* 48, 1225–1232.

Yang, L., Liu, H., Zhao, J., Pan, Y., Cheng, S., *et al.* (2018) *LITTLELEAF* (*LL*) encodes a WD40 repeat domain-containing protein associated with organ size variation in cucumber. *Plant Journal* 95, 834–847.

Yang, W. and Francis, D.M. (2005) Marker-assisted selection for combining resistance to bacterial spot and bacterial speck in tomato. *Journal of the American Society for Horticultural Science* 130, 716–721.

Yang, X., Caro, M., Hutton, S.F., Scott, J.W., Guo, Y., *et al.* (2014) Fine mapping of the tomato yellow leaf curl virus resistance gene *Ty-2* on chromosome11of tomato. *Molecular Breeding* 34, 749–760.

Ye, M., Peng, Z., Tang, D., Yang, Z., Li, D., *et al.* (2018) Generation of self-compatible diploid potato by knockout of S-RNase. *Nature Plants* 4, 651–654.

Yu, K., Park, S.J., and Poysa, V. (2000) Marker-assisted selection of common beans for resistance to common bacterial blight: efficacy and economics. *Plant Breeding* 119, 411–415.

Yu, Q., Wang, B., Li, N., Tang, Y., Yang, S., *et al.* (2017) CRISPR/Cas9-induced targeted mutagenesis and gene replacement to generate long-shelf life tomato lines. *Scientific Reports* 7: 11874.

Yu, S., Zhang, F., Yu, R., Zou, Y., Qi, J., *et al.* (2009) Genetic mapping and localization of a major QTL for seedling resistance to downy mildew in Chinese cabbage (*Brassica rapa* ssp. *pekinensis*). *Molecular Breeding* 23, 573–590.

Yu, S., Zhang, F., Zhao, X., Yu, Y., and Zhang, D. (2011) Sequence-characterized amplified region and simple sequence repeat markers for identifying the major quantitative trait locus responsible for seedling resistance to downy mildew in Chinese cabbage (*Brassica rapa* ssp. *pekinensis*). *Plant Breeding* 130, 580–583.

Yu, Y., Pan, Z., Wang, X., Bian, X., Wang, W., *et al.* (2021) Targeting of SPCSV-RNase3 via CRISPR-Cas13 confers resistance against sweet potato virus disease. *Molecular Plant Pathology* 23, 104–117.

Yu, Z.H., Wang, J.F., Stall, R.E., abd Vallejos, C.E. (1995) Genomic localization of tomato genes that control a hypersensitive reaction to *Xanthomonas campestris* pv. *vesicatoria* (Doidge) Dye. *Genetics* 141, 675–682.

Yuan, J., Bizimungu, B., de Koeyer, D., Rosyara, U., Wen, Z., and Lagüe, M. (2019) Genome-wide association study of resistance to potato common scab. *Potato Research* 63, 5005–5017.

Zamir, D., Ekstein-Michelson, I., Zakay, U., Navot, N., Zeidan, M., *et al.* (1994) Mapping and introgression of a tomato yellow leaf curl virus tolerance gene, *Ty-1*. *Theoretical and Applied Genetics* 88, 141–146.

Zhang, C., Ren, Y., Guo, S., Zhang, H., Gong, G., *et al.* (2013) Application of comparative genomics in developing markers tightly linked to the *Pm-2F* gene for powdery mildew resistance in melon (*Cucumis* melo L.). *Euphytica* 190, 157–168.

Zhang, C., Liu, L., Wang, X., Vossen, J., Li, G., *et al.* (2014) The *Ph-3* gene from *Solanum pimpinellifolium* encodes CC-NBS-LRR protein conferring resistance to *Phytophthora infestans*. *Theoretical and Applied Genetics* 127, 1353–1364.

Zhang, C.Y., Anarjan, M.B., Win, K.T., Begum, S., and Lee, S. (2021) QTL-Seq analysis of powdery mildew resistance in a Korean cucumber inbred line. *Theoretical and Applied Genetics* 134, 435–451.

Zhang, H., Yi, H., Wu, M., Zhang, Y., Zhang, X., *et al.* (2016) Mapping the flavor contributing traits on 'Fengwei Melon' (*Cucumis melo* L.) chromosomes using parent resequencing and super bulked-segregant analysis. *PLOS ONE* 11: e0148150.

Zhang, H., Si, X., Ji, X., Fan, R., Liu, J., *et al.* (2018) Genome editing of upstream open reading frames enables translational control in plants. *Nature Biotechnology* 36, 894–898.

Zhang, J. and Dilip, P.R. (2021) Development of codominant SCAR markers to detect the *Pto*, *Tm2-2*, *I3* and *Sw5* genes in tomato (*Solanum lycopersicum*). *Plant Breeding* 140, 342–348.

Zhang, J., Guo, S., Ji, G., Zhao, H., Sun, H., *et al.* (2020) Af unique chromosome translocation disrupting *ClWIP1* leads to gynoecy in watermelon. *Plant Journal* 101, 265–277.

Zhang, J., Zhao, J., Xu, Y., Liang, J., Chang, P., *et al.* (2015) Genome-wide association mapping for tomato volatiles positively contributing to tomato flavor. *Frontiers in Plant Science* 6: 1042.

Zhang, K., Wang, X., Zhu, W., Qin, X., Xu, J., *et al.* (2018) Complete resistance to powdery mildew and partial resistance to downy mildew in a *Cucumis hystrix* introgression line of cucumber were controlled by a co-localized locus. *Theoretical and Applied Genetics* 131, 2229–2243.

Zhang, K., Wei, Y., Njogu, M.K., Wang, X., Lou, Q., *et al.* (2019) Genetic mapping of angular leaf spot resistance to *Pseudomonas syringae* pv. *lachrymans* in a *Cucumis hystrix* introgression line of cucumber. *Euphytica* 215: 176.

Zhang, K., Li, Y., Zhu, W., Wei, Y., Njogu, M.K., *et al.* (2020) Fine mapping and transcriptome analysis of virescent leaf gene *v-2* in cucumber (*Cucumis sativus* L.). *Frontiers in Plant Science* 11: 570817.

Zhang, M., Liu, Q., Yang, X., Xu, J., Liu, G., *et al.* (2020) CRISPR/ Cas9-mediated mutagenesis of *Clpsk1* in watermelon to confer resistance to *Fusarium oxysporum* f.sp. *niveum*. *Plant Cell Reports* 39, 589–595.

Zhang, N., Xu, B.-H., Bi, Y.-F., Lou, Q.-F., Chen, J.-F., *et al.* (2017) Development of a musk-melon cultivar with improved resistance to gummy stem blight and desired agronomic traits using gene pyramiding. *Czech Journal of Genetics and Plant Breeding* 53, 23–29.

Zhang, R., Liu, C., Song, X., Sun, F., Xiao, D., *et al.* (2020) Genome-wide association study of turnip mosaic virus resistance in non-heading Chinese cabbage. *3 Biotech* 10: 363.

Zhang, S., Chen, X., Lu, C., Ye, J., Zou, M., *et al.* (2018) Genome-wide association studies of 11 agronomic traits in cassava (*Manihot esculenta* Crantz). *Frontiers in Plant Science* 9: 503.

Zhang, S.P., Miao, H., Yang, Y.H., Xie, B.Y., Wang, Y., and Gu, X.F. (2014) A major quantitative trait locus conferring resistance to *Fusarium* wilt was detected in cucumber by using recombinant inbred lines. *Molecular Breeding* 34, 1805–1815.

Zhang, X., Wang, G., Chen, B., Du, H., Zhang, F., *et al.* (2018) Candidate genes for first flower node identified in pepper using combined SLAF-seq and BSA. *PLOS ONE* 13: e0194071.

Zhang, X., Liu, T., Wang, J., Wang, P., Qiu, Y., *et al.* (2021) Pan-genome of *Raphanus* highlights genetic variation and introgression among domesticated, wild, and weedy radishes. *Molecular Plant* 14, 2032–2055.

Zhang, Z., Wang, B., Wang, S., Lin, T., Yang, L., *et al.* (2019) Genome-wide target mapping shows histone deacetylase complex 1 regulates cell proliferation in cucumber fruit. *Plant Physiology* 182, 167–184.

Zhao, J., Sauvage, C., Zhao, J., Bitton, F., Bauchet, G., *et al.* (2019) Meta-analysis of genome-wide association studies provides insights into genetic control of tomato flavor. *Nature Communications* 10: 1534.

Zhao, J., Xu, Y., Ding, Q., Huang, X., Zhang, Y., *et al.* (2016) Association mapping of main tomato fruit sugars and organic acids. *Frontiers in Plant Science* 7: 1286.

Zhao, X., Jayarathna, S., Turesson, H., Fält, A.S., Nestor, G., *et al.* (2021) Amylose starch with no detectable branching developed through DNA-free CRISPR-Cas9 mediated mutagenesis of two starch branching enzymes in potato. *Scientific Reports* 11: 4311.

Zhao, Y., Liu, Y., Zhang, Z., Cao, Y., Yu, H., *et al.* (2020) Fine mapping of the major anthracnose resistance QTL *AnRGO5* in *Capsicum chinense* 'PBC932'. *BMC Plant Biology* 20: 189.

Zhao, Z.Q., Sheng, X.G., Yu, H.F., Wang, J.S., Shen, Y.S., and Gu, H.H. (2020) Identification of QTLs associated with curd architecture in cauliflower. *BMC Plant Biology* 20: 177.

Zheng, Y., Wu, S., Bai, Y., Sun, H., Jiao, C., *et al.* (2019) Cucurbit Genomics Database (CuGenDB): a central portal for comparative and functional genomics of cucurbit crops. *Nucleic Acids Research* 47, D1128–D1136.

Zhi, X., Shu, J., Zheng, Z., Li, T., Sun, X., *et al.* (2021) Fine mapping of the *Ph-2* gene conferring resistance to late blight (*Phytophthora infestans*) in tomato. *Plant Disease* 105, 851–858.

Zhou, Q., Wang, S., Hu, B., Chen, H., Zhang, Z., and Huang, S. (2015) An *ACCUMULATION AND REPLICATION OF CHLOROPLASTS 5* gene mutation confers light green peel in cucumber. *Journal of Integrative Plant Biology* 57, 936–942.

Zhou, X., Zha, M., Huang, J., Li, L., Imran, M., and Zhang, C. (2017) StMYB44 negatively regulates phosphate transport by suppressing expression of *PHOSPHATE1* in potato. *Journal of Experimental Botany* 68, 1265–1281.

Zhu, W.Y., Huang, L., Chen, L., Yang, J.T., Wu, J.N., *et al.* (2016) A high-density genetic linkage map for cucumber (*Cucumis sativus* L.): based on specific length amplified fragment (SLAF) sequencing and QTL analysis of fruit traits in cucumber. *Frontiers in Plant Science* 7: 437.

Zia, M.A.B., Demirel, U., Nadeem, M.A., and Caliskan, M.E. (2020) Genome-wide association study identifies various loci underlying agronomic and morphological traits in diversified potato panel. *Physiology and Molecular Biology of Plants* 26, 1003–1020.

Zsögön, A., Čermák, T., Naves, E.R., Notini, M.M., Edel, K.H., *et al.* (2018) *De novo* domestication of wild tomato using genome editing. *Nature Biotechnology* 36, 1211–1216.

# 16 Genetic Engineering in Vegetable Crops: Present Scenario and Future Challenges

**Pallavi Mishra, Vidya Sagar and Shailesh K. Tiwari***
*Division of Vegetable Improvement, ICAR-Indian Institute of Vegetable Research, Varanasi, Uttar Pradesh, India*

**Abstract**
Vegetable crops include a large number of species from various plant families that grow in a variety of geographical locations and have a wide diversity of vegetative and reproductive characteristics. Vegetables, with their unique blend of taste, flavor, and freshness, are an important part of the human diet in terms of health and nutrition, and are consumed raw, cooked, and in processed form. Vegetable cultivation is an integral component of the agricultural economy, particularly in developing countries, and meeting rising global demand while ensuring the nutritional quality of vegetables is a significant challenge. Using cutting-edge biotechnological and genome-sequencing techniques, one can move closer to targeted breeding of vegetable crops for superior alleles and development of new cultivars fostered with precise genetic modifications. In this regard, *Agrobacterium*-mediated gene transfer remains the foundation for most vegetable transformation protocols; however, new emerging technologies such as transcriptome engineering and genome editing, particularly clustered regularly interspaced palindromic repeats (CRISPR)-CRISPR-associated protein 9 (Cas9), have had a significant impact on vegetable crop breeding. Using *Agrobacterium*-mediated gene transformation and genome-editing techniques, a variety of transgenic vegetables with economically important traits such as tolerance to abiotic/biotic stress, improved quality and shelf life, and expression of edible vaccines, antibodies, and other pharmaceutical proteins have been developed in recent years. However, in order to be used in a practical vegetable-breeding program, such techniques necessitate the use of a large number of genetic resources, efficient handling, and well-established transformation protocols. The potential applications of different transgenic techniques adopted for genetic improvement as well as the future implications of genetic modifications in various groups of vegetable crops are summarized in this chapter.

**Keywords:** abiotic stress, biopharming, CRISPR/Cas9, edible vaccine, genetically modified crops, plantibodies, transgene, vegetables

## 16.1 Introduction

As the global population quadruples over the next century, so will the demand for food, and the world is concomitantly facing problems such as hunger, health crises, and micronutrient deficiency. In 2018, about 821 million people were undernourished. According to the Food and Agriculture Organization of the United Nations (FAO) estimate in *The State of Food Security and Nutrition in the World* 2020 report, approximately 189.2 million people in India were malnourished. One out of every nine people does not get enough food to live an active and healthy life. An estimated

*Email: tiwarishailu@gmail.com


DOI: 10.1079/9781800627123.0016

52% of women of reproductive age are anemic and 35% of children have physical disabilities as a result of malnutrition and a lack of complete healthy food. To supply food to this burgeoning population, crop production on a global scale has to be increased in unprecedented ways. Because agriculture is the primary source of food, agricultural production must grow much faster than the growing population, which could be accomplished by increasing agricultural land, developing new crop improvement techniques, and adopting better farming practices.

As an exceptional source of minerals, vitamins, antioxidants, micronutrients, phytosterols, and dietary fibers, vegetables play a significant role in supplying complete nutrition and therefore comprise an important constituent of the human diet. Vegetable cultivation is important for the agricultural economy, and is a profitable business, especially in developing countries such as India, where the majority of the population is vegetarian. According to the latest data from the FAO, India is the world's second largest producer of vegetables after China, accounting for 61% of the world's cultivated area under vegetable cultivation and contributing 71% of total vegetable production (www.fao.org/faostat/en/, accessed 10 July 2024). Popular vegetable crops grown in India along with their major nutrients are shown in Table 16.1.

Biotic and abiotic stresses have a significant impact on vegetable productivity and quality, resulting in a destabilized agricultural economy in several countries. Pre- and postharvest losses, as well as a lack of proper storage facilities, also frequently result in a quantitative and qualitative diminution in vegetable production. This could be overcome by reducing abiotic/biotic stress losses and improving processing and storage facilities. Vegetable-breeding techniques used in the past have resulted in significant improvement in their quality, shelf life, yield, and abiotic/biotic stress tolerance. However, because it is a slow and labor-intensive method, it cannot solely help to achieve targeted goals such as improved nutritional and therapeutic value, abiotic/biotic stress tolerance, and surpassingly improved agronomic traits. Major constraints in conventional breeding approaches can be overcome by merging them with advanced biotechnological techniques. In this context, the use of genetic engineering in vegetables (despite serious controversy and opposition) has provided a greater contribution to revolutionizing the entire crop production system by providing crops with superior agronomic traits such as flavor, color, fruit shape and size, shelf life, disease resistance, and abiotic/biotic stress tolerance, while also focusing on economic and environmental sustainability. Genetic engineering has been successfully implemented in vegetable crops for their use in biopharming for production of therapeutics and industrial products, improvements in yield and quality parameters, storage life, and resistance. Several reports have highlighted that genetically engineered vegetables have improved agronomic traits while also being fortified with important nutraceuticals with human health benefits. In this chapter, we will summarize the major aspects, issues, and challenges associated with genetically engineered or genetically modified vegetable crops. Future perspectives on the use of genetic engineering and genome-editing techniques for vegetable improvement will also be discussed.

## 16.2 Present Scenario of Plant Genetic Engineering

Plant genetic engineering is a branch of biology that employs a variety of techniques for editing the plant genome, such as recombinant DNA

**Table 16.1.** Commonly occurring vegetables and their nutritional content.

| Vegetable group | Common examples | Major nutrients |
|---|---|---|
| Fruit vegetables | Tomato, cucumber, aubergine, pepper | Vitamin C, vitamin $B_1$, iodine |
| Leafy vegetables | Spinach, amaranth, cabbage | Iron, calcium, vitamin A |
| Root vegetables | Potato, sweet potato, radish, carrot | Carbohydrate, vitamin A, |
| Bulb vegetables | Onion, ginger and garlic | Vitamin C, calcium, potassium, sodium, iodine |
| Flower vegetables | Cauliflower, artichoke and broccoli | Calcium, iron, potassium |

technology and *in vitro* manipulations of plant tissues, as well as the most recent emerging techniques for *in silico* analysis of DNA/RNA via computational approaches. FlavrSavr, a transgenic tomato variety with a silenced glycoside hydrolase gene encoding the polygalacturonase enzyme responsible for fruit softening and early decay, was the first transgenic vegetable approved for commercial sale in the USA in 1994 (Bruening and Lyons, 2000). The next large-scale cultivation of genetically modified crops for herbicide tolerance and insect resistance began in 1996, with *Bacillus thuringiensis* (*Bt*) as the primary transforming agent. In the case of vegetables, large-scale commercialized vegetable crops include tomato and potato, which have been engineered for delayed ripening and virus resistance, respectively. FlavrSavr tomato and *Bt*-eggplant are the most well-known and successful transgenic stories in the history of horticultural crop breeding, but they faced significant opposition due to potential risks to human health and environmental safety. Besides these, many more genetically engineered vegetable crops have been developed so far (discussed in further sections) but unfortunately, as with FlavrSavr tomato and *Bt*-eggplant, socio-economic awareness about their possible benefits is under-rated.

In recent years, considerable progress has been made in introgression of quality traits, development of resistant cultivars, and production of industrial proteins in vegetable crops. Several studies have revealed that novel agronomic traits of interest can be introduced into vegetable crops. Genetic engineering has been successfully implemented to develop vegetables with desired phenotypic characteristics using the gene pool of wild species, as well as to reduce the expression of genes that produce naturally toxic compounds harmful to human health. For example, genes involved in the defense response against plant–pathogen interactions and in the response to various forms of stress (heat, cold, drought, salinity, and low oxygen) have been successfully engineered in vegetables.

## 16.3 Genetic Engineering in Vegetables for Quality Improvement

One of the major goals of introducing genetic engineering in vegetables is to improve qualitative traits such as nutritional content, organoleptic qualities such as flavor and aroma, and development of seedless vegetables.

### 16.3.1 Improvement of nutritional quality

Vegetables produce a variety of organic compounds that are essential to humans, such as vitamins, minerals, micronutrients, carotenoids, flavonoids, phytosterols, and dietary fibers. However, some vegetables are nutrient deficient and lack essential amino acids such as lysine and methionine. The essential amino acid content of such vegetables can be enhanced by modification of the amino acid sequence of the protein produced, overexpression of synthetically developed homologous proteins through genetic engineering, and/or metabolomics. Some of the different vegetables that have been subjected to genetic transformation are presented in Table 16.2 and are discussed below.

One of the best examples is potato, where the nutritive value of potato deficient in sulfur-containing amino acids such as methionine and cysteine and the essential amino acids lysine and tyrosine has been improved through the development of transgenic potatoes expressing the *Amaranthus hypochondriacus* seed albumin gene *AmA1* in their tubers. *AmA1* is a non-allergenic protein that contains all essential amino acids, and has been considered to be a complete protein by World Health Organization standards (Chakraborty *et al.*, 2000). Similarly, transgenic potatoes developed using an antisense threonine synthase (*TS*) gene exhibited up to a 30-fold increase in methionine content when compared with control plants (Zeh *et al.*, 2001). Transgenic potato with enhanced vitamin E (α-tocopherol) was developed by overexpressing homogentisate phytyltransferase (*AtHPT*) and γ-tocopherol-methyltransferase (*Atγ-TMT*) genes isolated from *Arabidopsis* in potato, which was shown to combat degenerative health problems, and decrease cancer risk and coronary heart diseases in humans (Upadhyaya *et al.*, 2020).

Another example is the development of tomato plants engineered with bacterial genes that produce phytoene desaturase enzyme, which converts phytoene to lycopene, which along with β-carotene serves as an important precursor for vitamin A biosynthesis through

**Table 16.2.** Genetic transformation for improvement of nutritional qualities in vegetables.

| Vegetable | Transgene/source/pathway | Transgene description | Reference |
|---|---|---|---|
| *Solanum tuberosum* L. | *AmA1* from *Amaranthus hypochondriacus* | Rich in essential amino acids lysine and tyrosine | Chakraborty *et al.* (2000) |
| | Threonine synthase (*TS*) gene | Antisense threonine synthase gene for increased methionine content | Zeh *et al.* (2001) |
| | Homogentisate phytyltransferase (*AtHPT*) and γ-tocopherol-methyltransferase (*Atγ-TMT*) genes from *Arabidopsis* | Enhanced vitamin E (α-tocopherol) content | Upadhyaya *et al.* (2020) |
| | *crtB* gene from *Erwinia uredovora* | Increase in β-carotene and lutein content in transgenic potato | Ducreux *et al.* (2005) |
| *Solanum lycopersicum* L. | Phytoene desaturase, converting phytoene to lycopene | Precursor for vitamin A biosynthesis through isoprenoid pathway, improved flavor | Römer *et al.* (2000) |
| | Genetically triggered carotenoid biosynthetic pathway for carotenoid accumulation | High lycopene and β-carotene contents | Fraser *et al.* (2009) |
| | Fruit-specific GTP cyclohydrolase I | Folate biofortification for increased folate and *p*-aminobenzoate content | Díaz de la Garza *et al.* (2004) |
| | *CHIA* gene from petunia encoding chalcone isomerase | Increased flavonoid content in tomato fruit peel | Muir *et al.* (2001) |
| | *LC* and *C1* transcription factors from corn | Up to 10-folds increase in flavonoid glycosides content | Le Gall *et al.* (2003) |
| | Hydroxycinnamoyl transferase (*HQT*) gene | Increased levels of chlorogenic acid | Niggeweg *et al.* (2004) |
| | Endogenous photomorphogenesis *DETI* gene | Increased carotenoid and flavonoid content | Davuluri *et al.* (2005) |
| | Lycopene β-cyclase (*tLcy-b*) | Conversion of lycopene to β-carotene | Giorio *et al.* (2007) |
| | *Del*/*Ros1* from *Antirrhinum majus* | Transcription factor, regulating anthocyanin production | Butelli *et al.* (2008) |
| *Phaseolus vulgaris* L. | *be2s2* from Brazil nut | Increased methionine content | Aragão *et al.* (1999) |

the isoprenoid pathway. Through this pathway, genetic manipulation of tomato cultivars resulted in a threefold increase in pro-vitamin A content and organoleptic qualities (Römer *et al.*, 2000). Similar genetic transformation has also been reported in potatoes engineered with *Erwinia uredovora crtB* gene, which expresses phytoene synthase, and reportedly exhibited a four- to sevenfold increase in β-carotene and lutein content in transgenic potato compared with control plants (Ducreux *et al.*, 2005). In sweet potato, to increase the levels of carotenoids, transgenic plants overexpressing *IbOr-Inscan* under the control of the cauliflower mosaic virus 35S promoter in an anthocyanin-rich purple-fleshed cultivar (referred to as IbOr plants) was developed. IbOr plants exhibited increased carotenoid levels (up to sevnfold) in their storage roots compared with wild-type plants. In tomato, the strategy adopted involved pathway extension beyond β-carotene through the expression of β-carotene hydroxylase (*CrtZ*) and oxyxgenase (*CrtW*) from *Brevundimonas* sp. in tomato fruit, followed by β-carotene enhancement through introgression of a lycopene β-cyclase (β-Cyc) allele from a *Solanum galapagense* background. Following successful use of bacterial genes in the development of Golden rice, the group of researchers successfully incorporated a similar system of carotenoid enhancement in potatoes (Diretto *et al.*, 2007). They introduced three *Erwinia*-derived genes encoding phytoene synthase (*CrtB*), phytoene desaturase (*CrtI*) and lycopene β-cyclase (*CrtY*) under the control of a constitutive promoter into potato. The carotenoid content increased approximately 20-fold, up to 114 μg $g^{-1}$ of dry weight and the β-carotene content by 3600-fold to 47 μg $g^{-1}$ of dry weight in the tubers of the transgenic plants ("golden" potatoes) (Diretto *et al.*, 2007). They further characterized the carotenoid metabolites and transcripts. Several endogenous genes have also been identified as key regulators for carotenoid biosynthesis (Diretto *et al.*, 2010).

In the case of tomatoes, genetic engineering has resulted in activation of the carotenoid biosynthetic pathway and carotenoid accumulation mechanisms, allowing the development of tomato lines supplemented with high lycopene and β-carotene content (Fraser *et al.*, 2009). Similarly, genetically engineered tomato plants with folate biofortification were created by overexpressing the fruit-specific GTP cyclohydrolase gene (*GCH1*), the first enzyme involved in pteridine synthesis (Díaz de la Garza *et al.*, 2004). Pteridine, along with *p*-aminobenzoate (PABA) and glutamate moieties, is known to be involved in folate biosynthesis in plants, and thus contributed to a significant increase in folate and PABA content in vegetables. The engineering of the metabolic pathway changed the folate content on average by twofold (Díaz de la Garza *et al.*, 2004). In addition, supplementation of PABA in these transformed plants increased the folate content tenfold (Díaz de la Garza *et al.*, 2004). In another study, Díaz de la Garza *et al.* (2007) showed that the combination of *GCH1* and aminodeoxychorismate synthase (*ADCS*), which catalyzes pteridine and PABA synthesis, increased the folate content of the ripe fruit by 25-fold compared with controls. An animal-based (chicken) *GCH1* was used to transform lettuce plants and a 2.1–8.5-fold increase in folate content was recorded (Nunes *et al.*, 2009). When modifying potatoes and *Arabidopsis* for higher folate content, Blancquaert *et al.* (2013) found that engineering of just the pterin and PABA pathways was insufficient to increase the folate content in the tubers. This example suggests that the two-gene system is insufficient to produce high folate plants in potato. We need to look for further steps in the folate biosynthesis pathway and this could be one of the reasons why there is no successful example of folate biofortification in many crops except for rice and tomato. Two mitochondrial channeling enzymes were proposed to limit the flux: first, the bifunctional hydroxymethyl dihydropterin pyrophosphokinase/dihydropteroate synthase (HPPK/DHPS), which performs condensation of both precursors in the mitochondria, and second, folylpolyglutamate synthase (FPGS) at the last step of the pathway, which could exert a pulling effect (Waller *et al.*, 2010; Gorelova *et al.*, 2017). By targeting a total of four genes, de Lepeleire *et al.* (2018) transformed potato with *GCH1*, *ADCS*, *HPPK/DHPS*, and *FPGS* genes and successfully achieved a satisfactory level of folate in the tubers. They were able to enhance the folate content of potato 12-fold (1925 mg 100 $g^{-1}$ dry weight), along with long-term storage stability. Flavonoids are an important group of secondary metabolites that are commonly used for their antioxidant properties, polyphenolic compounds, and role in fruit coloration. Intake of bioactive compounds, such as flavonoids, has been shown to reduce the incidence of chronic infections, inflammation, cancer and cardiovascular diseases in humans (Robak and Gryglewski, 1996; Yoon and Baek, 2005; García-Lafuente *et al.*, 2009). Because most of the flavonoids in tomato are limited to its peel, a transgenic approach to increase tomato flavonoid content was developed by overexpression of the enzymes involved in the flavonoid biosynthetic pathway. For this, the petunia *CHIA* gene, which encodes chalcone isomerase, was overexpressed in tomato, resulting in significantly higher levels of flavonols in tomato (Muir *et al.*, 2001). Similarly, transgenic tomato lines expressing the corn *LC* and *C1* transcription factors showed up to a tenfold increase in flavonoid glycoside content compared with the control plants (Le Gall *et al.*, 2003). Calystegines are glycosidic inhibitors that are found in vegetables such as potato, tomato, and cabbage. They are produced through the tropane alkaloid biosynthetic pathway. Calystegine synthesis has been shown to increase carbohydrate accumulation in root cultures and increase carbohydrate metabolism in transgenic potatoes (Richter *et al.*, 2007).

The selenocystine methyltransferase (*SMT*) gene was cloned from the selenium hyperaccumulator *Astragalus bisulcatus* (Neuhierl *et al.*, 1999) and transformed into *Arabidopsis* and *Brassica* plants (Pilon-Smits and LeDuc, 2009). Later, the same gene was transformed into *Nicotiana tabacum* and it was shown that this gene can be used for nutritional improvement purposes (McKenzie *et al.*, 2009). The same group also developed transgenic tomatoes for the production of methylselenocysteine (MeSeCy) by overexpression of the *SMT* gene. It was found that the accumulation of MeSeCy occurred in fruit but not in the leaves. It was also found that this MeSeCy was heat stable and can readily be used in tomato juices. Greater accumulation of MeSeCy was found when selenate was used (Brummell *et al.*, 2011).

Zinc is transported from root to shoot through the xylem via transporter proteins. The same transporters (HMA4 and $P_{1B}$-ATPase) can transport metals such as zinc (Zn), iron (Fe), and cadmium (Cd) to the xylem (Courbot *et al.*, 2007; Hanikenne *et al.*, 2008). High expression of HMA4 in the hyperaccumulator leads to a high concentration of Zn/Cd in the shoot (Hanikenne *et al.*, 2008). However, ectopic expression of *AtHMA4* under control of the CaMV promoter in tobacco resulted in increased levels of Zn in the shoots but decreased levels of Cd. This modification in transport may be suitable for the purpose of biofortification (Siemianowski *et al.*, 2011). This same construct was used to transform tomato plants to check its efficacy as a potential Zn biofortification candidate. It was found that the transformed tomato had a higher Zn content compared with wild-type plants (Kendziorek *et al.*, 2014). In a soil and hydroponics media study, Weremczuk *et al.* (2016) used transformed tomato carrying *AhHMA4p1::AhHMA4* genes to determine whether the mineral composition affects the translocation of these nutrients, especially Zn and Cd. They reported that the pattern of expression of the cross-homeostasis genes (*LeIRT1*, *LeChln*, and *LeNRAMP1*) changed in the transgenics in a medium-dependent fashion. Furthermore, when the plants were grown in soil with/without Cd, more efficient translocation of Zn was observed in the transgenic plants.

Short cation exchanger 1 (*sCAX1*), a vacuolar calcium/proton antiporter previously identified from *Arabidopsis thaliana* (Hirschi *et al.*, 1996), was used for transformation in carrot to increase calcium (Ca) levels. *sCAX1*-containing carrots were normal and had a yield equivalent to that of non-transformed carrots (S. Park *et al.*, 2004). The utility of these carrots was further proved by Morris *et al.* (2008) in feeding trials of mice and humans, where they raised the question of bioavailability of calcium and showed that the bioavailability in *sCAX1*-expressing carrots was lower than that of normal carrot. However, *sCAX1*-expressing carrots have been shown to be a better source of dietary calcium because the total amount of calcium absorbed per gram of carrot was significantly higher for *sCAX1*-expressing carrots than for control carrots (Morris *et al.*, 2008). Another approach to introducing variation in *CAX1* was achieved by the method of targeting induced local lesions in genome (TILLING) to produce plants that are efficient in metabolism of calcium. In *Brassica rapa* subsp. *trilocularis*, Navarro-León *et al.* (2018) used the TILLING approach to induce mutations in the *BraA.cax1* gene resulting in allelic variation. Three allelic variations were produced, *BraA.cax1a-4*, *BraA.cax1a-7*, and *BraA.cax1a-12*. Mutants accumulated more Ca and magnesium (Mg) in leaves under controlled and high doses of Ca and accumulated more Fe, independent of the level of Ca. The *BraA.cax1a-4* and *BraA.cax1a-7* mutants displayed lower overall chlorophyll, altered efficiency of photosynthesis, and higher reactive oxygen species (ROS) levels. In elevated Ca conditions, the *BraA.cax1a-12* mutant developed faster. The *BraA.cax1a-12* mutant is a good candidate for biofortification as it was able to accumulate high Ca, Mg, and Fe in the leaves together with 80% leaf yield advantage over the control (Navarro-León *et al.*, 2018).

### 16.3.2 Improvement of organoleptic qualities

The flavor and aroma of vegetables are caused by the presence of volatile compounds such as phenolics, ethers, aldehydes, and ketones, and have a major influence on consumer choices. The first transgenic tomato expressing the sweet protein E8-monellin from *Dioscoreophyllum cumminsii* was developed by Peñarrubia *et al.* (1992). The transgenic plants accumulated monellin protein at a significantly higher level in the leaves and tomato fruits, providing an alternative strategy for vegetable flavor enhancement. By introducing the thaumatin gene from *Thaumatococcus daniellii*, Bartoszewski *et al.* (2003) developed transgenic tomatoes with altered taste and sweetness levels. Similarly, *Arabidopsis AtbZIP11*, a class of leucine zipper (bZIP) transcription factor gene known to mediate the sucrose-induced repression of translation in *Arabidopsis*, has been used to create transgenic tomato plants with enhanced sweetness and sugar content (Sagor *et al.*, 2016). The FlavrSavr tomato, a transgenic tomato variety developed by silencing of polygalacturonase, a natural pectin-degrading enzyme in the plant cell wall, not only exhibits delayed fruit softening and decay, but also shows an increased development of flavor compounds for better taste in tomato (Clark *et al.*, 2004).

In the case of potato, flavor loss and browning are major factors that degrade their quality. To overcome such problems, transgenic potato with controlled enzymatic browning by inhibiting the synthesis of polyphenol oxidase has been developed using an antisense approach (Bachem *et al.*, 1994; Coetzer *et al.*, 2001; Chi *et al.*, 2014).

### 16.3.3 Seedless vegetables

Parthenocarpy, or the development of fruits without pollination/fertilization, results in the production of seedless vegetables. Seedless vegetables offer more choices to customers due to easier processing as well as improved quality of vegetables such as eggplant and tomato, where the seeds can make the taste bitter. Parthenocarpy has been linked to auxin biosynthesis in several studies (Pandolfini *et al.*, 2018), so efforts to engineer parthenocarpic fruits by triggering the auxin biosynthetic pathway for either increased auxin production or by increasing auxin sensitivity of ovaries have been made. Overexpression of *Antirrhinum majus* and *Pseudomonas syringae* pv. *savastanoi DefH9-iaaM* genes has been shown to confer parthenocarpy in transgenic eggplant and tomato fruits (Rotino *et al.*, 1997; Ficcadenti *et al.*, 1999). The *Agrobacterium rhizogenes*-derived *rolB* gene has also been shown to induce parthenocarpy in transgenic tomato by specifically controlling the *TPRP-F1* promoter (Carmi *et al.*, 2003). Parthenocarpy has also been reported to be induced in tomato by RNA interference (RNAi)-induced silencing of the chalcone synthase gene, which is the first step in the flavonoid biosynthesis pathway (Schijlen *et al.*, 2007). Recently, transient *SlIAA9* knockdown using novel tissue-specific promoters was used to develop genetically engineered parthenocarpic tomato plants (Kim *et al.*, 2019).

### 16.3.4 Fruit development and size

Plants synthesize certain compounds that not only act as an effective protectant against abiotic/biotic stresses but are also known to promote plant growth at vegetative and reproductive stages. The most common example is glycinebetaine, which is normally synthesized through pathways with two key enzymes, betaine aldehyde dehydrogenase (BADH) and choline oxidase (COD). Transgenic tomato lines expressing *codA* and *BADH* exhibited improved traits such as increased photosynthetic rate, higher chlorophyll content, and increased content of phytohormones (e.g. auxin, cytokinin, gibberellin, brassinolides) resulting in enlarged flowers and fruit development in transformed lines compared with the control plants (Zhang *et al.*, 2019). These and several other examples of genetic transformation for fruit quality traits, flavor, and aroma improvements in vegetables are summarized in Table 16.3.

## 16.4 Genetic Engineering of Vegetables for Abiotic Stress Tolerance

Abiotic stresses such as cold, drought, high temperature, and salinity have a significant impact on vegetable quality and productivity. Development of vegetables engineered with genes to withstand the damage caused by abiotic stress

**Table 16.3.** Improvement of fruit quality and characteristics of vegetables.

| Vegetable | Transgene/source | Transgene description | Reference |
|---|---|---|---|
| *Solanum lycopersicum* L. | E8-monellin gene from *Dioscoreophyllum cumminsii* | High accumulation of the sweet protein monellin in leaves and fruits | Peñarrubia *et al.* (1992) |
| | Δ-9 Desaturase gene from *Saccharomyces cerevisiae* | Improved profile of flavor compounds | Wang *et al.* (1996) |
| | *Adh2* using alcohol dehydrogenase cDNA | Improvement in flavor characteristics | Speirs *et al.* (1998) |
| | Thaumatin gene from *Thaumatococcus daniellii* | Modification in taste and sweetness levels | Bartoszewski *et al.* (2003) |
| | Gene silencing of polygalacturonase, a natural pectin-degrading enzyme in the cell wall | Delayed fruit decay and an increased profile of flavor compounds | Clark *et al.* (2004) |
| | *Arabidopsis AtbZIP11*, a class of bZIP transcription factor | For enhanced sweetness and sugar content in tomato fruits | Sagor *et al.* (2016) |
| *Solanum melongena* L., *Solanum lycopersicum* L. | *DefH9-iaaM* gene from *Antirrhinum majus* and *Pseudomonas syringae* pv. *savastanoi* | Induction of parthenocarpy in transgenic eggplant and tomato fruits | Rotino *et al.* (1997); Ficcadenti *et al.* (1999) |
| *Solanum lycopersicum* L. | *rolB* gene derived from *Agrobacterium rhizogenes* | Induction of parthenocarpy by controlling the *TPRP-F1* promoter in tomato | Carmi *et al.* (2003) |
| | RNA interference-induced silencing of chalcone synthase gene | Induction of parthenocarpy in tomato | Schijlen *et al.* (2007) |
| | *SlIAA9* knockdown approach | Induction of parthenocarpy in tomato | Kim *et al.* (2019) |
| | Betaine aldehyde dehydrogenase (*BADH*) and choline oxidase (*COD*) | Enlarged flowers and fruit development in transformed tomato lines | Zhang *et al.* (2019) |
| *Cucumis sativus* L. | *DefH9-iaaM* | Induction of parthenocarpy | Yin *et al.* (2006) |

could help to stabilize economic losses by enhancing vegetable production. Abiotic tolerance has been introduced into a variety of vegetable crops to improve antioxidant defense, ion homeostasis, water-use efficiency, and high temperature and salinity tolerance, summarized in Table 16.4.

### 16.4.1 Antioxidant defense

To combat the metabolic aberrations caused by abiotic stresses such as membrane and protein destability, plants possess various enzymatic (catalase, superoxide dismutase, glutathione reductase, glutathione peroxidase, glutathione *S*-transferase, ascorbate peroxidase, dehydroascorbate reductase, monodehydroascorbate reductase, guaiacol peroxidase) and non-enzymatic (phenolic compounds, glutathione, ascorbic acid and α-tocopherols) antioxidant defense mechanisms. These antioxidant enzymes protect plant cells from the oxidative damage caused by accumulation of ROS and form an important strategy for protecting plants from abiotic stresses (Møller, 2001; Zhu, 2002). Several attempts have been made to develop genetically engineered plants with enhanced ROS scavenging ability. In the

**Table 16.4.** Genetic engineering for development of abiotic stress tolerance in vegetables.

| Vegetable | Transgene/source | Transgene description | Reference |
|---|---|---|---|
| *Solanum lycopersicum* L. | C-repeat/dehydration responsive element binding factor 1 (*CBF1*) from *Arabidopsis* | Enhanced resistance to water-deficit stress | Hsieh *et al.* (2002a) |
| *Solanum lycopersicum* L. | Choline oxidase gene (*coda*) from *Arthrobacter globiformis* | Temperature and chilling tolerance | E.J. Park *et al.* (2004) |
| *Solanum lycopersicum* L. | *cAPX* | Heat stress and ultraviolet B tolerance | Wang *et al.* (2006) |
| *Solanum lycopersicum* L. | Overexpression of yeast *HAL1* | Enhanced $K^+$ uptake, improved tolerance to salinity stress | Gisbert *et al.* (2000); Rus *et al.* (2001) |
| *Solanum lycopersicum* L. | $Na^+/H^+$ antiporter gene (*AtNHX1*) from *Arabidopsis thaliana* | Ability to produce good-quality fruits during high salinity | Zhang and Blumwald (2001) |
| *Solanum lycopersicum* L. | *ABRC1/CBF1* | Chilling, salt and drought tolerance | Lee *et al.* (2003) |
| *Solanum lycopersicum* L. | *bspA*, a protein from *Populus tremula* | Drought stress | Roy *et al.* (2006) |
| *Solanum lycopersicum* L. | Betaine aldehyde dehydrogenase (*BADH*) | Salinity tolerance | Jia *et al.* (2002) |
| *Solanum tuberosum* L. | *TPS1* and *TPS2* | Drought tolerance | Kwon *et al.* (2004) |
| *Raphanus sativus* L. | Late embryogenesis abundant (*LEA*) gene from *Brassica napus* | Salinity and drought tolerance | Park *et al.* (2005b) |
| *Brassica oleracea* L. | *betA* gene for biosynthesis of glycinebetaine | Salinity tolerance | Bhattacharya *et al.* (2004) |
| *Brassica oleracea* L. | Overexpression of *vhb*, the *Vitreoscilla* hemoglobin gene | Submergence tolerance | Li *et al.* (2005) |
| *Brassica rapa* L. | *otsA/LEA* | Environmental stress tolerance | Park *et al.* (2003, 2005a) |
| *Brassica rapa* L. | Corn superoxide dismutase (*SOD*) and catalase (*CAT*) gene | Sulfur dioxide resistance | Tseng *et al.* (2007) |
| *Cucumis melo* L. | Yeast *HAL1* | Salinity tolerance | Bordas *et al.* (1997) |
| *Cucumis sativus* L. | Yeast *HAL1* | Salinity tolerance | Bordas *et al.* (1997) |
| *Phaseolus vulgaris* L. | *LEA* from *Brassica napus* | Drought resistance | Liu *et al.* (2005) |

case of vegetables, tomato plants have been genetically engineered to increase their susceptibility to oxidative stress and chilling injury by inhibition of catalase using antisense RNA approaches (Kerdnaimongkol and Woodson, 1999). Tolerance to oxidative and chilling stress has also been introduced into tomato by ectopic expression using the CaMV 35S promoter, which induced higher levels of catalase expression in transgenic tomato plants compared with the control plants (Hsieh *et al.*, 2002b). Interestingly, *CBF1*-overexpressing transgenic tomato plants also showed enhanced resistance to water-deficit stress (Hsieh *et al.*, 2002a). Virus-induced gene silencing of tomato *RBOH1* and *MPK1*, *MPK2*, and *MPK1/2* involved in mitogen-activated protein kinase cascades reduced the risks of damage caused by oxidative stress in transgenic tomato plants (Nie *et al.*, 2013). In an another study, it was discovered that the accumulation of hydrogen peroxide ($H_2O_2$) is mediated by brassinosteroid and abscisic acid (ABA), which together increase *RBOH1* expression and induce tolerance to heat and oxidative stress in tomato (Zhou *et al.*, 2014).

### 16.4.2 Ion homeostasis

Ion homeostasis is the process of maintaining a steady internal state of ions and is a fundamental process that all organisms require. It is also required for a variety of cellular and biological processes in plants, particularly when subjected to abiotic stress such as salinity. Plants use different minerals to carry out essential biochemical processes, but excessive mineral accumulation can be toxic. As a result, plants have evolved efficient mechanisms for maintaining balanced concentrations of these minerals in their cytosol and internal environment while also allowing dynamic physiological changes. Ion homeostasis is achieved by adopting different strategies such as excluding excess $Na^+$ via the plasma membrane through $Na^+/H^+$ antiporters, limiting $Na^+$ entry to the cell during $Na^+$ excess, $Na^+$ secretion outside the cell, and compartmentation of $Na^+$ into vacuoles to utilize $Na^+$ for osmotic adjustments (Zhu, 2002).

It is a well-known fact that plant cells maintain a high $K^+/Na^+$ level for their cellular and physiological metabolisms. Because $Na^+$ competes with $K^+$ for cellular uptake by plants during abiotic stress such as salinity, genetic manipulation of plants to improve the cation transport system by expressing promoters specific for $K^+$ uptake, accumulation, and transport would aid in achieving ionic changes inside the cell (Zhu, 2002). In the case of tomato, overexpression of the yeast *HAL1*, a regulator gene controlling $K^+$ transport, resulted in increased $K^+$ uptake and utilization, as well as improved tolerance to salinity stress compared with control plants (Gisbert *et al.*, 2000; Rus *et al.*, 2001). Genetic manipulation of tomato with the *A. thaliana* $Na^+/H^+$ antiporter *AtNHX1* gene resulted in the development of transgenic tomato plants that accumulated high $Na^+$ concentrations in leaves (but not fruits), and had the ability to produce good-quality fruits during high-salinity conditions (Zhang and Blumwald, 2001).

### 16.4.3 Improved water-use efficiency

Precise regulation of the stomatal opening and closing mechanism in plants can help to improve water-use efficiency and drought tolerance. During drought conditions, ABA-mediated stomatal closure reduces water loss, wherein the guard cells surrounding the stomatal pores act as a model for providing early signal transduction to regulate cellular and physiological processes such as hormone stimuli, gaseous exchange, water and carbon dioxide levels (Schroeder *et al.*, 2001; Kim *et al.*, 2010). Understanding the series of molecular mechanisms underlying these processes would help in the development of transgenic plants with increased water-use efficiency and drought tolerance. Drought tolerance in *Arabidopsis* has been achieved by the expression of ABA-INSENSITIVE 1 (ABI1-1R3) protein, which negatively regulates ABA signaling and improves drought tolerance in transgenic plants compared with wild-type plants (Gosti *et al.*, 1999). Another example is a mutation in the farnesyl transferase gene (*ERA1*), which causes hypersensitivity to ABA-mediated stomatal closure, resulting in increased stomatal closure and reduced wilting during drought conditions (Pei *et al.*, 1998). A deeper understanding of these signaling events would shed more light on how these genes can be used to improve drought tolerance in vegetable crops.

### 16.4.4 Accumulation of compatible osmolytes

Plants accumulate compatible osmolytes, which protect them in several ways, including turgor maintenance via osmotic adjustments, ROS detoxification, and protein secondary structure stabilization (Bohnert and Jensen, 1996), and are therefore related to yield improvement (Zhang *et al.*, 1999). These findings have stimulated the development of genetically engineered plants that can constitutively overproduce and accumulate compatible osmolytes in both normal as well as in physiological stress-challenged situations. Several transgenic plants overproducing osmoprotectants have been developed and have demonstrated improved tolerance to salinity and drought stress (Szegletes *et al.*, 2000; Aslam *et al.*, 2015; Singh *et al.*, 2015). Certain vegetable crops such as potato, tomato, and carrot have been genetically modified for overexpression of osmoprotectants such as trehalose, glycinebetaine, and proline. Trehalose is an important osmoprotectant that has been engineered to be overproduced in transgenic potatoes using the *Saccharomyces cerevisiae* trehalose-6-phosphate synthase (*TPS1*) gene that catalyzes trehalose synthesis. Transgenic potatoes overexpressing both *TPS1* and trehalose-6-phosphate phosphatase (*TPS2*) have been shown to be more drought tolerant (Yeo *et al.*, 2000; Kwon *et al.*, 2004). Glycinebetaine is yet another important compatible osmolyte that protects plants from abiotic stress-challenged situations. However, it cannot be accumulated by certain vegetables such as tomato. Hence, genetically manipulated tomato plants have been developed using a chloroplast-targeted *codA* gene produced by *Arthrobacter globiformis*, which encodes choline oxidase, which catalyzes the conversion of choline to glycinebetaine. Transgenic tomatoes expressing the *codA* gene showed improved tolerance to chilling stress in various developmental stages of the plant, as well as a 10–30% increase in fruit production compared with wild-type plants (E.J. Park *et al.*, 2004). Genetically modified carrots overexpressing glycinebetaine and *BADH* in their plastids exhibited very high tolerance to salinity (Kumar *et al.*, 2004).

## 16.5 Genetic Engineering of Vegetables for Disease Resistance

Plants have developed multi-layered defense mechanisms against disease-causing agents such as microbial pathogens and pests in the form of physiological and anatomical barriers. To combat this, pathogens have evolved various strategies to overcome the barriers created by their own host, and enter the host system by delivering elicitors and effectors into their cytoplasm to suppress the host defense system (Franceschetti and Maqbool, 2017). It has been observed that nearly all plant viruses produce viral suppressors of RNAi to hijack the host RNAi system and mute the host genes in order to promote viral pathogenicity (Zamore, 2004). Because pathogens have the ability to modify their genes over time, knowledge of only one aspect of plant–microbe interactions may not provide opportunities for genetic modification for disease resistance, and therefore genetic engineering may not be successful. Biotechnological advances and better understanding of the molecular mechanisms underlying plant–pathogen interactions have opened up new avenues for engineering resistance to microbial pathogens in plants in several ways. Host-plant genes encoding proteins required for mycotoxin breakdown or those inhibiting the activities of pathogen cell-wall-degrading enzymes can be genetically modified (Karlovsky, 2011). Similarly, to induce immunity against viral pathogens, targeted viral RNA degradation can be achieved by exploiting the RNAi machinery (Rosa *et al.*, 2018). Such methods can be used to modify the genomes of susceptible host organisms, reducing their ability to be recognised by microbial pathogens and thus lowering the risk of susceptibility. The same has been used to develop insect and disease-resistant vegetable crops (Table 16.5).

## 16.6 Pharmaceutical and Industrial Applications

Large-scale production of biomolecules and proteins of pharmaceutical and industrial interest is referred as biopharming. Genetic engineering has been exploited to transfer protein-producing genes of interest into different plant species and undertake agricultural-scale production for use as edible vaccines, enzymes, and/or plant-derived antibodies (plantibodies). Although many of the proteins are produced in fungal, bacterial, and animal systems, plants are equally preferred for large-scale production of desired proteins for a number of reasons such as simple biological mechanisms, cost-effectiveness, the possibility of huge biomass production with low-input requirements, and, most importantly, production of environmentally safe proteins. Another benefit is that the plants naturally do not contain pathogens that are harmful to humans, so the chances of allergenicity and cross-reactivity are quite low. Thus, genetic engineering of plants has wide applications in the therapeutics and pharmaceutical sectors. Potato and tomato are popular vegetable crops used as factories for large-scale production of recombinant enzymes, vaccines, and plantibodies, while other vegetables are also subjected to genetic transformations for industrial applications (Table 16.6).

**Table 16.5.** Genetic engineering for development of disease resistance in vegetables.

| Vegetable | Transgene/source | Transgene description | Reference |
|---|---|---|---|
| *Solanum lycopersicum* L. | Coat protein gene from tobacco mosaic virus (TMV) | Tolerance to tomato mosaic virus (ToMV) | Nelson *et al.* (1988) |
| | V1 capsid protein gene from tomato yellow leaf curl virus | For delayed disease symptoms | Kunik *et al.* (1994) |
| | Rep protein gene from tomato leaf curl virus (TLCV) | TLCV resistance | Praveen *et al.* (2005) |
| | *N* gene from *Nicotiana tabacum* | Resistance to TMV and ToMV | Whitham *et al.* (1996) |
| | Rep protein gene from cucumber mosaic virus (CMV) | Moderate resistance to CMV | Nunome *et al.* (2002) |
| | *NPR1* gene from *Arabidopsis* | Resistance to *Ralstonia solanacearum* | Lin *et al.* (2004) |
| | *Mi-1.2* gene from *Lycopersicon peruvianum* | Resistance to nematodes | Goggin *et al.* (2004) |
| *Capsicum annum* L. | Coat protein gene from CMV | Resistance to CMV | Zhu *et al.* (1996) |
| | Coat protein gene from TMV | Resistance to TMV | Lee *et al.* (2004) |
| *Solanum melongena* L. | *cry* gene from *Bacillus thuringiensis* | Resistance to eggplant fruit and shoot borer (*Leucinodes orbonalis*) | Kumar *et al.* (1998) |
| | Oryzacystatin (*OZC*) gene | Resistance to *Myzus persicae* and *Macrosiphum* spp. | Ribeiro *et al.* (2006) |
| | Δ-9 desaturase gene from yeast | Resistance to *Verticillium dahliae* | Xing and Chin (2000) |
| | *Mi-1.2* gene | Resistance to nematodes | Goggin *et al.* (2006) |
| *Brassica oleracea* L. | *B22IV* and *B22VI* from cauliflower mosaic virus | Resistance to cauliflower mosaic virus | Passelegue and Kerlan (1996) |
| | *Cry* gene from *Bacillus thuringiensis* | Resistance to *Plutella xylostella* | Jin *et al.* (2000); Bhattacharya *et al.* (2002) |
| | Glucose oxidase (*GO*) gene from *Aspergillus niger* | Resistance to *Xanthomonas campestris* | Lee *et al.* (2002) |
| | *CpTI*, cowpea trypsin inhibitor | Resistance to *Heliothis armigera* | Lv *et al.* (2005) |

### 16.6.1 Transgenic vegetables producing edible vaccines

Charles Arntzen coined the term edible vaccine in 1990. It refers to edible foods, typically plants, that produce proteins, vitamins, or other plant-derived products that, when consumed, stimulate the immune system and thus act as a vaccine against certain diseases. These vaccines are ingested by humans through the consumption of edible plant parts such as tubers, leaves, grains, and fruits. The first plant-derived edible vaccine was developed against the hepatitis B virus surface antigen (HBsAg) in transgenic tobacco plants transformed with the *HBsAg* gene (Mason *et al.*, 1992). Following this, subsequent attempts

**Table 16.6.** Production of pharmaceutical proteins, vaccines and plantibodies in vegetables.

| Vegetable | Transgene/source | Transgene description | Reference |
|---|---|---|---|
| *Solanum tuberosum* L. | Endoglucanase (*E1*) gene from *Acidothermus cellulolyticus* | Accumulation of high levels of E1 protein | Dai *et al.* (2000) |
| | Human lactoferrin and casein proteins | Expression of high levels of human lactoferrin and casein proteins | Chong and Langridge (2000) |
| | Sucrose isomerase enzyme (converting sucrose into palatinose) from *Erwinia rhapontici* | Expression of high levels of palatinose, a sweet protein, equivalent to monellin in tomato | Börnke *et al.* (2002) |
| | ADP-glucose pyrophosphorylase enzyme from barley endosperm | Increased starch content in potato | Regierer *et al.* (2002) |
| *Solanum lycopersicum* L. | Monellin protein from *Dioscoreophyllum cumminsii* | Development of tomato fruits expressing low-calorie monellin protein, having 100,000 times sweeter taste than sugar, preferable for diabetic patients | Peñarrubia *et al.* (1992) |
| *Solanum lycopersicum* L. | Miraculin gene from *Richardella dulcifica* | High miraculin content, a low-calorie sweet protein | Sun *et al.* (2007) |
| | Angiotensin I-converting enzyme inhibitor | Tomato fruits with anti-hypertensive properties | Hamamoto *et al.* (1993) |
| | Rabies glycoprotein | Oral immunization vaccine | McGarvey *et al.* (1995) |
| | Respiratory syncytial virus fusion gene | Vaccine | Sandhu *et al.* (2000) |
| | *ctxB* (cholera toxin B subunit) | Cholera vaccine | Jani *et al.* (2002); Jiang *et al.* (2007) |
| | *HEVE2* (partial gene from hepatitis E virus) | Vaccine | Ma *et al.* (2003) |
| | *AChE* (human acetylcholinesterase) | Organophosphate intoxication | Mor *et al.* (2001) |
| | IL-12, Mouse interleukin recombinant protein | Mucosal administration | Gutiérrez-Ortega *et al.* (2005) |
| *Brassica oleracea* L. | *B5* gene (smallpox) and spike protein gene from severe acute respiratory syndrome coronavirus (SARS-CoV) | Antigen production for vaccinia virus and human SARS-CoV | Pogrebnyak *et al.* (2006) |
| *Daucus carota* L. | *LTB* (a heat-labile enterotoxin) from *Escherichia coli* | Protection against diarrhea and cholera | Rosales-Mendoza *et al.* (2007, 2008) |
| | HepB (hepatitis B virus surface protein) | Protection against hepatitis B virus | Imani *et al.* (2002) |

have been made for large-scale vaccine production in common vegetables such as potato, tomato, carrot, and spinach.

### 16.6.2 Production of plantibodies

Vegetables have also been used as a suitable system for production of therapeutically valuable human proteins, recombinant enzymes to be used in genetic engineering, and plantibodies. For example, human α-interferon, a protein produced by the innate immune system in response to viral infections (mainly against chronic hepatitis C) has been successfully engineered for large-scale production in transgenic potato with transformed protoplasts, using lipofectin-mediated gene transformation (Sawahel, 2002). In the same way, plantibodies have extensive therapeutic uses due to the large-scale production potential of plants while requiring low cost and low inputs for growth (Zeitlin *et al.*, 1998; Ko *et al.*, 2003). Plantibody production has been manifested in vegetables such as potato, where they can be synthesized and stored in storage organs such as tubers, seeds, and leaves (Artsaenko *et al.*, 1998).

### 16.6.3 Industrial proteins and enzymes

A number of chemicals, enzymes, and raw materials of industrial use can be synthesized in plants in one of two ways: (i) by introducing a foreign gene encoding the desired protein so that plants can produce large quantities of the desired protein; or (ii) by modifying the plant metabolic pathways to produce large quantities of the final protein product. By introduction of a foreign gene in vegetables, large amounts of industrial enzymes such as cellulases, amylases, the sweet protein monellin, and silk proteins have been synthesized. Endoglucanase (*E1*), a cellulase gene from *Acidothermus cellulolyticus*, has been expressed in transgenic potato using a chloroplast *RbcS-2A* signal peptide and leaf-specific promoter, and large quantities of E1 protein were shown to be accumulated (Dai *et al.*, 2000). Human lactoferrin and casein proteins have also been expressed successfully in transgenic potato by inserting a cDNA fragment encoding human lactoferrin linked to a plant microsomal-retention signal peptide via *Agrobacterium tumefaciens*-mediated transformation (Chong and Langridge, 2000). Potato and tobacco plants were genetically modified using a recombinant *Nephila clavipes* dragline protein to obtain remarkable quantities of spider dragline silk, a proteinaceous and attractive fiber used for technical applications (Scheller *et al.*, 2001). Transformed potato and tobacco plants were able to express recombinant proteins and thus large quantities of the spider silk, which would otherwise have to be obtained from spiders. Monellin is a protein that is 100,000 times sweeter than sucrose, and has thus replaced it in confectionaries. It also has a low calorific value, making it popular among patients on a restrictive diet. Transgenic tomato plants expressing monellin from *Dioscoreophyllum cumminsii* have been developed, making these transgenic vegetables a suitable choice for diabetic patients (Peñarrubia *et al.*, 1992). However, due to taste and carcinogenicity limitations, monellin produced in tomatoes has frequently been replaced by palatinose produced in potato tubers (Börnke *et al.*, 2002), which is also a low-calorific-value artificial sweetener with a taste similar to sucrose. The enzyme sucrose isomerase, which converts sucrose to palatinose, was obtained from *Erwinia rhapontici* and cloned into potato using potato tuber-specific promoters. The transgenic potatoes accumulated 1.7–16 μmol $g^{-1}$ of palatinose in their tubers (Börnke *et al.*, 2002). Similarly, a transgenic potato overexpressing ADP-glucose pyrophosphorylase enzyme from barley endosperm has been developed for increased starch content in potato (Regierer *et al.*, 2002). Transgenic potato plants increased not only their soluble starch content but also their essential amino acid content by two- to fourfold (Regierer *et al.*, 2002). Thus, the production of industrial proteins and enzymes in vegetables fulfills nutritional requirements as well as improving the quality and flavor of fruits.

## 16.7 CRISPR/Cas9 in Genome Editing of Vegetable Crops

Genome engineering is an emerging technique that uses a variety of methods for the modification, removal, replacement, or insertion of unique genomic sequences of interest into a targeted site in the genome of a host organism (Gaj *et al.*, 2013; Ran *et al.*, 2013). It differs greatly from genetic transformation, in which a gene of interest is randomly inserted into the host genome at a site-specific location following genome-editing guides (Carr and Church, 2009). Scientists can now easily alter plant genomes directly at the desired location using sequence-specific endonucleases, thanks to the emergence of highly specific and cost-effective genome-editing techniques and the availability of sequenced genomes (Gaj *et al.*, 2013). A number of genome-editing techniques, such as meganucleases, zinc-finger nucleases (ZFNs), transcription activator-like effector nucleases (TALENs), and CRISPR/Cas9 have emerged in recent years, allowing more precise editing of genomic loci (Ran *et al.*, 2013). Meganucleases are sequence-specific endonucleases that recognize very large sequence sites (over 12 bp) and hence are less specific in their mode of action (Galetto *et al.*, 2009). However, they are an efficient tool for mammalian gene therapy in the treatment of diseases such as xeroderma pigmentosum, for the production of recombinant proteins, and for targeting virus-borne diseases (Pâques and Duchateau, 2007; Redondo *et al.*, 2008; Galetto *et al.*, 2009; Silva *et al.*, 2011). In contrast, ZFNs, TALENs, and CRISPR/Cas9 are more specific and accurate, and can precisely edit the target DNA site of the host genome (Hsu *et al.*, 2014). The CRISPR/Cas9 system is widely used in genome engineering in a variety of plants including popular solanaceous vegetables such as potato, tomato, and eggplant (Wang *et al.*, 2015; de Toledo Thomazella *et al.*, 2016; Ueta *et al.*, 2017; Andersson *et al.*, 2018; Gianoglio *et al.*, 2018; van Eck, 2018; Enciso-Rodriguez *et al.*, 2019; Maioli *et al.*, 2020; Zhi *et al.*, 2020).

## 16.8 Conclusion and Future Prospects

Genetic-engineering technology has obvious advantages over conventional breeding methods. It is possible to introduce, remove, and modify the specific gene of interest without affecting the rest of the genome. Transgenic crops with desired traits can be obtained in a relatively short number of generations compared with conventional breeding, saving a significant amount of time. Furthermore, the genetic materials used for transgenic development are not restricted within a particular species but can be exploited for interchange of desirable genes from different species. Significant progress in vegetable genetic improvement has been achieved in a relatively short timespan through the introduction of novel genes of agricultural, industrial, and pharmaceutical value. Despite the widespread applications of genetic engineering in crop improvement, commercialization of genetically modified crops still lags far behind, particularly in the case of vegetables. This is because of some limitations that have yet to be addressed. For starters, the presence of transgenes introgressed from wild relatives and sometimes from unrelated species raises serious concerns about public acceptance of transgenic crops. Second, there has been widespread speculation about the use of antibiotics and herbicide-resistant genes as selectable markers in transgenic development. Finally, genetic-modification techniques cannot modify all genes of interest at the same time, and the results in terms of toxicity assessments may not be equivalent in most cases. It has been observed that risk assessments of genetically modified crops are not proportionate to their substantial importance. Moreover, transgenic development is expensive and time consuming, resulting in abandonment of transgenic vegetable crop development and their decreased commercialization. Virus-resistant potato, *Bt*-eggppant, and FlavrSavr tomato with improved shelf life are some of the better known vegetables to face a ban on their global commercialization.

However, the importance of transgenic crops in developing countries burdened with growing populations cannot be overstated. It is critical to consider the importance and safety of transgenic crops in a timely manner in order to meet the growing demand for food. In addition to food requirements, the importance of transgenic crops should also be considered in terms of alleviating human suffering and battling life-threatening diseases such as hepatitis and cholera, which could possibly be achieved by consuming nutritionally richer transgenic vegetable supplemented with edible vaccines and/or therapeutic proteins. Malnutrition could also be overcome by consuming transgenic vegetables that have been engineered with specific traits such as essential amino acids, and high protein and mineral content. Transgenic technology holds great promise for the development of new varieties with novel agricultural traits that will help to improve food quality along with environmental and public safety in the near future.

## 16.9 References


Andersson, M., Turesson, H., Olsson, N., Fält, A.S., Ohlsson, P., *et al.* (2018) Genome editing in potato via CRISPR-Cas9 ribonucleoprotein delivery. *Physiologia Plantarum* 164, 378–384.

Aragão, F.J.L., Barros, L.M.G., de Sousa, M.V., Grossi de Sá, M.F., Almeida, E.R.P., *et al.* (1999) Expression of a methionine-rich storage albumin from the Brazil nut (*Bertholletia excelsa,* H.B.K., Lecythidaceae) in transgenic bean plants (*Phaseolus vulgaris* L., Fabaceae). *Genetics and Molecular Biology* 22, 445–449.

Artsaenko, O., Kettig, B., Fiedler, U., Conrad, U., and Düring, K. (1998) Potato tubers as a biofactory for recombinant antibodies. *Molecular Breeding* 4, 313–319.

Aslam, M., Maqbool, M.A., and Cengiz, R. (2015) Mechanisms of drought resistance. In: *Drought Stress in Maize (*Zea mays *L.)*. Springer, Cham, Switzerland, pp. 19–36.

Bachem, C.W., Speckmann, G.J., van der Linde, P.C., Verheggen, F.T., Hunt, M.D., *et al.* (1994) Antisense expression of polyphenol oxidase genes inhibits enzymatic browning in potato tubers. *Biotechnology* 12, 1101–1105.

Bartoszewski, G., Niedziela, A., Szwacka, M., and Niemirowicz-Szczytt, K. (2003) Modification of tomato taste in transgenic plants carrying a thaumatin gene from *Thaumatococcus daniellii* Benth. *Plant Breeding* 122, 347–351.

Bhattacharya, R.C., Viswakarma, N., Bhat, S.R., Kirti, P.B., and Chopra, V.L. (2002) Development of insect-resistant transgenic cabbage plants expressing a synthetic *cryIA(b)* gene from *Bacillus thuringiensis*. *Current Science* 83, 146–150.

Bhattacharya, R.C., Maheswari, M., Dineshkumar, V., Kirti, P.B., Bhat, S.R., and Chopra, V.L. (2004) Transformation of *Brassica oleracea* var. *capitata* with bacterial *bet*A gene enhances tolerance to salt stress. *Scientia Horticulturae* 100, 215–227.

Blancquaert, D., Storozhenko, S., van Daele, J., Stove, C., Visser, R.G., Lambert, W., and van der Straeten, D. (2013) Enhancing pterin and *para*-aminobenzoate content is not sufficient to successfully biofortify potato tubers and *Arabidopsis thaliana* plants with folate. *Journal of Experimental Botany* 64, 3899–3909.

Bohnert, H.J. and Jensen, R.G. (1996) Strategies for engineering water-stress tolerance in plants. *Trends in Biotechnology* 14, 89–97.

Bordas, M., Dabauza, M., Salvador, A., Roig, L.A., Serrano, R., and Moreno, V. (1997) Transfer of the yeast salt tolerance gene *HAL1* to *Cucumis melo* L. cultivars and *in vitro* evaluation of salt tolerance. *Transgenic Research* 6, 41–50.

Börnke, F., Hajirezaei, M., and Sonnewald, U. (2002) Potato tubers as bioreactors for palatinose production. *Journal of Biotechnology* 96, 119–124.

Bruening, G. and Lyons, J. (2000) The case of the FLAVR SAVR tomato. *California Agriculture* 54, 6–7.

Brummell, D.A., Watson, L.M., Pathirana, R., Joyce, N.I., West, P.J., *et al.* (2011) Biofortification of tomato (*Solanum lycopersicum*) fruit with the anticancer compound methylselenocysteine using a selenocysteine methyltransferase from a selenium hyperaccumulator. *Journal of Agricultural and Food Chemistry* 59, 10987–10994.

Butelli, E., Titta, L., Giorgio, M., Mock, H.P., Matros, A., *et al.* (2008) Enrichment of tomato fruit with health-promoting anthocyanins by expression of select transcription factors. *Nature Biotechnology* 26, 1301–1308.

Carmi, N., Salts, Y., Dedicova, B., Shabtai, S., and Barg, R. (2003) Induction of parthenocarpy in tomato via specific expression of the *rolB* gene in the ovary. *Planta* 217, 726–735.

Carr, P.A. and Church, G.M. (2009) Genome engineering. *Nature Biotechnology* 27: 1151.

Chakraborty, S., Chakraborty, N., and Datta, A. (2000) Increased nutritive value of transgenic potato by expressing a nonallergenic seed albumin gene from *Amaranthus hypochondriacus*. *Proceedings of the National Academy of Sciences USA* 97, 3724–3729.

Chi, M., Bhagwat, B., Lane, W.D., Tang, G., Su, Y., *et al.* (2014) Reduced polyphenol oxidase gene expression and enzymatic browning in potato (*Solanum tuberosum* L.) with artificial microRNAs. *BMC Plant Biology* 14:62.

Chong, D.K. and Langridge, W.H. (2000) Expression of full-length bioactive antimicrobial human lactoferrin in potato plants. *Transgenic Research* 9, 71–78.

Clark, D., Klee, H., and Dandekar, A. (2004) Despite benefits, commercialization of transgenic horticultural crops lags. *California Agriculture* 58, 89–98.

Coetzer, C., Corsini, D., Love, S., Pavek, J., and Tumer, N. (2001) Control of enzymatic browning in potato (*Solanum tuberosum* L.) by sense and antisense RNA from tomato polyphenol oxidase. *Journal of Agricultural and Food Chemistry* 49, 652–657.

Courbot, M., Willems, G., Motte, P., Arvidsson, S., and Roosens, N., *et al.* (2007) A major quantitative trait locus for cadmium tolerance in *Arabidopsis halleri* colocalizes with and a gene encoding a heavy metal ATPase. *Plant Physiology* 144, 1052–1065.

Dai, Z., Hooker, B.S., Anderson, D.B., and Thomas, S.R. (2000) Improved plant-based production of E1 endoglucanase using potato: expression optimization and tissue targeting. *Molecular Breeding* 6, 277–285.

Davuluri, G.R., van Tuinen, A., Fraser, P., Manfredonia, A., Newman, R., *et al.* (2005) Fruit-specific RNAi-mediated suppression of *DETI* enhances carotenoid and flavonoid content in tomatoes. *Nature Biotechnology* 23, 890–895.

de Lepeleire, J., Strobbe, S., Verstraete, J., Blancquaert, D., Ambach, L., *et al.* (2018) Folate biofortification of potato by tuber-specific expression of four folate biosynthesis genes. *Molecular Plant* 11, 175–188.

de Toledo Thomazella, D.P., Brail, Q., Dahlbeck, D., and Staskawicz, B. (2016) CRISPR-Cas9 mediated mutagenesis of a DMR6 ortholog in tomato confers broad-spectrum disease resistance. *Proceedings of the National Academy of Sciences USA* 118: e2026152118.

Díaz de la Garza, R., Quinlivan, E.P., Klaus, S.M., Basset, G.J., Gregory, J.F., and Hanson, A.D. (2004) Folate biofortification in tomatoes by engineering the pteridine branch of folate synthesis. *Proceedings of the National Academy of Sciences USA* 101, 13720–13725.

Díaz de la Garza, R.I.D., Gregory, J.F., and Hanson, A.D. (2007) Folate biofortification of tomato fruit. *Proceedings of the National Academy of Sciences USA* 104, 4218–4222.

Diretto, G., Al-Babili, S., Tavazza, R., Papacchioli, V., Beyer, P., and Giuliano, G. (2007) Metabolic engineering of potato carotenoid content through tuber-specific overexpression of a bacterial mini-pathway. *PLOS ONE* 2: e350.

Diretto, G., Al-Babili, S., Tavazza, R., Scossa, F., Papacchioli, V., *et al.* (2010) Transcriptional-metabolic networks in beta-carotene-enriched potato tubers: the long and winding road to the Golden phenotype. *Plant Physiology* 154, 899–912.

Ducreux, L.J., Morris, W.L., Hedley, P.E., Shepherd, T., Davies, H.V., *et al.* (2005) Metabolic engineering of high carotenoid potato tubers containing enhanced levels of β-carotene and lutein. *Journal of Experimental Botany* 56, 81–89.

Enciso-Rodriguez, F., Manrique-Carpintero, N.C., Nadakuduti, S.S., Buell, C.R., Zarka, D., and Douches, D. (2019) Overcoming self-incompatibility in diploid potato using CRISPR-Cas9. *Frontiers in Plant Science* 10: 376.

Ficcadenti, N., Sestili, S., Pandolfini, T., Cirillo, C., Rotino, G.L., and Spena, A. (1999) Genetic engineering of parthenocarpic fruit development in tomato. *Molecular Breeding* 5, 463–470.

Franceschetti, M. and Maqbool, A. (2017) Effectors of filamentous plant pathogens. *Commonalities amid Diversity* 81:e00066-16.

Fraser, P.D., Enfissi, E.M., and Bramley, P.M. (2009) Genetic engineering of carotenoid formation in tomato fruit and the potential application of systems and synthetic biology approaches. *Archives of Biochemistry and Biophysics* 483, 196–204.

Gaj, T., Gersbach, C.A., and Barbas, C.F., III (2013) ZFN, TALEN, and CRISPR/Cas-based methods for genome engineering. *Trends in Biotechnology* 31, 397–405.

Galetto, R., Duchateau, P., and Pâques, F. (2009) Targeted approaches for gene therapy and the emergence of engineered meganucleases. *Expert Opinion on Biological Therapy* 9, 1289–1303.

García-Lafuente, A., Guillamón, E., Villares, A., Rostagno, M.A., and Martínez, J.A. (2009) Flavonoids as anti-inflammatory agents: implications in cancer and cardiovascular disease. *Inflammation Research* 58, 537–552.

Gianoglio, S., Moglia, A., Comino, C., Acquadro, A., Jaime, P., *et al.* (2018) CRISPR/Cas9 knock-out of polyphenol oxidase genes for improving the quality of eggplant fruits. In: *Plant and Animal Genome XXVI, 13–17 January 2018, San Diego, California*, p. P1253.

Giorio, G., Stigliani, A.L., and d'Ambrosio, C. (2007) Agronomic performance and transcriptional analysis of carotenoid biosynthesis in fruits of transgenic HighCaro and control tomato lines under field conditions. *Transgenic Research* 16, 15–28.

Gisbert, C., Rus, A.M., Bolarín, M.C., López-Coronado, J.M., Arrillaga, I., *et al.* (2000) The yeast *HAL1* gene improves salt tolerance of transgenic tomato. *Plant Physiology* 123, 393–402.

Goggin, F.L., Shah, G., Williamson, V.M., and Ullman, D.E. (2004) Instability of *Mi*-mediated nematode resistance in transgenic tomato plants. *Molecular Breeding* 13, 357–364.

Goggin, F.L., Jia, L.L., Shah, G., Hebert, S., Williamson, V.M., and Ullman, D.E. (2006) Heterologous expression of the *Mi-1.2* gene from tomato confers resistance against nematodes but not aphids in eggplant. *Molecular Plant–Microbe Interactions* 19, 383–388.

Gorelova, V., de Lepeleire, J., van Daele, J., Pluim, D., Meï, C., *et al.* (2017) Dihydrofolate reductase/thymidylate synthase fine-tunes the folate status and controls redox homeostasis in plants. *Plant Cell* 29, 2831–2853.

Gosti, F., Beaudoin, N., Serizet, C., Webb, A.A., Vartanian, N., and Giraudat, J. (1999) ABI1 protein phosphatase 2C is a negative regulator of abscisic acid signaling. *Plant Cell* 11, 1897–1909.

Gutiérrez-Ortega, A., Sandoval-Montes, C., Olivera-Flores, T.J., Santos-Argumendo, L., and Gómez-Lim, M. (2005) Expression of functional interleukin-12 from mouse in transgenic tomato plants. *Transgenic Research* 14, 877–885.

Hamamoto, H., Suglyama, Y., Nakagawa, N., Hashida, Y., Matsunaga, Y., *et al.* (1993) A new tobacco mosaic virus vector and its use for the systemic production of angiotensin-I-converting enzyme inhibitor in transgenic tobacco and tomato. *Bio/Technology* 11, 930–932.

Hanikenne, M., Talke, I.N., Haydon, M.J., Lanz, C., Nolte, A., *et al.* (2008) Evolution of metal hyperaccumulation required *cis*-regulatory changes and triplication of HMA4. *Nature* 453, 391–395.

Hirschi, K.D., Zhen, R.G., Cunningham, K.W., Rea, P.A., and Fink, G.R. (1996) CAX1, an $H^+/Ca^{2+}$ antiporter from *Arabidopsis*. *Proceedings of the National Academy of Sciences USA* 93, 8782–8786.

Hsieh, T.H., Lee, J.T., Charng, Y.Y., and Chan, M.T. (2002a) Tomato plants ectopically expressing *Arabidopsis CBF1* show enhanced resistance to water deficit stress. *Plant Physiology* 130, 618–626.

Hsieh, T.H., Lee, J.T., Yang, P.T., Chiu, L.H., Charng, Y.Y., *et al.* (2002b) Heterology expression of the *Arabidopsis C-Repeat/Dehydration Response Element Binding Factor 1* gene confers elevated tolerance to chilling and oxidative stresses in transgenic tomato. *Plant Physiology* 129, 1086–1094.

Hsu, P.D., Lander, E.S., and Zhang, F. (2014) Development and applications of CRISPR-Cas9 for genome engineering. *Cells* 157, 1262–1278.

Imani, J., Berting, A., Nitsche, S., Schaefer, S., Gerlich, W.H., and Neumann, K.H. (2002) The integration of a major hepatitis B virus gene into cell-cycle synchronized carrot cell suspension cultures and its expression in regenerated carrot plants. *Plant Cell, Tissue and Organ Culture* 71, 157–164.

Jani, D., Meena, L.S., Rizwan-ul-Haq, Q.M., Singh, Y., Sharma, A.K., and Tyagi, A.K. (2002) Expression of cholera toxin B subunit in transgenic tomato plants. *Transgenic Research* 11, 447–454.

Jia, G.X., Zhu, Z.Q., Chang, F.Q., and Li, Y.X. (2002) Transformation of tomato with the BADH gene from *Atriplex* improves salt tolerance. *Plant Cell Reports* 21, 141–146.

Jiang, X.L., He, Z.M., Peng, Z.Q., Qi, Y., and Chen, Q. (2007) Cholera toxin B protein in transgenic tomato fruit induces systemic immune response in mice. *Transgenic Research* 16, 169–175.

Jin, R.G., Liu, Y.B., Tabashnik, B.E., and Borthakur, D. (2000) Development of transgenic cabbage (*Brassica oleracea* var. *capitata*) for insect resistance by *Agrobacterium tumefaciens*-mediated transformation. *In Vitro Cellular & Developmental Biology* 36, 231–237.

Karlovsky, P. (2011) Biological detoxification of the mycotoxin deoxynivalenol and its use in genetically engineered crops and feed additives. *Applied Microbiology and Biotechnology* 91, 491–504.

Kendziorek, M., Barabasz, A., Rudzka, J., Tracz, K., Mills, R.F., *et al.* (2014) Approach to engineer tomato by expression of *AtHMA4* to enhance Zn in the aerial parts. *Journal of Plant Physiology* 171, 1413–1422.

Kerdnaimongkol, K. and Woodson, W.R. (1999) Inhibition of catalase by antisense RNA increases susceptibility to oxidative stress and chilling injury in transgenic tomato plants. *Journal of the American Society for Horticultural Science* 124, 330–336.

Kim, J.S., Ezura, K., Lee, J., Ariizumi, T., and Ezura, H. (2019) Genetic engineering of parthenocarpic tomato plants using transient *SlIAA9* knockdown by novel tissue-specific promoters. *Scientific Reports* 9: 18871.

Kim, T.H., Böhmer, M., Hu, H., Nishimura, N., and Schroeder, J.I. (2010) Guard cell signal transduction network: advances in understanding abscisic acid, $CO_2$, and $Ca^{2+}$ signaling. *Annual Review of Plant Biology* 61, 561–591.

Ko, K., Tekoah, Y., Rudd, P.M., Harvey, D.J., Dwek, R.A., *et al.* (2003) Function and glycosylation of plant-derived antiviral monoclonal antibody. *Proceedings of the National Academy of Sciences USA* 100, 8013–8018.

Kumar, P.A., Mandaokar, A., Sreenivasu, K., Chakrabarti, S.K., Bisaria, S., *et al.* (1998) Insect-resistant transgenic brinjal plants. *Molecular Breeding* 4, 33–37.

Kumar, S., Dhingra, A., and Daniell, H. (2004) Plastid-expressed betaine aldehyde dehydrogenase gene in carrot cultured cells, roots, and leaves confers enhanced salt tolerance. *Plant Physiology* 136, 2843–2854.

Kunik, T., Salomon, R., Zamir, D., Navot, N., Zeidan, M., *et al.* (1994) Transgenic tomato plant expression the tomato yellow leaf curl virus capsid protein are resistant to the virus. *Bio/Technology* 12, 500–504.

Kwon, S.J., Hwang, E.W., and Kwon, H.B. (2004) Genetic engineering of drought resistant potato plants by co-introduction of genes encoding trehalose-6-phosphate synthase and trehalose-6-phosphate phosphatase of *Zygosaccharomyces rouxii*. *Genes and Genomics* 26, 199–206.

Le Gall, G., DuPont, M.S., Mellon, F.A., Davis, A.L., Collins, G.J., *et al.* (2003) Characterization and content of flavonoid glycosides in genetically modified tomato (*Lycopersicon esculentum*) fruits. *Journal of Agricultural and Food Chemistry* 51, 2438–2446.

Lee, J.T., Prasad, V., Yang, P.T., Wu, J.F., Ho, T.H.D., *et al.* (2003) Expression of *Arabidopsis* CBF1 regulated by an ABA/stress inducible promoter in transgenic tomato confers stress tolerance without affecting yield. *Plant, Cell & Environment* 26, 1181–1190.

Lee, Y.H., Yoon, I.S., Suh, S.C., and Kim, H.I. (2002) Enhanced disease resistance in transgenic cabbage and tobacco expressing a glucose oxidase gene from *Aspergillus niger*. *Plant Cell Reports* 20, 857–863.

Lee, Y.H., Kim, H.S., Kim, J.Y., Jung, M., Park, Y.S., *et al.* (2004) A new selection method for pepper transformation: callus-mediated shoot formation. *Plant Cell Reports* 23, 50–58.

Li, T., Liu, B., Spalding, M.H., Weeks, D.P., and Yang, B. (2012) High-efficiency TALEN-based gene editing produces disease-resistant rice. *Nature Biotechnology* 30, 390–392.

Li, X., Peng, R.H., Fan, H.Q., Xiong, A.S., Yao, Q.H., *et al.* (2005) *Vitreoscilla* hemoglobin overexpression increases submergence tolerance in cabbage. *Plant Cell Reports* 23, 710–715.

Lin, W.C., Lu, C.F., Wu, J.W., Cheng, M.L., Lin, Y.M., *et al.* (2004) Transgenic tomato plants expressing the *Arabidopsis NPR1* gene display enhanced resistance to a spectrum of fungal and bacterial diseases. *Transgenic Research* 13, 567–581.

Liu, Z.C., Park, B.J., Kanno, A., and Kameya, T. (2005) The novel use of a combination of sonication and vacuum infiltration in *Agrobacterium*-mediated transformation of kidney bean (*Phaseolus vulgaris* L.) with *lea* gene. *Molecular Breeding* 16, 189–197.

Lv, L.L., Lei, J.J., Song, M., Li, L.Y., and Cao, B.H. (2005) Study on transformation of cowpea trypsin inhibitor gene into cauliflower (*Brassica oleracea* L. var. *botrytis*). *African Journal of Biotechnology* 4, 45–49.

Ma, Y., Lin, S.Q., Gao, Y., Li, M., Luo, W.X., *et al.* (2003) Expression of ORF2 partial gene of hepatitis E virus in tomatoes and immunoactivity of expression products. *World Journal of Gastroenterology* 9, 2211–2215.

Maioli, A., Gianoglio, S., Moglia, A., Acquadro, A., Valentino, D., *et al.* (2020) Simultaneous CRISPR/Cas9 editing of three *PPO* genes reduces fruit flesh browning in *Solanum melongena* L. *Frontiers in Plant Science* 11: 1883.

Mason, H.S., Lam, D., and Arntzen, C.J. (1992) Expression of hepatitis B surface antigen in transgenic plants. *Proceedings of the National Academy of Sciences USA* 89, 11745–11749.

McGarvey, P.B., Hammond, J., Dienelt, M.M., Hooper, D.C., Fu, Z.F., *et al.* (1995) Expression of the rabies virus glycoprotein in transgenic tomatoes. *Bio/Technology* 13, 1484–1487.

McKenzie, M.J., Hunter, D.A., Pathirana, R., Watson, L.M., Joyce, N.I., *et al.* (2009) Accumulation of an organic anticancer selenium compound in a transgenic Solanaceous species shows wider applicability of the selenocysteine methyltransferase transgene from selenium hyperaccumulators. *Transgenic Research* 18, 407–424.

Møller, I.M. (2001) Plant mitochondria and oxidative stress: electron transport, NADPH turnover, and metabolism of reactive oxygen species. *Annual Review of Plant Biology* 52, 561–591.

Mor, T.S., Sternfeld, M., Soreq, H., Arntzen, C.J., and Masson, H.S. (2001) Expression of recombinant human acetylcholinesterase in transgenic tomato plants. *Biotechnology and Bioengineering* 75, 259–266.

Morris, J., Hawthorne, K.M., Hotze, T., Abrams, S.A., and Hirschi, K.D. (2008) Nutritional impact of elevated calcium transport activity in carrots. *Proceedings of the National Academy of Sciences USA* 105, 1431–1435.

Muir, S.R., Collins, G.J., Robinson, S., Hughes, S., Bovy, A., *et al.* (2001) Overexpression of petunia chalcone isomerase in tomato results in fruit containing increased levels of flavonols. *Nature Biotechnology* 19, 470–474.

Navarro-León, E., Ruiz, J.M., Graham, N., and Blasco, B. (2018) Physiological profile of CAX1a TILLING mutants of *Brassica rapa* exposed to different calcium doses. *Plant Science* 272, 164–172.

Nelson, R.S., McCormick, S.M., Delannay, X., Dube, P., Layton, J., *et al.* (1988) Virus tolerance, plant growth, and field performance of transgenic tomato plants expressing coat protein from tobacco mosaic virus. *Bio/Technology* 6, 403–409.

Neuhierl, B., Thanbichler, M., Lottspeich, F., and Böck, A. (1999) A family of *S*-methylmethionine-dependent thiol/selenol methyltransferases: role in selenium tolerance and evolutionary relation. *Journal of Biological Chemistry* 274, 5407–5414.

Nie, W.F., Wang, M.M., Xia, X.J., Zhou, Y.H., Shi, K., *et al.* (2013) Silencing of tomato *RBOH1* and *MPK2* abolishes brassinosteroid-induced $H_2O_2$ generation and stress tolerance. *Plant, Cell and Environment* 36, 789–803.

Niggeweg, R., Michael, A.J., and Martin, C. (2004) Engineering plants with increased levels of the antioxidant chlorogenic acid. *Nature Biotechnology* 22, 746–754.

Nunes, A.C.S., Kalkmann, D.C., and Aragão, F.J.L. (2009) Folate biofortification of lettuce by expression of a codon optimized chicken GTP cyclohydrolase I gene. *Transgenic Research* 18, 661–667.

Nunome, T., Fukumoto, F., Terami, F., Hanada, K., and Hirai, M. (2002) Development of breeding materials of transgenic tomato plants with a truncated replicase gene of cucumber mosaic virus for resistance to the virus. *Breeding Science* 52, 219–223.

Pandolfini, T., Molesini, B., and Spena, A. (2018) Parthenocarpy in crop plants. *Annual Plant Reviews Online* 38, 326–345.

Pâques, F. and Duchateau, P. (2007) Meganucleases and DNA double-strand break-induced recombination: perspectives for gene therapy. *Current Gene Therapy* 7, 49–66.

Park, B.J., Liu, Z.C., Kanno, A., and Kameya, T. (2005a) Genetic improvement of Chinese cabbage for salt and drought tolerance by constitutive expression of a *B. napus* LEA gene. *Plant Science* 169, 553–558.

Park, B.J., Liu, Z.C., Kanno, A., and Kameya, T. (2005b) Transformation of radish (*Raphanus sativus* L.) via sonication and vacuum infiltration of germinated seeds with *Agrobacterium* harboring a group 3 LEA gene from *B. napus*. *Plant Cell Reports* 24, 494–500.

Park, E.J., Jeknić, Z., Sakamoto, A., DeNoma, J., Yuwansiri, R., *et al.* (2004) Genetic engineering of glycinebetaine synthesis in tomato protects seeds, plants, and flowers from chilling damage. *Plant Journal* 40, 474–487.

Park, S., Kim, C.K., Pike, L.M., Smith, R.H., and Hirschi, K.D. (2004) Increased calcium in carrots by expression of an *Arabidopsis* $H^+/Ca^{2+}$ transporter. *Molecular Breeding* 14, 275–282.

Park, S.H., Jun, S.S., An, G.H., Hong, Y.N., and Park, M.C. (2003) A comparative study on the protective role of trehalose and LEA proteins against abiotic stresses in transgenic Chinese cabbage (*Brassica campestris*) overexpressing *CaLEA* or *otsA*. *Journal of Plant Biology* 46, 277–286.

Passelegue, E. and Kerlan, C. (1996) Transformation of cauliflower (*Brassica oleracea* var. *botrytis*) by transfer of cauliflower mosaic virus genes through combined cocultivation with virulent and avirulent strains of *Agrobacterium*. *Plant Science* 113, 79–89.

Pei, Z.M., Ghassemian, M., Kwak, C.M., McCourt, P., and Schroeder, J.I. (1998) Role of farnesyltransferase in ABA regulation of guard cell anion channels and plant water loss. *Science* 282, 287–290.

Peñarrubia, L., Kim, R., Giovannoni, J., Kim, S.H., and Fischer, R.L. (1992) Production of the sweet protein monellin in transgetic plants. *Bio/Technology* 10, 561–564.

Pilon-Smits, E.A.H. and LeDuc, D.L. (2009) Phytoremediation of selenium using transgenic plants. *Current Opinion in Biotechnology* 20, 207–212.

Pogrebnyak, N., Markley, K., Smirnov, Y., Brodzik, R., Bandurska, K., *et al.* (2006) Collard and cauliflower as a base for production of recombinant antigens. *Plant Science* 171, 677–685.

Praveen, S., Kushwaha, C.M., Mishra, A.K., Singh, V., Jain, R.K., and Varma, A. (2005) Engineering tomato for resistance to tomato leaf curl disease using viral *rep* gene sequences. *Plant Cell, Tissue and Organ Culture* 83, 311–318.

Ran, F.A., Hsu, P.D., Wright, J., Agarwala, V., Scott, D.A., and Zhang, F. (2013) Genome engineering using the CRISPR-Cas9 system. *Nature Protocols* 8, 2281–2308.

Redondo, P., Prieto, J., Munoz, I.G., Alibés, A., Stricher, F., *et al.* (2008) Molecular basis of xeroderma pigmentosum group C DNA recognition by engineered meganucleases. *Nature* 456, 107–111.

Regierer, B., Fernie, A.R., Springer, F., Perez-Melis, A., Leisse, A., *et al.* (2002) Starch content and yield increase as a result of altering adenylate pools in transgenic plants. *Nature Biotechnology* 20, 1256–1260.

Ribeiro, A.P.O., Pereira, E.J.G., Galvan, T.L., Picanco, M.C., Picoli, E.A.T., *et al.* (2006) Effect of eggplant transformed with oryzacystatin gene on *Myzus persicae* and *Macrosiphum euphorbiae*. *Journal of Applied Entomology* 130, 84–90.

Richter, U., Sonnewald, U., and Dräger, B. (2007) Calystegines in potatoes with genetically engineered carbohydrate metabolism. *Journal of Experimental Botany* 58, 1603–1615.

Robak, J. and Gryglewski, R. (1996) Bioactivity of flavonoids. *Polish Journal of Pharmacology* 48, 555–564.

Römer, S., Fraser, P.D., Kiano, J.W., Shipton, C.A., Misawa, N., *et al.* (2000) Elevation of the provitamin A content of transgenic tomato plants. *Nature Biotechnology* 18, 666–669.

Rosa, C., Kuo, Y.W., Wuriyanghan, H., and Falk, B.W. (2018) RNA interference mechanisms and applications in plant pathology. *Annual Review of Phytopathology* 56, 581–610.

Rosales-Mendoza, S., Soria-Guerra, R.E., Olivera-Flores, M.T.D., López-Revilla, R., Argüello-Astorga, G.R., *et al.* (2007) Expression of *Escherichia coli* heat-labile enterotoxin b subunit (LTB) in carrot (*Daucus carota* L.). *Plant Cell Reports* 26, 969–976.

Rosales-Mendoza, S., Soria-Guerra, R.E., López-Revilla, R., Moreno-Fierros, L., and Alpuche-Solís, A.G. (2008) Ingestion of transgenic carrots expressing the *Escherichia coli* heat-labile enterotoxin B subunit protects mice against cholera toxin challenge. *Plant Cell Reports* 27, 79–84.

Rotino, G.L., Perri, E., Zottini, M., Sommer, H., and Spena, A. (1997) Genetic engineering of parthenocarpic plants. *Nature Biotechnology* 15, 1398–1401.

Roy, R., Purty, R.S., Agrawal, V., and Gupta, S.C. (2006) Transformation of tomato cultivar 'Pusa Ruby' with *bspA* gene from *Populus tremula* for drought tolerance. *Plant Cell, Tissue and Organ Culture* 84, 55–67.

Rus, A., Estan, M., Gisbert, C., Garcia-Sogo, B., Serrano, R., *et al.* (2001) Expressing the yeast *HAL1* gene in tomato increases fruit yield and enhances $K^+/Na^+$ selectivity under salt stress. *Plant, Cell and Environment* 24, 875–880.

Sagor, G.H., Berberich, T., Tanaka, S., Nishiyama, M., Kanayama, Y., *et al.* (2016) A novel strategy to produce sweeter tomato fruits with high sugar contents by fruit-specific expression of a single bZIP transcription factor gene. *Plant Biotechnology Journal* 14, 1116–1126.

Sandhu, J.S., Krasnyanski, S.F., Domier, L.L., Korban, S.S., Osadjan, M.D., and Buetow, D.E. (2000) Oral immunization of mice with transgenic tomato fruit expressing respiratory syncytial virus-F protein induces a systemic immune response. *Transgenic Research* 9, 127–135.

Sawahel, W.A. (2002) The production of transgenic potato plants expressing human alpha-interferon using lipofectin-mediated transformation. *Cellular and Molecular Biology Letters* 7, 19–30.

Scheller, J., Gührs, K.H., Grosse, F., and Conrad, U. (2001) Production of spider silk proteins in tobacco and potato. *Nature Biotechnology* 19, 573–577.

Schijlen, E.G., de Vos, C.H., Martens, S., Jonker, H.H., Rosin, F.M., *et al.* (2007) RNA interference silencing of chalcone synthase, the first step in the flavonoid biosynthesis pathway, leads to parthenocarpic tomato fruits. *Plant Physiology* 144, 1520–1530.

Schroeder, J.I., Allen, G.J., Hugouvieux, V., Kwak, J.M., and Waner, D. (2001) Guard cell signal transduction. *Annual Review of Plant Biology* 52, 627–658.

Siemianowski, O., Mills, R.F., Williams, L.E., and Antosiewicz, D.M. (2011) Expression of the $P_{1B}$-type ATPase *AtHMA4* in tobacco modifies Zn and Cd root to shoot partitioning and metal tolerance. *Plant Biotechnology Journal* 9, 64–74.

Silva, G., Poirot, L., Galetto, R., Smith, J., Montoya, G., *et al.* (2011) Meganucleases and other tools for targeted genome engineering: perspectives and challenges for gene therapy. *Current Gene Therapy* 11, 11–27.

Singh, M., Kumar, J., Singh, S., Singh, V.P., and Prasad, S.M. (2015) Roles of osmoprotectants in improving salinity and drought tolerance in plants: a review. *Reviews in Environmental Science and Bio/Technology* 14, 407–426.

Speirs, J., Lee, E., Holt, K., Kim, Y.D., Scott, N.S., *et al.* (1998) Genetic manipulation of alcohol dehydrogenase levels in ripening tomato fruit affects the balance of some flavor aldehydes and alcohols. *Plant Physiology* 117, 1047–1058.

Sun, H.J., Kataoka, H., Yano, M., and Ezura, H. (2007) Genetically stable expression of functional miraculin, a new type of alternative sweetener, in transgenic tomato plants. *Plant Biotechnology Journal* 5, 786–777.

Szegletes, Z., Erdei, L., Tari, I., and Cseuz, L. (2000) Accumulation of osmoprotectants in wheat cultivars of different drought tolerance. *Cereal Research Communications* 28, 403–410.

Tseng, M.J., Liu, C.W., and Yiu, J.C. (2007) Enhanced tolerance to sulfur dioxide and salt stress of transgenic Chinese cabbage plants expressing both superoxide dismutase and catalase in chloroplasts. *Plant Physiology and Biochemistry* 45, 822–833.

Ueta, R., Abe, C., Watanabe, T., Sugano, S.S., Ishihara, R., *et al.* (2017) Rapid breeding of parthenocarpic tomato plants using CRISPR/Cas9. *Scientific Reports* 7: 507.

Upadhyaya, D.C., Bagri, D.S., Upadhyaya, C.P., Kumar, A., Thiruvengadam, M., and Jain, S.K. (2020) Genetic engineering of potato (*Solanum tuberosum* L.) for enhanced α-tocopherols and abiotic stress tolerance. *Physiologia Plantarum* 173, 116–128.

van Eck, J. (2018) Genome editing and plant transformation of solanaceous food crops. *Current Opinion in Biotechnology* 49, 35–41.

Waller, J.C., Akhtar, T.A., Lara-Núñez, A., Gregory, J.F. III, McQuinn, R.P., Giovannoni, J.J., and Hanson, A.D. (2010) Developmental and feedforward control of the expression of folate biosynthesis genes in tomato fruit. *Molecular Plant* 3, 66–77.

Wang, C., Chin, C.K., Ho, C.T., Hwang, C.F., Polashock, J.J., and Martin, C.E. (1996) Changes of fatty acids and fatty acid-derived flavor compounds by expressing the yeast α-9 desaturase gene in tomato. *Journal of Agricultural and Food Chemistry* 44, 3399–3402.

Wang, S., Zhang, S., Wang, W., Xiong, X., Meng, F., and Cui, X. (2015) Efficient targeted mutagenesis in potato by the CRISPR/Cas9 system. *Plant Cell Reports* 34, 1473–1476.

Wang, Y., Wisniewski, M., Meilan, R., Cui, M., and Fuchigami, L. (2006) Transgenic tomato (*Lycopersicon esculentum*) overexpressing *cAPX* exhibits enhanced tolerance to UV-B and heat stress. *Journal of Applied Horticulture* 8, 87–90.

Weremczuk, A., Barabasz, A., Ruszczyńska, A., Bulska, E., and Antosiewicz, D.M. (2016) Determination the usefulness of *AhHMA4p1::AhHMA4* expression in biofortification strategies. *Water, Air, and Soil Pollution* 227: 186.

Whitham, S., McCormick, S., and Baker, B. (1996) The *N* gene of tobacco confers resistance to tobacco mosaic virus in transgenic tomato. *Proceedings of the National Academy of Sciences USA* 93, 8776–8781.

Xing, J.S. and Chin, C.K. (2000) Modification of fatty acids in eggplant affects its resistance to *Verticillium dahliae*. *Physiological and Molecular Plant Pathology* 56, 217–225.

Yeo, E.T., Kwon, H.B., Han, S.E., Lee, J.T., Ryu, J.C., and Byu, M. (2000) Genetic engineering of drought resistant potato plants by introduction of the trehalose-6-phosphate synthase (*TPS1*) gene from *Saccharomyces cerevisiae*. *Molecules and Cells* 10, 263–268.

Yin, Z., Malinowski, R., Ziółkowska, A., Sommer, H., Pląder, W., and Malepszy, S. (2006) The *DefH9-iaaM*-containing construct efficiently induces parthenocarpy in cucumber. *Cellular & Molecular Biology Letters* 11, 279–290.

Yoon, J.H. and Baek, S.J. (2005) Molecular targets of dietary polyphenols with anti-inflammatory properties. *Yonsei Medical Journal* 46, 585–596.

Zamore, P.D. (2004) Plant RNAi: how a viral silencing suppressor inactivates siRNA. *Current Biology* 14, R198–R200.

Zeh, M., Casazza, A.P., Kreft, O., Roessner, U., Bieberich, K., *et al.* (2001) Antisense inhibition of threonine synthase leads to high methionine content in transgenic potato plants. *Plant Physiology* 127, 792–802.

Zeitlin, L., Olmsted, S.S., Moench, T.R., Co, M.S., Martinell, B.J., *et al.* (1998) A humanized monoclonal antibody produced in transgenic plants for immunoprotection of the vagina against genital herpes. *Nature Biotechnology* 16, 1361–1364.

Zhang, H.X. and Blumwald, E. (2001) Transgenic salt-tolerant tomato plants accumulate salt in foliage but not in fruit. *Nature Biotechnology* 19, 765–768.

Zhang, J., Nguyen, H.T., and Blum, A. (1999) Genetic analysis of osmotic adjustment in crop plants. *Journal of Experimental Botany* 50, 291–302.

Zhang, T., Liang, J., Wang, M., Li, D., Liu, Y., *et al.* (2019) Genetic engineering of the biosynthesis of glycinebetaine enhances the fruit development and size of tomato. *Plant Science* 280, 355–366.

Zhi, J., Liu, X., Li, D., Huang, Y., Yan, S., *et al.* (2020) CRISPR/Cas9-mediated *SlAN2* mutants reveal various regulatory models of anthocyanin biosynthesis in tomato plant. *Plant Cell Reports* 39, 799–809.

Zhou, J., Wang, J., Li, X., Xia, X.J., Zhou, Y.H., *et al.* (2014) $H_2O_2$ mediates the crosstalk of brassinosteroid and abscisic acid in tomato responses to heat and oxidative stresses. *Journal of Experimental Botany* 65, 4371–4383.

Zhu, J.K. (2002) Salt and drought stress signal transduction in plants. *Annual Review of Plant Biology* 53, 247–273.

Zhu, Y.X., Ouyang, W.J., Zhang, Y.F., and Chen, Z.L. (1996) Transgenic sweet pepper plants from *Agrobacterium* mediated transformation. *Plant Cell Reports* 16, 71–75.

# 17 Biotechnological Interventions for Vegetable Crop Improvement

**Manisha Mangal*, Arpita Srivastava, Nikita Baliyan, Priti Upadhyay and T.S. Aruna**

*Division of Vegetable Science, ICAR-Indian Agricultural Research Institute, New Delhi, India*

**Abstract**

Vegetable crops play a crucial role in global food security and nutrition, necessitating continuous efforts to enhance their productivity, quality, and resilience against biotic and abiotic stresses. Vegetable crop improvement relies heavily on the utilization of molecular markers, mapping populations, genome-wide association studies (GWAS), and quantitative trait locus (QTL) mapping. This chapter focuses on recent developments in these molecular approaches for enhancing vegetable crop breeding programs. Molecular markers have revolutionized genetic analysis by enabling the precise identification and tracking of desirable traits in vegetable crops. Their high-throughput nature and cost-effectiveness have facilitated marker-assisted selection (MAS), accelerating the breeding process for traits such as disease resistance, yield, and quality. Mapping populations, including biparental populations (e.g. recombinant inbred lines, $F_2$ populations) and multi-parental populations (e.g. nested association mapping populations), serve as valuable genetic resources for dissecting complex traits through linkage mapping and QTL analysis. These populations facilitate the identification of genomic regions associated with target traits, providing insights into the genetic architecture underlying trait variation. GWAS leverages natural allelic diversity across diverse germplasm collections to identify genomic regions associated with complex traits in vegetable crops. By analyzing large-scale single-nucleotide polymorphism data, GWAS enables the identification of candidate genes and molecular markers linked to important agronomic traits, offering valuable information for MAS and crop improvement. QTL mapping remains a powerful tool for dissecting the genetic basis of quantitative traits in vegetable crops. By utilizing mapping populations and molecular markers, QTL mapping enables the precise localization and validation of QTLs controlling traits such as yield, stress tolerance, and nutritional quality. Integration of QTL mapping with other omics approaches enhances our understanding of trait variation at the molecular level. In conclusion, molecular markers, mapping populations, GWAS, and QTL mapping constitute essential tools for accelerating vegetable crop improvement. Their integration with advanced genomic technologies and bioinformatics tools holds immense promise for enhancing breeding efficiency, accelerating trait introgression, and developing improved cultivars tailored to meet the demands of modern agriculture. Continued research efforts aimed at refining these molecular approaches and their integration into breeding pipelines are essential for sustaining global food security and agricultural sustainability in the face of evolving challenges.

**Keywords:** embryo rescue, genome sequencing, genome editing, haploids, molecular marker, QTL mapping, transgenics

*Email: manishamangal@rediffmail.com


DOI: 10.1079/9781800627123.0017

## 17.1 Introduction

The world's population has increased to approximately 7 billion in the last 50 years and the alarming rate at which it is growing indicates that it would reach 10 billion by 2050 (United Nations, n.d.). Therefore, the production of food needs to be enhanced to ensure global food and nutritional security. Vegetables are rich sources of vitamins, minerals, antioxidants, and dietary fibers, ensuring good health, and are a principal component of nutrition worldwide. At the global level, the area under vegetable cultivation is growing annually at a rate of 4.12%, while vegetable production is increasing at a rate of 6.48%. The mean productivity is 15.49 million t $ha^{-1}$, which is quite low (Swarnapriya, 2020). The major cause of low productivity is the various stresses that vegetables crops have to endure during their life cycle.

Stress in plants refers to external conditions that adversely affect growth, development, or productivity of plants (Verma *et al.*, 2013). In reaction to stress, a variety of biological processes are impacted, including changes in cellular metabolism, gene expression, growth rates, and crop yields. When there are abrupt changes in the environment, a plant becomes stressed. However, a species of plant that is tolerant to stress becomes accustomed to that particular stress after being exposed to it for a period of time.

Abiotic and biotic stresses are the two main classifications of plant stress. Biotic stress is brought on by attacks from living things—weeds, insects, nematodes, bacteria, fungi, viruses, and nematodes. These deprive the host plant of nutrients, which causes the host to die. Numerous biotic factors influence the host plant's ability to synthesize food. Many biotic agents have an impact on photosynthesis in the host plant, such as viral infections and chewing insects, which diminish the rate of photosynthesis per leaf area. One of the main reasons for pre- and postharvest losses is biological. Plants can withstand biotic stressors even if they lack an adaptive immune system by adapting to certain sophisticated mechanisms. In response to biotic stressors, the plant's genome encodes a number of resistance genes.

Abiotic stress is entirely distinct from biotic stress and is imposed on plants by non-living elements such as salinity, sunlight, temperature, cold, floods, and drought. These elements can be physical or chemical and have a negative impact on growth in terms of development, yield, and seed quality of crops and other plants. Stresses lead to a significant deviation from the plant's typical life cycle. Plants that are under stress go through three main response phases: the alarm phase (beginning of the stress), the resistance phase (activation of defense mechanisms), and the fatigue phase (loss caused by the stress) (Larcher, 2003). The ability of the plant to survive stress brought on by biotic or abiotic stimuli is referred to as resistance, which is a relative quality. Due to greater susceptibility of vegetable crops to these factors, it becomes hard to achieve the maximum yield potential. Indiscriminate use of pesticides for controlling diseases, insect pests, and nematodes is hazardous to health and the environment. Considering the growing awareness about residue-free vegetables, and reduced cost of production and export of fresh/canned vegetables, the current emphasis is on evolving varieties/hybrids that are resistant/tolerant to biotic and abiotic stresses so that the use of plant-protection chemicals is kept to a minimum. Moreover, the production of varieties/hybrids with inherent resistance/tolerance will always have a greater benefit over chemical control, as resistant varieties impede epidemics and maintain the natural balance. Therefore, there is a constant emphasis on breeding disease and pest-resistant cultivars in vegetables (Dhall, 2015).

## 17.2 Molecular Breeding Approaches to Combat Biotic and Abiotic Stresses

As conventional approaches for plant breeding are slow, costly, and labor-intensive, a number of tools and approaches have been developed and deployed in genetics and genomics that have enhanced the efficiency of plant-breeding programs. Thus, for a variety of crop species, including various vegetable species, whole-genome sequences, genetic linkage maps, marker assays, and molecular markers have been generated and published (Singh and Lebeda, 2007). Depending on the resources available and the complexity of the species genetics and breeding, a variety of new techniques are being used to improve vegetable crops, including marker-assisted selection (MAS) (Collard and Mackill, 2008), marker-assisted backcrossing (Collard and Mackill, 2008), marker-assisted recurrent selection (Charmet *et al.*, 1999), and genomic selection (Heffner *et al.*, 2009). Linkage mapping (Tanksley, 1993), genome-wide association studies (GWAS) (Thornsberry *et al.*, 2001), nested association mapping (NAM) (Tian *et al.*, 2011), and multi-parent advanced generation inter-cross (MAGIC) populations (Cavanagh *et al.*, 2008) are some of the mapping methods that have been developed for the detection of genetic variations. Until recently, major cereal crops were the main beneficiaries of these cutting-edge genetic and genomic tools and approaches, but they are now progressively making their way into vegetable genetics and breeding. In the sections that follow, efforts have been made to provide a thorough understanding of the main marker systems, as well as the creation of significant mapping populations for molecular breeding in vegetable crops.

### 17.2.1 Molecular markers

The foundation of traditional genetics and plant breeding is made up of molecular markers, which are crucial in selection and breeding for any given feature. Construction of genetic linkage maps, mapping of quantitative trait loci (QTLs), and fine mapping of the precise genes were all labor- and time-intensive operations as conventional markers were challenging to design and not very durable. The production of large-scale molecular markers such as simple sequence repeats (SSRs), insertions-deletions (indels), and single-nucleotide polymorphisms (SNPs) has become very straightforward with the introduction of modern sequencing tools. These markers not only made it easier to map target genes and create high-density genetic maps for crop breeding but also made it easier to find genetic variation linked to desirable traits.

The varieties and landraces of various vegetable crops are being characterized and fingerprinted using DNA molecular markers, and the accessions are being screened for specific biotic and abiotic stress tolerance and nutritional properties. Furthermore, the accurate mapping and tagging of the relevant gene with the appropriate molecular marker are exclusively responsible for the success of MAS and MAB in desired crops. For several vegetable crops, these molecular tags necessary for MAS and MAB have been produced utilizing various molecular markers.

MAS can be carried out in early-segregating populations and at early stages of plant growth after molecular markers closely associated with desirable features have been established. In order to create varieties with more desirable features, MAS can also be used to pyramid the main genes, including genes for biotic and abiotic stress resistance. The development of DNA markers has also sped up backcross breeding initiatives in a variety of vegetable crops. The sections that follow address several types of markers.

#### *Genetic markers*

A genetic marker is a gene or DNA sequence that controls a certain gene or trait and has a known chromosome position. Genetic markers serve as signs or flags, as they are closely related to the target gene. The two main categories of genetic marker are classical markers and DNA/molecular markers.

CLASSICAL MARKERS. There are three types of classical marker:

**1.** *Morphological markers.* These are related to phenotypic traits of plant/plant parts such as size, shape, color, surface of various plant parts, disease response, photoperiod sensitivity, and other visual traits.
**2.** *Cytological markers.* These are related to variations in the chromosome morphology such as chromosome size and number, meiotic behavior of chromosomes, etc.
**3.** *Biochemical markers.* These are related to variations in the amino acid and protein banding patterns, which includes allelic variants of enzymes called isozymes, monoterpenes, and protein-based markers.

DNA MARKERS. Molecular markers are sequences of nucleotide, and differences or variations in different individuals are reflected by variations or polymorphisms present between their nucleotide sequences. The basis of this variation or polymorphism can be insertions, deletions, point mutations, duplications, or translocations.

However, they do not necessarily affect the activity of the gene. The fact that morphological and biochemical markers are influenced by the environment and the developmental stage of the plant and are limited in number renders them unsuitable for practical breeding applications. In contrast, as DNA markers are unlimited in number and are not influenced by environmental factors, they are considered ideal for attaining breeding objectives and are the widely used type of marker. In general, an ideal DNA marker system should have the following properties: (i) easily available and identifiable in nature; (ii) evenly distributed throughout genome; (iii) high polymorphism and reproducibility; (iv) co-dominant in inheritance; and (v) cost effective and time saving.

CLASSIFICATION OF DNA MARKERS ACCORDING TO THEIR METHOD OF DETECTION. The plant species to be researched, the purpose of the research, and the availability of the necessary resources are taken into account when choosing a molecular marker to utilize (Amiteye, 2021). According to their technique of detection, the three main types of marker that are frequently used for molecular breeding are:

- hybridization-based markers, e.g. restriction fragment length polymorphism (RFLP);
- polymerase chain reaction (PCR)-based markers, e.g. random amplified polymorphic DNA (RAPD), amplified fragment length polymorphism (AFLP), sequence-characterized amplified region (SCAR), cleaved amplified polymorphic sequence (CAPS), SSRs, and inter-simple sequence repeats (ISSRs); and
- DNA sequence-based markers, e.g. SNPs.

*RFLP.* In RFLPs, the polymorphism is based on variation in DNA sequence length after the genomic DNA has been cut by a particular restriction enzyme at random but precise recognition sites, specific to that restriction enzyme, following hybridization with a DNA probe (Konzen *et al.*, 2017).

RFLPs are co-dominant markers. The restriction enzymes utilized will cut the DNA of one individual but not that of another if the sample has a SNP at the restriction site, leading to the production of restriction fragments with variable lengths. When the probes are radioactively or chemically tagged, autoradiography or chemiluminesence can used to identify the hybridized result (the target sample and probe). The key advantages of RFLP are that it is co-dominant in nature, does not require prior sequence-specific knowledge, and does not require PCR. The main drawbacks are that the technique is difficult and time consuming, requires a huge quantity of extremely pure DNA, and requires autoradiography skills and a steady supply of probes. Additionally, the very low assay efficiency of RFLP restricts the use of this technology in MAS.

*RAPD.* In this marker system, genomic DNA is amplified by PCR using a single, short decamer primer of arbitrary nucleotide sequence. Both of the primer annealing sites are similar to each other and in opposite directions. The RAPD profile varies due to changes in the primer-binding sites, which result in bands appearing or disappearing.

Arbitrarily-primed PCR, DNA amplification fingerprinting, and multiple arbitrary amplicon profiling are alternative names for RAPD. The advantages of this marker system are the tiny amount of DNA needed for the non-radioactive assay used and the lack of need for specialized probe libraries. Additionally, it does not really matter how good the DNA is as long as it can be efficiently and quickly screened for DNA sequences based on polymorphisms at numerous loci. However, the system's dominant nature, susceptibility to changes in PCR conditions, and potential for the formation of non-parental bands in the progeny of known pedigree make it necessary to utilize it with extreme caution.

*AFLP.* AFLP markers were designed to overcome the limitations of both RAPD and RFLP techniques (Vos *et al.*, 1995). This technique combines both RFLP and PCR techniques wherein the DNA is first digested and then amplified by PCR (Lynch and Walsh, 1998). In this marker system, two types of restriction enzyme, a frequent cutter and a rare cutter, are used for digesting the DNA. Each end of the resulting fragments is ligated with adapters in such a fashion that one adapter is specific for the rare cutter and the other for the frequent cutter. The primers used are complementary to the adaptor sequence, the restriction site sequence, and a few nucleotides inside the restriction site fragments.

The primers are designed such that only those fragments will be amplified that are cut at one end with the frequent cutter and at the other end with the rare cutter. Differences in the lengths of the PCR-amplified fragments are visualized on gel or capillary-based platforms. AFLP markers do not require any prior sequence information, are plentiful, and are cost effective. With the help of a gel scanner, AFLP markers can also discriminate between heterozygotes and homozygotes. Disadvantages include the high cost and requirement for sequencing experience.

*CAPS.* Target DNA in the CAPS marker system is digested using restriction enzymes following PCR amplification (Jarvis *et al.*, 1994; Michaels and Amasino, 1998). The constrained DNA samples are viewed on an acrylamide or agarose gel. Because of SNPs or indels, the target DNA's restriction enzyme recognition sequence changes or disappears, resulting in variations in the length of the digested DNA. The CAPS primers are created using either cloned RAPD bands, cDNA sequences, or sequence data found in genomic databases. CAPS are co-dominant markers (Weiland and Yu, 2003; Spaniolas *et al.*, 2006).

*SCAR.* By converting RAPD markers into SCAR markers, their dependability can be increased. Primer lengths for SCAR markers range from 15 to 30 nt and are created from cloned RAPD segments connected to an important characteristic. These markers are more repeatable than RAPDs as a longer PCR primer is used. These single-locus co-dominant markers are typically used for physical mapping.

*SSR.* SSRs, also called known as microsatellites (Litt and Luty, 1989) are short tandem repeats and simple sequence length polymorphisms (Schlotteroeretal, 1991). These are found abundantly as tandem repeat motifs of 1–6 nt in the genomes of various organisms (Beckmann and Weber, 1992). SSRs come in a variety of nucleotide forms, including mononucleotides (A), dinucleotides (GT), trinucleotides (ATT), tetranucleotides (ATCG), pentanucleotides (TAATC), and hexanucleotides (TGTGCA) (Weber, 1990). These are not only present throughout the nuclear genome but also in the chloroplast and mitochondria (Provan *et al.*, 2001; Rajendrakumar *et al.*, 2007). Microsatellite regions with different numbers of repeat units exhibit polymorphism, which can be identified using PCR (Kalia *et al.*, 2011). Because the flanking sequences of the SSR are preserved, they are used to create primers to amplify these markers. SSR markers are widely employed in plant mapping studies because they are co-dominant, highly repeatable, and dispersed widely throughout the genome (Kalia *et al.*, 2011). These markers are straightforward and simple to use, evenly distributed across the genome, extremely dependable and resilient, and have a lot of alleles per locus. Drawbacks include the fact that they are not present in all crop plants, the need for prior knowledge of DNA sequences, and the fact that their use is expensive and time consuming.

*ISSR.* ISSRs refer to the area between tandem repeat sequences or the hypervariable non-repeat region. Variation in the region between the repetitive sequences is the genetic basis for the polymorphism. For multilocus amplification, a single primer is used as both the forward and reverse primer. While SSRs are co-dominant markers, ISSRs are dominant markers.

*SNP.* Single base-pair changes present in the genome sequence of an individual are known as SNPs. Polymorphisms are observed among organisms due to addition, deletion, or substitution of nucleotide sequences, and are usually biallelic in nature (the presence of only two variants). In SNPs, nucleotides substitution can be in the form of transitions (C/T or G/A) or transversions (C/G, A/T, C/A, or T/G). The simplest and most common markers are provided by SNPs. These markers are widely distributed throughout the genome and can appear with varying frequencies in intergenic, non-coding, and coding regions (Xu, 2010). SNP genotyping is accomplished using a variety of allelic discrimination and detection platforms and methodologies. The simplest and most straightforward approach among these is RLFP (SNP-RFLP), while the CAPS marker technique can also be used for SNP discovery. After digestion, pieces of different lengths will be formed if one allele has a restriction enzyme recognition site while the other alleles do not. Sequence information from gene bank databases is analyzed to identify SNPs. Due to the abundance of SNPs and the fact that they are the most prevalent type of polymorphism and are biallelic in nature, this marker system provides numerous advantages. They can be used to combine physical maps with genetic maps and are ideal for creating high-density genetic maps as well as haplotype tags. The two main drawbacks are the high initial cost and the need for prior sequence knowledge.

### 17.2.2 Mapping populations

Over the last few decades, the development of molecular biology has given researchers new tools for investigating plant breeding. Genotype-based selection has grown in importance as a breeding strategy for agriculturally important plants. The limitations of morphological, chromosomal, and protein markers have been eliminated by molecular markers, and they also have special genetic characteristics that make them more valuable than other genetic markers. As discussed in the previous section, the types of molecular markers and their abundance and wide dispersion across the genome together with the fact that DNA-based indicators are highly heritable, relatively simple to assess, and unaffected by the environment gives them several benefits over phenotypic markers.

This section will discuss the various breeding populations that are being used for genetic improvement of crops, the principles of QTL mapping, and types of gene mapping.

Mapping populations are those that are used to map genes or to measure the genetic separation between genetic loci that have been created by managed hybridization between genetically varied lines. Choosing a parent is often an important phase in the growth of populations. In general, parents come from the same species; however, related species can also be used for population development if there is little diversity within the species. For the characteristic for which the mapping population is being produced, polymorphism between the parents is crucial. Additionally, the parents should have significant molecular polymorphisms and be entirely homozygous. The environment in which the parents and their offspring are to be phenotyped should also be one to which they are adapted. The size of the mapping population is another important aspect in the growth of the population. The genetics of the trait being mapped, the population type utilized for mapping, the goals of the experiment, the accessibility of resources for handling a sizable population, and the amount of time available for population development all affect the size of the mapping population. The optimum population size typically ranges from 100 to 3000 depending on the aforementioned variables. However, it is usually believed that a sample size of 200–300 is adequate for accurate trait mapping.

Depending on the number of parents utilized, the amount of time needed, and the availability of resources to increase the population, many kinds of population mapping have been proposed. A population that has two parents is known as a biparental mapping population, while populations that have more than two parents are known as multi-parental mapping populations.

#### *Biparental mapping populations*

Numerous biparental mapping populations are used across the world to map characteristics in various vegetable crops.

$F_2$ POPULATION. This population has been suggested solely for oligogenic characteristics and for preliminary mapping. By selfing $F_1$ offspring created from the hybridization of chosen parents, the population is created. As a result, the population is the outcome of a single meiotic cycle, and dominant markers exhibit a 3:1 segregation ratio while co-dominant markers exhibit a 1:2:1 ratio. As robust phenotyping and genotyping are both necessary for accurate gene mapping, the genetic makeup of $F_2$ populations prevents the collection of robust phenotypic data as the $F_2$ population is not entirely homozygous and multilocation phenotyping of the population is not feasible. The major benefits of the $F_2$ population are the little time and effort needed for creation. This is the most suitable population for identifying heterosis QTLs but is not suitable for fine mapping as it is immortal in nature.

To improve the reliability of results from $F_2$ populations, another type of population, namely $F_2$-derived $F_3$ populations ($F_2$:$F_3$), has been suggested. By selfing $F_2$ individuals for one generation to create $F_3$ families, the $F_2$:$F_3$ population is created. Every $F_3$ family consists of one individual

$F_2$ progeny. While $F_2$ individuals are being genotyped, $F_3$ families can be phenotyped.

DOUBLE HAPLOIDS. This population is created by doubling the number of haploid chromosomes. Consequently, there are two sets of chromosomes that are identical in each cell. There is only one allele accessible for every gene as they are entirely homozygous genetically. In addition to gynogenesis, androgenesis is the primary method of producing haploids (as in onions). This population's main benefits are its immortality, usefulness in mapping both quantitative and qualitative features, shorter development period, effective genetic analysis, and rapid generation of markers for important traits. However, the expense associated in establishing tissue culture and growth facilities is a major barrier to its development. In adddition, double haploid populations have weak linkage and just one cycle of recombination.

BACKCROSS POPULATION. To create a backcross population, $F_1$ plants are crossed with either of the $F_1$'s parents. This population has advantages because it develops more quickly, retains elite combinations, and is useful for resistance breeding. The mortality of this population, the inability to map quantitative features, and the preference for mapping of recessive traits are some of its major drawbacks.

BACKCROSS INBRED LINES OR ADVANCED BACKCROSS LINES. These lines are created by continuously selfing the $BC_1F_1$ progeny to produce homozygous lines, and then backcrossing the $F_1$ produced from a cross between two inbred lines to one of the parents. When backcrossing $F_1$ hybrids with parents who have greater values of the target trait, backcross inbred lines (BILs) should be developed. The population's strength comes from its eternal nature, which enables years of evaluation across many situations. It works well for mapping numerical qualities. However, it is challenging to create this population as it takes a lot of time, is challenging to examine the interactions of several unrelated genes from the donor parent, and the algorithms of the majority of QTL mapping tools are not made to function with this population type.

RECOMBINANT INBRED LINES. Recombinant inbred lines (RILs) are created by repeatedly crossing $F_2$ offspring of polymorphic homozygous parents. These are selfed for at least six generations, or until all of the $F_2$ individuals are homozygous and stable. Any characteristic that varies across the parental lines used to create it can be tentatively mapped using RILs. The longevity of RILs is a major benefit, and this population may be used again to map features that differ between the parents used to create RILs. These populations support the identification of numerous loci influencing every characteristic. However, because each RIL contains potentially confusing background genetic variation, these populations are less statistically strong for studying the impact of any one locus. These populations are difficult to generate in crops with inbreeding depression and require at least six generations for their development.

NEAR-ISOGENIC LINES. By repeatedly backcrossing the $F_1$ progeny to the recurrent parents and selecting for the desired characteristic in each generation, near-isogenic lines (NILs) are created. Except for the target gene, NILs produced by backcrossing resemble recurrent parents. Due to ongoing backcrossing, NILs are entirely homozygous, making them immortal mapping populations like RILs and BILs. Functional genomics is the primary application of NILs. Their primary drawbacks, however, include the need for several generations for development, higher costs, and linkage drag. These populations are only helpful for linkage mapping, precise mapping, and map-based cloning of genes.

### *Multi-parent mapping populations*

Recent studies on the limits of biparental populations have increased the relevance of multi-parent mapping populations. These are mostly accomplished by recombining the genomes of several parents rather than just two. Greater resolution capacity for mapping QTLs, increased genetic diversity, and minimum population structure are all benefits of including more than two parents. They offer germplasm for breeding projects in addition to being a tool for mapping QTLs. NAM and MAGIC are the top two methods used to produce multi-parent mapping populations.

NAM. The efficiency of both association and linkage mapping is combined in NAM. It is used to pinpoint and analyze the genetic layout of complex characteristics. Initially proposed in corn (Yu and Buckler, 2006; Yu *et al.*, 2008), approximately 5000 RILs were produced by crossing the homozygous founding parent (B73 line) with 25 lines, which covered a large portion of the domesticated corn gene pool (McMullen *et al.*, 2009). NAM involves phenotyping for a variety of desirable qualities, followed by thorough sequencing and genotyping of numerous founders and parents. A genome-wide analysis is then conducted to determine or comprehend the relationship between phenotypic and genotypic data.

MAGIC. Recently, the idea of MAGIC populations was suggested to enable the high-resolution mapping of quantitative features. MAGIC designs blend the genetic makeup of all the founders in a single line by repeatedly intercrossing numerous inbred founders (usually four, eight, or 16 parents) (Cavanagh *et al.*, 2008). By probing several alleles, this method offers a faster rate of recombination and improved mapping resolution compared with biparental mapping. These populations provide the aforementioned benefits mostly as a result of increased levels of intercrossing and opportunities for genomic shuffling. Genetic diversity and recombination rates in MAGIC populations are lower when compared with association mapping methods as these panels make use of a variety of varied breeding lines. However, confounding resulting from population structure based on sampling effects is a significant drawback of association mapping research (Caldwell *et al.*, 2006; Vilhjálmsson and Nordborg, 2013).

## 17.3 Applications of Molecular Markers

### 17.3.1 Mapping populations for QTL mapping

Understanding how many QTLs significantly contribute to the trait variation in a population is an important aspect for improving any defect in a crop. Quantitative trait values for a variety of crops and many kinds of polymorphic genetic markers make up the data for QTL mapping. The presence or absence of a specific molecular band at a specific marker can be expressed as a binary value of 1 or 0, as can the two marker genotypes (homozygote and heterozygote) for a backcross population derived from two inbred lines. Marker data can also be recorded digitally and are categorical. These markers can be arranged linearly or in linkage groups on chromosomes to form a genetic linkage map based on segregation analysis.

Quantitative trait data provide information on the variance of characteristics within a population, whereas marker data provide information on the segregation of a genome within a population. The QTL links the two data sets together. The segregation of QTLs, which are connected to some of the markers in the genome, is responsible for some of the phenotypic variation. As there are numerous genetic architectural characteristics, such as the number, locations, effects, and interactions of genes that impact the quantitative traits of interest, mapping QTLs is a statistical endeavor that relates variation in the quantitative trait to variation in genetic markers. The basic principles of QTL mapping will be discussed below.

#### *Principles of QTL mapping*

The basic idea behind figuring out whether a QTL is connected to a marker is to divide the mapping population into various genotypic classes based on genotypes at the marker locus and then use correlative statistics to see if individuals of one genotype differ noticeably from individuals of the other in terms of the trait being measured. In segregating mapping populations such as $F_2$-derived populations, RILs, NILs, double haploid lines, and backcross populations, QTL analysis is typically conducted. An appropriate mapping population derived from phenotypically different parents, a saturated linkage map based on molecular markers, reliable phenotypic screening of the mapping population, and an appropriate statistical package to analyze the genotypic information in conjunction with phenotypic information for QTL detection are essential components of QTL mapping. The experiment's aims, the amount of time available for population formation, and the available resources all influence the choice of mapping population. The kind of population used for QTL analysis, the genetic makeup of the target trait, and the availability of resources for managing a substantial mapping population will all influence the population's size. However, a mapping population of 200–300 individuals is sufficient to identify significant QTLs.

### Factors affecting the power of QTL mapping

The biological relevance of QTLs is not always apparent when they are identified using data from phenotyping and genotyping experiments. Major parameters influencing the effectiveness of QTL mapping include: (i) the number of genes regulating the target trait(s) and their position in the genome; (ii) the presence of genetic interactions between the QTLs found; (iii) the heritability of the trait; (iv) the type and size of the mapping population; (v) the number and coverage of markers in the linkage map; and (vi) the statistical approach utilized and significance level used for QTL mapping.

### Methods to detect QTLs

Regression analysis and the maximum likelihood approach are two statistical techniques that are frequently used for QTL determination. In QTL mapping, maximum likelihood techniques is often employed. The use of regression algorithms is simpler and the computing speed is substantially faster. However, because maximum probability takes greater computing power, it calls for specialized software. These techniques use a genetic map of the typed markers and, like analysis of variance, show if a single QTL is present close to a certain marker.

SINGLE-MARKER ANALYSIS. This approach seeks to identify the relationship between several marker classes and the phenotype of the quantitative trait. In this procedure, the population is divided into subpopulations based on the allelic segregation of certain loci and the phenotypic traits of each subpopulation are then measured. To ascertain whether or not the trait value differences between marker groups are statistically distinct, statistical techniques such as the *t*-test, analysis of variance, or simple linear regression are applied. However, single-marker analysis has limitations, including the inability to determine whether the markers are linked to one or more QTLs and the location of the QTLs. Furthermore, because they are mixed up with recombination frequencies, the impacts of the QTLs are likely to be overestimated.

SIMPLE INTERVAL MAPPING. Because this approach examines the intervals between neighboring markers rather than individual markers, it has an advantage over single-marker analysis. In simple interval mapping, each locus is taken into account individually and the model calculates the logarithm of the odds ratio to determine whether or not the particular locus is a real QTL. Therefore, this approach does away with the problem of marker and QTL recombination and subsequent QTL loss. When compared with single-marker analysis, this approach is statistically more effective.

COMPOSITE INTERVAL MAPPING. The likelihood of a single putative QTL at each site on the genome is evaluated using the simple interval mapping technique. The influence of QTLs situated elsewhere on the genome that may have interfered throughout the process is not seen. As a result, there is bias in the assessment of the position and effects of QTLs (Lander and Botstein, 1989). Ghost (non-existent) QTLs may also surface throughout this procedure (Haley and Knott, 1992; Martinez and Curnow, 1992). By scanning the complete genome at once, composite interval mapping is a common method that can be used to identify numerous QTLs that contribute to a characteristic (Lander and Botstein, 1989). Compared with single-QTL techniques, this method more correctly detects the site and impact of QTLs (Li *et al.*, 2007). Composite interval mapping utilizes a selection of marker loci as variables. In order to concentrate on the interval and remove confounding effects from other QTLs elsewhere in the genome, which is generally referred to as background noise, analysis is performed on the intervals between neighboring markers as well as extra markers unlinked to the interval markers (Li *et al.*, 2007). QTL Cartographer is the program most frequently used in composite interval mapping approaches.

### Methodological concerns in QTL analysis

The following methodological concerns apply to QTL analysis:

1. *Number of markers.* Approximately one polymorphic marker has to be found per 10 cM. As a result, the overall number of markers is influenced by the genome size.
2. *Common markers.* It is advised to have a number of shared markers to enable comparisons between various maps.
3. *Mapping population size.* Typically, larger populations are advised. However, a population of at least 150–200 is acceptable.

### 17.3.2 Association mapping

When compared with biparental mapping, association mapping (AM) saves time and offers a higher number of recombination events with better mapping resolution. As it uses natural genetic populations for study, AM makes it easier to identify more alleles. The availability of more genetic polymorphisms with a wider background allows the use of historical measured phenotypic data. The use of linkage disequilibrium-based AM can be used to circumvent the main drawbacks of QTL mapping in biparental populations. Candidate gene-based AM and GWAS are the two main subcategories of AM.

#### *Candidate gene-based AM*

This method investigates the relationship between an interesting characteristic and a DNA variation found in a potential gene. Candidate genes are those with recognized biological activities that have a direct or indirect impact on the characteristic of interest (Tabor *et al.*, 2002; Sehgal *et al.*, 2016). Based on the analysis of target traits from physiologic, chemical, and genetic research, candidate genes are chosen (Mackey, 2001). Identification of SNPs existing between lines and inside certain genes is necessary for candidate gene-based AM. The candidate gene is investigated by amplicon resequencing. Exon, promoter, and intron research is crucial to identify candidate gene SNPs. In recent years, the candidate gene method has been used successfully for the characterization and cloning of QTLs. Numerous tightly connected genes have been converted into functional indicators using this method (Lau *et al.*, 2015).

#### *GWAS*

The emergence of next-generation sequencing technology has enabled GWAS in a number of species. This potent method is mostly employed to investigate the genetics of desirable features seen in wild populations. GWAS is carried out using fixed homozygous lines that may be genotyped and phenotyped at several sites. A sizable population of up to tens of thousands of individuals is necessary for reliable QTL detection. GWAS aids in the identification of millions of SNPs and the technology-assisted increase in SNP number detection (Zargar *et al.*, 2015). Greater resolution, the capacity to examine haplotype blocks of a small size that are strongly connected with differences in quantitative traits, and high throughput are all made possible by this technology (Hindorff *et al.*, 2009; Atwell *et al.*, 2010).

### 17.3.3 Applications of molecular markers for biotic and abiotic stress tolerance genes in vegetable crops

DNA markers and QTL mapping help to identify stress tolerance genes at certain chromosomal loci (Younis *et al.*, 2020). The need for more accurate and focused research for bestowing biotic and abiotic stress tolerance with the aid of DNA marker technology, however, is driven by challenges brought on by climate change and the breakdown of resistance due to gene mutation in targeted insects/pests. We know that MAS is an efficient tool for improving biotic and abiotic resistance in plants, and a list of studies on diverse vegetable crops have been compiled in Tables 17.1 and 17.2 with details of the identified genes for various abiotic and biotic stresses in vegetable crops along with their source, molecular markers, and chromosomal locations. Figure 17.1 summarizes some of the applications of molecular markers for stress tolerance in vegetables.

## 17.4 Conclusion

Genomic technologies and methods such as next-generation sequencing, genotyping-by-sequencing, and GWAS have revolutionized plant genotyping and breeding. These technologies

**Table 17.1.** Genes for abiotic stress tolerance in various vegetable crops and their associated markers.

| Crop | Abiotic stress | Resistance genes | Source | Chromosome location | Molecular markers developed | Reference |
|---|---|---|---|---|---|---|
| Tomato | Drought | Gibberellin methyl transferase 1 | *Arabidopsis* | 5 | *MET 1* | Nir *et al.* (2014) |
| | | *ATHB* | *Arabidopsis* | 1 | ATHB-7 gene *Athb-8*, ATHB12 | Park *et al.* (2011); Mishra *et al.* (2012) |
| | | *HsfA1a* family | *Solanum lycopersicum* | 3, 4, 5, 6, 11 | *HsfA1a* | Wang *et al.* (2015) |
| | | *SlAREB1* | *S. lycopersicum* | Not known | Not known | Orellana *et al.* (2010) |
| | | Protease inhibitor gene families | *S. lycopersicum* | 3, 4, 9, 11 | Not known | Fan *et al.* (2019) |
| | Heat | *qCC-2-2*, *qHII-2-1*, *qREC-1-3*, *qCC-1-5*, *qHII-7-1*, *qHII-7-2*, *qHII-7-3*, *qHII-7-4*, *qHII-7-5*, *qHII-7-6*, *qHII-7-7*, and *qHII-7-8* | *S. lycopersicum* | 1, 2, 7 | *SSR96*, *C02M4005*, *SSR134-C01M86371*, *qCC-1-5* | Wen *et al.* (2019) |
| | | *Sp* genes | *Solanum pennellii* | 1, 3 | *qsp1* and *qsp3* | Xu *et al.* (2017); Egea *et al.* (2018) |
| | | *fln2.1_T1-2E*, *fln2.1_T2_2E fln11.1_T3_18*, *frn6.2_T1_2E frn6.2_T2_2E*, *frn2.1_T1_16*, *frn2.1_T1_17*, *frn2.1_T2_16*, *frn2.1_T2_17*, *frn2.2_T3_17*, *frs12.1_T3_18*, *fln11.2_T2*, and *frs7.1_T2_2E* | RIL population of *S. pimpinellifolium* and *S. lycopersicum* | 2, 4, 6, 7, 10, 11 | Gene-specific markers | Gonzalo *et al.* (2020) |
| | Salinity | *AlSOS1*, *SlNHX2*, *SlNHX4*, and *SlERF4* | Introgression lines of *S. lycopersicum* cv-6203 × *Solanum habrochaites* and *S. lycopersicum* M82 × *Solanum pennellii* | 2, 11 | Gene-specific markers | Ali *et al.* (2021) |
| | | *LeNHX3*, *HKT*, *SOS1*, and *NHX* genes | *F8 of S. lycopersicum and S.pimpinellifolium* L. | 1, 7 | Gene-specific markers | Asins *et al.* (2013) |

| | | | | | | |
|---|---|---|---|---|---|---|
| | Cold | *SlBBX* genes | *S. lycopersicum* | 2, 4, 5, 6, 7, 8, 9, 12 | Gene-specific markers | Bu *et al.* (2021) |
| | | Late embryogenesis abundant (LEA) protein 25 and glutaredoxin S17 | *S. lycopersicum* | 1, 2, 4, 9 | Gene-specific markers | Chand *et al.* (2021) |
| | | *LeCOR413PM2* | *S. lycopersicum* | 4, 7, 10 | SNP | Zhang *et al.* (2021) |
| Chili | Drought | *CaCIPK3* | *Capsicum annuum* L. | 1 | Gene-specific markers | Ma *et al.* (2021) |
| | | Aquaporins (CaAQPs) | *C. annuum* L. | 1 | Microsatellites | Sahitya *et al.* (2019) |
| | Heat | *Hsp70* | *C. annuum* L. | 3, 8 | *Hsp70-u2* and *Agi42* | Usman *et al.* (2019) |
| | | *CaWRKY40* | *C. annuum* L. | 1 | CAPS | Dang *et al.* (2013) |
| | Salinity | *AKT1*, *SOS1*, *SOS3*, ankyrin repeat gene family, and *CaDHN4* | *C. annuum* L. | 1, 2, 7, 8 | ESTs | Zhang *et al.* (2019); Lopez-Ortiz *et al.* (2020); Suarez *et al.* (2021) |
| | Cold | *EIN3.1* and *EIN3. 2*, *ICE1/2*, and *CaDHN4* | *C. annuum* L. | 1, 2, 6 | Gene-specific primers | Zhang *et al.* (2019); Tang *et al.* (2022) |
| Potato | Drought | Glyceraldehyde 3-phosphate dehydrogenase (GPD), dehydrins (TAS14 dehydrin and a 25 kDa protein dehydrin), 30S ribosomal protein S7 (RPS7), p23 co-chaperone, acyl-CoA-binding protein (ACBP), SUMO E2 conjugating enzyme SCE1, acid phosphatase, lipid transfer protein (LTP), RNA recognition motif (RRM)-containing protein, and two genes with unknown functions | *Solanum tuberosum* | 4, 5, 12 | Not known | Kappachery *et al.* (2013) |
| | | Mitogen-activated protein kinases (MAPKs) and LEA genes | *S. tuberosum* | 1, 2 | | Evers *et al.* (2010) |
| | | StGA2ox1 | *S. tuberosum* | Not known | Not known | Shi *et al.* (2019) |
| | Heat | HSc70, hexokinase 2, and hydrogen peroxide-producing germin-like protein (GLP) | *S. tuberosum* | 4, 6 | | Trapero-Mozos *et al.* (2018); Lehretz *et al.* (2021) |

*Continued*

**Table 17.1.** Continued.

| Crop | Abiotic stress | Resistance genes | Source | Chromosome location | Molecular markers developed | Reference |
|---|---|---|---|---|---|---|
| | Salinity | V-H+-ATPase genes, V-H$^+$-Ppase, *SlVP1* and *SlVP2* (VP1 and VP2), *SlNHX1–SlNHX4* (NHX1–NHX4), *AtHKT1*, and *StCYS1* | *S. tuberosum* | Not known | Gene-specific primers | Jaarsma *et al.* (2018); Wang *et al.* (2019); Liu *et al.* (2020) |
| | Cold | Protease inhibitors, patatins, heat-shock proteins, lipoxygenase, phospholipase A1, and leucine aminopeptidase (Lap) | *S. tuberosum* | Not known | SNP | Fischer *et al.* (2013) |
| Cucurbits | Drought | *CsATAF 1*, *CsGPA1*, and *CsVI2* | *Cucumis sativus* L. | Not known | Gene-specific primers | Wang *et al.* (2018); Yan *et al.* (2018); Chen *et al.* (2022) |
| | Heat | *qHT1.1*, *CsCaM3*, and *CsHSP70* | *C. sativus* L. | 1 | HT-1 | Li *et al.* (2016); Yu *et al.* (2018); Liu *et al.* (2021) |
| | Salinity | *qST6.2*, *Csa6G487740,Csa6G489940*, *M H$^+$-ATPase*, *SOS1*, and *SKOR* genes | *C. sativus* L. | 6 | SSR, gene-specific primers | Jiang *et al.* (2019); Liu *et al.* (2021) |
| | Cold | GRAS family, *Ch* gene, and *CsGPA1* | *C. sativus* L. | 1, 2, 3, 4, 5, 6, 7 | | Kozik (2008); Lu *et al.* (2020); Yan *et al.* (2022) |
| Brinjal | | ADC | *Solanum melongena* | Not known | Not known | |
| | Heat | *SmeHsf*, *Hsp21*, *Hsp23*, *Hsp70*, *Hsp90-5*, *Hsp18.2*, *Hsp17.9A*, *Hsp101*, *Hsp15.7*, *Hsp98.7*, *Hsp90-6*, *Hsp22.7*, *Hsp90-1*, *Hsp18.2*, *Hsp26.5*, *Hsp70-8*, *Hsp18.2*, *Hsp17.5-E*, *Hsp83A*, *Hsp18.1*, *Hsp70-7*, *Hsp18.1*, *HsfA3*, *HsfA2B*, *HsfA4A*, and HsfiA4C | *S. melongena* | Not known | Not known | Zhang *et al.* (2019); Wang *et al.* (2020) |

| | | | | | | |
|---|---|---|---|---|---|---|
| | Salinity | Halotolerance gene, *HAL1*, *AKT1*, *KAT1*, and *SOS1* | Yeast | Not known | Gene-specific primers | Kumar *et al.* (2014); Li *et al.* (2019) |
| | Cold | *SmICE1a*, miRNAs, and *SmCBF* | *S. melongena* L., *Solanum aculeatissimum* | Not known | Gene-specific primers | Yang *et al.* (2017); Zhou *et al.* (2020) |
| Cole crops | Drought | *BolTLP1* and dehydrin | *Brassica oleracea* L. var. *italica* (broccoli), *B. oleracea* | Not known | Gene-specific primers | He *et al.* (2021) |
| | Heat | *BnaCBLs*, 12 *BnaCIPKs*, and *GLYI* | *Brassica napus* | Not known | Gene-specific probes | Kourani *et al.* (2022) |
| | Salinity | *BnaCBLs*, 12 *BnaCIPKs*, and *GLYI* | *B. napus* | Not known | Gene-specific probes | Kourani *et al.* (2022) |
| | Cold | *BnaCBLs*, 12 *BnaCIPKs*, and *GLYI* | *B. napus* | Not known | Gene-specific probes | Kourani *et al.* (2022) |
| Okra | Heat | Genes encoding LEA proteins, heat-shock proteins (HSPs), and dehydrins | *Abelmoschus esculentus* L. | Not known | SNPs | Shi *et al.* (2020) |
| | Drought | Genes encoding LEA proteins and HSPs | *A. esculentus* L. | Not known | Not known | Razi *et al.* (2021) |

CAPS, cleaved amplified polymorphic sequences; EST, expressed sequence tag; RIL, recombinant inbred line; SSR, simple sequence repeat.

**Table 17.2.** Genes with a role in biotic stress tolerance in various vegetable crops with associated markers.

| Crop | Disease | | Resistance genes | Chromosomal location of the gene | Molecular markers developed | Reference |
|---|---|---|---|---|---|---|
| Tomato | Fungal | Late blight | *Ph1*, *Ph2*, *Ph3*, and *Ph5* | 7, 10, 9, 1 | SSR | Foolad and Panthee, (2012); Truong *et al.* (2013); Akbar *et al.* (2016) |
| | | *Verticillium* wilt | *Ve1*, and *Ve2* | 9 | SCAR | Kawchuk*et al.* (2001); Acciarri*et al.* (2007) |
| | | *Fusarium* wilt | *Fr1*, *I1*, *I2*, *I3*, and *I7* | 9, 7, 11, 7, 8 | CAPS, SCAR, SSR | Barillas *et al.* (2008); Mutlu *et al.* (2008); Catanzariti *et al.* (2015); Gonzalez-Cendales *et al.* (2016) |
| | | Powdery mildew | *ol-1* and *ol-2* | 6, 4 | SCAR, RAPD | Huang *et al.* (2000); de Giovanni *et al.* (2004) |
| | Bacterial | Bacterial wilt | *Bwr-6* and *Bwr-12* | 6, 12 | SNP | Kim *et al.* (2018) |
| | Viral | Tomato leaf curl | *Ty-1*, *Ty-2*, *Ty-3*, *Ty-4*, *Ty-5*, and *Ty-6* | 6, 11, 6, 3, 4, 10 | *Ty-1* (RAPD, SCAR, RFLP and CAPS), *Ty-2* (RFLP), *Ty-3* (RAPD, SCAR, RFLP and CAPS), *Ty-4* (RAPD), *Ty-5* (RAPD, SSRs), *Ty-6* (SNPs) | Zamir *et al.* (1994); Hanson *et al.* (2000); Ji *et al.* (2007a, 2008, 2009); Anbinder *et al.* (2009); Hutton and Scott (2013) |
| | | Cucumber mosaic virus | *Cmr* | 12 | RFLP | Stamova and Chetelat (2000) |
| | | Tomato spotted wilt virus | *Sw-5* and *Sw-7* | 9, 12 | *Sw-5* (RAPD, CAPS, SCAR, AFLP, indel), *Sw-7* (RFLP, SSR, SCAR) | Tanksley *et al.* (1992); Stevens *et al.* (1995); Brommonschenkel and Tanksley (1997); Garland *et al.* (2005); Dockter *et al.* (2009); Do Nascimento *et al.* (2009); Dianese *et al.* (2010); Padmanabhan *et al.* (2019) |
| | Nematodes | Root knot disease | *Mi-1* (*Mi-1.2*), *Mi-2*, *Mi-3*, *Mi-4*, *Mi-5*, *Mi-6*, *Mi-7*, *Mi-8*, *Mi-9*, and *Mi-HT* | 6 (*Mi-1*), 12 (*Mi-3*, *Mi-5*), 6 (*Mi-9*); *Mi-2*, *Mi-4*, *Mi-6*, *Mi-7*, and *Mi-8* not mapped | CAPS, SCAR, RAPD | Smith (1944); Yaghoobi *et al.* (1995); Veremis and Roberts (1996); Amiiraju *et al.* (2003); Seah *et al.* (2004); Yaghubi *et al.* (2005); Jablonska *et al.* (2007); Rani *et al.* (2009); Devran and Sogut (2010) |

| | | | | | | |
|---|---|---|---|---|---|---|
| Chili and sweet pepper | Fungal | Anthracnose | QTLs (*CaR12.2* and *CcR9*); pathogen pathotypes *Colletotrichum acutatum* and *Colletotrichum capsici* | LG12 (6.1 cM position), LG9 (10.0 cM position) | SSR | Lee *et al.* (2010) |
| | | | QTLs (*AnRGO5*, *AnRGT5*, *AnRGD5*, *AnRRO5*, *AnRRT5*, *AnRRD5*), pathogen pathotypes *Colletotrichum acutatum* | 5 | Indels, SSR, SCAR, CAPS | Sun *et al.* (2015) |
| | | | *co1*, *co2*, *co4*, and *Co5* | 2, 12, 9 | SNP | Mahasuk *et al.* (2009) |
| | | | Major QTLs (*An8.1*, *An9.1*), minor QTLs (*An7.3*, *An7.4*, *An4.1*, *An3.1*, *An3.2*), pathogen pathotype *Colletotrichum acutatum* | | | Kim *et al.* (2010) |
| | | *Phytophthora* blight | QTL and *Pc5.1* | 5 | SSR, SCAR | Mallard *et al.* (2013) |
| | | | *CaPhyto* | 5 | SSR | Wang *et al.* (2016) |
| | Bacterial | Bacterial wilt | QTL *Bw1* | 1 | SSR | Mimura *et al.* (2009) |
| | | Bacterial leaf spot | *Bs1*, *Bs2*, *Bs3*, *Bs4*, *bs5*, *bs6*, and *Bs7* | 6 (*Bs2*, *bs6*) | RAPD, SCAR, CAPS | Tai *et al.* (1999); Pierre *et al.* (2000); Csillery *et al.* (2004); Jones *et al.* (2004); Vallejos *et al.* (2010); Wai *et al.* (2015) |
| | Viral | Chili leaf curl disease | *ChiLCVD*-resistant gene | 6 | SSR | Thakur *et al.* (2020) |
| | | Cucumber mosaic virus | *Cmr1* and *Cmr2* | 2 | SNP | Kang *et al.* (2010) |
| | | Potyviruses | *pvr1* (eIF4E), *pvr2*, *pvr3*, *Pvr4* (CaNBARC322), *pvr5*, *pvr8*, *pvr6* (eIF(iso)4E), and *Pvr7* | 4 (*pvr1*, *pvr2*), 10 (*Pvr4*, *Pvr7* ), 3 (*pvr6*) | CAPS, SCAR, SSR, SNP | Kyle *et al.* (1997); Caranta *et al.* (1999); Arnedo-Andrés *et al.* (2002); Ruffel *et al.* (2002); Andrés *et al.* (2004); Yeam *et al.* (2005); Dato *et al.* (2015); Venkatesh *et al.* (2018) |

*Continued*

**Table 17.2.** Continued.

| Crop | Disease | | Resistance genes | Chromosomal location of the gene | Molecular markers developed | Reference |
|---|---|---|---|---|---|---|
| | Nematodes | Root knot disease | *Me* loci, *Me1*, *Mech2*, *Me3*, *Me4*, and *Me7* | 9 | CAPS, SCAR, SSCP | Djian-Caporalino *et al.* (2007); Uncu *et al.* (2015); Pinar *et al.* (2016); Guo *et al.* (2017); Wang *et al.* (2018); Changkwian *et al.* (2019) |
| | Major insect pests | Whitefly, thrips, mites, fruit borer, aphids | | | | |
| Brinjal | Fungal | *Fusarium* wilt | *Fomg* | – | $SCAR_{426}$ | Mutlu *et al.* (2008) |
| | | *Verticillium* wilt | *Ve* | 9 | $SIVR_{844}$ | Liu *et al.* (2014) |
| | Bacterial | Bacterial wilt | *ERs1* | Linkage group 2 | AFLP, SSR, SRAP | Lebeau et al. (2013) |
| | | | *ERPR1*, *-2*, *-3*, *-4*, *-6*, *-7*, *-8*, *-9* | 1, 2, 3, 4, 6, 7, 8, 9 | SNP | Salgon *et al.* (2018) |
| | | | *EBWR2*, *-9*, *-14* | 2, 9, 5 | SNP | Salgon *et al.* (2017) |
| Potato | Fungal | Late blight | *R1* | 5 | SCAR (R1-1205, BA47f2, CosA, GP76, GP179), CAPS (SPUD237, GP21(*Alu*I)) | Meksem *et al.* (1995); de Jong *et al.* (1997); Oberhagemenn *et al.* (1999); Gebhardt *et al.* (2004); Sokolova *et al.* (2011) |
| | | | *R3* (*R3a* and *R3b*) | 11 | SCAR (R31380) | Sokolova *et al.* (2011) |
| | | | *RB*/*Rpi-blbI* | 8 | SCAR (RB-629/638, RB1223) | Pankin *et al.* (2011); Sokolova *et al.* (2011) |
| | | | *Rpi-abpt* | 4 | CAPS (Th2) | Park et al. (2005) |
| | | | *Rpi-ber1* | 10 | CAPS (CT214 (*Dde*I), TG63), SCAR (Q133) | Tan *et al.* (2010) |
| | | | *Rpi-mcd1* | 4 | CAPS (TG339 (*Mnl*I)) | Tan *et al.* (2010) |
| | | | *Rpi-phu1* | 9 | RAPD (GP94) | Sliwka *et al.* (2006) |
| | | | *Rpi-sto1* | 8 | SCAR (Ssto-448) | Sokolova *et al.* (2011) |
| | | | *Rpi-cap1* | 11 | CAPS (CP58 (*Msp*I)) | Jacobs *et al.* (2010) |
| | | | *Rpi-dlc1* | 9 | CAPS (GP101 (*Alu*I)) | Golas *et al.* (2010) |
| | | | *Rpi-mcq1* | 9 | CAPS (TG328 (*Alu*I)) | Smilde *et al.* (2005) |
| | | | *Rpi-snk1.1* and *Rpi-snk 1.2* | 4 | CAPS (*Mbo*I) | Jacobs *et al.* (2010) |

| | | | | | | |
|---|---|---|---|---|---|---|
| | | Wart | *Sen 1* | 11 | SNP | Prodhomme *et al.* (2020) |
| | | Soft rot | *Eca1A* | 1 | AFLP (EM4-44) | Zimnoch-Guzowska *et al.* (2000) |
| | | | *Eca2A* | 2 | AFLP (HM4-14) | Zimnoch-Guzowska *et al.* (2000) |
| | | | *Eca6A* | 6 | AFLP (HM5-17) | Zimnoch-Guzowska *et al.* (2000) |
| | | | *Eca11A* | 11 | AFLP (EM1-17) | Zimnoch-Guzowska *et al.* (2000) |
| | | | *Eca11B* | 11 | AFLP (HM1-28) | Zimnoch-Guzowska *et al.* (2000) |
| | Viral | Potato leaf roll virus | *Rladg* | 5 | AFLP (E35M48.192), SCAR (RGASC850) | Velásquez *et al.* (2007); Mihovilovich *et al.* (2014) |
| | | Potato virus Y | *Rychc* | 9 | RFLP (CT220) | Sato *et al.* (2006) |
| | | | *Nytbr* | 4 | RFLP (TG506) | Celebi-Toprak *et al.* (2002) |
| | | | *Ryadg* | 11 | RFLP (TG508) | Hämäläinen *et al.* (1997) |
| | | | *Rysto* | 11 | AFLP (M45, M5) | Brigneti *et al.* (1997) |
| | | Potato virus X | *Nb* | 5 | CAPS (SPUD 237) | de Jong *et al.* (1997) |
| | | | *Rx1* | 12 | RFLP, CAPS (CP60) | Ritter *et al.* (1991); Bendahmane *et al.* (1997); Gebhardt *et al.* (2006) |
| | | | *Rx2* | 5 | RFLP (GP21) | Ritter *et al.* (1991) |
| | Nematodes | Potato cyst nematodes | *Gro1-4* | 7 | SCAR | Paal *et al.* (2004) |
| | | | *Gpa2* | 12 | RFLP | Roupee van der Voort *et al.* (1997); van der Vassen *et al.* (2000) |
| | | | *Hero* | 4 | RFLP | Ernst *et al.* (2002) |
| | | | *H1* | 5 | CAPS, AFLP | Bakker *et al.* (2004); Pajerowska-Mukhtar *et al.* (2009) |
| Onion | Fungal | Purple blotch | *ApR1* | Not known | STS (ApR-450), SSR(AcSSR7) | Chand *et al.* (2018) |
| | | Downy mildew | *Pd* | 3 | CAPS | Nickerson Zwaan B.V. (2010) |
| | | | *i25255* locus | 3 | Indel (DMR1) | Kim *et al.* (2016) |
| Cucumber | Fungal | Powdery mildew | *pm-s* | 5 | SSR | P.N. Liu *et al.* (2017) |
| | | | *CsLRR-RPK2* (*CsGy5G015660*) | 5 | CAPS | Badri Anarjan *et al.* (2021) |
| | | | *CsaMLO8* (*Csa5G623470*) | 5 | CAPS | Badri Anarjan *et al.* (2021) |
| | | Downy mildew | QTL *DM4.1* (RLK gene) | 4 | SNP | Berg *et al.* (2020) |
| | | Gummy stem blight | QTL *gsb5.1* | 5 | SSR (SSR15321 and SSR07711) | S. Liu *et al.* (2017) |

*Continued*

**Table 17.2.** Continued.

| Crop | Disease | | Resistance genes | Chromosomal location of the gene | Molecular markers developed | Reference |
|---|---|---|---|---|---|---|
| | Viral | Tomato leaf curl New Delhi virus (ToICNDV) causing leaf curl | Major QTL for *ToICNDV* | 8 | SNPs | Saez *et al.* (2020) |
| | | CMV | QTL *cmv6.1* | 6 | SSRs | Shi *et al.* (2018) |
| Musk melon | Fungal | Powdery mildew | QTL *qCmPMR-12* | 12 | SNPs (KASP markers) | Cao *et al.* (2021) |
| | | *Fusarium* wilt | *Fom-1* and *Fom-2* | 11 (*Fom-2*) | SCAR, CAPS | Wang *et al.* (2000); Brotman *et al.* (2005); Oumouloud *et al.* (2008); Tezuka *et al.* (2009); Wang *et al.* (2011) |
| | Bacterial | Bacterial fruit blotch | *MELO3C022157* | 9 | Indel marker MB157-2 | Islam *et al.* (2020) |
| | Viral | Cucumber mosaic virus | *cmv1* (*CmVPS4*) | 12 | – | Pascual *et al.* (2019) |
| Watermelon | Fungal | Powdery mildew | *Pm2* (*ClaPMR2*) | 2 | SNP, CAPS | Mandal *et al.* (2020) |
| | | Gummy stem blight | *Qgsb8.1* locus (*Cla001017* and *Cla001019*) | 8 | SNP | Ren *et al.* (2020) |
| Cabbage | Fungal | Black rot | *Bol031422* | 8 | Indel marker BR6-InDel | Hong *et al.* (2021) |
| | | Downy mildew | *BrRHP1* | 3 | RAPD (BrPERK15A and BrPERK15B) | Kim *et al.* (2011) |
| Cauliflower | Fungal | Downy mildew | *Ppa3* | – | OPC141186, OPE141881, ISSR-231103, SSR BoGMS0624 | Singh *et al.* (2012, 2015); Saha *et al.* (2021) |
| | | | *Ppa207* | 2 | BoGMS0486 and BoGMS0900 | Saha *et al.* (2020) |
| | | Black rot | *Xca1bo* | – | SCAR marker ScOPO-04833 | Saha *et al.* (2021) |
| Common beans | Fungal | Anthracnose and angular leaf spot | *CoPv01CDRK*/ *PhgPv01CDRK* | 01 | SNP | Goncalves *et al.* (2020) |
| | Viral | Common mosaic potyviruses | *bc-3* | – | RAPD, AFLP | Mukeshimana *et al.* (2005) |

| | | | | | | |
|---|---|---|---|---|---|---|
| Cowpea | Fungal | Anthracnose | *LRR-RLK* | – | ISSR-UBC811 | Dolagobinda *et al.* (2018) |
| | | *Cercospora* leaf spot | QTL *qCLS9.1* (*Vigun10g019300* and *Vigun10g019400)* | LG9 | CAPS | Heng *et al.* (2020) |
| | Bacterial | Bacterial blight | QTLs CoBB-1, CoBB-2 and CoBB-3 | LG3, LG5 and LG9 | SNP | Agbicodo *et al.* (2010) |
| Pea | Fungal | Powdery mildew | *er1*, *er2* and *Er3* | LG6, LG3, LG4 | RFLP, RAPD, SCAR, SSR | Devi *et al.* (2022) |
| | | *Ascochyta* blight | QTLs (*abIII-1* and *abI-IV-2*) | – | SNP | Jha Ambuj *et al.* (2017) |
| | | Root rot | QTL *WB.Fsp-Ps 5.1* | 5 | SNP | Williamson *et al.* (2021) |

AFLP, amplified fragment length polymorphism; CAPS, cleaved amplified polymorphic sequences; indel, insertion/deletion; KASP, kompetitive allele-specific PCR; RAPD, random amplified polymorphic DNA; RFLP, restriction fragment length polymorphism; SCAR, sequence-characterized amplified region; SNP, single-nucleotide polymorphism; SRAP, sequence-related amplified polymorphism; SSR, simple sequence repeats.

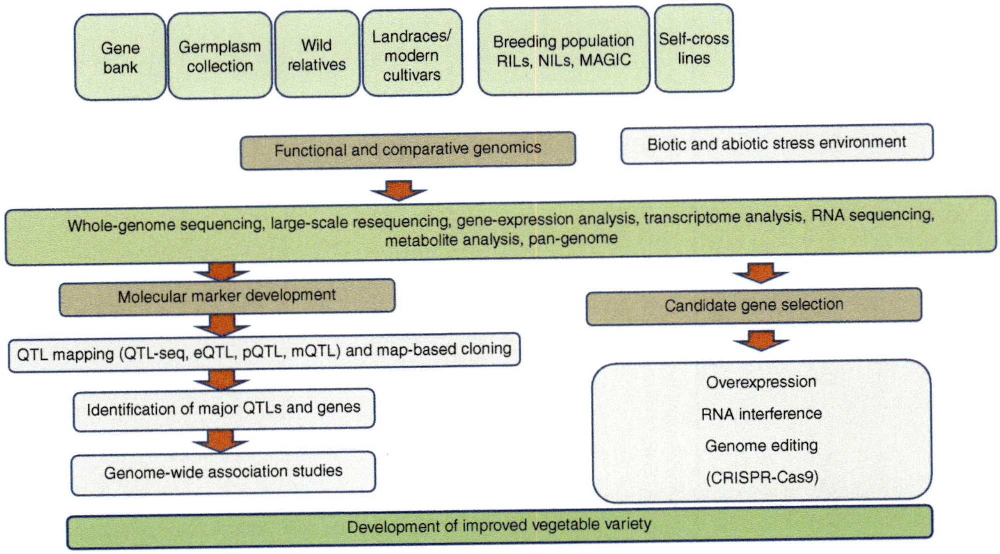


**Fig. 17.1.** Applications of molecular markers for biotic and abiotic stress tolerance genes in vegetable crops. MAGIC, multi-parent advanced generation inter-cross; NIL, near-isogenic line; QTL, quantitative trait locus; QTL-seq, QTL sequencing; eQTL, expression QTL; mQTL, methylation QTL; pQTL, protein QTL; RIL, recombinant inbred line.

allow easy identification of candidate genes and associated markers linked to traits of horticultural importance in vegetable crops. Vegetable crop breeding in the 21st century is broadly focused on the use of molecular markers and next-generation sequencing data, with a view to improving the efficiency of breeding programs. With the advent of new technologies, breeders can develop $F_1$ hybrids or inbreds in a short time with multiple gene/QTL combinations according to the requirements of consumers worldwide. MAS has been utilized extensively to confer biotic and abiotic tolerance in major vegetable crops such as tomato, potato, cucumber, chili, and cole crops, while some vegetable crops such as eggplant, spinach, chicory, and garlic lag behind in terms of MAS adoption. MAS is a popular and preferred technique for vegetable breeding owing to attributes such as cost-effectiveness, ease of marker generation, a shorter time requirement, and easy application. MAS should be implemented in breeding on a much larger scale to enhance productivity and sustainability with the changing climate. However, the full potential of this technology can only be exercised by overcoming the associated doubts and challenges. The efficiency of MAS for quantitative traits suffers from problems such as the inability to validate QTLs, exceptionally large QTL intervals (for most vegetable crops) that might include deleterious genes, identification of false negatives, unavailability of markers in strong linkage disequilibrium with the trait of interest, and QTLs from wild distant species. These factors have limited the application of MAS in vegetable crops. Thus, it is essential to generate fine maps in order to accurately locate and validate QTL/candidate genes, shorten the QTL interval, and select markers closely linked to the QTL/genes. If all the above key factors are taken into consideration, MAS will become a more popular, feasible, and frequently used approach for breeding in vegetable crops. New emerging genomic techniques such as SNP-chip genotyping and clustered regularly interspaced palindromic repeats (CRISPR)-CRISPR-associated protein 9 (Cas9) can drive the effort toward MAS, allowing gene-specific information can be correlated with phenotypic performance.

## 17.5 References

Acciarri, N., Rotino, G.L., Tamietti, G., Valentino, D., Voltattorni, S., and Sabatini, E. (2007) Molecular markers for *Ve1* and *Ve2 Verticillium* resistance genes from Italian tomato germplasm. *Plant Breeding* 126, 617–621.

Agbicodo, E.M., Fatokun, C.A., Bandyopadhyay, R., Wydra, K., Diop, N.N., *et al.* (2010) Identification of markers associated with bacterial blight resistance loci in cowpea [*Vigna unguiculata* (L.) Walp.]. *Euphytica* 175, 215–226.

Akbar, K., Abbasi, F.M., Sajid, M., Ahmad, M., Khan, Z.U., and Ali, H. (2016) Marker-assisted selection and pyramiding of *I1* and *Ph3* genes for multiple disease resistance in tomato through PCR analysis. *International Journal of Biosciences* 9, 108–113.

Ali, A.A.M., Romdhane, W.B., Tarroum, M., Al-Dakhil, M., Al-Doss, A., *et al.* (2021) Analysis of salinity tolerance in tomato introgression lines based on morpho-physiological and molecular traits. *Plants* 10: 2594.

Amiteye, S. (2021) Basic concepts and methodologies of DNA marker systems in plant molecular breeding. *Heliyon* 7: e08093.

Ammiraju, J.S.S., Veremis, J.C., Huang, X., Roberts, P.A., and Kaloshian, I. (2003) The heat-stable, root-knot nematode-resistance gene *Mi-9* from *Lycopersicon peruvianum* is localized on the short arm of chromosome 6. *Theoretical and Applied Genetics* 106, 478–484.

Anbinder, I., Reuveni, M., Azari, R., Ilan, P., Nahon, S., *et al.* (2009) Molecular dissection of tomato leaf curl virus resistance in tomato line Ty172 derived from *Solanum peruvianum*. *Theoretical and Applied Genetics* 119, 519–530.

Andrés, A., Luis Arteaga, M., and Gil Ortega, R. (2004) New genes related to PVY resistance in *C. annuum* L. 'Serrano Criollo de Morelos-334'. In: Voorrips, R.E. (ed.) *Proceedings of the XIIth EUCARPIA Meeting on Genetics and Breeding of Capsicum and Eggplant, 17–19 May 2004, Noordwijkerhout, Netherlands*. Plant Research International, Wageningen, Netherlands, pp. 134–138.

Arnedo-Andrés, M., Gil-Ortega, R., Luis-Arteaga, M., and Hormaza, I. (2002) Development of RAPD and SCAR markers linked to the *Pvr4* locus for resistance to PVY in pepper (*Capsicum annuum* L.). *Theoretical and Applied Genetics* 105, 1067–1074.

Asins, M.J., Villalta, I., Aly, M.M., Olías, R., Alvarez, D.E., *et al.* (2013) Two closely linked tomato *HKT* coding genes are positional candidates for the major tomato QTL involved in $Na^+/K^+$ homeostasis. *Plant, Cell & Environment* 36, 1171–1191.

Atwell, S., Huang, Y.S., Vilhjalmsson, B.J., Willems, G., Horton, M., *et al.* (2010) Genome-wide association study of 107 phenotypes in *Arabidopsis thaliana* inbred lines. *Nature* 465, 627–631.

Badri Anarjan, M., Bae, I., and Lee, S. (2021) Marker-assisted evaluation of two powdery mildew resistance candidate genes in Korean cucumber inbred lines. *Agronomy* 11: 2191.

Bakker, E., Achenbach, U., Bakker, J., van Vliet, J., Peleman, J., *et al.* (2004) A high-resolution map of the *H1* locus harbouring resistance to the potato cyst nematode *Globodera rostochiensis*. *Theoretical and Applied Genetics* 109, 146–152.

Barillas, A.C., Mejía, L., Sánchez-Pérez, A. and Maxwell, D.P. (2008) CAPS and SCAR markers for detection of *I-3* gene introgression for resistance to *Fusarium oxysporum* f. sp. *lycopersici* race 3. *Report of the Tomato Genetics Cooperative*, 58, 11–17.

Beckmann, J.S. and Weber, J.L. (1992) Survey of human and rat microsatellites. *Genomics* 12, 627–631.

Bendahmane, A., Kanyuka, K., and Baulcombe, D.C. (1997) High-resolution genetical and physical mapping of the *Rx* gene for extreme resistance to potato virus X in tetraploid potato. *Theoretical and Applied Genetics* 95, 153–162.

Berg, J.A., Hermans, F.W.K., Beenders, F., Lou, L., Vriezen, W.H., *et al.* (2020) Analysis of QTL DM4.1 for downy mildew resistance in cucumber reveals multiple subQTL: a novel *RLK* as candidate gene for the most important subQTL. *Frontiers in Plant Science* 11: 569876.

Brigneti, G., Garcia-Mas, J., and Baulcombe, D.C. (1997) Molecular mapping of the potato virus Y resistance gene *Rysto* in potato. *Theoretical and Applied Genetics* 94, 198–203.

Brommonschenkel, S.H. and Tanksley, S.D. (1997) Map-based cloning of the tomato genomic region that spans the *Sw-5* tospovirus resistance gene in tomato. *Molecular Genetics and Genomics* 256, 121–126.

Brotman, Y., Kovalski, I., Dogimont, C., Pitrat, M., Portnoy, V., and Perl-Treves, R. (2005) Molecular markers linked to papaya ring spot virus resistance and *Fusarium* race 2 resistance in melon. *Theoretical and Applied Genetics* 110, 337–345.

Bu, X., Wang, X., Yan, J., Zhang, Y., Zhou, S., *et al.* (2021) Genome-wide characterization of B-box gene family and its roles in responses to light quality and cold stress in tomato. *Frontiers in Plant Science* 12: 698525.

Caldwell, K.S., Russell, J., Langridge, P., and Powell, W. (2006) Extreme population-dependent linkage disequilibrium detected in an inbreeding plant species, *Hordeum vulgare*. *Genetics* 172, 557–567.

Cao, Y., Diao, Q., Chen, Y., Jin, H., Zhang, Y., and Zhang, H. (2021) Development of KASP markers and identification of a QTL underlying powdery mildew resistance in melon (*Cucumis melo* L.) by bulked segregant analysis and RNA-Seq. *Frontiers in Plant Science* 11: 593207.

Caranta, C., Thabuis, A., and Palloix, A. (1999) Development of a CAPS marker for the *Pvr4* locus: a tool for pyramiding potyvirus resistance genes in pepper. *Genome* 42, 1111–1116.

Catanzariti, A.M., Lim, G.T.T., and Jones, D.A. (2015) The tomato *I-3* gene: a novel gene for resistance to *Fusarium* wilt disease. *New Phytologist* 207, 106–118.

Cavanagh, C., Morell, M., Mackay, I., and Powell, W. (2008) From mutations to magic: resources for gene discovery, validation and delivery in crop plants. *Current Opinion in Plant Biology* 11, 215–221.

Celebi-Toprak, F., Slack, S.A., and Jahn, M.M. (2002) A new gene, *Nytbr*, for hypersensitivity to *Potato virus Y* from *Solanum tuberosum* maps to chromosome IV. *Theoretical and Applied Genetics* 104, 669–674.

Chand, J., Ahmad, R., Muhammad, S., Khan, M.S., Khan, N., *et al.*, (2021) Expression of antioxidant genes can increase cold resistance in tomato (*Solanum lycopersicum* L.). *Pakistan Journal of Botany* 53, 447–453.

Chand, S.K., Nanda, S., and Raj Kumar, J. (2018) Genetics and molecular mapping of a novel purple blotch-resistant gene *ApR1* in onion (*Allium cepa* L.) using STS and SSR markers. *Molecular Breeding* 38: 109.

Changkwian, A., Venkatesh, J., Lee, J.H., Han, J.W., Kwon, J.K., *et al.* (2019) Physical localization of the root-knot nematode (*Meloidogyne incognita*) resistance locus *Me7* in pepper (*Capsicum annuum*). *Frontiers in Plant Science* 10: 886.

Charmet, G., Robert, N., Perretant, M.R., Gay, G., Sourdille, P., *et al.* (1999) Marker-assisted recurrent selection for cumulative additive and interactive QTLs in recombinant inbred lines. *Theoretical and Applied Genetics* 99, 1143–1148.

Chen, L., Zheng, F., Feng, Z., Li, Y., Ma, M., *et al.* (2021) A vacuolar invertase CsVI2 regulates sucrose metabolism and increases drought tolerance in *Cucumis sativus* L. *International Journal of Molecular Sciences* 23: 176.

Collard, B.C.Y. and Mackill, D.J. (2008) Marker-assisted selection: an approach for precision plant breeding in the twenty-first century. *Philosophical Transactions of the Royal Society of London B; Biological Sciences* 363, 557–572.

Csillery, G., Szarka, E., Sardi, E., Mityko, J., Kapitany, J., *et al.* (2004) The unity of plant defense: genetics, breeding and physiology. In: *Proceedings of the 12th Eucarpia Meeting on Genetics and Breeding of Capsicum and Egg-Plant, 17–19 May 2004, Noordwijkerhout, Netherlands*, pp. 147–153.

Dang, F.F., Wang, Y.N., Yu, L., Eulgem, T., Lai, Y., *et al.* (2013) CaWRKY40, a WRKY protein of pepper, plays an important role in the regulation of tolerance to heat stress and resistance to *Ralstonia solanacearum* infection. *Plant, Cell & Environment* 36, 757–74.

de Giovanni, C., Dell'Orco, P., Bruno, A., Ciccarese, F., Lotti, C., and Ricciardi, L. (2004) Identification of PCR-based markers (RAPD, AFLP) linked to a novel powdery mildew resistance gene (*ol-2*) in tomato. *Plant Science* 166, 41–48.

de Jong, W., Forsyth, A., Leister, D., Gebhardt, C., and Baulcombe, D.C. (1997) A potato hypersensitive resistance gene against potato virus X maps to a resistance gene cluster on chromosome 5. *Theoretical and Applied Genetics* 95, 246–252.

Devi, J., Mishra, G.P., Sagar, V., Kaswan, V., Dubey, R.K., *et al.* (2022) Gene-based resistance to *Erysiphe* species causing powdery mildew disease in peas (*Pisum sativum* L.). *Genes* 13: 316.

Devran, Z. and Sogut, M.A. (2010) The occurrence of virulent root-knot nematode populations on tomatoes bearing the *Mi* gene in protected vegetable-growing areas of Turkey. *Phytoparasitica* 38, 245–251.

Dhall, R.K. (2015) Breeding for biotic stresses resistance in vegetable crops: a review. *Journal of Crop Science and Technology* 4, 13–27.

Di Dato, F., Parisi, M., Cardi, T., and Tripodi, P. (2015) Genetic diversity and assessment of markers linked to resistance and pungency genes in *Capsicum* germplasm. *Euphytica* 204, 103–119.

Dianese, E.C., de Fonseca, M.E.N., Goldbach, R., Kormelink, R., Inoue-Nagata, A.K., *et al.* (2010) Development of a locus-specific, co-dominant SCAR marker for assisted-selection of the *Sw-5* (*Tospovirus* resistance) gene cluster in a wide range of tomato accessions. *Molecular Breeding* 25, 133–142.

Djian-Caporalino, C., Fazari, A., Arguel, M.J., Vernie, T., VandeCasteele, C., *et al.* (2007) Root-knot nematode (*Meloidogyne* spp.) *Me* resistance genes in pepper (*Capsicum annuum* L.) are clustered on the P9 chromosome. *Theoretical and Applied Genetics* 114, 473–486.

Do Nascimento, I.R., Maluf, W.R., Figueira, A.R., Menezes, C.B., de Resende, J.T.V., *et al.* (2009) Marker assisted identification of tospovirus resistant tomato genotypes in segregating progenies. *Scientia Agricola* 66, 298–303.

Dockter, K.G., O'Neil, D.S., Price, D.L., Scott, J., and Stevens, M.R. (2009) Molecular mapping of the tomato spotted wilt virus resistance gene *Sw-7* in tomato. *Hortscience* 44: 1123.

Dolagobinda, P., Mathew, D., Mathew, S.K., and Nazeem, P.A. (2018) Identifying the markers and tagging a leucine-rich repeat receptor-like kinase gene for resistance to anthracnose disease in vegetable cowpea [*Vigna unguiculata* (L.) Walp.]. *Journal of Horticultural Science and Biotechnology* 93, 225–231.

Egea, I., Albaladejo, I., Meco, V., Morales, B., Sevilla, A., *et al.* 2018) The drought-tolerant *Solanum pennellii* regulates leaf water loss and induces genes involved in amino acid and ethylene/jasmonate metabolism under dehydration. *Scientific Reports* 8: 2791.

Ernst, K., Kumar, A., Kriseleit, D., Kloos, D.U., Phillips, M.S., and Ganal, M.W. (2002) The broad-spectrum potato cyst nematode resistance gene (*Hero*) from tomato is the only member of a large gene family of NBS-LRR genes with an unusual amino acid repeat in the LRR region. *Plant Journal* 31, 127–136.

Evers, D., Lefèvre, I., Legay, S., Lamoureux, D., Hausman, J.F., *et al.* (2010) Identification of drought-responsive compounds in potato through a combined transcriptomic and targeted metabolite approach. *Journal of Experimental Botany* 61, 2327–2343.

Fan, Y., Yang, W., Yan, Q., Chen, C., and Li J. (2019) Genome-wide identification and expression analysis of the protease inhibitor gene families in tomato. *Genes* 11: 1.

Fischer, M., Schreiber, L., Colby, T., Kuckenberg, M., Tacke, E., *et al.* (2013) Novel candidate genes influencing natural variation in potato tuber cold sweetening identified by comparative proteomics and association mapping. *BMC Plant Biology* 13: 113.

Foolad, M.R. and Panthee, D.R. (2012) Marker-assisted selection in tomato breeding. *Critical Reviews in Plant Sciences* 31, 93–123.

Garland, S., Sharman, M., Persley, D., and McGrath, D. (2005) The development of an improved PCR-based marker system for *Sw-5*, an important TSWV resistance gene of tomato. *Australian Journal of Agricultural Research* 56, 285–289.

Gebhardt, C., Ballvora, A., Walkemeier, B., Oberhagemann, P., and Schüler, K. (2004) Assessing genetic potential in germplasm collections of crop plants by marker-trait association:a case study for potatoes with quantitative variation of resistance to late blight and maturity type. *Molecular Breeding* 13, 93–102.

Gebhardt, C., Bellin, D., Henselewski, H., Lehmann, W., Schwarzfischer, J., and Valkonen, J. (2006) Marker-assisted combination of major genes for pathogen resistance in potato. *Theoretical and Applied Genetics* 112, 1458–1464.

Golas, T.M., Sikkema, A., Gros, J., Feron, R.M.C., van den Berg, R.G., *et al.* (2010) Identification of a resistance gene *Rpi-dlc1* to *Phytophthora infestans* in European accessions of *Solanum dulcamara*. *Theoretical and Applied Genetics* 120, 797–808.

Goncalves-Vidigal, M.C., Gilio, T.A.S., Valentini, G., Vaz-Bisneta, M., Vidigal Filho, P.S., *et al.* (2020) New Andean source of resistance to anthracnose and angular leaf spot: fine-mapping of disease-resistance genes in California Dark Red Kidney common bean cultivar. *PLOS ONE* 15: e0235215.

Gonzalez-Cendales, Y., Catanzariti, A.M., Baker, B., McGrath, D.J., and Jones, D.A. (2016) Identification of *I-7* expands the repertoire of genes for resistance to *Fusarium* wilt in tomato to three resistance gene classes. *Molecular Plant Pathology* 17, 448–463.

Gonzalo, M.J., Li, Y.C., Chen, K.Y., Gil, D., Montoro, T., *et al.* (2020) Genetic control of reproductive traits in tomatoes under high temperature. *Frontiers in Plant Science* 11: 326.

Guo, X., Yang, X.H., Yang, Y., Mao, Z.C., Liu, F., *et al.* (2017) Bacterial artificial chromosome library construction of root-knot nematode resistant pepper genotype HDA149 and identification of clones linked to *Me3* resistant locus. *Journal of Integrative Agriculture* 16, 57–64.

Haley, C.S. and Knott, S.A. (1992) A simple regression method for mapping quantitative trait loci in line crosses using flanking markers. *Heredity* 69, 315–324.

Hamalainen, J.H., Watanabe, K.N., Valkonen, J.P.T., Arihara, A., Plaisted, R.L., *et al.* (1997) Mapping and marker-assisted selection for a gene for extreme resistance to potato virus Y. *Theoretical and Applied Genetics* 94, 192–197.

Hanson, P., Bernacchi, M., Green, S., Tanksley, S.D., Muniyappa, V., and Padmaja, A.S. (2000) Mapping a wild tomato introgression associated with tomato yellow leaf curl virus resistance in a cultivated tomato line. *Journal of the American Society for Horticultural Science* 15, 15–20.

He, L., Li, L., Zhu, Y., Pan, Y., Zhang, X., *et al.* (2021) *BolTLP1*, a thaumatin-like protein gene, confers tolerance to salt and drought stresses in broccoli (*Brassica oleracea* L. var. *Italica*). *International Journal of Molecular Sciences* 22: 11132.

Heffner, E.L., Sorrells, M.E., and Jannink, J.L. (2009) Genomic selection for crop improvement. *Crop Science* 49, 1–12.

Heng, T., Kaga, A., Chen, X., and Somta, P. (2020) Two tightly linked genes coding for NAD-dependent malic enzyme and dynamin-related protein are associated with resistance to *Cercospora* leaf spot disease in cowpea (*Vigna unguiculata* (L.) Walp.). *Theoretical and Applied Genetics* 133, 395–407.

Hindorff, L.A., Sethupathy, P., Junkins, H.A., Ramos, E.M., Mehta, J.P., *et al.* (2009) Potential etiologic and functional implications of genome-wide association loci for human diseases and traits. *Proceedings of the National Academy of Sciences USA* 106, 9362–9367.

Hong, J.E., Afrin, K.S., Rahim, M.A., Jung, H.J., and Nou, I.S. (2021) Inheritance of black rot resistance and development of molecular marker linked to *Xcc* races 6 and 7 resistance in cabbage. *Plants* 10: 1940.

Huang, C.C., van de Putte, P.M., Haanstra-van der Meer, J.G., Meijer-Dekens, F., and Lindhout, P. (2000) Characterization and mapping of resistance to *Oidium lycopersicum* in two *Lycopersicon hirsutum* accessions: evidence for close linkage of two *Ol*-genes on chromosome 6. *Heredity* 85, 511–520.

Hutton, S.F. and Scott, J.W. (2013) Fine-mapping and cloning of *Ty-1* and *Ty-3*; and mapping of a new TYLCV resistance locus, "*Ty-6*". In: *Tomato Breeders Round Table Proceedings, 7 February 2013, Shangri La, Chiang Mai, Thailand*.

Islam, M.R., Hossain, M.R., Jesse, D.M.I., Jung, H.J., Kim, H.T., *et al.* (2020) Development of molecular marker linked with bacterial fruit blotch resistance in melon (*Cucumis melo* L.). *Genes* 11: 220.

Jaarsma, R. and de Boer, A.H. (2018) Salinity tolerance of two potato cultivars (*Solanum tuberosum*) correlates with differences in vacuolar transport activity. *Frontiers in Plant Science* 9: 737.

Jablonska, B., Ammiraju, J.S., Bhattarai, K.K., Mantelin, S., Martinez de Ilarduya, O., *et al.* (2007) The *Mi-9* gene from *Solanum arcanum* conferring heat-stable resistance to root-knot nematodes is a homolog of *Mi-1*. *Plant Physiology* 143, 1044–1054.

Jacobs, M.M.J., Vosman, B., Vleeshouwers, V.G.A.A., Visser, R.G.F., Henken, B., and van den Berg, R.G. (2010) A novel approach to locate *Phytophthora infestans* resistance genes on the potato genetic map. *Theoretical and Applied Genetics* 120, 785–796.

Jarvis, P., Lister, C., Szabo, V., and Dean, C. (1994) Integration of CAPS markers into the RFLP map generated using recombinant inbred lines of *Arabidopsis thaliana*. *Plant Molecular Biology* 24, 685–687.

Jha, A.B., Gali, K.K., Tar'an, B., and Warkentin, T.D. (2017) Fine mapping of QTLs for *Ascochyta* blight resistance in pea using heterogeneous inbred families. *Frontiers in Plant Science* 8: 765.

Ji, Y., Scott, J.W., and Maxwell, D.P. (2009) Molecular mapping of *Ty-4*, a new Tomato yellow leaf curl virus resistance locus on chromosome 3 of tomato. *Journal of the American Society for Horticultural Science* 134, 281–288.

Ji, Y., Scott, J.W., Hanson, P., Graham, E., and Maxwell, D.P. (2007) Sources of resistance, inheritance, and location of genetic loci conferring resistance to members of the tomato infecting begomoviruses. In: Czosnek H. (ed.) *Tomato Yellow Leaf Curl Virus Disease*. Springer, Dordrecht, Netherlands, pp. 343–362.

Ji, Y., Scott, J.W., Maxwell, D.P., and Schuster, D.J. (2008) Ty-4, a tomato yellow leaf curl virus resistance gene on chromosome 3 of tomato. *Report of the Tomato Genetics Cooperative* 58, 29–31.

Jiang, J.L., Tian, Y., Li, L., Yu, M., Hou, R.P., and Ren, X.M. (2019) $H_2S$ alleviates salinity stress in cucumber by maintaining the $Na^+/K^+$ balance and regulating $H_2S$ metabolism and oxidative stress response. *Frontiers in Plant Science* 10: 678.

Jones, J.B., Lacy, G.H., Bouzar, H., Stall, R.E., and Schaad, N.W. (2004) Reclassification of the xanthomonads associated with bacterial spot disease of tomato and pepper. *Systematic and Applied Microbiology* 27, 755–762.

Kalia, R.K., Rai, M.K., Kalia, S., Singh, R., and Dhawan, A.K. (2011) Microsatellite markers: an overview of the recent progress in plants. *Euphytica* 177, 309–334.

Kang, W.H., Hoang, N., Yang, H.B., Kwon, J.K., Jo, S.H., *et al.* (2010) Molecular mapping and characterization of a single dominant gene controlling CMV resistance in peppers (*Capsicum annuum* L.). *Theoretical and Applied Genetics* 120, 1587–1596.

Kappachery, S., Yu, J.W., Baniekal-Hiremath, G., and Park, S.W. (2013) Rapid identification of potential drought tolerance genes from *Solanum tuberosum* by using a yeast functional screening method. *Comptes Rendus Biologies* 336, 530–545.

Kawchuk, L.M., Hachey, J., Lynch, D.R., Kulcsar, F., van Rooijen, G., *et al.* (2001) Tomato *Ve* disease resistance genes encode cell surface-like receptors. *Proceedings of the National Academy of Sciences USA* 98, 6511–6515.

Kim, B., Hwang, I.S., Lee, H.J., Lee, J.M., Seo, E., *et al.* (2018) Identification of a molecular marker tightly linked to bacterial wilt resistance in tomato by genome-wide SNP analysis. *Theoretical and Applied Genetics* 131, 1017–1030.

Kim, S., Kim, K.T., Kim, D.H., Yang, E.Y., Cho, M.C., *et al.* (2010) Identification of quantitative trait loci associated with anthracnose resistance in chili pepper (*Capsicum* spp.). *Korean Journal of Horticultural Science and Technology* 28, 1014–1024.

Kim, S., Song, Y.H., Lee, J.Y., Choi, S.R., Dhandapani, V., *et al.* (2011) Identification of the *BrRHP1* locus that confers resistance to downy mildew in Chinese cabbage (*Brassica rapa* ssp. pekinensis) and development of linked molecular markers. *Theoretical and Applied Genetics* 123, 1183–1192.

Kim, S., Kim, C.-W., Choi, M.-S., and Kim, S. (2016) Development of a simple PCR marker tagging the *Allium roylei* fragment harboring resistance to downy mildew (*Peronospora destructor*) in onion (*Allium cepa* L.). *Euphytica* 208, 561–569.

Konzen, E., Peron, R., Ito, M.A., *et al.* (2017) Molecular identification of bamboo genera and species based on RAPD-RFLP markers. *Silva Fennica* 51: 1691.

Kourani, M., Mohareb, F., Rezwan, F.I., Anastasiadi, M., Hammond, J.P. (2022) Genetic and physiological responses to heat stress in *Brassica napus*. *Frontiers in Plant Science* 13: 832147.

Kozik, E.U. and Wehner, T.C. (2008) A single dominant gene *Ch* for chilling resistance in cucumber seedlings. *Journal of the American Society for Horticultural Science* 133, 225–227.

Kumar, S.K., Sivanesan, I., Murugesan, K., Jeong, B.R., and Hwang, S.J. (2014) Enhancing salt tolerance in eggplant by introduction of foreign halotolerance gene, *HAL1* isolated from yeast. *Horticulture, Environment, and Biotechnology* 55, 222–229.

Kyle, M.M. and Palloix, A. (1997) Proposed revision of nomenclature for potyvirus resistance genes in *Capsicum*. *Euphytica* 97, 183–188.

Lander, E.S. and Botstein, D. (1989) Mapping Mendelian factors underlying quantitative traits using RFLP linkage maps. *Genetics* 121, 185–199.

Larcher, W. (2003) Plants under stress. In: *Physiological Plant Ecology: Ecophysiology and Stress Physiology of Functional Groups*. Springer, Berlin/Heidelberg, Germany, pp. 345–437.

Lau, W.C., Rafii, M.Y., Ismail, M.R., Puteh, A., Latif, M.A., and Ramli, A. (2015) Review of functional markers for improving cooking, eating, and the nutritional qualities of rice. *Frontiers in Plant Science* 6: 832.

Lebeau, A., Gouy, M., Daunay, M.C., Wicker, E., Chiroleu, F., *et al.* (2013) Genetic mapping of a major dominant gene for resistance to *Ralstonia solanacearum* in eggplant. *Theoretical and Applied Genetics* 126, 143–158.

Lee, J., Hong, J., and Do, J.W. (2010) Identification of QTLs for resistance to anthracnose to two *Colletotrichum* species in pepper. *Journal of Crop Sciences Biotechnology* 13, 227–233.

Lehretz, G.G., Sonnewald, S., and Sonnewald, U. (2021) Assimilate highway to sink organs—physiological consequences of SP6A overexpression in transgenic potato (*Solanum tuberosum* L.). *Journal of Plant Physiology* 266: 153530.

Li, H., Wang, Y., Wang, Z., Guo, X., Wang, F., *et al.* (2016) Microarray and genetic analysis reveals that *csa-miR159b* plays a critical role in abscisic acid-mediated heat tolerance in grafted cucumber plants. *Plant, Cell & Environment* 39, 1790–804.

Li, H., Ye, G., and Wang, J.A. (2007) Modified algorithm for the improvement of composite interval mapping. *Genetics* 175, 361–374.

Li, J., Gao, Z., Zhou, L., Li, L., Zhang, J., *et al.* (2019) Comparative transcriptome analysis reveals $K^+$ transporter gene contributing to salt tolerance in eggplant. *BMC Plant Biology* 19: 67.

Litt, M. and Luty, J.A. (1989) A hypervariable microsatellite revealed by *in vitro* amplification of a dinucleotide repeat within the cardiac muscle actin gene. *American Journal of Human Genetics* 44, 397–401.

Liu, D., Dong, S., Bo, K., Miao, H., Li, C., *et al.* (2021) Identification of QTLs controlling salt tolerance in cucumber (*Cucumis sativus* L.) seedlings. *Plants* 10: 85.

Liu, J., Zheng, Z., Zhou, X., Feng, C., and Zhuang, Y. (2014) Improving the resistance of eggplant (*Solanum melongena*) to *Verticillium* wilt using wild species *Solanum linnaeanum*. *Euphytica* 201, 463–469.

Liu, M.-M., Li, Y.-I., Li, G., Dong, T.-T., Liu, S.-Y., *et al.* (2020) Overexpression of *StCYS1* gene enhances tolerance to salt stress in the transgenic potato (*Solanum tuberosum* L.) plant. *Journal of Integrative Agriculture* 19, 2239–2246.

Liu, P.N., Miao, H., Lu, H.W., Cui, J.Y., Tian, G.L., *et al.* (2017) Molecular mapping and candidate gene analysis for resistance to powdery mildew in *Cucumis sativus* stem. *Genetics and Molecular Research* 16: gmr16039680.

Liu, S., Shi, Y., Miao, H., Wang, M., Li, B., *et al.* (2017) Genetic analysis and QTL mapping of resistance to gummy stem blight in *Cucumis sativus* seedling stage. *Plant Disease* 101, 1145–1152.

Liu, Y., Dong, S., Wei, S., Wang, W., Miao, H., *et al.* (2021) QTL mapping of heat tolerance in cucumber (*Cucumis sativus* L.) at adult stage. *Plants* 10: 324.

Lopez-Ortiz, C., Peña-Garcia, Y., Natarajan, P., Bhandari, M., Abburi, V., *et al.* (2020) The ankyrin repeat gene family in *Capsicum* spp: genome-wide survey, characterization and gene expression profile. *Scientific Reports* 10: 4044.

Lu, X., Liu, W., Xiang, C., Li, X., Wang, Q., *et al.* (2020) Genome-wide characterization of GRAS family and their potential roles in cold tolerance of cucumber (*Cucumis sativus* L.). *International Journal of Molecular Sciences* 21: 3857.

Lynch, M. and Walsh, B. (1998) *Genetics and Analysis of Quantitative Traits*. Sinauer, Sunderland, Massachusetts.

Ma, X., Li, Y., Gai, W.X., Li, C. and, Gong, Z.H. (2021) The *CaCIPK3* gene positively regulates drought tolerance in pepper. *Horticulture Research* 8: 216.

Mackay, T.F. (2001) The genetic architecture of quantitative traits. *Annual Review of Genetics* 35, 303–339.

Mahasuk, P., Taylor, P.W.J., and Mongkolporn, O. (2009) Identification of two new genes conferring resistance to *Colletotrichum acutatum* in *Capsicum baccatum*. *Phytopathology* 99, 1100–1104.

Mallard, S., Cantet, S., Massire, A., Bachellez, A., Ewert, S., and Lefebvre, V. (2013) A key QTL cluster is conserved among accessions and exhibits broad-spectrum resistance to *Phytophthora capsici*: a valuable locus for pepper breeding. *Molecular Breeding* 32, 349–364.

Mandal, M.K., Suren, H., and Kousik, C. (2020) Elucidation of resistance signaling and identification of powdery mildew resistant mapping loci (*ClaPMR2*) during watermelon-*Podosphaera xanthii* interaction using RNA-Seq and whole-genome resequencing approach. *Scientific Reports* 10: 14038.

Martínez, O. and Curnow, R.N. (1992) Estimating the locations and the sizes of the effects of quantitative trait loci using flanking markers. *Theoretical and Applied Genetics* 85, 480–488.

McMullen, M.D., Kresovich, S., Villeda, H.S., Bradbury, P., Li, H., *et al.* (2009) Supporting online material for: genetic properties of the maize nested association mapping population. *Science* 325, 737–741.

Meksem, K., Leister, D., Salamini, F., and Gebhardt, C. (1995) A high resolution map of chromosome V of potato based on RFLP and AFLP markers in the vicinity of the *R1* locus. *Molecular Genetics and Genomics* 249, 74–81.

Michaels, S.D. and Amasino, R.M. (1998) A robust method for detecting single-nucleotide changes as polymorphic markers by PCR. *Plant Journal* 14, 381–385.

Mihovilovich, E., Aponte, M., Lindqvist-Kreuze, H., and Bonierbale, M. (2014) An RGA-derived SCAR marker linked to PLRV resistance from *Solanum tuberosum* ssp. *andigena*. *Plant Molecular Biology Reporter* 32, 117–128.

Mimura, Y., Kageyama, T., Minamiyama, Y., and Masashi, H. (2009) QTL analysis for resistance to *Ralstonia solanacearum* in *Capsicum* accession 'LS2341'. *Journal of the Japanese Society for Horticultural Science* 78, 307–313.

Mishra, K.B., Iannacone, R., Petrozza, A., Mishra, A., Armentano, N., *et al.* (2012) Engineered drought tolerance in tomato plants is reflected in chlorophyll fluorescence emission. *Plant Science* 182, 79–86.

Mukeshimana, G., Paneda, A., Rodriguez-Suarez, C., Ferreira, J.J., Giraldez, R., and Kelly, J.D. (2005) Markers linked to the *bc-3* gene conditioning resistance to bean common mosaic potyviruses in common bean. *Euphytica* 144, 291–299.

Mutlu, N., Boyacı, F.H., Gocmen, M., and Abak, K. (2008) Development of SRAP, SRAP-RGA, RAPD and SCAR markers linked with a *Fusarium* wilt resistance gene in eggplant. *Theoretical and Applied Genetics* 117, 1303–1312.

Nickerson Zwaan B.V. (2010) Resistance to downy mildew of onion caused by the fungus *Peronospora destructor*. Patent no. EP1668979B12004-12-08 2010-09-08.

Nir, I., Moshelion, M., and Weiss D. (2014) The *Arabidopsis* gibberellin methyl transferase 1 suppresses gibberellin activity, reduces whole-plant transpiration and promotes drought tolerance in transgenic tomato. *Plant, Cell & Environment* 37, 113–123.

Oberhagemann, P., Chatot-Balandras, C., Schäfer-Pregl, R., Wegener, D., Palomino, C., *et al.* (1999) A genetic analysis of quantitative resistance to late blight in potato: towards marker-assisted selection. *Molecular Breeding* 5, 399–415.

Orellana, S., Yañez, M., Espinoza, A., Verdugo, I., González, E., *et al.* (2010) The transcription factor SlAREB1 confers drought, salt stress tolerance and regulates biotic and abiotic stress-related genes in tomato. *Plant, Cell & Environment* 33, 2191–2208.

Oumouloud, A., Arnedo-Andrés, M.S., Gonzalez-Torres, R., and Alvarez, J.M. (2008) Development of molecular markers linked to the *Fom-1* locus for resistance to *Fusarium* race 2 in melon. *Euphytica* 164, 347–356.

Paal, J., Henselewski, H., Muth, J., Meksem, K., Menendez, C.M., *et al.* (2004) Molecular cloning of the potato *Gro1-4* gene conferring resistance to pathotype Ro1 of the root cyst nematode *Globodera rostochiensis*, based on a candidate gene approach. *Plant Journal* 38, 285–297.

Padmanabhan, C., Ma, Q., Shekasteband, R., Stewart, K.S., Hutton, S.F., *et al.* (2019) Comprehensive transcriptome analysis and functional characterization of PR-5 for its involvement in tomato *Sw-7* resistance to tomato spotted wilt tospovirus. *Scientific Reports* 9: 7673.

Pajerowska-Mukhtar, K., Stich, B., Achenbach, U., Ballvora, A., Lubeck, J., *et al.* (2009) Single nucleotide polymorphisms in the allene oxide synthase 2 gene are associated with field resistance to late blight in populations of tetraploid potato cultivars. *Genetics* 181, 1115–1127.

Pankin, A., Sokolova, E., Rogozina, E., Kuznetsova, M., Deahl, K., *et al.* (2011) Allele mining in the gene pool of wild *Solanum* species for homologues of late blight resistance gene *RB/Rpi-blb1*. *Plant Genetic Resources* 9, 305–308.

Park, J., Lee, H.-J., Cheon, C.-I., Kim, S.-H., *et al.* (2011) The *Arabidopsis thaliana* homeobox gene *ATHB12* is involved in symptom development caused by geminivirus infection. *PLOS ONE* 6: e20054.

Park, T.H., Vleeshouwers, V.G.A.A., Hutten, R.C.B., van Eck, H.J., van der Vossen, E., *et al.* (2005) High-resolution mapping and analysis of the resistance locus *Rpi-abpt* against *Phytophthora infestans* in potato. *Molecular Breeding* 16, 33–43.

Pascual, L., Yan, J., Pujol, M., Monforte, A.J., Picó, B., and Martín-Hernández, A.M. (2019) *CmVPS41* is a general gatekeeper for resistance to *Cucumber mosaic virus* phloem entry in melon. *Frontiers in Plant Science* 10: 1219.

Pierre, M., Noel, L., Lahaye, T., Ballvora, A., Veuskens, J., *et al.* (2000) High-resolution genetic mapping of the pepper resistance locus *Bs3* governing recognition of the *Xanthomonas campestris* pv *vesicatoria* AvrBs3 protein. *Theoretical and Applied Genetics* 101, 255–263.

Pinar, H., Mutlu, N., Ozaslandan, A., Argun, D., Keles, D., and Canhilal, R. (2016) Reliability assessment of molecular markers linked to resistance genes against *Meloidogyne* spp. in diverse peppers genotypes. *Egyptian Journal of Biological Pest Control* 26, 515–521.

Prodhomme, C., Vos, P.G., Paulo, M.J., Tammes, J.E., Visser, R.G.F., *et al.* (2020) Distribution of P1(D1) wart disease resistance in potato germplasm and GWAS identification of haplotype-specific SNP markers. *Theoretical and Applied Genetics* 133, 1859–1871.

Provan, J., Powell, W., and Hollingsworth, P.M. (2001) Chloroplast microsatellites: new tools for studies in plant ecology and evolution. *Trends in Ecology & Evolution* 16, 142–147.

Rajendrakumar, P., Biswal, A.K., Balachandran, S.M., Srinivasarao, K., and Sundaram, R.M. (2007) Simple sequence repeats in organellar genomes of rice: frequency and distribution in genic and intergenic regions. *Bioinformatics* 23, 1–4.

Rani, C.I., Muthuvel, I., and Veeraragavathatham, D. (2009) Evaluation of 14 tomato genotypes for yield and root knot nematode resistance parameters. *Pest Technology* 3, 76–80.

Razi, K., Bae, D.W., and Muneer, S. (2021) Target-based physiological modulations and chloroplast proteome reveals a drought resilient rootstock in okra (*Abelmoschus esculentus*) genotypes. *International Journal of Molecular Sciences* 22: 12996.

Rihan, H.Z., Al-Issawi, M., and Fuller, M.P. (2017) An analysis of the development of cauliflower seed as a model to improve the molecular mechanism of abiotic stress tolerance in cauliflower artificial seeds. *Plant Physiology and Biochemistry* 116, 91–105.

Ren, R., Xu, J., Zhang, M., Liu, G., Yao, X., *et al.* (2020) Identification and molecular mapping of a gummy stem blight resistance gene in wild watermelon (*Citrullus amarus*) germplasm PI 189225. *Plant Disease* 104, 16–24.

Ritter, E., Debener, T., Barone, A., Salamini, F., and Gebhardt, C. (1991) RFLP mapping on potato chromosomes of two genes controlling extreme resistance to potato virus X (PVX). *Molecular and General Genetics* 227, 81–85.

Rouppe van der Voort, J.R., Wolters, P., Folkertsma, R., Hutten, R., van Zandvoort, P., *et al.* (1997) Mapping of the cyst nematode resistance locus *Gpa2* in potato using a strategy based on comigrating AFLP markers. *Theoretical and Applied Genetics* 95, 874–880.

Ruffel, S., Dussault, M.H., Palloix, A., Moury, B., Bendahmane, A., *et al.* (2002) A natural recessive resistance gene against potato virus Y in pepper corresponds to the eukaryotic initiation factor 4E (eIF4E). *Plant Journal* 32, 1067–1075.

Saez, C., Martinez, C., Montero-Pau, J., Esteras, C., Sifres, A., *et al.* (2020) A major QTL located in chromosome 8 of *Cucurbita moschata* is responsible for resistance to *Tomato leaf curl New Delhi virus*. *Frontiers in Plant Science* 11: 207.

Saha, P., Ghoshal, C., Ray, S., Saha, N.D., Srivastava, M., *et al.* (2020) Genetic analysis of downy mildew resistance and identification of molecular markers linked to resistance gene *Ppa207* on chromosome 2 in cauliflower. *Euphytica* 216: 183.

Saha, P., Ghoshal, C., Saha, N.D., Verma, A., Srivastava, M., *et al.* (2021) Marker-assisted pyramiding of downy mildew-resistant gene *Ppa3* and black rot-resistant gene *Xca1bo* in popular early cauliflower variety Pusa Meghna. *Frontiers in Plant Science* 12, 603600.

Sahitya, U.L., Krishna, M.S.R., and Suneetha, P. (2019) Integrated approaches to study the drought tolerance mechanism in hot pepper (*Capsicum annuum* L.). *Physiology and Molecular Biology of Plants* 25, 637–647.

Salgon, S., Jourda, C., Sauvage, C., Daunay, M.C., Reynaud, B., *et al.* (2017) Eggplant resistance to the *Ralstonia solanacearum* species complex involves both broad-spectrum and strain-specific quantitative trait loci. *Plant Science* 8: 828.

Salgon, S., Raynal, M., Lebon, S., Baptiste, J.M., Daunay, M.C., *et al.* (2018) Genotyping by sequencing highlights a polygenic resistance to *Ralstonia solanacearum* in eggplant (*Solanum melongena* L.). *International Journal of Molecular Sciences* 19: 357.

Sato, M., Nishikawa, K., Komura, K., and Hosaka, K. (2006) Potato virus Y resistance gene, *Ry-chc*, mapped to the distal end of potato chromosome IX. *Euphytica* 149, 367–372.

Schlotteroer, C., Amos, B., and Tautz, D. (1991) Conservation of polymorphic simple sequence loci in cetacean species. *Nature* 354, 63–65.

Seah, S., Yaghoobi, J., Rossi, M., Gleason, C.A., and Williamson, V.M. (2004) The nematode-resistance gene, *Mi-1*, is associated with an inverted chromosomal segment in susceptible compared to resistant tomato. *Theoretical and Applied Genetics* 108, 1635–1642.

Sehgal, D., Singh, R., and Rajpal, V.R. (2016) Quantitative trait loci mapping in plants: concepts and approaches. In: Rajpal, V., Rao, S., and Raina, S. (eds.) *Molecular Breeding for Sustainable Crop Improvement, Vol. 2*. Springer, Cham, Switzerland, pp. 31–59.

Shi, D., Wang, J., Bai, Y., and Liu, Y. (2020) Transcriptome sequencing of okra (*Abelmoschus esculentus* L. Moench) uncovers differently expressed genes responding to drought stress. *Journal of Plant Biochemistry and Biotechnology* 29, 155–170.

Shi, L., Yang, Y., Xie, Q., Miao, H., Bo, K., *et al.* (2018) Inheritance and QTL mapping of cucumber mosaic virus resistance in cucumber (*Cucumis sativus* L.). *PLOS ONE* 13: e0200571.

Shi, J., Wang, J., Wang, N., Zhou, H., Xu, Q., and Yan G. (2019) Overexpression of *StGA2ox1* gene increases the tolerance to abiotic stress in transgenic potato (*Solanum tuberosum* L.) plants. *Applied Biochemistry and Biotechnology* 187, 1204–1219.

Singh, R.J. and Lebeda, A. (2007) Landmark research in vegetable crops. In: Singh, R.J. (ed.) *Genetic Resources, Chromosome Engineering, and Crop Improvement, Vol. 3: Vegetable Crops*. CRC Press, Boca Raton, Florida, pp. 1–15.

Singh, S., Sharma, S.R., Kalia, P., Deshmukh, R., Katara, J., *et al.* (2015) Identification of putative DNA markers for disease resistance breeding in Indian cauliflower (*Brassica oleracea* var. *botrytis* L.). *Indian Journal of Biotechnology* 14, 455–460.

Singh, S., Sharma, S.R., Kalia, P., Deshmukh, R., Kumar, V., *et al.* (2012) Molecular mapping of the downy mildew resistance gene *Ppa3* in cauliflower (*Brassica oleracea* var. *botrytis* L.). *Journal of Horticultural Science and Biotechnology* 87, 137–143.

Sliwka, J., Jakuczun, H., Lebecka, R., Marczewski, W., Gebhardt, C., and Zimnoch-Guzowska, E. (2006) The novel, major locus *Rpi-phu1* for late blight resistance maps to potato chromosome IX and is not correlated with long vegetation period. *Theoretical and Applied Genetics* 113, 685–695.

Smilde, W.D., Brigneti, G., Jagger, L., Perkins, S., and Jones, J.D. (2005) *Solanum mochiquense* chromosome IX carries a novel late blight resistance gene *Rpi-moc1*. *Theoretical and Applied Genetics* 110, 252–258.

Smith, P.G. (1944) Embryo culture of a tomato species hybrid. *Proceedings of the American Society for Horticultural Science* 44, 413–416.

Sokolova, E.A., Pankin, A., Beketova, M., Kuznetsova, M., Spiglazova, S., *et al.* (2011) SCAR markers of the *R*-genes and germplasm of wild *Solanum* species for breeding late blight-resistant potato cultivars. *Plant Genetic Resources* 9, 309–312.

Spaniolas, S., May, S.T., Bennett, M.J., and Tucker, G.A. (2006) Authentication of coffee by means of PCR-RFLP analysis and lab-on-a-chip capillary electrophoresis. *Journal of Agricultural and Food Chemistry* 54, 7466–7470.

Stamova, B.S. and Chetelat, R.T. (2000) Inheritance and genetic mapping of cucumber mosaic virus resistance introgressed from *Lycopersicon chilense* into tomato. *Theoretical and Applied Genetics* 101, 527–537.

Stevens, M.R., Lamb, E.M., and Rhoads, D.D. (1995) Mapping the *Sw-5* locus for tomato spotted wilt virus resistance in tomatoes using RAPD and RFLP analyses. *Theoretical and Applied Genetics* 90, 451–456.

Suarez, D.L., Celis, N., Ferreira, J.F.S., Reynolds, T., and Sandhu, D. (2021) Linking genetic determinants with salinity tolerance and ion relationships in eggplant, tomato and pepper. *Scientific Reports* 11: 16298.

Sun, C., Mao, S.L., Zhang, Z.H., Palloix, A., Wang, L.H., and Zhang, B.X. (2015) Resistances to anthracnose (*Colletotrichum acutatum*) of *Capsicum* mature green and ripe fruit are controlled by a major dominant cluster of QTLs on chromosome P5. *Scientia Horticulturae* 181, 81–88.

Swarnapriya, R. (2020) Breeding approaches for biotic stress resistance in vegetables. In: *Plant Breeding-Current and Future Views*. IntechOpen. Available at: www.intechopen.com/chapters/74317 (accessed 15 July 2024).

Tabor, H.K., Risch, N.J., and Myers, R.M. (2002) Candidate-gene approaches for studying complex genetic traits: practical considerations. *Nature Reviews Genetics* 3, 391–397.

Tai, T., Dahlbeck, D., Stall, R.E., Peleman, J., and Staskawicz, B.J. (1999) High-resolution genetic and physical mapping of the region containing the *Bs2* resistance gene of pepper. *Theoretical and Applied Genetics* 99, 1201–1206.

Tan, M.Y.A., Hutten, R.C.B., Visser, R.G.F., and van Eck, H.J. (2010) The effect of pyramiding *Phytophthora infestans* resistance genes $R_{pi\text{-}mcd1}$ and $R_{pi\text{-}ber}$ in potato. *Theoretical and Applied Genetics* 121, 117–125.

Tang, B., Xie, L., Yang, H., Li, X., Chen, Y., *et al.* (2022) Analysis of the expression and function of key genes in pepper under low-temperature stress. *Frontiers in Plant Science* 13: 852511.

Tanksley, S.D. (1993) Mapping polygenes. *Annual Review of Genetics* 27, 205–233.

Tanksley, S.D., Ganal, M.W., Prince, J.P., Devicente, M.C., Bonierbale, M.W., *et al.* (1992) High-density molecular linkage maps of the tomato and potato genomes. *Genetics* 132, 1141–1160.

Tezuka, T., Waki, K., Yashiro, K., Kuzuya, M., Ishikawa, T., *et al.* (2009) Construction of a linkage map and identification of DNA markers linked to *Fom-1*, a gene conferring resistance to *Fusarium oxysporum* f. sp. *melonis* race 2 in melon. *Euphytica* 168, 177–188.

Thakur, H., Jindal, S.K., Sharma, A., and Dhaliwal, M.S. (2020) Molecular mapping of dominant gene responsible for leaf curl virus resistance in chilli pepper (*Capsicum annuum* L.). *3 Biotech* 10: 182.

Thornsberry, J.M., Goodman, M.M., Doebley, J., Kresovich, S., Nielsen, D., and Buckler, E.S. (2001) *Dwarf8* polymorphisms associate with variation in flowering time. *Nature Genetics* 28, 286–289.

Tian, F., Bradbury, P., Brown, P., Hung, H., Sun, Q., *et al.* (2011) Genome-wide association study of leaf architecture in the maize nested association mapping population. *Nature Genetics* 43, 159–162.

Trapero-Mozos, A., Morris, W.L., Ducreux, L.J.M., McLean, K., Stephens, J., *et al.* (2018) Engineering heat tolerance in potato by temperature-dependent expression of a specific allele of *HEAT-SHOCK COGNATE 70*. *Plant Biotechnology Journal* 16, 197–207.

Truong, H.T.H., Tran, H.N., Choi, H.S., Park, P.H., and Lee, H.E. (2013) Development of a co-dominant SCAR marker linked to the *Ph-3* gene for *Phytophthora infestans* resistance in tomato (*Solanum lycopersicum*). *European Journal of Plant Pathology* 136, 237–245.

Uncu, A.T., Celik, I., Devran, Z., Ozkaynak, E., Frary, A., *et al.* (2015) Development of a SNP-based CAPS assay for the *Me1* gene conferring resistance to root knot nematode in pepper. *Euphytica* 206, 393–399.

United Nations (n.d.) Population. United Nations, New York. Available at: www.un.org/en/global-issues/population (accessed 11 July 2024).

Usman, M.G., Rafii, M.Y., Martini, M.Y., Yusuff, O.A., Ismail, M.R., and Miah, G. (2018) Introgression of heat shock protein (Hsp70 and sHsp) genes into the Malaysian elite chilli variety Kulai (*Capsicum annuum* L.) through the application of marker-assisted backcrossing (MAB). *Cell Stress and Chaperones* 23, 223–234.

Vallejos, C.E., Jones, V., Stall, R.E., Jones, J.B., Minsavage, G.V., *et al.* (2010) Characterization of two recessive genes controlling resistance to all races of bacterial spot in peppers. *Theoretical and Applied Genetics* 121, 37–46.

Van der Vossen, E.A., van der Voort, J.N., Kanyuka, K., Bendahmane, A., Sandbrink, H., *et al.* (2000) Homologues of a single resistance-gene cluster in potato confer resistance to distinct pathogens: a virus and a nematode. *Plant Journal* 23, 567–576.

Velasquez, A.C., Mihovilovich, E., and Bonierbale, M. (2007) Genetic characterization and mapping of major gene resistance to potato leafroll virus in *Solanum tuberosum* ssp. *andigena*. *Theoretical and Applied Genetics* 114, 1051–1058.

Venkatesh, J., An, J., Kang, W.-H., Jahn, M., and Kang, B.C. (2018) Fine mapping of the dominant potyvirus resistance gene *Pvr7* reveals a relationship with *Pvr4* in *Capsicum annuum*. *Phytopathology* 108, 142–148.

Veremis, J.C. and Roberts, P.A. (1996) Relationship between *Meloidogyne incognita* resistance gene in *Lycopersicon peruvianum* differentiated by heat sensitivity and nematode virulence. *Theoretical and Applied Genetics* 93, 950–959.

Verma, S., Nizam, S., and Verma, P.K. (2013) Biotic and abiotic stress signalling in plants. In: Sarwat, M., Ahmad, A., and Abdin, M.Z. (eds.) *Stress Signaling in Plants: Genomics and Proteomics Perspective*, Vol. 1. Springer, New York, pp. 25–49.

Vilhjálmsson, B.J. and Nordborg, M. (2013) The nature of confounding in genome-wide association studies. *Nature Reviews Genetics* 14, 1–2.

Vos, P., Hogers, R., Bleeker, M., Reijans, T., Tvandev, L., *et al.* (1995) AFLP: a new technique for DNA fingerprinting. *Nucleic Acids Research* 23, 4407–4414.

Wai, K.P., Siddique, M.I., Mo, H.S., Yoo, H.J., Byeon, S.E., *et al.* (2015) Pathotypes of bacterial spot pathogen infecting *Capsicum* peppers in Korea. *Plant Pathology Journal* 31, 428–432.

Wang, J., Hu, H., Wang, W., Wei, Q., Hu, T., and Bao, C. (2020) Genome-wide identification and functional characterization of the heat shock factor family in eggplant (*Solanum melongena* L.) under abiotic stress conditions. *Plants* 9: 915.

Wang, L., Liu, Y., Li, D., Feng, S., Yang, J., *et al.* (2019) Improving salt tolerance in potato through overexpression of *AtHKT1* gene. *BMC Plant Biology* 19: 357.

Wang, P., Wang, L., Guo, J., Yang, W., and Shen, H. (2016) Molecular mapping of a gene conferring resistance to *Phytophthora capsici* Leonian race 2 in pepper line PI201234 (*Capsicum annuum* L.). *Molecular Breeding* 36: 66.

Wang, S., Yang, J., and Zhang, M. (2011) Developments of functional markers for *Fom-2*-mediated *Fusarium* wilt resistance based on single nucleotide polymorphism in melon (*Cucumis melo* L.). *Molecular Breeding* 27, 385–393.

Wang, X., Fazari, A., Cao, Y., Zhang, Z., Palloix, A., *et al.* (2018) Fine mapping of the root-knot nematode resistance gene *Me1* in pepper (*Capsicum annuum* L.) and development of markers tightly linked to *Me1*. *Molecular Breeding* 38: 39.

Wang, Y., Cai, S., Yin, L., Shi, K., Xia, X., *et al.* (2015) Tomato HsfA1a plays a critical role in plant drought tolerance by activating *ATG* genes and inducing autophagy. *Autophagy* 11, 2033–2047.

Wang, Y.H., Thomas, C.E., and Dean, R.A. (2000) Genetic mapping of a *Fusarium* wilt resistance gene *Fom-2* in melon (*Cucumis melo* L.). *Molecular Breeding* 6, 379–389.

Weber, J.L. (1990) Informativeness of human (dC-dA) n.(dG-dT) n polymorphisms. *Genomics* 7, 524–530.

Weiland, J.J. and Yu, M.H. (2003) A cleaved amplified polymorphic sequence (CAPS) marker associated with root-knot nematode resistance in sugarbeet. *Crop Science* 43, 1814–188.

Wen, J., Jiang, F., Weng, Y., Sun, M., Shi, X., *et al.* (2019) Identification of heat-tolerance QTLs and high-temperature stress-responsive genes through conventional QTL mapping, QTL-seq and RNA-seq in tomato. *BMC Plant Biology* 19: 398.

Williamson-Benavides, B.A., Sharpe, R.M., Nelson, G., Bodah, E.T., Porter, L.D., and Dhingra, A. (2021) Identification of root rot resistance QTLs in pea using *Fusarium solani* f. sp. *pisi*-responsive differentially expressed genes. *Frontiers in Genetics* 12: 629267.

Xu, J., Driedonks, N., Rutten, M.J.M., Vriezen, W.H., de Boer, G.-J., and Rieu, I. (2017). Mapping quantitative trait loci for heat tolerance of reproductive traits in tomato (*Solanum lycopersicum*). *Molecular Breeding* 37: 58.

Xu, Y. (2010) *Molecular Plant Breeding*. CAB International, Wallingford, UK.
Yaghoobi, J., Kaloshian, I., Wen, Y., and Williamson, V.M. (1995) Mapping a new nematode resistance locus in *Lycopersicon peruvianum*. *Theoretical and Applied Genetics* 91, 457–464.
Yaghoobi, J., Yates, J.L., and Williamson, V.M. (2005) Fine mapping of the nematode resistance gene *Mi-3* in *Solanum peruvianum* and construction of an *S. lycopersicum* DNA contig spanning the locus. *Molecular Genetics and Genomics* 274, 60–69.
Yan, Y., Mintao, S., Si, M., Qian, F., Yijia, W., *et al.* (2022) Mechanism of *CsGPA1* in regulating cold tolerance of cucumber. *Horticulture Research* 9: uhac109.
Yan, Y., Zhang, W., Li, Y., He, C., Gao, L., and Yu, X. (2018) Functions of CsGPA1 on the hypocotyl elongation and root growth of cucumbers. *Scientific Reports* 8: 15583.
Yeam, I., Kang, B.C., Lindeman, W., Frantz, J.D., Faber, N., and Jahn, M.M. (2005) Allele-specific CAPS markers based on point mutations in resistance alleles at the *pvr1* locus encoding eIF4E in *Capsicum*. *Theoretical and Applied Genetics* 112, 178–186.
Younis, A., Ramzan, F., Ramzan, Y., Zulfiqar, F., Ahsan, M., and Lim, K.B. (2020) Molecular markers improve abiotic stress tolerance in crops: a review. *Plants* 9: 1374.
Yu, B., Yan, S., Zhou, H., Dong, R., Lei, J., *et al.* (2018) Overexpression of *CsCaM3* improves high temperature tolerance in cucumber. *Frontiers in Plant Science* 9: 797.
Yu, J. and Buckler, E.S. (2006) Genetic association mapping and genome organization of maize. *Current Opinion in Biotechnology* 17, 155–160.
Yu, J., Holland, J.B., McMullen, M.D., and Buckler, E.S. (2008) Genetic design and statistical power of nested association mapping in maize. *Genetics* 178, 539–551.
Zamir, D., Michelson, E.I., Zakay, Y., Novot, N., Zeidan, M., *et al.* (1994) Mapping and interrogation of a tomato yellow leaf curl virus tolerant gene, *Ty-1*. *Theoretical and Applied Genetics* 88, 141–146.
Zargar, S.M., Raatz, B., Sonah, H., Nazir, M., Bhat, J.A., *et al.* (2015) Recent advances in molecular marker techniques: Insight into QTL mapping, GWAS and genomic selection in plants. *Journal of Crop Science and Biotechnology* 18, 293–308.
Zhang, H.F., Liu, S.Y., Ma, J.H., Wang, X.K., Haq, S.U., *et al.* (2019) *CaDHN4*, a salt and cold stress-responsive dehydrin gene from pepper decreases abscisic acid sensitivity in *Arabidopsis*. *International Journal of Molecular Sciences* 21: 26.
Zhang, L., Guo, X., Zhang, Z., Wang, A., and Zhu, J. (2021) Cold-regulated gene *LeCOR413PM2* confers cold stress tolerance in tomato plants. *Gene* 764: 145097.
Zhang, S., Zhang, A., Wu, X., Zhu, Z., Yang, Z., *et al.* (2019) Transcriptome analysis revealed expression of genes related to anthocyanin biosynthesis in eggplant (*Solanum melongena* L.) under high-temperature stress. *BMC Plant Biology* 19: 387.
Zhou, L., He, Y.J., Li, J., Li, L.Z., Liu, Y., and Chen, H.Y. (2020) An eggplant *SmICE1a* gene encoding MYC-type ICE1-like transcription factor enhances freezing tolerance in transgenic *Arabidopsis thaliana*. *Plant Biology* 22, 450–458.
Zimnoch-Guzowska, E., Marczewski, W., Lebecka, R., Flis, B., SchäferPregl, R., *et al.* (2000) QTL analysis of new sources of resistance to *Erwinia carotovora* ssp. *atroseptica* in potato done by AFLP, RFLP, and resistance-gene-like markers. *Crop Science* 40: 1156.

# 18 Molecular Techniques for Biofortification in Vegetable Crops

**Feba Jacob[1], Rasmieh Hamid[2] and Deepu Mathew[1*]**
*[1]Centre for Plant Biotechnology and Molecular Biology, Kerala Agricultural University, Thrissur, Kerala, India; [2]Department of Plant Breeding, Cotton Research Institute of Iran (CRII), Agricultural Research, Education and Extension Organization (AREEO), Gorgān, Iran*

**Abstract**
Deficiency of vitamins and minerals results in "hidden hunger." Biofortification, the process of enhancing the bioavailability of vital elements in the edible portions of crops through agronomic methods or genetic improvement, could be an answer to hidden hunger or malnutrition. This instrumental intervention promises low-cost and long-term sustainability while giving nourishment to people who lack access to micronutrients. Vegetables are one of the most essential nutrient sources in the human diet and are considered protective foods, as they have the ability to shield the body from many diseases. Biofortification of vegetables is essential as more and more people are adopting plant-based diets. Strategies such as conventional breeding, transfer of mapped quantitative trait loci/genes into nutrient-deficient crops to make them nutrient-rich (e.g. zinc- and iron-rich tomato, vitamin A-enriched golden potato), and other molecular breeding processes have been useful for biofortification. More often, molecular breeding suffers from a lack of desired alleles or genes in the germplasm, and hence genetic engineering methods are resorted to for biofortification. Similarly, contemporary genome-sequencing technologies have facilitated the discovery of new regulatory genes and microRNAs, which may subsequently be improved using genome-editing techniques such as clustered regularly interspaced palindromic repeats (CRISPR)-CRISPR-associated protein 9 (Cas9). This chapter discusses molecular techniques that have been proven to be efficient and precise for crop biofortification.

**Keywords:** fortified foods, genetic engineering, genome editing, nutrient enrichment, RNAi, transgenics

## 18.1 Introduction

Tackling hunger in all its forms is the 21st century's most urgent wellness concern (Campisi *et al.*, 2022). In spite of having "enough to eat," the majority of people are not acquiring enough essential nutrients from food (Al-Hussaini *et al.*, 2022). According to the estimates of the World Health Organization, micronutrient deficiency, known as "hidden hunger," affects about 2 billion people worldwide (Monika *et al.*, 2023). A deficiency of vitamin A, iodine, iron and zinc is a key health problem in developing countries, where cereal-based diets are heavily relied on rather than fruits and vegetables (Vigneshwaran *et al.*, 2020). Thus, biofortification, the process of boosting bioavailable concentrations of essential nutrients in edible parts of plants through agronomic or crop-improvement strategies, could be a solution to malnutrition or hidden hunger (Gomathi *et al.*, 2017). *Biosis* is a Greek word meaning life, and *fortificare* in Latin means

*Email: deepu.mathew@kau.in


DOI: 10.1079/9781800627123.0018

to strengthen; thus, biofortification means to make life strong. Biofortification of regularly consumed food crops is the easiest answer to complex nutritional diseases. The biofortification program in general includes micronutrients such as iodine, iron, calcium, zinc, magnesium, selenium, and vitamins such as provitamin A and folate (Kumar and Sindhusha, 2021).

Three major techniques are used for biofortification of vegetables and other staple crops: conventional breeding, agronomic strategies using mineral fertilizers, and genetic engineering (Fig. 18.1) (Lal *et al.*, 2020). Agronomic biofortification could provide short-term micronutrient enhancement through fertilizer application. Parental lines with greater concentrations of vitamins or minerals can be crossed repeatedly in conventional plant breeding to generate plants with the essential nutritional requirements (Kumari *et al.*, 2022). Tissue culture-based somaclonal variations, and molecular methods such as marker-assisted breeding, genetic engineering, and gene editing offer effective ways to enhance the nutritional value of crops with greater certainty over a short time span compared with traditional methods (Kersey *et al.*, 2020; Kumar and Sindhusha, 2021; Lyzenga *et al.*, 2021; Kumar *et al.*, 2022).

Biofortification can be carried out by one of two strategies: (i) increasing the micronutrient or vitamin level in a staple crop or vegetable; or (ii) by reducing the level of anti-nutritional factors that either have negative health effects on consumption or hinder the digestion and absorption of proteins and nutrients from foods, thereby indirectly leading to malnutrition (Qaim, 2016). Golden rice is a genetically modified, biofortified crop with increased levels of β-carotene, the precursor of provitamin A. Anti-nutritional factors in crops such as alkaloids, lectins, phytic acid, gossypol, goitrogens, glucosinolates, oxalic acid, raffinose, saponins, enzyme inhibitors, tannins, hydrogen cyanide and β-*N*-oxalylaminoalanine have been successfully reduced through conventional breeding or genetic-engineering methods to enhance the nutritional value of crops and vegetables (Thakur *et al.*, 2019).

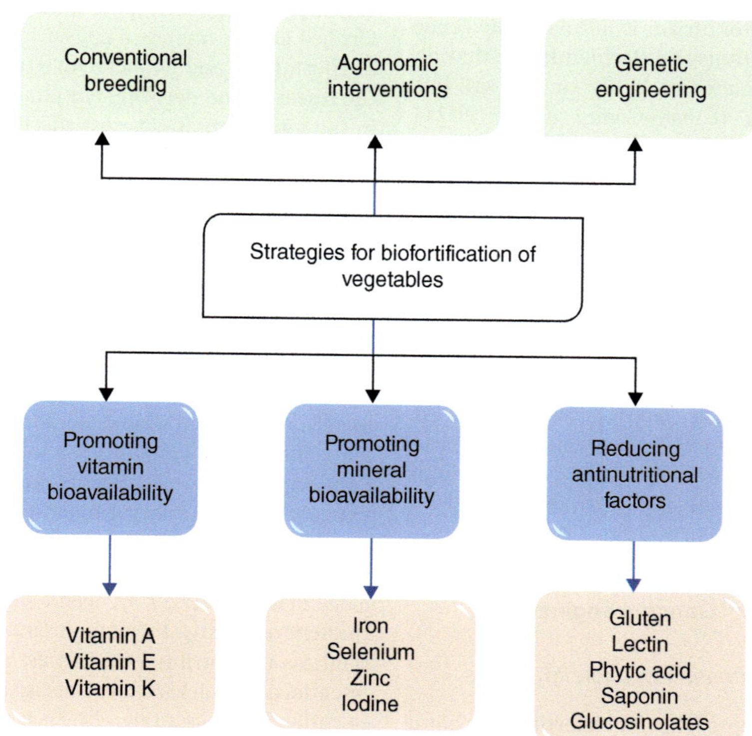


**Fig. 18.1.** Strategies for biofortification of vegetable crops.

With a major proportion of the population resorting to vegetarian or vegan food habits, biofortification of vegetables with essential nutrients is crucial for combating many health conditions, and hence this chapter discusses molecular methods for enhancing the nutrient availability in different vegetables. This will have substantial impact on increasing the availability of nutrient-rich foods for the targeted population all over the world, as well reducing malnutrition and ensuring overall food security; it will also be a vital tool for achieving the United Nations' Sustainable Development Goal (SDG) of eradicating hunger and malnutrition by 2030.

## 18.2 Biofortification in Vegetables through Molecular Approaches

Application of genetic engineering or molecular approaches enables the use of an infinite pool of genes to transfer and express desirable features from one organism to another that are evolutionarily and taxonomically different (Ahmar *et al.*, 2020; Basso *et al.*, 2020). Moreover, when a certain micronutrient is not naturally generated in crops, transgenic techniques are the only feasible alternative to fortify such crops with that nutrient (Lobato-Gómez *et al.*, 2021). Molecular approaches for biofortification include introducing new genes, overexpressing existing genes, downregulating the expression of certain genes, and disrupting the production pathway of genes of inhibitors. Homologous recombination, RNA interference (RNAi), transcription activator-like effector nucleases (TALENs), pentatricopeptide repeat proteins, zinc-finger nucleases (ZFNs), the clustered regularly interspaced palindromic repeats (CRISPR)-CRISPR-associated protein 9 (Cas9) system, intragenesis, and cisgenesis are among the genome-editing strategies that are used (Hanlon and Sewalt, 2021; Thakur *et al.*, 2022).

### 18.2.1 Genome engineering

*Transformation with foreign genes*

Transgenic approaches for biofortification entail transferring a foreign gene from one plant species to another, irrespective of the taxonomic or evolutionary position, or even introducing genes of interest from a completely different organism, such as bacteria. This can be a viable option for the production of biofortified crops when there is little to no genetic variability in the nutritional content of different plant species (Marques *et al.*, 2021). The capacity to discover and define gene function, followed by the use of these genes to modify plant metabolism, has been critical in producing transgenic crops. The metabolic pathways from bacteria and other organisms can also be mimicked in crops to leverage alternate metabolic processes (Dame *et al.*, 2021). The simultaneous insertion of genes involved in increasing the bioavailability and concentration of micronutrients and reducing the amount of anti-nutrients, which lower nutrient bioavailability in plants, is also possible using transgenic procedures. Moreover, genetic modifications can be utilized to reorganize the distribution of micronutrients in tissues, boost their concentration in edible plant parts, enhance the effectiveness of biochemical processes in edible tissues, and even recreate specific pathways (Jaiswal *et al.*, 2022). In contrast to agronomic biofortification programs, developing transgenically biofortified crops demands a considerable amount of effort, time, and financial investment during the research and development phase. However, in the long term, it is a cost-effective and productive approach (Meena *et al.*, 2022). Furthermore, there is no taxonomic limitation in genetic engineering, and even synthetic genes can be created and exploited. Transgenic crops with higher levels of micronutrients may reduce micronutrient deficiencies in their consumers, especially among the poor in developing countries (Buturi *et al.*, 2021).

Several crops have undergone genetic modification for enhanced micronutrient content. The nutritional status of food crops has been improved by using numerous genes from different sources to target micronutrients such essential amino acids, vitamins, minerals, and fatty acids. Biofortification targets include phytoene synthase (PSY), lycopene β-cyclase, and carotene desaturase for vitamins; nicotianamine synthase and ferritin for minerals; albumin for essential amino acids; and oleate desaturases for essential fatty acids (Fig. 18.2). Soybean with high-unsaturated fatty acid levels, maize with high lysine levels, and cassava rich in iron and

| Vitamins | Fatty acids | Minerals | Essential amino acids | Antioxidants |
|---|---|---|---|---|
| ▪ Carotene desaturase<br>▪ Phytoene synthase<br>▪ Lycopene β-cyclase<br>▪ DHR<br>▪ GTPCHI<br>▪ ADCS | ▪ Δ12 oleate desaturase<br>▪ Δ6 oleate desaturase | ▪ *OsIRT1*<br>▪ Nicotianamine synthase<br>▪ Ferritin | ▪ Albumin<br>▪ *Sb401*<br>▪ CgS<br>▪ DHPS | ▪ C1<br>▪ *LbMYB1* |

**Fig. 18.2.** Various targets for biofortification of vegetables. ADCS, amino deoxy chorismate synthase; C1: loci for carotenoid; CgS, cystathionine γ-synthase; DHPS, dihydropteroate synthase; DHR, dehydroascorbate reductase; GTPCHI, GTP cyclohydrolase I; *OsIRT1*, *Oryza sativa* Iron-regulated transporter1.

provitamin A are examples of biofortified crops. Reports on biofortified vegetables, fruits, legumes, grains, oilseeds, and fodder crops are available (Jaiswal *et al.*, 2022; Meena *et al.*, 2022).

### *Pathway manipulation by RNAi*

RNAi is a sequence-specific gene regulatory mechanism triggered by dsRNA, which in turn blocks the transcription or translation of a specific gene. Since its discovery, RNAi has revolutionized crop production (Solberg, 2021). It is a more accurate, reliable, and efficient technique than antisense technology. By biofortification and bioelimination, RNAi offers a platform for development of nutrient-dense vegetables, as well as the inclusion of biotic and abiotic stress tolerance. It is frequently used to improve crop nutritional quality and remove food allergens and pollutants (Rasool, 2021; Ratnaparkhe *et al.*, 2022). Applications of RNAi in vegetables include fruit quality enhancement including nutritional value, shelf life, seedlessness, and removal of allergens and anti-nutrients; development of tolerance to abiotic stresses such as salinity, drought, cold, and heat; development of tolerance to oxidative and biotic stresses such as viruses, bacteria, fungi, and insect pests; increased biomass in leafy vegetables and increased seed yield; and enhanced nutraceutical and pharmaceutical values (S. Singh *et al.*, 2020; Lata *et al.*, 2022). RNAi has been widely used for biofortification of vegetables in terms of improving vitamins, dietary fiber, minerals, and flavonoids, and as a protective food. Scientists have successfully used RNAi to generate tomato and potato varieties with higher quantities of carotene and vitamin C, and sweet potatoes with lower levels of amylose. It has also been used to reduce unwanted components such as toxins or anti-nutritional elements (Lata *et al.*, 2022).

### *Pathway manipulation by gene editing*

Sequence-specific nucleases are used in plant genome editing to create transgene-free plants through targeted and stably inherited genetic changes in the intended plant. A variety of sequence-specific nucleases such as ZFNs, TALENs, and the CRISPR-Cas system have been employed in plant genome editing (Kaul *et al.*, 2020; Cabrera-Ponce *et al.*, 2022). CRISPR/Cas9 is the most recent approach for genome editing to enhance desired features (Wada *et al.*, 2020; Gupta *et al.*, 2021; Alamillo *et al.*, 2023), and is a very effective and straightforward technique for initiating mutagenesis (Hunter, 2021). CRISPR/Cas9 has two components, single guide RNA (sgRNA) and the Cas9 protein (Riesenberg *et al.*, 2022). The complex of Cas9 and sgRNA uses the nucleases RuvC and HNH to cleave the target DNA, subsequently leading to either complementary or non-complementary breaks. The sgRNAs are intended to target particular DNA sequences located 3 bp upstream of the protospacer adjacent motif (PAM) (Gasiunas *et al.*, 2020; Chang, 2022). Moreover, Cas9 cleavage causes a double-strand break (DSB), which activates the DNA repair pathway. Non-homologous end-joining (NHEJ) or homology-directed repair (HDR) are used to repair DSBs. A DSB is more likely to activate NHEJ pathways, resulting in a variety of alterations, including insertions/deletions (indels) and substitutions (Xie *et al.*, 2020; Marr and Potter, 2021). With this method, the target genes are totally knocked out, and the genetic modifications are stably inherited from one generation to the next (Nasir *et al.*, 2022; Riesenberg *et al.*, 2022). The rapid development of next-generation sequencing (NGS) technologies, along with the accessibility of vegetables in the public domain, has enabled precise and accurate genome manipulation to obtain desired features. Genome editing of vegetable crops is currently restricted to a small number of vegetables (e.g. sweet potato, potato, squash, carrot, tomato, eggplant) owing to the dearth of regeneration methods and relevant genome sequence data (Das *et al.*, 2023). Table 18.1 lists the genes and traits that have been edited using CRISPR/Cas9 technology for biofortification of vegetable crops.

### 18.2.2 The role of NGS technologies in crop biofortification

Biofortification has been proven to be a beneficial alternative to tackle micronutrient deficiencies and human health issues in the last decade (Bouis *et al.*, 2017). Biofortification entails enhancing the bioavailability of micronutrients in specific crops that are usually consumed by low-income families around the world. The high-throughput deciphering of the genomes of multiple crops using NGS technology has the potential to change genetics, functional genomics, and breeding applications (N. Singh *et al.*, 2020). Numerous resequencing projects in various crops have greatly increased our knowledge of the structural variations throughout the genome. DNA markers identified through sequencing programs are enormously valuable in the development of nutritionally improved crop varieties and have shortened the time taken to release new varieties. NGS methods create huge amounts of sequence data in this setting, enabling low-cost genome-wide sequence reads that result in the detection of multiple single-nucleotide polymorphisms (SNPs). One of the most prominent advantages of NGS technology is SNP genotyping, which is a quick and low-cost technique for genotyping breeding populations. It enables breeders to implement genome-wide association studies (GWAS), genomic diversity studies, genomic selection, molecular marker discovery, haplotype-based gene analysis, and genetic linkage analysis (Chung *et al.*, 2017).

The current progress in sequencing technologies consists of first-generation sequencing technology (Sanger sequencing), second-generation sequencing technologies such as Genome Analyzer (Solexa, Illumina), SOLiD sequencer (Applied Biosystems), Roche/454 FLX Pyrosequencer, GS FLX Titanium/GS Junior, and third-generation sequencing technologies such as Oxford Nanopore Technology, PacBio, HiSeq/MiSeq (Illumina), and Ion Torrent PGM/Proton (Life Technologies). Due to their high throughput, long read lengths, and inexpensive sequencing, third-generation NGS technologies have become very common in recent years. NGS technologies are generally used for differential gene expression, transcriptomics, quantitative trait locus (QTL) mapping, targeting induced local lesions in genomes (TILLING), HapMap, epigenetic QTL analysis, GWAS, mutational mapping (MutMap), genomic selection, whole-genome resequencing, genotyping by sequencing, whole-genome bisulfite sequencing, whole-genome sequencing, exome sequencing, small RNA profiling, SHORE mapping, restriction site-associated DNA sequencing, and epigenetic methylated DNA immunoprecipitation and chromatin immunoprecipitation analysis (Babu *et al.*, 2020; Della Coletta *et al.*, 2021; Hussain *et al.*, 2022). In this regard, NGS technologies help us to look at pathways, accumulation processes, and nutritional traits. The application of quantitative genetics, pan-genomics, transcriptomics, metabolomics, ionomics, population genetics, proteomics, and nanotechnology methods increases our understanding of the exact selection of a particular trait of interest (Fig. 18.3).

### 18.2.3 GWAS and functional gene analysis

Comprehending the genetic basis of complex traits is essential for understanding developmental biology. Plant researchers have long been

**Table 18.1.** List of vegetables biofortified by genome editing.

| Vegetable crop (family) | Gene | Strategy | Mode of transformation | Trait improvement | Alteration in components | Reference |
|---|---|---|---|---|---|---|
| *Camelina sativa* (L.) Crantz (Brassicaceae) | *FAD2* | Knockout | *Agrobacterium tumefaciens*-mediated transformation | Quality improvement | Decreased linoleic acid (~16% to <4%), decreased linolenic acid (~35% to <10%) | Jiang *et al.* (2017) |
| | *FAD2* | Knockout | *A. tumefaciens*-mediated transformation | Quality improvement | Increased oleic acid (10% to 62%) | Morineau *et al.* (2017) |
| | *FAD2* | Knockout | *A. tumefaciens*-mediated transformation | Quality improvement | Increased oleic acid | Okuzaki *et al.* (2018) |
| | *FAE1* | Knockout | *A. tumefaciens*-mediated vacuum infiltration transformation | Quality improvement | Decrease in very long-chain fatty acids (60%) | Ozseyhan *et al.* (2018) |
| Tomato (*Solanum lycopersicum* L.) (Solanaceae) | *ANT1* | Knockout | *A. tumefaciens*-mediated transformation | Color | Increased anthocyanin | Karunarathna *et al.* (2020) |
| | *RIN* | Knockout | *A. tumefaciens*-mediated method | Fruit ripening | Decreased red color pigmentation in mutant | Ito *et al.* (2015) |
| | *PSY1* | – | *A. tumefaciens* strain GV3101 using cotyledon transformation | Color | – | Filler Hayut *et al.* (2017) |
| | *GAD* | Knockout | *A. tumefaciens*-mediated transformation | Quality improvement | Increased γ-aminobutyric acid | Nonaka *et al.* (2017) |
| | *SlALMT9* | Knockout | *A. tumefaciens* strain C58-mediated electroporation method | Quality improvement | Increased malate | Ye *et al.* (2017) |
| Sweet potato (*Ipomoea batatas* (L.) Lam.) (Convolvulaceae) | *GBSSI*, *SBEII* | Knockout | *A. tumefaciens* strain LB4404-mediated transformation | Quality improvement | Decreased amylose, decreased amylopectin | Wang *et al.* (2019) |

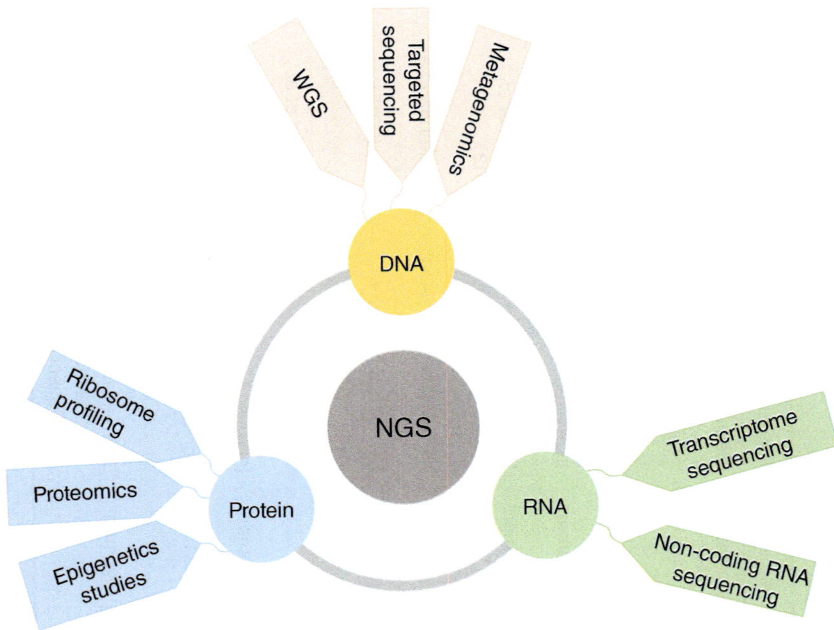


**Fig. 18.3.** Applications of next-generation sequencing (NGS) technologies that span the central dogma of molecular biology. WGS, whole-genome sequencing.

intrigued by phenotypic variation in distinct genetic backgrounds, which has paved the way for a number of association studies (Descalsota *et al.*, 2018). The development of high-density genotyping arrays, genome sequencing technology, and platforms for high-throughput phenotyping paved the path for GWAS (Varshney *et al.*, 2021). GWAS begins with genotyping using various NGS technologies, such as whole-genome sequencing or genotyping by sequencing, and then moves on to thorough phenotyping under a set of parameters. The data obtained from genotyping is then filtered out to acquire relevant SNPs that are subsequently subjected to trait-wise GWAS in order to generate SNP–trait relationships. As a result, potential genes for the traits of interest are identified (Alqudah *et al.*, 2020).

GWAS analysis performed using diploid potato germplasm from the Colombian *Solanum tuberosum* Group Phureja has identified SNPs and seven QTLs associated with anthocyanin content. The most consistent QTLs were discovered on chromosomes 1 and 10. Of the seven QTLs, three matched with previously identified genes implicated in anthocyanin regulation in potato tubers (Parra-Galindo *et al.*, 2019). The phenylalanine-ammonia lyase (*PAL*) gene was found in a broad set of indigenous landraces of diploid potatoes, along with other important SNPs and multiple anthocyanin homologs, and is connected with five anthocyanin compounds. *PAL* pleiotropy is an exceptional breeding target because it generates substantial variations in anthocyanins through recurrent selection. The multiple MYB transcription factors related to anthocyanin accumulation in potato, located in close proximity to *PAL*, showed that there may be multiple causal genes at the QTLs. At this cluster, most potatoes that have higher anthocyanin levels share the same genotype, indicating recurrent selection of the same alleles. As a result, this region has a high level of diversity, which is in line with balancing artificial selection to grow potatoes of various colors (Parra-Galindo *et al.*, 2021).

*ANTHOCYANIN1* (*StAN1*) on chromosome 10 is a vital gene regulating the biosynthesis of anthocyanin in potato tubers. The diagnostic markers available at present for finding functional *StAN1* alleles (*StAN1777* and *StAN1816*) are ineffective for predicting potato pigmentation patterns. As a result, it is necessary to develop new markers for detecting anthocyanin production, which can be employed effectively in potato breeding (Strygina *et al.*, 2019). In some potato genotypes, the absence of anthocyanins could be a result of a structural gene mutation, as demonstrated by *StF3′5′*, which interrupts anthocyanin synthesis partially and causes blue and purple pigmentation rather than red. As a result, it is an appealing subject for identification of purple- or red-fleshed potatoes using markers (Strygina and Khlestkina, 2017).

Of the 156 *MYB*-like genes, 22 in sweet potato are tightly linked, positively or negatively, to anthocyanin levels in storage roots or leaves. *IbMYB1*, one of the key genes governing anthocyanin biosynthesis, consists of three members *IbMYB1-1*, *IbMYB1-2a*, and *IbMYB1-2b*; the latter two are not required for the buildup of anthocyanin in cultivated species with purple rings, shoots, or spots on the flesh. Constant and robust *IbMYB1* expression is required to preserve the purple color of the storage roots and leaves. It is an anthocyanin biosynthetic early-response gene. *IbMYB2*s are similar to *IbMYB1* but not IbMYB1 members, hinting that there may be other *IbMYB1* members involved in anthocyanin biosynthesis control (D. Zhang *et al.*, 2021). A GWAS–expression QTL analysis showed that *IbMYB1-2* was a master regulator of the anthocyanin production in sweet potato storage roots (Zhang *et al.*, 2020).

A novel transcription factor complex (IbERF71–IbMYB340–IbbHLH2) coregulates the biosynthesis of anthocyanin in purple-fleshed sweet potatoes by binding to the promoter of *IbANS1*, as demonstrated by higher expression levels of *IbMYB340*, *IbbHLH2*, and *IbERF71*. The levels of expression were found to be favorably related to anthocyanin content. The co-transformation of *IbMYB340–IbbHLH2* led in anthocyanin accumulation in the leaves of tobacco and strawberry receptacles, with *IbERF71* significantly enhancing the color. *FvANS* and *FvGST*, which play a role in anthocyanin production and transport, respectively, in strawberries, were both significantly enhanced by transcription factors. Thus, a novel regulatory network involved in the biosynthesis of anthocyanin in purple-fleshed sweet potatoes was discovered (Ning *et al.*, 2021).

The increased accumulation of anthocyanin in cultivated tomato fruit was influenced by the *Solanum chilense Anthocyanin fruit* (*AftAft*/–) and *atroviolacium* (*atvatv*) loci (Cao *et al.*, 2017). The fruit color and anthocyanin concentration of the CRISPR/Cas9-mediated *Slan2* mutants were comparable to the wild-type cultivar 'Indigo Rose.' However, in comparison with the wild type, the *Slan2* mutant vegetative tissues had significantly lower anthocyanin content and lower relative expression levels of anthocyanin-related genes (Zhi *et al.*, 2020). *Slan2-like*, another potential candidate gene, in the *Aft* locus, was fine-mapped on chromosome 10 to a 145 kb region. Because of downregulation of numerous anthocyanin genes, CRISPR/Cas9-mediated *Slan2-like* mutants have considerably lower anthocyanin levels than the wild type. This suggests that *Slan2-like* is a promising candidate gene for boosting tomato anthocyanin concentration (Yan *et al.*, 2020). Alternative splicing is defined as "the process of generating different forms of mature mRNA from the primary gene transcript" (Mattoo *et al.*, 2022). A functional comparison of R2R3-MYB transcription factors in *Aft* and wild-type plants showed significant differences in *Slan2-like* expression and protein functioning. Splicing mutations rendered the wild-type protein useless, implying its role in tomato fruit anthocyanin pigmentation (Colanero *et al.*, 2020). Anthocyanin accumulation in tomato is regulated by photomorphogenesis-related transcription factor SlBBX20 (Xiong *et al.*, 2019). SlBBX20 increases anthocyanin biosynthesis by interacting with the SlDFR promoter, the gene for anthocyanin biosynthesis. It also interacts with the COP9 signalosome component, SlCSN5-2. *SlCSN5-2* silencing led to anthocyanin over-accumulation in the calluses and shoots of transgenic tomato, whereas overexpression of *SlCSN5-2* reduced the buildup of anthocyanin, implying that modulation of SlBBX20–SlCSN5 might represent an entirely new regulatory pathway for light-induced biosynthesis of anthocyanin in tomato (Luo *et al.*, 2021).

Noteworthy success has been made in finding the genes involved in the biosynthesis of anthocyanin in crops, a few of which have been characterized functionally (Table 18.2). An intensive strategy, however, is required for the functional characterization of other candidate

**Table 18.2.** Anthocyanin-rich potato, sweet potato and tomato cultivars/advanced breeding lines, and genetic stocks created through crossbreeding and selection.

| Vegetable crop | Cultivar(s) | Characteristics | Reference |
|---|---|---|---|
| Potato | 'Kufri Neelkanth' | Adapted for cultivation in northern Indian plains; much higher yield (35–38 t $ha^{-1}$) compared with national average of 23 t $ha^{-1}$; anthocyanin >100 µg 100 $g^{-1}$ FW | Sharma et al. (2009); https://icar.org.in/node/12076 (accessed 17 July 2024) |
| | 'Puma Makin,' 'Leona,' 'YawarManto,' 'Añil,' 'Sangre de Toro,' and 'Qequrani' | Native pigmented (red and purple) cultivars grown in Huancavelica region of Peru. Anthocyanin content: 'Puma Makin,' 74.3 mg $kg^{-1}$ SDW; 'Leona,' 6.32 mg $kg^{-1}$ FDW and 166.65 mg $kg^{-1}$ SDW; 'YawarManto,' 602.9 mg $kg^{-1}$ FDW and 709.4 mg $kg^{-1}$ SDW; 'Añil,' 104 mg $kg^{-1}$ FDW and 273.9 mg $kg^{-1}$ SDW; 'Sangre de Toro,' 27.5 mg $kg^{-1}$ FDW and 124.2 mg $kg^{-1}$ SDW; 'Qequrani,' 10.2 mg $kg^{-1}$ FDW | Bellumori *et al.* (2020) |
| | 'AmaRosa,' 'Purple Pelisse,' and 'Terra Rossa' | Released for cultivation in USA. 'AmaRosa,' TAC 18.2 mg compared with 13.8 mg 100 $g^{-1}$ FW in control 'All Blue'; 'Purple Pelisse,' TAC 34.2 mg compared with 12.6 mg 100 $g^{-1}$ FW in control 'All Blue' | Brown *et al.* (2012); Vales *et al.* (2012); Shock *et al.* (2018) |
| | 'Hongyoung' and 'Jayoung' | Released for cultivation in South Korea. 'Hongyoung,' TAC 31.8 mg 100 $g^{-1}$ FW, 3.6 times higher than control 'Jasim' | Park *et al.* (2009) |
| | 'HermannsBlaue,' 'Vitelotte,' 'Shetland Black,' and 'Valfi' | European, blue-fleshed potato cultivars | Qiu *et al.* (2018) |
| Sweet potato | 'Antin 1,' 'Antin 2,' and 'Antin 3' | Released in Indonesia. TAC 8.4, 130.2, and 150.7 mg 100 $g^{-1}$ FW | Ginting *et al.* (2020) |
| | 'Bhu Krishna' | Released in India. TAC 90.0 mg 100 $g^{-1}$ compared with zero in control | Parulekar *et al.* (2019) |
| | P4 | Released in USA. TAC up to 14 mg $g^{-1}$ DW | Xu *et al.* (2015) |
| | 'Borami,' 'Mokpo 62'; 'Shinzami,' and 'Zami' | Released in South Korea | Lee *et al.* (2013) |
| | 'Yamagawamurasaki,' 'Ayamurasaki,' 'ChiranMurasaki,' 'Tanegashima Murasaki,' 'Naka Murasaki,' and 'Purple Sweet' | Released for cultivation in Japan | Cuevas *et al.* (2011) |
| Tomato | Near-isogenic line (NIL) | Enhanced accumulation of anthocyanins and bioactive compounds with distinct changes in volatile compounds in NIL carrying *Aft*, *atv*, and *hp2* in the genetic background of red-fruited tomato | da Silva Souza *et al.* (2020) |

| | | | |
|---|---|---|---|
| | Breeding line | A purple-fruited line derived from a cross between ‘OSU Blue’ (blue fruit) and ‘Purple Mini’ (brown fruit) | Kang *et al.* (2018) |
| | ‘Sun Black’ | Deep purple skin but with a normal red-colored pulp and taste similar to traditional tomato released in Europe; TAC 7.1 mg 100 $g^{-1}$ FW, comparable to eggplant or red cherry | Mazzucato *et al.* (2013); Filler Hayut *et al.* (2017) |
| | ‘Indigo Rose’ | A blue-colored tomato released for cultivation in the USA. TAC up to 10 mg 100 $g^{-1}$ FW on a whole-fruit basis compared with normal tomato fruits devoid of anthocyanin | Myers (2012) |

FDW, flesh dry weight; FW, fresh weight; SDW, skin dry weight; TAC, total anthocyanin content.

genes to determine their contribution to the biosynthesis of anthocyanin and to create genetic tags for use as crossbreeding aids.

## 18.3 Biofortification in Solanaceous Vegetables

Solanaceous vegetables are common ingredients in dishes all over the world. They are commonly used in stews, soups, and sauces, as well as in medicinal concoctions. As a result, solanaceous vegetables have the potential to play critical roles in supplying dietary diversity, food and nutrition stability, and income generation for many households worldwide (Han *et al.*, 2021).

### 18.3.1 Tomato

Tomato is a rich source of antioxidants, minerals, and vitamins (Dorais *et al.*, 2008), and is a reservoir of carotenoids (e.g. β-carotene, lycopene), glycol-alkaloids such as tomatine, phenolic compounds (phenolic acids and flavonoids), and non-heme iron (0.3 mg 100 $g^{-1}$), the availability of which is enhanced by vitamin C (Porto *et al.*, 2016). The steps taken to increase the nutritional value of tomatoes are detailed below.

#### *Approaches based on genetic engineering*

Genetic engineering can be utilized to enhance the nutrient content, increase the efficiency of metabolic pathways, and redistribute the micronutrients in tissues. Genetic engineering is versatile because synthetic genes can be used and there are no genetic or taxonomic restrictions (Vats *et al.*, 2022). Many agricultural crops, including tomatoes, have been biofortified through genetic engineering. As the genome of tomato is diploid, it is relatively easy to carry out genetic modification (Kalloo, 2012). Tomato has many bioactive compounds (nutrients) that the human body cannot synthesize *de novo*, and thus increasing their concentrations will be advantageous.

Fruit-specific downregulation of the *DE-ETIOLATED1* (*DET1*) gene was found to enhance the levels of tocopherols, flavonoids, carotenoids, anthocyanidins, and phenylpropanoids in tomato. Metabolite profiling has verified these findings (Kaur *et al.*, 2017). Increases in lycopene and β-carotene levels in tomato mutants constitutively overexpressing the *SlPSY1* gene resulted in reduced pigmentation levels, fruits lacking carotenoids, and plant dwarfism. These results were most likely caused by geranylgeranyl diphosphate pool redirection for co-suppression or synthesis of gibberellin, with downregulation of *SlPSY* rather than overexpression (Vats *et al.*, 2022). Transgenic plants overexpressing the *SlPSY* gene had increased carotenoid content (Liu *et al.*, 2015). The pleiotropic effects were overcome through fruit-specific expression of the *Erwinia uredovora* PSY (*crtB*) gene in tomato and led to the development of plants with higher carotenoid levels (Farré *et al.*, 2014; Shwetha *et al.*, 2020). Bacterial *crtI*, which encodes the *Pds* gene, has been utilized to boost the β-carotene levels in transgenic tomato seedlings (Zhou *et al.*, 2008). *SlSGR1* has a direct negative correlation with *SlPSY1* (Luo *et al.*, 2013), which means that downregulating it through any strategy including RNAi could lead to an increase in the lycopene and β-carotene content in tomato (Luo *et al.*, 2013). It was also found that *SlSGR1*-repressed plants produce fruits with longer storage life and less chlorophyll breakdown. Multiple studies have found elevated β-carotene levels as a result of *CYC-B/LCY-B/LCY-1* overexpression (Vats *et al.*, 2022).

Carotenoid concentration is determined by the balance between catabolism and biosynthesis (Watkins and Pogson, 2020). Apocarotenoids are formed when carotenoids are degraded either enzymatically or non-enzymatically by reactive oxygen species. Differential expression of the enzymes that regulate carotenoid degradation is thus a practical method for increasing the amount of specific carotenoids (Joshi *et al.*, 2012). RNAi was used to target 9-*cis*-epoxycarotenoid dioxygenase 1 (*SlNCED1*), which resulted in deep red fruits with increased β-carotene and lycopene contents. The ectopic expression of *SlDXS* and 3-hydroxymethylglutaryl CoA (*hmgr-1*) genes, when metabolically engineered, increased the amount of β-carotene, phytoene, and sterol (Enfissi *et al.*, 2005). Plants with greater levels of β-carotene and xanthophyll (particularly zeaxanthin and β-cryptoxanthin) were obtained as a result of synchronous overexpression of β-carotene hydroxylase and *LCY-B* under *Pds*, a fruit-specific promoter (Dharmapuri *et al.*, 2002). Overexpression of β-carotene ketolase and hydroxylase resulted in the buildup of astaxanthin, a product obtained through hydroxylation of β-carotene (Dharmapuri *et al.*, 2002).

The upregulation of two genes from the GDP-mannose 3,5-epimerase (GME) family, *SlGME1* and *SlGME2*, catalyzes the conversion of GDP-D-mannose to GDP-L-galactose, an essential step in vitamin C biogenesis. Such plants were also more resistant to salt, oxidative, and temperature stress (Baldet *et al.*, 2013). Dehydroascorbate reductase and the concurrent expression of three genes, namely myo-inositol oxygenase, GDP-mannose pyrophosphorylase, and arabinono-1,4-lactone oxidase, were employed to enhance the vitamin C level of tomatoes (Cronjé, 2010).

The concentration of anthocyanin in tomatoes was improved by the overexpression of *Petunia* chalcone isomerase (Gonzali *et al.*, 2009) and *MYB75/PAP1* (*PRODUCTION OF ANTHOCYANINPIGMENT1*) (Zuluaga *et al.*, 2008), and two transcription factors, *DELILA* and *ROSEA1* (Naing *et al.*, 2017). The former showed a 78-fold rise in the flavonol content of tomato peel, primarily rutin. The expression of grapevine stilbene synthase increased *trans*-resveratrol levels (Verhoeyen *et al.*, 2002). Chlorogenic acid levels were enhanced by the overexpression of the genes hydroxycinnamoyltransferase (*HQT*) and hydroxycinnamoyl-CoA quinate (Niggeweg *et al.*, 2004). Overexpressing soybean isoflavone synthase (*GmIFS2*) enhanced genistin levels, whereas the overexpression of *AtMYB12* resulted in an increase of polyphenols (caffeoyl quinic acid) content in tomato (Liu *et al.*, 2019).

Thus, a significant amount of work has gone into the development of biofortified transgenics in tomato. However, due to widespread consumer skepticism and concern about transgenic crops, these initiatives have yet to bear fruit as commercial applications. As a result, agronomic and conventional breeding practices targeted at commercial exploitation continue to be preferred over transgenic approaches. Recent advances in genome editing technology have the ability to create plants that are transgene free and thus acceptable to the public.

### *Genome editing efforts and opportunities*

CRISPR/Cas-based genome editing is a simple technique involving only cloning of the constructs and direct transformation. CRISPR/Cas has already been used successfully and effectively in many products, including tomatoes (Zhu *et al.*, 2020). Undeniably, the simple diploid genome of tomato, the availability of a highly optimized protocol for *Agrobacterium*-mediated plant transformation (Sharma *et al.*, 2009; Sandhya *et al.*, 2022), and the ease of availability of genomic resources renders tomato a perfect candidate for accurate gene editing. Many attempts have already been made using gene editing to boost the level of nutrients in tomatoes, such as bioactive compounds and carbohydrates (Wan *et al.*, 2021). Studies have also been performed to investigate the roles of the regulatory genes that control the metabolism of these compounds.

Previously, map-based cloning was used to confirm the functions of genes identified through mutation or similar methods. In certain cases, pathways become complex and challenging, inhibiting the speedy identification of genes (Quinet *et al.*, 2019). A number of forward genetics screening techniques, including genome editing, are presently used to confirm the functions of putative genes. CRISPR/Cas was recently used to identify the role of mitogen-activated protein kinase 20 (MAPK-20), a group D MAPK, in the metabolism of carbohydrate and auxin in anthers, and SlMYB32 was identified as its target protein (Vats *et al.*, 2022). Using CRISPR/Cas, researchers discovered that aluminum-activated malate transporter 9 (ALMT9) plays an important part in determining tomato malate content (Xu *et al.*, 2018). The level of anthocyanin was altered in tomato plants to generate purple fruits using genome editing to overexpress an R2R3-MYB protein named SlANT2, which positively regulated production of anthocyanin content (Kiferle *et al.*, 2015). In at least two studies, γ-aminobutyric acid (GABA), a non-proteogenic amino acid with a hypotensive action in humans, was discovered to be upregulated in tomato. In the first investigation, the inhibitory domains of the glutamate decarboxylase (GAD) genes *SlGAD2* and *SlGAD3* were deleted (Takayama *et al.*, 2015), and in the second study, CRISPR/Cas9 was used to target five genes in the GABA shunt (Li *et al.*, 2018).

Targeting numerous genes involved in the carotenoid pathway, including lycopene β-cyclases, *SGR1* and *LCY-E*, has been shown to increase lycopene levels (Liu *et al.*, 2015; Luo *et al.*, 2023;

Yang *et al.*, 2023). A tomato-specific CRISPR kit was created using Golden Gate and BioBrick cloning principles. This binary vector assembly includes eight *Bsa*I-site primers and five *polIII* promoters (Hu *et al.*, 2019). In tomato, CRISPR/Cas has made multi-gene targeting easier (Hu *et al.*, 2019; Reem and van Eck, 2019; Chaudhuri *et al.*, 2022; Waltz, 2022) and platforms based on Csy4 (Wang *et al.*, 2021), tRNA (Hashimoto *et al.*, 2018), miRNA (Hong *et al.*, 2021), and ribozyme (Wada *et al.*, 2022). Li *et al.* (2016) used an intron-mediated site-specific gene replacement and insertion approach in rice, which can also be utilized for tomato.

By using the CRISPR-based multi-gene targeting method, Li *et al.* (2018) were able to promote lycopene biosynthesis while preventing its conversion to β- and α-carotene. Lycopene accumulation was aided by the deletion of numerous genes associated with the carotenoid metabolic pathway. The CRISPR/Cas9 technique was used to modify five specific genes in the genome.

The introduction of modified forms of the CRISPR/Cas technology, such as precise base editing employing the target AID system with cytidine deaminase (Hunziker *et al.*, 2020), geminiviral replicon-based HDR to enhance the repair template titer (Sun *et al.*, 2016), CRISPR-mediated methylome editing (Kang *et al.*, 2019), and other better CRISPR/Cas methods (Ryan *et al.*, 2018), has enhanced the prospects for the production of novel tomato varieties. Utilization of existing tomato crop information, as well as novel approaches and prospects offered by CRISPR/Cas, will be critical in the area of tomato biofortification.

### 18.3.2 Potato

Because potato is a short-term and simple-to-grow staple crop in many countries, it is a suitable target crop for biofortification (Singh *et al.*, 2022). Potato has very high iron bioavailability when compared with cereals (Shubham *et al.*, 2020). Vegetarians need 1.8 times the recommended dietary allowance (RDA) for iron compared with non-vegetarians (Gan *et al.*, 2018). A human gut model based on Caco2 cell lines and an *in vitro* gastrointestinal digestion approach have demonstrated that, on average, 70.6% of the iron in potato is liberated from its matrix and is thus available at the intestinal level (Boim *et al.*, 2020). As a result, potatoes could provide a significant percentage of the RDA. In addition, the existence of significant natural variation in potato germplasm could be employed to biofortify potatoes (Singh *et al.*, 2021). Numerous efforts have been made over the last decade to develop iron-rich biofortified potatoes through agronomic practices, conventional plant breeding, and advanced genetic engineering. However, our understanding of the various molecular processes that control iron uptake, translocation, and storage in potatoes remains limited.

#### *Potato biofortification by breeding*

The most crucial requirement for plant breeders is the availability of natural variation in a plant's germplasm. Fortunately, natural variations in skin texture, tuber form, size, color, and concentrations of different nutrients are abundant in potato germplasm (Blum, 2018; Zaidi *et al.*, 2019). For example, iron-rich potato accessions are found in *Solanum tuberosum* and *S. tuberosum* ssp. *andigena* (J. Sharma *et al.*, 2017). These accessions could be used as parental lines in breeding projects to generate iron-rich potato varieties, although more research into the trait's heritability is required. Knowledge of the genes involved in the absorption of iron and accumulation in tubers could aid potato-breeding programs targeted at increasing iron content. Various types of molecular markers associated with iron metabolism could be used in potatoes and other crops (Sen Gupta *et al.*, 2017). In addition, GWAS is now recognized as an effective method for identifying genes/markers linked to complex phenotypic traits such as mineral concentrations (Del Carpio *et al.*, 2019; Raggi *et al.*, 2019). The GWAS approach should be used to examine the marker–trait relationships linked to tuber iron content in various potato germplasm.

#### *Transgenesis*

Transgenic potato seedlings with the *AmA1* gene from *Amaranthus hypochondriacus*, for a non-allergenic protein, were developed. Compared with control plants, tubers from such modified potato plants had higher levels of all amino acids (Chakraborty *et al.*, 2000). Potatoes transformed with phosphofructokinase gene from the bacterium *Lactobacillus bulgaricus* were less preferred in the market as they had more simple sugars and become brown during frying. Transgenic potato had a lower sugar content and their chips were lighter in color than non-modified potato chips (Navrátil *et al.*, 2012).

As iron has lower phloem mobility, biofortification of potato with iron may be promising (Vos, 2009; Subramanian *et al.*, 2011). Iron biofortification could be enabled by the *IRON DEFICIENCY TOLERANT1* (*IDT1*) gene (Sharma and Yeh, 2020). Overexpression of ferritin (*FER3*) and iron-regulated transporter (*IRT1*) genes, according to one study, may help potato plants develop normally in calcareous soils without showing any iron-deficiency signs. (Challam *et al.*, 2021). Similarly, the expression levels of oligopeptide transporter, metal transporter, germin, and plasma membrane $H^+$-ATPase genes increased considerably in potato plants grown in soil with a lower iron content and complemented iron uptake (M. Zhang *et al.*, 2021). In contrast to other major crops, potatoes adopt a reduction-based method for iron uptake and hence the identification of genes involved in iron accumulation is challenging.

Transgenic potatoes with higher levels of provitamin A have been created by integrating *Erwinia herbicola* phytoene synthase (*CrtB*), lycopene β-cyclase (*CrtY*), and phytoene desaturase (*CrtI*) genes (Sharma and Yeh, 2020).

### 18.3.3 Cauliflower

Cauliflower (*Brassica oleracea* var. *botrytis* L.) is a major vegetable cultivated on approximately 9 million ha globally, with a combined output of approximately 27 million t (Natalini *et al.*, 2021; FAOSTAT, 2022). Cauliflower quality decided by biochemical traits such as vitamin C, glucosinolates, carotenoids, and polyphenols is emerging as a major factor in consumer preference (Habeeb Rahman, 2011). Along with yield, quality, and stress tolerance, their nutritional value as determined by phytochemicals and bioactive compounds is a major breeding objective in cauliflower (Singh and Kalia, 2021). Heterosis in cauliflower has been studied in terms of the presence of antioxidant pigments (Singh *et al.*, 2019; Natalini *et al.*, 2021).

Omics methods have mainly been used to study pathogen resistance and heat and salt stress tolerance in the genus *Brassica*. Cauliflower breeding efforts were focused on developing hybrids with a high carotene content (Ding and Jian, 2010). A crossbreed between an inbred line (orange mutant line producing small curds) and a cytoplasmic male sterile line (marketable white cauliflower) produced a saleable orange $F_1$ with good nutritional quality. The hybrid had 10.8 times higher carotene compared with the original white cauliflower (Rodriguez-Concepcion *et al.*, 2018).

Studies have also been carried out to phenotype and describe biochemical processes in cauliflower (Sandrini *et al.*, 2022). Metabolites, primarily glucosinolates, along with other primary and secondary metabolites have been studied extensively in cauliflower. A sulfur-linked d-glucopyranose and an amino acid-derived chain make up these molecules (Natalini *et al.*, 2021)

### 18.3.4 Carrot

Carrots are an excellent source of vitamins, β-carotene, and nutrients. There have been a few studies on the biofortification of carrots through transgenesis. Carrots have been transgenically modified to overexpress the calcium transporter gene *SCAX1* (Hawthorne *et al.*, 2009). Carrots thus engineered were found to have enhanced calcium uptake with the potential to alleviate diseases such as osteoporosis.

### 18.3.5 Cassava

Cassava is a staple vegetable in many diets around the world. It is an excellent source of nutrients but should not be eaten raw due to the presence of potentially hazardous quantities of the cyanogenic glucoside linamarin (Panghal *et al.*, 2019). The cyanogen levels in the leaves of many cassava cultivars (200–1300 mg cyanide (CN) equivalents $kg^{-1}$ dry weight (DW)) and roots (10–500 mg CN equivalents $kg^{-1}$ DW) exceed the FAO's highest recommended cyanide levels (10 mg CN equivalents $kg^{-1}$ DW) in foods. To create cyanogen-free cassava plants, researchers reduced expression of the *CYP79D1* and *CYP79D2* genes encoding the cytochrome P450s that catalyze the first stage in the synthesis of linamarin. As a consequence, linamarin levels in the leaves and roots of *CYP79D1* and *CYP79D2* antisense plants were reduced by up to 94% and 99%, respectively (Siritunga and Sayre, 2003). Furthermore, to improve the nutritional content of cassava roots, which have 85% starch by dry weight but lack protein, a synthetic *ASP1* gene expressing an important amino acid-rich storage protein (80%) was introduced by *Agrobacterium*-mediated gene transfer into embryogenic suspensions (Zidenga *et al.*, 2017).

### 18.3.6 Lettuce

Lettuce is commonly used as a salad foundation and is high in magnesium, phosphorus, and calcium. Lettuce is also high in fiber, and its iron level has recently been enhanced through transgenic biofortification. Because lettuce has a low iron content relative to other leafy vegetables, researchers have introduced the ferritin gene from soybean to boost its iron content (Goto *et al.*, 2000).

## 18.4 Challenges in Biofortification of Green Leafy Vegetables

Green leafy vegetables such as lettuce, amaranthus, cabbage, and spinach grow quicker than vegetable crops and have relatively higher carotenoid levels in their chloroplasts. Enhancing these levels or altering the carotenoid profiles, however, frequently involves photosynthesis interference and thus has an impact on plant physiology, fitness, and growth. To guarantee the correct assembly and performance of photosynthetic complexes, carotenoid levels in chloroplasts are carefully balanced with chlorophyll levels (Sandmann *et al.*, 2006; Morelli and Rodriguez-Concepcion, 2022). Alterations in the ratio of carotenoid:chlorophyll cause changes in the photosynthetic system conformation and disrupt the ability of chloroplasts to efficiently harness light energy. Variations in the carotenoid profile can also impair photoprotection and the plant's efficiency to absorb excess light energy as heat. Strategies for enhancing the accumulation of carotenoid in non-photosynthetic tissues, such as calluses or ripe fruit, have not been successful because the carotenoid pathway is strictly monitored in green tissues to prevent adverse effects on photosynthesis and photoprotection (Torres-Montilla and Rodriguez-Concepcion, 2021). Despite these constraints, there have been reports of successful carotenoid biofortification in plant leaves. Red plants with barely perceptible levels of endogenous carotenoids were produced by transformation of tobacco and lettuce with exogenous pathways for astaxanthin and other bacterial ketocarotenoids, which are the highly desired colors in feed additives (Bock, 2022). However, transplastomic plants lost their usual chloroplast characteristics such as grana and thylakoid membranes, and gained enormous plastoglobule-like osmophilic particles (Lu *et al.*, 2017). After transiently expressing the bacterial PSY-encoding gene *crtB*, it was found that leaves had a similar phenotype to plastids that resembled chromoplasts (Llorente *et al.*, 2020; Morelli and Rodriguez-Concepcion, 2022).The elevated phytoene production by the *crtB* gene localized in the plastid interfered with photosynthetic functionality, ultimately rendering the differentiation of chloroplasts into chromoplasts, as excess phytoene was gradually metabolized into carotenoids by plant enzymes produced endogenously (Llorente *et al.*, 2020). In artificial chromoplasts, by increasing the number of plastoglobules along with other storage structures, *crtB* triggered the buildup of isoprenoid compounds such as phylloquinone (vitamin K) and tocopherols (vitamin E) (Morelli and Rodriguez-Concepcion, 2022). This approach aids in the biofortification of lipophilic isoprenoid vitamins and carotenoids in a range of plant systems, including culinary crops such as lettuce and zucchini (Saini *et al.*, 2015; Rodriguez-Concepcion and Daròs, 2022). However, as the conversion from chloroplasts to chromoplasts leads to a loss of photosynthetic activity, this approach is only appropriate for plant biofortification when photosynthesis is not required (e.g. immediately before harvest).

Engineering the synthesis of carotenoids outside of the chloroplast is another method of supplying carotenoid-rich green tissues without hampering photosynthesis. Using viral vectors, crystals of lycopene were produced in the cytoplasm of cells by enhancing the endogenous mevalonate pathway (Howarth *et al.*, 2010) and directing the increased supply of dimethylalyl diphosphate and isopentenyl diphosphate to lycopene formation using bacterial genes (Zu *et al.*, 2020). Apart from avoiding direct interference with photosynthesis, using the cytosol as an alternative site for the production and storage of carotenoid has several benefits. Levels of production of carotenoid intermediates with health benefits (e.g. phytoene) can be enhanced in the cytosol, and as the plant enzymes associated with carotenoid metabolism exist only in plastids, such intermediates will not be transformed into downstream products.

## 18.5 Conclusion and Future Perspectives

Achieving its 17 established SDGs by 2030 is an established mandate of the United Nations. Of these, SDG3 is about guaranteeing healthy lifestyles and fostering well-being for people of all ages. Nutrition and the consumption of nutritious foods are inextricably related to good health. However, hidden hunger is a significant impediment to ensuring good health for the global population, especially in young children and women in developing nations.

Crop biofortification, in addition to providing balanced nutrition to the masses, assists governments in meeting SDG3 by creating crops that are nutritionally enriched. Effective biofortification programs should aim to create crop varieties with increased micronutrient content while not jeopardizing farmers' economic benefits (Lockyer *et al.*, 2018). The effectiveness of a biofortification breeding program depends heavily on the presence of genetic diversity, the decrease in anti-nutrients, and improved concentrations of amino acids (cysteine, lysine, and methionine) and ascorbic acid (vitamin C), which can enhance the absorption of essential elements (Garcia-Casal *et al.*, 2017).

Vitamins and minerals are necessary for human development and health (Thakur *et al.*, 2022). Vitamins A, C, E, and β-carotene are powerful antioxidants that also serve to improve health through other mechanisms (Cannon and Leitzmann, 2022). Vegetables are one of the most crucial dietary sources of nutrients and are referred to as "protective foods" as they can protect the body from a number of diseases. Biofortification of vegetables with various essential nutrients and reduction of anti-nutritional factors is essential, especially as a major proportion of the global population is moving toward a plant-based diet.

For biofortification of vegetables and other staple crops, three main strategies are used: conventional breeding, agronomic management, and genetic engineering. Although conventional breeding and agronomic strategies have been used for a long time, they have certain constraints such as the need for genetic diversity in the gene pool. To some extent, this can be mitigated through distant hybridization (P. Sharma *et al.*, 2017), but transgenic breeding can resolve this limitation.

Advanced genomics techniques, such as comparative transcriptomics studies and GWAS, can be used to identify key genes. The genes thus identified can be modified using RNAi, overexpression, or genome editing. Despite the tireless work undertaken for gene discovery, genetic modifications, its expression in the target organism, assessment of agronomical traits, and biosafety studies, at least in few countries, transgenesis for the release of new cultivars faces resistance from the public (Niazi and Brown, 2017). Most of the issues mentioned above can be solved through genome editing. It is highly accurate and effective in inducing specific changes in the genes of interest, and also is biologically harmless as the ultimate outcome is transgene free (Mushtaq *et al.*, 2021). The recent advances in genome editing are base editing (the irreversible change in a base at the location of interest without the use of DSBs, donor templates, or dependence on NHEJ and HDR), prime editing (the creation of indels and base-to-base conversions without DSBs, using prime editing guide RNA (pegRNA) to drive the Cas9 endonuclease), and genome editing via a rice zygote (which addresses the problem of macromolecule transfer to host cells and tissues, as well as the challenge of transformation and regeneration), which have created new opportunities for biotechnologists. Genome editing holds the key to the future of biofortification. Genetic engineering for improved absorption of soil micronutrients, maximal nutrient accumulation in the edible portions, and enhanced nutrient bioavailability should be prioritized.

## 18.6 References

Ahmar, S., Gill, R.A., Jung, K.H., Faheem, A., Qasim, M.U., *et al.* (2020) Conventional and molecular techniques from simple breeding to speed breeding in crop plants: recent advances and future outlook. *International Journal of Molecular Sciences* 21: 2590.

Alamillo, J.M., López, C.M., Rivas, F.J.M., Torralbo, F., Bulut, M., and Alseekh, S. (2023) Clustered regularly interspaced short palindromic repeats/CRISPR-associated protein and hairy roots: a perfect match for gene functional analysis and crop improvement. *Current Opinion in Biotechnology* 79: 102876.

Al-Hussaini, A.A., Alshehry, Z., AlDehaimi, A., and Bashir, M.S. (2022) Vitamin D and iron deficiencies among Saudi children and adolescents: a persistent problem in the 21st century. *Saudi Journal of Gastroenterology* 28: 157.

Alqudah, A.M., Sallam, A., Baenziger, P.S., and Börner, A. (2020) GWAS: fast-forwarding gene identification and characterization in temperate cereals: lessons from barley—a review. *Journal of Advanced Research* 22, 119–135.

Babu, P., Baranwal, D.K., Pal, D., Bharti, H., Joshi, P., *et al.* (2020) Application of genomics tools in wheat breeding to attain durable rust resistance. *Frontiers in Plant Science* 11: 567147.

Baldet, P., Bres, C., Okabe, Y., Mauxion, J.P., Just, D., *et al.* (2013) Investigating the role of vitamin C in tomato through TILLING identification of ascorbate-deficient tomato mutants. *Plant Biotechnology* 30, 309–314.

Basso, M.F., Arraes, F.B.M., Grossi-de-Sa, M., Moreira, V.J.V., Alves-Ferreira, M., and Grossi-de-Sa, M.F. (2020) Insights into genetic and molecular elements for transgenic crop development. *Frontiers in Plant Science* 11: 509.

Bellumori, M., Chasquibol Silva, N.A., Vilca, L., Andrenelli, L., Cecchi, L., *et al.* (2020) A Study on the biodiversity of pigmented Andean potatoes: nutritional profile and phenolic composition. *Molecules* 25: 3169.

Blum, A. (2018) *Plant Breeding for Stress Environments*. CRC Press, Boca Raton, Florida.

Bock, R. (2022) Transplastomic approaches for metabolic engineering. *Current Opinion in Plant Biology* 66: 102185.

Boim, A.G.F., Wragg, J., Canniatti-Brazaca, S.G., and Alleoni, L.R.F. (2020) Human intestinal Caco-2 cell line *in vitro* assay to evaluate the absorption of Cd, Cu, Mn and Zn from urban environmental matrices. *Environmental Geochemistry and Health*, 42, 601–615.

Bouis, H., Saltzman, A., Low, J., Ball, A., and Covic, N. (2017) An overview of the landscape and approach for biofortification in Africa. *African Journal of Food, Agriculture, Nutrition and Development* 17, 11848–11864.

Brown, C.R., Vales, I., Yilma, S., James, S., Charlton, B., *et al.* (2012) "AmaRosa," a red skinned, red fleshed fingerling with high phytonutrient value. *American Journal of Potato Research* 89, 249–254.

Buturi, C.V., Mauro, R.P., Fogliano, V., Leonardi, C., and Giuffrida, F. (2021) Mineral biofortification of vegetables as a tool to improve human diet. *Foods* 10: 223.

Cabrera-Ponce, J.L., Barraza, A., and Alvarez-Venegas, R. (2022) Cisgenic crops: major strategies to create cisgenic plants based on genome editing. In: Chaurasia, A. and Koole, C. (eds.) *Cisgenic Crops: Potential and Prospects*. Springer, Cham, Switzerland, pp. 213–235.

Campisi, S.C., Khan, A., Zasowski, C., and Bhutta, Z.A. (2022) Malnutrition. In: Guandalini, S., Dhawan, A., and Branski, D. (eds.) *Textbook of Pediatric Gastroenterology, Hepatology and Nutrition: A Comprehensive Guide to Practice*. Springer, Cham, Switzerland, pp. 609–623.

Cannon, G. and Leitzmann, C. (2022) Food and nutrition science: the new paradigm. *Asia Pacific Journal of Clinical Nutrition* 31, 1–15.

Cao, X., Qiu, Z., Wang, X., Giang, T.V., Liu, X., *et al.* (2017) A putative *R3 MYB* repressor is the candidate gene underlying atroviolacium, a locus for anthocyanin pigmentation in tomato fruit. *Journal of Experimental Botany* 68, 5745–5758.

Chakraborty, S., Chakraborty, N., and Datta, A. (2000) Increased nutritive value of transgenic potato by expressing a nonallergenic seed albumin gene from *Amaranthus hypochondriacus*. *Proceedings of the National Academy of Sciences USA* 97, 3724–3729.

Challam, C., Dutt, S., Sudhakar, D., Raveendran, M., Buckseth, T., and Singh, R.K. (2021) Increase in root branching enhanced ferric-chelate reductase activity under iron stress in potato (*Solanum tuberosum*). *Indian Journal of Agricultural Sciences* 91, 1646–1649.

Chang, A.Y. (2022) Single-guide RNAs: rationale and design. In: Tsang, S.H. (ed.) *CRISPR Genome Surgery in Stem Cells and Disease Tissues*. Academic Press, London, pp. 47–55.

Chaudhuri, A., Halder, K., and Datta, A. (2022) Classification of CRISPR/Cas system and its application in tomato breeding. *Theoretical and Applied Genetics* 135, 367–387.

Chung, Y.S., Choi, S.C., Jun, T.H., and Kim, C. (2017) Genotyping-by-sequencing: a promising tool for plant genetics research and breeding. *Horticulture, Environment, and Biotechnology* 58, 425–431.

Colanero, S., Perata, P., and Gonzali, S. (2020) What's behind purple tomatoes? Insight into the mechanisms of anthocyanin synthesis in tomato fruits. *Plant Physiology* 182, 1841–1853.

Cronjé, C. (2010) *Manipulation of Ascorbate Biosynthesis in Solanum lycopersicum* (cv Money maker). MSc thesis, University of Stellenbosch, Stellenbosch, South Africa.

Cuevas, E., Hillebrand, S., Winterhalter, P. (2011) Anthocyanins in purple sweet potato (*Ipomoea batatas* L.) varieties. *Fruit, Vegetable and Cereal Science and Biotechnology* 5, 19–24.

da Silva Souza, M.A., Peres, L.E., Freschi, J.R., Purgatto, E., Lajolo, F.M., and Hassimotto, N.M. (2020) Changes in flavonoid and carotenoid profiles alter volatile organic compounds in purple and orange cherry tomatoes obtained by allele introgression. *Journal of the Science of Food and Agriculture* 100, 1662–1670.

Dame, Z.T., Rahman, M., and Islam, T. (2021) Bacilli as sources of agrobiotechnology: recent advances and future directions. *Green Chemistry Letters and Reviews* 14, 246–271.

Das, T., Anand, U., Pal, T., Mandal, S., Kumar, M., *et al.* (2023) Exploring the potential of CRISPR/Cas genome editing for vegetable crop improvement: an overview of challenges and approaches. *Biotechnology and Bioengineering* 120, 1215–1228.

Del Carpio, P.D., Lozano, R., Wolfe, M.D., and Jannink, J.L. (2019) Genome-wide association studies and heritability estimation in the functional genomics era. In: Rajora, O.P. (ed.) *Population Genomics: Concepts, Approaches and Applications*, Springer Nature, Cham, Switzerland, pp. 361–425.

Della Coletta, R., Qiu, Y., Ou, S., Hufford, M.B., and Hirsch, C.N. (2021) How the pan-genome is changing crop genomics and improvement. *Genome Biology* 22: 3.

Descalsota, G.I.L., Swamy, B.M., Zaw, H., Inabangan-Asilo, M.A., Amparado, A., *et al.* (2018) Genome-wide association mapping in a rice MAGIC plus population detects QTLs and genes useful for biofortification. *Frontiers in Plant Science* 9: 1347.

Dharmapuri, S., Rosati, C., Pallara, P., Aquilani, R., Bouvier, F., *et al.* (2002) Metabolic engineering of xanthophyll content in tomato fruits. *FEBS Letters* 519, 30–34.

Ding, Y. and Jian, Y. (2010) An orange cauliflower hybrid 'Jinyu 60' with high β-carotene. *Acta Horticulturae* 856, 261–264.

Dorais, M., Ehret, D.L., and Papadopoulos, A.P. (2008) Tomato (*Solanum lycopersicum*) health components: from the seed to the consumer. *Phytochemistry Reviews* 7, 231–250.

Enfissi, E.M., Fraser, P.D., Lois, L.M., Boronat, A., Schuch, W., and Bramley, P.M. (2005) Metabolic engineering of the mevalonate and non-mevalonate isopentenyl diphosphate-forming pathways for the production of health-promoting isoprenoids in tomato. *Plant Biotechnology Journal* 3, 17–27.

FAOSTAT (2022) FAOSTAT. Food and Agriculture Organizatoin of the United Nations, Rome. Available at: www.fao.org/faostat/en/#home (accessed 17 July 2024).

Farré, G., Blancquaert, D., Capell, T., van der Straeten, D., Christou, P., and Zhu, C. (2014) Engineering complex metabolic pathways in plants. *Annual Review of Plant Biology* 65, 187–223.

Filler Hayut, S., Melamed Bessudo, C., and Levy, A.A. (2017) Targeted recombination between homologous chromosomes for precise breeding in tomato. *Nature Communications* 8: 15605.

Gan, W.Y., Boo, S., Seik, M.Y., and Khoo, H.E. (2018) Comparing the nutritional status of vegetarians and non-vegetarians from a Buddhist organisation in Kuala Lumpur, Malaysia. *Malaysian Journal of Nutrition* 24, 89–101.

Garcia-Casal, M.N., Peña-Rosas, J.P., Giyose, B., and Consultation Working Groups (2017) Staple crops biofortified with increased vitamins and minerals: considerations for a public health strategy. *Annals of the New York Academy of Sciences* 1390, 3–13.

Gasiunas, G., Young, J.K., Karvelis, T., Kazlauskas, D., Urbaitis, T., Jasnauskaite, M., *et al.* (2020) A catalogue of biochemically diverse CRISPR-Cas9 orthologs. *Nature Communications* 11:5512.

Ginting, E., Yulifianti, R., and Indriani, F. (2020) Selected purple-fleshed sweet potato genotypes with high anthocyanin contents. *IOP Conference Series: Earth and Environmental Science* 456: 012023.

Gomathi, M., Vethamoni, P.I., and Gopinath, P. (2017) Biofortification in vegetable crops—a review. *Chemical Science Review and Letters* 6, 1227–1237.

Gonzali, S., Mazzucato, A., and Perata, P. (2009) Purple as a tomato: towards high anthocyanin tomatoes. *Trends in Plant Science* 14, 237–241.

Goto, F., Yoshihara, T., and Saiki, H. (2000) Iron accumulation and enhanced growth in transgenic lettuce plants expressing the iron-binding protein ferritin. *Theoretical and Applied Genetics* 100, 658–664.

Gupta, S., Kumar, A., Patel, R., and Kumar, V. (2021) Genetically modified crop regulations: scope and opportunity using the CRISPR-Cas9 genome editing approach. *Molecular Biology Reports* 48, 4851–4863.

Habeeb Rahman, A. (2011) Quality aspects of cauliflower during storage. *International Food Research Journal* 18, 427–431.

Han, M., Opoku, K.N., Bissah, N.A., and Su, T. (2021) *Solanum aethiopicum*: the nutrient-rich vegetable crop with great economic, genetic biodiversity and pharmaceutical potential. *Horticulturae* 7: 126.

Hanlon, P. and Sewalt, V. (2021) GEMs: genetically engineered microorganisms and the regulatory oversight of their uses in modern food production. *Critical Reviews in Food Science and Nutrition* 61, 959–970.

Hashimoto, R., Ueta, R., Abe, C., Osakabe, Y., and Osakabe, K. (2018) Efficient multiplex genome editing induces precise, and self-ligated type mutations in tomato plants. *Frontiers in Plant Science* 9: 916.

Hawthorne, K.M., Morris, J., Hotze, T., Hirschi, K.D., and Abrams, S.A. (2009) Biotechnologically-modified carrots: calcium absorption relative to milk. *Journal of Bioequivalence and Bioavailability* 1, 34–38.

Hong, Y., Meng, J., He, X., Zhang, Y., Liu, Y., *et al.* (2021) Editing *miR482b* and miR482c simultaneously by CRISPR/Cas9 enhanced tomato resistance to *Phytophthora infestans*. *Phytopathology* 111, 1008–1016.

Howarth, J.L., Lee, Y.B., and Uney, J.B. (2010) Using viral vectors as gene transfer tools (Cell Biology and Toxicology Special Issue: ETCS-UK 1 day meeting on genetic manipulation of cells). *Cell Biology and Toxicology* 26, 1–20.

Hu, N., Xian, Z., Li, N., Liu, Y., Huang, W., *et al.* (2019) Rapid and user-friendly open-source CRISPR/Cas9 system for single-or multi-site editing of tomato genome. *Horticulture Research* 6: 7.

Hunter, C.T. (2021) CRISPR/Cas9 targeted mutagenesis for functional genetics in maize. *Plants* 10: 723.

Hunziker, J., Nishida, K., Kondo, A., Kishimoto, S., Ariizumi, T., and Ezura, H. (2020) Multiple gene substitution by target-AID base-editing technology in tomato. *Scientific Reports* 10: 20471.

Hussain, S., Habib, M., Ahmed, Z., Sadia, B., Bernardo, A., *et al.* (2022) Genotyping-by-sequencing based molecular genetic diversity of Pakistani bread wheat (*Triticum aestivum* L.) accessions. *Frontiers in Genetics* 13: 53.

Ito, Y., Nishizawa-Yokoi, A., Endo, M., Mikami, M., and Toki, S. (2015) CRISPR/Cas9-mediated mutagenesis of the RIN locus that regulates tomato fruit ripening. *Biochemical and Biophysical Research Communications* 467, 76–82.

Jaiswal, D.K., Krishna, R., Chouhan, G.K., de Araujo Pereira, A.P., Ade, A.B., *et al.* (2022) Bio-fortification of minerals in crops: current scenario and future prospects for sustainable agriculture and human health. *Plant Growth Regulation* 98, 5–22.

Jiang, W.Z., Henry, I.M., Lynagh, P.G., Comai, L., Cahoon, E.B., and Weeks, D.P. (2017) Significant enhancement of fatty acid composition in seeds of the allohexaploid, *Camelina sativa*, using CRISPR/Cas9 gene editing. *Plant Biotechnology Journal* 15, 648–657.

Joshi, V., Kumar, A., and Kumar, V. (2012) Antimicrobial, antioxidant and phyto-chemicals from fruit and vegetable wastes: a review. *International Journal of Food and Fermentation Technology* 2, 123–136.

Kalloo, G. (ed.) (2012) *Genetic Improvement of Tomato*. Springer Science and Business Media, Berlin/Heidelberg.

Kang, J.G., Park, J.S., Ko, J.H., and Kim, Y.S. (2019) Regulation of gene expression by altered promoter methylation using a CRISPR/Cas9-mediated epigenetic editing system. *Scientific Reports* 9: 11960.

Kang, S.I., Rahim, M.A., Afrin, K.S., Jung, H.J., Kim, H.T., *et al.* (2018) Expression of anthocyanin biosynthesis-related genes reflects the peel color in purple tomato. *Horticulture, Environment, and Biotechnology* 59, 435–445.

Karunarathna, N.L., Wang, H., Harloff, H.J., Jiang, L., and Jung, C. (2020) Elevating seed oil content in a polyploid crop by induced mutations in *SEED FATTY ACID REDUCER* genes. *Plant Biotechnology Journal* 18, 2251–2266.

Kaul, T., Sony, S.K., Verma, R., Motelb, K.F.A., Prakash, A.T., *et al.* (2020) Revisiting CRISPR/Cas-mediated crop improvement: special focus on nutrition. *Journal of Biosciences* 45: 137.

Kaur, B., Handa, A.K., and Mattoo, A.K. (2017) *Genetic Engineering of Tomato to Improve Nutritional Quality, Resistance to Abiotic and Biotic Stresses, and for Non-food Applications*. Burleigh Dodds Science Publishing, Cambridge, UK.

Kersey, P.J., Collemare, J., Cockel, C., Das, D., Dulloo, E.M., *et al.* (2020) Selecting for useful properties of plants and fungi—approaches, opportunities, and challenges. *Plants, People, Planet* 2, 409–420.

Kiferle, C., Fantini, E., Bassolino, L., Povero, G., Spelt, C., *et al.* (2015) Tomato R2R3-MYB proteins SlANT1 and SlAN2: same protein activity, different roles. *PLOS ONE* 10: e0136365.

Kumar, A. and Sindhusha, P. (2021) Biofortification: aAn imperative bond for nutritional quality enhancement in vegetables. *Journal of Food and Nutritional Disorders* 10: 7.

Kumar, P., Kumar, A., Dhiman, K., and Srivastava, D.K. (2022) Recent progress in cereals biofortification to alleviate malnutrition in India: an overview. In: Srivastava, D.K., Thakur, A.K., and Kumar, P. (eds.) *Agricultural Biotechnology: Latest Research and Trends*. Springer, Singapore, pp. 253–280.

Kumari, M., Sharma, D., and Sandeep, S. (2022) Biofortification of vegetable crops: an option for mitigating hidden hunger. *International Journal of Economic Plants* 9, 184–193.

Lal, M.K., Kumar, A., Kardile, H.B., Raigond, P., Changan, S.S., *et al.* (2020) Biofortification of vegetables. In: Sharma, T.R., Deshmukh, R., Sonah, H. (eds.) *Advances in Agri-food Biotechnology*. Springer, Singapore, pp. 105–129.

Lata, H., Sharma, A., Chadha, S., Kaur, M., and Kumar, P. (2022) RNA interference (RNAi) mechanism and application in vegetable crops. *Journal of Horticultural Science and Biotechnology* 97, 160–170.

Lee, M.J., Park, J.S., Choi, D.S., and Jung, M.Y. (2013) Characterization and quantitation of anthocyanins in purple-fleshed sweet potatoes cultivated in Korea by HPLC-DAD and HPLC-ESI-QTOF-MS/MS. *Journal of Agricultural and Food Chemistry* 61, 3148–3158.

Li, J., Meng, X., Zong, Y., Chen, K., Zhang, H., *et al.* (2016) Gene replacements and insertions in rice by intron targeting using CRISPR-Cas9. *Nature Plants* 2, 16139.

Li, R., Li, R., Li, X., Fu, D., Zhu, B., *et al.* (2018) Multiplexed CRISPR/Cas9-mediated metabolic engineering of γ-aminobutyric acid levels in *Solanum lycopersicum*. *Plant Biotechnology Journal* 16, 415–427.

Li, X., Wang, Y., Chen, S., Tian, H., Fu, D., *et al.* (2018) Lycopene is enriched in tomato fruit by CRISPR/Cas9-mediated multiplex genome editing. *Frontiers in Plant Science* 9: 559.

Liu, L., Shao, Z., Zhang, M., and Wang, Q. (2015) Regulation of carotenoid metabolism in tomato. *Molecular Plant* 8, 28–39.

Liu, Q., Yao, L., Xu, Y., Cheng, H., Wang, W., *et al.* (2019) *In vitro* evaluation of hydroxycinnamoyl CoA:quinatehydroxycinnamoyl transferase expression and regulation in *Taraxacum antungense* in relation to 5-caffeoylquinic acid production. *Phytochemistry* 162, 148–156.

Llorente, B., Torres-Montilla, S., Morelli, L., Florez-Sarasa, I., Matus, J.T., *et al.* (2020) Synthetic conversion of leaf chloroplasts into carotenoid-rich plastids reveals mechanistic basis of natural chromoplast development. *Proceedings of the National Academy of Sciences USA* 117, 21796–21803.

Lobato-Gómez, M., Hewitt, S., Capell, T., Christou, P., Dhingra, A., and Girón-Calva, P.S. (2021) Transgenic and genome-edited fruits: background, constraints, benefits, and commercial opportunities. *Horticulture Research* 8: 166.

Lockyer, S., White, A., and Buttriss, J. (2018) Biofortified crops for tackling micronutrient deficiencies—what impact are these having in developing countries and could they be of relevance within Europe? *Nutrition Bulletin* 43, 319–357.

Lu, Y., Stegemann, S., Agrawal, S., Karcher, D., Ruf, S., and Bock, R. (2017) Horizontal transfer of a synthetic metabolic pathway between plant species. *Current Biology* 27, 3034–3041.

Luo, D., Xiong, C., Lin, A., Zhang, C., Sun, W., *et al.* (2021) SlBBX20 interacts with the COP9 signalosome subunit SlCSN5-2 to regulate anthocyanin biosynthesis by activating *SlDFR* expression in tomato. *Horticulture Research* 8: 163.

Luo, J., Abid, M., Zhang, Y., Cai, X., and Tu, J., *et al.* (2023) Genome-wide identification of kiwifruit SGR family members and functional characterization of SGR2 protein for chlorophyll degradation. *International Journal of Molecular Sciences* 24: 1993.

Luo, Z., Zhang, J., Li, J., Yang, C., Wang, T., *et al.* (2013) A STAY-GREEN protein SlSGR1 regulates lycopene and β-carotene accumulation by interacting directly with SlPSY1 during ripening processes in tomato. *New Phytologist* 198, 442–452.

Lyzenga, W.J., Pozniak, C.J., and Kagale, S. (2021) Advanced domestication: harnessing the precision of gene editing in crop breeding. *Plant Biotechnology Journal* 19, 660–670.

Marques, E., Darby, H.M., and Kraft, J. (2021) Benefits and limitations of non-transgenic micronutrient biofortification approaches. *Agronomy* 11: 464.

Marr, E. and Potter, C.J. (2021) Base editing of somatic cells using CRISPR-Cas9 in *Drosophila*. *CRISPR Journal* 4, 836–845.

Mattoo, A.K., Dwivedi, S.L., Dutt, S., Singh, B., Garg, M., and Ortiz, R. (2022) Anthocyanin-rich vegetables for human consumption—focus on potato, sweetpotato and tomato. *International Journal of Molecular Sciences* 23: 2634.

Mazzucato, A., Willems, D., Bernini, R., Picarella, M.E., Santangelo, E., *et al.* (2013) Novel phenotypes related to the breeding of purple-fruited tomatoes and effect of peel extracts on human cancer cell proliferation. *Plant Physiology and Biochemistry* 72, 125–133.

Meena, R.K., Koli, D.K., and Koli, G.K. (2022) Transgenic approach for biofortification. *Biotica Research Today* 4, 388–391.

Monika, G., Kim, S.R.M., Kumar, P.S., Gayathri, K.V., Rangasamy, G., and Saravanan, A., (2023) Biofortification: a long-term solution to improve global health—a review. *Chemosphere* 314: 137713.

Morelli, L. and Rodriguez-Concepcion, M. (2022) Open avenues for carotenoid biofortification of plant tissues. *Plant Communications* 4: 100466.

Morineau, C., Bellec, Y., Tellier, F., Gissot, L., Kelemen, Z., *et al.* (2017) Selective gene dosage by CRISPR-Cas9 genome editing in hexaploid *Camelina sativa*. *Plant Biotechnology Journal* 15, 729–739.

Mushtaq, M., Ahmad Dar, A., Skalicky, M., Tyagi, A., Bhagat, N., *et al.* (2021) CRISPR-based genome editing tools: insights into technological breakthroughs and future challenges. *Genes* 12: 797.

Myers, J. (2012) Breeding tomatoes for increased flavonoids. In: *Strengthening Community Seed Systems Proceedings of the 6th Organic Seed Growers Conference, 19–21 January, Port Townsend, Washington*. Organic Seed Alliance, Port Townsend, Washington, pp. 50–51.

Naing, A.H., Park, K.I., Ai, T.N., Chung, M.Y., Han, J.S., *et al.* (2017) Overexpression of snapdragon *Delila* (*Del*) gene in tobacco enhances anthocyanin accumulation and abiotic stress tolerance. *BMC Plant Biology* 17: 65.

Nasir, M.A., Nawaz, S., and Huang, J. (2022) A review: computational approaches to design sgRNA of CRISPR-Cas9. *Current Bioinformatics* 17, 2–18.

Natalini, A., Acciarri, N., and Cardi, T., (2021) Breeding for nutritional and organoleptic quality in vegetable crops: the case of tomato and cauliflower. *Agriculture* 11: 606.

Navrátil, O., Bucher, P., and Vacek, J. (2012) Transgene coding of a key enzyme of the glycolytic pathway helps to decrease sugar content in potato tubers. *Czech Journal of Genetics and Plant Breeding* 48, 42–45.

Niazi, S.K. and Brown, J.L. (2017) *Fundamentals of Modern Bioprocessing*. CRC Press, Boca Raton, Florida.

Niggeweg, R., Michael, A.J., and Martin, C. (2004) Engineering plants with increased levels of the antioxidant chlorogenic acid. *Nature Biotechnology* 22, 746–754.

Ning, Z., Hu, K., Zhou, Z., Zhao, D., Tang, J., *et al.* (2021) *IbERF71*, with *IbMYB340* and *IbbHLH2*, coregulates anthocyanin accumulation by binding to the *IbANS1* promoter in purple-fleshed sweet potato (*Ipomoea batatas* L.). *Plant Cell Reports* 40, 157–169.

Nonaka, S., Arai, C., Takayama, M., Matsukura, C., and Ezura, H. (2017) Efficient increase of $\gamma$-aminobutyric acid (GABA) content in tomato fruits by targeted mutagenesis. *Scientific Reports* 7: 7057.

Okuzaki, A., Ogawa, T., Koizuka, C., Kaneko, K., Inaba, M., *et al.* (2018) CRISPR/Cas9-mediated genome editing of the fatty acid desaturase 2 gene in *Brassica napus*. *Plant Physiology and Biochemistry* 131, 63–69.

Ozseyhan, M.E., Kang, J., Mu, X., and Lu, C. (2018) Mutagenesis of the *FAE1* genes significantly changes fatty acid composition in seeds of *Camelina sativa*. *Plant Physiology and Biochemistry* 123, 1–7.

Panghal, A., Munezero, C., Sharma, P., and Chhikara, N. (2019) Cassava toxicity, detoxification and its food applications: a review. *Toxin Reviews* 40, 1–16.

Park, Y.E., Cho, J.H., Cho, H.M., Yi, J.Y., Seo, H., and Choung, M.G. (2009) A new potato cultivar 'Hongyoung', with red skin and flesh color, and high concentrations of anthocyanins. *Korean Journal of Breeding Science* 41: 87.

Parra-Galindo, M.A., Piñeros-Niño, C., Soto-Sedano, J.C., and Mosquera-Vasquez, T. (2019) Chromosomes I and X harbor consistent genetic factors associated with the anthocyanin variation in potato. *Agronomy* 9: 366.

Parra-Galindo, M.A., Soto-Sedano, J.C., Mosquera-Vásquez, T., and Roda, F. (2021) Pathway-based analysis of anthocyanin diversity in diploid potato. *PLOS ONE* 16: e0250861.

Parulekar, Y., Haldankar, P., Dalvi, N., Salvi, B., and Bhattacharyya, T. (2019) Nutraceuticals and their biofortification in vegetable crops: a review. *Advanced Agricultural Research and Technology Journal* 3, 219–229.

Porto, J.S., Rebouças, T.N.H., Moraes, M.O.B., Bomfim, M.P., Lemos, O.L., *et al.* (2016) Quality and antioxidant activity of tomato cultivated under different sources and doses of nitrogen. *Revista Caatinga* 29, 780–788.

Qaim, M. (2016) New and future GM crop applications. In: *Genetically Modified Crops and Agricultural Development*, Springer, New York, pp. 85–107.

Qiu, G., Wang, D., Song, X., Deng, Y., and Zhao, Y. (2018) Degradation kinetics and antioxidant capacity of anthocyanins in air-impingement jet dried purple potato slices. *Food Research International* 105, 121–128.

Quinet, M., Angosto, T., Yuste-Lisbona, F.J., Blanchard-Gros, R., Bigot, S., *et al.* (2019) Tomato fruit development and metabolism. *Frontiers in Plant Science* 10: 1554.

Raggi, L., Caproni, L., Carboni, A., and Negri, V. (2019) Genome-wide association study reveals candidate genes for flowering time variation in common bean (*Phaseolus vulgaris* L.). *Frontiers in Plant Science* 10: 962.

Rasool, N. (2021) New plant breeding techniques: problems and prospects. In: Bhat, T.A. (ed.) *Cytotoxicity and Crop Improvement: Revolutionizing Food Science*. Cambridge Scholars Publishing, Newcastle upon Tyne, UK, pp. 70–72.

Ratnaparkhe, M.B., Satpute, G.K., Kumawat, G., Chandra, S., Kamble, V.G., *et al.* (2022) Genomic designing for abiotic stress tolerant soybean. In: Kole, C. (ed.) *Genomic Designing for Abiotic Stress Resistant Oilseed Crops*. Springer, Cham, Switzerland, pp. 1–73.

Reem, N.T. and van Eck, J. (2019) Application of CRISPR/Cas9-mediated gene editing in tomato. In: Qi, Y. (ed.) *Plant Genome Editing with CRISPR Systems: Methods and Protocols*. Springer, New York, pp. 171–182.

Riesenberg, S., Helmbrecht, N., Kanis, P., Maricic, T., and Pääbo, S. (2022) Improved gRNA secondary structures allow editing of target sites resistant to CRISPR-Cas9 cleavage. *Nature Communications* 13: 489.

Rodriguez-Concepcion, M., and Daròs, J.A. (2022) Transient expression systems to rewire plant carotenoid metabolism. *Current Opinion in Plant Biology* 66: 102190.

Rodriguez-Concepcion, M., Avalos, J., Bonet, M.L., Boronat, A., Gomez-Gomez, L., *et al.* (2018) A global perspective on carotenoids: metabolism, biotechnology, and benefits for nutrition and health. *Progress in Lipid Research* 70, 62–93.

Ryan, D.E., Taussig, D., Steinfeld, I., Phadnis, S.M., Lunstad, B.D., *et al.* (2018) Improving CRISPR-Cas specificity with chemical modifications in single-guide RNAs. *Nucleic Acids Research* 46, 792–803.

Saini, R.K., Nile, S.H., and Park, S.W. (2015) Carotenoids from fruits and vegetables: chemistry, analysis, occurrence, bioavailability and biological activities. *Food Research International* 76, 735–750.

Sandhya, D., Jogam, P., Venkatapuram, A.K., Savitikadi, P., Peddaboina, V., *et al.* (2022) Highly efficient *Agrobacterium*-mediated transformation and plant regeneration system for genome engineering in tomato. *Saudi Journal of Biological Sciences* 29: 103292.

Sandmann, G., Römer, S., and Fraser, P.D. (2006) Understanding carotenoid metabolism as a necessity for genetic engineering of crop plants. *Metabolic Engineering* 8, 291–302.

Sandrini, M., Nerva, L., Sillo, F., Balestrini, R., Chitarra, W., and Zampieri, E. (2022) Abiotic stress and belowground microbiome: the potential of omics approaches. *International Journal of Molecular Sciences* 23: 1091.

Sen Gupta, D., McPhee, K., and Kumar, S. (2017) Development of molecular markers for iron metabolism related genes in lentil and their expression analysis under excess iron stress. *Frontiers in Plant Science* 8: 579.

Sharma, J., Sharma, V., Dua, V., Kumar, V., and Singh, B. (2017) Evaluation of Indian potato germplasm for iron and zinc content. *Indian Journal of Plant Genetic Resources* 30, 232–236.

Sharma, M.K., Solanke, A.U., Jani, D., Singh, Y., and Sharma, A.K. (2009) A simple and efficient *Agrobacterium*-mediated procedure for transformation of tomato. *Journal of Biosciences* 34: 423–433.

Sharma, P., Aggarwal, P., and Kaur, A. (2017) Biofortification: a new approach to eradicate hidden hunger. *Food Reviews International* 33, 1–21.

Sharma, R. and Yeh, K.C. (2020) The dual benefit of a dominant mutation in Arabidopsis *IRON DEFICIENCY TOLERANT1* for iron biofortification and heavy metal phytoremediation. *Plant Biotechnology Journal* 18, 1200–1210.

Shock, C., Brown, C., Sathuvalli, V., Charlton, B., Yilma, S., *et al.* (2018) TerraRossa: a mid-season specialty potato with red flesh and skin and resistance to common scab and golden cyst nematode. *American Journal of Potato Research* 95, 597–605.

Shubham, K., Anukiruthika, T., Dutta, S., Kashyap, A., Moses, J.A., and Anandharamakrishnan, C. (2020) Iron deficiency anemia: a comprehensive review on iron absorption, bioavailability and emerging food fortification approaches. *Trends in Food Science and Technology* 99, 58–75.

Shwetha, H.J., Shilpa, S., Arathi, B.P., Raju, M., and Lakshminarayana, R. (2020) Biofortification of carotenoids in agricultural and horticultural crops: a promising strategy to target vitamin A malnutrition. In: Benkeblia, N. (ed.) *Vitamins and Minerals Biofortification of Edible Plants*. Wiley, New York, pp. 123–161.

Singh, B., Goutam, U., Kukreja, S., Sharma, J., Sood, S., and Bhardwaj, V. (2021) Potato biofortification: an effective way to fight global hidden hunger. *Physiology and Molecular Biology of Plants* 27, 2297–2313.

Singh, B., Goutam, U., Kukreja, S., Siddappa, S., Sood, S., *et al.* (2022) Biofortification strategies to improve iron concentrations in potato tubers: lessons and future opportunities. *Potato Research* 65, 51–64.

Singh, N., Bhatt, V., Rana, N., and Shivaraj, S. (2020) Advances of next-generation sequencing (NGS) technologies to enhance the biofortifications in crops. In: Sharma, T.R., Deshmukh, R., and Sonah, H. (eds.) *Advances in Agri-Food Biotechnology*. Springer, Singapore, pp. 427–450.

Singh, S. and Kalia, P. (2021) Advances in cauliflower (*Brassica oleracea* var. *botrytis* L.) breeding, with emphasis on India. In: Al-Khayri, J.M., Jain, S.M., and Johnson, D.V. (eds.) *Advances in Plant Breeding Strategies: Vegetable Crops, Vol. 10: Leaves, Flowerheads, Green Pods, Mushrooms and Truffles*. Springer, Cham, Switzerland, pp. 247–301.

Singh, S., Dey, S., Bhatia, R., Kumar, R., Sharma, K., and Behera, T. (2019) Heterosis and combining ability in cytoplasmic male sterile and doubled haploid based *Brassica oleracea* progenies and prediction of heterosis using microsatellites. *PLOS ONE* 14: e0210772.

Singh, S., Selvakumar, R., Mangal, M., and Kalia, P. (2020) Breeding and genomic investigations for quality and nutraceutical traits in vegetable crops—a review. *Indian Journal of Horticulture* 77, 1–40.

Siritunga, D. and Sayre, R.T. (2003) Generation of cyanogen-free transgenic cassava. *Planta* 217, 367–373.

Solberg, M. (2021) RNAi: an instrument for exploratory experimentation. In: Solberg, M. (ed.) *A Cognitive Ethnography of Knowledge and Material Culture: Cognition, Experiment, and the Science of Salmon Lice*. Springer, Cham, Switzerland, pp. 169–217.

Strygina, K. and Khlestkina, E. (2017) Anthocyanins synthesis in potato (*Solanum tuberosum* L.): genetic markers for smart breeding. *Sel'skokhozyaistvennaya Biologiya* 52, 37–49.

Strygina, K.V., Kochetov, A.V., and Khlestkina, E.K. (2019) Genetic control of anthocyanin pigmentation of potato tissues. *BMC Genetics* 20, 35–43.

Subramanian, N.K., White, P.J., Broadley, M.R., and Ramsay, G. (2011) The three-dimensional distribution of minerals in potato tubers. *Annals of Botany* 107, 681–691.

Sun, Y., Li, J., and Xia, L. (2016) Precise genome modification via sequence-specific nucleases-mediated gene targeting for crop improvement. *Frontiers in Plant Science* 7: 1928.

Takayama, M., Koike, S., Kusano, M., Matsukura, C., Saito, K., *et al.* (2015) Tomato glutamate decarboxylase genes *SlGAD2* and *SlGAD3* play key roles in regulating γ-aminobutyric acid levels in tomato (*Solanum lycopersicum*). *Plant and Cell Physiology* 56, 1533–1545.

Thakur, A., Sharma, V., and Thakur, A. (2019) An overview of anti-nutritional factors in food. *International Journal of Chemical Studies* 7, 2472–2479.

Thakur, V., Sharma, A., Sharma, P., Kumar, P., and Shilpa (2022) Biofortification of vegetable crops for vitamins, mineral and other quality traits. *Journal of Horticultural Science and Biotechnology* 97, 417–428.

Torres-Montilla, S., and Rodriguez-Concepcion, M. (2021) Making extra room for carotenoids in plant cells: new opportunities for biofortification. *Progress in Lipid Research* 84: 101128.

Vales, M., Brown, C., Yilma, S., Hane, D., James, S., *et al.* (2012) Purple pelisse: a specialty 'fingerling' potato with purple skin and flesh and medium specific gravity. *American Journal of Potato Research* 89, 306–314.

Varshney, R.K., Bohra, A., Roorkiwal, M., Barmukh, R., Cowling, W.A., *et al.* (2021) Fast-forward breeding for a food-secure world. *Trends in Genetics* 37, 1124–1136.

Vats, S., Bansal, R., Rana, N., Kumawat, S., Bhatt, V., *et al.* (2022) Unexplored nutritive potential of tomato to combat global malnutrition. *Critical Reviews in Food Science and Nutrition* 62, 1003–1034.

Verhoeyen, M., Bovy, A., Collins, G., Muir, S., Robinson, S., *et al.* (2002) Increasing antioxidant levels in tomatoes through modification of the flavonoid biosynthetic pathway. *Journal of Experimental Botany* 53, 2099–2106.

Vigneshwaran, M., Saud, R.K., Borah, K., and Bora, S.S. (2020) Biofortification in cereals is a promising approach to improve nutrition. *Journal of Pharmacognosy and Phytochemistry* 9, 1222–1229.

Vos, J. (2009) Nitrogen responses and nitrogen management in potato. *Potato Research* 52, 305–317.

Wada, N., Ueta, R., Osakabe, Y., and Osakabe, K. (2020) Precision genome editing in plants: state-of-the-art in CRISPR/Cas9-based genome engineering. *BMC Plant Biology* 20: 234.

Wada, N., Miyaji, T., Abe-Hara, C., Osakabe, K., and Osakabe, Y. (2022) CRISPR/Cas9 tools for multiplex genome editing in crops. In: Chen, J.-T. and Ahmar, S. (eds.) *Genome Editing Technologies for Crop Improvement.* Springer, Singapore, pp. 95–107.

Waltz, E. (2022) GABA-enriched tomato is first CRISPR-edited food to enter market. *Nature Biotechnology* 40, 9–11.

Wan, L., Wang, Z., Tang, M., Hong, D., Sun, Y., *et al.* (2021) CRISPR-Cas9 gene editing for fruit and vegetable crops: strategies and prospects. *Horticulturae* 7: 193.

Wang, H., Wu, Y., Zhang, Y., Yang, J., Fan, W., *et al.* (2019) CRISPR/Cas9-based mutagenesis of starch biosynthetic genes in sweet potato (*Ipomoea batatas*) for the improvement of starch quality. *International Journal of Molecular Sciences* 20: 4702.

Wang, Z., Wan, L., Xin, Q., Zhang, X., Song, Y., *et al.* (2021) Optimizing glyphosate tolerance in rapeseed by CRISPR/Cas9-based geminiviral donor DNA replicon system with Csy4-based single-guide RNA processing. *Journal of Experimental Botany* 72, 4796–4808.

Watkins, J.L. and Pogson, B.J. (2020) Prospects for carotenoid biofortification targeting retention and catabolism. *Trends in Plant Science* 25, 501–512.

Xie, F., Zhou, X., Lin, T., Wang, L., Liu, C., *et al.* (2020) Production of gene-edited pigs harboring orthologous human mutations via double cutting by CRISPR/Cas9 with long single-stranded DNAs as homology-directed repair templates by zygote injection. *Transgenic Research* 29, 587–598.

Xiong, C., Luo, D., Lin, A., Zhang, C., Shan, L., *et al.* (2019) A tomato B-box protein *Sl*BBX 20 modulates carotenoid biosynthesis by directly activating *PHYTOENE SYNTHASE 1*, and is targeted for 26S proteasome-mediated degradation. *New Phytologist* 221, 279–294.

Xu, J., Su, X., Lim, S., Griffin, J., Carey, E., *et al.* (2015) Characterisation and stability of anthocyanins in purple-fleshed sweet potato P40. *Food Chemistry* 186, 90–96.

Xu, L., Qiao, X., Zhang, M., and Zhang, S. (2018) Genome-wide analysis of aluminum-activated malate transporter family genes in six Rosaceae species, and expression analysis and functional characterization on malate accumulation in Chinese white pear. *Plant Science* 274, 451–465.

Yan, S., Chen, N., Huang, Z., Li, D., Zhi, J., *et al.* (2020) *Anthocyanin Fruit* encodes an R2R3-MYB transcription factor, SlAN2-like, activating the transcription of *SlMYBATV* to fine-tune anthocyanin content in tomato fruit. *New Phytologist* 225, 2048–2063.

Yang, T., Ali, M., Lin, L., Li, P., He, H., *et al.* (2023) Recoloring tomato fruit by CRISPR/Cas9-mediated multiplex gene editing. *Horticulture Research* 10: uhac214.

Ye, J., Wang, X., Hu, T., Zhang, F., Wang, B., *et al.* (2017) An InDel in the promoter of *Al-ACTIVATED MALATE TRANSPORTER9* selected during tomato domestication determines fruit malate contents and aluminum tolerance. *Plant Cell* 29, 2249–2268.

Zaidi, S.S.E.A., Vanderschuren, H., Qaim, M., Mahfouz, M.M., Kohli, A., *et al.* (2019) New plant breeding technologies for food security. *Science* 363, 1390–1391.

Zhang, D., Tan, Y., Dong, F., Zhang, Y., Huang, Y., *et al.* (2021) The expression of *IbMYB1* is essential to maintain the purple color of leaf and storage root in sweet potato [*Ipomoea batatas* (L.) Lam]. *Frontiers in Plant Science* 12: 688707.

Zhang, L., Yu, Y., Shi, T., Kou, M., Sun, J., *et al.* (2020) Genome-wide analysis of expression quantitative trait loci (eQTLs) reveals the regulatory architecture of gene expression variation in the storage roots of sweet potato. *Horticulture Research* 7: 90.

Zhang, M., Wang, Y., Chen, X., Xu, F., Ding, M., *et al.* (2021) Plasma membrane $H^+$-ATPase overexpression increases rice yield via simultaneous enhancement of nutrient uptake and photosynthesis. *Nature Communications* 12:735.

Zhi, J., Liu, X., Li, D., Huang, Y., Yan, S., *et al.* (2020) CRISPR/Cas9-mediated *SlAN2* mutants reveal various regulatory models of anthocyanin biosynthesis in tomato plant. *Plant Cell Reports* 39, 799–809.

Zhou, X., van Eck, J., and Li, L. (2008) Use of the cauliflower *Or* gene for improving crop nutritional quality. *Biotechnology Annual Review* 14, 171–190.

Zhu, H., Li, C., and Gao, C. (2020) Applications of CRISPR-Cas in agriculture and plant biotechnology. *Nature Reviews Molecular Cell Biology* 21, 661–677.

Zidenga, T., Siritunga, D., and Sayre, R.T. (2017) Cyanogen metabolism in cassava roots: impact on protein synthesis and root development. *Frontiers in Plant Science* 8: 220.

Zu, Y., Prather, K.L., and Stephanopoulos, G. (2020) Metabolic engineering strategies to overcome precursor limitations in isoprenoid biosynthesis. *Current Opinion in Biotechnology* 66, 171–178.

Zuluaga, D.L., Gonzali, S., Loreti, E., Pucciariello, C., Degl'Innocenti, E., *et al.* (2008) *Arabidopsis thaliana* MYB75/PAP1 transcription factor induces anthocyanin production in transgenic tomato plants. *Functional Plant Biology* 35, 606–618.

# 19 RNA Interference Strategies for Vegetable Crop Improvement

**Deepu Mathew***
*Centre for Plant Biotechnology and Molecular Biology, Kerala Agricultural University, Thrissur, Kerala, India*

**Abstract**

RNA interference (RNAi) is an antiviral mechanism in plants involving the production of dsRNA and its cleavage by Dicer enzyme to small interfering RNAs, which specifically direct the RNA-induced silencing complex (RISC) to degrade the homologous RNAs. Due to its specific gene-silencing capability, this strategy has been extensively employed in crop plants including vegetables in various breeding programs. RNAi has been useful in modification of plant architecture, enhancing the quality of the harvested product including biofortification, imparting tolerance to biotic and abiotic stress tolerance, and various other traits. Above all, RNAi has emerged as a precise functional genomics strategy to reveal the various roles played by putative genes and pathways. This chapter discusses the improvements made in the leading vegetable crops through RNAi.

**Keywords:** cloning, gene silencing, genetic engineering, hairpin construct, molecular techniques, RNAi, transgenesis

## 19.1 Introduction

Crop improvement has been a highly dynamic process, passing through natural and manual sections, artificial induction of new alleles through mutations, marker-assisted selection, transgenesis for new traits, overexpression and downregulation of genes, gene and quantitative trait locus mapping, association mapping, genome editing, and more. Throughout this journey, the quest to identify the genes and pathways responsible for each trait has been evident. Functional genomics and reverse genetics tools have been used extensively to answer this query. Downregulation or knockdown of genes was the classical and direct approach to show the immediate function of a gene. Mutations using chemical or physical agents, site-directed mutagenesis, RNA interference (RNAi), and gene editing have been the strategies followed for gene silencing.

As vegetable crops are the most important component of human food, there has been a substantial effort in breeding to improve them. Each of the breeding strategies has been equally good and applicable to vegetable breeding. As they are a diverse category of crops, each approach has been selectively employed in vegetables and RNAi is no exception. RNAi has been used extensively in vegetable crops for yield improvement, enhancing product quality, enhancing biotic and abiotic stress tolerance, and altering the plant architecture. The improvements made in vegetable crops through RNAi are discussed in this chapter.

*Email: deepu.mathew@kau.in

 *Genetic Engineering of Vegetable Crops* (Pritam Kalia ed.)
DOI: 10.1079/9781800627123.0019

## 19.2 Mechanism of Gene Silencing by RNA Interference

RNAi is a natural antiviral defense mechanism reported in plants (Baulcombe 1996; Prins and Goldbach 1996). Following the discovery of loss of petunia flower pigmentation (Jorgensen, 1990) and dsRNA-triggered gene silencing (Fire *et al.*, 1998), RNAi emerged as a widely employed tool in functional genomics. The molecular mechanisms behind RNAi have been detailed previously by Milhavet *et al.* (2003), López-Fraga *et al.* (2009), Shan (2010) and many others. RNAi involves the production of dsRNA. Dicer enzymes convert/cleave this dsRNA into 21 nt RNAs known as small interfering RNAs (siRNAs). Dicer-like protein 4 (DCL4) is involved in siRNA production. These siRNAs then provide specificity to an endonuclease-containing RNA-induced silencing complex (RISC), which targets the homologous RNA for degradation.

For regulation of plant gene expression, vectors harboring intron-containing hairpin RNA (ihpRNA) for the gene of interest are inserted into the plant DNA. Ligase-based vectors were used initially to generate ihpRNA constructs (Smith *et al.*, 2000), and were subsequently substituted with ligase-independent vectors (Xu *et al.*, 2010; Ganesh *et al.*, 2022). These expression constructs code for dsRNAs and later the siRNAs, triggering the homology-based targeting of cognate transcripts and generating a pre-programmed silencing of target genes.

## 19.3 RNAi in Improvement of Solanaceous Vegetables

### 19.3.1 Tomato

Tomato is the most important vegetable in the world. RNAi has been extensively employed to improve the plant and fruit traits and tolerance to abiotic and biotic stresses in this crop.

*Yield improvement*

RNAi strategies have been used to understand the functional roles of major genes in fruit and yield traits in tomato. The C class flowering gene *AGAMOUS* (*AG*) encodes a transcription factor protein that regulates flowering and fruiting in angiosperms. The RNAi approach was used to generate transgenic tomato lines with suppressed expression of *TOMATO AGAMOUS* (*SlTAG1*) or *TOMATO AGAMOUS-LIKE1* (*SlTAGL1*) genes. The *SlTAGL1* RNAi lines were defective in fruit ripening, whereas *SlTAG1* RNAi lines had defects with stamen and carpel formation (Pan *et al.*, 2010). This study revealed the multiple roles of *AG* in flowering and fruit development in tomato. Similarly, downregulation of the transcription factor gene *WUSCHEL* (*SlWUS*) by an RNAi approach has developed tomato plants with smaller flowers and smaller fruit with lower locule number (Li *et al.*, 2017). The study provided evidence that *SlWUS* is a candidate gene for the *lc* locus controlling locule number in tomato fruit, which has a profound role in flower development.

Fruit development is a function of balanced action of plant growth regulators. To reveal the significance of endogenous abscisic acid (ABA) balance in tomato fruit set, Wang *et al.* (2023) developed lines transformed with a fruit-specific RNAi construct, controlled by a fruit-specific *E8* promoter and targeting the *SlNCED1* gene. The *SlNCED1* gene encodes 9-*cis*-epoxycarotenoid dioxygenase (NCED), an important enzyme in ABA biosynthesis. The *SlNCED1*-RNAi downregulated ABA signaling pathway genes such as *SlPYL3*, *SlPP2C1/2/5* and *SnRK2.6*, strongly repressing ABA signaling transduction. The pistil showed aberrant development, fruit setting was drastically reduced, and the fruit were parthenocarpic with an abnormal shape. This study showed the significance of ABA balance in fruit set in tomato. In a parallel study using RNAi, Zhang *et al.* (2019) showed that the MADS-box protein of an *AG* gene of tomato, *SlMBP3*, has roles in seed formation in tomato. The transgenic lines had fleshy placenta with no locular gel and highly malformed seeds with no seed-coat. RNAi-based experiments have also proven the significance of the *SlNAC3* gene in seed development in tomato (Han *et al.*, 2014).

The enzyme *N*-acetyl-glucosaminyltransferase I (GNTI), which initiates a complex *N*-glycan maturation in the Golgi apparatus, was shown to be essential for normal development and seed set in tomato (Kaulfürst-Soboll *et al.*, 2021). The *GNTI*-RNAi plants had necrotic fruit stalks, patchy incomplete fruit ripening, early fruit drop, and higher susceptibility to infections. Their continued attempts to produce low-allergenic fruit (reduced levels of core α-1,3-fucose and β-1,2-xylose) by downregulating the Golgi α-mannosidase II (*MANII*) gene resulted in normal fruit that remained attached for a longer time with no necrosis, but fruit had no or few enlarged seeds.

### Fruit quality improvement

Suppression of the expression of *DET1*, an endogenous photomorphogenesis regulatory gene using RNAi technology along with fruit-specific promoters has improved the carotenoid and flavonoid content in tomato (Davuluri *et al.*, 2005). Other fruit quality parameters remained unchanged, showing the potential of this technology in biofortification. In a similar way, fruit-specific RNAi of the *SlNCED1* gene was shown to enhance the lycopene and β-carotene contents in the fruit (Sun *et al.*, 2012). A fruit-specific *E8* promoter used in the RNAi construct reduced the ABA accumulation and *SlNCED1* transcript levels by up to 50%. Reduced NCED activity blocked the carbon, which normally gets used up in the synthesis of free ABA and ABA metabolites in the ripening process, and it was diverted to the carotenoid pathway, leading to the accumulation of lycopene and β-carotene, which are upstream compounds in the pathway.

Enzyme ζ-carotene desaturase (ZDS) catalyzes the conversion of ζ-carotene to lycopene. RNAi-mediated silencing of the ζ-carotene desaturase gene was found to cause an increase in phytoene content in tomato leaves and fruit (Babu *et al.*, 2021). Additionally, the transgenic lines had whitish-green fruit and photobleached and larger leaves.

Storage parenchyma cells of tomato fruit usually show significant accumulation of hexoses diverted from the fruit apoplasm. The hexose transporters LeHT1, LeHT2, and LeHT3 in the fruit storage parenchyma cells are thought to be responsible for hexose uptake. Using an RNAi strategy, McCurdy *et al.* (2010) confirmed the role of these transporters in hexose accumulation in the fruit. Knockdown of fruit expression of the *LeHT* genes by 80–90% reduced the levels of fruit hexose by 55%.

The *Lycopersicum Invertase5* (*LIN5*) gene encoding a cell-wall invertase enzyme regulates the total soluble solids in tomato fruit. Zanor *et al.* (2009) developed transgenic lines with *LIN5* silenced by RNAi. Metabolome characterization of ovaries, unripe fruits, and red-ripe fruits showed significant changes in sugar metabolism. Hormone and transcript profiling showed changes in the hormones and transcripts encoding the enzymes involved in hormone biosynthesis and their response elements.

In the ascorbate metabolism pathway, reduced L-ascorbic acid is oxidized by ascorbate peroxidase (APX) enzyme, leading to a reduction in the vitamin C content in the fruit. Suppression of the mitochondrial *APX* gene through RNAi decreased the cytosolic APX activity, leading to up to a 2.2-fold increase in ascorbic acid content in tomato fruit (Zhang *et al.*, 2011b). A parallel study by Cai *et al.* (2016) showed that suppression of *SlDof22* transcription factor gene by RNAi enhanced ascorbic acid accumulation in tomato fruit. Additionally, expression of the genes in the D-mannose/L-galactose pathway and ascorbic acid recycling were affected. As the transcription factor protein SlDof22 binds to the promoter sequence of *SlSOS1*, a key gene in the salt-stress response, the transgenic plants lost salt tolerance. Another RNAi study in tomato showed that downregulation of the ascorbate oxidase gene enhanced ascorbic acid levels in tomato fruit (Zhang *et al.*, 2011a).

In an attempt to reduce the allerginicity of tomato fruits, Le *et al.* (2006a) silenced the *Lyc e 1.01* and *Lyc e 1.02* genes in tomato through an RNAi approach. These genes encode profilin, a small actin-binding protein that contributes to the allergenicity in many fruits and vegetables. The resulting fruit had a tenfold reduced level of profiling, and skin prick tests showed reduced hypoallergenic responses in patients with a tomato allergy. The same group has later showed that inhibition of *Lyc e 3* also reduced the allergenicity in tomato fruit (Le *et al.*, 2006b).

### Resistance to biotic stresses

Tomato lines transformed with RNAi constructs for the conserved sequences of 21–200 nt of the *AC4* gene of tomato leaf curl virus and expressing dsRNA from the sense and antisense inserts of these targets were tested for their resistance to the virus (Praveen *et al.*, 2010). Silencing of the viral target gene was more effective with dsRNA with a longer arm length and coding for a pool of diced siRNAs. Similarly, Fuentes *et al.* (2016) developed lines resistant to tomato yellow leaf curl virus (TYLCV) through RNAi targeting of the *C1* gene. The $F_6$ and $F_4$ progenies of two homozygous transgenic lines were found to stably express the respective siRNAs and showed consistent field resistance. A common set of differentially expressed genes found through the transcriptome profiling has suggested that alteration of the transcriptional pattern of non-target genes is a limitation of this technology. Development of TYLCV-resistant tomato lines using an RNAi strategy was also reported by Ammara *et al.* (2015). Similarly, development of plants with resistance to tomato spotted wilt virus (TSWV) was reported by Tabein *et al.* (2020).

Tomato lines transformed with an RNAi construct harboring an inverted repeat of a partial replicase gene of cucumber mosaic virus strain O (CMV-O) and expressing dsRNA specific to the gene were found to be resistant to CMV (Ntui *et al.*, 2014). The resistant transgenic lines were also shown to be resistant to the related virus strain CMV-Y. A similar study showed that transgenic tomato plants expressing a viroid hairpin RNA (hpRNA) construct and accumulating high levels of siRNAs acquired resistance to potato spindle tuber viroid (Schwind *et al.*, 2009).

Silencing of the *Pbs2*, *Hog1*, and *Fmk1* genes involved in mitogen-activated protein kinase signaling in the fungus *Fusarium oxysporum* was found to reduce infection in tomato (Pareek and Rajam, 2017). When the spores were transformed using RNAi constructs, *Fmk1* transformants lost surface hydrophobicity coupled with lower invasion of fruit and reduced virulence toward seedlings. The *Hog1* and *Pbs2* transformants showed altered conidial size with reduced invasion and pathogenesis.

The dsRNAs targeting genes that are sites of action of fungicides, *FoCYP51*, *FoChs1*, and *FoEF2*, in *Fusarium oxysporum* f. sp. *radicis-lycopersici* causing crown and root rot in tomato, were developed (Mosa and Youssef, 2021). Delivery of naked dsRNAs on layered double hydroxide nanosheets ensured their stability for long-term sustained release in plant cells. The treated plants shown reduced invasion by the fungus.

RNAi strategies have been used extensively to reveal the functional roles of genes and pathways in pathogen resistance in tomato. Genes such as tomato receptor-like cytoplasmic kinase (*TRK1*) (Jaiswal *et al.*, 2022), *PEPR1 ORTHOLOG RECEPTOR-LIKE KINASE1* (Xu *et al.*, 2018), and protein kinase 1b (AbuQamar *et al.*, 2008) have been shown to be important in resistance mechanisms as their knockdown lines were found to be susceptible to multiple fungi. Some other targeted genes are described in Table 19.1.

RNAi has also been shown to be effective in developing plants with resistance to pests. Transgenic plants expressing dsRNAs targeting vital insect genes will silence the genes in the feeding larvae. Tomato lines transferred with the dsRNA for vacuolar ATPase-A and arginine kinase genes, were found to be resistant to leafminer (*Tuta absoluta*) (Camargo *et al.*, 2016). The dsRNAs were delivered into tomato leaflets through direct uptake by the leaflets and through an *in planta* strategy. Larvae that fed on the leaves containing dsRNA showed nearly a 60% reduction in target gene transcript accumulation, higher mortality, and less damage. In another similar study, Bento *et al.* (2020) attempted the management of *T. absoluta* by RNAi-mediated silencing of the target insect genes juvenile hormone inducible protein (*JHP*), juvenile hormone epoxide hydrolase protein (*JHEH*), ecdysteroid 25-hydroxylase (*PHM*), chitin synthase A (*CHI*), carboxylesterase (*COE*) and arginine kinase (*AK*). The larvae were fed an artificial diet containing *Escherichia coli* cells expressing dsRNAs targeting *JHP*, *CHI*, *COE*, and *AK*, which resulted in significantly reduced transcript accumulation and enhanced larval mortality. In a similar study, Ramkumar *et al.* (2021) topically applied dsRNAs targeting the v-ATPase B and juvenile hormone-binding protein genes of *T. absoluta* on tomato leaves to successfully control these insects through an RNAi strategy. In yet another parallel study, Majidiani *et al.* (2019) attempted the injection and root delivery of dsRNAs targeting acetylcholinesterase, nicotinic acetylcholine α-6 receptor, and ryanodine receptor protein genes to manage *T. absoluta* in tomato. Injection with 2–5 μg dsRNA resulted in up to 92.59% mortality, whereas the root feeding with 5 μg dsRNAs resulted in up to 80.5% mortality. Camargo *et al.* (2015) sequenced and *de novo* assembled the transcriptomes of healthy and

**Table 19.1.** Studies on the management of tomato diseases using RNA interference (RNAi) strategies.

| Disease/pathogen | Target gene(s) | Reference |
|---|---|---|
| Pepino mosaic virus | Conserved sequences from three virus strains | Leibman *et al.* (2021) |
| Monopartite begomovirus tomato leaf curl Taiwan virus (ToLCTWV) and bipartite tomato yellow leaf curl Thailand virus (TYLCTHV) | Fusions of *C1*, *C2*, and *C3* genes of ToLCTWV and the corresponding genes of the TYLCTHV DNA-A genome | H.M. Chen *et al.* (2016) |
| Tomato leaf curl New Delhi virus and tomato leaf curl Gujarat virus | RNAi suppressor proteins AC2 and AC4 of the geminivirus | Singh *et al.* (2015) |
| Tomato leaf curl virus | Viral *AC1* and *AC4* genes | Ramesh *et al.* (2007) |
| Tomato chlorosis virus (ToCV) and tomato yellow leaf curl virus (TYLCV) | Coat protein genes of ToCV and TYLCV | Jin *et al.* (2020) |
| Potyviruses | Eukaryotic translation initiation factor genes (*eIF4E1* and *eIF4E2*) | Mazier *et al.* (2011) |
| Cucumber mosaic virus | *1a*, *2a*, *2b*, *3a* and coat protein genes | Miao *et al.* (2016) |
| *Fusarium oxysporum* and *F. solani* | Pathogenicity genes of the fungi: Zn(II)2Cys6-type transcription regulator (*FOW2*) and class V chitin synthase (*chsV*) | Shanmugam *et al.* (2017) |
| *Fusarium oxysporum* f. sp. *lycopersici* | Ornithine decarboxylase, a key polyamine (putrescine, spermidine, and spermine) biosynthesis gene of the fungus | Singh *et al.* (2020) |
| Tomato wilt caused by *Fusarium oxysporum* f. sp. *lycopersici* | Peroxisomal biogenesis factor (*PEX6*) and β-1,3-glucanosyltransferase (*GAS1*) genes of the fungus | Tetorya and Rajam (2021) |
| *Verticillium dahliae* | RNAi suppression of *WALLS ARE THIN 1* (*SlWAT1*) gene, a susceptibility (*S*) gene in tomato for *V. dahliae* | Hanika *et al.* (2021) |

leafminer-infected leaves. Differentially expressed genes were identified as targets for RNAi-mediated control of this insect.

Wuriyanghan *et al.* (2011) used the expressed sequence tag sequence information from Asian citrus psyllid (*Diaphorina citri*) to identify the RNAi targets in tomato psyllid (*Bactericera cockerelli*). The mRNAs abundant in the gut were targeted by injection and oral acquisition of dsRNAs and siRNAs. This induced mortality in the recipient psyllids. Gogoi *et al.* (2017) showed that, following exogenous application of dsRNA molecules on tomato plants, the dsRNAs moved rapidly within the plant and were taken up by agricultural pests. The dsRNA molecules were derived from the zucchini yellow mosaic virus HC-Pro region, and after application to the plants, dsRNAs were detected in aphids, whiteflies, and mites that fed on the leaves. The siRNAs were also detected in the treated plants and insects. Some of the other studies on insect pests are given in Table 19.2.

### Resistance to abiotic stresses

Compared with the research carried out to manage biotic stresses in tomato, fewer studies have been done on abiotic stresses. RNAi-mediated downregulation of the transcription factor gene *SlDREB2* demonstrated its role in drought tolerance in tomato (Tao *et al.*, 2022). The *SlDREB2*-RNAi lines were drought susceptible with lower levels of proline and soluble sugars.

Downregulation of a GRAS transcription factor gene in tomato, *SlGRAS10*, by RNAi resulted in dwarf plants with smaller leaves and internodes and with higher flavonoid accumulation (Habib *et al.*, 2021). The *SlGRAS10*-RNAi plants had levels of higher superoxide dismutase, peroxidase, catalase, malondialdehyde, and

**Table 19.2.** Studies on the management of pests of tomato using RNA interference strategies.

| Pest | Target gene(s) | Reference |
|---|---|---|
| *Helicoverpa armigera* | Chitinase gene of the insect | Mamta and Rajam (2016) |
| Western flower thrips (*Frankliniella occidentalis*) | Coatomer protein subunit epsilon (*CopE*) and Toll-like receptor 6 (*TLR6*) genes of the thrips | Venkatesh *et al.* (2023) |
| Whitefly (*Bemisia tabaci*) | v-ATPase gene of the whiteflies | Pizetta *et al.* (2022) |
| Aphids (*Myzus persicae*) | Acetylcholinesterase 1 (*Ace 1*) gene | Faisal *et al.* (2019) |
| Leaf-chewing and lacerate-and-flush feeding insects | Plastid transformation with RNAi construct targeting v-ATPase-A gene from cotton mealybug (*Phenacoccus solenopsis*) | Kaplanoglu *et al.* (2022) |
| *Tetranychus evansi*, *T. truncatus*, and *T. cinnabarinus* | Transplastomic tomato lines carrying RNAi construct targeting spider mite β-actin gene | Wu *et al.* (2023) |

proline, leading to enhanced salt, drought, and ABA tolerances. The *SlMX1* gene encoding a MIXTA-like MYB transcription factor is yet another gene identified to impart drought tolerance in tomato (Ewas *et al.*, 2016). In a similar study, Guo *et al.* (2023) demonstrated the role of the histone deacetylase gene *SlHDA3* in drought and salt tolerance in tomato. The RNAi approach has been successful in a number of studies, including: revealing the role of SlWHY2 protein in maintaining mitochondrial function under drought tolerance (Meng *et al.*, 2020); enhancement of drought tolerance by the SlNAC6 NAC transcription factor protein (Jian *et al.*, 2021); promotion of reactive oxygen species (ROS) scavenging and chlorophyll accumulation under drought conditions by methionine sulfoxide reductase B2 (Cui *et al.*, 2022); upregulation of defense pathways under drought and salt tolerance conditions by SlMBP8 MADS-box transcription factor protein (Yin *et al.*, 2017); heat tolerance by the SUMO E3 ligase enzyme SlSIZ1 (Zhang *et al.*, 2018); enhancement of the expression of the *hsp21.5a* gene and promotion of high-temperature tolerance by the WHIRLY1 protein (Zhuang *et al.*, 2020); upregulation of the tolerance pathways under cold stress by LeCOR413PM2 (Zhang *et al.*, 2021); cold storage tolerance of tomato fruits by monodehydroascorbate reductase (El Airaj *et al.*, 2013); enhancement of tolerance to drought, heat, and combined stresses by the downregulation of a cytokinin receptor gene *SlHK2* (Mushtaq *et al.*, 2022); and increased high-temperature stress tolerance in tomato through downregulation of the mitogen-activated protein kinase gene *SlMPK1* (Ding *et al.*, 2018). Suppression of ascorbate oxidase activity was reported to increase tomato yield under drought stress (Garchery *et al.*, 2013).

### Other traits

Development of the late-ripening tomato cultivar Flvr Savr in 1994, as the first genetically engineered whole food, was perhaps the initial demonstration of the potential of RNAi in crop improvement. The tomato polygalacturonase (*PG*) gene was suppressed by transformation of an antisense expression cassette of the *PG* cDNA (pCGN1436) (Sheehy *et al.*, 1988). The *PG* suppression was attributed to the synthesis of 21 nt siRNAs (Krieger *et al.*, 2008). In a similar study, tomato lines showing delayed ripening and better processing qualities were developed by suppressing the expression of 1-aminopropane-1-carboxylate (ACC) synthase (*ACS*) gene using RNAi (Gupta *et al.*, 2013). Three *ACS* homologs were silenced to develop the lines with repressed ethylene production in the fruits, extending the shelf life to around 45 days. Ethylene suppression upregulated the polyamine biosynthesis and ascorbic acid metabolism genes and downregulated the cell-wall-hydrolyzing enzyme genes. The ACC oxidase enzyme catalyzes the oxidation of ACC to ethylene, and transgenic tomato lines in which the *ACC oxidase* gene was downregulated by RNAi produced fruits with a shelf life of more than 120 days (Xiong *et al.*, 2005). The efficiency of longer and shorter linker DNAs to achieve complete interference was also tested in this study. A shorter linker (7 nt) used to link the sense and antisense configurations was found to be more efficient compared with longer (1002 bp) linker to produce dsRNAs of the ACC oxidase gene and to achieve full RNAi.

Ji *et al.* (2014) demonstrated significant roles played by the *SlNCED1* gene encoding 9-*cis*-epoxycarotenoid dioxygenase involved in ABA biosynthesis and the *SlCYP707A2* gene encoding ABA 8′-hydroxylase involved in ABA oxidative catabolism in fruit ripening of tomato. The *SlNCED1*-RNAi plants produced slow-ripening orange fruits, whereas ripening was promoted in *SlCYP707A2*-silenced fruits. Similarly, the histone deacetylase *SlHDA1* gene was also identified as a repressor of fruit ripening and carotenoid accumulation in tomato fruits (Guo *et al.*, 2017). Using RNAi, Karlova *et al.* (2011) showed that the transcription factor gene *APETALA2a* (*AP2a*) has a significant role in fruit ripening, as the transgenic lines had ripe fruits with altered shapes and carotenoid accumulation. The *AP2a*-RNAi fruits also showed changes in expression pattern of genes involved in chromoplast differentiation, ripening-associated processes, and phenylpropanoid and carotenoid pathways, as well as in hormone synthesis and perception. A similar study by Dong *et al.* (2013) showed that the *SlMADS1* gene of tomato, encoding a MADS-box transcription factor, is a negative regulator of fruit ripening.

Polyamines have important roles in flower induction and fertility regulation in plants, and *S*-adenosylmethionine decarboxylase (*SAMDC*) is a key gene in polyamine biosynthesis. Silencing of this gene in the tapetal tissues by RNAi was shown to cause male sterility in tomato (Sinha and Rajam, 2013). Coding regions from three *SAMDC* gene homologs were cloned in a hairpin RNA construct harboring a tapetum-specific *A9* promoter and used for transformation. The siRNAs were expressed in the anther and resulted in aborted and sterile pollen, but female fertility remained unaffected, leading to fruit set and seed development through cross-pollination. In a similar study, Schijlen *et al.* (2007) showed that RNAi suppression of the chalcone synthase gene led to the development of parthenocarpic fruits in tomato.

RNAi also holds enormous potential in functional genomics. The roles of three members of the GA 20-oxidase gene family involved in gibberellin biosynthesis, *GA20ox1*, *GA20ox2*, and *GA20ox3*, were investigated using an RNAi strategy (Xiao *et al.*, 2006). Three transgenic tomato lines, each with one of the genes downregulated, demonstrated that *GA20ox1* or *GA20ox2* suppression led to shorter internodes and smaller dark green leaves while *GA20ox3* suppression resulted in no visible changes.

### 19.3.2 Potato

#### *Tuber quality improvement*

RNAi silencing of the β-carotene hydroxylase (*bch*) gene, which converts β-carotene to zeaxanthin, through RNAi was found to enhance the β-carotene content in tubers. Use of a tuber-specific, granule-bound starch synthase promoter in the construct enhanced the β-carotene content up to 331 μg 100 $g^{-1}$ fresh weight. A few transformants were also found to have a higher level of lutein accumulation (van Eck *et al.*, 2007).

To make the potato tubers a more efficient protein expression system and to produce human therapeutic recombinant glycoproteins, Kim *et al.* (2008) demonstrated the knockdown of patatin, the major storage glycoproteins in potato tubers, through RNAi. A conserved target sequence was amplified from the potato patatin gene (*pat3-k1*) and a cauliflower mosaic virus 35S promoter-driven hpRNAi vector was prepared. The *pat3-k1*-RNAi tubers resulted in rapid purification of other potato glycoproteins with less contamination.

#### *Resistance to biotic stresses*

The dsRNA targeting a replicase gene of CMV-O and -Y showed promise in developing resistant potato lines (Ntui *et al.*, 2013). The pEKH2IN2C-MVai construct harboring an inverted repeat of a defective CMV replicase gene fragment from RNA2 of CMV-O was transformed into a susceptible potato cultivar. Transgenic lines constitutively expressed the siRNAs, offering complete resistance. Another study showed that transformation of potato with an RNAi construct carrying a concatenation of viral coat protein sequences from Potato virus X, Potato virus Y, and Potato virus S as a 600 bp inverted repeat, and its expression using a constitutive 35S promoter resulted in combined resistance to these RNA viruses (Hameed *et al.*, 2017). A similar study by Petrov *et al.* (2015) showed that potato plants transformed with an RNAi construct targeting the HC-Pro gene of potato virus Y were completely resistant to the virus.

Plants transformed with RNAi constructs and expressing the dsRNA targeting the genes of Colorado potato beetle, as well as feeding the beetles with the dsRNAs expressed in bacteria, were effective in killing this insect (Palli, 2014; Koo *et al.*, 2023). A similar study by Zhu *et al.* (2011) also showed the potential of feeding dsRNA expressed in bacteria or synthesized *in vitro* to manage Colorado potato beetle. Using this method, five target genes (actin, *Sec23*, *vATPaseE*, *vATPaseB*, and *COPβ*), were suppressed, resulting in beetle mortality.

Chen *et al.* (2005) used RNAi to show that the protein coded by *gr-ams-1* and secreted by the sense organs of the potato cyst nematode is essential for host location and invasion. RNAi-mediated knockout of the β-1,4-endoglucanase gene, encoding a pathogenicity protein of this nematode, reduced the ability of the nematodes to invade roots. This study suggested that research programs aiming to develop potato cultivars with nematode resistance should target molecules interacting with these proteins secreted by the nematode.

#### *Resistance to abiotic stresses*

Antioxidant tocopherols protect chloroplast membranes and the photosynthetic machinery from photo-oxidative damage, especially under abiotic stress. The sucrose export defective1 (*sxd1*) gene encodes tocopherol cyclase. *StSXD1*-RNAi potato plants had defective tocopherol cyclase activity, leading to tocopherol deficiency, causing a photoassimilate export-defective phenotype (Hofius *et al.*, 2004). This study suggested a role of tocopherols in maintaining photosynthetic activity under abiotic stress.

### 19.3.3 Bell peppers and hot peppers

#### *Fruit quality improvement*

*N*-glycan processing enzymes have significant roles in the softening of climacteric fruits. Through RNAi of the genes encoding the *N*-glycan processing enzymes (α-mannosidase and β-D-*N*-acetylhexosaminidase), Ghosh *et al.* (2011) demonstrated their role in the softening of *Capsicum* fruits. The α-mannosidase- and β-D-*N*-acetylhexosaminidase-RNAi fruits were two times firmer compared with wild fruits, and their deterioration was delayed by nearly a week.

#### *Resistance to biotic stresses*

Kim *et al.* (2021) found that at higher temperatures, when resistance (*R*) gene-mediated resistance is suppressed, production of virus-derived siRNAs was significantly enhanced in peppers, leading to RNAi-mediated viral resistance. Thus, plants retain their viral resistance by activating an RNAi mechanism through transcription of temperature-specific virus-derived siRNAs. Similarly, biogenesis and accumulation of siRNAs specific to tomato leaf curl New Delhi virus (ToLCNDV) during infection was reported in solanaceous crops including peppers (Kushwaha *et al.*, 2015). Fletcher *et al.* (2022) once again showed that RNAi is a robust defense mechanism in plants when they demonstrated the presence of abundant virus-derived siRNAs, which targeted viral transcripts for degradation, through small RNA sequencing of peppers infected by TSWV or *Capsicum* chlorosis virus. The host plants were also found to abundantly generate virus-activated siRNAs against many endogenous transcripts.

The TOM1 and TOM3 proteins of the tonoplast membrane are necessary for the multiplication of tobamoviruses. *CaTOM1*- and *CaTOM3*-RNAi lines of pepper were free from the mosaic symptoms caused by tobacco mosaic virus (TMV) infection (Kumar *et al.*, 2012). Hence, siRNA-mediated silencing of these genes was identified as an efficient method for developing TMV-resistant pepper plants. Similarly, an hpRNAi construct in a binary vector based on the overlapping and highly conserved regions of the *Rep* and *TrAP* genes of pepper leaf curl virus was successful in blocking infection by this virus and DNA-B of ToLCNDV (Shafiq, 2014). Another RNAi study demonstrated the profound role of the *CaBtf3* gene, encoding the β-nascent polypeptide-associated complex subunit of the NAC transcription factor, in regulating the transcription of pathogenesis-related (PR) proteins during TMV infection in pepper (Huh *et al.*, 2012). The *CaBtf3*-RNAi lines had lower cell death resulting from the hypersensitive response (HR), decreased expression of HR-associated genes, and higher accumulation of TMV coat proteins.

RNA vaccination of *Capsicum chinense* 'Bhut Jolokia' through exogenous application with a crude extract of bacterially expressed dsRNA targeting the CMV *2b* gene was shown to successful manage CMV infection (Borah *et al.*, 2018). A comparable result was later obtained by Bhorali (2021) in the same cultivar by transforming the plants with an hpRNAi construct targeting the replicase gene of CMV. Although peppers in general are highly recalcitrant to *in vitro* regeneration, Bhorali (2021) optimized multiple shoot induction from calluses in Murashige and Skoog (MS) medium with 8.5 mg $l^{-1}$ kinetin, 0.5 mg $l^{-1}$ thidiazuron and 5 mg $l^{-1}$ silver nitrate, and root induction in half-strength MS medium with 1.5 mg $l^{-1}$ 1-naphthalene-acetic acid.

Using RNAi experiments, H. Li *et al.* (2023) have shown that infection of pepper plants with vector-borne viruses rendered the plants able to release volatiles that are more attractive to insect vectors. The volatile *cis*-3-hexenal emitted by tomato zonate spot virus-infected pepper plants, recognized by chemosensory protein 1 of thrips, was more attractive to the virus vector. RNAi with the gene encoding the chemosensory protein of the thrips drastically decreased the electroantennogram responses in thrips' antennae to *cis*-3-hexenal and impaired the insects' responses.

In a study to optimize the methods to manage TSWV, *Capsicum* plants were given a topical spray of dsRNAs cultured in bacterial cells and targeting the *N* gene of the virus. Challenging the plants with virus by mechanical inoculation, 1 day after the spray, resulted in a significant reduction in viral load in the treated plants (Nilon *et al.*, 2021).

Downregulation of fungal RXLR effectors through RNAi was found to be a promising strategy to protect pepper plants from *Phytophthora capsici* infection (Cheng *et al.*, 2022). This study showed that *in planta* expression of RNAi constructs as well as exogenously applied siRNAs targeting the *RXLR1* or *RXLR4* gene could restrict growth of the fungus.

Injection with dsRNAs targeting the *Snf7* and *v-ATPase-A* genes of pepper weevil (*Anthonomus eugenii*) completely silenced these genes and within 12 days, more than 90% mortality of the weevils was obtained (Wu *et al.*, 2019). A similar study showed that topical spray application of *Capsicum* leaves with dsRNAs targeting an aquaporin homolog gene of thrips, cultured in bacterial cells, was effective in managing the thrips (Nilon *et al.*, 2021).

### 19.3.4 Eggplant

#### *Yield improvement*

SmARF8, a nuclear protein and active transcriptional regulator, has been shown to be involved in parthenocarpic fruit development in eggplant. *SmARF8*-RNAi lines showed parthenocarpy in unfertilized flowers, and it was concluded that this protein affects dimerization of auxin/indole acetic acid repressors and thus induces fruit development (Du *et al.*, 2016).

#### *Fruit quality improvement*

In eggplant, silencing of *SmCIP7*, encoding a protein that interacts with *COP1* (*CONSTITUTIVELY PHOTOMORPHOGENIC1*), through RNAi reduced the fruit anthocyanin and chlorophyll levels, fruit size, flesh browning, and seed yield. The reduction in fruit size in *SmCIP7*-RNAi lines was proposed to be the result of upregulation of the *SmYABBY1* gene (Y. Li *et al.*, 2023).

#### *Resistance to biotic stresses*

Downregulation of *MYB44*, an R2R3-MYB transcription factor gene, by an RNAi strategy demonstrated a role in bacterial wilt resistance in eggplant (Qiu *et al.*, 2019). The SmMYB44 protein binds directly to the promoter of the spermidine synthase (*SmSPDS*) gene and initiates its transcription. Following *Ralstonia solanacearum* infection, *SmSPDS* is induced and spermidine content increases, especially in resistant lines. The *SmMYB44*-RNAi lines had a reduced abundance of *SmSPDS* transcripts and were more susceptibility to wilt. In a similar study, RNAi was used to confirm the role of *RE-bw* with NB-ACR and

WRKY domains and involvement in protein binding and defense response progress in bacterial wilt resistance in eggplant. The *RE-bw*-RNAi lines had reduced salicylic acid levels, ROS-scavenging enzymatic activities, cell-wall thickness, and lignin content, resulting in higher levels of bacterial colonization. The RE-bw transcription factor protein interacts with *EDS1*, *PAD4*, *NPR1*, *SGT1*, and *WRKY70* genes and the avirulence effector PopP2 of the bacterium (Xu *et al.*, 2015).

A study by N. Chen *et al.* (2016) demonstrated the role of an NAC transcription factor gene in bacterial wilt susceptibility. *SmNAC*-RNAi lines had enhanced bacterial wilt resistance resulting from higher accumulation of salicylic acid and higher expression of the isochorismate synthase 1 gene, which is involved in salicylic acid biosynthesis. Recognition of bacterial type III effectors triggers the specific resistance mechanisms in plants. Thus, ripAX2, a type III effector of *R. solanacearum*, interacts with resistance genes at the *EBWR9* locus and makes a wild species of eggplant resistant. A bacterial strain mutated at the *ripAX2* locus can cause disease (Morel *et al.*, 2018).

The ecdysone-induced protein 93F (E93) has profound roles in the development of eggplant ladybeetle at pre-pupal and pupal stages. Downregulation of *E93* through RNAi at the pupal stage affected adult eclosion, and injection of fourth-instar larvae and pupae with dsE93 resulted in abnormal development of the insect (Wu *et al.*, 2022). This study suggested that *E93*, which is a repressor of larval characteristics and a specifier of adult structures, could be a target for the management of this insect. The jasmonic acid–coumarin pathway is enhanced during infestation of the mite *Tetranychus cinnabarinus* in eggplant and this safeguards the plants due to reduced fecundity of the insect. This pathway is known to downregulate the expression of the vitellogenin gene in mites. Xu *et al.* (2023) showed that *TcVg6*-RNAi eggplant lines had significantly reduced egg production and hatching rate of these mites.

The *Mi-msp-1* effector gene of the root-knot nematode *Meloidogyne incognita* is important for the parasitization process. *Mi-msp-1*-RNAi eggplant lines had long-term effects on enhanced resistance to nematode penetration, development, and reproduction (Chaudhary *et al.*, 2019). Similarly, the effector proteins Mi-msp-18 and Mi-msp-20 are involved in the late and early stages of root-knot nematode parasitism, respectively, and are known to interact with cell-wall-modifying enzymes. Shivakumara *et al.* (2017) showed that *msp-18*-RNAi and *msp-20*-RNAi eggplant lines had enhanced nematode resistance by affecting the transcription of genes related to cell-wall modification in the nematodes.

## 19.4 RNAi in Improvement of Other Vegetables

### 19.4.1 Cucumber

*Plant growth and fruit development*

RNAi has been used to decipher the functions of various genes in cucumber (*Cucumis sativus*) plant growth and development. The HANABA TARANU (HAN) transcription factor interacts with multiple genes including *CsWUS* and *CsSTM*, and regulates the organization of shoot apical meristem and flower organ development. *CsHAN1*-RNAi lines had delayed development of shoot apical meristem (Ding *et al.*, 2015). Similarly, the transcription factor protein encoded by the *LEAFY* (*LFY*) gene regulates the flowering and branching processes in plants. *CsLFY*-RNAi lines were found to have adventitious shoots with fewer leaves. Bimolecular fluorescence complementation assays revealed the interaction between the CsLFY transcription factor and *CsWUS*, controlling shoot meristem development in cucumber (Zhao *et al.*, 2018).

The *TEOSINTE BRANCHED1/CYCLOIDEA/PCF* (*TCP*) family gene *BRANCHED1* (*BRC1*) is a repressor of shoot branching in plants. *CsBRC1* binds the promoter of the *CsPIN3* gene encoding an auxin efflux carrier protein, PIN-FORMED, and downregulates it. Thus, auxin accumulation is enhanced in the cells, leading to suppression of branching. *CsBRC1*-RNAi cucumber lines showed reduced auxin accumulation in lateral buds and a correspondingly enhanced number of branches (Shen *et al.*, 2019). A similar study showed that *CsTFL1* is a suppressor of determinate growth habit in cucumber, and a single mutation in this gene leads to determinate growth (Wen *et al.*, 2019). The CsTFL1 protein competes with the CsFT protein to interact with the CsNOT2a (negative on TATA-less 2a)–CsFDP (FD PARALOG) complex to suppress floral meristem identity genes in the shoot tip and leads to indeterminate growth. *CsTFL1*-RNAi lines showed terminal flowers at the shoot apex.

The role of the *CsSPL* gene, a *SPOROCYTELESS* (*SPL*)/*NOZZLE* (*NZZ*) homolog in cucumber, in anther and ovule development was investigated by Liu *et al.* (2018). *CsSPL*-RNAi lines had malformed pollen, poorly developed ovules, and reduced male and female fertility. *CsSPL* was shown to be a linker between *CsPHB*, an important regulatory gene in the nucellus, and *CsWUS*. Similarly, the development of male flowers in cucumbers is regulated by the homeodomain-leucine zipper type IV transcription factor gene *GL2-LIKE* through interaction with a jasmonate ZIM (zinc-finger inflorescence meristem) domain protein, CsJAZ1. Downregulation of this gene caused delayed male flower initiation, reduced pollen and seed vigor, and partial male sterility (Cai *et al.*, 2020).

### *Fruit quality improvement*

The cucumber *FRUITFULL-like* MADS-box transcription factor gene *CsFUL1*$^{A}$ downregulates *CsSUP*-mediated cell division and *CsPIN*-mediated auxin transport. *CsFUL1*$^{A}$-RNAi lines had longer fruits, whereas the overexpression lines had shorter fruits. A Q/K single-nucleotide polymorphism in this gene was found to be associated with fruit length (Zhao *et al.*, 2019).

RNAi-mediated downregulation of a gibberellin receptor gene, *CsGID1a*, demponstrated its role in carpel and locule formation in cucumber (Liu *et al.*, 2016). Similarly, *CLV3* and *WUS* are major genes regulating the number of locules in fruits. *CsCLV3*-RNAi and *CsWUS*-overexpression lines had higher numbers of petals and carpels, showing that *CsCLV3* is a negative regulator and *CsWUS* is a positive regulator of florogenesis in cucumber (Che *et al.*, 2020).

Raffinose family oligosaccharides (RFOs), the major translocated sugars in cucumber vascular bundles, can be detected only in traces in its fruits. α-Galactosidase enzymes catalyze the first step in the catabolism of RFOs in vascular bundles. Although six α-galactosidase genes have been identified in the cucumber genome through RNAi, *CsAGA1* and *CsAGA2* have been shown to be important in hydrolysis. *CsAGA1*-RNAi plants accumulated high raffinose and stachyose levels, whereas *CsAGA2*-RNAi plants accumulated high stachyose levels (Hua *et al.*, 2021). A similar RNAi study showed that the CsSWEET7a protein acts as a hexose transporter and regulates phloem unloading during cucumber fruit development. *CsSWEET7a*-RNAi lines had smaller fruits with low total soluble solids (Li *et al.*, 2021). Another study revealed that the source–sink carbon partitioning in cucumber fruits is determined by the balanced expression of stachyose synthase (*CsSTS*) and its antisense RNA (Dai *et al.*, 2023). RNAi studies have shown that *CsSTS* is negatively regulated by antisense *CsSTS*.

The WD-repeat gene in cucumber, *CsTTG1*, expressed in the ovary epidermis regulates trichome differentiation and fruit warts. *CsTTG1*-RNAi lines had reduced fruit spines, and protein from this gene was shown to interact with the trichome formation regulator *Mict*/*CsGL1* (C.H. Chen *et al.*, 2016). Another RNAi-based study demonstrated the role of the spine base size 1(*CsSBS1*) gene, encoding a $C_2H_2$ zinc-finger transcription factor, in deciding the size of the spine base in cucumber fruits. Expansion of the fruit spine base was inhibited in *CsSBS1*-RNAi lines. *CsSBS1* along with *CsTTG1* and homeodomain *GLABROUS1* (*CsGL1*), two positive regulators, decide the spine base size through ethylene signaling (Yang *et al.*, 2022). The *TS1* gene encoding oleosin proteins and *CsTS1* regulate fruit tubercule development. *CsTS1*-RNAi lines were found to have smaller tubercules. Further studies showed that CsTu transcription factors act as positive regulators of *CsTS1*, leading to fruit tubercule development (Yang *et al.*, 2019).

### *Resistance to biotic stresses*

Downregulation of the basic helix–loop–helix transcription factor *IRREGULAR VASCULATURE PATTERNING* (*CsIVP*) gene through RNAi has resulted in cucumber plants with higher resistance to downy mildew (S. Yan *et al.*, 2020). As this protein interacts with *NIM1-INTERACTING1* (*CsNIMIN1*), which negatively regulates the salicylic acid signaling pathway, *CsIVP*-RNAi plants have shown better resistance through enhanced accumulation of salicylic acid.

#### Resistance to abiotic stresses

NAC transcription factors have important roles in drought tolerance in plants and *CsATAF1* enhances drought tolerance in cucumber by upregulation of ABA and scavenging of ROS. *CsATAF1*-RNAi plants were drought susceptible owing to the higher water loss from leaves due to the reduced response of stomata to ABA, higher level of hydrogen peroxide, superoxide radicals and malondialdehyde, and reduced antioxidant enzyme activity (Wang *et al.*, 2018). The stress-responsive genes *CsABI5*, *CsCu-ZnSOD*, and *CsDREB2C* were identified to be the direct target genes of the CsATAF1 protein. Similarly, the role of G-proteins in drought tolerance in cucumber was demonstrated by Liu *et al.* (2021) through RNAi-mediated silencing of the *CsGPA1* gene, encoding the α-submit of G-protein. The transgenic seedlings showed higher drought susceptibility.

Exogenous application with melatonin enhances abiotic stress tolerance in plants. Cucumber plants in which the serotonin *N*-acetyltransferase (*CsSNAT*) gene was silenced by RNAi lost salt tolerance (Qi *et al.*, 2020). RNAi lines have shown reduced expression of the stress-tolerance genes *CsSOD*, *CsCAT*, *CsP5CS1*, *CsP5CS2*, *CsPIP*, and *CsHSP90*, and seedling germination and growth were impaired. Similarly, RNAi downregulation of the G-protein gene *CsGPA1* was shown to result in the loss of salt tolerance in cucumber seedlings (Y. Yan *et al.*, 2020). Under salt stress, *CsGPA1* reversely regulated the aquaporin genes (*CsAQPs*) in cucumber leaves and roots, and hence the water content in RNAi seedlings was reduced substantially, leading to rapid wilting.

The *IRREGULAR VASCULATURE PATTERNING* (*IVP*) gene changes the leaf and vascular patterns in plants. Under low nitrogen, *CsIVP* downregulates the high nitrate affinity transporters and reallocation transporters, and hence *CsIVP*-RNAi lines were found to better tolerate nitrogen deficiency. Additionally, this gene interacts with *CsDREB2C* and downregulates high-temperature responses in plants. Thus, with enhanced expression of heat-shock transcription factors, heat-shock proteins, DREB2C, MBF1b, and WRKY33, *CsIVP*-RNAi lines were also tolerant to high temperature (S. Yan *et al.*, 2022).

Through RNAi, the role of the *CsSTS* gene in cold tolerance in cucumber was demonstrated (Lü *et al.*, 2017). The low stachyose and starch levels in the leaves of *CsSTS*-RNAi lines have shown the role of this gene in phloem loading. Transgenic lines were susceptible to low temperature stress and showed enhanced malondialdehyde content and electrical conductivity. A similar study showed that upregulation of a heat-shock transcription factor gene, *CsHSFA1d*, following heat shock led to activation of the jasmonic acid signaling pathway, leading to cold acclimation in cucumber. *CsHSFA1d*-RNAi lines were cold sensitive. A higher jasmonic acid level is proposed to degrade the CsJAZ5 protein, a repressor of the jasmonic acid pathway, and upregulate *CsICE1*, which activates the ICE–CBF–COR pathway (Qi *et al.*, 2022).

RNAi was also useful to show the role of the G-protein gene *CsGPA1* in cold tolerance in cucumber. This G-protein interacts with a positive brassinolide signal gene, *CsCDL1*, in the cell membrane and nucleus and with *CsCOR413PM2* in the cell membrane. Under cold stress, *CsGPA1*-RNAi lines showed reduced expression of most of brassinolide-related genes. Both *CsGPA1*-RNAi and *CsCOR413PM2*-RNAi lines had decreased calcium influx at low temperatures, which subsequently reduced the expression of *CsICE–CsCBF* (Y. Yan *et al.*, 2022).

### 19.4.2 Onion

#### Bulb quality improvement

Propanthial *S*-oxide is the lachrymatory factor that causes the burning sensation of the eyes when cutting onion bulbs. Catabolism of *trans*-1-propenyl-L-cysteine sulfoxide by the enzyme alliinase produces 1-propenyl sulfenic acid, which is converted to lachrymatory factor by the enzyme lachrymatory factor synthase (LFS). Kamoi (2008) produced tearless onion through an RNAi strategy targeting the *LFS* gene. In the transgenic lines, LFS enzyme activity was downregulated 1573-fold and lachrymatory factor production upon tissue rupture was reduced by up to 67-fold. A related study showed that suppressing the *LFS* gene changed the levels of sulfur secondary metabolites in onion. By *LFS* downregulation, the *trans*-*S*-1-propenyl-L-cysteine sulfoxide breakdown pathway was shifted, resulting in the conversion of more 1-propenyl sulfenic acid to di-1-propenyl thiosulfinate. Enhanced thiosulfinate levels increased the production of zwiebelane isomer and other volatile sulfur compounds, di-1-propenyl disulfide, and 2-mercapto-3,4-dimethyl-2,3-dihydrothiophene (Eady *et al.*, 2008). The same laboratory has previously reported the generation of onions with reduced allinase activity (Eady *et al.*, 2004).

Flavonoids anthocyanins, flavonols, and chalcones impart the red, pale yellow, and bright yellow colors, respectively, to onion bulbs. RNAi-mediated silencing of the R2R3-MYB transcription factor gene *MYB1* has revealed the role played by this gene in anthocyanin accumulation (Schwinn *et al.*, 2016).

#### *Resistance to biotic stresses*

The RNA sequencing data from the onion thrips *T. tabaci* was used to identify the RNAi pathway genes. Feeding the insects with *dsSNF7* and *dsAQP* caused reduced gene expression and 62% and 72% mortality, respectively (Singh *et al.*, 2019).

### 19.4.3 French bean

#### *Resistance to biotic stresses*

Silencing the coding region of the *AC1* gene of bean golden mosaic virus through RNAi was found to generate French bean (*Phaseolus vulgaris*) plants highly resistant to golden mosaic disease (Bonfim *et al.*, 2007).

Another RNAi strategy in French bean was successful to silence the histidine kinase gene *slnCl1*, confirming it as the pathogenicity factor of the fungal pathogen *Colletotrichum lindemuthianum* (Nogueira *et al.*, 2019).

Transformation of French bean with an ihpRNAi construct targeting the *Bemisia tabaci* v-ATPase gene *BtvATPase* through particle bombardment resulted in the development of lines with stable expression of siRNAs and strong resistance to whiteflies (Ferreira *et al.*, 2022).

#### *Other traits*

RNAi has been useful for the functional characterization of genes in French bean. Secretion of Nod factors (lipochito-oligosaccharide signals) by *Rhizobium* initiates its symbiotic association with plants. These signals enhance calcium levels and ROS production in the roots and start the transcription of nodulin genes. RNAi studies by Montiel *et al.* (2012) and Fonseca-García *et al.* (2021) showed the essential role of the NADPH oxidase gene *PvRbohB* (*Respiratory burst oxidase homolog B*) in carbon metabolism and root infection by *Rhizobium* in French bean. The roots of *PvRbohB*-RNAi lines had lower ROS production and lateral root density, and impaired nodulation. The role of this gene in colonization of the arbuscular mycorrhizal fungus *Rhizophagus irregularis* in French bean was subsequently demonstrated by Arthikala *et al.* (2013) through RNAi-mediated downregulation. However, RNAi studies by Fonseca-García *et al.* (2019) showed that *PvRbohB* enhances *Rhizobium tropici* infection and inhibits *Rhizophagus irregularis* colonization in *P. vulgaris*. In addition to root colonization by rhizobia and arbuscular mycorrhizal fungus, RNAi studies have shown that *PvRbohB* is involved in lateral root development and ROS activities in French bean (Montiel *et al.*, 2013).

The annexin gene *PvAnn1* encoding a calcium- and membrane-binding protein, is also essential in this crop for rhizobial infection and nodulation (Carrasco-Castilla *et al.*, 2018). Inoculation of *PvAnn1*-RNAi roots with rhizobia showed a reduced number of infection threads compared with the wild type. Similarly, using RNAi, a cysteine-rich receptor-like kinase gene, *CRK12*, was shown to be an important gene for the successful colonization of *R. tropici* in French bean (Lecona *et al.*, 2023). Downregulation of *CRK12* through RNAi in the *P. vulgaris* hairy root system resulted in a reduction in nodule numbers. RNAi studies have shown that an isoflavone reductase gene, *PvIFR1*, also has a significant role in root nodulation in common bean. Isoflavone reductase is directly involved in the biosynthesis of isoflavonoids, which act as signals from the roots to the rhizobia and also inhibit auxin transportation, leading to accelerated cortical cell division during nodulation. *PvIFR1* is expressed in the roots, under nitrogen deficiency and in response to rhizobial inoculation. *PvIFR1*-RNAi lines of common bean had reduced shoot and root length and reduced nodulation with different strains of rhizobia (Rípodas *et al.*, 2013). Using RNAi, other genes encoding metallothionein 1A (*PvMT1A*; Fonseca-García *et al.*, 2022), calcineurin B-like protein-interacting protein kinase (*PvCIPK7*; Arthikala *et al.*, 2023), class II trehalose-6-phosphate synthase (*PvTPS9*; Barraza *et al.*, 2016), receptor for activated C kinase (*PvRACK1*; Islas-Flores *et al.*, 2011), BYPASS1 (*BPS1*; Arthikala *et al.*, 2018), and phosphatidylinositol 3-kinase (*PvPI3K*; Estrada-Navarrete *et al.*, 2016) were also shown to have significant roles in rhizobial nodulation in common bean.

RNAi-mediated knockdown of the target of rapamycin gene, *PvTOR*, has shown that it regulates arbuscule development during mycorrhizal symbiosis in common bean (Arthikala *et al.*, 2021). *PvTOR*-RNAi roots had poor root length and colonization and stunted arbuscules. Expression of arbuscular mycorrhizal symbiosis-induced genes was also repressed.

### 19.4.4 Cabbage

*Resistance to biotic stresses*

Downregulation of the *Sar1*, *Sec23*, *Sec24*, *Sec13*, and *Sec31* genes encoding coat protein II (COPII), a primary coated vesicle that exports molecules from the endoplasmic reticulum, using RNAi by injecting dsRNA for each gene inhibited feeding and reproductive development by cabbage beetle (Tian *et al.*, 2022). Silencing any one of these genes repressed the juvenile hormone and ecdysone signaling. It was also shown that RNAi with any one of the 18 enzymes involved in the mevalonate pathway in cabbage beetle reduced expression of the juvenile hormone-responsive gene *Krüppel homolog 1* (*Kr-h1*). Subsequent biochemical changes induced diapause-associated traits in female beetles (Tian *et al.*, 2021). In a similar study, Huang *et al.* (2022) showed that downregulation of the genes for MAPK-ERK kinase, extracellular regulated protein kinase, kinase suppressor of RAS 2, and P38 through RNAi impaired reproductive development and longevity in cabbage beetle.

Cedden *et al.* (2023) achieved up to 76% mortality of cabbage stem flea beetles by oral delivery of dsRNAs targeting the *Sec23* and *VatpG* (vacuolar adenosine triphosphatase subunit G) genes. Similarly, Howell *et al.* (2020) showed that RNAi with genes for inhibitor of apoptosis, ATPase N2B, serine/threonine protein phosphatase PP1-β catalytic subunit, signal recognition particle 54 kDa protein, and G-protein-coupled receptor 161-like protein in adults and nymphs through injection or feeding with the respective dsRNAs led to up to 75% mortality in harlequin bug, a major pest of all cabbage family plants. A parallel study showed that feeding diamondback moth larvae with disks of cabbage leaf coated with dsRNA targeting the tyrosine hydroxylase gene caused larval mortality (Ellango *et al.*, 2018). The tyrosine hydroxylase gene regulates the biosynthesis of 3,4-dihydroxyphenylalanine, a precursor of many proteins involved in vital physiological functions in insects.

*Resistance to abiotic stresses*

Translationally controlled tumor protein (TCTP) acts in the target of rapamycin (TOR) signaling pathway, regulates cell growth, and its expression is enhanced under high temperature and salt stresses. The role of this gene in abiotic stress tolerance was proven when *BoTCTP*-RNAi cabbage plants showed reduced growth and tolerance to cold and high temperatures and salt stress (Cao *et al.*, 2010). Using RNAi studies, Zhang *et al.* (2011) showed that the proline dehydrogenase (*ProDH*) gene, which suppresses the degradation of proline, has a significant role in drought- and salt-stress tolerance in cabbage.

### 19.4.5 Chinese cabbage

*Resistance to abiotic stresses*

RNAi with the *BrDST71* gene is reported to enhance drought tolerance in Chinese cabbage (*Brassica rapa*) (Kim *et al.*, 2020). Another *B. rapa* gene, *Drought Stress Resistance* (*BrDSR*), is also reported to impart drought resistance in Chinese cabbage (Yu *et al.*, 2016).

*Resistance to biotic stress*

Using RNAi, Shin *et al.* (2017) developed Chinese cabbage lines with resistance to mites.

*Other traits*

Downregulation of the type 2A serine/threonine protein phosphatase gene was found to restore self-compatibility in Chinese cabbage (Shin *et al.*, 2019).

### 19.4.6 Carrot

*Tuber quality improvement*

The ζ-carotene desaturase genes *ZDS1* and *ZDS2* contribute to β-carotene and α-carotene synthesis in carrot. *ZDS1* contributes to leaf development, while *ZDS2* is expressed in the leaves and storage roots of a mature plant. *ZDS1*-RNAi lines had reduced carotene content and lower expression of *ZDS2* and *PSY2* genes, whereas *ZDS2*-RNAi lines maintained the normal activity of the other two genes as well as the carotene content, indicating that carrot plant development and carotene accumulation are determined by *ZDS1*, and *ZDS2* has only a complementary role (Flores-Ortiz *et al.*, 2020).

Under microbial invasion or abiotic stress, the salicylic acid signaling pathway induces two highly similar genes, *Dau c 1.01* and *Dau c 1.02*, coding for pathogenesis-related protein 10 (PR10). PR10 is known to increase the allergenic potential of carrot. Silencing these genes through RNAi has produced lines with reduced allergenicity, as determined through skin prick tests (Peters *et al.*, 2011).

*Resistance to biotic stresses*

RNAi with a gene encoding the nuclear inclusion protein (NIa) of the potyvirus carrot virus Y was proposed to develop resistant lines (McCormick *et al.*, 2004). Similarly, Li *et al.* (2010) showed that carrot plants transformed with an RNAi construct targeting the cathepsin B gene (*Rscb1*) of the nematode *Radopholus similis* led to reduced fecundity of the nematodes. In a similar study, hairy root lines of carrot transformed with RNAi constructs targeting the actin (*act4*) or proteasomal alpha subunit 4 (*pas4*) genes and expressing the respective dsRNAs were found to downregulate these genes in *R. similis* feeding on them, and suppressed nematode multiplication (Roderick *et al.*, 2018). Silencing the *pat10* and *unc87* genes of the root lesion nematodes *Pratylenchus thornei* and *P. zeae* through dsRNA treatment drastically reduced nematode reproduction on carrot mini disks (Tan *et al.*, 2013).

*Resistance to abiotic stresses*

The phenylalanine ammonia-lyase (*PAL*) gene in carrot (*DcPAL1*) is activated under ultraviolet-B irradiation, fungal elicitation or other stresses. Box-L-like sequences in its promoter are the binding site for the MYB transcription factor DcMYB1. *DcMYB1* is expressed 0.5 h after stress induction and *DcPAL1* is expressed 1 h after *DcMYB1* expression. *DcMYB1*-RNAi showed no upregulation of *DcPAL1* under stress, demonstrating the roles of these genes in stress-tolerance mechanisms (Maeda *et al.*, 2005).

### 19.4.7 Beetroot

*Resistance to biotic stresses*

Plants transformed with an RNAi construct targeting the coat protein gene of beet necrotic yellow vein virus were found to be resistant (Andika *et al.*, 2005).

Injection with siRNAs of the genes for chitinase 1 and 7, plasma glutamate carboxypeptidase, ATPase, tubulin 1 and 2, and ADP-ribosylation factor 1 and 2 (*arf1* and *arf2*) of the beet army worm *Spodoptera exigua*, identified through transcriptome sequencing at various stages of insect development, into fourth-instar larvae knocked down the respective genes and caused insect mortality (Li *et al.*, 2013). Similarly, Park and Kim (2013) showed that RNAi with glycerol-3-phosphate dehydrogenase and glycerol kinase genes associated with glycerol biosynthesis in *S. exigua* led to the loss of rapid cold hardening in this insect, making it cold susceptible. Similarly, silencing the chitinase genes *SeChi* and *SeChi-h* (Zhang *et al.*, 2012) and the β1-integrin subunit gene *βSe1* (Surakasi *et al.*, 2011) of *S. exigua* through dsRNA injection resulted developmental deformities at different stages.

### 19.4.8 Okra

*Yield improvement*

Hyperoside is an important flavonoid that promotes flowering and seed set in okra. The gene *AeUF3GaT1*, coding for the enzyme that catalyzes the last step of hyperoside biosynthesis, was found to be strongly induced during external hyperoside application on okra plants. The transcription factor protein AeMYB30 regulates the transcription of *AeUFG3aT1*. Thus, external application of hyperoside activates transcription of *AeMYB30* followed by *AeUFG3aT1*, leading to hyperoside biosynthesis and better seed set. *AeUF3GaT1*-RNAi lines had an aberrant phenotype with impaired seed set, providing proof for the function of this gene (Yang *et al.*, 2021).

*Resistance to biotic stresses*

Our laboratory at Kerala Agricultural University, India, has developed yellow vein mosaic virus-resistant okra plants by transforming a highly susceptible variety, Salkeerthi, with an ihpRNA construct designed using the pRNAi-LIC plasmid and targeting the *βC1* open reading frame of the β-satellite of this begomovirus (Ganesh *et al.*, 2022). For transformation with the ihpRNA-βC1 cassette, an *in planta* strategy was followed, with a success rate of 11.42%. In a similar work, okra plants transformed with anRNAi construct targeting the *AC1*, *AC2*, and *AC4* genes encoding the replication-associated protein, transcriptional activator protein, and suppressor of post-transcriptional gene silencing, respectively, of the virus, through *Agrobacterium*-mediated tissue-culture-dependent regeneration, were found to be virus resistant (Sharma *et al.*, 2023).

## 19.5 Conclusion

RNAi is a powerful tool for downregulating or silencing genes in a target-specific manner. This strategy has been proven efficient to: (i) silence susceptibility pathways in plants to make them resistant; (ii) silence pathways that synthesize anti-nutrients resulting in biofortification; (iii) develop plants harboring the hairpin constructs targeting important genes of viruses, fungi, bacteria, and insects, making the plants resistant to these pests; and (iv) downregulate susceptibility factors for environmental stresses in plants. Above all, this is an important functional genomics approach for proving the important physiological roles played by each gene. As vegetable crops are very important in human food and in agriculture as a whole, this technique has attracted much research effort. This chapter has discussed the important RNAi research studies reported in vegetable crops.

## 19.6 References


AbuQamar, S., Chai, M.F., Luo, H., Song, F., and Mengiste, T. (2008) Tomato protein kinase 1b mediates signaling of plant responses to necrotrophic fungi and insect herbivory. *Plant Cell* 20, 1964–1983.

Ammara, U.E., Mansoor, S., Saeed, M., Amin, I., Briddon, R.W., and Al-Sadi, A.M. (2015) RNA interference-based resistance in transgenic tomato plants against *Tomato yellow leaf curl virus*-Oman (TYLCV-OM) and its associated betasatellite. *Journal of Virology* 12: 38.

Andika, I.B., Kondo, H., and Tamada, T. (2005) Evidence that RNA silencing-mediated resistance to *Beet necrotic yellow vein virus* is less effective in roots than in leaves. *Molecular Plant–Microbe Interactions* 18, 194–204.

Arthikala, M.K., Montiel, J., Nava, N., Santana, O., Sánchez-López, R., *et al.* (2013) *PvRbohB* negatively regulates *Rhizophagus irregularis* colonization in *Phaseolus vulgaris*. *Plant and Cell Physiology* 54, 1391–1402.

Arthikala, M.K., Nanjareddy, K., and Lara, M. (2018) In *BPS1* downregulated roots, the BYPASS1 signal disrupts the induction of cortical cell divisions in bean-*Rhizobium* symbiosis. *Genes* 9: 11.

Arthikala, M.K., Nanjareddy, K., Blanco, L., Alvarado-Affantranger, X., and Lara, M. (2021) Target of rapamycin, *PvTOR*, is a key regulator of arbuscule development during mycorrhizal symbiosis in *Phaseolus*. *Scientific Reports* 11: 11319.

Arthikala, M.K., Blanco, L., Alvarado-Affantranger, X., Márquez-Guzmán, J., Lara, M., and Nanjareddy, K. (2023) Identification of *CBL* and *CIPK* gene families and functional characterization of *PvCIPK7* as an essential regulator of root nodule development and nitrogen fixation in *Phaseolus vulgaris*. *Journal of Plant Biology* 66, 535–549.

Babu, M.A., Srinivasan, R., Subramanian, P., and Muthukaliannan, K.G. (2021) RNAi silenced ζ-carotene desaturase developed variegated tomato transformants with increased phytoene content. *Plant Growth Regulation* 93, 189–201.

Barraza, A., Contreras-Cubas, C., Estrada-Navarrete, G., Reyes, J.L., Juárez-Verdayes, M.A., *et al.* (2016) The class II trehalose 6-phosphate synthase gene *PvTPS9* modulates trehalose metabolism in *Phaseolus vulgaris* nodules. *Frontiers in Plant Science* 7: 1589.

Baulcombe, D.C. (1996) Mechanisms of pathogen-derived resistance to viruses in transgenic plants. *Plant Cell* 8, 1833–1844.

Bento, F.M., Marques, R.N., Campana, F.B., Demétrio, C.G., Leandro, R.A., *et al.* (2020) Gene silencing by RNAi via oral delivery of dsRNA by bacteria in the South American tomato pinworm, *Tuta absoluta*. *Pest Management Science* 76, 287–295.

Bhorali, P. (2021) Optimization of *in vitro* transformation protocol and RNAi based gene silencing for viral cucumber mosaic virus resistance in Bhut jolokia *Capsicum chinense* Jacq. PhD thesis, Assam Agricultural University, Jorhat, India.

Bonfim, K., Faria, J.C., Nogueira, E.O., Mendes, É.A., and Aragão, F.J. (2007) RNAi-mediated resistance to *Bean golden mosaic virus* in genetically engineered common bean (*Phaseolus vulgaris*). *Molecular Plant–Microbe Interactions* 20, 717–726.

Borah, M., Berbati, M., Reppa, C., Holeva, M., Nath, P.D., and Voloudakis, A. (2018) RNA-based vaccination of Bhut Jolokia pepper (*Capsicum chinense* Jacq.) against cucumber mosaic virus. *Virus Disease* 29, 207–211.

Cai, X., Zhang, C., Shu, W., Ye, Z., Li, H., and Zhang, Y. (2016) The transcription factor SlDof22 involved in ascorbate accumulation and salinity stress in tomato. *Biochemical and Biophysical Research Communications* 474, 736–741.

Cai, Y., Bartholomew, E.S., Dong, M.M., Zhai, X.L., Yin, S., *et al.* (2020) The HD-ZIP IV transcription factor *GL2-LIKE* regulates male flowering time and fertility in cucumber. *Journal of Experimental Botany* 71, 5425–5437.

Camargo, R.D.A., Herai, R.H., Santos, L.N., Bento, F.M., Lima, J.E., *et al.* (2015) *De novo* transcriptome assembly and analysis to identify potential gene targets for RNAi-mediated control of the tomato leafminer (*Tuta absoluta*). *BMC Genomics* 16: 635.

Camargo, R.D.A., Barbosa, G.O., Possignolo, I.P., Peres, L.E., Lam, E., *et al.* (2016) RNA interference as a gene silencing tool to control *Tuta absoluta* in tomato (*Solanum lycopersicum*). *PeerJ* 4: e2673.

Cao, B., Lu, Y., Chen, G., and Lei, J. (2010) Functional characterization of the translationally controlled tumor protein (*TCTP*) gene associated with growth and defense response in cabbage. *Plant Cell, Tissue and Organ Culture* 103, 217–226.

Carrasco-Castilla, J., Ortega-Ortega, Y., Jáuregui-Zúñiga, D., Juárez-Verdayes, M.A., Arthikala, M.K., *et al.* (2018) Down-regulation of a *Phaseolus vulgaris* annexin impairs rhizobial infection and nodulation. *Environmental and Experimental Botany* 153, 108–119.

Cedden, D., Güney, G., Scholten, S., and Rostás, M. (2023) Lethal and sublethal effects of orally delivered double-stranded RNA on the cabbage stem flea beetle, *Psylliodes chrysocephala*. *Pest Management Science* 80, 2282–2293.

Chaudhary, S., Dutta, T.K., Tyagi, N., Shivakumara, T.N., Papolu, P.K., *et al.* (2019) Host-induced silencing of *Mi-msp-1* confers resistance to root-knot nematode *Meloidogyne incognita* in eggplant. *Transgenic Research* 28, 327–340.

Che, G., Gu, R., Zhao, J.Y., Liu, X.F., Song, X.F., *et al.* (2020) Gene regulatory network controlling carpel number variation in cucumber. *Development* 147: dev184788.

Chen, C.H., Yin, S., Li, X.W., Liu, B., Yang, S., *et al.* (2016) The WD-repeat protein CsTTG1 regulates fruit wart formation through interaction with the homeodomain-leucine zipper I protein Mict. *Plant Physiology* 171, 1156–1168.

Chen, H.M., Lin, C.Y., Tsai, W.S., Kenyon, L., Chan, M.T., *et al.* (2016) Resistance to viral yellow leaf curl in tomato through RNAi targeting two *Begomovirus* species strains. *Journal of Plant Biochemistry and Biotechnology* 25, 199–207.

Chen, N., Wu, S., Fu, J., Cao, B., Lei, J., *et al.* (2016) Overexpression of the eggplant (*Solanum melongena*) NAC family transcription factor *SmNAC* suppresses resistance to bacterial wilt. *Scientific Reports* 6: 31568.

Chen, Q., Rehman, S., Smant, G., and Jones, J.T. (2005) Functional analysis of pathogenicity proteins of the potato cyst nematode *Globodera rostochiensis* using RNAi. *Molecular Plant–Microbe Interactions* 18, 621–625.

Cheng, W., Lin, M., Chu, M., Xiang, G., Guo, J., *et al.* (2022) RNAi-based gene silencing of RXLR effectors protects plants against the oomycete pathogen *Phytophthora capsici*. *Molecular Plant–Microbe Interactions* 35, 440–449.

Cui, L., Zheng, F., Zhang, D., Li, C., Li, M., *et al.* (2022) Tomato methionine sulfoxide reductase B2 functions in drought tolerance by promoting ROS scavenging and chlorophyll accumulation through interaction with catalase 2 and RBCS3B. *Plant Science* 318: 111206.

Dai, H., Zhang, W., Hua, B., Zhu, Z., Zhang, J., *et al.* (2023) Cucumber *STACHYOSE SYNTHASE* is regulated by its *cis*-antisense RNA *asCsSTS* to balance source–sink carbon partitioning. *Plant Cell* 35, 435–452.

Davuluri, G.R., van Tuinen, A., Fraser, P.D., Manfredonia, A., Newman, R., *et al.* (2005) Fruit-specific RNAi-mediated suppression of *DET1* enhances carotenoid and flavonoid content in tomatoes. *Nature Biotechnology* 23, 890–895.

Ding, H., He, J., Wu, Y., Wu, X., Ge, C., *et al.* (2018) The tomato mitogen-activated protein kinase SlMPK1 is as a negative regulator of the high-temperature stress response. *Plant Physiology* 177, 633–651.

Ding, L., Yan, S.S., Jiang, L., and Liu, M. (2015) *HANABA TARANU* regulates the shoot apical meristem and leaf development in cucumber (*Cucumis sativus* L.). *Journal of Experimental Botany* 66, 7075–7087.

Dong, T., Hu, Z., Deng, L., Wang, Y., Zhu, M., *et al.* (2013) A tomato MADS-box transcription factor, *Sl-MADS1*, acts as a negative regulator of fruit ripening. *Plant Physiology* 163, 1026–1036.

Du, L., Bao, C., Hu, T., Zhu, Q., Hu, H., He, Q., and Mao, W. (2016) SmARF8, a transcription factor involved in parthenocarpy in eggplant. *Molecular Genetics and Genomics* 291, 93–105.

Eady, C.C., Trueman, L., McCallum, J., Shaw, M., Pither-Joyce, M., *et al.* (2004) Transgenic onions with reduced alliinase activity: biochemical and molecular assessment. *Acta Horticulturae* 688, 181–190.

Eady, C.C., Kamoi, T., Kato, M., Porter, N.G., Davis, S., *et al.* (2008) Silencing onion lachrymatory factor synthase causes a significant change in the sulfur secondary metabolite profile. *Plant Physiology* 147, 2096–2106.

El Airaj, H., Gest, N., Truffault, V., Garchery, C., Riqueau, G., *et al.* (2013) Decreased monodehydroascorbate reductase activity reduces tolerance to cold storage in tomato and affects fruit antioxidant levels. *Postharvest Biology and Technology* 86, 502–510.

Ellango, R., Asokan, R., Chandra, G.S., Kumar, N.K., Mahmood, R., and Ramamurthy, V.V. (2018) Tyrosine hydroxylase, a potential target for the RNAi-mediated management of diamondback moth (Lepidoptera: Plutellidae). *Florida Entomologist* 101, 1–5.

Estrada-Navarrete, G., Cruz-Mireles, N., Lascano, R., Alvarado-Affantranger, X., Hernández-Barrera, A., *et al.* (2016) An autophagy-related kinase is essential for the symbiotic relationship between *Phaseolus vulgaris* and both rhizobia and arbuscular mycorrhizal fungi. *Plant Cell* 28, 2326–2341.

Ewas, M., Gao, Y., Wang, S., Liu, X., Zhang, H., *et al.* (2016) Manipulation of *SlMXl* for enhanced carotenoids accumulation and drought resistance in tomato. *Science Bulletin* 61, 1413–1418.

Faisal, M., Abdel-Salam, E.M., Alatar, A.A., Saquib, Q., Alwathnani, H.A., and Canto, T. (2019) Genetic transformation and siRNA-mediated gene silencing for aphid resistance in tomato. *Agronomy* 9: 893.

Ferreira, A.L., Faria, J.C.D., da Costa Moura, M., de Mendonça Zaidem, A.L., Pizetta, C.S.R., *et al.* (2022) Whitefly-tolerant transgenic common bean (*Phaseolus vulgaris*) line. *Frontiers in Plant Science* 13: 984804.

Fire, A., Xu, S., Montgomery, M., Kostas, S., Driver, S., and Mello, C. (1998) Potent and specific genetic interference by double-stranded RNA in *Caenorhabditis elegans*. *Nature* 391, 806–811.

Fletcher, S.J., Peters, J.R., Olaya, C., Persley, D.M., Dietzgen, R.G., *et al.* (2022) Tospoviruses induce small interfering RNAs targeting viral sequences and endogenous transcripts in solanaceous plants. *Pathogens* 11: 745.

Flores-Ortiz, C., Alvarez, L.M., Undurraga, A., Arias, D., Durán, F., *et al.* (2020) Differential role of the two ζ-carotene desaturase paralogs in carrot (*Daucus carota*): *ZDS1* is a functional gene essential for plant development and carotenoid synthesis. *Plant Science* 291: 110327.

Fonseca-García, C., Zayas, A.E., Montiel, J., Nava, N., Sánchez, F., and Quinto, C. (2019) Transcriptome analysis of the differential effect of the NADPH oxidase gene *RbohB* in *Phaseolus vulgaris* roots following *Rhizobium tropici* and *Rhizophagus irregularis* inoculation. *BMC Genomics* 20: 800.

Fonseca-García, C., Nava, N., Lara, M., and Quinto, C. (2021) An NADPH oxidase regulates carbon metabolism and the cell cycle during root nodule symbiosis in common bean (*Phaseolus vulgaris*). *BMC Plant Biology* 21: 274.

Fonseca-García, C., López-García, C.M., Pacheco, R., Armada, E., Nava, N., *et al.* (2022) Metallothionein1A regulates rhizobial infection and nodulation in *Phaseolus vulgaris*. *International Journal of Molecular Sciences* 23: 1491.

Fuentes, A., Carlos, N., Ruiz, Y., Callard, D., Sánchez, Y., *et al.* (2016) Field trial and molecular characterization of RNAi-transgenic tomato plants that exhibit resistance to tomato yellow leaf curl geminivirus. *Molecular Plant–Microbe Interactions* 29, 197–209.

Ganesh, K.V., Mathew, D., Augustine, R., Soni, K.B., Alex, S., Shylaja, M.R., and Cherian, K.A. (2022) Development of transgenic okra (*Abelmoschus esculentus* L. Moench) lines having RNA mediated resistance to *Yellow vein mosaic virus* (Geminiviridae). *Journal of Virological Methods* 301: 114457.

Garchery, C., Gest, N., Do, P.T., Alhagdow, M., Baldet, P., *et al.* (2013) A diminution in ascorbate oxidase activity affects carbon allocation and improves yield in tomato under water deficit. *Plant, Cell & Environment* 36, 159–175.

Ghosh, S., Meli, V.S., Kumar, A., Thakur, A., Chakraborty, N., *et al.* (2011) The *N*-glycan processing enzymes α-mannosidase and β-D-*N*-acetylhexosaminidase are involved in ripening-associated softening in the non-climacteric fruits of capsicum. *Journal of Experimental Botany* 62, 571–582.

Gogoi, A., Sarmah, N., Kaldis, A., Perdikis, D., and Voloudakis, A. (2017) Plant insects and mites uptake double-stranded RNA upon its exogenous application on tomato leaves. *Planta* 246, 1233–1241.

Guo, J.E., Hu, Z., Zhu, M., Li, F., Zhu, Z., *et al.* (2017) The tomato histone deacetylase *SlHDA1* contributes to the repression of fruit ripening and carotenoid accumulation. *Scientific Reports* 7: 7930.

Guo, J.E., Wang, H., Yang, Y., Li, J., and Zhu, Z. (2023) Histone deacetylase gene *SlHDA3* is involved in drought and salt response in tomato. *Plant Growth Regulation* 99, 359–372.

Gupta, A., Pal, R.K., and Rajam, M.V. (2013) Delayed ripening and improved fruit processing quality in tomato by RNAi-mediated silencing of three homologs of 1-aminopropane-1-carboxylate synthase gene. *Journal of Plant Physiology* 170, 987–995.

Habib, S., Lwin, Y.Y., and Li, N. (2021) Down-regulation of *SlGRAS10* in tomato confers abiotic stress tolerance. *Genes* 12: 623.

Hameed, A., Tahir, M.N., Asad, S., Bilal, R., van Eck, J., *et al.* (2017) RNAi-mediated simultaneous resistance against three RNA viruses in potato. *Molecular Biotechnology* 59, 73–83.

Han, Q.Q., Song, Y.Z., Zhang, J.Y., and Liu, L.F. (2014) Studies on the role of the *SlNAC3* gene in regulating seed development in tomato (*Solanum lycopersicum*). *Journal of Horticultural Science and Biotechnology* 89, 423–429.

Hanika, K., Schipper, D., Chinnappa, S., Oortwijn, M., Schouten, H.J., *et al.* (2021) Impairment of tomato *WAT1* enhances resistance to vascular wilt fungi despite severe growth defects. *Frontiers in Plant Science* 12: 721674.

Hofius, D., Hajirezaei, M.R., Geiger, M., Tschiersch, H., Melzer, M., and Sonnewald, U. (2004) RNAi-mediated tocopherol deficiency impairs photoassimilate export in transgenic potato plants. *Plant Physiology* 135, 1256–1268.

Howell, J.L., Mogilicherla, K., Gurusamy, D., and Palli, S.R. (2020) Development of RNAi methods to control the harlequin bug, *Murgantia histrionica*. *Archives of Insect Biochemistry and Physiology* 104: e21690.

Hua, B., Zhang, M., Zhang, J., Dai, H., Zhang, Z., and Miao, M. (2021) *CsAGA1* and *CsAGA2* mediate RFO hydrolysis in partially distinct manner in cucumber fruits. *International Journal of Molecular Science* 22: 13285.

Huang, Z., Tian, Z., Zhao, Y., Zhu, F., Liu, W., and Wang, X. (2022) MAPK signaling pathway is essential for female reproductive regulation in the cabbage beetle, *Colaphellus bowringi*. *Cells* 11: 1602.

Huh, S.U., Kim, K.J., and Paek, K.H. (2012) *Capsicum annuum* basic transcription factor 3 (CaBtf3) regulates transcription of pathogenesis-related genes during hypersensitive response upon *Tobacco mosaic virus* infection. *Biochemical and Biophysical Research Communications* 417, 910–917.

Islas-Flores, T., Guillén, G., Alvarado-Affantranger, X., Lara-Flores, M., Sánchez, F., and Villanueva, M.A. (2011) *PvRACK1* loss-of-function impairs cell expansion and morphogenesis in *Phaseolus vulgaris* L. root nodules. *Molecular Plant–Microbe Interactions* 24, 819–826.

Jaiswal, N., Liao, C.J., Mengesha, B., Han, H., Lee, S., *et al.* (2022) Regulation of plant immunity and growth by tomato receptor-like cytoplasmic kinase TRK1. *New Phytologist* 233, 458–478.

Ji, K., Kai, W., Zhao, B., Sun, Y., Yuan, B., *et al.* (2014) *SlNCED1* and *SlCYP707A2*: key genes involved in ABA metabolism during tomato fruit ripening. *Journal of Experimental Botany* 65, 5243–5255.

Jian, W., Zheng, Y., Yu, T., Cao, H., Chen, Y., *et al.* (2021) SlNAC6, a NAC transcription factor, is involved in drought stress response and reproductive process in tomato. *Journal of Plant Physiology* 264: 153483.

Jin, F.M., Song, J., Xue, J., Sun, H.B., Zhang, Y., *et al.* (2020) Successful generation of anti-ToCV and TYLCV transgenic tomato plants by RNAi. *Plant Biology* 64, 490–496.

Jorgensen, R. (1990) Altered gene expression in plants due to trans interactions between homologous genes. *Trends in Biotechnology* 8, 340–344.

Kamoi, T. (2008) Developing an optimal method for producing a tearless onion. PhD thesis, Lincoln University, Oakland, California.

Kaplanoglu, E., Kolotilin, I., Menassa, R., and Donly, C. (2022) Plastid transformation of Micro-tom tomato with a Hemipteran double-stranded RNA results in RNA interference in multiple insect species. *International Journal of Molecular Sciences* 23: 3918.

Karlova, R., Rosin, F.M., Busscher-Lange, J., Parapunova, V., Do, P.T., *et al.* (2011) Transcriptome and metabolite profiling show that *APETALA2a* is a major regulator of tomato fruit ripening. *Plant Cell* 23, 923–941.

Kaulfürst-Soboll, H., Mertens-Beer, M., Brehler, R., Albert, M., and von Schaewen, A. (2021) Complex *N*-glycans are important for normal fruit ripening and seed development in tomato. *Frontiers in Plant Science* 12: 635962.

Kim, H., Choi, Y., Park, J.S., Shin, Y.H., Park, K.M., and Park, Y.D. (2020) Evaluation of gene expression level and phenotype for drought-stress tolerant Chinese cabbage with RNA interference *BrDST71* gene. *Horticultural Science and Technology* 38, 269–270.

Kim, Y., Kim, Y.J., and Paek, K.H. (2021) Temperature-specific vsiRNA confers RNAi-mediated viral resistance at elevated temperature in *Capsicum annuum*. *Journal of Experimental Botany* 72, 1432–1448.

Kim, Y.S., Lee, Y.H., Kim, H.S., Kim, M.S., Hahn, K.W., *et al.* (2008) Development of patatin knockdown potato tubers using RNA interference (RNAi) technology, for the production of human-therapeutic glycoproteins. *BMC Biotechnology* 8: 36.

Koo, J., Gurusamy, D., and Palli, S.R. (2023) Inefficient uptake of small interfering RNAs is responsible for their inability to trigger RNA interference in Colorado potato beetle cells. *Archives of Insect Biochemistry and Physiology* 114: e22036.

Krieger, E.K., Allen, E., Gilbertson, L.A., Roberts, J.K., Hiatt, W., and Sanders, R.A. (2008) The Flavr Savr tomato, an early example of RNAi technology. *HortScience* 43, 962–964.

Kumar, S., Dubey, A.K., Karmakar, R., Kini, K.R., Mathew, M.K., and Prakash, H.S. (2012) Inhibition of TMV multiplication by siRNA constructs against *TOM1* and *TOM3* genes of *Capsicum annuum*. *Journal of Virological Methods* 186, 78–85.

Kushwaha, N., Singh, A.K., Basu, S., and Chakraborty, S. (2015) Differential response of diverse solanaceous hosts to *Tomato leaf curl New Delhi virus* infection indicates coordinated action of NBS-LRR and RNAi-mediated host defense. *Archives of Virology* 160, 1499–1509.

Le, L.Q., Mahler, V., Lorenz, Y., Scheurer, S., Biemelt, S., *et al.* (2006a) Reduced allergenicity of tomato fruits harvested from Lyc e 1-silenced transgenic tomato plants. *Journal of Allergy and Clinical Immunology* 118, 1176–1183.

Le, L.Q., Lorenz, Y., Scheurer, S., Fötisch, K., Enrique, E., *et al.* (2006b) Design of tomato fruits with reduced allergenicity by dsRNAi-mediated inhibition of ns-LTP (*Lyc e 3*) expression. *Plant Biotechnology Journal* 4, 231–242.

Lecona, A.M., Nanjareddy, K., Blanco, L., Piazza, V., Vera-Núñez, J.A., *et al.* (2023) CRK12: a key player in regulating the *Phaseolus vulgaris–Rhizobium tropici* symbiotic interaction. *International Journal of Molecular Sciences* 24: 11720.

Leibman, D., Ortega-Parra, N., Wolf, D., Shterkman, M., Hanssen, I., and Gal-On, A. (2021) A transgenic RNAi approach for developing tomato plants immune to *Pepino mosaic virus*. *Plant Pathology* 70, 1003–1012.

Li, H., Jiang, W., Zhang, Z., Xing, Y., and Li, F. (2013) Transcriptome analysis and screening for potential target genes for RNAi-mediated pest control of the beet armyworm, *Spodoptera exigua*. *PLOS ONE* 8: e65931.

Li, H., Qi, M., Sun, M., Liu, Y., Liu, Y., *et al.* (2017) Tomato transcription factor *SlWUS* plays an important role in tomato flower and locule development. *Frontiers in Plant Science* 8: 457.

Li, H., Chen, Y., Lu, C., Tian, H., Lin, S., *et al.* (2023) Chemosensory protein regulates the behavioural response of *Frankliniella intonsa* and *Frankliniella occidentalis* to tomato zonate spot virus-infected pepper (*Capsicum annuum*). *PLOS Pathogens* 19: e1011380.

Li, Y., Xie, H., Xu, C., Li, D., and Zhang, C. (2010) RNAi effect of cathepsin B gene on reproduction of *Radopholus similis*. *Scientia Agricultura Sinica* 43, 1608–1616.

Li, Y., Liu, H., Yao, X., Wang, J., Feng, S., *et al.* (2021) Hexose transporter CsSWEET7a in cucumber mediates phloem unloading in companion cells for fruit development. *Plant Physiology* 186, 640–654.

Li, Y., Xing, M., Yang, Q., Wang, Y., Jiang, J., *et al.* (2023) SmCIP7, a COP1 interactive protein, positively regulates anthocyanin accumulation and fruit size in eggplant. *International Journal of Biological Macromolecules* 234: 123729.

Liu, B., Liu, X., Yang, S., Chen, C., Xue, S., *et al.* (2016) Silencing of the gibberellin receptor homolog, *CsGID1a*, affects locule formation in cucumber (*Cucumis sativus*) fruit. *New Phytologist* 210, 551–563.

Liu, X.F., Ning, K., Che, G., Yan, S.S., Han, L.G., *et al.* (2018) *CsSPL* functions as an adaptor between HD-ZIP III and CsWUS transcription factors regulating anther and ovule development in *Cucumis sativus* (cucumber). *Plant Journal* 94, 535–547.

Liu, Y., Du, Q., Bai, L., Sun, M., Li, Y., *et al.* (2021) Interference of *CsGPA1*, the α-submit of G protein, reduces drought tolerance in cucumber seedlings. *Horticultural Plant Journal* 7, 209–220.

López-Fraga, M., Martinez, T., and Jiménez, A. (2009) RNA interference technologies and therapeutics: from basic research to products. *BioDrugs* 23, 305–332.

Lü, J., Sui, X., Ma, S., Li, X., Liu, H., and Zhang, Z. (2017) Suppression of cucumber stachyose synthase gene (*CsSTS*) inhibits phloem loading and reduces low temperature stress tolerance. *Plant Molecular Biology* 95, 1–15.

Maeda, K., Kimura, S., Demura, T., Takeda, J., and Ozeki, Y. (2005) DcMYB1 acts as a transcriptional activator of the carrot phenylalanine ammonia-lyase gene (*DcPAL1*) in response to elicitor treatment, UV-B irradiation and the dilution effect. *Plant Molecular Biology* 59, 739–752.

Majidiani, S., PourAbad, R.F., Laudani, F., Campolo, O., Zappalà, L., *et al.* (2019) RNAi in *Tuta absoluta* management: effects of injection and root delivery of dsRNAs. *Journal of Pest Science* 92, 1409–1419.

Mamta, R.K.R.K. and Rajam, M.V. (2016) Targeting chitinase gene of *Helicoverpa armigera* by host-induced RNA interference confers insect resistance in tobacco and tomato. *Plant Molecular Biology* 90, 281–292.

Mazier, M., Flamain, F., Nicolaï, M., Sarnette, V., and Caranta, C. (2011) Knock-down of both *eIF4E1* and *eIF4E2* genes confers broad-spectrum resistance against potyviruses in tomato. *PLOS ONE* 6: e29595.

McCormick, N.L., Ford, R., Taylor, P.W., and Rodoni, B. (2004) The development of virus resistant carrot genotypes using RNAi technology. In: Fischer, T., Turner, N., Angus, J., McIntyre, L., Robertson, M., *et al.* (eds.) *New Directions for a Diverse Planet: Proceedings of the 4th International Crop Science Congress, 26 September–1 October 2004, Brisbane, Australia* (poster).

McCurdy, D.W., Dibley, S., Cahyanegara, R., Martin, A., and Patrick, J.W. (2010) Functional characterization and RNAi-mediated suppression reveals roles for hexose transporters in sugar accumulation by tomato fruit. *Molecular Plant* 3, 1049–1063.

Meng, C., Yang, M., Wang, Y., Chen, C., Sui, N., *et al.* (2020) SlWHY2 interacts with SlRECA2 to maintain mitochondrial function under drought stress in tomato. *Plant Science* 301: 110674.

Miao, B.A.I., Chen, W.T., Xie, B.Y., and Yang, G.S. (2016) A novel strategy to enhance resistance to *Cucumber mosaic virus* in tomato by grafting to transgenic rootstocks. *Journal of Integrative Agriculture* 15, 2040–2048.

Milhavet, O., Gary, D.S., and Mattson, M.P. (2003) RNA interference in biology and medicine. *Pharmacological Reviews* 55, 629–648.

Montiel, J., Nava, N., Cárdenas, L., Sánchez-López, R., Arthikala, M.K., *et al.* (2012) A *Phaseolus vulgaris* NADPH oxidase gene is required for root infection by *Rhizobia*. *Plant and Cell Physiology* 53, 1751–1767.

Montiel, J., Arthikala, M.K., and Quinto, C. (2013) *Phaseolus vulgaris RbohB* functions in lateral root development. *Plant Signaling & Behavior* 8: e22694.

Morel, A., Guinard, J., Lonjon, F., Sujeeun, L., Barberis, P., *et al.* (2018) The eggplant AG91-25 recognizes the type III-secreted effector RipAX2 to trigger resistance to bacterial wilt (*Ralstonia solanacearum* species complex). *Molecular Plant Pathology* 19, 2459–2472.

Mosa, M.A. and Youssef, K. (2021) Topical delivery of host induced RNAi silencing by layered double hydroxide nanosheets: an efficient tool to decipher pathogenicity gene function of *Fusarium* crown and root rot in tomato. *Physiological and Molecular Plant Pathology* 115: 101684.

Mushtaq, N., Wang, Y., Fan, J., Li, Y., and Ding, J. (2022) Down-regulation of cytokinin receptor gene *SlHK2* improves plant tolerance to drought, heat, and combined stresses in tomato. *Plants* 11: 154.

Nilon, A., Robinson, K., Pappu, H.R., and Mitter, N. (2021) Current status and potential of RNA interference for the management of tomato spotted wilt virus and thrips vectors. *Pathogens* 10: 320.

Nogueira, G.B., Dos Santos, L.V., de Queiroz, C.B., Corrêa, T.L.R., Menicucci, R.P., *et al.* (2019) The histidine kinase *slnCl1* of *Colletotrichum lindemuthianum* as a pathogenicity factor against *Phaseolus vulgaris* L. *Microbiological Research* 219, 110–122.

Ntui, V.O., Kynet, K., Azadi, P., Khan, R.S., Chin, D.P., *et al.* (2013) Transgenic accumulation of a defective cucumber mosaic virus (CMV) replicase derived double stranded RNA modulates plant defence against CMV strains O and Y in potato. *Transgenic Research* 22, 1191–1205.

Ntui, V.O., Kong, K., Azadi, P., Khan, R.S., Chin, D.P., *et al.* (2014) RNAi-mediated resistance to cucumber mosaic virus (CMV) in genetically engineered tomato. *American Journal of Plant Sciences* 5: 43606.

Palli, S.R. (2014) RNA interference in Colorado potato beetle: steps toward development of dsRNA as a commercial insecticide. *Current Opinion in Insect Science* 6, 1–8.

Pan, I.L., McQuinn, R., Giovannoni, J.J., and Irish, V.F. (2010) Functional diversification of *AGAMOUS* lineage genes in regulating tomato flower and fruit development. *Journal of Experimental Botany* 61, 1795–1806.

Pareek, M. and Rajam, M.V. (2017) RNAi-mediated silencing of MAP kinase signalling genes (*Fmk1*, *Hog1*, and *Pbs2*) in *Fusarium oxysporum* reduces pathogenesis on tomato plants. *Fungal Biology* 121, 775–784.

Park, Y. and Kim, Y. (2013) RNA interference of glycerol biosynthesis suppresses rapid cold hardening of the beet armyworm, *Spodoptera exigua*. *Journal of Experimental Biology* 216, 4196–4203.

Peters, S., Imani, J., Mahler, V., Foetisch, K., Kaul, S., *et al.* (2011) *Dau c 1.01* and *Dau c 1.02*-silenced transgenic carrot plants show reduced allergenicity to patients with carrot allergy. *Transgenic Research* 20, 547–556.

Petrov, N., Stoyanova, M., Andonova, R., and Teneva, A. (2015) Induction of resistance to potato virus Y strain NTN in potato plants through RNAi. *Biotechnology & Biotechnological Equipment* 29, 21–26.

Pizetta, C.S.R., Ribeiro, W.R., Ferreira, A.L., da Costa Moura, M., Bonfim, K., *et al.* (2022) RNA interference-mediated tolerance to whitefly (*Bemisia tabaci*) in genetically engineered tomato. *Plant Cell, Tissue and Organ Culture* 148, 281–291.

Praveen, S., Ramesh, S.V., Mishra, A.K., Koundal, V., and Palukaitis, P. (2010) Silencing potential of viral derived RNAi constructs in *Tomato leaf curl virus*-AC4 gene suppression in tomato. *Transgenic Research* 19, 45–55.

Prins, M. and Goldbach, R. (1996) RNA-mediated virus resistance in transgenic plants. *Archives of Virology* 141, 2259–2276.

Qi, C., Zhang, H., Liu, Y., Wang, X., Dong, D., *et al.* (2020) CsSNAT positively regulates salt tolerance and growth of cucumber by promoting melatonin biosynthesis. *Environmental and Experimental Botany* 175: 104036.

Qi, C., Dong, D., Li, Y., Wang, X., Guo, L., *et al.* (2022) Heat shock-induced cold acclimation in cucumber through CsHSFA1d-activated JA biosynthesis and signaling. *Plant Journal* 111, 85–102.

Qiu, Z., Yan, S., Xia, B., Jiang, J., Yu, B., *et al.* (2019) The eggplant transcription factor MYB44 enhances resistance to bacterial wilt by activating the expression of *spermidine synthase*. *Journal of Experimental Botany* 70, 5343–5354.

Ramesh, S.V., Mishra, A.K., and Praveen, S. (2007) Hairpin RNA-mediated strategies for silencing of tomato leaf curl virus *AC1* and *AC4* genes for effective resistance in plants. *Oligonucleotides* 17, 251–257.

Ramkumar, G., Asokan, R., Prasannakumar, N.R., Kariyanna, B., Karthi, S., *et al.* (2021) RNA interference suppression of *v-ATPase B* and juvenile hormone binding protein genes through topically applied dsRNA on tomato leaves: developing biopesticides to control the South American pinworm, *Tuta absoluta* (Lepidoptera: Gelechiidae). *Frontiers in Physiology* 12: 742871.

Rípodas, C., Dalla Via, V., Aguilar, O.M., Zanetti, M.E., and Blanco, F.A. (2013) Knock-down of a member of the isoflavone reductase gene family impairs plant growth and nodulation in *Phaseolus vulgaris*. *Plant Physiology and Biochemistry* 68, 81–89.

Roderick, H., Urwin, P.E., and Atkinson, H.J. (2018) Rational design of biosafe crop resistance to a range of nematodes using RNA interference. *Plant Biotechnology Journal* 16, 520–529.

Schijlen, E.G.W.M., Ric de Vos, C.R., Martens, S., Jonker, H.H., Rosin, F.M., *et al.* (2007) RNA interference silencing of chalcone synthase, the first step in the flavonoid biosynthesis pathway, leads to parthenocarpic tomato fruits. *Plant Physiology* 144, 1520–1530.

Schwind, N., Zwiebel, M., Itaya, A., Ding, B., Wang, M.B., *et al.* (2009) RNAi-mediated resistance to *Potato spindle tuber viroid* in transgenic tomato expressing a viroid hairpin RNA construct. *Molecular Plant Pathology* 10, 459–469.

Schwinn, K.E., Ngo, H., Kenel, F., Brummell, D.A., Albert, N.W., *et al.* (2016) The onion (*Allium cepa* L.) R2R3-MYB gene *MYB1* regulates anthocyanin biosynthesis. *Frontiers in Plant Science* 7: 1865.

Shafiq, M. (2014) RNA interference based resistance against chilli leaf curl disease complex. Thesis, Quaid-i-Azam University, Islamabad, Pakistan.

Shan, G. (2010) RNA interference as a gene knockdown technique. *International Journal of Biochemistry & Cell Biology* 42, 1243–1251.

Shanmugam, V., Sharma, V., Bharti, P., Jyoti, P., Yadav, S.K., *et al.* (2017) RNAi induced silencing of pathogenicity genes of *Fusarium* spp. for vascular wilt management in tomato. *Annals of Microbiology* 67, 359–369.

Sharma, N., Sarao, N.K., Mohanpuria, P., and Sharma, A. (2023) Effective *Agrobacterium*-mediated genetic transformation of okra (*Abelmoschus esculentus* L.) and generation of RNAi plants resistant to *Begomovirus* infecting okra. *Journal of Horticultural Science and Biotechnology* 99, 146–159.

Sheehy, R.E., Kramer, M., and Hiatt, W.R. (1988) Reduction of polygalacturonase activity in tomato fruit by antisense RNA. *Proceedings of the National Academy of Sciences USA* 85, 8805–8809.

Shen, J., Zhang, Y., Ge, D., Wang, Z., Song, W., *et al.* (2019) *CsBRC1* inhibits axillary bud outgrowth by directly repressing the auxin efflux carrier CsPIN3 in cucumber. *Proceedings of the National Academy of Sciences USA* 116, 17105–17114.

Shin, N.R., Shin, Y.H., Jeon, Y., and Park, Y.D. (2019) Identification of self-compatible transgenic Chinese cabbage with RNA interference Type 2A serine/threonine protein phosphatase gene. *Journal of Horticultural Science and Technology* 37, 200–201.

Shin, Y.H., Park, J.S., Woo, E.T., and Park, Y.D. (2017) Development of transgenic Chinese cabbage using RNA interference for mites control. *Journal of Horticultural Science and Technology* 35, 119–120.

Shivakumara, T.N., Chaudhary, S., Kamaraju, D., Dutta, T.K., Papolu, P.K., *et al.* (2017) Host-induced silencing of two pharyngeal gland genes conferred transcriptional alteration of cell wall-modifying enzymes of *Meloidogyne incognita vis-à-vis* perturbed nematode infectivity in eggplant. *Frontiers in Plant Science* 8: 473.

Singh, A., Taneja, J., Dasgupta, I., and Mukherjee, S.K. (2015) Development of plants resistant to tomato geminiviruses using artificial trans-acting small interfering RNA. *Molecular Plant Pathology* 16, 724–734.

Singh, N., Mukherjee, S.K., and Rajam, M.V. (2020) Silencing of the *ornithine decarboxylase* gene of *Fusarium oxysporum* f. sp. *lycopersici* by host-induced RNAi confers resistance to *Fusarium* wilt in tomato. *Plant Molecular Biology Reporter* 38, 419–429.

Singh, S., Gupta, M., Pandher, S., Kaur, G., Goel, N., *et al.* (2019) RNA sequencing, selection of reference genes and demonstration of feeding RNAi in *Thrips tabaci* (Lind.) (Thysanoptera: Thripidae). *BMC Molecular and Cell Biology* 20: 6.

Sinha, R. and Rajam, M.V. (2013) RNAi silencing of three homologues of *S*-adenosylmethionine decarboxylase gene in tapetal tissue of tomato results in male sterility. *Plant Molecular Biology* 82, 169–180.

Smith, N., Singh, S., Wang, M., Stoutjesdijk, P., Green, A., and Waterhouse, P. (2000) Total silencing by intron-spliced hairpin RNAs. *Nature* 407, 319–320.

Sun, L., Yuan, B., Zhang, M., Wang, L., Cui, M., *et al.* (2012) Fruit-specific RNAi-mediated suppression of *SlNCED1* increases both lycopene and β-carotene contents in tomato fruit. *Journal of Experimental Botany* 63, 3097–3108.

Surakasi, V.P., Mohamed, A.A., and Kim, Y. (2011) RNA interference of β1 integrin subunit impairs development and immune responses of the beet armyworm, *Spodoptera exigua*. *Journal of Insect Physiology* 57, 1537–1544.

Tabein, S., Jansen, M., Noris, E., Vaira, A.M., Marian, D., *et al.* (2020) The induction of an effective dsRNA-mediated resistance against *Tomato spotted wilt virus* by exogenous application of double-stranded RNA largely depends on the selection of the viral RNA target region. *Frontiers in Plant Science* 11: 533338.

Tan, J.A.C., Jones, M.G., and Fosu-Nyarko, J. (2013) Gene silencing in root lesion nematodes (*Pratylenchus* spp.) significantly reduces reproduction in a plant host. *Experimental Parasitology* 133, 166–178.

Tao, L., Yu, G., Chen, H., Wang, B., Jiang, L., *et al.* (2022) *SlDREB2* gene specifically recognizing to the universal DRE elements is a transcriptional activator improving drought tolerance in tomato. *Scientia Horticulturae* 295: 110887.

Tetorya, M. and Rajam, M.V. (2021) RNAi-mediated silencing of *PEX6* and *GAS1* genes of *Fusarium oxysporum* f. sp. *lycopersici* confers resistance against *Fusarium* wilt in tomato. *3Biotech* 11: 443.

Tian, Z., Guo, S., Li, J.X., Zhu, F., Liu, W., and Wang, X.P. (2021) Juvenile hormone biosynthetic genes are critical for regulating reproductive diapause in the cabbage beetle. *Insect Biochemistry and Molecular Biology* 139: 103654.

Tian, Z., Guo, S., Zhu, F., Liu, W., and Wang, X.P. (2022) Targeting coat protein II complex genes via RNA interference inhibits female adult feeding and reproductive development in the cabbage beetle *Colaphellus bowringi*. *Pest Management Science* 78, 2141–2150.

van Eck, J., Conlin, B., Garvin, D.F., Mason, H., Navarre, D.A., and Brown, C.R. (2007) Enhancing beta-carotene content in potato by RNAi-mediated silencing of the beta-carotene hydroxylase gene. *American Journal of Potato Research* 84, 331–342.

Venkatesh, J., Kim, S.J., Siddique, M.I., Kim, J.H., Lee, S.H., and Kang, B.-C. (2023) *CopE* and *TLR6* RNAi-mediated tomato resistance to western flower thrips. *Journal of Integrative Agriculture* 22, 471–480.

Wang, J., Zhang, L., Cao, Y., Qi, C., Li, S., *et al.* (2018) *CsATAF1* positively regulates drought stress tolerance by an ABA-dependent pathway and by promoting ROS scavenging in cucumber. *Plant and Cell Physiology* 59, 930–945.

Wang, X., Cui, X., Shang, S., Zhang, X., Wang, H., *et al.* (2023) The destruction of endogenous ABA balance caused by fruit-specific *SlNCED1*-RNAi affects tomato fruit set. *Journal of Plant Growth Regulation* 42, 7200–7214.

Wen, C., Zhao, W., Liu, W., Yang, L., Wang, Y., *et al.* (2019) *CsTFL1* inhibits determinate growth and terminal flower formation through interaction with *CsNOT2a* in cucumber. *Development* 146: dev180166.

Wu, J.J., Chen, F., Yang, R., Shen, C.H., Ze, L.J., *et al.* (2022) Knockdown of *Ecdysone-Induced Protein 93F* causes abnormal pupae and adults in the eggplant lady beetle. *Biology* 11: 1640.

Wu, K., Taylor, C.E., Pinheiro, D.H., Skelley, L.H., McAuslane, H.J., and Siegfried, B.D. (2019) Lethal RNA interference response in the pepper weevil. *Journal of Applied Entomology* 143, 699–705.

Wu, M., Zhang, Q., Dong, Y., Wang, Z., Zhan, W., *et al.* (2023) Transplastomic tomatoes expressing double-stranded RNA against a conserved gene are efficiently protected from multiple spider mites. *New Phytologist* 237, 1363–1373.

Wuriyanghan, H., Rosa, C., and Falk, B.W. (2011) Oral delivery of double-stranded RNAs and siRNAs induces RNAi effects in the potato/tomato psyllid, *Bactericerca cockerelli*. *PLOS ONE* 6: e27736.

Xiao, J., Li, H., Zhang, J., Chen, R., Zhang, Y., Ouyang, B., Wang, T., and Ye, Z. (2006) Dissection of GA 20-oxidase members affecting tomato morphology by RNAi-mediated silencing. *Plant Growth Regulation* 50, 179–189.

Xiong, A.S., Yao, Q.H., Peng, R.H., Li, X., Han, P.L., and Fan, H.Q. (2005) Different effects on ACC oxidase gene silencing triggered by RNA interference in transgenic tomato. *Plant Cell Reports* 23, 639–646.

Xu, G., Sui, N., Tang, Y., Xie, K., Lai, Y., and Liu, Y. (2010) One-step, zero-background ligation-independent cloning intron-containing hairpin RNA constructs for RNAi in plants. *New Phytologist* 187, 240–250.

Xu, S., Liao, C.J., Jaiswal, N., Lee, S., Yun, D.J., *et al.* (2018) Tomato *PEPR1 ORTHOLOG RECEPTOR-LIKE KINASE1* regulates responses to systemin, necrotrophic fungi, and insect herbivory. *Plant Cell* 30, 2214–2229.

Xu, X., Cao, B., Li, G., Lei, J., Chen, Q., *et al.* (2015) Functional characterization of a putative bacterial wilt resistance gene (*RE-bw*) in eggplant. *Plant Molecular Biology Reporter* 33, 1058–1073.

Xu, Z., Cheng, J., Wang, T., Huang, Q., Liu, P., *et al.* (2023) Novel jasmonic acid–coumarin pathway in the eggplant that inhibits *vitellogenin* gene expression to prevent mite reproduction. *Journal of Agricultural and Food Chemistry* 71, 13979–13987.

Yan, S., Ning, K., Wang, Z., Liu, X., Zhong, Y., *et al.* (2020) CsIVP functions in vasculature development and downy mildew resistance in cucumber. *PLOS Biology* 18: e3000671.

Yan, S., Yu, B., Ming, F., Liang, Y., Zhong, Y., *et al.* (2022) CsIVP modulates low nitrogen and high-temperature resistance in cucumber. *Plant and Cell Physiology* 63, 605–617.

Yan, Y., Mintao, S., Si, M., Qian, F., Yijia, W., *et al.* (2022) Mechanism of *CsGPA1* in regulating cold tolerance of cucumber. *Horticulture Research* 9: uhac109.

Yan, Y., Sun, M., Li, Y., Wang, J., He, C., and Yu, X. (2020) The *CsGPA1-CsAQPs* module is essential for salt tolerance of cucumber seedlings. *Plant Cell Reports* 39, 1301–1316.

Yang, Q., Song, Z., Dong, B., Niu, L., Cao, H., *et al.* (2021) Hyperoside regulates its own biosynthesis via MYB30 in promoting reproductive development and seed set in okra. *Plant Physiology* 185, 951–968.

Yang, S., Wen, C.L., Liu, B., Cai, Y.L., Xue, S.D., *et al.* (2019) A *CsTu-TS1* regulatory module promotes fruit tubercule formation in cucumber. *Plant Biotechnology Journal* 17, 289–301.

Yang, S., Wang, Y., Zhu, H., Zhang, M., Wang, D., *et al.* (2022) A novel HD-Zip I/C2H2-ZFP/WD-repeat complex regulates the size of spine base in cucumber. *New Phytologist* 233, 2643–2658.

Yin, W., Hu, Z., Hu, J., Zhu, Z., Yu, X., *et al.* (2017) Tomato (*Solanum lycopersicum*) MADS-box transcription factor *SlMBP8* regulates drought, salt tolerance and stress-related genes. *Plant Growth Regulation* 83, 55–68.

Yu, J.-G., Lee, G.-H, and Park, Y.-D. (2016) Characterization of a drought-tolerance gene, *BrDSR*, in Chinese cabbage. *Korean Journal of Horticultural Science and Technology* 34, 102–111.

Zanor, M.I., Osorio, S., Nunes-Nesi, A., Carrari, F., Lohse, M., *et al.* (2009) RNA interference of *LIN5* in tomato confirms its role in controlling Brix content, uncovers the influence of sugars on the levels of fruit hormones, and demonstrates the importance of sucrose cleavage for normal fruit development and fertility. *Plant Physiology* 150, 1204–1218.

Zhang, D., Chen, J., Yao, Q., Pan, Z., Chen, J., and Zhang, W. (2012) Functional analysis of two chitinase genes during the pupation and eclosion stages of the beet armyworm *Spodoptera exigua* by RNA interference. *Archives of Insect Biochemistry and Physiology* 79, 220–234.

Zhang, J., Wang, Y., Naeem, M., Zhu, M., Li, J., *et al.* (2019) An *AGAMOUS* MADS-box protein, SlMBP3, regulates the speed of placenta liquefaction and controls seed formation in tomato. *Journal of Experimental Botany* 70, 909–924.

Zhang, L., Guo, X., Zhang, Z., Wang, A., and Zhu, J. (2021) Cold-regulated gene *LeCOR413PM2* confers cold stress tolerance in tomato plants. *Gene* 764: 145097.

Zhang, N., Huang, Y.-Y., Feng, J., Liu, L.-S., Peng, Y., *et al.* (2011) Cloning of *ProDH* gene in cabbage and construction of the RNAi vector. *Journal of China Agricultural University* 16, 87–94.

Zhang, S., Wang, S., Lv, J., Liu, Z., Wang, Y., *et al.* (2018) SUMO E3 ligase SlSIZ1 facilitates heat tolerance in tomato. *Plant and Cell Physiology* 59, 58–71.

Zhang, Y.Y., Li, H.X., Shu, W.B., Zhang, C.J., Zhang, W., and Ye, Z.B. (2011a) Suppressed expression of ascorbate oxidase gene promotes ascorbic acid accumulation in tomato fruit. *Plant Molecular Biology Reporter* 29, 638–645.

Zhang, Y.Y., Li, H.X., Shu, W.B., Zhang, C.J., and Ye, Z.B. (2011b) RNA interference of a mitochondrial *APX* gene improves vitamin C accumulation in tomato fruit. *Scientia Horticulturae* 129, 220–226.

Zhao, J.Y., Jiang, L., Che, G., Pan, Y.P., Li, Y.Q., *et al.* (2019) A functional allele of *CsFUL1* regulates fruit length through repressing *CsSUP* and inhibiting auxin transport in cucumber. *Plant Cell* 31, 1289–1307.

Zhao, W.S., Chen, Z.J., Liu, X.F., Che, G., Gu, R., *et al.* (2018) *CsLFY* is required for shoot meristem maintenance via interaction with *WUSCHEL* in cucumber (*Cucumis sativus*). *New Phytologist* 218, 344–356.

Zhu, F., Xu, J., Palli, R., Ferguson, J., and Palli, S.R. (2011) Ingested RNA interference for managing the populations of the Colorado potato beetle, *Leptinotarsa decemlineata*. *Pest Management Science* 67, 175–182.

Zhuang, K., Gao, Y., Liu, Z., Diao, P., Sui, N., *et al.* (2020) WHIRLY1 regulates HSP21.5A expression to promote thermotolerance in tomato. *Plant and Cell Physiology* 61, 169–177.

# 20 Designing Futuristic Vegetable Varieties for Multiple Purposes

**Pritam Kalia***
*Division of Vegetable Science, ICAR-Indian Agricultural Research Institute, New Delhi, India*

**Abstract**

Vegetables are considered an essential protective food for well-balanced nutritional diets. The production and marketing of vegetables crops are undergoing continuous change globally. This is mainly due to the growing demands of consumers for safe and healthy vegetables, increasing urbanization, and the growth in scale and influence of supermarket chains. The general objectives for growers are high yield, disease and insect pest resistance, uniformity, and abiotic stress resistance, including adaptation to changing climatic conditions. The objectives for consumers are quality, appearance, shelf life, taste, and nutritional value. For growers to survive, cultivars must be accepted by the market. A diet rich in vegetables provides micronutrients and health-promoting phytochemicals that alleviate malnutrition. The beneficial health effects are attributed mainly to diverse antioxidant compounds such as vitamins, carotenoids, phenolics, alkaloids, nitrogen-containing compounds, and organosulfur compounds. Although the main long-term breeding objective will continue to be increasing yield to meet the food requirements of an ever-increasing population, in order to ensure health security, the multi-purpose utility of varieties for the fresh market, and industry suitability, it is imperative that nutraceutical, edible color, and bioactive compound-rich vegetable varieties are bred, ensuring high remuneration to farmers. Conventional breeding in conjunction with molecular biology has bright prospects for developing vegetable varieties high in nutraceuticals, edible colors and bioactive compounds suitable for these purposes. Although, it is possible to develop new vegetable varieties or integrate favorable genes for nutraceuticals, bioactive compounds, and edible color into cultivated varieties through conventional breeding, the process is slow and time consuming. Advances in molecular biology and recombinant technology have paved the way to enhance the pace of special trait variety development using marker-assisted breeding and designing new vegetable crop plants using transgenic and genome-editing approaches.

**Keywords:** biotechnological approaches, conventional breeding, multi-nutrients, nutraceuticals, nutritious vegetable varieties

## 20.1 Introduction

There are around 795 million undernourished people around the world today (FAO, 2015). That means that one in nine people do not get enough food to be healthy and lead an active life. Vegetables are considered an essential protective food for a well-balanced nutritional diet. The production and marketing of vegetables crops are undergoing continuous changes worldwide, mainly due to growing demands of consumers for safe and healthy vegetables,

*Email: pritam.kalia@gmail.com


DOI: 10.1079/9781800627123.0020

increasing urbanization, and the growth in scale and influence of supermarket chains. The final goal of vegetable breeding programs is to develop new improved varieties with elite combinations of desirable horticultural characteristics. Consumers (and therefore supermarkets) are seeking nutritionally rich vegetable varieties/hybrids for new nutrient-rich cuisines, different quality attributes, and improved convenience, and are prepared to pay a premium for such produce if their expectations are met.

Vegetable breeding has to address and satisfy the needs of both the consumer and the grower. The general objectives for growers are high yield, disease and insect pest resistance, uniformity, and abiotic stress resistance, including adaptation to changing climatic conditions. The objectives for consumers are quality, appearance, shelf life, taste, and nutritional value. For growers to survive, cultivars must be accepted by the market. Thus, color, appearance, taste, and shape are usually more important than productivity. Vegetables are valued for their extrinsic and intrinsic quality traits. A diet rich in vegetables provides micronutrients and health-promoting phytochemicals that alleviate malnutrition. The beneficial health effects are attributed mainly to diverse antioxidant compounds such as vitamins, carotenoids, phenolics, alkaloids, nitrogen-containing compounds, and organosulfur compounds. Although the main long-term breeding objective will continue to be increasing yield to meet the food requirements of the ever-increasing population, in order to ensure health security and multi-purpose utility of varieties for the fresh market and industry suitability, it is imperative that nutraceutical, edible color, and bioactive compound-rich vegetable varieties are bred, ensuring high remuneration to farmers. Quality in vegetables is a complex characteristic influenced by both genetic and environmental factors. Breeding for quality has been unsystematic and often empirical, but significant progress has been made in a number of vegetable crops. Conventional breeding in conjunction with molecular biology has good prospects of developing vegetable varieties high in nutraceuticals, edible colors, and bioactive compounds suitable for the fresh market as well as for the developing functional/fusion food industry (Rajasree and Pugalendhi, 2021).

## 20.2 Nutraceutical/Bioactive Compounds

These are chemical compounds found in plants that may be considered a food or part of a food that not only provides nutrition for body building but also helps in prevention and treatment of diseases. They are also referred to as products isolated or purified from foods and generally sold in medicinal forms not usually associated with food and demonstrated to have a physiological benefit or to provide protection against chronic diseases. Some of the nutraceutical/bioactive compounds found in different vegetable crops are shown in Table 20.1.

**Table 20.1.** Vegetable sources of nutraceutical/bioactive compounds.

| Vegetables | Nutraceutical/bioactive compounds |
|---|---|
| *Allium* vegetables (garlic, onions, chives, leeks) | Allyl sulfides |
| Cruciferous vegetables (broccoli, cauliflower, cabbage, Brussels sprouts, kale, turnips, kohlrabi) | Indoles/glucosinolates, sulfaforaphane, isothiocyanates/thiocyanates, thiols |
| Solanaceous vegetables (tomatoes, peppers) | Lycopene |
| Umbelliferous vegetables (carrots, celery, parsley, parsnips) | Carotenoids, phthalides, polyacetylenes |
| Compositae plants (artichoke) | Silymarin |
| Beans | Flavonoids (isoflavones) |
| Carrots, squash, broccoli, sweet potatoes, tomatoes, kale, collards, cantaloupe, pumpkin | Vitamin A (retinol) |
| Green peppers, broccoli, green leafy vegetables, cabbage, tomatoes | Vitamin C (ascorbic acid) |
| Green leafy vegetables | Vitamin E |

### 20.2.1 Edible colors

Edible colors are natural pigments found in plant tissues. They are chemical compounds produced by several biochemical pathways and give color to the food. These colors change according to the growth stage of the plant part or vegetable product and include anthocyanins, betalains, carotenoids, and chlorophylls. These pigments play important ecological and metabolic functions in the plant (Grotewold, 2006), and are frequently exploited as a source of nutraceuticals to address a number of human ailments. These pigments have been implicated in regimes to maintain human health, to protect against chronic disease, and to restore wellness by repairing tissues after disease has been established. A wide range of bioassays and tests have been used to establish the biological efficacy of natural pigments in human health interventions, including *in vitro* bioassays, and *in vivo*, epidemiological, and clinical trials (Lila, 2004). The major color pigments and their vegetable sources are given in Table 20.2, and their impacts on human health are described below.

#### *Anthocyanins*

Anthocyanins are natural pigments belonging to the flavonoid family. They are responsible for the blue, purple, red, and orange color of many fruits and vegetables. Anthocyanins are capable of acting on different cells involved in the development of atherosclerosis, one of the leading causes of cardiovascular dysfunction. Anthocyanins and the aglycone cyanidin were found to inhibit cyclo-oxygenase enzymes, which can be a marker for the initiation stage of carcinogenesis. Recently, both anthocyanins and cyanidin from tart cherries were found to reduce cell growth of human colon cancer cell lines (Kang *et al.*, 2003). On the one hand, they can interfere with glucose absorption, while on the other, they may have a protective effect on pancreatic cells. The most extensively documented phytomedicinal role of anthocyanin pigments is in improving eyesight, including night vision. Anthocyanins exert significant antimicrobial properties, and (in association with other flavonoids) have demonstrated effective inhibition of aflatoxin biosynthesis (Norton, 1999).

#### *Betalains*

Betalains (also known as betacyanins) have been widely used as natural colorants for many centuries, but their attractiveness for use as colorants of foods (or drugs and cosmetics) has increased recently due to their reportedly high anti-oxidative, free-radical-scavenging activities and concerns about the use of various synthetic alternatives. Beetroot is a natural source of betacyanins.

#### *Carotenoids*

Carotenoids are the second most abundant pigment in nature, and comprise more than 700 types (Britton, 1998). Carotenoids play an important role in plant reproduction through their role in attracting pollinators and in seed dispersal, and are essential components of the human diet. β-Carotene provides protection to vision and eye function, and protects against macular degeneration and cataracts. Carotenoids are credited with biological promotion of the immune system response. They are associated with inhibition of several types of cancer including cervical, esophageal, pancreatic, lung, prostate, colorectal, and stomach.

**Table 20.2.** Color-rich vegetables and their bioactive compounds.

| Vegetables | Color | Pigment/bioactive compound |
|---|---|---|
| Tomatoes, carrot, watermelon | Red | Lycopene |
| Beetroot | Red/crimson | Betacyanins |
| Carrot, cantaloupe, pumpkin, sweet potato, cassava | Orange | Beta-carotene |
| Baby corn, yellow corn, carrot, watermelon | Yellow | Lutein |
| Broccoli, kale, spinach, cabbage, asparagus | Green | Chlorophyll |
| Carrot, eggplant, broccoli, cauliflower, lettuce, okra | Black/purple | Anthocyanins |

### *Chlorophylls*

Chlorophyll is the most important plant pigment, giving plants their green color. Chlorophylls are typically consumed in much higher doses in a diet that incorporates green and leafy vegetables (Delgado-Vargas and Paredes-Lopez, 2003). The anti-mutagenic properties of chlorophylls have been demonstrated in various assays, and an intake of chlorophyll has the potential to act as a chemopreventive compound in humans.

## 20.2.2 Edible colors in carrots

### *Orange-colored carrots*

The color of orange carrots is due to carotenoids. β-Carotene (provitamin A) is a precursor of vitamin A in the human body. Four carotenoids, β-carotene, α-carotene, γ-carotene and β-cryptoxanthin, have vitamin A activity in humans, which means they can be converted into the visual pigment retinol and are classified as essential nutrients. The predominant carotenoids are the provitamin A carotenes, α- and β-carotene, accounting for 13–40% and 45–80% of the carotenoids in orange carrots, respectively. The dietary reference intake for vitamin A is expressed as the retinol activity equivalent (RAE), which takes bioavailability into account. The recommended dietary reference intake for males and females is 900 and 700 RAE, respectively (higher in pregnancy and when lactating), and for children is 400–500 RAE. One RAE is equivalent to 1 mg of pure retinol, 2 mg of pure β-carotene dissolved in oil, or 12 mg of β-carotene in food. More than 4 million children, mostly from developing countries, exhibit clinical symptoms of severe vitamin A deficiency, including poor immunity, loss of vision in low light conditions (night blindness) and in extreme cases an irreversible form of blindness called xerophthalmia, and loss of healthy epithelial cell differentiation, normal reproductive performance, and visual function. The dark orange varieties have a β-carotene concentration of up to 130–500 ppm (Simon *et al.*, 2009).

### *Yellow-colored carrots*

The yellow color of the roots of these carrots is due to xanthophylls (lutein and zeaxanthin), which is collectively referred to as macular pigment. It is found in the macular region of the eye in humans and may have importance in eye health and protection from age-related macular degeneration. Lutein is an important pigment in yellow carrots, found in the range of 1–5 ppm. Xanthophyll-rich carrots are associated with a significant reduction in the risk for cataracts and for age-related macular degeneration.

### *Red-colored carrots*

Red carrots contain lycopene in addition to α- and β-carotene. The utility of red carrot as a functional food depends on the bioavailability of its constituent carotenoids. Lycopene is the carotenoid primarily responsible for the red color. It is an excellent dietary antioxidant and plays a role in reducing the risk of a number of cancers and coronary heart diseases. The lycopene content in carrot varies from 85 to 100 ppm among different varieties. Red carrots are commonly consumed as a salad and gajar halwa during the winter months (November–January) in northern India.

### *Purple-colored carrots*

A group of polyphenolic pigments called anthocyanins are responsible for the color of purple carrots. Current cultivars of purple varieties include solid purple carrots, often referred to as black carrot. The primary anthocyanins found in purple carrots are derivatives of cyanidin, but pelargonidin and peonidin glycosides have also been identified. Dietary anthocyanins may play a role in health promotion and protection from cardiovascular disease and anti-cancer activity, and have beneficial effects on diabetes. The total anthocyanin content in dark purple carrots ranges from 300 to 350 mg 100 $g^{-1}$ of fresh weight in dark purple carrots.

### *Multi-colored carrots*

Colorful specialty carrots have been "rediscovered" by modern plant breeders interested in improving and diversifying the nutritional quality of the food supply. Carrot has gained popularity in recent times due to increased awareness of its nutritional value, as well the physical appeal of the root color. Orange carrots rewarded color as "good for the eyes" due to their high amount of carotenoids especially β-carotene. Breeding single-colored varieties/$F_1$ hybrids of orange, yellow, red, and purple provides a single source of pigments. However, the use of novel breeding strategies has combined purple and orange (anthocyanins + β-carotene), purple and white (anthocyanins + lutein), and purple and red (anthocyanins + lycopene) carrots to provide increased anticancer potential. In an effort to develop pigment-rich tropical carrot varieties to gain public awareness, a rainbow mix of novel seeds from biofortified varieties is distributed to farmers and kitchen garden growers. Purple carrots with a white core contain very low levels of carotene (4–6 ppm), whereas purple–orange carrots can contain as much or more total carotene as typical orange carrots (38–130 ppm). The continued research in carrot cultivar development has produced a novel purple–orange–red cultivar that contains approximately 40 ppm carotene and 62 ppm lycopene (Arscott and Tanumihardjo, 2010).

### 20.2.3 Glucosinolates present in cole crops

Nearly 120 different glucosinolates are known to occur naturally in plants. They are a class of plant secondary metabolite, such as β-thioglucoside *N*-hydroxysulfates (also known as (Z)-(or *cis*)-*N*-hydroximinosulfate esters or *S*-glucopyranosyl thiohydroximates), with a side chain (R) and a sulfur-linked β-D-glucopyranose moiety. The common structural feature of intact glucosinolates is shown in Fig. 20.1 (Fahey *et al.*, 2001).

Glucosinolate content in some tissues of the *Brassica* vegetables is approximately 1% of dry weight, although the content has been found to be as high as 10% in the seeds of some plants. Among the cultivated brassicas, broccoli has attracted attention after the discovery that it contains high levels of isothiocyanate sulforaphane (4-(methylsulfinyl)butyl isothiocyanate), produced from the breakdown of the glucosinolate glucoraphanin (4-(methylsulfinyl)butyl glucosinolate) and showed significant potential anti-carcinogenic properties. Sulforaphane is the most potent naturally occurring phase II enzyme inducer and is considered to play an important role in the detoxification of xenobiotic compounds when assimilated into the diet (Rosa *et al.*, 1997; Fahey *et al.*, 2001).

Plants contain the enzyme myrosinase, which, in the presence of water, cleaves the glucose group from a glucosinolate. The remaining molecule then quickly converts to an isothiocyanate, a nitrite, or a thiocyanate; these are the active substances (Cartea *et al.*, 2008). The hydrolytic product of glucosinolate and the metabolic products act as chemoprotective agents against chemically induced carcinogens by blocking the initiation of tumors in a variety of rodent tissues, such as the liver, colon, mammary gland, and pancreas. The glucosinolates found in cole crops are isothiocyanate (cauliflower, cabbage, kohlrabi), sulforaphane (broccoli; most important in preventing cancer), glucobrassicin (cabbage and broccoli), gluconasturtiin (cabbage), indole-3-carbinol (cabbage), and sinigrin (Brussels sprout and broccoli). While research continues into the health-promoting mechanisms of *Brassica* isothiocyanates, others have developed high-glucosinolate broccoli germplasm that results in plants that produce mainly isothiocyanates, compared with standard broccoli cultivars that also produce nitriles (Sarikamis *et al.*, 2006). Introgressed chromosome segments from a wild ancestor, *Brassica villosa*, were used to enhance glucosinolate levels. These authors

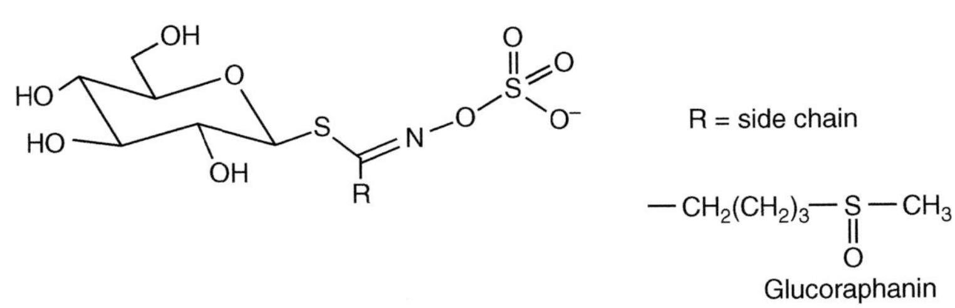


**Fig. 20.1.** General structure of intact glucosinolates.

also found *B. villosa* alleles that determine whether hydrolysis generates iberin or sulphoraphane. These findings suggest that high-glucosinolate broccoli might be suitable to increase the amount of sulforaphane in the diet. The major glucosinolates found in cole crops and their concentrations (Verkerk *et al.*, 2009) are given in Table 20.3.

### 20.2.4 Prebiotic-rich nutritious vegetables

Prebiotics have been characterized as a group of carbohydrates that resist digestion and absorption in the gastrointestinal tract, and are fermented by the gut microbiota. They selectively promote the growth and activity of a limited number of colonic bacteria, and alter the colonic microflora balance toward a healthier composition (Dwivedi *et al.*, 2014). The consumption of prebiotics may enhance immune function, improve colonic integrity, decrease the incidence and duration of intestinal infections, downegulate allergenic responses, and improve digestion and elimination.

To date, the best-known sources of the prebiotic carbohydrates fructo-oligosaccharides, inulin, and galacto-oligosaccharides are plants, especially vegetables and root and tuber crops. Fructan-rich vegetables include chicory, onion, and artichoke, while inulin-rich vegetables include garlic, yam, and Jerusalem artichoke.

### 20.2.5 Anti-obesity vegetables

Obesity is a cause of 3.4 million deaths, 3.9% of years of life lost, and 3.8% of disability adjusted life-years worldwide. Populations with a body mass index (BMI) of 25 or greater are more in the developed (up to 38%) than in the developing (up 13.4%) world. Overweight and obesity have therefore become a major global health challenge. Individuals with a BMI above 25 are at increased risk of diabetes mellitus, cardiovascular diseases, fatty liver (non-alcoholic), and hypertension (Williams *et al.*, 2013).

A number of vegetables are a source of phytochemicals that can help in the fight against obesity. Some of these are listed in Table 20.4.

## 20.3 Breeding for Nutraceutical/Bioactive Compounds and Edible Colors

Nutraceutical/bioactive and edible colors are natural compounds that are regulated by several biochemical pathways and controlled by genetic and environmental factors. From early times, people knowingly or unknowingly selected

**Table 20.3.** Major glucosinolates in commercial crucifer vegetables.

| Vegetable | Glucosinolate groups | Aliphatic glucosinolates (mg 100 $g^{-1}$ fresh weight) |
|---|---|---|
| Green broccoli | Glucoraphanin | 11.6–34 |
| Purple broccoli | Glucoraphanin | 6.7 |
| White cauliflower | Glucoiberin | 0.5–6.6 |
| Green cauliflower | Glucoiberin | 1.2–27.7 |
| Purple cauliflower | Glucoraphanin | 11.6 |
| | Glucoiberin | 4.6 |
| Red cabbage | Glucoraphanin | 4–18.2 |
| | Glucoiberin | 4–13.6 |
| | Sinigrin | 3–16.7 |
| Savoy cabbage | Glucoiberin | 10.4–21.2 |
| | Sinigrin | 15.5–18.6 |
| Brussels sprouts | Sinigrin | 22–25.3 |
| | Glucoiberin | 6.4–13.9 |
| Kale | Sinigrin | 2.2–22.7 |
| | Glucoiberin | 13.4–16 |

**Table 20.4.** Major vegetables as a source of anti-obesity phytochemicals.

| Vegetables | Phytochemicals |
|---|---|
| Potato, sweet potato, carrot, lettuce, globe artichoke | Cholorogenic acids |
| Onion, lettuce, broccoli | Flavonol: quercetin, kaempferol |
| Capsicum, broccoli, celery, parsely | Flavones: luteolin, apigenin |
| Allium group vegetables | Organosulfurs |

several vegetable crops to use as food. Vegetables contain nutritional and phytochemical compounds, and there are many wild or cultivated vegetables that are rich in these beneficial compounds. The biochemical pathways and synthesis of these compounds are controlled by one or many genes, which are scattered in the available or unknown germplasm of a particular vegetable crop. India is endowed with diverse agro-climatic regions ranging from tropical to temperate, making it possible to grow all kinds of vegetable crops. In addition, there is plenty of diversity in different vegetable crops that can be exploited for the development of varieties with special traits. Through conventional breeding, it is possible to develop new vegetable varieties, or integrate favorable genes for nutraceuticals, bioactive compounds, and edible color into cultivated varieties. However, advances in molecular biology and recombinant technology have paved the way for enhancing the pace of special trait variety development using marker-assisted breeding and transgenic and genome-editing approaches.

### 20.3.1 Breeding techniques

Breeding plants for improved nutritional quality is an old idea. One can imagine early gatherers and farmers avoiding eating certain plants that made them ill. As research in the nutritional sciences continues to demonstrate the importance to our health and well-being of plant-derived nutrients in foods, more breeding is under way to increase the type and amount of nutrients in cereals, legumes, vegetables, and fruits. The human diet consists of a diverse array of crops. Because specific nutrient levels in many of these crops are often under genetic control, the prospects for improving the nutritional quality of food crops including vegetable crops are enormous.

Breeding methods in any crop depend on the breeding system and genetic architecture resulting from natural selection as well as human selection during the course of cultivation. The genetic architecture or the pattern of inheritance of characteristics is another important consideration when determining the most appropriate breeding procedure applicable to any particular crop. The choice of breeding method is largely guided by the nature of gene action and the relative magnitude of additive genetic variance, dominance variance, and epistasis in a breeding population. An efficient breeding procedure should be effective in manipulation and selection of favorable gene combinations, additive genetic variance, exploitation of dominance variance, and achieving a close relationship between expected genetic gain and realized progress from selection.

### 20.3.2 Self-pollinated vegetable crops

#### *Pure line selection*

A pure line is the progeny of single self-fertilized homozygous individuals. In this process, a large number of plants are initially selected on the basis of phenotypic performance and their individual progenies are evaluated until homozygosity is attained. The best progeny is then developed as a pure line variety. This procedure may be applied to improve any local variety or exotic collection. If there is variation within the progeny plants, single-plant selection is followed until complete homozygosity is attained.

#### *Single-plant selection*

This method is used to select chance variants, off types, or superior types from a mixed homozygous population in self-pollinated vegetable crops.

*Hybridization*

Hybridization is used to generate variability in segregating progenies, using procedures such as pedigree breeding, bulk breeding, single-seed descent, backcrossing, and their combinations to develop pure line cultivars. In the hybridization program, parents are selected on the basis of phenotypic performance, morphological characteristics, and the general combining ability (GCA) of the varieties. Generally, parents exhibiting high GCA, where there is involvement of additive genetic variance, are selected in self-pollinated crops. Two, three, four, or more plants may be selected, depending on the situation. After selecting parents, a single cross, double cross, or three-way cross is attempted. Sometimes $F_1$ progeny are backcrossed to their respective parents for the traits of interest.

PEDIGREE METHOD. This method has been adopted to develop pure-line varieties with different nutritional qualities. In several vegetable crops, improvements have been accomplished through this method, resulting in the development of a large number of varieties. In this method, single-plant selection is carried out up to generations $F_5$ or $F_6$, and in the advanced generations, families are selected on the basis of their phenotypic performance. Pedigree records of the selected plants and families are maintained. Initially, the $F_2$ population, which is highly heterozygous, is grown in rows and single-plant selection is made. The larger the population, the more chance there is of obtaining superior segregants. In subsequent generations, within-family variation is reduced, with an increase in homozygosity. The ratio of populations grown and plants selected in the $F_3$ generation may vary from 20:1 to 100:1. It is essential to grow checks after every fifth or sixth row. In generations $F_3$ and $F_4$, there will still be within-family variation and therefore emphasis is laid on within-family selection. As visual selection is applied, the intensity of selection should be considerably high and the heritability of characteristics should be taken into account. In this method, selection is more effective for nutraceuticals and edible colors with a high heritability.

BULK METHOD. This is one of the most economical methods of handling segregating populations based on natural selection. In natural selection, poorly adapted types are eliminated and superior types exist. This method consists of growing large populations in each generation, harvesting the seeds in bulk, and planting a sample of seeds in the next generations. No selection is made until generation $F_5$ or $F_6$, so only natural selection operates. Generally, there is no selection of plants on the basis of phenotypic performance, but sometimes superior plants are selected and their seeds are combined. This type of mass selection will increase the chance of obtaining superior plants and their seeds up to $F_3$, after which the pedigree method is very effective. In $F_6$, single plants are selected and their progeny are grown in the next generations. Superior families are selected on the basis of phenotypic performance, and are further evaluated against standard checks.

SINGLE-SEED DESCENT METHOD. This method is based on the assumption of additive genetic variance, presumably a rule in self-fertilizing crops. In this method, all plants in the $F_2$ generation are advanced to the next generations by harvesting and growing on seed from each plant. This process is continued up to generations $F_5/F_6$ in which superior individual plants are selected and tested in multi-location trials.

BACKCROSS METHOD. This method is used to transfer characteristics controlled by a single gene. It is more convenient to execute this method in a self-fertilizing crop. The backcross method is based on the principle of quick and efficient transfer of a gene through repeated backcrossing and then selection in the $F_2$ and subsequent generations. With continuous backcrossing with the recurrent parent, the backcross population becomes more similar to the recurrent parent with the additional trait of the donor parent.

### 20.3.3 Cross-pollinated vegetable crops

The principal methods for cross-pollinated vegetable crops are: introduction, selection (mass selection, line breeding, family breeding, recurrent selection), hybridization, and heterosis breeding. The base materials for improvement may be open-pollinated cultivars coming from introduction, populations derived through hybridization, synthetics, or genetically broad-based stocks. The selection methods commonly applied in cross-pollinated vegetable crops are always based on a population concept. Selfing is the most efficient means of increasing homozygosity. Vegetable crops sensitive to inbreeding depression, in descending order, are: carrot, onion, cabbage, and cauliflower (this varies with variety depending on the proportion of self-compatible and self-incompatible plants occurring in the variety). Therefore, techniques such as pure-line breeding are never undertaken to develop a variety except as a means of developing inbred lines for producing hybrids in strictly allogamous crops such as cole crops, cucurbits, bulb crops, root crops, leafy vegetables, and sweet corn. The selection methods followed are: mass selection, recurrent selection, and disruptive selection.

### Introduction

Introduced materials may be valuable in selecting desirable plants. In the case of non-uniform introductions, the desirable plants can be selected, and their progeny increased, purified, and later tested against standard or local varieties to select the most promising lines. It is equally important to collect and domesticate the wild relatives of cultivated species with desirable genes for nutraceuticals and edible colors. The chances of an introduction depend to a great extent on the relationship between the agroclimates, particularly temperature and day length, of the donor and receptor areas.

### Selection

MASS SELECTION. In this method, the best individual plants are selected in the population and their seeds are combined to raise the following generation. As the mass selection is based exclusively on the phenotype of the plants without any progeny testing, the success of selection depends on the heritability of the characteristics under selection.

LINE BREEDING. In this method, superior lines are selected on the basis of progeny testing and their seeds are combined. Progeny for testing can be developed by various methods such as open pollination, selfing, crossing, pair crossing, diallel crossing, and poly-crossing.

FAMILY BREEDING. This involves elaborate testing of the progeny not only in the $F_1$ generation but also in later $F_2$ or $F_3$ generations, and usually more than one cycle of selection. This method is practiced in beet and can also be adopted in radish, carrot, and cauliflower.

RECURRENT SELECTION. The source population for recurrent selection is heterozygous. It may be an open-pollinated variety, a single-cross or double-cross $F_1$ hybrid, intercrossed progeny of selected inbreed lines, a synthetic variety, or a composite. The different types of recurrent selection are as follows:

**1.** *Simple recurrent selection.* In this method the best individual plants are selected, their selfed progenies are intercrossed, and further selections are made from the intercross progeny. The self-progeny of the selected individuals are again intercrossed the following year to complete the first cycle of selection. As no test cross is carried out in this method, it is effective in the improvement of highly heritable characteristics, which can be assessed by a simple test or phenotypic evaluation.

**2.** *Recurrent selection for general (GCA) and specific (SCA) combining ability.* Recurrent selection for GCA and SCA is almost the same except for the tester stock used for the test cross. A heterozygous base tester is used for GCA while a narrow base tester is used for SCA. On the basis of the result of the test crosses for GCA and SCA, the procuring families can be selected for intercrossing and further selection in an intercross composite population.

**3.** *Reciprocal recurrent selection.* This method is useful in simultaneous selection for both GCA and SCA in two heterozygous and preferably genetically unrelated source populations.

### Heterosis breeding

Hybrid varieties are developed by exploiting the dominance variance in heterosis breeding. Hybrids are available for commercial cultivation in all cross-pollinated crops such as cole crops, cucurbits, and bulb crops. Development of $F_1$ hybrids is suitable to enhance nutraceuticals and edible colors. For example, the β-carotene content in muskmelon was significantly increased in an $F_1$ hybrid (Moon *et al.*, 2002).

### 20.3.4 Advanced breeding techniques

#### *Mutation breeding*

Put simply, a mutation is a random or directed change in the structure of the DNA or chromosome, which often results in a visible or detectable chance in a specific characteristic or trait. In self-pollinated crops, mutation breeding is well known and is based on self-fertilization of the mutant plants until the desired trait is stably expressed in the population; however, in cross-pollinated crops, its application is more difficult and identification of the origin of the desirable genotype is difficult. Both physical (X-rays, ultraviolet light) and chemical (ethyl ethane sulfonate, ethyl methane sulfonate, 5-bromouracil) mutagens cause macro-mutations (major, mono-, or oligogenic qualitative mutation) and/or micro-mutations (minor, polygenic, or quantitative changes) in the genotypes under treatment. Sapir *et al.* (2008) reported the *high pigment-1* (*hp-1*) mutation in tomatoes, which is known to increase flavonoid content in fruits.

#### *Polyploidy breeding*

Polyploidy (having more than two complete sets of chromosomes) can be induced due to aberrations in cell division. This may occur in both mitosis and meiosis. This method can be used successfully in vegetable breeding as a means of enhancing nutraceuticals and color traits in vegetables. For example, tetraploids of radish, pumpkin, and watermelon are highly productive and have improved quality.

Polyploidy can be induced in plants by colchicine treatment, which prevents the chromosomes from separating, thus transforming a diploid plant ($2n$) to a tetraploid one ($4n$). Following selfing and selection, a stable tetraploid plant is established, which can then undergo normal meiosis and perpetuate the tetraploidy through the sexual cycle,

Zhang *et al.* (2010) developed tetraploid muskmelon rich in vitamin C, with a higher level than that found in the diploid fruit.

#### *Haploid breeding*

The development of haploids in a number of plant species is now recognized as the most rapid route to the achievement of homozygosity and production of pure lines. Haploids occur in a number of families and genera of both angiosperms and gymnosperms. There are several ways in which haploidy can be achieved:

1. Spontaneous occurrence in plant populations where they are recognized by small narrow leaves, male sterility, and abnormal physical features. Such haploids occur at very low frequency and may be of maternal or paternal origin.
2. Anther and microspore culture where the male gamete develops into an embryo by a process called androgenesis.
3. Culture of unfertilized ovules and ovaries leading to embryo development from one or more of the haploid cells within the unfertilized embryo sac. This process is called gynogenesis.

Haploids should be produced efficiently to be useful in crop improvement. With anther and microspore culture, the potential exists for the recovery of large numbers of haploid plants because of the large number of microspores/pollen grains in each anther. As a consequence, this is the method of choice for haploid production in monocots and dicots. In *Brassica* and related species, this is the most widely used method of haploid plant production, and although there are variations, such plants can be produced in a wide variety of genotypes, some of which are of commercial importance. With available methods for chromosome doubling, doubled haploid plants can be efficiently recovered in most cases.

#### *Double-haploid breeding*

Conventional plant breeding methods require six to seven cycles of selection for desirable traits in order to get the required homozygosity level. This takes 11–13 years to release a cultivar, whereas double-haploid production, by chromosome doubling, enables complete homozygous individuals to be produced in a single generation, thus speeding up breeding programs. In crucifer plants, double-haploid breeding is used mainly for construction of linkage maps, genetic analysis of quantitative traits, and capturing genetic variation.

Currently, little breeding effort is ongoing for improvement of *Brassica* spp. for nutraceuticals, and there are very few successful double-haploid protocols. To explore the potential application of doubled-haploid technology for improvement of these species, Ferrie and Keller (1995) assessed the screening of several species for their microspore culture response using the *Brassica napus* microspore culture protocol.

## 20.4 Biotechnological Approaches

### 20.4.1 Molecular markers and marker-assisted selection (MAS)

Molecular markers such as random amplified polymorphic DNA (RAPD), inter simple sequence repeats (ISSRs), simple sequence repeats (SSRs), sequence-characterized amplified regions (SCAR), and cleaved amplified polymorphic sequences (CAPS) are used to study linkage of genes responsible for nutraceuticals, bioactive compounds, and edible colors using mapping populations. Of late, the use of single-nucleotide polymorphism (SNP) markers is becoming more common. In marker-assisted selection, a marker (morphological, biochemical, or DNA) is used for indirect selection of a trait of interest. Mapping populations such as near-isogenic lines (NILs) and recombinant inbred lines (RILs) are used to identify the molecular markers linked to genes of interest.

### 20.4.2 Development of NILs

Development of NILs involves several generations of backcrossing. NILs are developed by crossing the source line (non-recurrent parent) with the recipient line (recurrent parent). The progenies are again crossed with the recurrent parent, followed by phenotypic selection. Generally, it is assumed that if two NILs differ in phenotypic performance, it might be the effect of the alleles carried by the introgressed DNA fragment in the given NIL. The positive lines are forwarded for back-crossing. Generally, six to eight back-crosses are made to recover the recurrent parent genome.

### 20.4.3 Development of RILs

The parents are crossed to produce the $F_1$ population. The $F_1$ is selfed to produce the $F_2$ population. The $F_2$ population is selfed or intermated among different $F_2$s and subsequent generations. This is followed up to the $F_7/F_8$ generation to produce RILs. In RILs, alleles derived from either of the parents are arranged in an alternative way along each chromosome. In each generation, meiotic events lead to further recombination and reduce heterozygosity until completely homozygous RILs with fragments of either parental genome are achieved. Because of this, RILs are considered a permanent resource that can be replicated indefinitely and can be shared by many groups of researchers. In RILs, a population size of 50–250 is generally better for high-resolution mapping.

### 20.4.4 Bulked segregant analysis

Bulked segregant analysis (BSA) is frequently used in gene tagging or identifying major quantitative trait loci (QTLs). BSA is based on the principle of isogenic lines. It has been the most frequently used and effective method for assigning mutant loci to a specific chromosome. BSA measures allele frequencies in pools of segregants that have been sorted according to phenotype. Two $F_2$ wild-type pools derived from the $F_1$ of two different strains have the same allele frequency throughout the genome. In BSA, two parents (e.g. resistant and susceptible) showing a high degree of molecular polymorphism and that contrast for the target trait are crossed and the $F_1$ is selfed to generate the $F_2$ population. In the $F_2$, individual plants are phenotyped for resistance and susceptibility. Usually, the DNA isolated from ten plants in each group is pooled to constitute resistant and susceptible bulks. The resistant parent, susceptible parent, resistant bulk, and susceptible bulk are surveyed for polymorphisms using molecular markers. A marker showing a polymorphism between the parents as well as the bulks is considered to be putatively linked to the target trait and is further used for mapping using individual $F_2$ plants.

### 20.4.5 QTL analysis

Quantitative traits are defined as traits that have a continuous phenotypic distribution. Variance of these traits is often controlled by the segregation of many loci, called QTLs. Therefore, quantitative traits are often synonymously called polygenic traits. Another characteristic of quantitative traits is that environmental variants can play a large role in determining the phenotypic variance. The polygenic nature and the ability to be modified by the environment make studies of the genetic basis of quantitative traits more difficult than that for monogenic traits.

QTL analysis is the study of the alleles that occur at a locus and the phenotypes that they produce. Most traits of interest are governed by more than one gene, and defining and studying the entire locus of genes related to a trait aims to understand the effect of the genotype of an individual. The advent of molecular maps and the derived QTL mapping technology has provided strong evidence that, despite the inferior phenotype, exotic germplasm is likely to contain QTLs that can increase the quality of elite breeding lines. These results have motivated the development of a new molecular breeding strategy, referred to as advanced backcross (AB)-QTL method, which integrates QTL analysis with variety development, by simultaneously identifying and transferring favorable QTL alleles from unadapted to cultivated germplasm. Using bin mapping, bin 3C has previously been described as harboring a single gene mutation, *r*, for yellow flesh in tomato (Fray and Grierson, 1993).

### 20.4.6 Advanced backcross QTL analysis

The AB-QTL strategy has been applied commercially in tomato and pepper. The most extensive experiments have been conducted in tomato, where populations involving crosses with five wild *Solanum* spp. have been genotyped and field tested in a number of locations around the world for numerous traits important for the tomato processing industry. Through the application of marker and phenotypic analysis of segregating generations of the cultivated tomato and wild *Solanum* spp., it was possible to reveal QTLs that improved fruit color. These QTLs originating from red-fruited (*Solanum pimpinellifolium*) and green-fruited (*Solanum peruvianum*, *S. habrochaites*, *S. parviflorum*) wild relatives had been detected in segregating populations of crosses of these species and the cultivated tomato.

### 20.4.7 Prebreed populations

The exploitation of wild tomato relatives has traditionally been carried out through the initial development of an interspecific hybrid and subsequent backcrossing to the cultivated species. This approach is costly in terms of both time and money. The construction of a prebreed population makes use of the effort made during this process. Development of these populations starts with the initial cross between a donor parent (the wild relative accession of interest) and a recurrent parent (the cultivated tomato accession). There are different types of prebreed populations depending on the crosses developed. They all have in common the fact that they consist of a set of lines, each of which contains a fraction of the wild species genome in the cultivated species genetic background. Obtaining these populations is also a long process. The more advanced prebreed populations are RILs, backcross RILs (BCRILs), and introgression lines (ILs). These populations are obtained through successive selfings of the initial interspecific cross until a set of homozygous plants is obtained (Fig. 20.2). Each of the plants differs from the others in the combination of alleles. One disadvantage of RILs is the fact that each plant includes a high proportion of the wild genome, which can cause fertility problems and make the evaluation of some traits related to yield or fruit quality difficult. Moreover, the development of RILs is not always possible, given that for some crosses, selfing of the interspecific hybrid and the first generations is not possible. In such cases, a pseudo-$F_2$ can be obtained by crossing two $F_1$ plants. Starting with the interspecific hybrid, BCRILs are obtained by subsequent and repeated backcrosses and selfings. After several generations of backcrossing, the plant set obtained has the whole genome of the donor parent divided into small fragments in the genetic background of the cultivated species. Moreover,

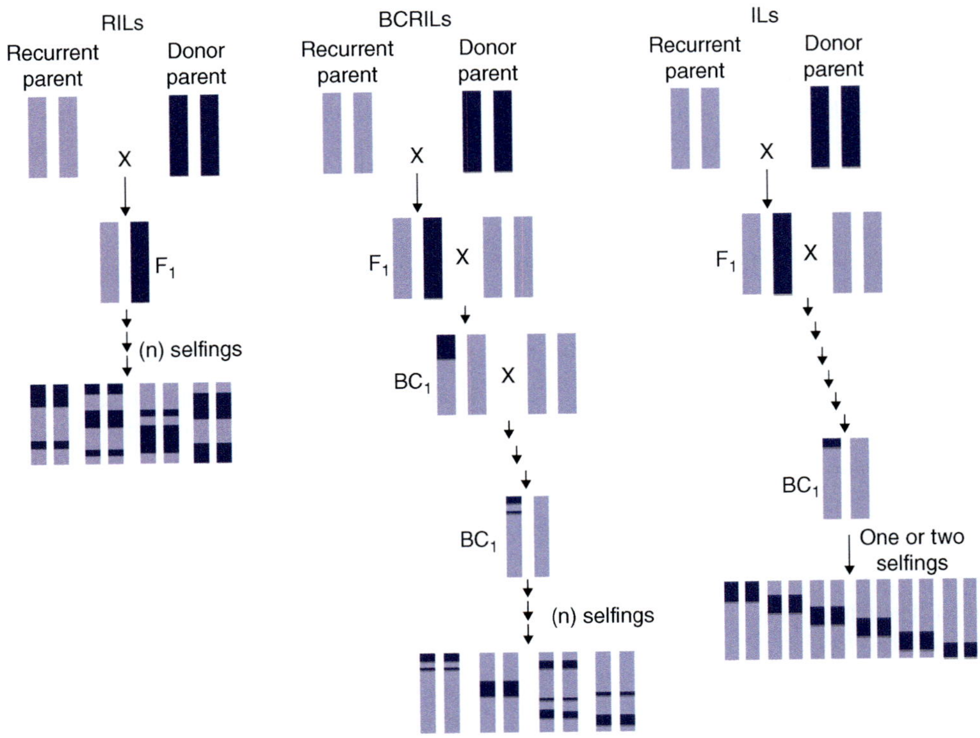


**Fig. 20.2.** Diagrammatic representation of the development of prebreeding populations. BCRILs, backcross recombinant inbred lines; ILs, introgression lines; RILs, recombinant inbred lines.

the successive selfings have increasing homozygosity levels (Julián Rodríguez, 2014).

ILs consist of a set of lines, each of which contains only one fragment of the wild species genome in the cultivated species genetic background in such a way that, as a whole, all of the wild species genome is represented in the introgressed fragments. These populations are more advanced than BCRILs as they have more backcross generations. Therefore, each line technically contains only a single fragment from the donor parent, and the introgressed fragments are also smaller than in BCRILs. An IL is an NIL with respect to their recurrent parent, given that they differ only in a very small region of the genome. For that reason, phenotypic differences among each line and the recurrent parent can only be due to the introgressed fragment. In addition, these populations are highly repeatable as they are homozygous. Another advantage of ILs is the elimination of possible interference between genes or QTLs, as a consequence of the dissection of the donor parent genome into very small fragments. However, this also represents a disadvantage, as interactions cannot be detected. In these cases, crosses between ILs that contain the putative interacting QTLs would allow the study of the interaction in the progeny. All these characteristics make ILs a powerful tool for gene mapping, superior to the other types of populations. The main disadvantage is the long time required for their development. The currently available introgression libraries in tomato are listed in Table 20.5.

### 20.4.8 Transgenic approaches

*β-Carotene enhancement*

Three genes, encoding phytoene synthase (*CrtB*), phytoene desaturase (*CrtI*), and lycopene β-cyclase (*CrtY*), from *Erwinia* have been introduced in potato to produce β-carotene. Romer

**Table 20.5.** Currently available introgression libraries in tomato.

| Wild species | Accessions | References |
|---|---|---|
| *Solanum pennellii* | LA0716 | Eshed and Zamir (1994) |
| | LA4028, LA4103 | Dani Zamir's group |
| *Solanum habrochaites* | LA1777, LA407 | Monforte and Tanksley (2000); Francis *et al.* (2001) |
| | LYC4 | Finkers *et al.* (2007) |
| *Solanum chmielewskii* | LA1840 | Peleman and van der Voort (2003); Prudent *et al.* (2009) |
| *Solanum pimpinellifolium* | LA 1589 | Tanksley *et al.* (1996) |
| | acc.TO-937 | Chetelat and Meglic (2000); Barrante *et al.* (2014) |
| *Solanum lycopersicoides* | LA2951 | Canady *et al.* (2005) |

*et al.* (2000) developed transgenic tomato to enhance the carotenoid content and profile of tomato fruit, and produced transgenic lines containing a bacterial carotenoid gene (*crtI*) encoding the enzyme phytoene desaturase, which converts phytoene into lycopene. Gerjets and Sandmann (2006) developed genetically engineered potato for the production of commercially important ketocarotenoids including astaxanthin (3,3′-dihydroxy 4,4′-diketo-β-carotene). Lu *et al.* (2006) suggested that transgenic cauliflower with the *Or* transgene is associated with a cellular process that triggers the differentiation of proplastids or other non-colored plastids into chromoplasts for carotenoid accumulation, and also reported that *Or* can be used as a novel genetic tool to induce carotenoid accumulation in a major staple food crop.

### *Tearless onion*

When onions are cut, two compounds are formed: propanethial sulfoxide, also known as lachrymatory factor, and 1-propanesulfenic acid. Lachrymatory factor reacts with nerve-cell membranes in the eye to produce tears, causing the familiar crying when cutting onions. In normal conditions, levels of 1-propanesulphenic acid are low because it is rapidly converted to lachrymatory factor. Recently, Eady *et al.* (2008) silenced the gene for the lachrymatory factor enzyme using RNA interference to produce "tearless onions": 1-propanesulphenic acid self-condenses to 1-propenyl 1-propenethiosulfinate, which then undergoes further reactions (Block, 2010). This reduces the levels of lachrymatory factor by up to 30-fold but does not diminish the overall levels of organosulfur compounds in the bulb. These tearless onions have potential health benefits for consumers as they do not induce tears but retain their health-promoting properties (Goldman, 2011).

### *Vegetable varieties rich in natural pigment/ bioactive compounds*

Pigments present in vegetable varieties can be extracted and used as natural dyes, edible food coloring, and bioactive health compounds for functional food development. Some of these are listed in Table 20.6.

### *Genome editing*

Clustered regularly interspaced palindromic repeats (CRISPR)-CRISPR-associated protein 9 (Cas9) editing is a precise method used to mutate key genes, and it is possible to rapidly generate new germplasm resources for the promotion of desirable horticultural traits. This is made possible by the availability of whole-genome sequencing data and information on the function of genes responsible for important desirable traits.

Genome-editing technologies such as zinc-finger nucleases (ZFNs), transcription activator-like effector nucleases (TALENs), and CRISPR/Cas systems or RNA-guided engineered nucleases can be effectively employed to harness the plant's native genomic repertoire by precise alterations in the DNA sequences. While ZFNs and TALENs are limited by technical complexity and limited efficiency, the CRISPR/Cas9 is simple with high efficiency. CRISPR/Cas9 is a part of the adaptive immune system in prokaryotes, which protects them against invading DNA such as viruses by cleaving the DNA in a sequence-dependent manner. CRISPR/Cas9 is an advanced

**Table 20.6.** Edible color-rich varieties of vegetable crops.

| Vegetables | Variety | Pigment |
|---|---|---|
| Carrot | Pusa Asita | Anthocyanin |
| Paprika | KTPL-19 | Capsanthin |
| Amaranth | Pusa Lal Chaulai | Anthocyanin |
| Beetroot | Detroit Dark Red | Betalain |
| Red cabbage | Kinner Red | Anthocyanin |
| Purple-headed broccoli | Palam Vichitra | Anthocyanin |
| Carrot | Pusa Rudhira | Lycopene |
| | Pusa Vrishti | Lycopene |
| | Pusa Yamdagini | Carotene |
| | Pusa Nayanjyoti | Carotene |
| Tomato | Pusa Rohini | Lycopene |

genome-editing technique that enables researchers to change, add, or remove specific DNA sequences to modify specific regions of the genome.

Among vegetables, tomato is the species with the largest number of studies using the CRISPR/Cas9 system, both because of the economic importance of the crop and because of the ease of genetic transformation using *Agrobacterium*. In tomato, parthenocarpy is potentially a desirable trait because of consumer preference and for processing purposes (Ito *et al.*, 2015). The CRISPR/Cas9 system can be used as a breeding tool for the development of parthenocarpic tomato plants (Nekrasov *et al.*, 2017).

γ-Aminobutyric acid (GABA) is regarded as a health-promoting functional compound and has received much attention in classical tomato breeding studies (Li *et al.*, 2018). Editing several genes in the GABA pathway resulted in a 19-fold higher increase in GABA content in tomato genome-edited plants (Bertier *et al.*, 2018). In potato, reducing steroidal glycoalkaloids (SGAs) such as α-solanine and α-chaconine content in tubers is a key requisite for breeding superior potatoes, as the presence at high levels may confer a bitter taste and potential undesirable effects on humans. Thus, the CRISPR/Cas9 system was used to target a steroid 16α-hydroxylase (*St16DOX*) in SGA biosynthesis in potato. Two SGA-free potato lines carrying some deletions in the *St16DOX* gene were generated in this study (Soyk *et al.*, 2017).

The CRISPR/Cas9 system is a promising, revolutionary technology for crop breeding and biological research through directed and controlled changes in the genome. To date, most CRISPR/Cas9 studies in horticultural crops have involved non-homologous end joining-mediated gene repair to create precise mutations to knock out or modify the target gene's function.

Thus, genetic engineering has been a useful technique to supplement crop breeding to date with foreign DNA including selectable marker genes being introduced into transgenic plants. The integration of the exogenous DNA in a random fashion may also interrupt endogenous genes or modify the expression of neighboring genes (Hunter, 2014; Pauwels *et al.*, 2014). One major advantage of CRISPR/Cas9 over transgenic approaches is the possibility of generating non-transgenic plants.

## 20.5 References

Arscott, S.A. and Tanumihardjo, S.A. (2010) Carrots of many colors provide basic nutrition and bioavailable phytochemicals acting as a functional food. *Comprehensive Reviews in Food Science and Food Safety* 9, 223–239.

Barrantes, W., Fernández-del-Carmen, A., López-Casado, G., González-Sánchez, M.A., Fernández-Muñoz, R., *et al.* (2014) Highly efficient genomics-assisted development of a library of introgression lines of *Solanum pimpinellifolium*. *Molecular Breeding* 34, 1817–1831.

Block, E. (2010) *Garlic and Other Alliums: The Lore and the Science*. RSC, Cambridge, UK.

Britton, G. (1998) Overview of carotenoid biosynthesis. In: Britton, G., Liaaen-Jensen, S., and Pfander, H. (eds.) *Carotenoids.* Birkhauser, Basel, Switzerland, pp. 13–147.

Canady, M.A., Meglic, V., and Chetelat, R.T. (2005) A library of *Solanum lycopersicoides* introgression lines in cultivated tomato. *Genome* 48, 685–697.

Cartea, M.A. and Velasco, P. (2008) Glucosinolates in *Brassica* foods: bioavailability in food and significance for human health. *Phytochemistry Reviews* 2, 213–229.

Chetelat, R.T. and Meglic, V. (2000) Molecular mapping of chromosome segments introgressed from *Solanum lycopersicoides* into cultivated tomato (*Lycopersicon esculentum*). *Theoretical and Applied Genetics* 100, 232–241.

Delgado-Vargas, F., and Paredes-Lopez, O. (eds.) (2003) *Natural Colorants for Food and Nutraceutical Uses*. CRC Press, Boca Raton, Florida.

Dwivedi, S., Sahrawat, K., Puppala, N., and Ortiz, R. (2014) Plant prebiotics and human health: biotechnology to breed prebiotic-rich nutritious food crops. *Electronic Journal of Biotechnology* 17, 238–245.

Erpen-Dalla Corte, L., Mahmoud, L.M., Moraes, T.S., Mou, Z., Grosser, J.W., and Dutt, M. (2019) Development of improved fruit, vegetable, and ornamental crops using the CRISPR/Cas9 genome editing technique. *Plants* 8: 601.

Eshed, Y. and Zamir, D. (1994) Introgressions from *Lycopersicon pennellii* can improve the soluble solids yield of tomato hybrids. *Theoretical and Applied Genetics* 88, 891–897.

Fahey, J.W., Zalcmann, A.T., and Talalay, P. (2001) The chemical diversity and distribution of glucosinolates and isothiocyanates among plants. *Phytochemistry* 56, 5–51.

FAO (Food and Agriculture Organization of the United Nations) (2015) *The State of Food Insecurity in the World*. FAO, Rome.

Ferrie, A.M.R. and Keller, W.A. (1995) Microspore culture for haploid plant production. In: Gamborg, O.L. and Phillips, G.C. (eds.) *Plant Cell, Tissue and Organ Culture: Fundamental Methods*. Springer, Berlin, pp. 155–164.

Finkers, R., van Heusden, A.W., Meijer-Dekens, F., van Kan, J.A., Maris, P., and Lindhout, P. (2007) The construction of a *Solanum habrochaites* LYC4 introgression line population and the identification of QTLs for resistance to *Botrytis cinerea*. *Theoretical and Applied Genetics* 114, 1071–1080.

Francis, D.M., Kabelka, E., Bell, J., Franchino, B., and St. Clair, D. (2001) Resistance to bacterial canker in tomato (*Lycopersicon hirsutum* LA407) and its progeny derived from crosses to *L. esculentum*. *Plant Disease* 85, 1171–1176

Fray, R.G. and Grierson, D. (1993) Molecular genetics of tomato fruit ripening. *Trends in Genetics* 9, 438–443.

Gerjets, T. and Sandmann, G. (2006) Keto-carotenoid formation in transgenic potato. *Journal of Experimental Botany* 57, 3639–3645.

Goldman, I.L. (2011) Molecular breeding of healthy vegetables. *EMBO Reports* 12, 96–102.

Grotewold, E. (2006) The genetics and biochemistry of floral pigments. *Annual Review of Plant Biology* 57, 761–780.

Hunter, P. (2014) "Genetically Modified Lite" placates public but not activists. *EMBO Reports* 15, 138–141.

Ito, Y., Nishizawa-Yokoi, A., Endo, M., Mikami, M., and Toki, S. (2015) CRISPR/Cas9-mediated mutagenesis of the *RIN* locus that regulates tomato fruit ripening. *Biochemical and Biophysical Research Communications* 467, 76–82.

Julián Rodríguez, O. (2014) Exploitation of *Solanum chilense* and *Solanum peruvianum* in tomato breeding for resistance to Tomato yellow leaf curl disease. PhD thesis, Universitat Politècnica de València, Valencia, Spain.

Kang, S., Seeram, N., Nair, M., and Bourquin, L. (2003) Tart cherry anthocyanins inhibit tumor development in $Apc^{Min}$ mice and reduce proliferation of human colon cancer cells. *Cancer Letters* 194, 13–19.

Li, R., Fu, D., Zhu, B., Luo, Y., and Zhu, H. (2018) CRISPR/Cas9-mediated mutagenesis of *lncRNA1459* alters tomato fruit ripening. *Plant Journal* 94, 513–524.

Lila, M.A. (2004) Plant pigments and human health. In: Davis, K. (ed.) *Plant Pigments and their Manipulation. Annual Plant Reviews, Vol. 14*. CRC Press, Boca Raton, Florida, pp. 248–271.

Lu, S., Eck, J.V., Zhou, X., Lopez, A.B., Halloran, D.M., *et al.* (2006) The cauliflower *Or* Gene encodes a DnaJ cysteine-rich domain-containing protein that mediates high levels of β-carotene accumulation. *Plant Cell* 18, 3594–3605.

Monforte, A.J. and Tanksley, S.D. (2000a) Development of a set of near isogenic and backcross recombinant inbred lines containing most of the *Lycopersicon hirsutum* genome in a *L esculentum* genetic background: a tool for gene mapping and gene discovery. *Genome* 43, 803–813.

Moon, S.S., Verma, V.K., and Munshi, A.D. (2002) Gene action of quality traits in muskmelon (*Cucumis melo* L.). *Vegetable Science* 29, 134–136.

Nekrasov, V., Wang, C., Win, J., Lanz, C., Weigel, D., and Kamoun, S. (2017) Rapid generation of a transgene-free powdery mildew resistant tomato by genome deletion. *Scientific Reports* 7: 482.

Norton, R.A. (1999) Inhibition of aflatoxin $B_1$ biosynthesis in *Aspergillus flavus* by anthocyanidins and related flavonoids. *Journal of Agricultural and Food Chemistry* 47, 1230–1235.

Pauwels, K., Podevin, N., Breyer, D., Carroll, D., and Herman, P. (2014) Engineering nucleases for gene targeting: safety and regulatory considerations. *New Biotechnology* 31, 18–27.

Peleman, J.D. and van der Voort, J.R. (2003) Breeding by design. *Trends in Plant Science* 8, 330–334.

Prudent, M., Causse, M., Génard, M., *et al.* (2009) Genetic and physiological analysis of tomato fruit weight and composition: influence of carbon availability on QTL detection. *Journal of Experimental Botany* 60, 923–937.

Rajasree, V. and Pugalendhi, L. (2021) Breeding vegetables for nutritional security. In: Jovandaric, M.Z. (ed.) Veganism—A Fashion Trend or Food as a Medicine. IntechOpen – a United Nations Sustainable Development Goals Publishers Compact. DOI: http://dx.doi.org/10.5772/intechopen.95349.

Romer, S., Fraser, P.D., Kiano, J.W., Shipton, C.A., Misawa, N., *et al.* (2000) Elevation of the pro-vitamin A content of transgenic tomato plants. *Nature Biotechnology* 18, 666–669.

Rosa, E.A.S., Hesney, R.K., Fenwick, G.R., and Portas, C. (1997) Glucosinolates in crop plants. *Horticultural Sciences* 19, 99–215.

Sarikamis, G., Marquez, J., Maccormack, R., Bennett, R.N., Roberts, J., and Mithen, R.(2006) High glucosinolate broccoli: a delivery system for sulforaphane. *Molecular Breeding* 18, 219–228.

Sapir, M., Shamir, M.O., Ovadia, R., Reuveni, M., Evenor, D., *et al.* (2008) Molecular aspects of *Anthocyanin fruit* tomato in relation to *high pigment-1*. *Journal of Heredity* 99, 292–303.

Simon, P., Pollak, L., Clevidence, B., Holden, J., and Haytowitz, B. (2009) Plant breeding for human nutritional quality. *Plant Breeding Reviews* 31, 325–392.

Soyk, S., Müller, N., Park, S.J., Schmalenbach, I., Jiang, K., *et al.* (2017) Variation in the flowering gene *SELF PRUNING 5G* promotes day-neutrality and early yield in tomato. *Nature Genetics* 49, 162–168.

Tanksley, S.D., Grandillo, S., Fulton, T.M., Zamir, D., Eshed, Y., *et al.* (1996) Advanced backcross QTL analysis in a cross between an elite processing line of tomato and its wild relative *L. pimpinellifolium*. *Theoretical and Applied Genetics* 92, 213–224.

Verkerk, R., Schreiner, M., Krumbein, A., Ciska, E., Holst, B., *et al.* (2009) Glucosinolates in *Brassica* vegetables: the influence of the food supply chain on intake, bioavailability and human health. *Molecular Nutrition & Food Research* 53 (Suppl. 2): S219.

Williams, D.J., Edwards, D., Hamernig, I., Jian, L., James, A.P., *et al.* (2013) Vegetables containing phytochemicals with potential anti-obesity properties: a review. *Food Research International* 52, 323–333.

Zhang, W., Hao, H., Ma, L., Zhao, C., and Yu, X. (2010) Tetraploid muskmelon alters morphological characteristics and improves fruit quality. *Scientia Horticulturae* 125, 396–400.

# Index

Note: The page numbers in italics and bold represents figures and tables respectively.

Printed and bound by CPI Group (UK) Ltd, Croydon, CR0 4YY

13/09/2026

14961116-0004